z Values for
Rejection Regions for
Large-Sample Tests
of Hypotheses

	Type of Alternative Hypothesis, H_a		
Value of α	$>$	$<$	$\neq$
$\alpha = .10$	Reject H_0 if $z > 1.28$	Reject H_0 if $z < -1.28$	Reject H_0 if $z > 1.65$ or $z < -1.65$
$\alpha = .05$	Reject H_0 if $z > 1.65$	Reject H_0 if $z < -1.65$	Reject H_0 if $z > 1.96$ or $z < -1.96$
$\alpha = .01$	Reject H_0 if $z > 2.33$	Reject H_0 if $z < -2.33$	Reject H_0 if $z > 2.58$ or $z < -2.58$

z-Scores for
Large-Sample
Confidence Intervals

	Confidence Desired		
	90%	*95%*	*99%*
Value of z	1.65	1.96	2.58

D1451638

BASIC STATISTICS
AN INFERENTIAL APPROACH

THIRD EDITION

BASIC STATISTICS
AN INFERENTIAL APPROACH

THIRD EDITION

FRANK H. DIETRICH II
THOMAS J. KEARNS

Northern Kentucky University

DELLEN PUBLISHING COMPANY
San Francisco

COLLIER MACMILLAN PUBLISHERS
London

DIVISIONS OF MACMILLAN, INC.

ON THE COVER:

Vessel of Light No. 23, Lullaby for the Careful Cradle Watch is an oil on canvas painting executed by Martin Facey in 1988. The work measures 32″ × 32″. Facey was awarded the prestigious Guggenheim Fellowship in painting in 1982. While teaching painting at the University of New Mexico, he was influenced by the colors employed in American Indian artifacts. Facey is represented by Tortue Gallery in Santa Monica, California.

Permissions: Dellen Publishing Company
              400 Pacific Avenue
              San Francisco, California 94133

Orders: Dellen Publishing Company
         c/o Macmillan Publishing Company
         Front and Brown Streets
         Riverside, New Jersey 08075

Collier Macmillan Canada, Inc.

LIBRARY OF CONGRESS CATALOGING-IN-PUBLICATION DATA
Dietrich, Frank H.
  Basic statistics.

  Includes index.
  1. Statistics. I. Kearns, Thomas J. II. Title.
QA276.12.D53 1989   519.5   88-28370

Printing: 4 5 6 7   Year: 2 3

ISBN 0-02-328801-9

CONTENTS

Chapter 8

INFERENCES COMPARING TWO POPULATIONS 403

Chapter 9

REGRESSION AND CORRELATION:
A STRAIGHT-LINE RELATIONSHIP 491

Chapter 10

ANALYSIS OF VARIANCE: COMPARING MORE THAN
TWO POPULATION MEANS 567

ANSWERS TO SELECTED EXERCISES   723

INDEX   741

PREFACE

This text is designed for an introductory course in statistics. It is intended for students who do not have a strong mathematical background; a second-year high school algebra course should be sufficient.

This text is written to provide students not only with the basic methods of elementary statistics but also with some sense of the subject as a coherent whole rather than as a disjoint collection of seemingly unrelated procedures. For this reason, an inferential approach has been adopted, and the text's organization, examples, applications, and interpretations are focused on the theme of statistical inference.

AN INFERENTIAL APPROACH

One of the most distinguishing features of this text, compared to other introductory statistics texts, is the continuous reference to inference-making. Throughout the first six chapters, the reasoning behind hypothesis testing is developed gradually and intuitively. The treatment of such topics as numerical descriptive measures, probability, and sampling distributions emphasizes how they are related to making inferences. Because students can see how each topic is integrated into the inferential process, they do not wonder why these topics are studied.

Chapters 1 and 2 focus on the two main areas of the science of statistics—inferential and descriptive statistics. Measures of central tendency and relative standing are discussed in Chapter 3. The idea that most data are within 2 standard deviations of the mean is stressed, and the idea of a rare event is developed. It is at this point that the basic logic of hypothesis testing is intuitively introduced.

Probability is discussed in a very concise way in Chapter 4 to provide the language and a bit more precision to the hypothesis testing logic. The notion of the sampling distribution of a statistic is introduced in Chapter 5. The sampling distributions for binomial experiments are used to reinforce the previously developed logic behind statistical inferences.

The Central Limit Theorem is used in Chapter 6 to extend the ideas of sampling distributions to include the normal distribution. Through z-scores, the sample mean or sample proportion for large samples is used to test hypotheses, again informally.

Finally, in the beginning of Chapter 7, the testing of statistical hypotheses is formally introduced. By the gradual development of the logic behind hypothesis testing and the reinforcement of these ideas, students readily understand such concepts as a null hypothesis, alternative hypothesis, Type I error, Type II error, etc. In the remainder of the text, the same type of logic is used to discuss confidence intervals and other statistical methods in a variety of situations.

PRINCIPAL CHANGES FOR THE THIRD EDITION

The principal change in this edition is the introduction of material involving computers. Examples of computer solutions and exercises suitable to computer analysis are included throughout the text. Emphasis is then upon analysis of the results. The widespread availability of computers and statistical software packages makes the inclusion of this material possible for most users of the text. The simple fact that most real users of statistics utilize the computer for analysis makes it desirable.

The material is not dependent upon any particular hardware or software. What is intended is that students understand what a typical computer statistical package can provide and how the output provided by the computer can be analyzed and utilized. The particular output for this text was generated using Minitab. Minitab was chosen because of its wide availability, ease of use, and the fact that readers using other software should find the output comparable. For users of Minitab, a supplement detailing Minitab instructions and keyed to this text is available.

FEATURES RETAINED FROM THE PREVIOUS EDITIONS

1. **Introductory Motivating Examples**   The introduction to each chapter is a brief summary of an interesting, real-life application of a statistical concept. These applications serve as motivating examples and help to answer by example the often-asked questions, "Why should I study statistics? Of what value is statistics to my program?" These applications represent a wide range of disciplines but require no specialized knowledge to appreciate them.

2. **The Use of Examples as a Teaching Device**   Every important concept is introduced by one or more examples. It is our belief that students generally do not understand definitions, generalizations, and abstractions until an application is provided.

3. **A Simple, Clear Style**   We have written this book in a simple, clear style. Topics that are tangential to our objective have been avoided, and we have not taken an

encyclopedic approach in the presentation of material. The logic behind the concepts and statistical tests is emphasized where appropriate. However, in many instances—for example, the derivations of test statistics—the mathematics and detailed reasoning are left for more advanced courses.

4. **An Emphasis on Inference Throughout**   As outlined above, each of the first six chapters contains a section relating the topics discussed in that chapter to inference-making. This emphasis provides continuity and serves as a constant reminder that the primary objective is to learn about statistical inference. The final six chapters then deal with specific inferential methods.

5. **A Great Number and Variety of Exercises**   Exercises are provided at the end of most sections to give the student immediate reinforcement of the material. The exercises are divided into the following types:

   a. **Learning the Language**   A list of important terms and symbols that were introduced is provided at the end of each section (where appropriate). The student can quickly ascertain whether these important items have been learned.

   b. **Learning the Mechanics**   These exercises are intended to be straightforward, and often are designed to strengthen the computational skills of the student. They are introduced in a few words and are unhampered by background information, which often detracts from instructional objectives.

   c. **Using the Tables**   These exercises are also mechanical, but are specifically designed to help the student learn to use the statistical tables provided in the Appendix.

   d. **Applying the Concepts**   These realistic exercises are designed to exhibit the broad spectrum of applications of statistics. After the mechanics are mastered, these exercises develop the student's skill at analyzing realistic problems.

   e. **Supplementary Exercises**   Many supplementary exercises are provided at the end of each chapter for review and practice.

6. **Treatment of Probability**   The treatment of probability is minimal, covering in three sections only those topics that are essential to an understanding of the role probability plays in making inferences. This includes the concepts of sample space, probability assignment, and events, but counting is restricted to ordered samples, and emphasis is placed on random samples and equal likelihood. This is done to allow a more complete treatment of inferential statistics and to minimize material extraneous to this theme that can cause confusion for students.

   Additional topics in probability, including compound and complementary events, conditional probability, and independence, are covered in two optional sections. Classes with the time and desire to cover these sections would, of course, profit from them, but they are not essential for the material that follows.

   Instructors requiring a more complete treatment of probability within the context of a beginning statistics course may want to consider *Statistics*, by James T. McClave and Frank H. Dietrich II.*

* James T. McClave and Frank H. Dietrich II, *Statistics*, 4th ed. (San Francisco: Dellen Publishing Company, 1988).

7. Chapter Sequence The topics have been arranged in a sequence that seems to be preferred by most instructors. Chapters 1–8, the core of the text, are logically dependent and should be covered in that order. The remaining four chapters are independent of each other and can be covered in any order, after Chapter 8.

STUDENT AIDS

1. Helpful Hints Hints are included at appropriate places in the text to provide help with difficult or confusing points.

2. Boxes Boxes highlight important definitions, rules, and helpful hints.

3. Second Color A second color is used throughout to make reading easier, especially in tables, graphs, and charts.

4. Chapter Summaries Each chapter includes a summary of the important concepts and methods introduced in the chapter. These provide the student with a compact overview of the chapter for reference and review.

5. Chapter Quizzes A chapter quiz follows the supplementary exercises. This provides the student with a self-testing device and still another method of assessing whether the chapter material has been learned.

6. Cumulative Quizzes After the chapter quiz, a cumulative quiz covering material in the current and preceding chapters is given. This provides a check on the retention of previous material, as well as on the student's ability to assimilate the past and current material. Since the course examinations are likely to be cumulative in nature, these cumulative quizzes should be most helpful in preparing for exams.

7. On Your Owns Each chapter ends with an exercise entitled "On Your Own." The intent of this exercise is to give the student some "hands-on" experience with an application of the statistical concepts introduced. In most cases the student is required to collect, analyze, and interpret data relating to some real applications of the chapter's concepts.

8. Inside Covers Commonly used tables are placed inside the front and back covers of the text for convenient reference.

9. Supplements A solutions manual, Minitab computer supplement, and a statistical package to accompany this text are described below.

SUPPLEMENTS

1. Solutions Manual (by Nancy J. Shafer). The solutions manual presents the solutions to most odd-numbered exercises in the text. Many points are clarified and expanded to provide maximum insight into and benefit from each exercise.

2. **Minitab Computer Supplement** (by Ruth K. Meyer and David D. Krueger). The Minitab computer supplement was developed to be used with Minitab Release 5.1, a general-purpose statistical computing system. The supplement, which was written especially for the student with no previous experience with computers, provides step-by-step descriptions of how to use Minitab effectively as an aid in data analysis. Each chapter begins with a list of new commands introduced in the chapter. Brief examples are then given to explain new commands, followed by examples from the text illustrating the new and previously learned commands. Where appropriate, simulation examples are included. Exercises, many of which are drawn from the text, conclude each chapter.

A special feature of the supplement is a chapter describing a survey sampling project. The objectives of the project are to illustrate the evaluation of a questionnaire, provide a review of statistical techniques, and illustrate the use of Minitab for questionnaire evaluation.

3. **DellenStat: Statistical Software Manual** (by Michael Conlon). DellenStat is an integrated statistics package consisting of a workbook and an IBM PC floppy diskette with software and example sets of data. The system contains a file creation and management facility, a statistics facility, and a presentation facility. The software is menu-driven and has an extensive help facility. It is completely compatible with the text.

The DellenStat workbook describes the operation of the software and uses examples from the text. After an introductory chapter for new computer users, the remaining chapters follow the outline of the text. Additional chapters show how to create new sets of data. Technical appendices cover material for advanced users and programmers.

DellenStat runs on any IBM PC or close compatible with at least 256K of memory and at least one floppy disk drive.

4. **DellenTest.** This unique computer-generated random test system is available to instructors without cost. Utilizing an IBM PC computer and a number of commonly used dot-matrix printers, the system will generate an almost unlimited number of quizzes, chapter tests, final examinations, and drill exercises. At the same time, the system produces an answer key and student worksheet with an answer column that exactly matches the column on the answer key.

ACKNOWLEDGMENTS

The authors wish to thank users of previous editions who provided helpful comments and suggestions: William Applebaugh, Indiana University South East; Carole Bernett, William Rainey Harper College; John Kellermeier, Plattsburg State University of New York; Bert Liberi, Westchester County Community College; Douglas Nychka, North Carolina State University; David Trunnell, Xavier University; Coburn Ward, University of the Pacific. We are particularly grateful to Susan Hardy, University of California–Berkeley, for her detailed review of the second edition.

BASIC STATISTICS
AN INFERENTIAL APPROACH

THIRD EDITION

Chapter 1

WHAT IS STATISTICS?

THE SALK POLIO VACCINE EXPERIMENT

The largest and most expensive medical experiment in history was carried out in 1954. Well over a million young children participated, and the immediate direct costs were over 5 million dollars. The experiment was carried out to assess the effectiveness, if any, of the Salk vaccine as a protection against paralysis or death from poliomyelitis.[*]

How could the information collected from over a million young children be summarized? How could such information be used to determine the effectiveness of the vaccine? A knowledge of statistics will help answer such questions. In this chapter we will begin discussing the meanings, uses, and values of statistics. We will also introduce some very basic terminology.

[*] From an article by Paul Meier in *Statistics: A Guide to the Unknown*, 2nd ed. (1978), p. 3. (References are listed at the end of each chapter.)

1.1

STATISTICS: WHAT IS IT?

Statistics. Is it a field of study? Is it a collection of numbers that summarize phenomena such as the state of our national economy, the performance of a football team, or the social conditions in a particular city? Or, as the title of a popular book (Tanur et al., 1978) suggests, is it "a guide to the unknown"?

The following two meanings of the word *statistics* will be used in this text:

Definition 1.1

a. Statistics are numbers calculated from a collection of data.
b. Statistics is the science of collecting and interpreting data.

When people hear the word *statistics*, they usually think of the first meaning—numbers that summarize and describe a collection of data. We will regularly use that meaning of statistics throughout this book, but the overall purpose of the book is to acquaint you with the second meaning—statistics as a science. The science of statistics can be divided into two broad areas: **descriptive statistics** and **inferential statistics.**

Definition 1.2

Descriptive statistics is the science of summarizing or describing data.

Definition 1.3

Inferential statistics is the science of interpreting data in order to make predictions, estimates, or decisions.

The following two examples illustrate these two areas of statistics.

EXAMPLE 1.1   Figure 1.1, which appeared in the January 11–13, 1985, issue of *USA Today*, indicates
which agencies and organizations Americans were most satisfied with in 1984. The
graph summarizes the opinions obtained in a poll of 2,000 adults. The figure indicates
that 67% of those polled were completely satisfied with the airlines, while only 38%
were completely satisfied with the IRS. Since the purpose of this figure is to summarize
and describe the opinions of the 2,000 adults who were polled, it provides an example
of *descriptive statistics*.   ■

Figure 1.1
Percentages of people
completely satisfied with
agencies and organizations
in 1984
Source: USA Today,
January 11–13, 1985, p. 1A.

Willing to help

Most helpful agencies and organizations of 1984:

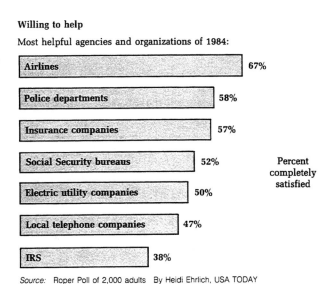

Percent
completely
satisfied

Source: Roper Poll of 2,000 adults By Heidi Ehrlich, USA TODAY

EXAMPLE 1.2   In the introduction to this chapter we mentioned the Salk polio vaccine experiment
of 1954. As stated in the quote, the purpose of the experiment was to investigate the
effectiveness of the vaccine. The data obtained from the experiment allowed the investigators to compare the rate of occurrence of various types of polio among children
who had received the vaccine as opposed to children who had not. The investigators
then had to make a decision about whether the vaccine was or was not effective in
reducing the incidence of polio. This is an example of the science of *inferential
statistics*.   ■

In this section we have examined the meanings of the word *statistics*. In the next
section we will introduce some commonly used statistical terms.

1.2

THE LANGUAGE OF STATISTICS

Certain elements are common to all statistical problems, regardless of the area in which the problems arise. To proceed further with a discussion of statistics, you must become familiar with the terminology for these common elements. The first element, a **population,** is used in both the descriptive and inferential areas of statistics. This term has two meanings.

Definition 1.4

a. A **population** is a set of data that characterizes some phenomenon of interest.

b. A **population** is the collection of all objects on which these data are obtained.

A set of data may be either numerical or nonnumerical in nature. Examples of non-numerical data are political party affiliation, religious preference, sex, highest educational degree received, and month of the year. Examples of numerical data are age, weight, test score, length of time, temperature, and pressure. Thus, we may think of **data** as observations or measurements of a phenomenon of interest.

EXAMPLE 1.3   The personnel department of a college is interested in the length of time each of the 573 full-time faculty members has been employed by the college. By examining the records on file for each faculty member, the desired information is obtained. Explain how the word *population* might be used in this example.

Solution   The population as a set of measurements would be 573 numbers, where each number represents the length of employment for one of the 573 full-time faculty members.

The population as a collection of objects is the group of 573 full-time faculty members.

In the context of this example, there is no ambiguity in thinking of the population as the group of 573 faculty members. It is understood that the length of employment for each member of the faculty is the measurement being observed. ∎

EXAMPLE 1.4   A political candidate is naturally interested in all registered voters in the district. What difficulties arise if we define "all the voters in the district" as a population?

Solution The major difficulty is that we have not specified what set of data is of interest to the political candidate. Many characteristics of the voters—such as age, political party affiliation, marital status, and income—could be recorded for each registered voter, and each would define a population of data. ■

When defining a population as a collection of objects, it must be done in the context of a specific problem. Otherwise, it is not clear what the statistical population (set of data) is.

The second element common to statistical problems, the **sample,** also has two meanings.

Definition 1.5

a. A **sample** is a set of data taken from the population of interest.
b. A **sample** is a collection of objects chosen from all possible objects in the population of interest.

The sample is a subset (part) of the population. From a statistical viewpoint it is preferable to consider the sample as a set of data. However, in the context of a specific example, we may want to use the word *sample* to represent a collection of objects.

In the Salk polio vaccine experiment, over 1.8 million children in the first, second, and third grades were studied. Records were kept, indicating whether each child had been vaccinated and what type of polio, if any, was diagnosed. This information was the data of interest. In the sense of Definition 1.5**b**, the children who were studied were a sample from the population of all children in the first, second, and third grades. In the sense of Definition 1.5**a**, the recorded data—vaccination status and type of polio— were the sample.

EXAMPLE 1.5 The map shown in Figure 1.2 (page 6) appeared in the journal *Environmental Comment* (September 1980). It shows the distribution of approximately 100 hazardous waste management facilities in the United States. Suppose an environmentalist is interested in determining the percentage of these facilities that use recovery methods to recycle potentially dangerous, but very valuable wastes.

a. What is the population of interest?
b. Suppose the environmentalist cannot survey every facility, but has the resources to survey only thirty facilities. Give both meanings of the term *sample* in this case.

Solution **a.** The population is the collection of all the hazardous waste management facilities shown in Figure 1.2. For each facility the observation of interest is whether recovery methods are or are not used.

Figure 1.2
Geographical distribution of hazardous waste management facilities
Source: Hazardous Waste Management Facilities in the United States—1977. Cincinnati, Ohio:
U.S. Environmental Protection Agency, Office of Solid Waste, 1977.

b. The sample would be the thirty facilities that are surveyed, and it would be noted
whether each facility does or does not use recovery methods. Figure 1.3 illustrates
a possible sample of thirty facilities that could be obtained.   ■

A comparison of Figures 1.2 and 1.3 clearly shows that the sample is only a part
of the population. Although there are instances where an entire population is examined, we would not consider this to be a sample. For our purposes, a sample must
be smaller than the population to satisfy the definition.

Helpful Hint
███████████████████

A statistical sample is always smaller than the population.

Descriptive statistics are used to summarize the information (data) in both samples
and populations. The foundation of every application of inferential statistics is a pop-

Figure 1.3
Possible sample of thirty hazardous waste management facilities

ulation from which a sample is obtained. To clarify the connection between the sample and the population and to continue the discussion of inferential statistics, two more elements must be considered. The first of these is the **inference.**

Definition 1.6

An **inference** is a decision, estimate, prediction, or generalization about a population based on information contained in a sample from that population.

In the Salk vaccine experiment, the children studied were only a portion—a sample—of all the children in the country. For these children, the investigators found that the frequency of diagnosed polio cases among those who had been vaccinated was only about half the frequency among those who had not been vaccinated. The investigators inferred that this was representative of what would be found for the whole population and, therefore, that the vaccine was effective.

EXAMPLE 1.6   A sociologist is interested in the average number of murders committed last month in all cities in the United States containing fewer than 50,000 people. Rather than examine the police records for each of these cities, the sociologist samples twenty-five of the cities and finds that the average for these twenty-five cities was 2.6 murders during the last month. Using this information, the sociologist concludes that the average for the collection of all such cities in the United States is approximately 2.6 murders per city during the last month.

    a. Identify the population.
    b. Identify the sample.
    c. Identify the inference that was made.

Solution   **a.** The population is the collection of all cities in the United States with fewer than 50,000 residents.
    b. The sample is the collection of twenty-five cities observed. This sample had an average of 2.6 murders per city last month.
    c. The sociologist made the inference that the average number of murders per city for all cities in the United States containing fewer than 50,000 people was approximately 2.6 last month. Using the sample information, the sociologist has estimated the average for the population. ■

The next, and perhaps most important, element of a statistical problem is a **measure of reliability** for an inference.

Definition 1.7

A **measure of reliability** is a statement that indicates how certain we are that an inference is correct.

Since all statistical inferences are based upon only the partial information about a population contained in a sample, there is always a chance that an inference will be incorrect. The science of statistics recognizes this fact, and requires that every inference be accompanied by a measure of reliability.

EXAMPLE 1.7   After examining the medical records of 1,000 patients of a general practitioner, a psychologist estimates that 60% of all illnesses treated by the doctor are psychosomatic in nature. The psychologist is 99% sure that the true percentage of all the disorders treated by the doctor that are psychosomatic is within 4% of the estimate of 60%.

    a. Identify the population and sample in this example.
    b. Identify the inference made.
    c. Identify the statement of reliability.

Solution   **a.** The population of objects on which observations are taken is the collection of all the doctor's patients. The sample is the collection of 1,000 patients whose records were examined. The observation taken on each patient was whether the patient's illness was or was not psychosomatic in nature.

b. The inference is that "60% of all illnesses treated by the doctor are psychosomatic in nature."

c. The statement of reliability tells us how certain the psychologist is that the inference is correct. For this example, it is that the psychologist is 99% certain that the percentage of psychosomatic disorders for the population is within 4% of the estimate. The 4% leeway means that the percentage of psychosomatic disorders for the population might be as small as 56% (60% − 4%) or as large as 64% (60% + 4%). With this leeway, there is a 99% certainty that the inference is correct. There is a small chance (1%) that the inference is incorrect. ■

We conclude this section with a summary of the elements common to statistical problems and an example to illustrate all four elements.

Four Elements Common to Statistical Problems

1. Population of interest
2. Sample from this population
3. Inference about the population based on information contained in the sample
4. Measure of the reliability of the inference

EXAMPLE 1.8   In certain types of industries, some byproducts are mildly radioactive, and these by-products can get into our freshwater supply. In 1976, the Environmental Protection Agency (EPA) issued regulations limiting the amount of radioactivity in supplies of drinking water and setting the maximum for naturally occurring radiation at 5 pico-curies per liter of water. (A picocurie is a unit of radiation.) In order to evaluate the quality of a particular city's water supply, the number of picocuries per liter was measured for each of twenty water specimens taken from the supply. Identify the four elements of this statistical problem.

Solution   **1.** The *population* is most conveniently thought of as the city's entire water supply. In a statistical sense, it is the number of picocuries per liter of water in each possible specimen that could be taken from the water supply.

2. The *sample* is the collection of twenty water specimens actually selected from the water supply, or the number of picocuries per liter measured for each of the twenty water specimens.

3. Although an *inference* is not explicitly stated, an obvious question of interest is, "Does the city's water supply meet the EPA's guidelines?" Thus, depending on the data in the sample, one of two inferences could be made: It might be inferred that the city's water supply is safe (is within the maximum of 5 picocuries per liter), or it might be inferred that the entire water supply is radioactively contaminated.

4. Whichever inference is actually made, we need a *measure of its reliability*. The inference itself is of little worth unless we can be fairly certain that it is correct. Suppose the twenty water specimens have an average of 4.9 picocuries of radiation per -liter, and it is inferred that the city's water supply is safe. Unless we are very sure that the inference is correct, we should be cautious about using the water. Since the inference is based on a sample, which is only part of the population, we must recognize the possibility that the sample data may not accurately reflect the population data. A measure of reliability will help us assess the possibility of error. ∎

1.3

WHY STUDY STATISTICS?

Why study statistics? There are many reasons for studying statistics, but we will mention only two of the more important ones. First, the growth in data collection associated with business, government, science, public health, the environment, etc., has been astounding over the past several decades. Even people who are not involved in scientific research are exposed to data published in newspapers and magazines and broadcast on radio and television virtually every day. An examination of advertisements alone would prove this point. A basic knowledge of statistics will enable you to evaluate numerical data and claims made about such data. You may be called upon to use this ability to make intelligent decisions, inferences, and generalizations in both your professional and personal life. For this reason, the study of statistics is an essential preparation for a role in modern society.

Second, since you will have to interpret data and make inferences so frequently, another—and perhaps more important—reason for studying statistics becomes evident. The measure of reliability that accompanies every inference separates the science of statistics from the art of fortune-telling. A palm reader, like a statistician, may examine a sample (your hand) and make inferences about the population (your life). However, no meaningful measure of reliability can be attached to the palm reader's inferences. On the other hand, we will always be sure to assess the reliability of our statistical inferences. An understanding of statistical methods will allow you to better judge the reliability of inferences.

In brief, here are the two most important reasons for learning about the science of statistics:

Two Reasons for Studying Statistics

1. To be able to evaluate data intelligently
2. To be better able to evaluate statements of reliability

Chapter Summary

■■■■■■■■

Statistics is the science of collecting and interpreting data. The word *statistics* also refers to numbers calculated from a collection of data.

The science of statistics can be **descriptive,** when data are being summarized or described, or **inferential,** when data are being interpreted in order to make predictions, estimates, or decisions.

The data that characterize the phenomenon of interest are called the **population.** The word *population* also refers to the objects on which data are obtained. A subset of a population is called a **sample.**

When information from a sample is used to make a decision, estimate, prediction, or generalization about the entire population, the result is called an **inference** about the population. Connected with any inference should be some **measure of reliability,** a statement indicating how certain we can be that the inference is correct.

In statistical problems we should look for a population, a sample drawn from that population, an inference based upon the sample, and some measure of reliability for the inference.

Exercises (1.1–1.11)

■■■■■■■■

Learning the Language

1.1 Define the following terms:
a. Descriptive statistics   e. Inferential statistics
b. Population   f. Statistics
c. Inference   g. Measure of reliability
d. Sample

Applying the Concepts

1.2 Suppose that to evaluate the current status of the dental health of school children, the American Dental Association conducted a survey to estimate the average number of cavities per child in grade school in the United States. One thousand school children from across the country were selected and the number of cavities for each child was recorded.
a. Describe the population of interest to the American Dental Association.
b. Describe the sample.

1.3 Suppose you work for a major public opinion pollster and you want to estimate the proportion of adult citizens who think the president is doing a good job in handling the nation's economy. Clearly define the population you want to sample.

1.4 A first-year chemistry student conducts an experiment to determine the amount of hydrochloric acid necessary to neutralize 2 milliliters of a basic solution. The student prepares five 2-milliliter portions of the solution and adds enough hydrochloric acid to each portion to neutralize the solution. The amount of acid necessary to achieve neutrality of the solution is recorded for each of the five portions.
a. Describe the population of measurements of interest to the student.
b. Describe the sample.

1.5   A manufacturer of vacuum cleaners has decided that an assembly line is operating
satisfactorily if less than 2% of the vacuum cleaners produced per day are defective.
If 2% or more of the vacuum cleaners are defective, the assembly line must be shut
down and adjustments made. Checking every vacuum cleaner as it comes off the
line would be costly and time-consuming, so the manufacturer decides to choose
thirty vacuum cleaners at random from a specific day's production and test them for
defects.
   a.   Describe the population of interest to the manufacturer.
   b.   Describe the sample.
   c.   Give an example of an inference that might be made.

1.6   A drug company advertises a new drug that is reported to be effective in treating a
particular disease, but also produces undesirable side effects in certain patients. To
determine the proportion of treated patients who develop the undesirable side effects,
research physicians administer the drug to forty-five carefully selected volunteer patients and monitor their responses.
   a.   Identify the population of interest to the research physicians.
   b.   Identify the sample.
   c.   Give an example of an inference the research physicians might make as a result
        of this experiment.

1.7   An insurance company would like to determine the proportion of all medical doctors
who have been involved in one or more malpractice suits. The company selects 500
doctors at random and determines how many in the sample were ever involved in
a malpractice suit.
   a.   Describe the population of interest to the insurance company.
   b.   Describe the sample.
   c.   Give an example of an inference the insurance company might make.

1.8   A new teaching method has been used with twenty-five students in the third grade
at a local elementary school. At the end of the school year, the twenty-five students
will take a standardized test to measure their progress. If these students have an
average score of 80 or more on the standardized test, the teacher will conclude that
the new method should be used with all third-grade students at this school.
   a.   Describe the population of interest.
   b.   Describe the sample.
   c.   Give an example of an inference that might be made.
   d.   What should accompany any inference that might be made?

1.9   A manufacturer of car batteries wants to revise the length of the battery guarantee
so that only 5% of the batteries sold must be-replaced while under warranty. Using
a sample of their records of battery sales and replacements, the manufacturer decides
to guarantee the batteries for 3 years.
   a.   Identify the population of interest to the manufacturer.
   b.   What inference was made about this population?
   c.   What information about the sample do you think would be useful in measuring
        the reliability of this inference?

1.10 Each month the Gallup Poll in Princeton, New Jersey, publishes results of its surveys in *The Gallup Report.* To help explain how the surveys are conducted and how they should be interpreted, the following is stated inside the cover:

The Sample: The sampling procedure of the Gallup Poll is designed to produce samples which are representative of the U.S. civilian adult population. National survey results are based on interviews with a minimum of 1,500 adults.

Sampling Tolerances: In interpreting the survey findings, it should be borne in mind that all sample surveys are subject to "sampling error," a statistical estimate of the extent to which the survey results might differ from results that would be obtained if the entire population in question were interviewed.

The size of the sampling error is mainly dependent upon the number of interviews in each population sub-group. For example, samples of 1,500 have a possible sampling error of plus or minus 3 percentage points 95 out of 100 times, and samples of 1,000 may have an error of 4 percent.

The number of interviews for each sub-sample is shown on the tables to assist readers in determining the appropriate sampling error.*

The July 1984 issue of *The Gallup Report* contained the results shown in the accompanying table. They are from a national survey of people who answered the question:

Would you favor or oppose a national law that would raise the legal drinking age in all states to 21?

Favor	Oppose	No Opinion	Number of Interviews
79%	18%	3%	1,522

Source: "National Drinking Age of 21" survey. *The Gallup Report,* No. 226, July 1984, p. 3.

a. What was the population of interest?
b. How large a sample was taken?
c. What percentage of the population would you infer favors raising the legal drinking age?
d. According to *The Gallup Report,* approximately how large is the "sampling error" associated with this inference? (In Chapter 7 we will discuss how this value was found.)
e. Using the terminology of this chapter, what does the "sampling error" provide?

1.11 The July 1984 issue of *The Gallup Report* described the outcome of another national survey. The question of interest this time was:

Suppose the presidential election were being held today. If President Reagan were the Republican candidate and Walter Mondale were the Democratic candidate, which would you like to see win?

* From "Note to Reader," *The Gallup Report,* No. 226, July 1984.

The responses to this question are summarized in the table.

Reagan	Mondale	Other	Undecided	Number of Interviews
54%	37%	2%	7%	1,190

Source: "Reagan vs. Mondale Trial Heat" survey. *The Gallup Report*, No. 226, July 1984, p. 14.

a. What was the population of interest?
b. How large a sample was taken?
c. What percentage of the population would you infer would like to see Reagan win? (Note that this survey was taken in July 1984.)
d. Approximately how large is the "sampling error" associated with this inference?
e. Compare your answer in part *d* of this exercise to that of part *d* of Exercise 1.10. Why do you think the sampling error is larger in this survey?

CHAPTER 1 QUIZ

1. Identify the two broad areas of the science of statistics.
2. What is a statistical inference?
3. Define the terms *population* and *sample*.
4. How will learning statistics help you cope with today's world?
5. A veterinarian was interested in the average number of times each day that the clinic personnel would be required to make a "house call," that is, travel away from the clinic to treat an animal. For a total of 14 days, the veterinarian recorded the number of house calls made each day. The average number of daily visits was 3.1.
   a. Identify the population of interest to the veterinarian.
   b. Identify the sample.
   c. Give an example of an inference the veterinarian might make.
   d. What should accompany any inference the veterinarian makes?

On Your Own

Get a copy of your local daily newspaper and look for articles that contain numerical data. The data might be a summary of the results of a public opinion poll, the results of a vote by the United States Senate, crime rates, birth or death rates, an election result, etc. For each article containing data that you find, answer the following questions:

a. Do the data constitute a sample or an entire population? If a sample has been taken, clearly identify both the sample and the population; otherwise, identify the population.

b. If a sample has been observed, does the article present an explicit (or implied) inference about the population of interest? If so, state the inference made in the article.

c. If an inference has been made, has a measure of reliability been included? What is it?

Reference

Meier, P. "The Biggest Public Health Experiment Ever: The 1954 Field Trial of the Salk Poliomyelitis Vaccine." In Tanur, J. M., Mosteller, F., Kruskal, W. H., Link, R. F., Pieters, R. S., & Rising, G. R. (eds.), *Statistics: A Guide to the Unknown*, 2nd ed. E. L. Lehmann, special editor. San Francisco: Holden-Day, 1978.

USING GRAPHS TO DESCRIBE DATA

PUBLIC OPINION POLL

Introduction

The graphs in Figure 2.1 (page 18) summarize some opinions of married couples about questions related to work. Newspapers, magazines, and television stations often use graphs similar to these to describe data. How are such graphs constructed? What information do they provide? How can they be distorted to mislead the viewer? In this chapter we will answer these questions.

Graphs are intuitively appealing descriptive devices that may be used to describe samples or populations. This is important because being able to describe data is the first step in learning the process of making inferences. In this chapter we will first characterize data as being one of two types. Since different kinds of graphs are appropriate depending on their purpose and the type of data, several types of graphs will be introduced. Guidelines will be provided for constructing the graphs, and pitfalls that distort information will be noted.

AT HOME AND AT WORK

a. Question: Would you be willing to leave your job and relocate if your spouse was offered a better position?

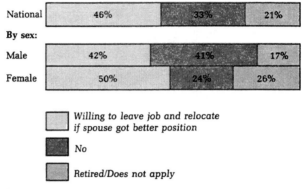

Source: Survey by Mark Clements Research, Inc. for *Parade* magazine, November 21–December 12, 1983.

b. Question: In families where both husband and wife work full time, whose job demands should take priority?

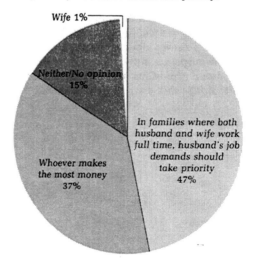

Note: Sample size = 1,207.
Source: Survey by Audits and Surveys for the Merit Report, May 10–15, 1983.

Figure 2.1
Graphs summarizing married couples' opinions about jobs
Source: Adapted from *Public Opinion*, Vol. 7, No. 4, August/September 1984, p. 35.

2.1

TWO TYPES OF DATA

Remember that data are observations or measurements of a phenomenon of interest. Data can generally be classified as one of two types: **qualitative** or **quantitative.**

Definition 2.1

Qualitative data are observations that can be classified into nonnumerical categories. An individual observation belongs to exactly one category.

Examples of qualitative data are:

1. The major area of study for each of the seniors at the University of Hawaii
2. The political affiliation of each person in a group of registered voters
3. The favorite color of each person in a group of people

An individual observation in each of these examples can be classified into one and only one category according to an attribute or quality.

Definition 2.2

Quantitative data are observations or measurements that are numerical.

Examples of quantitative data are:

1. The IQ score of each student in a group of college freshmen
2. The number of patients a crisis center will serve daily during January
3. How long it will take each student to finish the first exam in this course

An individual observation in each of these examples is a number that measures some quantity and thus results in quantitative data.

As you would expect, the methods used to summarize and analyze data depend on whether the data are qualitative or quantitative. Thus, it is important that you be able to identify these types of data.

Exercises (2.1–2.5)

Learning the Language

2.1 Define the following terms:
a. Qualitative data   b. Quantitative data

Applying the Concepts

2.2   State whether the following types of data are qualitative or quantitative:

  a. The amount of personal life insurance held by each professor at your university
  b. The flavor of ice cream each of 200 shoppers buys at a supermarket
  c. The marital status of each person living on a city block
  d. The number of years of experience of each salesperson at a car dealership
  e. The price of each new house listed for sale in the classified section of a newspaper
  f. The length of time it takes each of fifty migraine headache sufferers to obtain relief from pain
  g. The eye color of each student in your class
  h. The health insurance plan held by each patient in a hospital
  i. The monthly long-distance telephone bill for each of twenty-seven small businesses
  j. The greatest fear of each of fifty physicians

2.3   State whether the following types of data are qualitative or quantitative:

  a. The hair color of each student in your class
  b. The length of time each of thirty patients must stay in a hospital
  c. The sex of each United States senator
  d. The religious affiliation of each of a psychologist's patients
  e. The drugs used to treat a person with arthritis
  f. The length of each bolt produced daily by a machine
  g. Seventy responses to the question: Do you think life begins at conception?
  h. The amount of soft drink actually dispensed into each of twenty-four 12-ounce cans
  i. The number of questions each student answers correctly on a quiz
  j. The amount of money stolen in each of twenty-eight robberies

2.4   The 1985 *World Almanac & Book of Facts* provides information about major U.S. dams and reservoirs.* State whether the type of data associated with each of the following characteristics is qualitative or quantitative:

  a. The name of the dam
  b. The state in which the dam is located
  c. The river on which the dam is located
  d. The height of the dam (in feet)
  e. The length of the dam (in feet)
  f. The total volume (in cubic yards) of the dam
  g. The purpose of the dam (e.g., irrigation, flood control, hydroelectricity, etc.)
  h. The year in which the structure was completed

2.5   The 1985 *World Almanac & Book of Facts* also provides information about the nations of the world.† Consider the sets of data that contain information about every nation for each category in the following list. Classify each set of data as being either qualitative or quantitative.

---

* *The World Almanac & Book of Facts 1985* (New York: Newspaper Enterprise Association, Inc., 1985), p. 178.
† Ibid., pp. 515–600.

a. The size of the population
b. The languages spoken
c. The geographic area in square miles
d. The type of government
e. The chief crops
f. The number of televisions in use
g. The literacy rate

2.2

BAR GRAPHS AND PIE CHARTS

Both qualitative and quantitative data can be presented in graphical form to provide quick, visual summaries of information about the phenomenon of interest. Two of the most commonly used graphical methods are **bar graphs** and **pie charts.** These methods are usually used with qualitative data, which naturally fall into categories. They may also be used with quantitative data when there are relatively few different values that the numbers in the data set may assume. (Each value may then serve as a category to group the data.) The following two definitions will be useful when constructing a bar graph or pie chart to summarize a sample:*

Definition 2.3

The **frequency** of a category is the total number of observations in a sample that fall in that category.

Definition 2.4

The **relative frequency** of a category is the frequency of that category divided by the total number of observations. That is,

$$\text{Relative frequency} = \frac{\text{Frequency}}{n}$$

where n is the total number of observations (n is often called the **sample size**).

---

* Since we will be primarily interested in describing samples, we will state definitions in terms of samples and describe how to construct graphs for samples. If we wanted to construct graphs for populations, the same methods would be used.

EXAMPLE 2.1   A city is considering lowering property taxes even though this might reduce some
police and fire department services. A city commissioner, wondering how residents of
the city view this proposal, samples fifty residents. Each resident is interviewed and
classified into one of three categories:

F: Favor lower property taxes

O: Oppose lower property taxes

N: No opinion

The results of the survey are shown in Table 2.1. We want to determine the frequency
and relative frequency for each of these categories.

Table 2.1

Opinions of lower
property taxes

Resident	Category	Resident	Category	Resident	Category
1	O	18	F	35	N
2	F	19	F	36	O
3	F	20	F	37	F
4	N	21	O	38	N
5	N	22	F	39	O
6	F	23	F	40	F
7	O	24	N	41	F
8	F	25	N	42	N
9	F	26	O	43	O
10	N	27	O	44	O
11	O	28	F	45	F
12	F	29	N	46	O
13	O	30	O	47	F
14	F	31	F	48	F
15	O	32	F	49	O
16	F	33	N	50	N
17	N	34	O		

Solution   Counting the number of observations in each category in Table 2.1 provides the
frequencies shown in Table 2.2. The sum of the frequencies for the three categories
is equal to 50, the total number of observations in the sample. The sum of the
category frequencies for any sample will always equal n, the sample size.

Table 2.2

Frequencies and relative
frequencies for tax
opinion data

Category	Frequency	Relative Frequency
F	22	$22/50 = .44$
O	16	$16/50 = .32$
N	12	$12/50 = .24$
Totals	50	$50/50$ 1.00

The relative frequency for each category in Table 2.2 was found by dividing each
frequency by 50, the total number of observations. The sum of the relative frequencies
will always equal 1 (except for possible rounding errors). ∎

A **bar graph,** similar to the graph in Figure 2.1**a**, can be constructed from frequencies or relative frequencies using the steps listed in the box.

How to Construct a Bar Graph

1. List categories for the data set of interest. If the categories are quantitative, order them from the smallest value to the largest value.
2. Compute the frequency (or relative frequency) of the observations in each category.
3. Label each category listed in step 1 on the horizontal axis of the graph. Use the vertical axis for the frequency (or relative frequency).
4. Plot each frequency (or relative frequency) as a bar (rectangle) over the corresponding category.

The height and width of a bar graph should be approximately equal. This is for aesthetic reasons and to ensure that the visual representation of the data is not distorted. If either the height or the width of a graph is exaggerated, it may provide a false impression. We will discuss graphical distortions in more detail in Section 2.5.

EXAMPLE 2.2   Refer to Example 2.1. Draw a bar graph for the tax opinion data of Example 2.1.

Solution   The categories, frequencies, and relative frequencies for the tax opinion data are shown in Table 2.2. Thus, we have already completed the first two steps required to construct a bar graph. Both a frequency bar graph and a relative frequency bar graph for the tax opinion data are shown in Figure 2.2. The two graphs are identical in appearance and provide essentially the same information. The choice of whether to use frequencies or relative frequencies on the vertical scale is a matter of personal preference.

Figure 2.2
Bar graphs for the tax opinion example

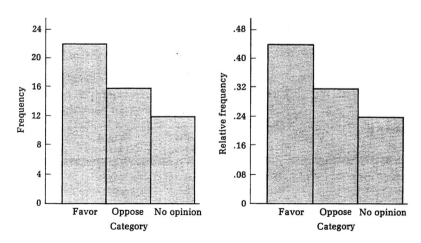

Both bar graphs in Figure 2.2 provide a useful summary of the tax opinion data. For example, the most popular response (22 out of 50, or .44) favored lower property taxes, while a substantial number (12 out of 50, or .24) had no opinion on the issue. ■

Another method commonly used to summarize qualitative data is the **pie chart.** Steps for constructing a pie chart are listed in the box.

How to Construct a Pie Chart

1. List the categories for the data set of interest.
2. Calculate the relative frequency for each category.
3. Construct a circle (pie) and allocate one portion (slice) of the pie to each category. The size of a slice constructed for a category is proportional to the fraction of observations in that category. Thus, the central angle of the corresponding slice should be the relative frequency for the category (from step 2) times 360°.
4. Compute the percentage of observations in each category, and label each section of the pie by the appropriate category and percentage. The percentage is calculated from the relative frequency:

$$\text{Percentage} = 100 \times (\text{Relative frequency})$$

For example, consider a category with a relative frequency of .20. The corresponding slice of the pie should have a central angle equal to .20 × 360° = 72°. This slice of the pie would be labeled as containing 100 × (.20) = 20% of the observations. Precise pie charts require the use of a protractor to measure the central angles, but for quick visualizations of the frequencies of categories we can simply approximate the size of each slice.

EXAMPLE 2.3   Construct a pie chart for the tax opinion data discussed in Examples 2.1 and 2.2.

Solution   We must compute the size of the central angle and the percentage of observations corresponding to each category given in Table 2.2. This is done in Table 2.3. Using this information, we draw a pie chart, as shown in Figure 2.3.

Table 2.3

Calculations for tax opinion data

Category	Relative Frequency	Size of Central Angle	Percentage
Favor	.44	.44 × 360.0° = 158.4°	100 × .44 = 44%
Oppose	.32	.32 × 360.0° = 115.2°	100 × .32 = 32%
No opinion	.24	.24 × 360.0° = 86.4°	100 × .24 = 24%
Totals	1.00	360.0°	100%

Figure 2.3
Pie chart for tax
opinion data

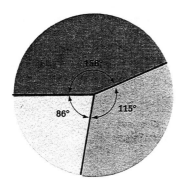

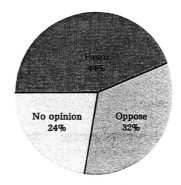

a. Central angles   **b.** Categories and percentages

The pie chart provides essentially the same information as the bar graph. We can see that nearly half the residents favor lower taxes and about one-fourth of the residents have no opinion. ■

In the next example, we will construct a bar graph for a quantitative data set. The purpose again will be to provide a visual summary of the data.

EXAMPLE 2.4   A rehabilitation officer who works with juvenile offenders is concerned about the number of youngsters with multiple convictions. Forty-two juvenile offenders are questioned to determine how many convictions each had prior to the current offense. The results of this survey are shown in Table 2.4.

Table 2.4

Juvenile Offender	Number of Previous Convictions	Juvenile Offender	Number of Previous Convictions	Juvenile Offender	Number of Previous Convictions
1	0	15	1	29	1
2	1	16	2	30	1
3	1	17	0	31	3
4	3	18	1	32	0
5	0	19	2	33	0
6	1	20	1	34	1
7	2	21	0	35	2
8	1	22	1	36	1
9	4	23	3	37	1
10	1	24	1	38	1
11	0	25	0	39	2
12	1	26	4	40	0
13	1	27	2	41	4
14	0	28	1	42	3

Record of juveniles'
previous convictions

a. Draw a bar graph for the data using frequencies.
b. Interpret the information the graph provides.

Solution **a.** We must first list the different values that appear in the data set. Since the data are quantitative, we list the values from the smallest to the largest and record the frequency of each. This is done in Table 2.5. From the information in Table 2.5, we can construct the bar graph shown in Figure 2.4, following the guidelines given earlier.

Table 2.5

Frequency table for juvenile offender data

Number of Previous Convictions	Frequency
0	10
1	19
2	6
3	4
4	3
Total	42

Figure 2.4
Bar graph for the juvenile offender data

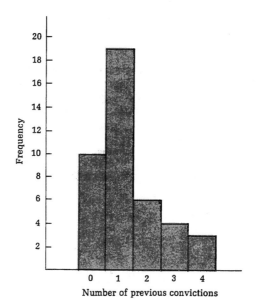

b. The message provided by the graph in Figure 2.4 is quite clear. Many of the juveniles have had one prior conviction and, although the next most frequent category is none at all, an alarming number of juveniles had been convicted more than once before. The graph clearly demonstrates that the rehabilitation officer is justified in being concerned about how many youngsters have multiple convictions.

Keep the following rules in mind when constructing bar graphs or pie charts:

Helpful Hints

1. The sum of the frequencies must always equal n, the sample size.
2. The sum of the relative frequencies must always equal 1, except for possible rounding errors.
3. The sum of the percentages must always equal 100%, except for possible rounding errors.

Bar graphs and pie charts are most appropriate when the data of interest fall naturally into relatively few categories. These graphs provide an excellent way to present such a set of data to a group of people. In Section 2.3 we will discuss how to summarize data quickly for one's personal use. When a quantitative data set contains many different numerical values, a bar graph is inappropriate since there would be too many categories. We will discuss how to construct a graph for such data in Section 2.4.

Exercises (2.6–2.19)

Learning the Language

2.6   Define the following terms:
a. Frequency   b. Relative frequency

2.7   What is denoted by *n*?

Learning the Mechanics

2.8   a. Complete the following table:

Category	Frequency	Relative Frequency	Percentage
Red	5		
Green	20		
Yellow	50		
Blue	25		
Totals			

b. Draw a bar graph for the data using the frequencies.
c. Draw a pie chart for the data.

2.9 For a sample containing $n = 200$ observations, the following information is given:

Category	Frequency	Relative Frequency	Percentage
A	20		
B	40		
C	20		
D	40		
E	80	_____	_____
Totals	200		

a. Complete the table.
b. Draw a bar graph for the data using the relative frequencies.
c. Draw a pie chart for the data.

2.10 The following is a sample of thirty numbers:

1	4	6	6	1	2	6	1	4	1
4	6	4	2	5	3	6	2	1	6
5	1	2	2	6	2	3	6	4	6

a. Find the frequency for each number in this data set.
b. Use the results of part a to draw a bar graph for the data.

2.11 Consider the following sample of forty letters:

D	D	S	D	T	S	T	D	S	D
D	D	T	D	D	T	D	S	T	S
S	R	S	D	R	R	D	D	S	D
T	S	D	D	T	S	S	D	S	S

a. List each letter (category) that appears.
b. Find the frequency for each category.
c. Find the relative frequency for each category.
d. Find the percentage of observations in each category.
e. Draw a bar graph for the data using the relative frequencies.
f. Draw a pie chart for the data.

2.12 Consider the following sample of fifty numbers:

3	0	3	1	2	4	2	1	1	4
3	2	1	1	1	2	4	4	5	4
0	1	2	2	1	4	2	0	5	1
0	2	1	1	0	3	1	4	3	4
1	1	4	1	2	2	4	4	4	3

a. Find the frequency of each number in this sample.
b. Draw a bar graph using the frequencies found in part a.

Applying the Concepts

2.13 The directors of a hospital are interested in the types of medical insurance their patients have. The directors have defined four categories of interest:

M: Medicare

G: Group plans

I: Individual, private policies

N: No coverage

A survey of 627 patients produced the following results:

M	G	I	N	Total
252	288	37	50	627

a. Construct a bar graph for the data using frequencies.

b. Construct a pie chart.

c. Interpret the information provided by the graphs.

2.14 The following twenty-five scores are the quiz grades for an introductory statistics class:

8	9	9	9	10
10	10	5	10	10
10	9	10	9	9
9	8	10	10	8
10	10	10	10	10

a. Construct a bar graph for the data using frequencies.

b. How many of the students scored 9 or 10?

c. If the highest possible score on the quiz is 10, did the class tend to do well or poorly on the quiz? Explain.

2.15 After a campaign encouraging car pools, a university monitored traffic arriving on campus one morning. The number of occupants in each of 723 passenger cars was recorded, and the results are shown in the table.

Number of Occupants	1	2	3	4	5	6
Frequency	421	130	119	42	8	3

a. Construct a bar graph using this information.

b. Does the bar graph provide sufficient information to determine whether the campaign to encourage car pools was successful? Is there other information you might need in order to assess the campaign's effectiveness? Explain.

2.16 Refer to Figure 2.1 on page 18.

a. Interpret the graphs that are drawn below the question, "Would you be willing to leave your job and relocate if your spouse was offered a better position?"

b. Interpret the graph that is drawn below the question, "In families where both husband and wife work full time, whose job demands should take priority?"

c. The graphs shown in Figure 2.1 summarized opinions from surveys taken in 1983. If a similar study were conducted today, in what way, if any, do you think the graphs would change?

2.17 The U.S. Department of Commerce periodically asks Americans to evaluate their housing units. The data shown in the table indicate owners' opinions about their structures in 1977.

Evaluation	Number (thousands)	Percentage
Excellent	20,515	42.3
Good	22,790	47.0
Fair	4,774	9.8
Poor	460	0.9
Totals	48,539	100.0

Source: From Table 3/1, "Evaluation of Housing Units by Location and Tenure: 1974 and 1977," in Center for Demographic Studies, U.S. Bureau of the Census, Social Indicators III (Washington, D.C.: Government Printing Office, 1980), p. 140.

a. Draw a bar graph or pie chart for the data.
b. What percentage of owners considered their houses to be fair or poor?
c. Interpret the graph drawn in part a.

2.18 Refer to Exercise 2.17. The data in the accompanying table indicate how renters evaluated their housing in 1977.

Evaluation	Number (thousands)	Percentage
Excellent	5,149	19.5
Good	12,576	47.7
Fair	6,898	26.2
Poor	1,721	6.5
Totals	26,344	99.9

Source: From Table 3/1, "Evaluation of Housing Units by Location and Tenure: 1974 and 1977," in Center for Demographic Studies, U.S. Bureau of the Census, Social Indicators III (Washington, D.C.: Government Printing Office, 1980), p. 140.

a. Summarize the data using a figure like that drawn in part a of Exercise 2.17.
b. What percentage of renters considered their dwellings to be fair or poor?
c. Compare the two figures drawn for owners and renters.

2.19 Many American adults participate in basic and secondary education. The accompanying pie chart summarizes the percentages of racial/ethnic groups involved in such studies in the mid 1970's.

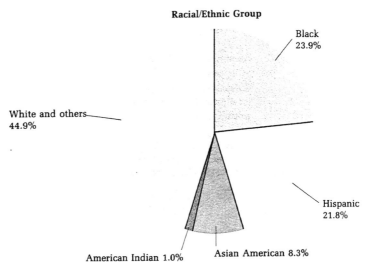

Racial/Ethnic Group

Black 23.9%

White and others 44.9%

Hispanic 21.8%

American Indian 1.0%   Asian American 8.3%

Source: Adapted from Chart 6/23, "Participation in Basic and Secondary Adult Education, by Age, Race, and Ethnicity: 1972–1976," in Center for Demographic Studies, U.S. Bureau of the Census, *Social Indicators III* (Washington, D.C.: Government Printing Office, 1980), p. 283.

a. What group had the largest percentage of adults participating in basic or secondary education?

b. What percentage of participants were not Hispanic?

2.3

EXPLORATORY DATA ANALYSIS: STEM AND LEAF PLOTS

The ideas we will introduce in this section found popularity in 1977 when *Exploratory Data Analysis*, by John W. Tukey, was published. We will discuss only the most basic types of exploratory data analysis, concentrating on simple, easy-to-draw graphs. Such graphs are often the first step in an analysis. They help the analyst to make a preliminary summary of the data and can lead to more formal methods of data description or analysis.

The first type of data summary we will discuss is a stem and leaf plot. We will illustrate the construction of a stem and leaf plot in Example 2.5.

EXAMPLE 2.5   The accompanying data set contains twenty-three midterm exam scores in an introductory statistics course. Construct a stem and leaf plot for the data.

83	78	61	100	91	55
34	59	58	72	68	71
86	86	83	85	79	72
78	88	94	97	96	

Solution   We will designate the last digit of each test score as its **leaf,** and the first digit, or digits,
as its **stem.** The stems and leaves of the first five test scores are shown in Table 2.6.
Notice that the stem indicates whether a test score is in the 60's, 70's, 80's, etc.—the
most important information about that number. The leaf provides the finer detail of
whether a score in the 60's is, for example, a 61 or a 68.

Table 2.6

Stems and leaves for five
test scores

Test Score	Stem	Leaf
83	8	3
78	7	8
61	6	1
100	10	0
91	9	1

The first step in forming a stem and leaf plot is to list all possible stem values in
a column. We begin with the smallest value (3, corresponding to the test score 34)
and end with the largest (10, corresponding to the test score 100). The stem and leaf
plot with only the stem completed is shown in Figure 2.5.

Figure 2.5
The stem for a stem and
leaf plot

```
STEM | LEAF
 3   |
 4   |
 5   |
 6   |
 7   |
 8   |
 9   |
10   |
```

Next, the leaf of each test score is placed in the row corresponding to that number's
stem. After the leaf of each test score has been placed by the appropriate stem, the
stem and leaf plot is complete. This display is shown in Figure 2.6 (page 33).

Figure 2.6
The completed stem and
leaf plot

STEM	LEAF
3	4
4	
5	5 9 8
6	1 8
7	8 2 1 9 2 8
8	3 6 6 3 5 8
9	1 4 7 6
10	0

■

A natural analogy can be made between a stem and leaf plot and a plant. The stem of a number (or a plant) is the main portion of the number (plant), and the leaf is an outgrowth of the stem.

Notice that if you turn the stem and leaf plot on its side, you obtain a figure that is similar to a bar graph. One advantage of the stem and leaf plot is that it is easy to construct. It is not as attractive as a bar graph but can be made more quickly.

Stem and leaf plots have other nice features besides being easy to draw. For example, the values in the original data set can be retrieved from the plot. There is no loss of information. Also, a stem and leaf plot arranges the data in an orderly manner and makes it easy to determine certain characteristics, such as the smallest number, the largest number, or what the middle number in the data set is.

The steps shown in the box will be useful when you are constructing stem and leaf plots.

How to Construct a Stem and Leaf Plot

1. Decide which part of each number is the stem and which is the leaf. Usually the leaf is the last digit in the number and the rest is the stem.
2. List the stems in order in a column. Start with the smallest stem and continue with all possible values, ending with the largest.
3. For each number in the data set, place the leaf for that number in the row of the corresponding stem.

EXAMPLE 2.6   The accompanying data are the lengths of time (in years) that twenty employees have been working at a medical clinic. Construct a stem and leaf plot for the data.

6	12	5	16
10	11	15	14
22	19	5	31
12	8	17	2
17	27	21	13

Solution   Since these are one- and two-digit numbers, we will view the one-digit numbers as being preceded by a 0. For example, 6 is considered to be 06, etc. We can now envision all the numbers as having two digits. We will let the first digit be the stem and the second digit be the leaf. The only values for the stem are 0, 1, 2, and 3, and the completed stem and leaf plot is shown in Figure 2.7.

Figure 2.7
Stem and leaf plot for employment data

STEM	LEAF
0	6 5 5 8 2
1	2 6 0 1 5 4 9 2 7 7 3
2	2 7 1
3	1

     Stem and leaf plots are used to provide quick summaries of relatively small sets of quantitative data. Qualitative data sets may also be summarized quickly without drawing actual bar graphs. We will demonstrate this procedure with an example.

EXAMPLE 2.7   A sample of twenty-five voters is selected, and each voter is classified into one of three categories:

     D: Democrat

     R: Republican

     O: Other

The results of the sample are shown in Table 2.7. Construct a graph that quickly summarizes the data.

Table 2.7

Party affiliations of twenty-five voters

Voter	Affiliation	Voter	Affiliation
1	D	14	R
2	D	15	D
3	R	16	D
4	D	17	D
5	R	18	O
6	R	19	D
7	R	20	R
8	D	21	D
9	O	22	D
10	D	23	R
11	D	24	R
12	R	25	R
13	O		

Solution   We will use techniques that are similar to those used in drawing a stem and leaf plot. The final graph is shown in Figure 2.8. Since the data fall naturally into qualitative categories, we use these categories as "stems" when drawing the graph. Next to each stem, we record a count of how many voters fall into each category. This can be

Figure 2.8
Graph for voter affiliation
data

PARTY AFFILIATION	COUNT
D	1 2 3 4 5 6 7 8 9 0 1 2
R	1 2 3 4 5 6 7 8 9 0
O	1 2 3

done quickly by going through the data, recording a 1 for the first observation in a category, and increasing the count by one digit each time an additional observation fits that category.

Note that when the tenth Democrat (the nineteenth voter) is reached, a 0 is recorded. Similarly, a 1 is recorded for the eleventh Democrat (the twenty-first voter). This is done to ensure that the relative weight given to every count is the same. Otherwise, if double digits were used, the graph would appear distorted.

As before, if this graph is turned on its side, it resembles a bar graph. If the data were to be summarized and presented to a group of people, the actual bars would be drawn. For a preliminary analysis, however, the graph shown in Figure 2.8 provides a nice summary. ■

Since the primary purpose of the graphs presented in this section is to summarize data quickly for one's personal use, there is quite a bit of room for flexibility. You may expand upon the ideas we have presented when forming your own graphs. However, be sure that any graphs you develop provide an accurate summary of the data.

Throughout the remainder of the text we will, where appropriate, conclude sections with examples demonstrating the use of the computer. There are many options available if one wants to use a computer to help with statistical analyses. Most often, what is called a *statistical package* is used. A **statistical package** is essentially a set of computer programs that have been written to perform frequently used analyses of data. Generally, to use a statistical package, basic steps similar to the ones listed here are followed:

1. Obtain access to the computer to be used. That is, "turn the computer on."
2. Give the computer the appropriate commands to allow use of the desired statistical package.
3. Enter the data that are to be analyzed into the computer. This step is often referred to as the "input" of the data.
4. Give the commands that will instruct the statistical package to perform whatever statistical analyses are of interest. These commands will depend on the specific package you are using.
5. Get the results of the analyses printed in a usable form. This is called a "computer printout."
6. Let the computer know that the statistical package is no longer being used.
7. Turn off the computer, or "log off," as it is more often called.

In the examples in this text, we are going to present only the results of statistical analyses—namely, computer printouts. We will indicate what types of analyses can be performed and interpret the results. We will not discuss how to obtain the results or in general how to use a computer. We will use this approach because many different computers and statistical packages are available. It would be virtually impossible to cover all the possible ways in which analyses could be obtained.

The computer printouts in this book were obtained using the general-purpose statistical computing package Minitab. This particular package was originally developed to teach statistical analysis, and it is one of the easiest to use and most widely available statistical packages on the market today. For those of you who want to learn how to use Minitab to do statistical analyses, a supplement has been written to accompany this text.

COMPUTER EXAMPLE 2.1   In this example, we will examine a stem and leaf plot that was constructed by Minitab. The data in Table 2.8 give the percentages of alcohol in fifty-two wine coolers. Figure 2.9 displays a stem and leaf plot that was created by Minitab from these data.

Figure 2.9
Computer-generated stem and leaf plot for percentages of alcohol in wine coolers

```
Stem-and-leaf of %ALCOHOL   N  = 52
Leaf Unit = 0.10

     1      2 9
     1      3
     3      3 56
     5      4 23
     8      4 599
    20      5 122333333444
   (17)     5 55566677778889999
    15      6 000012244
     6      6 55569
     1      7 0
```

The stem of the graph is the second column of the computer printout. The numbers in the third and succeeding columns are the leaves. At the top of the printout we are informed that there are N = 52 numbers in this graph and that the leaf unit is 0.10.

Table 2.8

Percentages of alcohol in fifty-two wine coolers

Brand/Bottler or Vendor	% Alcohol	Brand/Bottler or Vendor	% Alcohol
Bartles and Jaymes WC		Cape Cod Cranberry Cooler	
E & J Gallo	5.2	Cape Cod Wine	3.6
California Cooler—Tropical		Citronet WC	
California Cooler	6.0	Heublein Wines	4.3
Calvin Cooler—Chablis		Coastal Valley Cooler	
Calvin Cooler Wines	6.5	Glunz Cellars	5.4
Calvin Cooler—Citrus		Creative Cooler—Wine Tea	
Calvin Cooler Wines	5.7	Creative Cooler	6.5
Calvin Cooler—Pina Pineapple		Creative WC—Margarita	
Calvin Cooler Wines	5.9	Creative Cooler	5.7

Table 2.8 *continued*

Brand/Bottler or Vendor	% Alcohol	Brand/Bottler or Vendor	% Alcohol
Dewey Stevens Premium Light WC		Royal Dutch Cooler—Banana Cream	
Creative Cooler	4.5	Marquis B. V., Holland	6.6
Diamond Citrus Twist Cooler		Royal Dutch Cooler—Strawberry Cream	
Diamond Island Cellars	6.0	Marquis B. V., Holland	6.4
Diamond Orange Squeeze Cooler		Seagram's WC—Citrus & White Wine	
Diamond Island Cellars	5.3	Joseph E. Seagram & Sons	4.2
Diamond Passion Punch Cooler		Seagram's WC—Golden	
Diamond Island Cellars	5.8	Joseph E. Seagram & Sons	5.2
Diamond Wild Berry Cooler		Seagram's WC—Peach Flavored	
Diamond Island Cellars	5.8	Joseph E. Seagram & Sons	5.7
Florida WC—Apple		Seagram's WC—Premium	
Florida Wine	6.1	Joseph E. Seagram & Sons	5.4
Florida WC—Cool White		Steidl's WC—Red	
Florida Wine	6.2	Steidl Wine	5.7
Florida WC—Orange		Steidl's WC—White	
Florida Wine	6.5	Steidl Wine	5.8
Florida WC—Peach		Sun Country Tropical Cooler	
Florida Wine	6.4	Sun Country Cellars	5.5
Florida WC—Strawberry		The Grape Vine WC—Rose	
Florida Wine	7.0	Monarch Wine	4.9
Franzia WC—White Zinfandel		The Grape Vine WC—White	
Franzia Winery	5.3	Upper Bay Wine Cellars	5.6
HI 5 Cooler		20/20 WC—Citrus	
Tom Pree Wine	5.1	20/20 Wine	5.3
Manischewitz WC—Berry; reported sample #1		20/20 WC—Lambrusco	
Manischewitz Wine	5.3	20/20 Wine	6.9
Manischewitz WC—Berry; reported sample #2		20/20 WC—Peach	
Manischewitz Wine	5.3	20/20 Wine	5.6
Manischewitz WC—Cream White		20/20 WC—Raspberry	
Manischewitz Wine	5.3	20/20 Wine	5.6
Manischewitz WC—Pina Coconetta;		20/20 WC—Strawberry	
reported sample #1		20/20 Wine	5.9
Manischewitz Wine	5.4	20/20 WC—Tropical	
Manischewitz WC—Pina Coconetta;		20/20 Wine	5.5
reported sample #2		Widmer Niagara WC	
Manischewitz Wine	4.9	Widmer's Wine Cellars	2.9
Manischewitz WC—Lemonade		Wild Irish Rose WC—Citrus Rose	
Manischewitz Wine	5.5	Richards Wine	5.9
Peach-A-Roo WC		Wild Irish Rose WC—Orange Rose	
Central Vineyards	5.9	Richards Wine	6.0
Quenchette French Cooler—Raspberry		Wild Irish Rose WC—Tropical Rose	
Les Grands Chais France	3.5	Richards Wine	6.0
Quinn's Cooler—White Wine and Citrus			
R. R. Quinn and Co.	6.2		

Source: "Calories and Alcohol in Wine and Beer Coolers." *Consumers' Research*, Vol. 70, No. 3, March 1987, pp. 30–32. Connecticut Agricultural Experiment Station, New Haven CT. Bulletin 840. Analysis of Wine and Beer Coolers. October 1986.

Thus, the first number given in the graph is 2.9 and the last is 7.0. Minitab also orders the numbers in each leaf from the smallest to the largest. In the first column of the printout, Minitab provides the cumulative number of measurements from either the smallest or the largest measurement, whichever is closer. Thus, we can observe that 4.3 is the fifth smallest measurement. Similarly, there are fifteen numbers larger than or equal to 6.0. Finally, for the stem that includes the number(s) in the middle of the data set, the number of leaves on that stem is indicated in parentheses in the first column. Thus, there are seventeen leaves on the stem that includes the twenty-sixth and twenty-seventh numbers (the middle) of the data set.   ∎

Exercises (2.20–2.27)

Learning the Mechanics

2.20   Consider the following sample data:

189	258	192	212
205	209	229	233
224	226	216	211
200	197	231	205
227	224	211	200

a. Using the first two digits of each number as its stem, list all possible stem values from the smallest to the largest.

b. Complete the stem and leaf plot by adding the leaf for each number to the appropriate stem.

2.21   Consider the following sample data:

9.2	12.7	7.9	8.8	8.2
10.6	6.4	11.5	9.1	10.3
8.5	9.0	8.4	7.7	7.0
11.3	5.0	12.2	10.9	5.6
7.1	10.1	6.3	5.8	8.4

a. Using the decimal point and any digits preceding it as a stem, form a stem and leaf plot for the data.

b. What is the value of the smallest number? The largest number?

c. What is the value of the fifth smallest number? The fifth largest?

2.22   Consider the following sample of qualitative data:

A	D	E	A	E	C	D
C	B	C	D	B	A	C
A	B	C	C	C	C	B
C	C	B	D	C	B	C

a. Form a quick graph to summarize the data.

b. What is the total number of A's and B's?

Applying the Concepts

2.23 At a small college, many undergraduate students attend school part-time while working toward their degrees. The following data set shows how many semesters a sample of thirty of these students were enrolled before graduating.

8	9	10	12	8	8
8	10	8	10	10	9
9	8	8	9	10	8
8	16	9	12	11	12
11	12	9	8	13	11

a. Construct a stem and leaf plot of the data using the digits 0 and 1 as stems.

b. Construct a quick graph of the data by using the numbers 8, 9, 10, 11, 12, 13, 14, 15, and 16 as stems. Record the count for each stem value to complete the graph.

c. Compare the two graphs drawn in parts a and b. Which provides a better description of the data? Why?

2.24 Twenty-five people, living in a neighborhood where unemployment is low, are polled and asked if they believe the president is doing a good job of handling the nation's economy. They respond in one of the following ways:

Y: Yes

- N: No

U: Unsure

The twenty-five responses are shown below:

Y	N	N	Y	U
Y	U	Y	Y	Y
N	Y	Y	Y	N
N	Y	U	Y	Y
Y	Y	Y	N	Y

a. Construct a quick graph to summarize the data.

b. How would you describe the opinions of the people polled?

c. If the same poll were conducted in a neighborhood where unemployment is high, do you think the results would be similar? Explain.

2.25 The accompanying data set shows the number of unemployed persons 16 years old and over for the years 1948–1979. (The numbers have been rounded and are presented in terms of millions of people unemployed per year.)

2.3	1.9	2.8	3.9	3.8	2.8	4.8	7.3
3.6	1.8	2.9	4.7	3.4	2.8	4.3	6.9
3.3	3.5	4.6	3.9	2.9	4.1	5.1	6.0
2.1	2.9	3.7	4.1	3.0	5.0	7.8	6.0

a. Construct a stem and leaf plot for the data.

b. What was the least number of people unemployed in any year?

2.26   The data in the accompanying table give the percentages of the total number of college or university student loans that are in default, by state (and the District of Columbia). Minitab was used to produce the accompanying stem and leaf plot of these data.

a.   Discuss this graph.

*b.   Use a computer package to produce a similar graph.

State	%	State	%	State	%
Ala.	12.0	Ky.	10.3	N. Dak.	4.8
Alaska	19.7	La.	13.5	Ohio	10.4
Ariz.	12.1	Maine	9.7	Okla.	11.2
Ark.	12.9	Md.	16.6	Oreg.	7.9
Calif.	11.4	Mass.	8.3	Pa.	8.7
Colo.	9.5	Mich.	11.4	R. I.	8.8
Conn.	8.8	Minn.	6.6	S. C.	14.1
Del.	10.9	Miss.	15.6	S. Dak.	5.5
D.C.	14.7	Mo.	8.8	Tenn.	12.3
Fla.	11.8	Mont.	6.4	Tex.	15.2
Ga.	14.8	Nebr.	4.9	Utah	6.0
Hawaii	12.8	Nev.	10.1	Vt.	8.3
Idaho	7.1	N.H.	7.9	Va.	14.4
Ill.	9.3	N.J.	12.0	Wash.	8.4
Ind.	6.7	N. Mex.	7.5	W. Va.	9.5
Iowa	6.2	N.Y.	11.3	Wis.	9.0
Kansas	5.7	N.C.	15.5	Wyo.	2.7

Source: National Direct Student Loan Program.

Graph for Exercise 2.26

```
Stem-and-leaf of DEFAULT    N  = 51
Leaf Unit = 0.10

      1      2 7
      1      3
      3      4 89
      5      5 57
     10      6 02467
     14      7 1599
     21      8 3347888
     (5)     9 03557
     25     10 1349
     21     11 23448
     16     12 001389
     10     13 5
      9     14 1478
      5     15 256
      2     16 6
      1     17
      1     18
      1     19 7
```

* Exercises, or parts of exercises, that require the use of a computer will be marked with an asterisk (*).

*2.27 According to the U.S. Department of Education, the national dropout rate for high school students fell by more than 1%, from 30.3% to 29.1%, between 1982 and 1984. The accompanying table shows the dropout rate, defined as the percentage of ninth graders who do not graduate, for each state (and the District of Columbia) in 1982 and 1984.

State	**1982**	**1984**	**State**	**1982**	**1984**	**State**	**1982**	**1984**
Ala.	36.6	37.9	Ky.	34.1	31.6	N.Dak.	16.1	13.7
Alaska	35.7	25.3	La.	38.5	43.3	Ohio	22.5	20.0
Ariz.	36.6	35.4	Maine	27.9	22.8	Okla.	29.2	26.9
Ark.	26.6	24.8	Md.	25.2	22.2	Oreg.	27.6	26.1
Calif.	39.9	36.8	Mass.	23.6	25.7	Pa.	24.0	22.8
Colo.	29.1	24.6	Mich.	28.4	27.8	R.I.	27.3	31.3
Conn.	29.4	20.9	Minn.	11.8	10.7	S.C.	37.4	35.5
Del.	25.3	28.9	Miss.	38.7	37.6	S.Dak.	17.3	14.5
D.C.	43.1	44.8	Mo.	25.8	23.8	Tenn.	32.2	29.5
Fla.	39.8	37.8	Mont.	21.3	17.9	Tex.	36.4	35.4
Ga.	35.0	36.9	Nebr.	18.1	13.7	Utah	25.0	21.3
Hawaii	25.1	26.8	Nev.	35.2	33.5	Vt.	20.4	16.9
Idaho	25.6	24.2	N.H.	23.0	24.8	Va.	26.2	25.3
Ill.	23.9	25.5	N.J.	23.5	22.3	Wash.	23.9	24.9
Ind.	28.3	23.0	N.Mex.	30.6	29.0	W.Va.	33.7	26.9
Iowa	15.9	14.0	N.Y.	36.6	37.8	Wis.	16.9	15.5
Kansas	19.3	18.3	N.C.	32.9	30.7	Wyo.	27.6	24.0

a. Use a computer package to generate separate stem and leaf plots for the dropout rates in 1982 and 1984.

b. Use the graphs created in part *a* to find the highest and lowest dropout rates in each of the two years. With which states are they associated? Do the states having the extreme rates change from 1982 to 1984?

c. Compare the two plots generated in part *a* and discuss any similarities or differences you find.

2.4

HISTOGRAMS

In the preceding sections we constructed graphs for data that fell naturally into specific categories. But quantitative data often assume many different values, so we cannot proceed directly to a bar graph (or pie chart). Before graphing such data, we must limit the number of categories by grouping the data into **intervals.** The resulting bar graph is called a **histogram.** Thus, a histogram is nothing more than a bar graph for grouped data.

How to Construct a Histogram

1. Choose categories (intervals) to satisfy the following conditions:
   a. All the categories are of equal width.
   b. Each measurement falls into one and only one category.
   c. The number of categories is reasonable, given the number of measurements in the sample. This number is arbitrary, but generally about ten categories will provide a useful histogram. You could use as few as five categories for a small number of measurements, or as many as fifteen for a large number of measurements.
2. Construct a bar graph for these categories.

EXAMPLE 2.8   Table 2.9 contains EPA mileage ratings for 100 experimental cars. Construct a histogram for the data.

Table 2.9

EPA mileage ratings of 100 cars

36.3	41.0	36.9	37.1	44.9	36.8	30.0	37.2	42.1	36.7
32.7	37.3	41.2	36.6	32.9	36.5	33.2	37.4	37.5	33.6
40.5	36.5	37.6	33.9	40.2	36.4	37.7	37.7	40.0	34.2
36.2	37.9	36.0	37.9	35.9	38.2	38.3	35.7	35.6	35.1
38.5	39.0	35.5	34.8	38.6	39.4	35.3	34.4	38.8	39.7
36.3	36.8	32.5	36.4	40.5	36.6	36.1	38.2	38.4	39.3
41.0	31.8	37.3	33.1	37.0	37.6	37.0	38.7	39.0	35.8
37.0	37.2	40.7	37.4	37.1	37.8	35.9	35.6	36.7	34.5
37.1	40.3	36.7	37.0	33.9	40.1	38.0	35.2	34.8	39.5
39.9	36.9	32.9	33.8	39.8	34.0	36.8	35.0	38.1	36.9

Solution   We must first decide what the categories should be. Remember that the number of categories is arbitrary, but we will generally choose around ten. If we look through the listing of mileage ratings, we find that the smallest value is 30.0 and that the numbers have one decimal place. Thus, to be sure to include all the measurements, and to ensure that each measurement falls into one and only one category, we will begin the first category at 29.95. This endpoint has one more decimal place than the original data.

To find the width of each category, we begin by finding the range of the data, that is, the largest value minus the smallest value. The largest mileage rating listed in Table 2.8 is 44.9, so the range is $44.9 - 30.0 = 14.9$. If we wanted twelve categories, each one would have to be $^{14.9}/_{12} = 1.24$ units wide. But we want a width with only one decimal place, because that will ensure that the endpoints of every category will have two decimal places (one more than the original data) and end in a 5 (since we are starting at 29.95). So, we round 1.24 up or down to 1.2 or 1.3. Similarly, if we wanted ten categories, we would divide the range by 10, obtaining 1.49, and then round up to a category width of 1.5.

Let us now arbitrarily decide to use ten categories. The first category begins at 29.95 and ends at $29.95 + 1.5 = 31.45$. The second category will begin at 31.45 and end at $31.45 + 1.5 = 32.95$. We continue in this manner until we reach the category that includes the largest measurement. Notice that no measurement can fall on a boundary, and each measurement will fall into one and only one category. These categories are shown in the first column of Table 2.10, and the number of measurements in each category—the frequency—is shown in the second column. We are now ready to construct the histogram shown in Figure 2.10. Notice that the upper and lower boundaries of the categories are marked on the horizontal axis.

Table 2.10

Categories and frequencies
for the car mileage data

Category	Frequency
29.95 to 31.45	1
31.45 to 32.95	5
32.95 to 34.45	9
34.45 to 35.95	14
35.95 to 37.45	33
37.45 to 38.95	18
38.95 to 40.45	12
40.45 to 41.95	6
41.95 to 43.45	1
43.45 to 44.95	1
Total	100

Figure 2.10
Histogram for car
mileage data

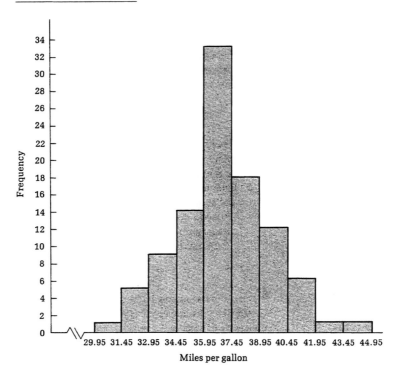

The histogram in Figure 2.10 provides a good summary of the car mileage data. We can see that thirty-three of the cars had a mileage rating between 35.95 and 37.45 and that this category has the greatest frequency. The categories tend to contain a smaller fraction of the measurements as the mileage ratings move away from (above or below) this category.

We can obtain other information from the table or the graph. By adding frequencies, we see that sixty-five of the 100 mileage ratings are in the three categories that lie between 34.45 and 38.95. Similarly, only two of the cars obtained a mileage rating over 41.95. Many other summary statements can be made by further study of the histogram. However, the principal use of histograms is to provide a *visual* representation of data. Numerical information is more easily extracted from frequency tables.

∎

Frequency histograms and bar graphs for quantitative data are often called **frequency distributions** because they show how the data are distributed in the categories that are marked off on the horizontal axis. Because the categories are ordered according to the size of the measurements, you can see at a glance the approximate proportion of the measurements that exceed or fall short of some value.

When constructing histograms, remember the following:

Helpful Hints

1. The approximate width of each category (interval) may be found by dividing the range (the largest value minus the smallest value) by the number of categories. That is,

$$\text{Width of category} = \frac{\text{Range}}{\text{Number of categories}}$$

Usually, ten categories will provide a useful graph. Round the width found from the above equation to the same number of decimal places as the data.

2. Each category should begin and end with numbers that end in a 5, and should have one more decimal place than the data being graphed. This will ensure that each measurement falls into one and only one category.

EXAMPLE 2.9   An article in the *Journal of the American Dental Association* (Miller et al., 1974) discusses mercury vapor contamination levels in the air of dental offices and the resulting danger to dentists and dental assistants. Such contamination might occur due to spills, open storage of scrap amalgam, or aerosols created when high-speed drills are

Figure 2.11
Histogram for mercury
vapor levels

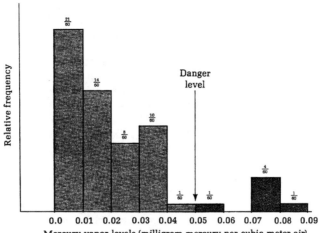

used to remove old amalgam fillings. The cumulative absorption of small quantities of mercury can result in serious medical problems. Therefore, the constant daily exposure of dentists and their assistants to possible mercury contamination is an important concern. A level of 0.05 milligram of mercury per cubic meter of air is the largest concentration considered safe for those working a 40-hour work week.

Mercury vapor levels were measured in sixty dental offices in San Antonio, Texas. A histogram summarizing the resulting information is given in Figure 2.11. This histogram clearly shows that an alarming fraction ($\frac{6}{60}$, or $\frac{1}{10}$) were above the danger level of 0.05 milligram. You can see that the data have been effectively summarized and clearly indicate a need for strict policing of mercury vapor levels in dental offices.

In this histogram the relative frequency of each category is given at the top of the bar for each category, and the value of the danger level has been labeled. Adding details such as these enhances the information a graph provides. ■

COMPUTER
EXAMPLE 2.2

Refer to the EPA mileage ratings for 100 experimental cars that were given in Table 2.9. If we use Minitab to generate a histogram of this data set, the graph shown in Figure 2.12 results.

Figure 2.12
Computer-generated
histogram for car mileage
data

```
Histogram of MILEAGES    N = 100

Midpoint    Count
       30       1    *
       32       5    *****
       34      12    ************
       36      31    *******************************
       38      31    *******************************
       40      15    ***************
       42       4    ****
       44       1    *
```

Notice that the categories formed by the Minitab program are identified by their midpoints rather than by their endpoints. Thus, the first category has a midpoint of 30, the second of 32, and so forth. The corresponding endpoints are obvious from the graph. Roughly speaking, the categories are from 29 to 31, 31 to 33, and so forth. Actually, adjacent categories cannot share the same endpoint, of course. Minitab considers the lower endpoint to be part of a category, whereas the upper endpoint is not a part of that category. For example, the first category includes all numbers greater than or equal to 29, but less than 31. (We could, therefore, represent the endpoints of the categories as 28.95 to 30.95, 30.95 to 32.95, etc.)

Minitab labels the number of mileages in each category as "count" rather than frequency. We find 1 mileage in the first category, 5 in the second, and so forth. Finally, asterisks "draw" the bars of the histogram. Since the bars run horizontally, this is called a **horizontal histogram,** whereas the histogram in Figure 2.10 is a **vertical histogram.**

Both histograms provide a similar summary of the data. They both indicate that most of the mileages are near the middle of the data set, so that the categories tend to contain a smaller fraction of the measurements as the mileage ratings become larger or smaller.   ■

Exercises (2.28–2.40)

Learning the Mechanics

2.28 The following table summarizes a sample of forty observations:

Category	Frequency
0.5 to 1.5	2
1.5 to 2.5	3
2.5 to 3.5	12
3.5 to 4.5	10
4.5 to 5.5	6
5.5 to 6.5	0
6.5 to 7.5	7
Total	40

Use this information to draw a histogram.

2.29 Here is a sample of twenty-four numbers:

8.4	5.3	5.6	7.8	7.0	5.6	9.1	8.8
3.7	3.1	6.5	4.4	5.5	5.4	6.9	8.2
8.2	1.2	8.6	9.4	8.5	2.8	5.7	3.1

a. Let the first category be 1.15 to 2.15. Specify the remaining categories.
b. Find the frequency of measurements for each category in part a.
c. Using the results of part b, draw a histogram for the data.
d. Could you allow the first category to be 1.0 to 2.0? Explain. [*Hint:* What category would 7.0 be in?]

2.30 Construct a histogram for the data summarized in the following table:

Category	Relative Frequency
0.5 to 2.5	.10
2.5 to 4.5	.15
4.5 to 6.5	.25
6.5 to 8.5	.20
8.5 to 10.5	.05
10.5 to 12.5	.10
12.5 to 14.5	.10
14.5 to 16.5	.05

2.31 Here is a sample of twenty measurements:

26	22	34	12	21	26	32	39	32	25
36	31	28	30	38	23	17	27	29	19

    a. Give the upper and lower boundaries for six categories, starting the lower boundary of the first category at 10.5 and using a width of 5. Find the relative frequency for each of the six categories.

    b. Construct a histogram using the results of part a.

2.32 Use the data below to construct a histogram. Use ten intervals.

5.9	5.3	1.6	7.4	9.8	1.7	8.6	1.2	2.1
4.0	6.5	7.2	7.3	8.4	8.9	6.7	9.2	2.8
4.5	6.3	7.6	9.7	9.4	8.8	3.5	1.1	4.3
3.3	3.1	1.3	8.4	1.6	8.2	6.5	4.1	3.1
1.1	5.0	9.4	6.4	7.7	2.7	0.2	9.6	3.4
2.0	3.3	8.9	5.1	5.5	9.4	1.6	6.2	3.8
8.3	1.4	8.3	7.3	6.8	3.9	7.3	8.1	4.4
4.8	9.4	7.9	4.8	6.2	7.7	6.6	6.5	1.2
3.9	1.1	8.3	9.4	9.1	7.7			

2.33 Construct a histogram for the following data:

23	39	12	99	82	88	12	24	67	30
52	12	24	19	17	53	15	50	60	18
32	16	49	40	37	51	55	61	81	35

Applying the Concepts

2.34 The graph at the top of page 48 summarizes the scores of 100 students on a questionnaire designed to measure aggressiveness. Scores are integer values ranging from 0 to 20. A high score indicates a high level of aggression.

    a. Which category contains the highest proportion of test scores?

    b. What proportion of the scores lie between 3.5 and 5.5?

    c. What proportion of the scores exceed 11.5?

    d. How many students scored less than 5.5?

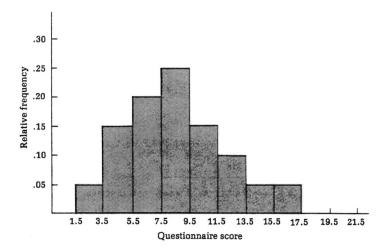

2.35   The number of years of experience (to the nearest half year) for each of the teachers in the Statistics Department of a college was recorded, and the results are listed below.

5.0	4.5	19.5	1.0	6.5
0.5	4.0	1.0	2.5	9.0
5.5	5.0	3.5	4.5	1.5
8.0	21.0	7.0	13.5	3.5

a. Using six categories, beginning with 0.45 and ending with 21.45, find the relative frequency for each category.

b. Using the answer to part a, construct a histogram for the data.

c. To qualify for certain benefits, a teacher must have over 13 years of experience. What proportion of the teachers in the Statistics Department are eligible?

2.36   The following is a sample of the prices (in cents) charged for regular gasoline at twenty-five self-serve or full-serve gas stations in the Cincinnati, Ohio, area during January 1985:

114.9	125.9	96.9	97.9	131.9
96.9	101.9	98.9	118.9	97.9
96.9	123.9	112.9	108.9	123.9
96.9	159.9	96.9	96.9	101.9
121.9	98.9	134.9	106.9	107.9

a. Construct a histogram for the data.

b. Suppose you sampled twenty-five gas stations from another area of the country—say, New York City or rural Kansas. Do you think a histogram for those twenty-five measurements would resemble the one you constructed in part a? Explain.

c. Many of the gasoline prices are near one dollar per gallon, but some are quite a bit higher. What explanations can you provide for these large differences in prices?

2.37 On January 26, 1985, *The Cincinnati Enquirer* reported temperatures for the coldest recorded months of January in Cincinnati. The official daily low temperatures for January 1940 and January 1977 are given here.*

JANUARY 1940: DAILY LOW TEMPERATURES

7	−1	32	17	4	−4	20
3	12	33	16	9	5	3
19	32	−3	21	6	17	11
28	−12	10	15	−1	23	21
3	−2	24				

JANUARY 1977: DAILY LOW TEMPERATURES

2	9	−10	−21	9	10	−5
2	−6	−1	−24	−5	−1	23
−5	−4	−25	−4	−11	26	17
24	−4	23	−11	14	0	10
11	23	−1				

a. Draw two histograms, one for each month's data.

b. Compare the two histograms constructed in part a.

2.38 The *Handbook of Airline Statistics* publishes data on such information as intercity distances of flights and number of passengers carried. The following data are intercity distances (in miles) for fifty of the most frequented U.S. flights in 1972.[†]

191	410	1,188	358	197
1,092	329	256	222	1,017
721	227	216	100	273
215	406	252	312	2,288
2,453	238	102	839	907
355	755	1,003	1,363	1,416
1,071	289	274	671	407
489	344	675	882	2,399
2,574	591	860	959	1,246
1,740	2,556	1,853	1,258	1,030

a. Construct a histogram for the data.

b. What fraction of the flights were more than 1,000 miles?

2.39 Refer to Computer Example 2.1, where a stem and leaf plot was generated to summarize the percentages of alcohol in 52 wine coolers. The histogram at the top of page 50 summarizes the same data.

* From "Our Coldest Januarys," *The Cincinnati Enquirer*, January 26, 1985, p. A-1.

† From Table 6A in Civil Aeronautics Board, *Handbook of Airline Statistics* (Washington, D.C.: Government Printing Office, 1973), p. 433.

```
Histogram of %ALCOHOL    N = 52

Midpoint    Count
    3.0       1   *
    3.5       2   **
    4.0       1   *
    4.5       2   **
    5.0       5   *****
    5.5      19   *******************
    6.0      14   **************
    6.5       6   ******
    7.0       2   **
```

a. Compare this histogram to the stem and leaf plot shown in Figure 2.9. Examine the visual images of both of these graphs. Do they provide essentially the same image?

b. Which, if either, graph do you prefer? Give reasons for your answer.

*2.40 Refer to the data in Exercise 2.27 on the high school dropout rates in 1982 and 1984. These data for each state and the District of Columbia are reproduced here.

State	1982	1984	State	1982	1984	State	1982	1984
Ala.	36.6	37.9	Ky.	34.1	31.6	N.Dak.	16.1	13.7
Alaska	35.7	25.3	La.	38.5	43.3	Ohio	22.5	20.0
Ariz.	36.6	35.4	Maine	27.9	22.8	Okla.	29.2	26.9
Ark.	26.6	24.8	Md.	25.2	22.2	Oreg.	27.6	26.1
Calif.	39.9	36.8	Mass.	23.6	25.7	Pa.	24.0	22.8
Colo.	29.1	24.6	Mich.	28.4	27.8	R.I.	27.3	31.3
Conn.	29.4	20.9	Minn.	11.8	10.7	S.C.	37.4	35.5
Del.	25.3	28.9	Miss.	38.7	37.6	S.Dak.	17.3	14.5
D.C.	43.1	44.8	Mo.	25.8	23.8	Tenn.	32.2	29.5
Fla.	39.8	37.8	Mont.	21.3	17.9	Tex.	36.4	35.4
Ga.	35.0	36.9	Nebr.	18.1	13.7	Utah	25.0	21.3
Hawaii	25.1	26.8	Nev.	35.2	33.5	Vt.	20.4	16.9
Idaho	25.6	24.2	N.H.	23.0	24.8	Va.	26.2	25.3
Ill.	23.9	25.5	N.J.	23.5	22.3	Wash.	23.9	24.9
Ind.	28.3	23.0	N.Mex.	30.6	29.0	W.Va.	33.7	26.9
Iowa	15.9	14.0	N.Y.	36.6	37.8	Wis.	16.9	15.5
Kansas	19.3	18.3	N.C.	32.9	30.7	Wyo.	27.6	24.0

a. Using a computer package, construct a relative frequency histogram for the dropout rates in each of the two years.

b. Can you perceive a shift in the distribution of dropout rates from 1982 to 1984? If so, in which direction is the shift?

c. Compare the relative frequency histograms to the stem and leaf displays in Exercise 2.27. Which do you prefer as a descriptive display of these data?

2.5

DISTORTING THE TRUTH WITH GRAPHS

Even though there is much truth in the adage "a picture is worth a thousand words," it is also true that pictures can be used to convey an exaggerated or distorted message to the viewer. So another adage applies: "Let the buyer (reader) beware." You should examine all graphical descriptions of data with extreme caution. The following example demonstrates a few pitfalls to watch for when analyzing any bar graph or histogram.

Suppose we want to analyze the frequency of visits by members of minority groups to a community mental health clinic. The clinic serves four minority groups, which we will call A, B, C, and D. A total of 300 visits by members of these minority groups is examined, and Table 2.11 summarizes the data.

We first present a bar graph in Figure 2.13, which gives a graphical description of the data. The vertical axis for relative frequencies is approximately equal in length to the horizontal axis, and the differences between the relative frequencies are honestly portrayed. We will now consider various ways that we can alter the graph to change our interpretation of the data.

Table 2.11

Data for minority use of community mental health clinic

Minority Group	Number of Visits (Frequency)	Relative Frequency
A	66	.22
B	96	.32
C	54	.18
D	84	.28
Totals	300	1.00

Figure 2.13
Minority use of community mental health clinic—proper display of information

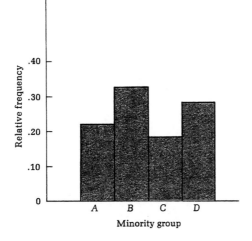

Figure 2.14
Minority use of community
mental health clinic—packed
vertical axis

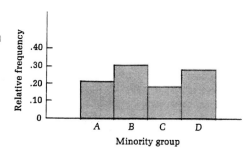

In Figure 2.14, the *differences* between the relative frequencies for the four minority groups appear smaller. This is because more units per inch have been "packed in" on the vertical axis. That is, the vertical axis has been shrunk, and you can see that this makes the differences between the relative frequencies appear moderate. This same effect can be produced by stretching the horizontal axis (increasing the width of each bar). Even rather large differences between relative frequencies can be made to appear moderate by simultaneously shrinking the vertical axis and stretching the horizontal axis.

With the same data, you can magnify the differences between the relative frequencies by using several tricks. In Figure 2.15 the vertical axis has been stretched. This increases the distance between units on the vertical axis (with only a few units per inch), which gives the appearance of enlarging the differences between the relative frequencies for the minority groups. Compare Figure 2.15 to the original graph in Figure 2.13.

Figure 2.15
Minority use of community
mental health clinic—
stretched vertical axis

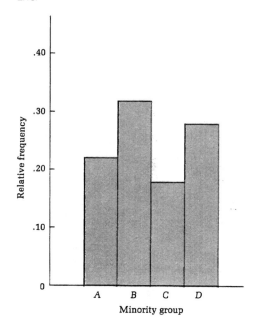

Figure 2.16
Minority use of community mental health clinic—vertical axis started at a point greater than 0

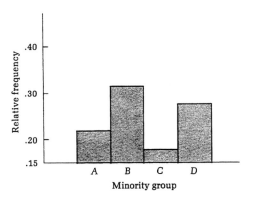

The telltale sign of a stretched vertical axis is that it is much longer than the horizontal axis, but this can be disguised by starting the vertical axis at some point above 0. This has been done in Figure 2.16, where the differences between the relative frequencies again seem to be quite large.

A particularly effective way to distort the differences in a bar graph is to make the width of each bar proportional to its height. For example, look at the bar graph in Figure 2.17, which again depicts the mental health clinic data. The minority groups are rearranged in ascending order by relative frequency. This is a common practice when using graphical techniques to distort the truth. Both the width and the height of the bars grow as the relative frequency grows. The eye tends to equate the area of each bar with the relative frequency, when in fact the true relative frequency is proportional only to the height of the bar.

You have probably seen graphs with figures, rather than bars, that increase (or decrease) in area. For example, to depict the decrease in oil consumption from 1981 to 1983, a graph might show barrels diminishing in size, as shown in Figure 2.18 (page 54). A graph showing an increase in U.S. health expenditures might use dollar signs as in Figure 2.19. Finally, to show an increase in the number of U.S. airline pas-

Figure 2.17
Minority use of community mental health clinic—width of bars grows with height

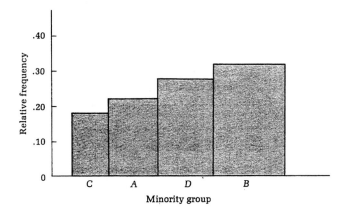

Figure 2.18
Misleading graph of oil
consumption

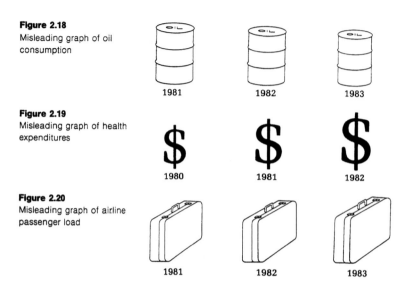

Figure 2.19
Misleading graph of health
expenditures

Figure 2.20
Misleading graph of airline
passenger load

sengers, a graph like the one shown in Figure 2.20 might be seen. Graphs such as these give an exaggerated image of the actual changes being reported.

We have presented only a few of the ways that bar graphs can be used to convey misleading pictures of qualitative data. However, these same types of distortions can be used with any graph, and the lesson is clear. Examine all graphical descriptions of data with care. Ignore apparent visual differences (or lack of differences) and concentrate on the actual numerical differences shown by the graph. In particular, the following should be remembered:

Aids for Detecting Graphical Distortions

1. Check both the vertical axis and the horizontal axis to be sure neither has been shrunk or stretched. The length of the vertical axis should be approximately equal to the length of the horizontal axis.
2. Check the vertical axis to determine whether it starts at 0.
3. Check the widths of the bars to determine whether they are equal.
4. Concentrate on actual numerical differences in the graph, not apparent visual differences.

Exercises (2.41–2.43)

Learning the Mechanics

2.41 Identify the methods used in graphs b and c to distort the interpretation of the data accurately portrayed in graph a.

a.

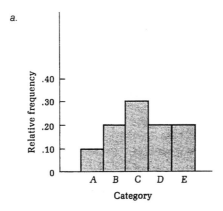

b.

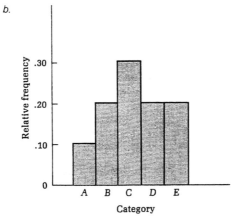

c.

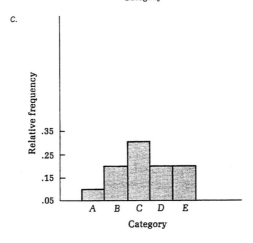

2.42 What is the main graphical distortion that would result from each of the following?
a. Stretched vertical axis
b. Stretched horizontal axis
c. Shrunk horizontal axis
d. Shrunk vertical axis
e. Vertical axis starting above 0

2.43 A study was conducted to examine whether pregnant women's attitudes toward pregnancy relate to miscarriages. Ten thousand pregnant women who received the same basic care had their attitudes evaluated. Seven hundred of them miscarried. The attitudes toward pregnancy expressed by these 700 women are shown in the table.

Pregnant Woman's Attitude toward Pregnancy	Frequency of Miscarriage
Very negative	461
Slightly negative	152
Positive	87

a. Construct a bar graph for the data.
b. What changes could be made in the graph in part a to make the differences between the frequencies seem smaller? Larger?

2.6

USING GRAPHS TO MAKE INFERENCES

We will use the following example to discuss the use of graphs to make inferences.

EXAMPLE 2.10   In Example 2.9 we presented a histogram that summarized the mercury vapor levels measured in sixty dental offices in San Antonio, Texas. That histogram is repeated in Figure 2.21 for your convenience.

a. Based on Figure 2.21, how would you visualize (infer) a graph for the population of all dental offices in the United States?
b. Can you give a measure of reliability for this inference?

Solution   a. We can expect the population histogram to be somewhat similar to the sample histogram shown in Figure 2.21. Examining the histogram in Figure 2.21, we see that most offices sampled had a very small mercury vapor level, and the relative frequencies tend to decrease as the mercury vapor levels increase. Thus, the population histogram might look like Figure 2.22a. Alternatively, since Figure 2.21 shows a slight increase in the relative frequency for the category 0.07 to 0.08, the population histogram might appear as shown in Figure 2.22b. Which graph you prefer

Figure 2.21
Histogram for mercury
vapor levels

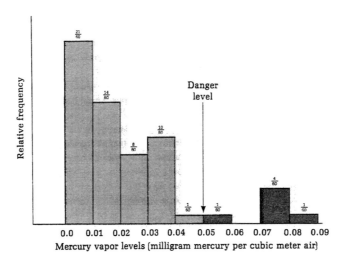

Figure 2.22
Possible histograms for
mercury vapor levels of
all dental offices

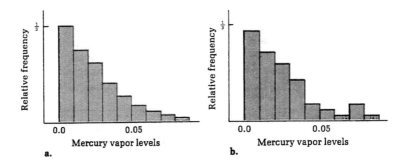

is not really important, since the inferences about the populations are essentially the same. In particular, we can infer that most of the dental offices are below the danger level of 0.05 milligram of mercury per cubic meter of air, although some exceed this level.

b. It is true that the sample and population graphs will be similar. But how can you explain how close the two figures will be? In other words, how can you measure the reliability of the inference? It is virtually impossible to answer this question. ∎

Example 2.10 illustrates a shortcoming of using graphs for the purpose of making inferences—namely, graphs do not lend themselves to providing a measure of reliability. In later chapters, we will develop other descriptive measures using numbers, and these will allow us to measure the reliability of our inferences.

Chapter Summary

Graphs are intuitively appealing descriptive techniques that can be used to give visual descriptions of samples or populations. The type of graph used depends on the type of data being summarized.

Qualitative data consist of observations that can be classified into exactly one of a set of nonnumerical categories. **Quantitative data** are numerical observations or measurements.

Either form of data can be pictured in **bar graphs** or **pie charts.** These graphs usually display frequencies or relative frequencies for categories of data. The **frequency** of a category is the actual number of observations in a sample that fall within the category. If a sample consists of n observations (**sample size** $= n$), then the **relative frequency** of a category is the frequency divided by n.

Stem and leaf plots can be used to summarize and analyze relatively small sets of either quantitative or qualitative data.

When quantitative data assume a large number of values, we can group the data into intervals before graphing them. A bar graph resulting from such an approach is called a **histogram.**

The rules for constructing bar graphs, pie charts, stem and leaf plots, and histograms can be reviewed on pages 23, 24, 33, and 42. Bar graphs and histograms are easily distorted to give a false impression of the data. In general, a fair presentation has horizontal and vertical axes about equal in length; only the heights of the bars, not their areas, reflect the frequencies. Since graphs are meant to be only a descriptive technique, they are not appropriate for making inferences.

SUPPLEMENTARY EXERCISES (2.44–2.62)

Learning the Mechanics

2.44   Construct a histogram for the data summarized in the following table:

Category	Relative Frequency
0.00 to 0.75	.03
0.75 to 1.50	.02
1.50 to 2.25	.04
2.25 to 3.00	.06
3.00 to 3.75	.11
3.75 to 4.50	.15
4.50 to 5.25	.20
5.25 to 6.00	.16
6.00 to 6.75	.13
6.75 to 7.50	.10

2.45 *a.* Complete the following table:

Category	Frequency	Relative Frequency	Percentage
X	90		
Y	50		
Z	60	_____	_____
Totals	200		

b. Draw a bar graph for the data using the relative frequencies.

c. Draw a pie chart for the data.

2.46 *a.* Draw a graph to summarize the following data properly:

85	80	89	85	89	81	86
90	85	85	88	86	81	89
87	86	84	81	88	84	87
88	84	82	86	87	86	85
81	83	80	87	80	85	88
90	88	85	88	84	85	87

b. Double the length of the vertical axis of the graph you drew in part *a*. (Do not change the horizontal axis.) Compare this graph to the one in part *a*.

Applying the Concepts

2.47 A survey was taken in rural Ohio to compare the rates of crime for burglary, larceny, and vandalism. For 500 crimes investigated, the following frequencies of occurrence were found:

Type of Crime	Burglary	Larceny	Vandalism
Frequency	35	155	310

a. Calculate the relative frequency and percentage for each type of crime.

b. Using your answers to part *a*, draw a bar graph and a pie chart.

c. Which of the three types of crime seems to occur most often?

2.48 Most companies require managers to give their employees annual performance reviews. One company surveyed 300 employees and asked what bothered them most about their manager's evaluation. A summary of the complaints is given in the table.

Complaint	Frequency
Rated only on a boss's whim	40
Judged by irrelevant standards	55
Supervisors do not want to do the review	88
Evaluations are superficial	103
Evaluations are threatening	14

a. Draw a bar graph for the data.

b. Using the graph from part a, what recommendations would you make to the personnel department of this company?

2.49 Automobile manufacturers in the United States invest heavily in research to improve their products' gas mileage. One manufacturer, hoping to achieve 30 miles per gallon with one of its full-size models, measured the mileage obtained by thirty test versions of the model. The results follow (rounded to the nearest mile for convenience):

30	30	32	31	30	27	28	31	33	28
30	31	29	31	27	33	31	30	35	29
32	27	31	34	30	27	26	29	28	30

a. Draw a bar graph to summarize the data.

b. What fraction of the test versions got 30 or more miles per gallon?

c. If the data had been rounded to the nearest tenth of a mile, instead of the nearest mile, would a bar graph still be appropriate to summarize the data? Explain.

2.50 Utility companies are concerned about the temperatures that might occur during various times of the year. To get an idea of the temperatures in Florida during the month of January, a company obtained the average monthly temperatures for that month from 1951 through 1980. These temperatures (in °F) are given below.*

58.9	54.3	55.1	55.8	58.5	54.7
62.4	63.3	58.4	60.7	65.0	49.9
59.8	52.5	57.0	57.7	59.3	52.5
61.0	57.1	57.1	57.8	69.1	54.9
57.3	58.4	58.7	53.1	63.0	58.7

a. Draw a stem and leaf plot for the data.

b. Construct a histogram for the data.

c. Would you use the graph in part a or the one from part b to summarize the data? Would each graph be appropriate for certain purposes? Explain.

d. In how many years was the average temperature above 60°F?

2.51 A pharmacologist studied a sample of 100 women through pregnancy and delivery. The number of drugs each used during this period is given below:

4	5	5	7	6	5	4	4	4	6
1	6	5	4	0	3	4	0	4	2
5	2	4	5	7	2	1	9	5	4
6	5	0	8	3	4	6	5	2	7
3	3	5	1	4	5	5	3	3	5
6	2	3	3	4	6	2	5	7	7
4	4	5	4	5	4	5	5	5	4
3	0	4	6	6	5	7	6	4	8
5	5	5	6	3	6	3	6	6	4
2	5	1	2	0	2	4	6	4	2

a. Draw a bar graph for the data.

* From Table 15, *American Statistics Index 1982* (Bethesda, Md.: Congressional Information Service, Inc., 1983), p. 24.

b. How many of the women used more than four drugs?

c. How many of the women used no drugs?

2.52 After purchasing a policy from a life insurance company, a person has a specified period of time to cancel the policy without financial obligation. An insurance company is interested in the value of the policies that are cancelled. The values (in thousands of dollars) for seventy cancelled policies are shown below:

3	20	15	8	7	15	20	8	20	50
15	15	3	25	35	5	10	35	15	10
25	10	15	30	10	30	40	3	15	15
10	20	25	25	15	20	25	20	45	30
15	25	15	20	15	25	40	10	25	25
25	15	30	27	15	40	25	50	50	40
5	30	10	20	25	20	40	6	25	20

a. Draw a bar graph for the data.

b. Were more policies cancelled at values below 15 thousand dollars than at values above 30 thousand dollars?

c. Which policy was the most frequently cancelled?

2.53 In the game of chess, the first few moves are crucial to the outcome. Five opening strategies are highly favored by chess experts. A sample of 100 Grand Masters is taken, and each is asked which of these strategies he or she prefers. A summary of their responses is shown in the table.

Strategy	A	B	C	D	E
Frequency	17	23	22	18	20

a. Draw a pie chart for the data.

b. Interpret the pie chart from part a.

2.54 A geneticist crossed red and white snapdragons and produced 1,000 offspring. The bar graph shown here summarizes the results of this experiment. What was the relative frequency for each color offspring?

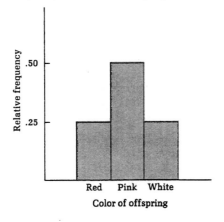

2.55   Sixty fields of corn are treated with a new fertilizer, and the yield of each field is measured. The yields of corn, in bushels per acre, are shown below:

67.0	69.8	69.4	66.3	71.2	67.8	69.9	71.0	70.8	71.1
72.8	73.7	72.8	75.0	76.8	67.3	68.2	76.5	71.8	71.5
68.5	68.8	69.6	65.8	68.8	74.2	67.1	65.4	70.2	69.4
73.3	68.1	70.8	72.4	66.8	74.8	71.6	72.5	72.2	70.5
70.5	70.7	70.2	63.2	69.4	72.0	69.7	72.5	70.3	69.1
70.9	74.0	71.4	71.5	66.5	70.0	73.8	70.6	71.3	67.4

a. Draw a histogram for the data. Use a category width of 1.5.
b. How many fields yielded more than 65 bushels of corn per acre?
c. How many fields yielded less than 70 bushels of corn per acre?

2.56   The ages (in years) of a sample of fifty students attending a community college are shown below:

20	18	19	19	21	18	24	22	24	18
25	20	21	18	22	20	21	22	20	19
18	19	19	18	18	25	19	18	20	20
22	18	19	23	23	19	26	18	19	21
25	35	18	24	23	18	21	25	21	19

a. Draw an appropriate graph for the data.
b. Interpret the graph of part a.

2.57   What is the major shortcoming associated with using graphs to make inferences?

2.58   The graph shown on page 63 appeared January 11, 1985, in the *Los Angeles Times*. The graph contains two pie charts—one describing the revenues of the proposed 1985–1986 California state budget, and the other showing the expenditures. Answer the following questions concerning these graphs.
a. What are the largest and second largest proposed sources of revenue?
b. What are the largest and second largest proposed expenditures?

2.59   The following data are the numbers of passengers (in thousands) carried by twenty-nine major and regional airlines in the United States from January to September 1984.*

3,057	924	5,426	10,149	5,900	31,060
1,922	7,831	2,531	3,805	11,817	12,908
1,753	28,191	1,048	10,897	9,080	8,090
25,137	28,581	1,509	8,510	662	927
1,712	73	1,482	10,467	13,947	

a. What kind of graph is most appropriate to summarize the data?
b. Construct the graph you chose in part a.

* Adapted from table, "U.S. Airline System Traffic: January–September 1984," *Air Transport World*, Vol. 21, No. 12, December 1984, p. 106.

Proposed California Budget

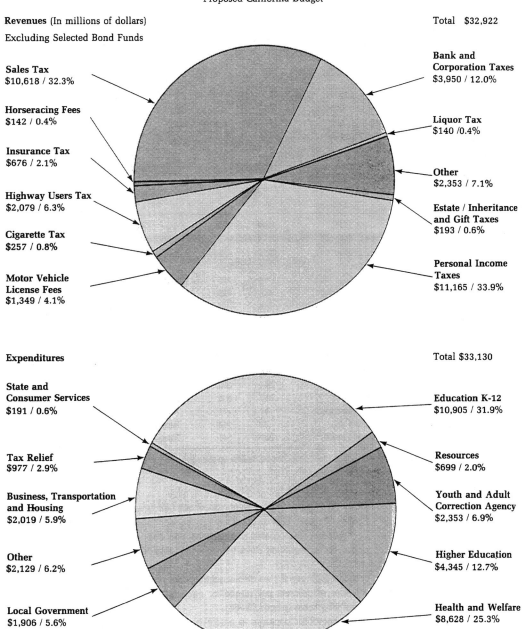

Revenues (In millions of dollars)   Total $32,922

Excluding Selected Bond Funds

Sales Tax
$10,618 / 32.3%

Horseracing Fees
$142 / 0.4%

Insurance Tax
$676 / 2.1%

Highway Users Tax
$2,079 / 6.3%

Cigarette Tax
$257 / 0.8%

**Motor Vehicle
License Fees**
$1,349 / 4.1%

**Bank and
Corporation Taxes**
$3,950 / 12.0%

Liquor Tax
$140 /0.4%

Other
$2,353 / 7.1%

**Estate / Inheritance
and Gift Taxes**
$193 / 0.6%

**Personal Income
Taxes**
$11,165 / 33.9%

Expenditures   Total $33,130

**State and
Consumer Services**
$191 / 0.6%

Tax Relief
$977 / 2.9%

**Business, Transportation
and Housing**
$2,019 / 5.9%

Other
$2,129 / 6.2%

Local Government
$1,906 / 5.6%

Education K-12
$10,905 / 31.9%

Resources
$699 / 2.0%

**Youth and Adult
Correction Agency**
$2,353 / 6.9%

Higher Education
$4,345 / 12.7%

Health and Welfare
$8,628 / 25.3%

Source: David Puckett, "Proposed California Budget," *Los Angeles Times*, January 11, 1985, Part I, p. 28.

2.60 The following data are the approximate per capita expenditures (in dollars) in 1974 for police protection in twenty-six selected U.S. cities. Draw a stem and leaf plot for the data.

77	50	52	60	50	44	46
124	27	74	67	42	45	64
25	86	74	43	52	71	50
44	123	35	41	78		

*2.61 The table contains the populations (in thousands) of 30 sunbelt cities in 1960 and 1980.

City	1960	1980	City	1960	1980
Albuquerque, NM	201	332	Las Vegas, NV	64	165
Anaheim, CA	104	219	Lubbock, TX	129	174
Arlington, TX	45	160	Miami, FL	292	347
Atlanta, GA	487	425	New Orleans, LA	628	558
Austin, TX	187	345	Oklahoma City, OK	324	403
Charlotte, NC	202	314	Orlando, FL	88	128
Dallas, TX	680	904	Phoenix, AZ	439	790
Fort Lauderdale, FL	84	153	St. Petersburg, FL	181	239
Fort Worth, TX	356	385	San Antonio, TX	588	786
Fresno, CA	134	218	San Diego, CA	573	876
Honolulu, HI	294	365	San Francisco, CA	740	679
Houston, TX	938	1,595	San Jose, CA	204	629
Huntington Beach, CA	11	171	Tampa, FL	275	272
Huntsville, AL	72	143	Tucson, AZ	213	331
Jacksonville, FL	201	541	Tulsa, OK	262	361

Source: U.S. Bureau of the Census, *Statistical Abstract of the United States: 1981* (pp. 21–23), *1985* (pp. 23–25).

  a. Using a computer program, construct two histograms, one for the 1960 data set and one for the 1980 data set.

  b. Compare your histograms and describe what they reveal about the change in the populations of sunbelt cities between 1960 and 1980.

*2.62 In Computer Example 2.1 we discussed the percentages of alcohol in 52 wine coolers. The number of calories per ounce for each of these wine coolers is given in the accompanying table (pages 65–66).

  a. Use a computer package to obtain both a stem and leaf plot and a histogram for these data.

  b. Discuss both the graphs obtained in part *a*. Which graph do you think provides a better summary of the data? Why?

Brand/Bottler or Vendor	Calories/oz.	Brand/Bottler or Vendor	Calories/oz.
Bartles and Jaymes WC		Manischewitz WC—Berry;	
E & J Gallo	15	reported sample #1	
California Cooler—Tropical		Manischewitz Wine	20
California Cooler	19	Manischewitz WC—Berry;	
Calvin Cooler—Chablis		reported sample #2	
Calvin Cooler Wines	19	Manischewitz Wine	20
Calvin Cooler—Citrus		Manischewitz WC—Cream White	
Calvin Cooler Wines	20	Manischewitz Wine	20
Calvin Cooler—Pina Pineapple		Manischewitz WC—Pina Coconetta;	
Calvin Cooler Wines	20	reported sample #1	
Cape Cod Cranberry Cooler		Manischewitz Wine	20
Cape Cod Wine	13	Manischewitz WC—Pina Coconetta;	
Citronet WC		reported sample #2	
Heublein Wines	13	Manischewitz Wine	20
Coastal Valley Cooler		Manischewitz WC—Lemonade	
Glunz Cellars	18	Manischewitz Wine	20
Creative Cooler—Wine Tea		Peach-A-Roo WC	
Creative Cooler	20	Central Vineyards	26
Creative WC—Margarita		Quenchette French Cooler—Raspberry	
Creative Cooler	18	Les Grands Chais France	18
Dewey Stevens Premium Light WC		Quinn's Cooler—White Wine and Citrus	
Creative Cooler	14	R. R. Quinn and Co.	18
Diamond Citrus Twist Cooler		Royal Dutch Cooler—Banana Cream	
Diamond Island Cellars	20	Marquis B. V., Holland	29
Diamond Orange Squeeze Cooler		Royal Dutch Cooler—Strawberry Cream	
Diamond Island Cellars	17	Marquis B. V., Holland	25
Diamond Passion Punch Cooler		Seagram's WC—Citrus & White Wine	
Diamond Island Cellars	19	Joseph E. Seagram & Sons	14
Diamond Wild Berry Cooler		Seagram's WC—Golden	
Diamond Island Cellars	19	Joseph E. Seagram & Sons	17
Florida WC—Apple		Seagram's WC—Peach Flavored	
Florida Wine	21	Joseph E. Seagram & Sons	18
Florida WC—Cool White		Seagram's WC—Premium	
Florida Wine	20	Joseph E. Seagram & Sons	18
Florida WC—Orange		Steidl's WC—Red	
Florida Wine	19	Steidl Wine	18
Florida WC—Peach		Steidl's WC—White	
Florida Wine	23	Steidl Wine	18
Florida WC—Strawberry		Sun Country Tropical Cooler	
Florida Wine	19	Sun Country Cellars	18
Franzia WC—White Zinfandel		The Grape Vine WC—Rose	
Franzia Winery	16	Monarch Wine	18
HI 5 Cooler		The Grape Vine WC—White	
Tom Pree Wine	17	Upper Bay Wine Cellars	19

(continued)

Brand/Bottler or Vendor	Calories/oz.	Brand/Bottler or Vendor	Calories/oz.
20/20 WC—Citrus		20/20 WC—Tropical	
20/20 Wine	22	20/20 Wine	23
20/20 WC—Lambrusco		Widmer Niagara WC	
20/20 Wine	20	Widmer's Wine Cellars	13
20/20 WC—Peach		Wild Irish Rose WC—Citrus Rose	
20/20 Wine	18	Richards Wine	19
20/20 WC—Raspberry		Wild Irish Rose WC—Orange Rose	
20/20 Wine	20	Richards Wine	20
20/20 WC—Strawberry		Wild Irish Rose WC—Tropical Rose	
20/20 Wine	17	Richards Wine	20

Source: "Calories and Alcohol in Wine and Beer Coolers." *Consumers' Research*, Vol. 70, No. 3, March 1987, pp. 30–32. Connecticut Agricultural Experiment Station, New Haven, CT. Bulletin 840. Analysis of Wine and Beer Coolers. October 1986.

CHAPTER 2 QUIZ

1. State whether the following types of data are qualitative or quantitative:
   a. The years of teaching experience of each professor at a university
   b. The price per pound of ground beef at your favorite supermarket
   c. The bus route assigned for next Monday morning to each driver of a city transit company

2. What kinds of graphs are generally used to summarize qualitative data?

3. Define the following terms:
   a. Qualitative data   b. Relative frequency

4. Why are graphs rarely used when making inferences?

5. A grocery store manager wonders which advertising method is most effective. A sample of 125 customers is asked what influenced them to shop at this store. The results are summarized in the table.

Type of Influence	Television ad	Newspaper ad	Radio ad	Word of mouth	Other
Frequency	43	27	15	33	7

   a. Draw a bar graph for the data.
   b. Interpret the graph drawn in part a.

CHAPTERS 1 AND 2 CUMULATIVE QUIZ

1. Are graphs more useful in the area of descriptive statistics or inferential statistics? Why?

2. The manager of a high school cafeteria wondered what the students preferred to drink with their lunches. The manager polled 100 of the students and obtained the responses given in the table.

Drink	Iced tea	Milk	Soft drink	Other
Frequency	19	23	48	10

    a. Identify the population of interest to the cafeteria manager.
    b. Identify the sample.
    c. Give an example of an inference the manager might make.
    d. What should accompany any inference that is made?

3. Refer to Question 2.
    a. Are the data of interest qualitative or quantitative?
    b. Sketch a pie chart to summarize the data for the 100 students who were polled.
    c. Interpret the pie chart drawn in part *b*.

4. *a.* When is a bar graph used to summarize a set of quantitative data?
    b. When is a histogram used to summarize a set of quantitative data?

5. The histogram shown here summarizes 100 measurements.

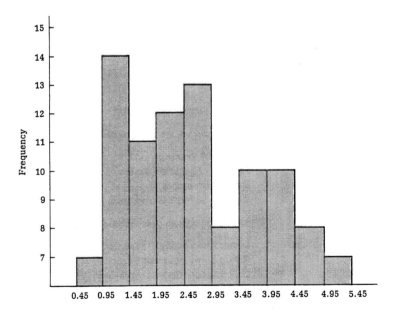

    a. How does the frequency of measurements between 0.45 and 0.95 compare to the frequency of measurements between 0.95 and 1.45?

b. Does the graph provide an accurate visual comparison of the two categories compared in part *a*? Explain.

c. What distortion does this graph contain, and what causes the distortion?

On Your Own

We list below several sources of real-life data sets that have been obtained from *Statistics Sources* (Wasserman & O'Brien, 1984). This index of data sources is very complete and is a useful reference for anyone interested in finding almost any type of data. First, we list some almanacs:

CBS News Almanac
Information Please Almanac
The World Almanac & Book of Facts

United States government publications are also rich sources of data:

Agricultural Statistics
Digest of Educational Statistics
Handbook of Labor Statistics
Housing and Urban Development Yearbook
Social Indicators
Uniform Crime Reports for the United States
Vital Statistics of the United States
Business Conditions Digest
Economic Indicators
Monthly Labor Review
Survey of Current Business
Bureau of the Census Catalog

Many data sources are published on an annual basis:

Commodity Yearbook
Facts and Figures on Government Finance
Municipal Yearbook
Standard & Poor's Corporation, Trade and Securities: Statistics

Some sources contain data that are international in scope:

Compendium of Social Statistics
Demographic Yearbook
United Nations Statistical Yearbook
World Handbook of Political and Social Indicators

Using the data sources listed above, sources suggested by your instructor, or your own resourcefulness, find two real-life data sets—one qualitative and one quantitative—from areas of particular interest to you.

a. Construct appropriate graphs for the two data sets you found.

b. What information do these graphs provide?

References

Miller, S. L., Domey, R. G., Elston, S. F. A., & Milligan, G. "Mercury Vapor Levels in the Dental Office: A Survey." *Journal of the American Dental Association*, November 1974, *89*, 1084–1091.

Tukey, J. W. *Exploratory Data Analysis*. Reading, Mass.: Addison-Wesley Publishing Company, 1977.

Wasserman, P. & O'Brien, J. *Statistics Sources*, 9th ed. Detroit: Gale Research Company, 1984.

Chapter 3

USING NUMBERS TO DESCRIBE DATA

THE QUALIFYING EXAM FOR REGISTERED NURSES

In order to become a registered nurse in the United States, several requirements must be met. As part of the process of licensing, candidates take the State Board Test Pool Examination for Registered Nurse Licensure (SBTPE). The examination is designed to test essential knowledge of nursing and to ensure the candidate's ability to apply that knowledge to clinical situations.

In July 1982, a new version of the SBTPE was introduced, and the scoring system for this test was discussed in a book designed to help people prepare for the examination (McQuaid & Kane, 1981, p. 20):

> For the comprehensive SBTPE to be introduced in July of 1982, there will be a new scoring system. The new examination will have a single score rather than the five separate scores currently being used, and if you fail one part you must re-take the entire exam in order to pass. The new mean will be 2,000 with a standard deviation of 400. Using this scale, scores will range from about 800 up to about 3,200, and there will be no overlap with the range of scores for the current SBTPE. This will eliminate any chance for confusion between the old scoring system and the new scoring system.
>
> To keep the difficulty level of the examination the same, the passing score for the comprehensive examination will be 1,600, or one standard deviation below the mean. With 1,600 as the passing score, the pass rate for the first time candidates will remain at about 84 or 85 percent.*

In this chapter we will discuss how numbers such as the *mean* and *standard deviation* are used to describe sets of data. For instance, we will consider how they are related to the highest and lowest measurements. We will calculate numerical descriptive measures for sample data, interpret these numbers, and begin to consider how such numbers are used to make inferences.

* Reprinted by permission from *The State Board Test Pool Examination for Registered Nurse Licensure*, published by Chicago Review Press.

3.1

MEASURES OF CENTRAL TENDENCY

In Chapter 2 we presented graphical methods for summarizing and describing sets of data. We mentioned that graphs for quantitative data are often called *frequency distributions* because they show how the data are distributed—that is, they show the general pattern or arrangement of the data.

When we consider a distribution of data, we might be interested in what numbers occur in the data set; how they are clustered with, or isolated from, each other; and so forth. Small sets of data can be pictured in a graph called a **dot diagram** by using dots and a number line. For example, a sample of nine values, 0, 0, 1, 4, 4, 4, 5, 5, and 7, is pictured in a dot diagram in Figure 3.1. In this figure we can see immediately that the data range from 0 to 7, that there are two main clusters (one at 0 and 1, the other at 4 and 5), and that the data are centered somehow about 3 or 4.

Figure 3.1
A dot diagram

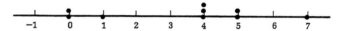

Large data sets can be pictured by histograms, or similar types of graphs, as we discussed in Chapter 2. However, none of these graphs enable us to make inferences. In order to make an inference, we must supplement the pictorial representation of a data set with numerical descriptions that reveal the essential characteristics of the distribution. The first kind of numerical descriptions we will discuss are **measures of central tendency.**

Definition 3.1

A **measure of central tendency** is a number that locates the approximate center of a distribution of data.

One of the most popular and best understood measures of central tendency is the **mean,** or in everyday terms, the **average,** of a data set.

EXAMPLE 3.1   Calculate the mean of the following sample of five measurements:

5, 3, 8, 5, 6

Solution   We must first find the sum of all the measurements:

Sum = 5 + 3 + 8 + 5 + 6 = 27

The mean is this sum (27) divided by the number of measurements (5):

$$\text{Mean} = \frac{27}{5} = 5.4$$ ■

Definition 3.2

The **mean** of a set of data is the sum of all the measurements divided by the total number of measurements in the data set.

At this point, it will be very helpful to present some shorthand notation that will simplify instructions for calculating the mean and other numerical descriptive measures. The notation is given in the box.

---

x: The letter x is used to represent an arbitrary measurement of a sample.

n: A lowercase n is used to represent the number of measurements in a sample.

$\sum x$: The summation symbol (Greek capital letter sigma), followed by the letter x, is used to mean "add all the measurements of a sample."

$\bar{x}$: The letter x with a bar over it (read "x bar") is used to denote the sample mean.

Thus, with this notation, the sample mean is given by the formula

$$\bar{x} = \frac{\sum x}{n} = \frac{\text{Sum of all the sample measurements}}{\text{Number of measurements in the sample}}$$

EXAMPLE 3.2   The times required by a sample of students to complete an introductory statistics test are given below. (The times were rounded to the nearest minute.)

37, 56, 43, 49, 52, 33, 45

a. Give the value of n.
b. Compute $\sum x$.
c. Compute $\bar{x}$.

Solution   **a.** The number of measurements in a sample is denoted by n. Thus,

$n = 7$

b. When you see $\sum x$, think "add all the sample measurements." We obtain

$\sum x = 37 + 56 + 43 + 49 + 52 + 33 + 45 = 315$

c. The symbol $\bar{x}$ represents the sample mean, or average. Using our notation, we have

$$\bar{x} = \frac{\sum x}{n} = \frac{315}{7} = 45$$ ∎

The **median** is another commonly used measure of central tendency. It is located in the middle of a data set. The method of determining the median depends on the number of measurements in the data set.

Definition 3.3

a. The **median** of a data set with an *odd* number of measurements is the middle number when the measurements are arranged from the smallest to the largest (or largest to smallest).

b. The **median** of a data set with an *even* number of measurements is the mean of the two middle numbers when the measurements are arranged from the smallest to the largest (or largest to smallest).

When a set of data is arranged in increasing order, half the data are above the median and half are below it. The median exactly splits the entire data set in half for an even number of measurements. For an odd number of measurements the median is a number in the data set, and this number splits the remaining data exactly in half.

EXAMPLE 3.3   Consider the following sample of $n = 7$ measurements:

    5, 7, 4, 9, 5, 6, 2

a. Calculate the median of this sample.

b. Eliminate the number 2, and calculate the median of the remaining $n = 6$ measurements.

Solution   **a.** Since 7 is an odd number, the median is the middle number, as shown below:

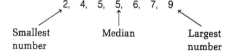

Note that *every* measurement in the sample is listed from the smallest to the largest. This is important to remember when numbers are repeated in the sample, such as the two 5's in this example.

b. After removing the 2, there are $n = 6$ observations, which is an even number. We again list the measurements in increasing order. Then, the median is the mean of the two middle numbers:

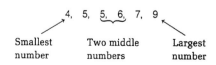

Thus, the median is

$$\text{Median} = \frac{5 + 6}{2} = \frac{11}{2} = 5.5$$ ■

A third measure of central tendency is the **mode** of a set of measurements.

Definition 3.4

The **mode** of a set of data is the measurement that occurs most often.

EXAMPLE 3.4   Determine the mode for the following sample of $n = 10$ measurements:

8, 7, 9, 4, 8, 10, 9, 9, 3, 5

Solution   Since 9 occurs most often, the mode is 9.   ■

EXAMPLE 3.5   Refer to the sample of 10 measurements given in Example 3.4.

a. Calculate the median.
b. Calculate the mean.
c. Draw a dot diagram of the data, and mark the mean, median, and mode on this diagram. Discuss their values.

Solution   **a.** To find the median, we first arrange the 10 measurements in order from the smallest to largest:

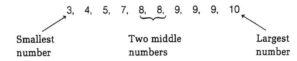

Since there is an even number of measurements (10), the median is the average of the two middle numbers, which are both 8's. Thus, the median is

$$\text{Median} = \frac{8 + 8}{2} = 8$$

b. The sample mean $\bar{x}$ is computed from the formula

$$\bar{x} = \frac{\sum x}{n}$$

First, we find $\sum x$:

$$\sum x = 3 + 4 + 5 + 7 + 8 + 8 + 9 + 9 + 9 + 10 = 72$$

Then, since we have $n = 10$, the mean is

$$\bar{x} = \frac{72}{10} = 7.2$$

c. The dot diagram showing the locations of the three measures of central tendency is given in Figure 3.2.

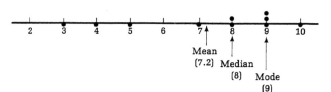

The mean, median, and mode found in Examples 3.4 and 3.5 are quite different. Each gives slightly different weight to the characteristics of the data set, and consequently, they usually differ in the place where they locate the center of the distribution. For example, the mode of 9 would create a different mental image of the location of the data than would the mean of 7.2.

The relative positions of the mean, median, and mode depend on the shape of the frequency distribution. The following definitions will be useful for discussing distributions of data.

Definition 3.5

A distribution of data in which measurements above the mean occur less frequently than measurements below the mean is said to be **skewed to the right.** See Figure 3.3a.

Definition 3.6

A distribution of data in which measurements below the mean occur less frequently than measurements above the mean is said to be **skewed to the left.** See Figure 3.3b.

Definition 3.7

A distribution of data that is neither skewed to the right nor skewed to the left is called **symmetric.** Measurements at equal distances from the center of the distribution occur with the same frequency. See Figure 3.3c.

Figure 3.3
Comparison of the mean, median, and mode for three types of distributions

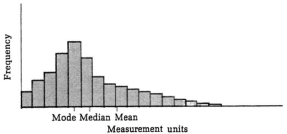

a. Distribution that is skewed to the right

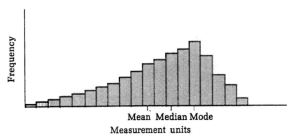

b. Distribution that is skewed to the left

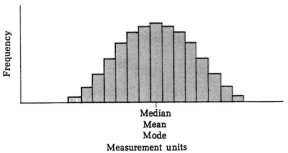

c. Distribution that is symmetric

Figure 3.3 shows the relative positions of the mean, median, and mode for the three types of data just defined. If a data set with one mode is symmetric, all three measures of central tendency are equal. However, when a data set is skewed to the left or right, the three measures of central tendency may be quite different. Note that the direction in which a distribution is skewed refers to the direction of the "tail" of the distribution, not the direction in which it is concentrated.

Histograms for very large data sets almost appear to form smooth curves. As can be seen in Figure 3.4a (page 78), the width of the rectangles may be quite small for a very large data set and this results in a smooth-looking outline (graph). Thus, to graph a very large data set, we may approximate the histogram by a smooth curve, as shown in Figure 3.4b. It is easier to sketch the smooth curve, and the two graphs give us approximately the same information.

Figure 3.4
Histograms for very large
data sets

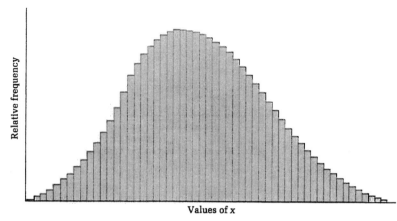

a. Actual histogram for a very large data set

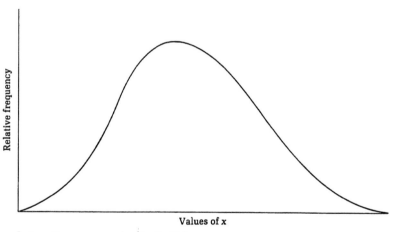

b. Smooth curve approximating the histogram

EXAMPLE 3.6   Suppose you are a newly graduated psychiatrist and you are considering working
for a small psychiatric clinic. The present employees at the clinic are the founding
member and three associates who were hired by the founder. You ask what salary you
will receive if you join the firm. Unfortunately, you get two answers:

> *Answer A:* The founding member tells you that an "average employee" earns
> $37,500.
>
> *Answer B:* One of the associates later tells you that an "average employee" earns
> $25,000.

Which answer can you believe?

Solution   The confusion exists because the phrase "average employee" has not been clearly defined. Suppose that the four salaries are $25,000 for each of the three associates and $75,000 for the founding member. Then,

$$\text{Mean} = \frac{\$150,000}{4} = \$37,500$$

Median = $25,000

Mode = $25,000

You can now see how the two answers were obtained. The founding member reported the mean of the four salaries, and the associate could have reported either the median or the mode since they are equal. The two answers were misleading because neither person stated which measure of central tendency was being used. ■

As you can see from Example 3.6, when reporting an observation "typical" of a data set, it is important to state which measure of central tendency was used. Always keep in mind the three measures of central tendency introduced in this section:

Three Measures of Central Tendency

1. **Mean**—the average of a data set
2. **Median**—the middle of a data set
3. **Mode**—the number that occurs most frequently in a data set

---

COMPUTER
EXAMPLE 3.1

In Example 2.8 we constructed a histogram for the EPA mileage ratings of 100 cars. The data were given in Table 2.9 (page 42). We will now use Minitab to order the 100 mileage ratings from the smallest to the largest, construct a histogram of the data, calculate the median, and calculate the mean.

After the data shown in Table 2.9 are entered into the computer, Minitab can be used to obtain the ordered measurements shown in Figure 3.5 and the histogram in Figure 3.6 (page 80).

MILEAGES

30.0	31.8	32.5	32.7	32.9	32.9	33.1	33.2	33.6	33.8	33.9
33.9	34.0	34.2	34.4	34.5	34.8	34.8	35.0	35.1	35.2	35.3
35.5	35.6	35.6	35.7	35.8	35.9	35.9	36.0	36.1	36.2	36.3
36.3	36.4	36.4	36.5	36.5	36.6	36.6	36.7	36.7	36.7	36.8
36.8	36.8	36.9	36.9	36.9	37.0	37.0	37.0	37.0	37.1	37.1
37.1	37.2	37.2	37.3	37.3	37.4	37.4	37.5	37.6	37.6	37.7
37.7	37.8	37.9	37.9	38.0	38.1	38.2	38.2	38.3	38.4	38.5
38.6	38.7	38.8	39.0	39.0	39.3	39.4	39.5	39.7	39.8	39.9
40.0	40.1	40.2	40.3	40.5	40.5	40.7	41.0	41.0	41.2	42.1
44.9										

Figure 3.5
EPA mileage ratings ordered from smallest to largest

Figure 3.6
Computer-generated
histogram of mileage ratings

```
Histogram of MILEAGES    N = 100

Midpoint    Count
      30      1    *
      32      5    *****
      34     12    ************
      36     31    *******************************
      38     31    *******************************
      40     15    ***************
      42      4    ****
      44      1    *
```

The median and the mean of the 100 ratings were also found using Minitab. These values are

```
MEDIAN =        37.000
MEAN    =       36.994
```

Since there are 100 mileage ratings, an even number, the median falls between the fiftieth and fifty-first values. From the ordered data in Figure 3.5, we can see that both of these values are 37.0. This verifies that the median is in fact 37.0, as calculated by Minitab.

A look at the histogram in Figure 3.6 indicates that this set of data is fairly symmetric. Thus, the median and the mean are very nearly equal; that is, 36.994 is approximately equal to 37.0.   ■

Exercises (3.1–3.17)

Learning the Language

3.1 Define the following terms:
a. Measures of central tendency   c. Median   e. Skewed
b. Mean   d. Mode   f. Symmetric

3.2 What is denoted by the following symbols?
a. x b. n c. $\sum x$ d. $\bar{x}$

Learning the Mechanics

3.3 Consider the following sample: 2, 0, 4, 3, 5
a. Find the value of n.   b. Calculate $\sum x$.   c. Calculate $\bar{x}$.

3.4 Calculate the mean, median, and mode for each of the following samples:
a. 4, 2, 6, 5, 6, 6   b. 0, 1, 0, 1, 2, 4, 0, 2
c. −3, 0, −2, 1, 4, 1, 1, 3, −2, 3, 2, 1

3.5 Find the sample mean, $\bar{x}$, in each of the following situations:
a. $\sum x = 35$ and $n = 7$ c. $\sum x = 100$ and $n = 20$
b. $\sum x = 35$ and $n = 5$ d. $\sum x = 20$ and $n = 100$

3.6 Calculate the mean, median, and mode for each of the following samples:
a. −1, −3, −2, −2, −2, 0, 7   b. 2, 9, 8, 3, 7, 7, 6, 7, 5
c. 1, 9, 9, 9, 9, 9, 100

3.7 In each of the following graphs, indicate the relative positions of the mean, median, and mode:
a.

Units of measurement

b.

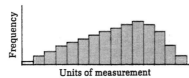

Units of measurement

c.

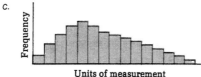

Units of measurement

d.

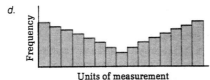

Units of measurement

Applying the Concepts

3.8 A psychologist has developed a new technique intended to improve rote memory. To test the method against other standard methods, twenty high school students are selected at random and each is taught the new technique. The students are then asked to memorize a list of 100 word phrases using the technique. The following are the numbers of word phrases memorized correctly by the students:

91	64	98	66	83	87	83	86	80	93
83	75	72	79	90	80	90	71	84	68

a. Define the terms mean, median, and mode in the context of this problem.
b. Construct a relative frequency histogram for the data.
c. Compute the mean, median, and mode for the above data set and locate them on the histogram. Do these measures of central tendency appear to locate the center of the distribution of data?

3.9   Would you expect the data sets described below to be symmetric, skewed to the right, or skewed to the left? Explain.
  a. Salaries of all persons employed by a large university
  b. Grades on an easy test
  c. Grades on a difficult test
  d. Amounts of time students in your class studied last week
  e. Ages of automobiles on a used car lot
  f. Amounts of time spent by students on a difficult exam (maximum time is 50 minutes)

3.10 One index used by social scientists to measure socioeconomic status is personal income. The yearly incomes (in dollars) of twenty residents of a certain community are given below:

16,000	16,000	16,000	90,800	19,800
16,000	15,600	19,000	19,000	10,200
15,000	17,000	11,500	13,000	18,000
19,500	13,500	12,200	17,000	11,400

  a. Compute the mean, median, and mode for this sample.
  b. Now, eliminate the largest income (90,800) and repeat part a. What effect does this have on the measures of central tendency computed in part a? Which measure of central tendency seems to be most sensitive to extreme values?

3.11 The scores for a class of twenty students on a statistics test are as follows:

| 87 | 76 | 96 | 77 | 94 | 92 | 88 | 85 | 66 | 89 |
| 79 | 95 | 50 | 91 | 83 | 88 | 82 | 58 | 18 | 69 |

  a. Compute the mean, median, and mode for the data.
  b. Which of the three measures of central tendency do you think best represents the achievement of the class?
  c. Eliminate the two lowest scores, and again compute the mean, median, and mode. Which measure of central tendency do you think is most affected by extremely low (or extremely high) scores?

3.12 Ten trained rats were released in a maze. Their times to escape (in seconds) are recorded below. The N's represent two rats that had still not escaped at the time of the termination of the experiment.

| 100 | 38 | N | 122 | 95 | 116 | 56 | 135 | 104 | N |

  a. Can you calculate the mean for the data? Explain.
  b. Is the median a meaningful measure of central tendency for the data? Explain. Calculate the median.

3.13 In the January 28, 1985, issue of The Sporting News, the salaries of players in the National Basketball Association (NBA) were reported. These salaries, listed by team, are shown on pages 83–84. An examination of the data indicates that a few players earn over $1 million annually—and a couple even earn over $2 million. Most players, however, earn much less.
  a. Which measure of central tendency, the mean or the median, would be larger? Why?

Player salaries for Exercise 3.13

Atlanta

Player	Salary
Rollins	$678,000
Johnson	491,000
Wilkins	460,000
Williams	450,000
*Carr	250,000
*Willis	250,000
*Glenn	225,000
Levingston	200,000
Wittman	170,000
Hastings	160,000
Brown	160,000
Rivers	105,000

Dallas

Player	Salary
*Aguirre	$800,000
*Perkins	500,000
*Blackman	255,000
Vincent	250,000
Nimphius	250,000
Davis	210,000
Harper	200,000
Ellis	200,000
Bryant	130,000
*Sitton	75,000
*Sluby	75,000

Houston

Player	Salary
Sampson	$1,300,000
*Olajuwon	780,000
McCray	360,000
*Reid	320,000
Lloyd	150,000
*Hollins	135,000
*Leavell	130,000
Wiggins	130,000
*Micheaux	100,000
*Petersen	100,000
Ehlo	80,000
*McDowell	65,000

L.A. Lakers

Player	Salary
Johnson	$2,500,000
Abdul-Jabbar	1,530,000
Kupchak	1,150,000
McAdoo	923,000
Wilkes	860,000
Worthy	400,000
Scott	350,000
Cooper	336,000
*Rambis	325,000
McGee	200,000
Lester	120,000
Spriggs	90,000
*Jones	90,000

Philadelphia

Player	Salary
Malone	$2,125,000
Erving	1,054,000
Jones	500,000
C. Johnson	406,000
Cheeks	350,000
Toney	310,000
*Barkley	307,000
*Wood	215,000
Richardson	170,000
*Williams	150,000
Threatt	78,000
*G. Johnson	75,000

Seattle

Player	Salary
Sikma	$1,149,000
King	500,000
Wood	450,000
Vranes	392,000
Chambers	374,000
*Henderson	325,000
Sobers	277,000
*McCormick	185,000
Sundvold	158,000
*Blackwell	80,000
*Brickowski	65,000
*Schweitz	65,000

Boston

Player	Salary
Bird	$1,800,000
McHale	1,000,000
*Maxwell	830,000
Parish	700,000
Johnson	405,000
Wedman	400,000
Ainge	400,000
Buckner	239,000
Carr	175,000
Kite	120,000
Clark	65,000
*Carlisle	65,000

Denver

Player	Salary
English	$790,000
Natt	708,000
Issel	614,000
Cooper	280,000
Schayes	225,000
*Dunn	200,000
*Hanzlik	200,000
Lever	185,000
*Turner	100,000
Kopicki	80,000
*White	75,000
Evans	65,000

Indiana

Player	Salary
Stipanovich	$420,000
Kellogg	406,000
Garnett	269,000
Williams	200,000
*Stansbury	190,000
*Fleming	175,000
Sichting	135,000
*Durrant	100,000
Waiters	100,000
*Gray	80,000
Thomas	80,000
*Brown	65,000

Milwaukee

Player	Salary
Moncrief	$884,000
Cummings	410,000
Grevey	383,000
*Mokeski	345,000
Breuer	167,000
Pressey	152,000
*Dunleavy	150,000
Lister	150,000
Pierce	140,000
Hodges	100,000
*Fields	100,000
*Davis	80,000

Phoenix

Player	Salary
Davis	$670,000
*Lucas	650,000
Adams	500,000
Robey	425,000
Edwards	406,000
Nance	400,000
Macy	305,000
*Humphries	150,000
Foster	150,000
Scott	125,000
*Jones	120,000
Pittman	95,000
*Holton	80,000
Sanders	80,000
	70,000

Utah

Player	Salary
Dantley	$515,000
Griffith	363,000
Green	255,000
*Wilkins	250,000
Bailey	225,000
*Kelley	215,000
*Paultz	180,000
Eaton	133,000
*Stockton	125,000
Roberts	125,000
Hansen	86,000
*Anderson	65,000
*Mannion	65,000
	65,000

Chicago

Player	Salary
Corzine	$660,000
*Greenwood	550,000
*Jordan	550,000
Jones	500,000
Johnson	369,000
Woolridge	300,000
Green	275,000
Dailey	250,000
Whatley	200,000
Higgins	160,000
*Oldham	125,000
*Matthews	65,000

Cleveland

Player	Salary
Free	$675,000
Davis	475,000
Shelton	395,000
Poquette	284,000
*Turpin	260,000
Bagley	209,000
Jones	205,000
Hubbard	200,000
Hinson	150,000
*Anderson	100,000
Thompson	90,000
West	80,000
*Williams	65,000

Detroit

Player	Salary
Thomas	$750,000
*Tripucka	700,000
Roundfield	674,000
*Johnson	625,000
*Laimbeer	600,000
Long	330,000
Benson	270,000
Tyler	225,000
Cureton	200,000
*Campbell	110,000
*Steppe	100,000
*Jones	65,000

Golden State

Player	Salary
Short	$495,000
Johnson	363,000
Smith	300,000
Whitehead	225,000
Floyd	200,000
Conner	160,000
*Bratz	135,000
*Aleksinas	100,000
*Burtt	70,000
*Plummer	70,000
*Wilson	70,000
*Thibeaux	70,000

Kansas City

Player	Salary
Drew	$700,000
Theus	435,000
Olberding	375,000
Woodson	375,000
Thompson	277,000
Johnson	225,000
*Meriweather	225,000
*Thorpe	225,000
McNamara	165,000
*Buse	130,000
*Verhoeven	65,000
*Pope	65,000

L.A. Clippers

Player	Salary
Walton	$1,350,000
Johnson	900,000
Donaldson	470,000
Nixon	408,000
*Smith	400,000
Bridgeman	350,000
Catchings	262,000
*Cage	215,000
*Gordon	175,000
Warrick	110,000
*White	90,000
*Murphy	65,000

New Jersey

Player	Salary
Birdsong	$1,075,000
Dawkins	768,000
Williams	513,000
Richardson	427,000
Gminski	350,000
King	275,000
*Ransey	275,000
Cook	255,000
O'Koren	244,000
*Turner	75,000
*Johnson	65,000
*Sappleton	65,000

New York

Player	Salary
King	$874,000
*Cummings	800,000
Cartwright	600,000
Robinson	540,000
Sparrow	500,000
Webster	450,000
*Bailey	425,000
Orr	325,000
Tucker	273,000
Grunfeld	250,000
Walker	230,000
Carter	160,000
*Wilkins	65,000
*Bannister	65,000
*Cavenall	65,000

Portland

Player	Salary
*Paxson	$800,000
Vandeweghe	755,500
Thompson	750,000
*Bowie	600,000
Carr	465,000
Valentine	210,000.
Drexler	175,000
Norris	110,000
*Thompson	105,000
*Scheffler	90,000
*Colter	70,000
*Kersey	65,000

San Antonio

Player	Salary
Gervin	$831,000
Mitchell	790,000
Gilmore	600,000
Knight	433,000
Banks	265,000
*Robertson	210,000
Moore	200,000
Cook	142,000
Paxson	102,000
Iavaroni	100,000
*Jones	65,000

Washington

Player	Salary
Gus Williams	$730,000
Ruland	660,000
Robinson	450,000
Mahorn	361,000
*Ballard	360,000
*McMillen	250,000
Johnson	232,000
Malone	211,000
*Sewell	100,000
Daye	95,000
*Guy Williams	75,000
*Bradley	75,000

* Current salaries are estimates because they were not included in list distributed by NBA to its teams. Player is either a rookie, or a veteran who negotiated a new contract after the 1983–1984 season. Salaries include bonuses.

Source: Table, "Player Payrolls," The Sporting News, January 28, 1985, p. 29.

b. If the players' association were negotiating with the owners for more fringe benefits, should the players' association use the mean or the median to describe players' salaries? Why?

3.14 To demonstrate the idea of the mean of a set of data, a professor calculates the average height of the ten students attending class that day. The mean (or average) height of these ten students is 68 inches. Suppose that the center on the girls' basketball team walks into class right after this mean has been calculated. If she is 79 inches tall, what is the mean height of the eleven students who are now in the classroom?

3.15 In 1975 and again in 1978 surveys asked people the question: "On an average day, about how many hours do you personally watch television?" The responses to these surveys appear in the accompanying table.

Responses	1975	1978
0 hours a day	57	91
1 hour a day	256	316
2 hours a day	395	418
3 hours a day	291	287
4 or more hours a day	484	416
Number responding	1,483	1,528

Source: Adapted from Table 11/17, "Frequency of ... Time Spent Viewing Television, Selected Years: 1972–1978," in Center for Demographic Studies, U.S. Bureau of the Census, Social Indicators III (Washington, D.C.: Government Printing Office, 1980), p. 561.

a. Calculate the median of the responses for 1975.
b. Calculate the median of the responses for 1978.
c. Compare the medians calculated in parts a and b. What does this seem to indicate?
d. Could you calculate the mean response for 1975 or 1978? Explain.

3.16 The data shown in the accompanying table are the median prices of existing homes in the years 1981–1986. If the average prices of existing homes were calculated for

Year	Median-Priced Existing Home
1981	$66,460
1982	67,800
1983	70,300
1984	72,400
1985	75,500
1986	80,300

Source: National Association of Realtors.

each of these years, how do you think these values would compare to the median prices given? That is, would the averages be higher, lower, or equal in value to the medians? Why?

3.17　　Refer to Exercise 3.13. Given the player salaries shown on pages 83 and 84, Minitab was used to find the means and medians of the salaries for the Philadelphia and Utah teams. The results are listed in the table.

Philadelphia	Utah
Mean = $478,333	Mean = $195,538
Median = $308,500	Median = $180,000

a. Compare the two means. What does this indicate about the salaries of Philadelphia players compared with those of Utah players?

b. Compare the two medians. What does this indicate about the salaries of Philadelphia players compared with those of Utah players?

c. Compare the mean and median for each team. Explain the difference between the values of the mean and median for each team.

3.2

MEASURES OF VARIABILITY

Measures of central tendency provide only a partial description of a quantitative data set. The description is incomplete without a **measure of variability.**

Definition 3.8

A **measure of variability** is a number that describes the spread, or variation, in a distribution of data.

Perhaps the simplest measure of the variability of a quantitative data set is its **range.** The range was introduced briefly in the discussion of histograms in Section 2.4. We now define it formally:

Definition 3.9

The **range** of a set of data is equal to the largest measurement minus the smallest measurement.

EXAMPLE 3.7   Calculate the range of the following data set: 2, 7, 15, 4, 6, 10

Solution   The largest number is 15, and the smallest is 2. Thus, the range is

Range = 15 − 2 = 13   ∎

EXAMPLE 3.8   Consider the histograms for two different data sets shown in Figure 3.7.

Figure 3.7
Histograms for two data sets:
Example 3.8

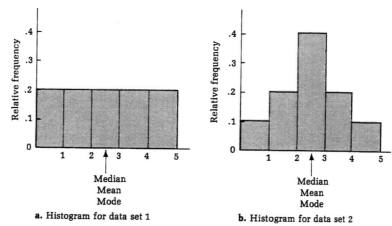

a. Histogram for data set 1   b. Histogram for data set 2

a. Calculate the range for data set 1.
b. Calculate the range for data set 2.
c. Compare the two values found in parts a and b. Discuss possible shortcomings of using the range as a measure of variability.

Solution   a. Range = 5 − 0 = 5.
b. Range = 5 − 0 = 5.
c. Both sets of data have a range of 5, but more of the measurements in data set 2 tend to concentrate near the center of its distribution than those in data set 1. Consequently, data set 2 is not as variable as data set 1. Although the range is easy to compute and easy to understand, the range is not a very sensitive measure of the variability in a data set.   ∎

A measure of variability that is more sensitive than the range is the **sample variance.**

Definition 3.10

The **sample variance** is equal to the sum of the squared differences between each of the n sample measurements and the sample mean, divided by $(n − 1)$. In symbols, using s^2 to represent the sample variance, this translates into the following formula:

$$s^2 = \frac{\sum (x − \bar{x})^2}{n − 1}$$

Notice that in the formula for the sample variance we use the *squared* differences—that is, we use $(x - \bar{x})^2$. If we calculate the *difference* $(x - \bar{x})$ between each measurement in a sample and the sample mean, some of these differences will be negative and some will be positive (since some measurements are smaller than $\bar{x}$ and some are larger). As a matter of fact, the sum of these differences will be exactly 0 for *any* sample of measurements. To eliminate the cancelling effect of the positive and negative values, the differences are squared when calculating the sample variance s^2.

You may wonder why we use the divisor $(n - 1)$ instead of n when calculating s^2. The reason is that by using the divisor $(n - 1)$, we obtain a quantity that is better suited to making inferences. Since we will ultimately want to use the sample variance s^2 to make inferences, $(n - 1)$ is preferred to n.*

If the sample mean, $\bar{x}$, is not an integer, calculating s^2 using the formula of Definition 3.10 is difficult. An equivalent formula that makes computations easier is shown in the box.

Calculating Formula for the Sample Variance

$$s^2 = \frac{\left(\begin{array}{c}\text{n times the sum of}\\ \text{the squares of}\\ \text{the sample measurements}\end{array}\right) - \left(\begin{array}{c}\text{Sum of}\\ \text{the sample}\\ \text{measurements}\end{array}\right)^2}{n(n - 1)}$$

Using our notation, we can write the calculating formula as

$$s^2 = \frac{n \sum x^2 - \left(\sum x\right)^2}{n(n - 1)}$$

EXAMPLE 3.9   Consider the two samples shown in the two dot diagrams in Figure 3.8.

Figure 3.8
Dot diagrams for
Example 3.9

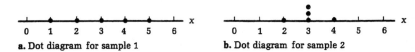

a. Dot diagram for sample 1   **b.** Dot diagram for sample 2

a. Calculate the sample mean, $\bar{x}$, for each sample.
b. Intuitively, which sample seems more variable?
c. Calculate the sample variance, s^2, for each sample.
d. Do the values calculated in part **c** agree with the answer to part **b**?

* If one were to divide by n rather than by $(n - 1)$, the resulting quantity would tend to underestimate the population variance. Dividing by $(n - 1)$ eliminates this tendency, and s^2 does not tend to underestimate or overestimate the population variance.

Solution **a.** The mean for sample 1 is

$$\bar{x} = \frac{1 + 2 + 3 + 4 + 5}{5} = \frac{15}{5} = 3$$

The mean for sample 2 is

$$\bar{x} = \frac{2 + 3 + 3 + 3 + 4}{5} = \frac{15}{5} = 3$$

b. Sample 1 seems to be more variable since its measurements do not concentrate near the mean as much as those in sample 2.

c. We will use sample 1 to illustrate how to calculate a sample variance using the calculating formula given in the preceding box. First, we construct Table 3.1. The sample measurements are listed in the first column under the heading x. The sum of the measurements is 15; that is, $\sum x = 15$. The square of each measurement is given in the second column. The sum of these numbers is $\sum x^2 = 55$.

Table 3.1

Table for calculating s^2: Sample 1

	x	x^2
	1	1
	2	4
	3	9
	4	16
	5	25
Sums	15	55

We can now calculate s^2. Since $n = 5$, the equation is

$$s^2 = \frac{n\sum x^2 - (\sum x)^2}{n(n-1)} = \frac{5(55) - (15)^2}{5(5-1)}$$

$$= \frac{275 - 225}{5(4)} = \frac{50}{20} = 2.5$$

We will now find s^2, the sample variance, for sample 2. The calculations are shown in Table 3.2.

Table 3.2

Table for calculating s^2: Sample 2

	x	x^2
	2	4
	3	9
	3	9
	3	9
	4	16
Sums	15	47

From the table, we have $\sum x = 15$ and $\sum x^2 = 47$. Using the calculating formula for s^2, we obtain the sample variance for sample 2:

$$s^2 = \frac{n\sum x^2 - (\sum x)^2}{n(n-1)} = \frac{5(47) - (15)^2}{5(5-1)}$$

$$= \frac{235 - 225}{5(4)} = \frac{10}{20} = 0.5$$

d. The variance of sample 1, $s^2 = 2.5$, is larger than the variance of sample 2, $s^2 = 0.5$. This indicates that the measurements in sample 1 are spread farther from the mean than those in sample 2. This result agrees with the answer to part **b.**   ■

The sample variance, s^2, is always expressed in squared units rather than the original units of measurement. A measure of variability that eliminates this awkward situation and is easier to interpret is the **sample standard deviation.**

Definition 3.11

The **sample standard deviation** is the positive square root of the sample variance. In symbols, using s to represent the sample standard deviation, this translates into the following formula:

$$s = \sqrt{s^2}$$

EXAMPLE 3.10   Refer to Example 3.9.

a. Calculate the standard deviation for sample 1. (Recall that $s^2 = 2.5$.)
b. Calculate the standard deviation for sample 2. (Recall that $s^2 = 0.5$.)
c. Compare the answers you get in parts **a** and **b.**

Solution   **a.** The sample standard deviation for sample 1 is

$$s = \sqrt{s^2} = \sqrt{2.5} \approx 1.6^*$$

b. For sample 2, the standard deviation is

$$s = \sqrt{s^2} = \sqrt{0.5} \approx 0.7$$

c. The standard deviation for sample 1 is larger than that for sample 2. This indicates that sample 1 is more variable, which, of course, agrees with our comparison of the two variances. A more significant application of the sample standard deviation will be provided in Section 3.4.   ■

* The symbol "$\approx$" means "approximately equal to." Square roots can be obtained from most electronic calculators, and throughout this text we will assume students have access to such instruments.

When calculating descriptive measures such as the mean, variance, and standard deviation, we will often have to approximate, or round off, an answer. We will use the following rules when rounding off.

Rules for Rounding Off

1. Carry a large number of decimal places in intermediate calculations. Round only in the final answer.
2. Use one more decimal place in the answer than appears in the original data.
3. If only a 5 follows the number in the decimal place of interest, round down if the number is even and round up if it is odd.

EXAMPLE 3.11   The following is a sample of $n = 9$ measurements:

8,   2,   2,   7,   4,   6,   5,   3,   4

a. Calculate the sample variance, s^2, using the calculating formula.
b. Calculate the standard deviation, s.

Solution   **a.** To use the calculating formula, we must first find the values of $\sum x$ and $\sum x^2$. These values are conveniently found in the following table:

x	x^2
8	64
2	4
2	4
7	49
4	16
6	36
5	25
3	9
4	16
Sums 41	223

Looking at the table, we see that

$$\sum x = 41 \quad \text{and} \quad \sum x^2 = 223$$

Since $n = 9$, we calculate

$$s^2 = \frac{n\sum x^2 - (\sum x)^2}{n(n-1)} = \frac{9(223) - 41^2}{9(8)} = \frac{2{,}007 - 1{,}681}{72}$$

$$= \frac{326}{72} = 4.527778 \approx 4.5$$

Rounding to one more decimal place than the original data contained, we get $s^2 = 4.5$.

b. To calculate the standard deviation, s, we use the value of s^2 that has not been rounded. Thus,

$$s = \sqrt{s^2} = \sqrt{4.527778} \approx 2.127858$$

We now round this answer to one decimal place and obtain

$$s = 2.1. \qquad \blacksquare$$

The mean, variance, and standard deviation are descriptive measures that can be calculated for either a sample of data or an entire population of data. In this section we calculated the sample mean, $\bar{x}$, the sample variance, s^2, and the sample standard deviation, s. Although we will not calculate the corresponding population descriptive measures at this time (these calculations will be discussed in Chapter 5), we are ready to introduce symbols for these quantities. The symbols μ (mu) and σ (sigma), which are lowercase Greek letters, are used to represent the mean and standard deviation, respectively, of a population. Since we will want to refer to both sample and population descriptive measures, the notation given in the box will help differentiate between the two.

Population Descriptive Measures

μ = Population mean

σ^2 = Population variance

σ = Population standard deviation

Sample Descriptive Measures

$\bar{x}$ = Sample mean

s^2 = Sample variance

s = Sample standard deviation

COMPUTER EXAMPLE 3.2   There is some concern about the nutritional value of the fast foods American consume. Table 3.3 lists several items from the fast-food world that have markedly high levels of sodium. We will use Minitab to calculate the mean and standard deviation of the given values of sodium (in milligrams).

Table 3.3

High-sodium fast food items

Food Item	Chain	Sodium (mg)
Large roast beef w/cheese	Roy Rogers	1,953
Pancake breakfast	Jack-in-the-Box	1,815
Hot Ham 'n Cheese	Arby's	1,655
Chicken Supreme	Jack-in-the-Box	1,582
Pasta seafood salad	Jack-in-the-Box	1,570
Specialty ham & cheese	Burger King	1,550
Ham biscuit	Hardee's	1,415
Pancake platter	Roy Rogers	1,264
Potato w/bacon & cheese	Wendy's	1,180
Whaler sandwich	Burger King	1,013

Source: "The Fast-Food Eating Guide," Business and Society Review, Vol. 59, Feb. 1986, p. 57.

The sodium levels shown in Table 3.3 were input into the computer, and Minitab calculated the following values:

```
MEAN    =    1499.7
ST.DEV. =    287.73
```

The mean milligrams of sodium for these fast-food items is 1,499.7 and the standard deviation is 287.73 (which could be rounded to 287.7). Note that Minitab uses the abbreviation "ST.DEV." for standard deviation. ∎

Exercises (3.18–3.31)

Learning the Language

3.18 Define the following terms:
a. Measure of variability   c. Sample variance
b. Range   d. Sample standard deviation

3.19 What is denoted by the following symbols?
a. μ b. x^2 c. $\sum x^2$ d. s e. s^2 f. σ g. σ^2

Learning the Mechanics

3.20 Consider the following sample: 2, 0, 4, 3, 5
a. Calculate $\sum x$.   c. Find the value of n.   e. Calculate s.
b. Calculate $\sum x^2$.   d. Calculate s^2.

3.21 Calculate the range, variance, and standard deviation for each of the following samples:
a. 4, 2, 6, 5, 6, 6   c. $-3, 0, -2, 1, 4, 1, 1, 3, 7$
b. 0, 4, 0, 2, 3, 4, 0, 3

3.22 Find the sample variance, s^2, using the information given in each of the following:
a. $\sum x = 10$, $\sum x^2 = 500$, $n = 50$ c. $\sum x = 10$, $\sum x^2 = 500$, $n = 20$
b. $\sum x = 50$, $\sum x^2 = 10$, $n = 500$ d. $\sum x = 20$, $\sum x^2 = 500$, $n = 10$

3.23 Calculate the range, variance, and standard deviation for each of the following samples:
  a.   −1, −3, −2, −2, −2, 0, 7   c.   1, 9, 9, 9, 9, 9, 100
  b.   2, 9, 8, 3, 7, 7, 6, 7, 5

3.24 Three sets of data are shown in the accompanying graphs. All three data sets have the same mean, $\bar{x} = 50$, and the same range, 100, but they have different variances. Which data set is the most variable? Which is the least variable?

a.

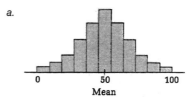

c.

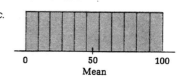

b.

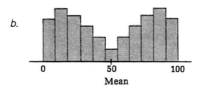

3.25 The mean and variance of a sample are calculated to be the following:

$$\bar{x} = 23.64516 \qquad s^2 = 9.575$$

  a.   Assuming the observations in the sample are integers, round off the values of $\bar{x}$ and s^2.
  b.   Assuming the sample has values with one decimal place, round off the values of $\bar{x}$ and s^2.
  c.   Calculate the value of s and round it off for the sample as described in part b.

Applying the Concepts

3.26 A group of twenty economists was asked to project what the prime interest rate of a well-known New York bank would be at the start of the second quarter next year. The following are their projections:

| 0.11 | 0.13 | 0.13 | 0.12 | 0.09 | 0.10 | 0.12 | 0.10 | 0.10 | 0.09 |
| 0.11 | 0.11 | 0.09 | 0.11 | 0.15 | 0.09 | 0.11 | 0.11 | 0.09 | 0.13 |

Compute $\bar{x}$, s^2, and s for the data.

3.27 A research cardiologist is interested in knowing the age when adult male heart attack victims suffer their first heart attack. The cardiologist takes a random sample of the medical records of thirty coronary patients and obtains the following results (recorded in years):

51	64	43	54	52	38	45	70	75	71
49	42	62	55	65	63	40	61	49	57
58	67	53	54	44	59	54	42	60	50

a. Calculate $\bar{x}$, s^2, and s for this data set.

b. Add two standard deviations to the sample mean, and subtract two standard deviations from the sample mean. How many of the original thirty measurements fall within the interval $(\bar{x} - 2s)$ to $(\bar{x} + 2s)$?

c. Repeat part b using three standard deviations. How many of the original thirty measurements fall within the interval $(\bar{x} - 3s)$ to $(\bar{x} + 3s)$?

3.28 The final grades given by two professors in an introductory statistics course have been carefully examined. Students in the first professor's class had grades with an average of 3.0 and a standard deviation of 0.2. Those in the second professor's class had grades with an average of 3.0 and a standard deviation of 1.0. If you had a choice, which professor would you take for this course? Explain.

3.29 Consider the following two samples:

Sample 1:   10, 0, 1, 9, 10, 0

Sample 2:   0, 5, 10, 5, 5, 5

a. Examine both samples and identify the one that you believe has greater variability.

b. Calculate the range for each sample. Does this agree with your answer to part a? Explain.

c. Calculate the variance for each sample. Does this agree with your answer to part a? Explain.

d. Which of the two—the range or the variance—provides a better measure of variability? Why?

3.30 The accompanying table shows the 1975 estimated daily production of crude oil in the United States, given in thousands of barrels (1 barrel equals 42 U.S. gallons).

State	Estimated Production	State	Estimated Production
Alabama	30	Montana	95
Alaska	198	Nebraska	17
Arkansas	42	New Mexico	261
California	1,055	New York	2
Colorado	100	North Dakota	55
Florida	113	Ohio	25
Illinois	75	Oklahoma	438
Indiana	12	Pennsylvania	9
Kansas	164	Texas	3,813
Kentucky	20	Utah	102
Louisiana	1,670	West Virginia	7
Michigan	60	Wyoming	377
Mississippi	131	Miscellaneous	6

Total—U.S. 8,877

Source: Douglas M. Considine (ed.), *Energy Technology Handbook* (New York: McGraw-Hill, 1977), Table 8, pp. 3-60.

*a.   Use a computer package to calculate the mean and standard deviation for this data set. Order the measurements as well (this will help answer parts b and c).

b.   What fraction of the measurements fall within two standard deviations of the mean?

c.   What fraction of the measurements fall within three standard deviations of the mean?

3.31   The data shown in the accompanying table are the mine and alluvial production figures for uncut diamonds in 1976. Both gem and industrial diamonds are included in the total production. Industrial diamonds are small and impure diamonds that are usable only as abrasives. It should also be noted that the figures given refer to legal production of diamonds. There is a large illicit trade in illegally produced diamonds.

Country	Production (Thousands of carats)
Zaire	11,820
Soviet Union	9,900
South Africa	7,022
Botswana	2,361
Ghana	2,343
Namibia	1,694
Sierra Leone	1,500
Venezuela	833
Angola	660
Tanzania	450
Central African Empire	405
Liberia	400
Brazil	270
Guinea	80
Ivory Coast	60
India	20
Indonesia	15
Guyana	14
Lesotho	3

Source: George T. Kurian, The Book of World Rankings
(New York: Facts on File, Inc., 1979), p. 190.

*a.   Use a computer package to calculate $\bar{x}$ and s for the data.

b.   Calculate $\bar{x} - 2s$ and $\bar{x} + 2s$ to indicate where most of the measurements fall.

c.   How many of the nineteen measurements actually fall in this interval?

3.3

MEASURES OF RELATIVE STANDING

We will now consider a third category of numerical descriptive measures—those that describe the relative position of an observation within a data set.

Definition 3.12

A **measure of relative standing** is a number that indicates the position of an individual measurement within a data set in relation to the remaining measurements.

A measure of relative standing commonly used to evaluate performance on standardized educational tests is the **percentile.**

Definition 3.13

The **pth percentile** of a data set is a number such that p% of the measurements in the data set are smaller than or equal to it and the remaining measurements are greater than or equal to it.

A percentile is best used to locate a measurement within a very large data set. Since we will not present the actual data for such large data sets, we will not calculate percentiles directly from the data, but instead will calculate them using a graph called a **cumulative relative frequency distribution.** In the following example we will discuss what a cumulative frequency distribution is, and how it may be used to find percentiles.

EXAMPLE 3.12   The graph in Figure 3.9 is a cumulative relative frequency distribution. The set of data used to construct this graph consists of the ages of all registered voters in a city in

Figure 3.9
Cumulative relative frequency distribution for the ages of registered voters in a city in the United States

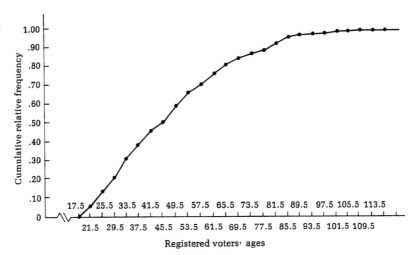

the United States. The graph was constructed by first plotting a point above each value on the horizontal axis at a height equal to the fraction of registered voters whose ages are equal to or less than that value. Successive points were then connected by straight lines to form the graph. The graph is called a *cumulative relative frequency distribution* because each plotted point, the fraction of registered voters less than or equal to an age, represents the total, or accumulation, of the relative frequencies for the younger age categories.

The youngest age at which a voter may register is 18, so the cumulative relative frequency is 0 until the category beginning with 17.5. The rate at which the graph rises decreases for the older age categories because fewer registered voters are in these categories. Finally, at an age of about 110 the graph attains a value of 1.00, indicating all registered voters in this city are younger than 110 years of age.

We can find the approximate fraction of voters less than or equal to any age directly from the graph. For example, at an age of 57.5 years, the cumulative relative frequency is .70. Thus, .70 of all registered voters in the city are 57.5 years old or younger; this means that 57.5 is the 70th percentile.

Now, use the cumulative relative frequency distribution in Figure 3.9 to find the 20th, 50th, and 99th percentiles.

Solution   The calculation of the 20th, 50th, and 99th percentiles is demonstrated in Figure 3.10 (a copy of the graph in Figure 3.9). To find the 20th percentile, draw a horizontal line from the cumulative relative frequency of .20 to the line connecting the points. Then draw a vertical line from there to the horizontal axis, and the value of 29.5 on the horizontal axis is the 20th percentile. Twenty percent of all the registered voters are approximately 29.5 years old or younger. Similarly, we find that the 50th percentile

Figure 3.10
Cumulative relative frequency distribution for the ages of registered voters in a city in the United States

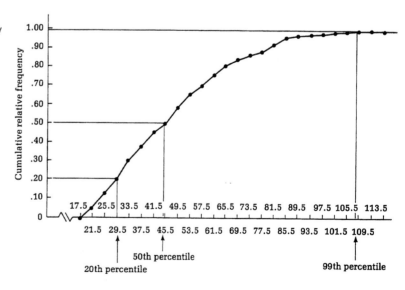

(the median) is approximately 45 years, and the 99th percentile is approximately 108 years. ∎

A cumulative relative frequency distribution is very useful for finding percentiles. Since we will concentrate on interpreting percentiles, rather than calculating them, we will not discuss how to construct cumulative relative frequency distributions any further.* The graph will be provided whenever it is needed.

EXAMPLE 3.13   Suppose you score 670 on the verbal portion of the Scholastic Aptitude Test (SAT) and the testing service reports that a score of 670 is at the 98th percentile.

a. What percentage of people taking the test scored less than or equal to 670?
b. What percentage scored higher than 670?

Solution   **a.** Since a score of 670 is at the 98th percentile, 98% of the examination scores were less than or equal to 670.
b. Consequently, 100% − 98% = 2% of the examination scores were higher than 670. ∎

You may sometimes hear or see the term **quartile.** Actually, quartiles are names given to certain percentiles.

Definition 3.14

The **first quartile** is the 25th percentile.
The **second quartile** is the 50th percentile.
The **third quartile** is the 75th percentile.

Thus, the three quartiles separate the data into four equal parts, as shown in Figure 3.11.

Figure 3.11
A display of the first, second, and third quartiles

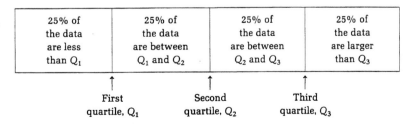

| 25% of the data are less than Q_1 | 25% of the data are between Q_1 and Q_2 | 25% of the data are between Q_2 and Q_3 | 25% of the data are larger than Q_3 |

First quartile, Q_1 Second quartile, Q_2 Third quartile, Q_3

---

* If you are interested in constructing a cumulative relative frequency distribution, refer to Chapter 3 of *Statistics: A Tool for the Social Sciences* (Ott, Larson, & Mendenhall, 1983).

Another popular measure of relative standing uses the mean and standard deviation of a data set and is called a **z-score.** We will calculate z-scores for both samples and populations.

Definition 3.15

A **z-score** is a number calculated for an individual measurement in a data set using the following rule:

$$z = \frac{\text{(Individual measurement)} - \text{(Mean of the data set)}}{\text{Standard deviation of the data set}}$$

Denoting an individual measurement by the symbol x, and using our standard notation, the z-scores for a measurement are given by the following formulas:

SAMPLE z-SCORE   POPULATION z-SCORE

$$z = \frac{x - \bar{x}}{s} \qquad\qquad z = \frac{x - \mu}{\sigma}$$

A z-score is the distance between an individual measurement, x, and the mean, expressed in standard deviations. That is, the z-score gives the number of standard deviations from the mean for the individual measurement.

EXAMPLE 3.14   The scores on the mathematics portion of the March 1985 Scholastic Aptitude Test (SAT) had a mean of 455 and a standard deviation of 112. Suppose a student scored 575 on this test.

a. What is the z-score for this student's test score?
b. Interpret this z-score.

Solution   **a.** Since we are discussing *all* the scores on this particular test, we will consider this to be a population of data. Therefore, we denote the mean and standard deviation by

$$\mu = 455 \quad \text{and} \quad \sigma = 112$$

Since we are interested in the score x = 575, the z-score is

$$z = \frac{x - \mu}{\sigma} = \frac{575 - 455}{112} \approx 1.07$$

b. This means that the student's test score of 575 is 1.07 standard deviations above the mean test score. Figure 3.12 shows the relative position of this student's test score.

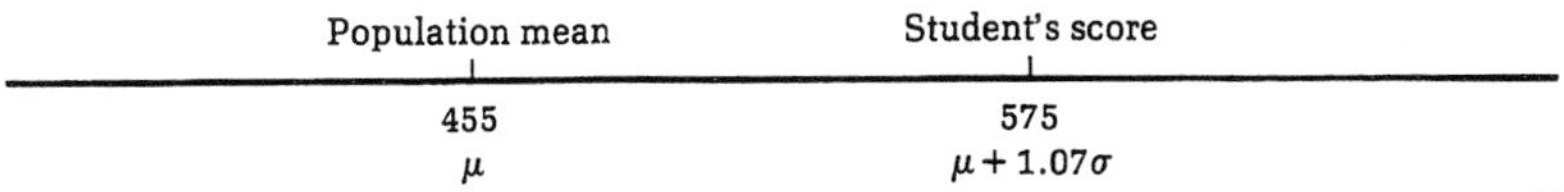

Figure 3.12
Relative position of $x = 575$

Population mean	Student's score
455	575
μ	$\mu + 1.07\sigma$

∎

The following facts are useful when discussing z-scores:

Facts About z-Scores

1. A z-score will rarely be smaller than -3.0 or larger than $+3.0$.
2. A large positive z-score indicates that a measurement is larger than most other measurements.
3. A "large" negative z-score indicates that a measurement is smaller than most other measurements.
4. A z-score near 0 indicates that a measurement is near the mean of a set of data.

We may also use z-scores to compare measurements from two different sets of data.

EXAMPLE 3.15   Refer to Example 3.14. Suppose another student took the mathematics portion of the American College Test (ACT) in 1985 and scored 30. The mean score on this part of the ACT was 17.3 and the standard deviation was 7.9.

a. What is the z-score for a test score of 30 on the ACT?

b. Relative to the two sets of test scores, which test score is higher, 575 on the SAT or 30 on the ACT?

Solution   **a.** We are considering all the scores on the ACT in 1985, and thus we will denote the mean and standard deviation by

$$\mu = 17.3 \quad \text{and} \quad \sigma = 7.9$$

The z-score for $x = 30$ is

$$z = \frac{x - \mu}{\sigma} = \frac{30 - 17.3}{7.9} \approx 1.61$$

b. Recall that the z-score for an SAT score of 575 was 1.07. Since the ACT score of 30 has a larger z-score (1.61), we see that the score on the ACT is higher than the score on the SAT in a relative sense. That is, in reference to the two populations of test scores, the ACT score is 1.61 standard deviations above the mean, while the SAT score is only 1.07 standard deviations above the mean. Remember, this comparison is relative to the different means and standard deviations of the two sets of test scores.   ∎

Exercises (3.32–3.43)

Learning the Language

3.32 Define the following terms:
  a. Measure of relative standing
  b. Percentile
  c. Quartile
  d. z-score

Learning the Mechanics

3.33 Calculate the z-score for the x-value given in each of the following situations:
  a. $x = 10,$ $\bar{x} = 5,$ $s = 5$
  b. $x = 10,$ $\bar{x} = 0,$ $s = 5$
  c. $x = 0,$ $\bar{x} = 5,$ $s = 10$
  d. $x = 50,$ $\mu = 30,$ $\sigma = 15$
  e. $x = 4{,}000,$ $\mu = 5{,}000,$ $\sigma = 500$

3.34 The accompanying graph is a smooth curve in the approximate shape of a histogram for a very large set of data. The median and several percentiles are shown on the x-axis of the graph. Using this figure, answer the following questions.

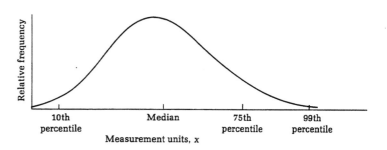

  a. What percentage of measurements are less than the 10th percentile? The median?
  b. What percentage of measurements are greater than the 10th percentile? The 99th percentile?
  c. What percentage of measurements are between the median and the 75th percentile? Between the 10th and 99th percentiles?

3.35 In which of the following situations is the x-value largest in relation to the data set from which it comes?
  a. $x = 37,$ $\bar{x} = 20,$ $s = 10$
  b. $x = 500,$ $\bar{x} = 200,$ $s = 250$
  c. $x = 3.0,$ $\bar{x} = 1.0,$ $s = 0.7$

3.36   A cumulative relative frequency distribution is shown below:

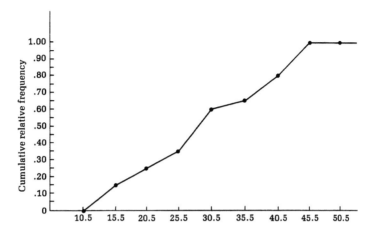

a.   Find the first, second, and third quartiles.
b.   Find the 10th, 25th, 60th, and 75th percentiles.

3.37   In which of the following situations is the x-value smallest in relation to the data set from which it comes?

a. $x = 6$, $\mu = 20$, $\sigma = 9$
b. $x = 350$, $\mu = 400$, $\sigma = 20$
c. $x = 1.6$, $\mu = 2.0$, $\sigma = 0.3$

3.38   In which of the following situations is the x-value farthest from the mean of the data set from which it comes?

a. $x = 20$, $\bar{x} = 47$, $s = 10$
b. $x = 402$, $\mu = 350$, $\sigma = 20$
c. $x = 1.68$, $\mu = 2.00$, $\sigma = 0.20$

3.39   In the exercises for Section 3.2 (Exercises 3.21 and 3.23), you calculated the standard deviation, s, for each of the samples listed below. For convenience, we give this value and the value of the sample mean, $\bar{x}$, with each sample:

(i) 4, 2, 6, 5, 6, 6	$\bar{x} = 4.8$	$s = 1.6$
(ii) 0, 4, 0, 2, 3, 4, 0, 3	$\bar{x} = 2.0$	$s = 1.8$
(iii) -3, 0, -2, 1, 4, 1, 1, 3, 7	$\bar{x} = 1.3$	$s = 3.0$
(iv) -1, -3, -2, -2, -2, 0, 7	$\bar{x} = -0.4$	$s = 3.4$
(v) 2, 9, 8, 3, 7, 7, 6, 7, 5	$\bar{x} = 6.0$	$s = 2.3$
(vi) 1, 9, 9, 9, 9, 9, 100	$\bar{x} = 20.9$	$s = 35.0$

a.   Calculate the largest and smallest z-score for each sample.
b.   Examine the answers to part a. Do these agree with the statement that "a z-score will rarely be smaller than -3.0 or larger than $+3.0$"? Explain.

Applying the Concepts

3.40 The numbers of revenue passengers enplaned in 1983 by the major domestic airlines are shown in the table. The figures given are in millions and have been rounded off to the nearest hundred thousand. The mean and standard deviation of these figures are 20.39 and 11.8, respectively.

Airline	Revenue Passengers Enplaned
American	29.8
Continental	9.3
Delta	36.1
Eastern	33.9
Northwest	10.4
Pan Am	6.2
Republic	17.8
TWA	15.6
United	38.1
USAir	16.4
Western	10.7

Source: Table, "Summary of Major Carriers Domestic Traffic for 1983," *World Aviation Directory,* No. 89, Winter 1984–1985, p. 92.

a.   Find the *z*-score for the number of revenue passengers enplaned for each airline.

b.   How many standard deviations away from the mean is each of the figures for revenue passengers enplaned?

3.41 The composite scores on an ACT had a mean of 18.5 and a standard deviation of 5.8. The composite scores on an SAT given the same year had a mean of 440 and a standard deviation of 110.

a.   Suppose you are told that a person scored at the 90th percentile on the ACT. Interpret this statement.

b.   Suppose that one student scored a composite 25 on the ACT and a second student scored a 680 on the SAT. Which student has a higher score relative to his or her individual test? Why?

3.42 In 1974 there were 160 inhabitants per nurse in the United States. Relative to countries worldwide, this placed the United States in the 6th percentile.* Explain how this locates the United States in the inhabitants-per-nurse distribution for countries worldwide.

3.43 Many firms use on-the-job training to teach their employees computer programming. Suppose you work in the Personnel Department of a firm that just finished training a group of its employees to program, and you have been requested to review the performance of one of the trainees on the final test that was given to all trainees. The

* *Source:*   George T. Kurian, *The Book of World Rankings* (New York: Facts on File, Inc., 1979), p. 292.

mean and standard deviation of the test scores are 80 and 5, respectively, and the distribution of scores is symmetric.

a. The employee in question scored 65 on the final test. Compute the employee's z-score.

b. If a trainee were arbitrarily selected from those who had taken the final test, is it more likely that he or she would score 90 or above, or 65 or below?

3.4

USING NUMERICAL MEASURES TO DESCRIBE DATA SETS

Now that we have calculated some numerical descriptive measures, we will discuss how they can be used to construct a visual picture of the frequency distribution for a data set. The rule of thumb given in the next box will help us use the mean and standard deviation to describe a data set and ultimately to make inferences.

Rule of Thumb

For sets of data that result from real-life experiments, the following statements are generally true:

1. Most of the measurements will be within 2 standard deviations of the mean. Equivalently, few of the measurements will lie more than 2 standard deviations from the mean.
2. All or almost all of the measurements will be within 3 standard deviations of the mean. Equivalently, none or almost none of the measurements will lie more than 3 standard deviations from the mean.

We have used the rather vague term *most* to describe the number of measurements that are within 2 standard deviations of the mean, and you might be wondering what we mean by *most*. A Russian mathematician named Tchebysheff proved that for *any* set of data, *at least* 75% of the measurements will lie within 2 standard deviations of the mean.* However, most data sets contain a much higher percentage of measurements within 2 standard deviations of the mean. As a matter of fact, for the 75% figure to be appropriate, a data set would have to have a rather unique distribution—one almost identical to that shown in Figure 3.13 (page 106). If a set of data has any other distribution, the percentage of measurements within 2 standard deviations of the mean is higher than 75%.

* If you are interested in more details concerning Tchebysheff's result, consult *Introduction to Probability and Statistics* (Mendenhall, 1983).

Figure 3.13
Distribution of a data set
containing 75% of the
measurements within
2 standard deviations of
the mean

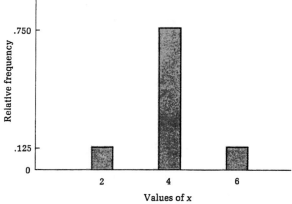

Figure 3.14
Six distributions of data in
which over 90% of the
measurements are within
2 standard deviations of
the mean

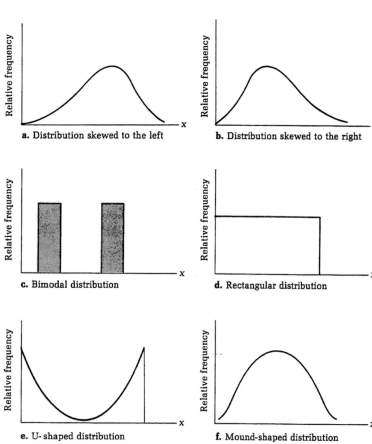

Practical, real-life experiments have shown that for most sets of data, 90% or more of the measurements are within 2 standard deviations of the mean. Since the actual percentage will vary somewhat from data set to data set, and the *exact* percentage is not important to us at this time, we have chosen to use the descriptive term *most* in the rules of thumb given above. When making statistical inferences, the fact that is important is that there are few measurements more than 2 standard deviations from the mean and none or almost none more than 3 standard deviations from the mean for practical sets of data.

In Figure 3.14 we have sketched the graphs of six different distributions of data. The shapes of these distributions are quite different, but in each case *more than 90% of the measurements in each set of data are within 2 standard deviations of the mean.* Thus, although Tchebysheff's result guarantees only that at least 75% of the measurements are within 2 standard deviations of the mean for any set of data, the actual percentage is usually much larger. As long as the distribution of a set of data does not have the unique shape shown in Figure 3.13, we can expect over 90% of the measurements to be within 2 standard deviations of the mean.

As we progress in this text, we will study specific distributions of data and calculate the exact percentage of measurements within 2 or 3 standard deviations of the mean. As you will see, the results will always agree with the rule of thumb given above. Exercises 3.44, 3.45, and 3.46 are designed to convince you that the rule of thumb may be used to describe sets of data with various distributions. We will demonstrate the mechanisms of the rule of thumb with the following example.

EXAMPLE 3.16   A sample of 100 measurements contains one 1, two 2's, three 3's, four 4's, five 5's, ten 6's, fifteen 7's, etc., as listed in the following table:

Sample Value, x	1	2	3	4	5	6	7	8	9	10
Frequency of Occurrence	1	2	3	4	5	10	15	25	20	15

We have calculated the sample mean and sample standard deviation to be

$$\bar{x} = 7.5 \quad \text{and} \quad s = 2.0$$

a. Draw a bar graph for this sample.

b. Find the number of sample measurements within 2 standard deviations of the mean. Does this agree with the rule of thumb?

c. Find the number of measurements more than 2 standard deviations from the mean. Does this agree with the rule of thumb?

d. Find the number of measurements within 3 standard deviations of the mean. Does this agree with the rule of thumb?

e. Find the number of measurements more than 3 standard deviations from the mean. Does this agree with the rule of thumb?

Solution   **a.** A bar graph for the 100 measurements is shown in Figure 3.15 (page 108). Notice that the data are distinctly skewed to the left.

Figure 3.15
Bar graph for Example 3.16

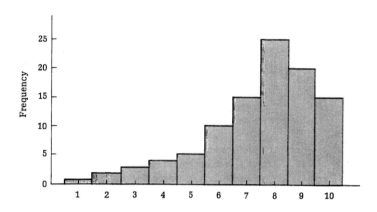

b. We must first find the values that are 2 standard deviations above and below the mean. These are

$$\bar{x} + 2s = 7.5 + 2(2.0) = 7.5 + 4.0 = 11.5$$

and

$$\bar{x} - 2s = 7.5 - 2(2.0) = 7.5 - 4.0 = 3.5$$

Thus, all the sample measurements between 3.5 and 11.5 are within 2 standard deviations of the mean, and these are the measurements that take values 4 through 10, inclusive. Referring to the table given at the beginning of the example, we see that there are $4 + 5 + 10 + 15 + 25 + 20 + 15 = 94$ such measurements. Of the 100 measurements, 94 are within 2 standard deviations of the mean. This certainly agrees with the rule of thumb which states: *Most of the measurements will be within 2 standard deviations of the mean.*

c. Since 94 measurements are within 2 standard deviations of the mean, only 6 $(100 - 94 = 6)$ are more than 2 standard deviations from the mean. Again, this agrees with the rule of thumb: *Few of the measurements will lie more than 2 standard deviations from the mean.*

d. We now calculate

$$\bar{x} + 3s = 7.5 + 3(2.0) = 7.5 + 6.0 = 13.5$$

and

$$\bar{x} - 3s = 7.5 - 3(2.0) = 7.5 - 6.0 = 1.5$$

All the measurements between 1.5 and 13.5 are within 3 standard deviations of the mean, and these are the values 2 through 10, inclusive. Referring to the table, we see that there are

$$2 + 3 + 4 + 5 + 10 + 15 + 25 + 20 + 15 = 99$$

such measurements. Of the 100 measurements, 99 are within 3 standard deviations of the mean. This agrees very nicely with the rule of thumb: *All or almost all of the measurements will be within 3 standard deviations of the mean.*

e. Since 99 measurements are within 3 standard deviations of the mean, only 1 $(100 - 99 = 1)$ measurement is more than 3 standard deviations from the mean. Of course, this agrees with the rule of thumb: *None or almost none of the measurements will lie more than 3 standard deviations from the mean.* ■

EXAMPLE 3.17 To help consumers assess the risks they are taking, the Food and Drug Administration (FDA) publishes information on the amount of nicotine found in all commercial brands of cigarettes. Suppose a new cigarette has recently been marketed, and the FDA tests this cigarette for nicotine content. For 1,000 cigarettes tested, the mean nicotine content was 26.4 milligrams and the standard deviation was 2.0 milligrams. Use the rule of thumb to describe the sample of nicotine contents.

Solution The rule of thumb tells us that most of the measurements will lie within 2 standard deviations of the mean. First, we should calculate $(\bar{x} - 2s)$ and $(\bar{x} + 2s)$. The FDA has found that $\bar{x} = 26.4$ and $s = 2.0$, so,

$$\bar{x} - 2s = 26.4 - 2(2.0) = 26.4 - 4.0 = 22.4$$
$$\bar{x} + 2s = 26.4 + 2(2.0) = 26.4 + 4.0 = 30.4$$

Thus, most of the cigarettes sampled should contain between 22.4 and 30.4 milligrams of nicotine.

The rule of thumb also states that almost all of the measurements are within 3 standard deviations of the mean. Thus, since

$$\bar{x} - 3s = 26.4 - 3(2.0) = 26.4 - 6.0 = 20.4$$
$$\bar{x} + 3s = 26.4 + 3(2.0) = 26.4 + 6.0 = 32.4$$

very few, if any, sampled cigarettes will contain less than 20.4 milligrams of nicotine or more than 32.4 milligrams of nicotine. ■

EXAMPLE 3.18 The rule of thumb for interpreting the value of a standard deviation can be put to a practical use as a check on the calculation of a standard deviation. Suppose you have a data set for which the smallest measurement is 20 and the largest is 80. You have calculated the standard deviation of the data set to be

$$s = 190$$

How can you use the rule of thumb to provide a rough check on your calculated value of s?

Solution Most of the measurements will be within 2 standard deviations of the mean, and approximately all of them will fall within 3 standard deviations of the mean. Consequently, we expect the range of the measurements to be equal to somewhere between 4 and 6 standard deviations—that is, between $4s$ and $6s$ (see Figure 3.16 on page 110). For the given data set, the range is

Range = Largest measurement − Smallest measurement
$$= 80 - 20 = 60$$

Then, if we let the range equal approximately 6s, we obtain

$$6s \approx \text{Range}$$

$$s \approx \frac{\text{Range}}{6} = \frac{60}{6} = 10$$

Or, if we let the range equal approximately 4s (see Figure 3.16), we obtain a larger value for s:

$$4s \approx \text{Range}$$

$$s \approx \frac{\text{Range}}{4} = \frac{60}{4} = 15$$

Figure 3.16
The relationship between the range and the standard deviation

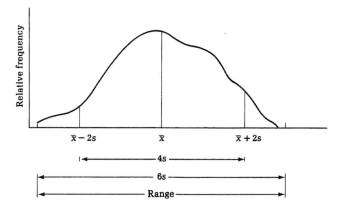

Now, you can see that in this example it does not make much difference whether you let the range equal 4s or 6s.* In either case, it is clear that the value you have supposedly calculated, s = 190, is too large, and you should check your calculations.   ∎

EXAMPLE 3.19   The SBTPE considered in the introduction to this chapter has a mean of 2,000 and a standard deviation of 400. Use the rule of thumb to describe the results and to estimate the highest and lowest scores on this test.

Solution   Most measurements will lie within 2 standard deviations of the mean. Since $\mu = 2{,}000$ and $\sigma = 400$, most measurements will be between

$$\mu - 2\sigma = 2{,}000 - (2)(400) = 1{,}200$$

and

$$\mu + 2\sigma = 2{,}000 + (2)(400) = 2{,}800$$

That is, most scores will be between 1,200 and 2,800.

* Note that $s \approx \text{Range}/4$ always provides a larger—and thus more conservative—estimate for s than $s \approx \text{Range}/6$. Therefore, we will use $s \approx \text{Range}/4$ throughout the remainder of this text.

Almost all measurements will lie within 3 standard deviations of the mean, that is, between

$$\mu - 3\sigma = 2{,}000 - (3)(400) = 800$$

and

$$\mu + 3\sigma = 2{,}000 + (3)(400) = 3{,}200$$

Thus, almost all scores will be between 800 and 3,200. This corresponds exactly to the claim made in the introduction and gives estimates for the highest and lowest test scores. ∎

COMPUTER
EXAMPLE 3.3

In Computer Example 3.1, we used Minitab to order 100 mileage ratings from the smallest to the largest. The results of that analysis are repeated in Figure 3.17. We also calculated the mean of the 100 mileage ratings to be 36.994, which we will round to 37.0. We will now use Minitab to find the standard deviation of the mileage ratings.

MILEAGES

30.0	31.8	32.5	32.7	32.9	32.9	33.1	33.2	33.6	33.8	33.9
33.9	34.0	34.2	34.4	34.5	34.8	34.8	35.0	35.1	35.2	35.3
35.5	35.6	35.6	35.7	35.8	35.9	35.9	36.0	36.1	36.2	36.3
36.3	36.4	36.4	36.5	36.5	36.6	36.6	36.7	36.7	36.7	36.8
36.8	36.8	36.9	36.9	36.9	37.0	37.0	37.0	37.0	37.1	37.1
37.1	37.2	37.2	37.3	37.3	37.4	37.4	37.5	37.6	37.6	37.7
37.7	37.8	37.9	37.9	38.0	38.1	38.2	38.2	38.3	38.4	38.5
38.6	38.7	38.8	39.0	39.0	39.3	39.4	39.5	39.7	39.8	39.9
40.0	40.1	40.2	40.3	40.5	40.5	40.7	41.0	41.0	41.2	42.1
44.9										

Figure 3.17
EPA mileage ratings ordered from smallest to largest

Minitab calculated the standard deviation of the 100 mileage ratings to be

ST.DEV. =   2.4179

We will round this to 2.4.

The values that are 2 standard deviations above and below the mean are

$$\bar{x} + 2s = 37.0 + 2(2.4) = 37.0 + 4.8 = 41.8$$

and

$$\bar{x} - 2s = 37.0 - 2(2.4) = 37.0 - 4.8 = 32.2$$

From Figure 3.17 we can see that all but four of the mileage ratings are between 32.2 and 41.8. Thus, there are 96 measurements within 2 standard deviations of the mean. This agrees very nicely with the rule of thumb, which states that most of the measurements will be within 2 standard deviations of the mean.

We now find the values 3 standard deviations from the mean:

$$\bar{x} + 3s = 37.0 + 3(2.4) = 37.0 + 7.2 = 44.2$$

and

$$\bar{x} - 3s = 37.0 - 3(2.4) = 37.0 - 7.2 = 29.8$$

Examining Figure 3.17, we find that only the value 44.9 is not in this interval, and therefore 99 of the 100 mileage ratings are between 29.8 and 44.2. This agrees with the rule of thumb: All or almost all of the measurements will be within 3 standard deviations of the mean. ∎

Exercises (3.44–3.56)

Learning the Mechanics

Exercises 3.44, 3.45, and 3.46 are designed to help convince you that the rule of thumb works reasonably well for data sets with different types of distributions.

3.44 The following table summarizes a sample of 100 measurements:

Sample Value, x	0	1	2	3	4	5	6	7	8	9	10
Frequency of Occurrence	1	2	4	8	10	50	10	8	4	2	1

Using the data and the fact that $\bar{x} = 5.0$ and $s = 1.7$:
a. Draw a bar graph for this sample.
b. Find the number of measurements within 2 standard deviations of the mean.
c. Find the number of measurements that lie more than 3 standard deviations from the mean.
d. Compare the answers to parts b and c with the rule of thumb.

3.45 A sample of 2,000 measurements is summarized in the following table:

Sample Value, x	0	1	2	3	4	5	6	7	8	9	10
Frequency of Occurrence	1	2	4	8	20	35	60	120	250	500	1,000

The mean and standard deviation for these data are $\bar{x} = 9.0$ and $s = 1.4$.
a. Draw a bar graph to describe the data.
b. Find the number of measurements that lie more than 2 standard deviations from the mean.
c. Find the number of measurements that lie within 3 standard deviations of the mean.
d. Compare the answers to parts b and c with the rule of thumb.

3.46 A sample of 70 measurements is summarized in the following table:

Sample Value, x	1	2	3	4	5	6	7	8	9	10
Frequency of Occurrence	1	3	4	10	5	2	10	20	10	5

a. Draw a bar graph to describe the data.
b. Calculate $\bar{x}$ and s.
c. Find the number of measurements that lie within 2 standard deviations of the mean.

    d. Find the number of measurements that lie within 3 standard deviations of the mean.

    e. Compare the answers to parts *c* and *d* with the rule of thumb.

3.47 Consider the set of $n = 10$ measurements: 3, 1, 0, 3, 2, 5, 3, 6, 3, 4

    a. Give the range for this data set.

    b. Use the range to calculate an approximate value of *s* for the data.

    c. Calculate the variance and standard deviation for the data. Use your answer from part *b* as a rough check on your arithmetic.

3.48 Consider the set of $n = 15$ measurements:

      3, 1, −1, 3, 0, −2, 1, 3, 1, 0, 0, 3, 4, 1, −1

    a. Give the range for this set of data.

    b. Use the range to calculate an approximate value of *s* for the data.

    c. Calculate the variance and standard deviation for the data. Use your answer from part *b* as a rough check on your arithmetic.

3.49 What does the rule of thumb say about the number of measurements in a data set that lie within 2 standard deviations of the mean? 3 standard deviations?

Applying the Concepts

3.50 A buyer for a lumber company must determine whether to buy a piece of land containing 5,000 pine trees. If 1,000 of the trees are at least 40 feet tall, the buyer will purchase the land; otherwise, he will not. The owner of the land reports that the heights of the trees have a mean of 30 feet and a standard deviation of 3 feet. Based on this information, what is the buyer's decision?

3.51 For a sample of 50 days, the number of vehicles using a certain road was counted by a city engineer. The mean was 385, and the standard deviation was 15. Suppose you are interested in the number of days that between 340 and 430 vehicles used the road. What would you say about this number?

3.52 Suppose the distribution of IQ's has a mean of 100 and a variance of 225.

    a. Are there many people with an IQ higher than 145? Explain.

    b. Within what limits do most IQ's fall?

3.53 The following quotation briefly describes the purpose of the article, "Physicians' Self-Reports of Reactions to Malpractice Litigation":

> The authors devised a survey as a first step in assessing physicians' perceptions of the impact of medical malpractice litigation on their professional practice and personal lives. Subjects were a sample of physicians in Cook County, Ill., who had been sued during 1977–1981.*

Pertinent characteristics of the doctors surveyed were reported, including "the mean ($\pm$SD) number of years since earning their medical degree," which was 26.9 ± 10.8 years.

    a. For the doctors surveyed, what is the mean number of years since earning their medical degree?

* S. C. Charles, J. R. Wilbert, and E. C. Kennedy, *The American Journal of Psychiatry*, Vol. 141, No. 4, April 1984, p. 563.

b. The abbreviation SD stands for *standard deviation*. What is the standard deviation of the number of years since earning their medical degrees?

c. What do you think is the smallest number of years since one of these doctors earned a medical degree? Why?

d. What do you think is the largest number of years since one of these doctors earned a medical degree? Why?

3.54 In September 1974, there were seventeen countries worldwide that had at least one nuclear power reactor installed and operating.* The mean and standard deviation of the numbers of nuclear power reactors for the seventeen countries are 7.5 and 11.1, respectively.

a. The United States had 42 operating nuclear power reactors. Do you think any country had more? Why?

b. Do you think the data are symmetric or skewed? Why?

*3.55 According to *Consumers' Digest* (Nov.–Dec. 1980), sugar is the leading food additive in the U.S. food supply. Sugar may be listed more than once on a product's ingredient list since it goes by different names depending on its source (e.g., sucrose, corn sweetener, fructose, and dextrose). Thus, when you read a product's label you may have to total up the sugar in the product to see how much sweetener it contains. The table gives a list of candy bars and the percentage of sugar they contain relative to their weight.

Brand	Percentage of Sugar by Weight	Brand	Percentage of Sugar by Weight
Baby Ruth	23.7	Power House	30.6
Butterfinger	29.5	Bit-O-Honey	23.5
Mr. Goodbar	34.2	Chunky	38.4
Milk Duds	36.0	Milk Chocolate covered	
Mello Mint	79.6	Raisinettes	24.7
M & M Plain Chocolate		Oh Henry!	31.2
Candies	52.2	Borden Cracker Jack	14.7
Mars Chocolate Almond	36.4	Good & Plenty Licorice	28.2
Milky Way	26.8	Nestle's Crunch	43.5
Marathon	36.7	Planter Jumbo Block	
Snickers	28.0	Peanut Candy	21.5
3 Musketeers	36.1	Switzer Licorice	8.4
Junior Mints	45.3	Switzer Red Licorice	2.8
Pom Poms	29.5	Tootsie Pop Drops	54.1
Sugar Babies	41.0	Tootsie Roll	21.1
Sugar Daddy	22.0	Fancy Fruit Lifesavers	77.6
Almond Joy	20.0	Spear-O-Mint Lifesavers	67.6

Source: National Confectioners Association Brand Name Guide to Sugar (Nelson Hall Paperback). *Secondary Source: Consumers' Digest,* Nov.–Dec. 1980, p. 11.

* *Source:* Douglas M. Considine (ed.), *Energy Technology Handbook* (New York: McGraw-Hill, 1977), p. 5-36.

a. Use a computer package to calculate the mean and standard deviation of these data. Also, order the data from the smallest to the largest.

b. Find the number of measurements within 2 standard deviations of the mean. Does this result agree with the rule of thumb?

*3.56   During the 1980's the pharmaceutical industry has placed an increased emphasis on producing revolutionary new products. As a result, research and development (R&D) costs have increased, and companies are taking a greater interest in R&D management. The table lists the 1984 R&D expenditures (in millions of dollars) of the world's largest pharmaceutical manufacturers.

Company	R&D Expenditures	Company	R&D Expenditures
Merck	$290	Upjohn	200
American Home	90	Takeda	125
Bristol–Myers	162	Hoffman–LaRoche	363
Pfizer	159	Sandoz	181
Ciba–Geigy	230	Johnson & Johnson	187
Hoechst	274	Boehringer Ingelheim	176
Warner–Lambert	162	Squibb	114
Abbott	110	Schering–Plough	129
Smith Kline–Beckman	158	Rhone–Poulenc	110
Bayer	200		

Source: Business Quarterly, Fall 1984, p. 81.

a. Use a computer package to calculate the mean and standard deviation of the R&D expenditures.

b. Find the number of measurements within 3 standard deviations of the mean. Compare your answer to the rule of thumb.

3.5

USING NUMBERS TO MAKE INFERENCES

Since z-scores measure the distance from the mean in units of standard deviations, we can restate the rule of thumb for z-scores.

The Rule of Thumb for z-Scores

The z-scores associated with the measurements from real-life experiments satisfy these rules:

1. Most z-scores are between -2 and $+2$. Few z-scores are less than -2 or greater than $+2$.

2. All or almost all z-scores are between -3 and $+3$. No (or almost no) z-scores will be less than -3 or greater than $+3$.

Since few sampled measurements are more than 2 standard deviations from the mean, we will define such measurements to be **rare events.**

Definition 3.16

If a measurement is sampled from a set of data and that measurement is more than 2 standard deviations from the mean, the measurement is considered to be a **rare event.** That is, a measurement with a z-score less than -2 or greater than $+2$ is considered to be a rare event.

This definition will be modified in Chapter 4 to make it slightly more restrictive, but we can apply it now so that we can begin using numbers to make inferences.

EXAMPLE 3.20   Suppose a female bank executive believes that her salary is low as a result of sex discrimination. To try to substantiate her belief, she collects information on the salaries of her male counterparts in the banking business. The men selected are also bank executives holding a position similar to hers. She finds that the distribution of the males' salaries has a mean of $38,000 and a standard deviation of $1,400. Her salary is $33,240. Does this information help support her claim of sex discrimination?

Solution   Recall that a z-score counts the number of standard deviations between a measurement and the mean. So, first we calculate the z-score for the woman's salary with respect to the data sampled for her male counterparts:

$$z = \frac{x - \bar{x}}{s} = \frac{\$33{,}240 - \$38{,}000}{\$1{,}400} = -3.4$$

This tells us that the woman's salary is 3.4 standard deviations *below* (because the z-score is negative) the mean of the male salary distribution, and her salary of $33,240 is a very rare event in reference to the male salary data.

There are two possible interpretations of this information.

1. The female bank executive's salary represents a very unusual measurement (a rare event) for the male salary distribution.
2. The female bank executive's salary represents a measurement from a distribution different from the male salary distribution.

Well, which of the two interpretations do you think is correct? Since we do not expect rare events to occur with great frequency, we conclude that the female executive's salary does not come from the male salary distribution. This helps to support her claim of sex discrimination. However, the careful investigator should require more information before inferring sex discrimination as the cause. We would want to know more about the sample collection technique the woman used, and more about her competence at her job. Other factors, such as the length of employment, should also be considered in the analysis.   ■

Example 3.20 illustrates an approach to statistical inference that we will call the **rare event approach.** An experimenter hypothesizes a specific frequency distribution to describe a population of measurements. Then a sample of measurements is drawn from the population. If the experimenter finds it unlikely that the sample came from the hypothesized distribution, the hypothesis is concluded to be false. Thus, in Example 3.20, the woman believes her salary reflects sex discrimination. She hypothesizes that her salary could be just another measurement in the distribution of her male counter-parts' salaries if no discrimination exists. However, it was so unlikely that the sample (in this case, her salary) came from the male frequency distribution that she rejects that hypothesis, concluding that the distribution from which her salary is drawn is different from that for the males.

The fact that we will make such an inference only if a rare event is observed provides a measure of reliability for the inference. There will be few times that we make a mistake using this decision-making procedure.

EXAMPLE 3.21   Refer to Example 3.20. Suppose the woman earning $33,240 had found that the salaries of her male counterparts in banking had a mean of $37,000 and a standard deviation of $2,600. Would this information substantiate her claim of sex discrimination?

Solution   Following the same line of reasoning used in Example 3.20, we first calculate a z-score for the woman's salary with respect to the sample of her male counterparts:

$$z = \frac{\$33{,}240 - \$37{,}000}{\$2{,}600} \approx -1.45$$

The woman's salary is now only about 1.45 standard deviations below the mean for the male salary distribution, and this is not a rare event. Based on this information, there is little support for her claim of sex discrimination. She may still be correct, but more information is needed to substantiate her claim.   ■

At this stage we have used only one sampled observation to make an inference about a population. Naturally, in practical problems we would like to use more observations (that is, larger samples) to make inferences, and we will do so in later chapters. Our immediate objective is to introduce the reasoning behind statistical inferences. As we proceed, we will keep refining the rare event approach to make it more precise.

The hints given in the box on page 118 should help you use the rare event approach as we now know it.

Helpful Hints

When using the rare event approach to make inferences:

1. Calculate the z-score for the sampled observation, based on the hypothesized values of the mean and standard deviation for the population of interest.
2. If the z-score is smaller than -2 or larger than $+2$, then infer that the sampled observation came from a population that somehow differs from the population of interest. If the z-score is between -2 and $+2$, there is insufficient evidence to infer that the population of interest has a mean or standard deviation different from those hypothesized.
3. The reliability of the inference, roughly provided by the magnitude of the z-score, is very important; its calculation will be discussed in more precise terms in later chapters.

Exercises (3.57–3.63)

Learning the Language

3.57 Define the following term: rare event

Learning the Mechanics

3.58 It is hypothesized that $\mu = 50$ and $\sigma = 10$ for a certain population of interest, and one observation is chosen from the population. For each of the following observations, calculate a z-score and state whether it represents a rare event:

a. 15   b. 85   c. 33   d. 55   e. 0

Applying the Concepts

3.59 A chemical company produces a substance composed of 98% cracked corn particles and 2% zinc phosphide for use in controlling rat populations in sugar cane fields. Production must be carefully controlled to maintain the 2% zinc phosphide because too much zinc phosphide will cause damage to the sugar cane and too little will be ineffective in controlling the rat population. Records from past production indicate that the distribution of the actual percentage of zinc phosphide present in the substance has a mean of 2.0% and a standard deviation of 0.08%. If the production line is operating correctly and a batch is chosen at random from a day's production, what is the z-score for a measurement of 1.80% zinc phosphide? Suppose the batch chosen actually contains 1.80% zinc phosphide. What would you conclude? Explain your reasoning.

3.60 Solar energy is considered by many to be the energy of the future. A survey was taken to compare the cost of solar energy to the cost of gas or electric energy. Its results revealed that the average monthly utility bill of a three-bedroom house using gas or electricity was $165 and the standard deviation was $25.

   a. What can you say about the number of all three-bedroom houses using gas or electric energy having bills between $90 and $240?

   b. Suppose a house with solar energy units had a utility bill of $90. Does this suggest that solar energy units might result in lower utility bills? Explain.

   c. Suppose that a house with solar energy units had a utility bill of $130. Compute the z-score for this amount and interpret it.

3.61 Suppose a light bulb manufacturer claims that the mean lifetime of its bulbs is 750 hours. Assume you have prior knowledge that bulb lifetimes have a standard deviation of 80 hours.

   a. If the manufacturer's claim is true, approximately what number of light bulbs will burn out in less than 510 hours?

   b. Suppose you randomly select one of the bulbs and it burns out in less than 510 hours. What would you conclude about the manufacturer's claim? Explain your reasoning.

   c. Suppose you randomly select one of the bulbs and it burns out at approximately 650 hours. How do you interpret this result?

3.62 In the fall of 1988, freshmen entering a midwestern university had composite ACT scores with a mean of 17.6 and a standard deviation of 5.0.

   a. Within what limits do most of these composite ACT scores fall?

   b. Suppose a 1988 college freshman is randomly selected, and that student has a composite ACT score of 25. Could a student with this score be attending this university? Explain.

3.63 In 1976 the top twenty-five wine-producing countries had a mean of 11.62 gallons per capita and a standard deviation of 9.57 gallons per capita.*

   a. In 1976 the Soviet Union produced 3.24 gallons of wine per capita. Is it possible that the Soviet Union was among the top twenty-five wine-producing countries? Why?

   b. If a country claimed to produce 39 gallons of wine per capita in 1976, would you be suspicious of the claim? Why?

3.6

EXPLORATORY DATA ANALYSIS: BOX AND WHISKER PLOTS

The mean and standard deviation are numbers that are often used in a formal description or analysis of a data set. The standard deviation is probably the most sensitive and most often used measure of variability. As you may have noticed, however, the calculation of the standard deviation can be quite tedious, particularly for

* George T. Kurian, *The Book of World Rankings* (New York: Facts on File, Inc., 1979), p. 182.

a large data set. In this section we will discuss a quick and easy method to describe a set of data. The numbers used in this description are easier to calculate than the mean and standard deviation, but are not as useful for making inferences. As with stem and leaf plots, the methods discussed in this section are used mostly in preliminary data analyses.

In Section 3.3 we introduced three measures of relative standing: Q_1, the first quartile; Q_2, the second quartile (or median); and Q_3, the third quartile. Recall that the three quartiles separate a data set into four equal parts. Thus, 25% of the data are less than Q_1, 25% of the data are between Q_1 and Q_2, 25% of the data are between Q_2 and Q_3, and 25% of the data are larger than Q_3.* The methods of data description used in this section will be based upon the three quartiles, Q_1, Q_2, and Q_3.

We have discussed calculating the median, Q_2, in Section 3.1. The calculations of the first and third quartiles are similar to those for the median. The method we will use is given in the box. We will illustrate the method with two examples.

Steps for Calculating Q_1, Q_2, and Q_3

1. Arrange the measurements from the smallest to the largest.
2. Calculate the second quartile (the median), Q_2. When the data set contains an odd number of measurements, Q_2 is the middle number. For an even number of measurements, Q_2 is the mean of the two middle numbers.
3. The median splits the data in half. Calculate the median of the half of the data set containing the smallest observations. The calculated value is Q_1.
4. Calculate the median of the half of the data set containing the largest observations. The calculated value is Q_3.

Note: The value of the median is not included when calculating Q_1 or Q_3.

EXAMPLE 3.22   Consider the following ten test scores:

76, 93, 84, 79, 76, 35, 61, 70, 78, 96

Calculate the first quartile, Q_1, the second quartile, Q_2, and the third quartile, Q_3.

Solution   The first step is to arrange the test scores from the smallest to the largest, as given below:

35, 61, 70, 76, 76, 78, 79, 84, 93, 96

---

* For the sets of data we will discuss, these percentages will be met approximately. The larger the data set, the closer the percentages will be to 25%.

We will first calculate Q_2. Since there is an even number of measurements, its value is

$$Q_2 = \frac{76 + 78}{2} = 77$$

To calculate Q_1, we use the lower half of the data:

35, 61, 70, 76, 76

The value of Q_1 is 70, since there is an odd number of measurements.
To calculate Q_3 we use the upper half of the data:

78, 79, 84, 93, 96

The value of Q_3 is 84. ■

EXAMPLE 3.23   Refer to Example 3.22. Eliminate the test score of 70. Recalculate the values of Q_1, Q_2, and Q_3 for the nine test scores that remain.

Solution   The nine test scores arranged in order are

35, 61, 76, 76, 78, 79, 84, 93, 96

Again, we calculate Q_2 first. Now there is an odd number of measurements, and the value of Q_2 is 78.
The lower half of the data set, not including $Q_2 = 78$, is

35, 61, 76, 76

Since there is an even number of measurements, the value of Q_3 is

$$\frac{61 + 76}{2} = 68.5$$

The upper half of the data set is

79, 84, 93, 96

The value of Q_3 is

$$\frac{84 + 93}{2} = 88.5$$ ■

Now that we have discussed how to calculate the three quartiles, we will introduce a useful way to summarize such numbers. A **box and whisker plot** will be used to show the relative positions of the three quartiles as well as the smallest and largest measurements. To demonstrate what a box and whisker plot is, we will use the 1983–1984 salaries of the Boston Celtics. The data were given on page 83 and are repeated in Table 3.4 (page 122).

Table 3.4

Player	Salary	Player	Salary
Bird	$1,800,000	Ainge	$400,000
McHale	1,000,000	Buckner	239,000
Maxwell	830,000	Carr	175,000
Parish	700,000	Kite	120,000
Johnson	405,000	Clark	65,000
Wedman	400,000	Carlisle	65,000

Boston Celtics' salaries

The following are easily found from the table:

Smallest measurement = $65,000

First quartile, Q_1 = $147,500

Second quartile, Q_2 = $400,000

Third quartile, Q_3 = $765,000

Largest measurement = $1,800,000

A box and whisker plot for these numbers is shown in Figure 3.18. The smallest measurement, the three quartiles, and the largest measurement are marked on the horizontal axis. To emphasize where the quartiles are located, a rectangle is drawn such that the ends are vertical lines directly above the first and third quartiles. A vertical line is drawn inside the rectangle to indicate the position of the second quartile. This rectangle forms the "box" part of the diagram.

Figure 3.18
Box and whisker plot for
Boston Celtics' salaries

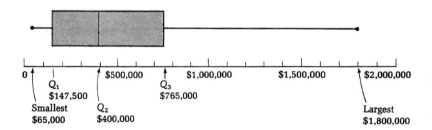

Horizontal lines are drawn connecting the ends of the rectangle to the smallest and largest measurements. These lines are the "whisker" part of the diagram, and together with the rectangle complete the box and whisker plot.

The diagram in Figure 3.18 provides a great deal of information about the Boston Celtics' salary data. The box indicates where 50% of the salaries fall—between $147,500 and $765,000. The whiskers provide information about the skewness of the data. The smallest salary, $65,000, is much nearer to $Q_1 = \$147,500$ than the largest salary, $1,800,000, is to $Q_3 = \$765,000$. The box and whisker plot clearly shows this difference and indicates that the data are skewed to the right. The measurements are less concentrated in the upper part of the data set as compared to the lower part.

General guidelines to follow when constructing a box and whisker plot for a set of data are given in the accompanying box.

How to Construct a Box and Whisker Plot

1. Determine the values of the smallest measurement; the first quartile, Q_1; the second quartile, Q_2; the third quartile, Q_3; and the largest measurement.
2. Mark the five values determined in step 1 on a horizontal line (axis).
3. Draw a rectangle slightly above the horizontal axis, running lengthwise from Q_1 to Q_3. The width of the rectangle is arbitrary but is often chosen to be about three-quarters of an inch.
4. Draw a vertical line inside the rectangle above the value of Q_2.
5. Complete the diagram by drawing a horizontal line from the middle of the left end of the rectangle to a point above the smallest measurement and a second horizontal line from the middle of the right end of the rectangle to a point above the largest measurement.

Box and whisker plots are especially useful for comparing two sets of data. In the next example we will compare the salaries of the Boston Celtics to the salaries of the Indiana Pacers.

EXAMPLE 3.24   The 1983–1984 salaries of the Indiana Pacers are given in Table 3.5. Construct a box and whisker plot for these salaries and compare it to the box and whisker plot for the Boston Celtics' salaries.

Table 3.5

Indiana Pacers' salaries

Player	Salary	Player	Salary
Stipanovich	$420,000	Sichting	$135,000
Kellogg	406,000	Durrant	100,000
Garnett	269,000	Waiters	100,000
Williams	200,000	Gray	80,000
Stansbury	190,000	Thomas	80,000
Fleming	175,000	Brown	65,000

Solution   First, the following values are calculated for the data given in Table 3.5:

Smallest measurement = $65,000

First quartile, Q_1 = $90,000

Second quartile, Q_2 = $155,000

Third quartile, Q_3 = $234,500

Largest measurement = $420,000

a. Indiana Pacers

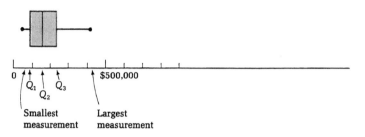

b. Boston Celtics

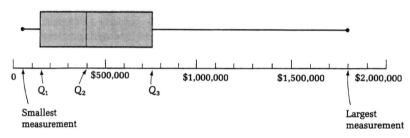

Figure 3.19
Box and whisker plots for NBA salaries

Following the steps given in the preceding box, we constructed the box and whisker plot shown in Figure 3.19**a** using the five numbers just calculated. The box and whisker plot for the Boston Celtics was first shown in Figure 3.18, and we repeat it in modified form in Figure 3.19**b** for easy comparison.

The two diagrams in Figure 3.19 provide a striking comparison between the salaries of the Boston Celtics and the Indiana Pacers. About the only point of similarity between the two sets of salaries is that the smallest salary is the same on both teams. The data for the Celtics are quite skewed, while the data for the Pacers are nearly symmetric. The salaries of the Boston team are much larger than those of the Indiana team. The median Celtic salary is almost as large as the largest Pacer salary! The Celtics' salaries are also much more variable than the Pacers' salaries. This is indicated by the relative differences in the lengths of both the boxes and the whiskers. ■

Box and whisker plots provide an easy way to describe sets of data. They give a quick way of checking to see if a data set is skewed, where most of the measurements fall, where the middle of the data is, and how variable the data are. As an initial step in data analysis, they provide much information. A more complete analysis—for example, calculating the mean and standard deviation—could of course provide additional information about the data.

COMPUTER
EXAMPLE 3.4

Refer to Computer Example 3.2, in which we presented some fast foods that have extremely high levels of sodium. These data are given again in Table 3.6.

Table 3.6

High-sodium fast food items

Food Item	Chain	Sodium (mg)
Large roast beef w/cheese	Roy Rogers	1,953
Pancake breakfast	Jack-in-the-Box	1,815
Hot Ham 'n Cheese	Arby's	1,655
Chicken Supreme	Jack-in-the-Box	1,582
Pasta seafood salad	Jack-in-the-Box	1,570
Specialty ham & cheese	Burger King	1,550
Ham biscuit	Hardee's	1,415
Pancake platter	Roy Rogers	1,264
Potato w/bacon & cheese	Wendy's	1,180
Whaler sandwich	Burger King	1,013

Source: "The Fast-Food Eating Guide," *Business and Society Review,*
Vol. 59, Feb. 1986, p. 57.

The box and whisker plot shown in Figure 3.20 was produced by Minitab using the sodium data. From Table 3.6 we can find the following values:

Smallest measurement = 1.013

First quartile, Q_1 = 1.264

Second quartile, Q_2 = 1,560

Third quartile, Q_3 = 1,655

Largest measurement = 1,953

Figure 3.20
Computer-generated box
and whisker plot of high
sodium levels

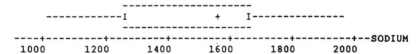

The whiskers of the graph in Figure 3.20 end at 1,013 and 1,953, the smallest and largest measurements, respectively. The left end of the box begins at Q_1 = 1,264 and the right end stops at Q_3 = 1,655. The box and whisker plot constructed by Minitab is very similar to the ones we constructed in this section. The biggest difference is how the second quartile is denoted in the box. We drew a vertical line, but it is difficult to denote vertical lines on many printers. Therefore, Minitab denotes the second quartile, Q_2 = 1,560, by "+"—the plus symbol.   ∎

Exercises (3.64–3.71)

Learning the Mechanics

3.64   Consider the following sample of fourteen numbers:

49, 100, 55, 47, 50, 71, 70, 36, 78, 41, 34, 90, 79, 50

a. What are the smallest and largest numbers?

b. Calculate Q_1, Q_2, and Q_3.

c. Draw a box and whisker plot for the data.

3.65 Construct a box and whisker plot for the following data:

8, 15, 6, 23, 17, 28, 73, 10, 14, 23, 20, 19, 26

3.66 Refer to Exercise 3.65.

a. Delete the value 73, then draw a box and whisker plot for the remaining twelve numbers.

b. Compare the diagram drawn in part a with that drawn in Exercise 3.65.

Applying the Concepts

3.67 The accompanying table shows the number of aircraft that each of the major U.S. airlines had in operation in 1983 and 1982.

Airline	1983	1982
American	234	231
Continental	69	112
Delta	226	213
Eastern	278	253
Northwest	111	110
Pan Am	134	130
Republic	163	160
TWA	154	179
United	316	342
USAir	126	108
Western	73	70

Source: *World Aviation Directory*, No. 89, Winter 1984–1985, p. 109.

a. Construct a box and whisker plot for the 1983 data.

b. Construct a box and whisker plot for the 1982 data.

c. Compare the two plots.

3.68 In a discussion of the average size of farms, *The Book of World Rankings* states:

Next to land tenure, the most important factor in agricultural economics is the average size of farms. The size of farms is determined by land colonization practices. In Latin America, for example, the ease with which the conquistadores could establish large estates called latifundios or haciendas led to a system in which large farms were the rule rather than the exception. On the other hand, in countries where Muslim or Hindu inheritance laws prevail, the division of farms among all male heirs has led to fragmentation of holdings into minute parcels.*

The average number of acres per farm (rounded to the nearest acre) for forty-one countries is shown in the table.

* *Source:* George T. Kurian, *The Book of World Rankings* (New York: Facts on File, Inc., 1979), p. 135.

Country	Average	Country	Average
Argentina	667	Botswana	12
Uruguay	516	Morocco	11
United States	390	Mali	11
Mexico	306	Kenya	10
Chile	293	Senegal	9
Paraguay	269	Philippines	9
Venezuela	201	Thailand	9
Brazil	196	Uganda	8
Costa Rica	101	Greece	8
Nicaragua	92	Togo	6
United Kingdom	61	Pakistan	6
Colombia	56	Sri Lanka	4
Peru	50	Egypt	4
Tunisia	38	Vietnam	3
Spain	37	China	3
Dominican Republic	21	Nepal	3
Guatemala	20	Japan	3
El Salvador	17	Indonesia	3
India	16	Madagascar	3
Iran	15	South Korea	2
Turkey	12		

a. Construct a box and whisker plot for the data.

b. Is the data set symmetric? Explain.

c. What other information does the box and whisker plot provide?

3.69 Shown in the accompanying table are the NBA salaries for the Chicago Bulls, Cleveland Cavaliers, and Los Angeles Lakers. Construct a box and whisker plot for each set of salaries, and compare the three diagrams.

Bulls	Cavaliers	Lakers
$660,000	$675,000	$2,500,000
550,000	475,000	1,530,000
550,000	395,000	1,150,000
500,000	284,000	923,000
369,000	260,000	860,000
300,000	209,000	400,000
275,000	205,000	350,000
250,000	200,000	336,000
200,000	150,000	325,000
160,000	100,000	200,000
125,000	90,000	120,000
65,000	80,000	90,000
	65,000	90,000

Source: Table, "Player Payrolls," *The Sporting News,* January 28, 1985, p. 29.

*3.70 The table shows the average Scholastic Aptitude Test scores for 21 states and the District of Columbia in 1982 and 1985.

State	1982	1985	State	1982	1985
California	889	904	New Hampshire	925	939
Connecticut	896	915	New Jersey	869	889
Delaware	897	918	New York	896	900
D.C.	821	844	North Carolina	827	833
Florida	889	884	Oregon	908	928
Georgia	823	837	Pennsylvania	885	893
Hawaii	857	877	Rhode Island	877	895
Indiana	860	875	South Carolina	790	815
Maine	890	898	Texas	868	878
Maryland	889	910	Vermont	904	919
Massachusetts	888	906	Virginia	888	908

Source: U.S. Department of Education.

a. Use a computer package to construct a box and whisker plot for each set of SAT scores.

b. Did the scores tend to be higher in 1982 or 1985?

c. What other information do the box plots provide?

*3.71 How much does a new 1,600-square-foot house with 1½ or 2 baths in a desirable neighborhood cost? To find out, Better Homes and Gardens® Real Estate Service

City	Sale Price	City	Sale Price
Albany, N.Y.	$ 72,500	Lake Tahoe, Nev.	$190,000
Allentown, Pa.	82,500	Little Rock, Ark.	85,000
Amarillo, Tex.	75,000	Manhattan, Kans.	85,000
Baltimore, Md.	100,000	Montgomery, Ala.	65,000
Baton Rouge, La.	100,000	Nashville, Tenn.	84,000
Biloxi, Miss.	78,500	Olympia, Wash.	75,000
Bowling Green, Ky.	67,400	Owensboro, Ky.	72,500
Cheyenne, Wyo.	82,000	Pittsburgh, Pa.	72,500
Cincinnati, Ohio	85,000	Pleasanton, Calif.	185,000
Columbus, Ga.	69,000	Raleigh, N.C.	94,000
Cranston, R.I.	98,500	Reno, Nev.	93,950
Darien, Conn.	290,000	Rockford, Ill.	76,000
Des Moines, Iowa	93,900	San Diego, Calif.	140,000
Fargo, N.D.	100,000	Sheboygan, Wis.	72,000
Greenville, S.C.	76,800	South Bend, Ind.	71,300
Honolulu, Hawaii	165,000	Springfield, Mass.	75,000
Jacksonville, Fla.	77,000	Trenton, N.J.	100,000
Joplin, Mo.	65,000	Vancouver, Wash.	68,000

Source: Consumers' Research, February 1985, p. 19.

conducted a nationwide survey of real estate agents. The table contains a portion of the resulting sample data.

a.   Construct a box and whisker plot for the sale price data using a computer package.
b.   Are the data symmetric, skewed to the right, or skewed to the left?
c.   What are the largest and smallest sale prices?
d.   What is the median sale price?

Chapter Summary

Numerical descriptions can summarize the important characteristics of a distribution of data. **Measures of central tendency** locate the approximate center of a distribution. Three common measures of central tendency are the mean, the median, and the mode. The **mean** is the sum of all the measurements divided by the total number of measurements in the data set. When measurements are arranged in increasing or decreasing order, the **median** is the middle number if there is an odd number of measurements, or the mean of the two middle numbers if there is an even number of measurements. The **mode** is the measurement that occurs most often.

The general shape of a distribution of data can be described by the direction in which the distribution is skewed. If measurements above the mean occur less frequently than measurements below the mean, the distribution is **skewed to the right;** if they occur more frequently, it is **skewed to the left;** and if they occur with the same frequency, it is **symmetric.**

Measures of variability describe the spread, or variation, of a distribution of data. The **range** is the difference between the largest and smallest measurements in the data set. The **sample variance** is the sum of the squared differences between each of the n sample measurements and the sample mean, divided by $(n - 1)$. The **sample standard deviation** is the square root of the sample variance.

Measures of relative standing indicate the position of individual measurements in a data set with respect to all other measurements in the set. The **pth percentile** is a number such that $p\%$ of the measurements are less than or equal to it. The **first quartile, second quartile,** and **third quartile** are the 25th, 50th, and 75th percentiles, respectively. A **z-score** for a measurement in a data set is obtained by subtracting the mean of the data set from the individual measurement and then dividing this difference by the standard deviation.

Most measurements will be within 2 standard deviations of the mean—that is, most will have z-scores between -2 and $+2$. Almost all measurements will be within 3 standard deviations of the mean—that is, almost all will have z-scores between -3 and $+3$. For real-life data, *most* means approximately 90% and *almost all* means approximately 99%. A measurement more than 2 standard deviations from the mean is called a **rare event.**

The **rare event approach** to making inferences can be described roughly as follows: Hypothesize values for the mean and standard deviation of the population; calculate the z-score for a sampled observation; if the z-score is less than -2 or greater than $+2$, it is unlikely that the sampled observation came from a population with the hypothesized mean and standard deviation. This approach will be refined in subsequent chapters.

To simplify calculations, these symbols are used:

x: Individual measurement

n: Number of measurements in a sample

$\bar{x}$: Sample mean

μ: Population mean

s^2: Sample variance

σ^2: Population variance

s: Sample standard deviation

σ: Population standard deviation

The symbol $\sum$ indicates that a sum is to be taken; for example, $\sum x$ means the sum of the individual measurements. The sample mean, variance, and standard deviation for a sample with n measurements can be defined in symbols, as follows:

$$\bar{x} = \frac{\sum x}{n} \qquad s^2 = \frac{\sum (x - \bar{x})^2}{n - 1} \qquad s = \sqrt{s^2}$$

$\left(\sum \text{ means sum over all } n \text{ measurements in the sample.}\right)$ An alternative formulation of the sample variance for computational purposes is

$$s^2 = \frac{n\left(\sum x^2\right) - \left(\sum x\right)^2}{n(n - 1)}$$

The z-scores for an individual measurement x are

$$z = \frac{x - \mu}{\sigma} \qquad \text{and} \qquad z = \frac{x - \bar{x}}{s}$$

for a population and sample, respectively.

Box and whisker plots can be used as an initial step in data analysis to see if the data set is skewed, where most measurements fall, where the middle of the data is, and how variable the data are.

SUPPLEMENTARY EXERCISES (3.72–3.103)

Learning the Mechanics

3.72 Calculate the mean, median, range, variance, and standard deviation for each of the following samples:

a. 1, 5, 0, 2, 5, 7, 1   b. 10, 8, 12, 2   c. 1, 2, 0, 0, 5, 4   d. 3, 4, 10, 2

3.73 Calculate $\bar{x}$, s^2, and s for each of the following:

a. $\sum x^2 = 246$, $\sum x = 45$, $n = 15$ c. $\sum x^2 = 65$, $\sum x = 9$, $n = 7$

b. $\sum x^2 = 543$, $\sum x = 106$, $n = 25$

3.74 Calculate a z-score for the measurement x in each of the following situations:

a. $x = 10$, $\mu = 20$, $\sigma = 5$ c. $x = 0$, $\bar{x} = 50$, $s^2 = 400$

b. $x = 15$, $\bar{x} = 5$, $s = 5$ d. $x = 100$, $\mu = 50$, $\sigma^2 = 900$

3.75 Suppose a sample of measurements has a mean of 75 and a standard deviation of 5. Answer the following questions:

a. What can you say about the number of sample measurements between 65 and 85?

b. What can you say about the number of sample measurements between 60 and 90?

c. If a measurement chosen from this sample was equal to 92, would it be a rare event? Explain.

Applying the Concepts

3.76 The owner of a service station decided to conduct a survey of service records to determine the length of time (in months) between customer oil changes. A random sample of the station records produced the following times between oil changes:

6	6	24	8	6
6	6	16	6	12
18	8	4	12	12

Compute the sample mean, variance, and standard deviation.

3.77 In Exercise 3.76, if the customer whose oil was changed every 24 months had instead changed it every 18 months, would the sample variance increase or decrease? Why? If instead of every 24 or 18 months, the customer had changed the oil every 6 months, how would the resulting sample variance compare to the sample variances when the oil was changed every 24 months? Every 18 months?

3.78 Automobile manufacturers in the United States are deeply involved in research to improve their products' gas mileage. One manufacturer, hoping to achieve 30 miles per gallon with one of its full-size models, measured the mileage obtained by thirty test versions of the model. The results follow (rounded to the nearest mile for convenience):

30	30	32	31	30	27	28	31	33	28
30	31	29	31	27	33	31	30	35	29
32	27	31	34	30	27	26	29	28	30

(The data given here were presented in Exercise 2.49. The purpose of that exercise was to summarize the data using a bar graph.)

a. Compute $\bar{x}$, s^2, and s for the data set.

b. If the manufacturer will be satisfied with a (population) mean of 30 miles per gallon, how should it react to the above test data?

3.79 As part of an agricultural experiment, a new type of soybean was planted in five different fields of equal area. The fields had the following yields at harvest (in hundreds of bushels):   16, 14, 11, 13, 11

a. Calculate the mean yield for this sample.

b. Calculate the variance and standard deviation of the sample measurements.

3.80   In a study of response to pain, subjects were exposed to a pain-producing stimulus. The object of the study was to see whether each subject showed consistency in pain response. The following represent a sample of pain responsiveness scores from the study:

10	13	20	15	13
16	13	21	19	11
12	16	11	15	16

a. Calculate the range for the data, and use the range to calculate an estimate of the sample standard deviation.

b. Find the mean, variance, and standard deviation for the sample. Use the estimate of s from part a to check your calculation of s.

c. How many observations in the data set fall within 2 standard deviations of the mean?

3.81   A severe drought affected several western states for 3 years in a row. A Christmas tree farmer, who is concerned about the drought's effect on the size of the present crop of trees, decides to take a sample of the heights of twenty-five trees and obtains the following results (recorded in inches):

60	57	62	69	46
54	64	60	59	58
75	51	49	67	65
44	58	55	48	62
63	73	52	55	50

a. Compute $\bar{x}$, s^2, and s for the data.

b. How many of the tree heights would you expect to find in the intervals $(\bar{x} - 2s)$ to $(\bar{x} + 2s)$ and $(\bar{x} - 3s)$ to $(\bar{x} + 3s)$?

c. Count the number of measurements that actually fall within each interval of part b. Compare these numbers to the results of part b.

3.82   The Community Attitude Assessment Scale (CAAS) measures citizens' attitudes toward fifteen life areas (for example, education, employment, health) in four respects— importance, influence, equality of opportunity, and satisfaction. In order to develop the CAAS, a number of households in each of twenty-five communities were randomly selected and sent questionnaires. Because relatively low response rates suggest that there could be a substantial, but unknown, opinion bias in the reported data, the percentage of the sample responding to the survey was determined in each community. The results (in percentages) are given below:

21	14	18	20	14
16	6	22	28	16
26	14	13	15	25
21	14	7	12	8
15	14	21	22	10

a. Find the range for the data, and use it to estimate a value for *s*.

b. Calculate the variance and standard deviation for the data. Use the estimate of *s* from part *a* to check your calculation of *s*.

c. Find the number of measurements that fall in the interval $(\bar{x} - 2s)$ to $(\bar{x} + 2s)$.

3.83 A small computing center has found that the number of jobs submitted to its computers per day has a distribution with a mean of 83 jobs and a variance of 100. What does the rule of thumb say about the number of days that would have between 53 and 113 jobs submitted?

3.84 A professor thinks that if a class is allowed to work on an examination as long as desired, the times spent by the students would have a mean of 40 minutes and a standard deviation of 6 minutes. Approximately how long should be allotted for the examination if the professor wants all or almost all of the class to finish?

3.85 A veterinarian was interested in determining how many animals were treated in the clinic each day. A random sample of 20 days' records produced the following results:

15	17	24	16	18	15	19	22	25	21
18	17	20	20	20	18	10	16	12	21

a. Calculate the sample mean, variance, and standard deviation.

b. Suppose the veterinarian had seen two additional animals on each of the 20 days. Again, calculate the sample mean, variance, and standard deviation. Compare these values with your results in part *a*. What is the effect on the sample mean, variance, and standard deviation of adding a constant to each measurement in the sample?

3.86 A coach of distance runners was interested in knowing how many miles America's top distance runners usually run in a week. The coach surveyed fifteen of the best distance runners and obtained the following information (in miles):

120	95	110	95	70
90	80	100	125	75
85	100	115	130	90

a. Find the sample mean, variance, and standard deviation for the data.

b. Suppose each runner cut the weekly mileage in half. Again calculate the mean, variance, and standard deviation for the data. Compare these results with those obtained in part *a*. What is the effect on the sample mean, variance, and standard deviation of multiplying each measurement in the sample by a constant?

3.87 By law, a box of cereal labeled as containing 16 ounces must contain at least 16 ounces of cereal. It is known that the machine filling the boxes produces a distribution of fill weights with a mean equal to the setting on the machine and with a standard deviation equal to 0.05 ounce. If a box of cereal is selected at random and the machine is set at 16.15, do you think the box will contain more than 16.25 ounces? Explain.

3.88 The United States surgeon general has warned that the two most dangerous ingredients contained in cigarettes are tar and nicotine. The amount of tar (in milligrams) present in ten selected brands is recorded in the table (page 134). All ten brands chosen are kingsize, filter-tipped cigarettes. Calculate the mean, range, variance, and standard deviation for the data shown.

Brand	Tar	Brand	Tar
Merit	9	Pall Mall	17
Marlboro	17	True	5
Merit Ultra Lights	4	Bull Durham	28
Marlboro Lights	11	Decade	6
L & M	14	Winston	15

Source: U.S. Federal Trade Commission, *Federal Trade Commission Report: "Tar," Nicotine and Carbon Monoxide of the Smoke of 207 Varieties of Domestic Cigarettes* (Washington, D.C.: Federal Trade Commission, 1984), pp. 1–5.

3.89   Most people living in metropolitan areas receive impressions of what is happening in their area primarily through their major newspapers. A study was conducted to determine whether the Uniform Crime Report, compiled by the United States Federal Bureau of Investigation, and the daily newspaper gave consistent information about the trend and distribution of crime in a metropolitan area. An attention score, based on the amount of space devoted to a story, was calculated for each paper's coverage of murders, assaults, robberies, etc. Suppose the average murder attention score of metropolitan newspapers across the country, μ, is 54 and $\sigma = 6$. If the St. Louis *Globe-Democrat* has a murder attention score of 69, do many newspapers have a murder attention score higher than the *Globe-Democrat*? Explain.

3.90   Audience sizes at concerts given by a city's symphony over the past 2 years were recorded and found to have a sample mean of 3,125 and a sample standard deviation of 25. Calculate the intervals $(\bar{x} - 2s)$ to $(\bar{x} + 2s)$ and $(\bar{x} - 3s)$ to $(\bar{x} + 3s)$. What number of the recorded audience sizes of the past 2 years would be expected to fall in each of these intervals?

3.91   A producer of alkaline batteries was interested in obtaining a statistical description of the shelf-life of the battery it manufactures. Twenty-five batteries were selected at random as they came off the assembly line and their shelf-lives were tested. The following are the lifetimes of the twenty-five batteries rounded to the nearest month:

24	21	24	20	19
25	27	24	30	21
23	24	24	19	24
26	22	25	24	24
25	23	28	23	25

a.   Compute $\bar{x}$, s^2, and s for this data set.

b.   How many of the measurements would you expect to find in the intervals $(\bar{x} - 2s)$ to $(\bar{x} + 2s)$ and $(\bar{x} - 3s)$ to $(\bar{x} + 3s)$?

c.   Count the number of measurements that actually fall within the intervals of part b. Compare these results with the results of part b.

3.92 Most students wonder how their test scores compare to those of the rest of the class. Suppose that you and five friends are taking a psychology course, and the six of you received the following grades: 83, 70, 74, 89, 97, 85. The instructor in the class tells you that the average and standard deviation for all the test scores are 85 and 5, respectively.
  a. Calculate the z-score for each of the six test scores.
  b. State whether each of the test scores lies above or below the mean and by how many standard deviations.
  c. Do you think any of these test scores is the highest or the lowest in the class? Explain.

3.93 A city librarian claims that books have been checked out an average of 7 (or more) times in the last year. You suspect the librarian has exaggerated the checkout rate (book usage) and that, in fact, the mean number of checkouts per book per year is less than 7. Assume that from prior knowledge we know the standard deviation of the number of checkouts per book per year is 1. Using the card catalog, we randomly sample one book and find that it has been checked out 4 times in the last year.
  a. If the mean number of checkouts per book per year really is 7, what is the z-score corresponding to 4 checkouts?
  b. Considering your answer to part a, is there evidence to indicate that the librarian's claim is wrong? Explain.
  c. If the standard deviation of the number of checkouts per book per year were 2 (instead of 1), would your answer to part b change? Explain.

3.94 A radio station claims that the number of minutes of advertising per hour of broadcasting time has an average of 3 and a standard deviation equal to 2.1. You listen to the radio station for 1 randomly selected hour and observe that the amount of advertising time is equal to 7 minutes. Does this observation appear to disagree with the radio station's claim? Explain.

3.95 On the English portion of the ACT, a midwestern university's incoming freshmen in Fall 1988 had scores with a mean of 17.3 and a standard deviation of 4.9. Their scores on the mathematics portion of the ACT had a mean of 15.4 and a standard deviation of 7.2.
  a. Were the test scores more variable on the English portion or on the mathematics portion?
  b. Did students have a better average score on the English or the mathematics portion of the test?
  c. Considering all these entering freshmen, do you think there were more scores above 27 on the English or on the mathematics portion? Why?

3.96 The table at the top of page 136 lists the percentages of registered voters for thirty countries during the period 1975–1978.
  a. Calculate the mean, variance, and standard deviation for the data.
  b. How many of the thirty percentages fall within 2 standard deviations of the mean?

Country	Percentage of Registered Voters	Country	Percentage of Registered Voters
Finland	80.6	Malta	64.2
United Kingdom	72.0	Canada	63.6
West Germany	71.7	Luxembourg	61.7
Norway	71.1	Israel	61.4
Greece	70.7	France	59.3
Japan	70.6	Iceland	59.1
Sweden	69.9	Ireland	58.5
Netherlands	69.3	Switzerland	58.2
United States	69.2	India	54.8
Denmark	68.3	Mexico	48.2
Italy	68.3	Chile	47.7
East Germany	68.0	Turkey	45.4
Austria	66.7	Venezuela	41.2
Belgium	65.0	Brazil	35.3
Soviet Union	64.4	South Africa	9.7

Source: George T. Kurian, *The Book of World Rankings* (New York: Facts on File, Inc., 1979), p. 52.

3.97 Refer to Exercise 3.96.
a. Find the first, second, and third quartiles for the thirty percentages.
b. Draw a box and whisker plot for the data.

3.98 The 1976 reserves of uranium for countries around the world are given in the table.

Country	Reserve in Tons	Country	Reserve in Tons
United States	523,000	Spain	6,800
South Africa	306,000	Greenland	5,800
Australia	289,000	Mexico	4,700
Canada	167,000	Yugoslavia	4,500
Niger	74,000	Turkey	4,100
France	37,000	South Korea	3,000
India	29,800	Finland	1,900
Algeria	28,000	Austria	1,800
Gabon	20,000	United Kingdom	1,800
Brazil	18,200	Zaire	1,800
Argentina	17,800	West Germany	1,500
Central African Empire	8,000	Italy	1,200
Japan	7,700	Sweden	1,000
Portugal	6,800		

Source: George T. Kurian, *The Book of World Rankings* (New York: Facts on File, Inc., 1979), p. 197.

a. Calculate the mean, variance, and standard deviation for this set of data.

b. Calculate the z-score for the largest and smallest measurements.

c. Find the median and calculate the z-score for this value. Interpret this z-score.

3.99 In the mid-1970's the United States labor force was 37% female. Worldwide, women represented a mean percentage of 29.2% of the labor force, with a standard deviation of 11.9%.

a. Calculate the z-score for the United States' figure of 37% and interpret this value.

b. Considering the values of the mean and standard deviation, what is the largest percentage of women you would expect in any country's work force? Why?

3.100 *The Book of World Rankings* states:

> Despite the importance of food production in a hungry world, agriculture remains labor-intensive, agricultural products are subject to price fluctuations, and agricultural pursuits are characterized by low productivity. As a result, agriculture's share of the GDP is highest in the least-developed countries, and in most countries of the world it has been declining in relation to other sectors. National policy in almost all countries deliberately favors industry over agriculture and even short-term growth prospects in this sector are depressed by the irreversible migration of manpower to the towns and cities and consequent depletion of the agricultural labor force.*

According to data contained in this publication, the mean percentage of agriculture's share of the GDP (gross domestic product) was found to be 24% for countries worldwide. The standard deviation was 16%.

a. Within what limits do you expect to find most countries' agricultural percentage of the GDP? Why?

b. Do you think there is a country with an agricultural share of the GDP equal to 80%? Explain.

*3.101 The table contains the price per acre of farmland for a sample of states that includes nine eastern states and eleven western states (i.e., west of the Mississippi River).

State	Price per Acre (1984)	State	Price per Acre (1984)
Arizona	$ 265	Nebraska	$ 444
California	1,726	Nevada	229
Colorado	435	New Hampshire	1,419
Connecticut	3,208	New Jersey	3,525
Delaware	1,642	New Mexico	163
Florida	1,527	North Dakota	360
Kansas	466	Pennsylvania	1,510
Massachusetts	2,372	Rhode Island	3,335
Maryland	2,097	South Dakota	250
Montana	222	Wyoming	177

Data: U.S. Department of Agriculture.
Source: Business and Society Review, Fall 1985, No. 55, p. 84.

* Source: George T. Kurian, *The Book of World Rankings* (New York: Facts on File, Inc., 1979), p. 211.

a. Use a computer package to find the mean and median price per acre for the sample of twenty states.

b. Use a computer package to find the mean price per acre for the eastern states and for the western states.

c. Compare the means found in part *b* to the mean you found in part *a*. What do the comparisons reveal about the value of farmland in the United States?

d. Use a computer package to order the twenty prices per acre. Circle the prices per acre that correspond to the nine eastern states. Does this representation of the data help support your answer to part *c*?

*3.102 The data in the table are the 1960 and 1980 population sizes of thirty sunbelt cities, expressed in thousands. For example, in 1960 the population of Albuquerque, New Mexico, was approximately 201,000.

City	1960	1980	City	1960	1980
Albuquerque, NM	201	332	Las Vegas, NV	64	165
Anaheim, CA	104	219	Lubbock, TX	129	174
Arlington, TX	45	160	Miami, FL	292	347
Atlanta, GA	487	425	New Orleans, LA	628	558
Austin, TX	187	345	Oklahoma City, OK	324	403
Charlotte, NC	202	314	Orlando, FL	88	128
Dallas, TX	680	904	Phoenix, AZ	439	790
Fort Lauderdale, FL	84	153	St. Petersburg, FL	181	239
Fort Worth, TX	356	385	San Antonio, TX	588	786
Fresno, CA	134	218	San Diego, CA	573	876
Honolulu, HI	294	365	San Francisco, CA	740	679
Houston, TX	938	1,595	San Jose, CA	204	629
Huntington Beach, CA	11	171	Tampa, FL	275	272
Huntsville, AL	72	143	Tucson, AZ	213	331
Jacksonville, FL	201	541	Tulsa, OK	262	361

Source: U.S. Bureau of the Census, *Statistical Abstract of the United States: 1981* (pp. 21–23), *1985* (pp. 23–25).

a. Use a computer package to construct box and whisker plots for these two data sets.

b. How do the two distributions compare? Include a discussion of central tendency, variability, and skewness in your comparison.

c. Using only these box plots, can you comment on the growth of these cities during this 20-year period compared with cities outside the sunbelt? Explain.

*3.103 "In most industries, goods are shipped, the product works, and the customer pays. But in high-technology [industries], confusion abounds over when a sale is really a sale. Sometimes, the product is shipped, and a sale is recorded—but the product has bugs, and customers balk at paying" ("High-Tech Sales: Now You See Them, Now

You Don't," *Business Week*, Nov. 18, 1985, p. 106). The table, also from *Business Week*, lists a sample of high-technology companies and the number of days (on average) it takes each to collect payment for sales made.

Company	Average No. of Days to Collect on Sale
CPT	155
Data Switch	143
C3	231
Scan-Optics	144
Aydin	178
Computer Entry Systems	137
UTL	242
Seagate Technology	143
DSC Communications	217
Computervision	132
Ungermann-Bass	131
Radiation Systems	184
Porta Systems	141
Silicon General	140
Network Systems	181
Masstor Systems	163
Intecom	131
Microdyne	154
T-Bar	144
Applied Data Research	126

Data: Standard & Poor's Compustat Services, Inc.

a. Use a computer package to order these data from the smallest to the largest.
b. Use a computer package to calculate the mean and standard deviation of these data.
c. Find the number of measurements within 2 standard deviations of the mean. Compare this result with the rule of thumb.

CHAPTER 3 QUIZ

1.   For the sample of five numbers,

3, 0, 0, 6, 7

calculate each of the following:

a. Median	c. Mean	e. Mode
b. Range	d. Variance	f. Standard deviation

2. What can be said about a measurement x in relation to the rest of the data set if:
   a. x is the 15th percentile?
   b. x has a z-score of -3?
   c. x is the third quartile?
   d. x has a z-score of $+2$?
   e. x has a z-score of 0?

3. The daily high temperature of a Florida city has an average of 78°F and a standard deviation of 7°F.
   a. What is the hottest you would expect the temperature to be in the city?
   b. What is the coldest?
   c. Would the temperature very often get above 92°F or below 64°F? Explain.

4. A typist claims that when typing for a long period of time, the number of words typed per minute has an average of 80 and a standard deviation of 3.
   a. This typist was observed while typing for 1 minute, and was found to have typed 70 words in that minute. How do you interpret this information? Explain.
   b. Suppose the typist had been observed to type 90 words during that minute. Would your interpretation change? Explain.
   c. Suppose the typist had been observed to type 75 words during that minute. How would you interpret this? Explain.

CHAPTERS 1–3 CUMULATIVE QUIZ

1. What advantage is there to using numbers to make inferences, as compared to using graphs?

2. An English literature class has twenty-three students. The following measurements are the ages (in years) of the students who sit in the front row: 19, 30, 19, 20, 22
   a. Are these five measurements a population or a sample? Explain.
   b. Calculate the mean, median, and mode for the data.
   c. Calculate the range, variance, and standard deviation for the data.

3. The percentage of family income allocated to rent was studied in a community. It was found that the mean percentage allocated to rent was 17.2 and the standard deviation was 0.1.
   a. Calculate the z-score for a family that allocates 16.9% of its income to rent.
   b. Interpret the z-score calculated in part a.

4. Fifty people were given a test for social awareness. Based on the results of the test, each individual was given a classification: Unaware, Moderately aware, or Very aware. The table summarizes the classifications of the fifty people tested.

Classification	Unaware	Moderately aware	Very aware
Frequency	7	31	12

a. Sketch a bar graph for the data.

b. Interpret the graph drawn in part *a.*

5. The president of a large bank claims that the number of years the employees of the bank have attended school has a mean of 17 and a standard deviation of 1.7. One employee is randomly chosen and her years of education are noted.

a. If the chosen employee attended school for 12 years, would you infer that the president's claim is incorrect? Why?

b. If the chosen employee attended school for 15 years, would you infer that the president's claim is incorrect? Why?

On Your Own

In the On Your Own at the end of Chapter 2, we listed several sources of real-life data sets. Using those data sources, sources suggested by your instructor, or your own resourcefulness, find one real-life quantitative data set from an area of particular interest to you. If you found a quantitative data set in Chapter 2, you may use it again.

a. Find the mean, median, variance, and standard deviation for this data set.

b. Count the actual number of measurements that fall within 2 and 3 standard deviations of the mean for this data set.

c. Compare your answers to part *b* with the rule of thumb.

References

McQuaid, E. A. & Kane, M. *The State Board Test Pool Examination for Registered Nurse Licensure.* Chicago: Chicago Review Press, 1981.

Mendenhall, W. *Introduction to Probability and Statistics.* 6th ed. Boston: Duxbury, 1983.

Ott, L., Larson, R. F., & Mendenhall, W. *Statistics: A Tool for the Social Sciences.* 3rd ed. Boston: Duxbury, 1983.

Chapter 4

PROBABILITY: A MEASURE OF RELIABILITY

THE MEISSEN MONKEY BAND CASE

The following burglary case description was reported by Chief Richard R. Anderson, Omaha Police Department:

> A man was shown a display of thirty-one (31) pieces of Meissen in an attempt to identify some as belonging to him. Within the display of thirty-one (31) pieces, there were eleven (11) that were accounted for; however, the man was unaware of this. Without the knowledge that eleven (11) were accounted for, the man selected nineteen (19) and stated these were his. Out of the nineteen selected, none were any of the eleven (11) that we had verified were not his. The one piece that was left could have been his; however, he said he was unsure and would not select it just to say it was his. All items on display are porcelain-type figurines, all painted differently with different musical instruments. The collection is referred to as a Meissen Monkey Band [Stephens, 1980].*

The fact that the man selected nineteen pieces and none of them were from the group that was accounted for seemed to be strong evidence that the pieces did, in fact, belong to him. Probability theory provided a method for measuring how strong the evidence was.

What is probability theory? How does probability theory provide a measure of the strength of sample evidence? In the first three sections of this chapter, we will introduce probability and discuss its role in making statistical inferences. In the optional sections that follow, additional topics in probability will be discussed. These topics may be useful in subsequent studies in statistics, but they are *not* required to continue through this text.

* Reprinted by permission of the *Journal of Police Science and Administration*, copyright 1980 by the International Association of Chiefs of Police, Inc., Vol. 8, No. 3, p. 353.

4.1

EXPERIMENTS AND SAMPLE SPACES

The word *probability* is familiar to most of you. Probability is used when discussing the likelihood, or chance, that an uncertain event will occur when some experiment is performed. The "experiment" can be as simple as observing whether it rains on a given day, or it can be extremely complex. When discussing probability, the term *experiment* is defined in a very broad manner. Namely, an **experiment** is any process that produces observations or measurements.* Further examples of experiments are listed below.

1. Recording 1,000 United States citizens' opinions about the president's economic policy
2. Tossing a coin three times and observing the upward face on each toss
3. Choosing a chairperson and a secretary from a committee of five people
4. Treating ten cancer patients with a new drug and observing how each responds
5. Drawing two marbles from a box containing six marbles
6. Measuring the amount of dissolved oxygen in a polluted river
7. Observing the fraction of insects killed by a new insecticide
8. Giving a test to twenty-three students and noting each student's grade
9. Choosing two bottles of wine from a case of twelve bottles and noting whether each chosen bottle has fermented properly
10. Choosing nineteen figurines from a set of thirty-one

As you can see from these examples, the notion of an experiment is very comprehensive: To qualify as an experiment, a process need only produce observations or measurements. Naturally, the possible outcomes of an experiment are of interest; the collection of all possible outcomes is called the **sample space.**

Definition 4.1

The **sample space** is the collection of all the possible outcomes of an experiment.[†]

EXAMPLE 4.1   One of the examples of experiments given above is "Choosing a chairperson and a secretary from a committee of five people." Suppose the committee consists of Jim, Kathy, Steve, Carol, and Bart. Give the sample space for this experiment.

Solution   The sample space is the collection of all possible outcomes to this experiment. It is a list of all the different ways a chairperson and a secretary could be chosen from the five people on the committee. If you examine the sample space given in Table 4.1, you can see that the first listed outcome is that Jim is chairperson and Kathy is secretary; the second is that Jim is chairperson and Steve is secretary; and so on.

---

* An experiment could also result in a single observation or measurement.
[†] Some texts refer to the outcomes in a sample space as **simple events.**

Table 4.1

Sample space for
Example 4.1

	Chairperson	Secretary		Chairperson	Secretary
1.	Jim	Kathy	11.	Steve	Carol
2.	Jim	Steve	12.	Steve	Bart
3.	Jim	Carol	13.	Carol	Jim
4.	Jim	Bart	14.	Carol	Kathy
5.	Kathy	Jim	15.	Carol	Steve
6.	Kathy	Steve	16.	Carol	Bart
7.	Kathy	Carol	17.	Bart	Jim
8.	Kathy	Bart	18.	Bart	Kathy
9.	Steve	Jim	19.	Bart	Steve
10.	Steve	Kathy	20.	Bart	Carol

For experiments such as the one in Example 4.1, it is fairly easy to identify all the possible outcomes in the sample space. But for other experiments, this task may be formidable. Since our main objective in this chapter is to introduce the basic concepts of probability and the role of probability in making inferences, we will restrict our discussion to experiments of the less formidable type.

In statistics, we are interested in making an inference about a population using the information contained in a sample. The process of sampling from a population is thus the experiment of interest to us, and the outcomes are samples. In this chapter we will consider the particular experiment of sampling n items (objects or measurements) from a population containing a specific number, N, of items.* This type of experiment is illustrated in the following example.

EXAMPLE 4.2 A population contains the $N = 4$ measurements 1, 2, 3, and 4. A sample of $n = 2$ measurements is to be taken from this population by choosing one number from the four in the population, recording its value, and then choosing a second number from the remaining three and recording its value. Give the sample space for this experiment.

Solution We can consider the experiment to consist of choosing a first and second number from a total of four numbers. An outcome of the experiment must then specify the first number chosen and the second number chosen. The collection of all the possible outcomes is given in Table 4.2.

Table 4.2

Sample space for
Example 4.2

	First Number	Second Number		First Number	Second Number
1.	1	2	7.	3	1
2.	1	3	8.	3	2
3.	1	4	9.	3	4
4.	2	1	10.	4	1
5.	2	3	11.	4	2
6.	2	4	12.	4	3

* A population that contains a specific number of items is called **finite.**

The outcomes for the experiment of Example 4.2 are the different possible samples of $n = 2$ measurements that can be selected from a population of $N = 4$ measurements. When sampling from a population, the experimental outcomes are samples, and thus *we may consider the sample space to be the collection of all the possible samples the experiment could produce.*

Our approach to Examples 4.1 and 4.2 illustrates the important characteristics of all the experiments we will consider in this chapter. In each experiment, *n objects are sampled one by one without replacement from a population of N objects.* In other words, the first object is selected from the N objects in the population. This first sampled item is not replaced, and a second item is sampled from the remaining $(N - 1)$ objects in the population. A third object is sampled from the remaining $(N - 2)$ objects in the population, and this process is continued until a sample of n objects is obtained.

Note that our approach to Example 4.2 distinguishes between drawing a 1 first and 2 second (outcome 1) and drawing a 2 first and 1 second (outcome 4). In one sense, of course, the end result is the same: The numbers chosen are 1 and 2. But we will consider our samples to be obtained in order, because counting and listing the sample space will then be easier in the long run.

EXAMPLE 4.3   In Example 4.1 we presented the experiment of choosing a chairperson and a secretary from a committee of five people. We may consider this to be an experiment of obtaining a sample from a population. To demonstrate this:

  a. Describe the experiment.
  b. Describe the population.
  c. Give the samples in the sample space.

Solution   **a.** The experiment consists of sampling $n = 2$ people from a population of $N = 5$ people. The first person in the sample will be the chairperson and the second person will be the secretary.
  b. The population is the collection (committee) of five people: Jim, Kathy, Steve, Carol, and Bart.
  c. The sample space is identical to the one given in Example 4.1. Whether you say the experiment produces *outcomes* or *samples* is not really important; these are two different terms for the same thing in statistical experiments. The sample space is given in Table 4.1. ∎

For the experiment of sampling n objects one by one without replacement from N objects, we can count the total number of samples (outcomes) in the sample space by multiplying together n numbers: The first number is N, the second is $(N - 1)$, the third is $(N - 2)$, etc. This follows from the general principle that if a first thing can be done in p ways and a second thing can be done in q ways, then the two things can be done in order in $(p \times q)$ ways.

In Example 4.2, $N = 4$ and $n = 2$. The total number of samples in the sample space can be found by multiplying

$$N \times (N - 1) = 4 \times 3 = 12$$

If you refer to the sample space shown in Table 4.2, you can see that there are twelve different samples.

In Example 4.3, a sample of $n = 2$ people was selected from a population of $N = 5$ people. The total number of samples in the sample space is

$$N \times (N - 1) = 5 \times 4 = 20$$

The sample space of Example 4.3, which is given in Table 4.1, contains twenty samples.

Since we will be referring frequently to the total number of samples in a sample space, it is convenient to use a shorthand notation to represent this value. The notation is given in the box.

Notation

The **total number of samples in a sample space** will be denoted by $\# S$.

The next box summarizes how to find the value of $\# S$ for the experiments we will discuss.

Calculating $\# S$

Consider an experiment that consists of randomly sampling n objects from a population of N objects. To count the total number of samples in the sample space, multiply together n numbers: The first number is N, the second is $(N - 1)$, the third is $(N - 2)$, and so forth. The number yielded by this multiplication is $\# S$.

EXAMPLE 4.4   An experiment consists of randomly sampling three letters from the alphabet. Find the value of $\# S$.

Solution   The population contains $N = 26$ letters. A sample of $n = 3$ letters is to be obtained. Thus, to find $\# S$, the total number of samples in the sample space, we multiply $n = 3$ numbers together. These three numbers are $N = 26$, $N - 1 = 25$, and $N - 2 = 24$. We obtain

$$\# S = 25 \times 24 \times 23 = 13,800$$

samples in the sample space. Typical samples are ABC, ABD, ABE, ABF, and so on. ∎

In general, to list the samples in a sample space, it is helpful to follow the steps given in the box on page 148.

Finding the Sample Space

1. *Describe the experiment.* State the number of objects to be sampled, n, and the number of objects in the population, N.
2. *Describe the population.* Identify each of the N objects in the population.
3. *Calculate #S.* Find the total number of samples in the sample space by multiplying the n terms N, (N − 1), (N − 2), etc.
4. *Give the sample space.* For small sample spaces, list the samples in the sample space. For very large sample spaces, list typical samples to obtain an understanding of their nature.

We conclude this section with some examples.

EXAMPLE 4.5   A jar contains two pennies, two nickels, a dime, and a quarter. Two coins are sampled from the jar and the denomination of each is noted. Give the sample space for this experiment.

Solution   We will follow the four steps given in the box.

1. *Describe the experiment.* A sample of $n = 2$ coins is to be selected from a population of $N = 6$ coins.
2. *Describe the population.* The population contains two pennies, two nickels, a dime, and a quarter. To identify each of the coins, we will denote the two pennies by P1 and P2, the two nickels by N1 and N2, the dime by D, and the quarter by Q. Note that we have identified *each* coin in the population; P1 is not the same as P2 and N1 is not the same as N2.
3. *Calculate #S.* Since $N = 6$ and $n = 2$, $\#S = 6 \times 5 = 30$.
4. *Give the sample space.* The thirty samples in the sample space are listed in Table 4.3.

Table 4.3

Sample space for
Example 4.5

	First Coin	Second Coin		First Coin	Second Coin		First Coin	Second Coin
1.	P1	P2	11.	N1	P1	21.	D	P1
2.	P1	N1	12.	N1	P2	22.	D	P2
3.	P1	N2	13.	N1	N2	23.	D	N1
4.	P1	D	14.	N1	D	24.	D	N2
5.	P1	Q	15.	N1	Q	25.	D	Q
6.	P2	P1	16.	N2	P1	26.	Q	P1
7.	P2	N1	17.	N2	P2	27.	Q	P2
8.	P2	N2	18.	N2	N1	28.	Q	N1
9.	P2	D	19.	N2	D	29.	Q	N2
10.	P2	Q	20.	N2	Q	30.	Q	D

Note that samples 1 and 6 both contain the two pennies, P1 and P2, and thus are very similar. However, since we are considering experiments in which objects are sampled one by one without replacement, the order in which the coins are drawn is relevant and we consider samples 1 and 6 to be two different samples. Drawing P1 first and P2 second is physically different from drawing P2 first and P1 second. ■

EXAMPLE 4.6   A part-time college student wants to take two evening classes. Five courses are available—one each in English, French, mathematics, psychology, and sociology. Give the sample space for this experiment.

Solution   We will follow the usual steps to find the sample space.

1. *Describe the experiment.* The experiment consists of choosing $n = 2$ courses from a total of $N = 5$ courses.
2. *Describe the population.* The population is the collection of the five courses in English, French, mathematics, psychology, and sociology. We will denote these by their first letters: E, F, M, P, and S.
3. *Calculate* $\#S$. Since $N = 5$ and $n = 2$, $\#S = 5 \times 4 = 20$.
4. *Give the sample space.* The sample space is given in Table 4.4.

Table 4.4

Sample space for Example 4.6

	First Course	Second Course		First Course	Second Course
1.	E	F	11.	M	P
2.	E	M	12.	M	S
3.	E	P	13.	P	E
4.	E	S	14.	P	F
5.	F	E	15.	P	M
6.	F	M	16.	P	S
7.	F	P	17.	S	E
8.	F	S	18.	S	F
9.	M	E	19.	S	M
10.	M	F	20.	S	P

■

EXAMPLE 4.7   An employer has interviewed five applicants for three jobs. Three of the applicants are women and two are men. We are interested in the sex of each of the three people whom the employer hires. Give the sample space for this experiment.

Solution   1. *Describe the experiment.* The employer chooses $n = 3$ people from a total of $N = 5$ applicants.

2. *Describe the population.* The population is the collection of five applicants, two men and three women. Since we are interested in their sex, we will denote the objects in the population by M1, M2, W1, W2, and W3.
3. *Calculate* $\#S$. Since $N = 5$ and $n = 3$, $\#S = 5 \times 4 \times 3 = 60$.

4. *Give the sample space.* Since the sample space has sixty samples, we will not list all of them. Typical samples in the sample space are

> M1 gets job 1, M2 gets job 2, and W1 gets job 3.
>
> M1 gets job 1, M2 gets job 2, and W2 gets job 3.
>
> M1 gets job 1, M2 gets job 2, and W3 gets job 3.
>
> M1 gets job 1, W1 gets job 2, and W2 gets job 3.   ■

Exercises (4.1–4.13)

Learning the Language

4.1 Define the term *sample space*.

4.2 What is denoted by each of the following?
a. N   *b. n*   *c.* #**S**

Learning the Mechanics

4.3 Calculate #**S**, the number of samples in the sample space for each of the following experiments in which *n* objects are sampled from *N* objects:
a. $n = 4$, $N = 7$ *d.* $n = 3$, $N = 6$
b. $n = 3$, $N = 50$ *e.* $n = 5$, $N = 10$
c. $n = 2$, $N = 10$ *f.* $n = 1$, $N = 25$

4.4 A box contains slips of paper with the numbers 1, 2, 3, 4, 5, and 6 written on them. One slip of paper is drawn and its number is noted, and then a second slip of paper is drawn and its number is noted.
a. Describe the experiment.
b. Describe the population.
c. Calculate #**S**.
d. Give the sample space.

4.5 A sample of $n = 3$ letters is chosen from the letters A, B, C, and D.
a. Describe the experiment.
b. Describe the population.
c. Calculate #**S**.
d. Give the sample space.

4.6 A bag contains two red marbles, two green marbles, and one orange marble. Two marbles are chosen from the bag.
a. Describe the experiment.
b. Describe the population.
c. Calculate #**S**.
d. Give the sample space.

Applying the Concepts

4.7   The companies in the highly competitive razor blade industry do a tremendous amount of advertising each year. Company G gave a supply of the three top-selling brands, G, S, and W, to a customer and asked him to pick the brand he liked best and the brand he liked second best. The company was, of course, hoping the customer would state a preference for its brand. Give the sample space for this experiment.

4.8   One regulation that concerns high school students who participate in varsity track is that they may enter at most three events during a meet. Suppose a track star at a certain school is qualified to compete in the following four events: high jump, triple jump, long jump, and 100-yard dash. The track star is going to choose three of these events in which to participate at an upcoming track meet. Give the sample space for this experiment.

4.9   A student is going to buy two nonfiction books from the top five on the list of best sellers. Give the sample space for this experiment.

4.10 The hostess of a dinner party is going to choose two of the following six vegetables to serve at the dinner: asparagus, broccoli, cauliflower, peas, spinach, and string beans. Give the sample space for this experiment.

4.11 Among weekly work schedules, the 40-hour work week is often considered to be the norm. In some countries, however, the average work week exceeds 40 hours. Shown in the table is the average number of working hours per week in manufacturing for seventeen countries with long work weeks.

Country	Average Hours Per Week	Country	Average Hours Per Week
Cyprus	45.0	Peru	48.1
Fiji	45.1	Thailand	48.1
Burma	45.6	Singapore	48.4
Gibraltar	45.6	Guyana	48.8
Mexico	45.6	Brunei	50.7
South Africa	46.1	Ecuador	51.0
Panama	47.1	South Korea	52.5
Syria	47.1	Egypt	57.0
Guatemala	47.2		

Source: George T. Kurian, *The Book of World Rankings* (New York: Facts on File, Inc., 1979), p. 214.

   a. If we want to sample five of these seventeen countries to study their working habits more thoroughly, how many possible samples can we take?

   b. List a few of the samples of five countries that could be obtained.

4.12 At Po Folks, a family restaurant, the Po Plate is a meal consisting of four vegetables served with cornbread or biscuits. The list of vegetables from which to choose, as provided in the menu, is reproduced below:*

Turnip greens	Baked Po-taters	Sliced tomaters
Corn on the cob	Fried Po-taters	Cottage cheese
Green beans	Po-tater salad	Red beans & rice
Fried okra	Coleslaw	Rice & gravy
New Po-taters & carrots		Seafood gumbo

a. How many samples of four vegetables can be selected from this list?

b. Give an example of a sample that excludes both corn and potatoes.

4.13 Out of 100 million or so individuals who file federal income tax returns each year, more than 1 million will be audited by the Internal Revenue Service. The chance of being audited depends on factors such as the filer's income and the complexity of the return. The accompanying table contains the number of returns, number of audits, and percentage audited for various income levels and types of return for 1984. Consider the last line of the table, which describes the returns of Schedule C filers who earned over $100,000.

Your Chances of Getting Audited			
Income & Return	# of Returns	# of Audits	% Audited
Less than $10,000, Form 1040A	20,806,000	73,814	.35%
Less than $10,000, Non-1040A	9,980,000	43,531	.44
$10,000–$25,000, Simple returns	20,622,000	132,737	.64
$10,000–$25,000, Complex returns	10,025,000	167,808	1.67
$25,000–$50,000	22,409,900	451,862	2.02
$50,000+	6,874,000	242,757	3.53
Filers of Schedule C with Incomes of:			
Less than $25,000	1,873,000	27,097	1.45
$25,000–$100,000	1,909,000	48,627	2.55
$100,000+	1,004,000	54,179	5.40

Source: Joseph A. Harb, "How to Survive a Tax Audit," Consumers' Research, Vol. 70, No. 3, March 1987, p. 22.

a. If a random sample of one such return is selected, how many possible samples are there?

b. If a random sample of two such returns is selected, how many possible samples are there?

* Po Folks, Inc., 1983.

4.2

EVENTS AND PROBABILITY

In Section 4.1 we defined the sample space to be the collection of all possible outcomes of an experiment. For statistical experiments, the outcomes are samples, and we are now ready to discuss the **probability** of observing a sample.

Definition 4.2

The **probability** of a sample (an outcome) is a number that measures the likelihood that the sample (outcome) will occur when the experiment is performed.

The probability of an outcome is what we would expect the relative frequency of the outcome to be in a very long series of repetitions of an experiment. When we do not know the relative frequency, we must assign the probability based upon other considerations. The following example will give you some insight into this process and help you understand some properties of probabilities.

EXAMPLE 4.8   A bag contains a red ball, a green ball, and a blue ball. The balls differ only in color; they are identical in all other aspects such as size, weight, texture, etc. After the balls are well mixed, two balls are drawn from the bag.

a. Give the sample space for this experiment.
b. What is the probability of observing each of the samples in the sample space?

Solution   **a.** We will follow the four steps given in Section 4.1 for finding the sample space.

1. *Describe the experiment.* The experiment consists of selecting $n = 2$ balls from $N = 3$ balls. The three balls are well mixed before the two are drawn.
2. *Describe the population.* The population is a collection of three balls that are identical except for their colors. There is a red ball, a green ball, and a blue ball. We will denote these by R, G, and B. respectively.
3. *Calculate $\#S$.* Since $n = 2$ and $N = 3$, $\#S = 3 \times 2 = 6$.
4. *Give the sample space.* The sample space is given in Table 4.5.

Table 4.5

Sample space for Example 4.8

	First Ball Drawn	Second Ball Drawn		First Ball Drawn	Second Ball Drawn
1.	R	G	4.	G	B
2.	R	B	5.	B	R
3.	G	R	6.	B	G

b. Since the three balls are identical except for their colors and are well mixed before the two balls are drawn, we would expect each possible sample to occur with approximately the same relative frequency if the experiment were repeated a very large number of times. Since there are six possible samples, each should occur approximately $\frac{1}{6}$ of the time. We will choose this number as the probability for each sample and say the probability that each of the samples will occur is $\frac{1}{6}$.   ■

Notice in Example 4.8 that the probability of each sample occurring is a fraction ($\frac{1}{6}$) and that the sum of the probabilities of all the samples in the sample space is 1; that is,

$$\frac{1}{6} + \frac{1}{6} + \frac{1}{6} + \frac{1}{6} + \frac{1}{6} + \frac{1}{6} = 1$$

This helps us identify two rules that must be followed when selecting the probabilities of outcomes for any experiment.

Rules of Probabilities

1. The probability of any sample (outcome) must be between 0 and 1.
2. The sum of the probabilities of all the samples (outcomes) in a sample space must equal 1.

In order to choose the probability of occurrence for each sample in a sample space, we must know exactly how the experiment producing that sample space is performed. We need this information to judge how frequently we would expect each sample to occur in many performances of the experiment. A common method of obtaining a sample from a population is **random sampling.**

Definition 4.3

If a sample of size n is selected from a population of size N in such a way that every possible sample of size n has the same probability of occurring, the sampling procedure (experiment) is called **random sampling.***

A sample produced by random sampling is called a **random sample.** Experiments that produce random samples are often described as follows: "n objects are randomly chosen," "All objects are well mixed and n are drawn," "A random selection of n objects is taken," "We have randomly sampled," etc. All the inferential procedures we will discuss in this text rely upon random sampling, and all the experiments we will discuss in this chapter will produce random samples.

* Some texts refer to this sampling procedure as **simple random sampling.**

EXAMPLE 4.9   A group of five cards contains an ace, a king, a queen, a jack, and a ten. The cards are well shuffled, and two cards are selected.

a. Give the sample space for this experiment.
b. Assign a probability for each possible sample.

Solution   **a.** We will follow the usual four steps to find the sample space.

1. *Describe the experiment.* The experiment consists of sampling $n = 2$ cards from a population of $N = 5$ cards. Since the cards are well shuffled, random sampling is being used.
2. *Describe the population.* The population consists of the five cards: ace, king, queen, jack, and ten. We will denote them by A, K, Q, J, and T, respectively.
3. *Calculate #S.* Since $n = 2$ and $N = 5$, $\#S = N \times (N - 1) = 5 \times 4 = 20$.
4. *Give the sample space.* The sample space is given in Table 4.6.

Table 4.6

Sample space for
Example 4.9

	First Card	Second Card		First Card	Second Card
1.	A	K	11.	Q	J
2.	A	Q	12.	Q	T
3.	A	J	13.	J	A
4.	A	T	14.	J	K
5.	K	A	15.	J	Q
6.	K	Q	16.	J	T
7.	K	J	17.	T	A
8.	K	T	18.	T	K
9.	Q	A	19.	T	Q
10.	Q	K	20.	T	J

b. Since random sampling is being used, each sample has the same probability of occurrence. Since there are twenty samples, each sample must have a probability of $\frac{1}{20}$ to satisfy the two rules of probabilities given earlier in the box. If the experiment were repeated a very large number of times, we would expect to observe each sample approximately $\frac{1}{20}$ of the time.   ■

In Example 4.9 there are twenty samples in the sample space, and each sample has a probability of $\frac{1}{20}$. This relationship may be generalized for random sampling.

Rule for Assigning Probabilities to Random Samples

For random sampling, the probability of observing each sample is

$$\frac{1}{\text{Number of samples in the sample space}} = \frac{1}{\#S}$$

EXAMPLE 4.10   A random sample of three objects is to be selected from a population of ten objects. What is the probability of observing each sample in the sample space?

Solution   We begin by following the four steps for finding the sample space.

1. *Describe the experiment.* The experiment consists of randomly sampling $n = 3$ objects from a population of $N = 10$ objects.
2. *Describe the population.* The population contains ten objects which we shall denote by $1, 2, 3, \ldots, 10$.
3. *Calculate* $\#S$. Since $n = 3$ and $N = 10$, $\#S = 10 \times 9 \times 8 = 720$.
4. *Give the sample space.* Since there are 720 samples in the sample space, we will not list them all, but some typical samples are 1, 2, 3; 1, 2, 4; 1, 2, 5; and so on.

Fortunately, to answer the question of interest, we do not have to list all the samples in the sample space. Since there are 720 samples, the probability of observing each sample is $\frac{1}{720}$. Thus, for random sampling, we need to know only the total number of samples in a sample space in order to determine the probability that each sample occurs.   ■

When an experiment is performed, we may not be interested in the probability that a particular sample occurs. Instead, we may want to know the probability that any one sample in a collection of several samples occurs. This leads to the concept of an **event,** which is defined in the next box.

An event is often denoted by a capital letter such as *A, B, C,* etc. Example 4.11 will illustrate how to find the probability of an event.

Definition 4.4

An **event** is a specific collection of samples (outcomes) in the sample space of an experiment; that is, an event is a subset of the sample space.

EXAMPLE 4.11   In Example 4.6 (Section 4.1), we considered the experiment in which a part-time college student was to choose two courses from five: English, French, mathematics, psychology, and sociology. These were denoted by E, F, M, P, and S, respectively. The sample space is repeated in Table 4.7 for convenience.

a. List the samples in each of the following events:

    A: Student does not take mathematics.

    B: Student takes at least one language.

    C: Student takes psychology and does not take sociology.

b. If the two courses are randomly selected from the five available, what is the probability of observing each sample in the sample space?

c. Using the answers to parts **a** and **b**, compute the probability of observing each of the events given in part **a**.

Table 4.7

Sample space for
Example 4.11

	First Course	Second Course		First Course	Second Course
1.	E	F	11.	M	P
2.	E	M	12.	M	S
3.	E	P	13.	P	E
4.	E	S	14.	P	F
5.	F	E	15.	P	M
6.	F	M	16.	P	S
7.	F	P	17.	S	E
8.	F	S	18.	S	F
9.	M	E	19.	S	M
10.	M	F	20.	S	P

Solution

a. The samples in event A are all those listed in Table 4.7 that do not contain M: 1, 3, 4, 5, 7, 8, 13, 14, 16, 17, 18, and 20.

The samples in event B are all those listed in Table 4.7 that contain E or F (or both): 1, 2, 3, 4, 5, 6, 7, 8, 9, 10, 13, 14, 17, and 18.

The samples in event C are all those listed in Table 4.7 that contain P but do not contain S: 3, 7, 11, 13, 14, and 15.

b. Since random sampling is being used and there are twenty samples in the sample space, the probability of observing each sample is $\frac{1}{20}$.

c. Using the answers to parts **a** and **b**, we can obtain the probability of observing event A. Since the probability of observing each sample is $\frac{1}{20}$, each sample would occur approximately $\frac{1}{20}$ of the time if the experiment were repeated a very large number of times. Since event A contains twelve samples, each with a probability of $\frac{1}{20}$, we would expect event A to occur approximately $\frac{12}{20}$ of the time if the experiment were repeated a very large number of times. Thus, the probability that event A occurs is $\frac{12}{20}$. That is, if two courses are randomly selected from the five available, the probability is $\frac{12}{20} = \frac{3}{5}$ that the student will not take mathematics. Since event B contains fourteen samples, the probability of event B occurring is $\frac{14}{20} = \frac{7}{10}$.

Event C contains six samples, so the probability that event C occurs is $\frac{6}{20} = \frac{3}{10}$. ■

Before proceeding further, it would be advantageous to introduce the following notation.

Notation

Let A be any event of interest.

The number of samples in event A will be denoted by #A.

The probability of event A will be denoted by P(A).

Similarly, for an event D,

$\#D$ = The number of samples in event D

and

$P(D)$ = The probability of event D

Examining part **c** of Example 4.11, we see that the probability of each event was found by dividing the number of samples in the event of interest by the total number of samples in the sample space. This may be generalized to the following rule.

Rule for Calculating Probabilities of Events for Random Sampling*

If n objects are randomly sampled from N objects, and A is any event of interest, then

$$P(A) = \frac{\text{Number of samples in event } A}{\text{Number of samples in the sample space}} = \frac{\#A}{\#S}$$

A summary of the procedure that should be followed to find the probability of an event for an experiment that employs random sampling is given in the box.

Steps for Finding the Probability of an Event A When Random Sampling Is Used

1. Describe the experiment.
2. Describe the population.
3. Calculate $\#S$, the number of samples in the sample space, and list them if it is feasible (that is, if there are not too many).
4. Calculate $\#A$.
5. Find the probability of event A,

$$P(A) = \frac{\text{Number of samples in event } A}{\text{Number of samples in the sample space}} = \frac{\#A}{\#S}$$

---

* Note: This rule is appropriate only for random sampling.

We conclude this section with several examples.

EXAMPLE 4.12   Five tickets have been sold for a lottery, and you have bought ticket number 1. A first and second prize are to be awarded based on a random selection of two tickets from the five. Find the probability of the following events:

     F: You win first prize.

     L: You do not win a prize.

Solution   We will follow the five steps given in the box.

1. *Describe the experiment.* The experiment consists of randomly sampling $n = 2$ tickets from $N = 5$ tickets.
2. *Describe the population.* The population is the collection of the five tickets that were sold. We will denote these by 1, 2, 3, 4, and 5. (You bought ticket 1.)
3. *Calculate #S and list the samples, if feasible.* Since $n = 2$ and $N = 5$, $\#S = 5 \times 4 = 20$. This is a relatively small number, so we give the sample space in Table 4.8.

Table 4.8

Sample space for
Example 4.12

	First Prize	Second Prize		First Prize	Second Prize
1.	1	2	11.	3	4
2.	1	3	12.	3	5
3.	1	4	13.	4	1
4.	1	5	14.	4	2
5.	2	1	15.	4	3
6.	2	3	16.	4	5
7.	2	4	17.	5	1
8.	2	5	18.	5	2
9.	3	1	19.	5	3
10.	3	2	20.	5	4

4. *Calculate #F and #L.* A sample is in event *F* if you win first prize. Thus, the four samples 1, 2, 3, and 4 are in event *F*, and $\#F = 4$. A sample is in event *L* if you do not win a prize. The twelve samples 6, 7, 8, 10, 11, 12, 14, 15, 16, 18, 19, and 20 are in event *L*, and $\#L = 12$.
5. *Find the probabilities of the events of interest.* Since $\#F = 4$ and $\#S = 20$,

$$P(F) = \frac{\#F}{\#S} = \frac{4}{20} = \frac{1}{5}$$

Similarly,

$$P(L) = \frac{\#L}{\#S} = \frac{12}{20} = \frac{3}{5}$$ ∎

EXAMPLE 4.13   In Example 4.5 (Section 4.1), the experiment consisted of selecting two coins from a jar that contained two pennies, two nickels, a dime, and a quarter. The sample space is repeated in Table 4.9 for convenience. Note that $\#S = 30$.

Table 4.9

Sample space for
Example 4.13

	First Coin	Second Coin		First Coin	Second Coin		First Coin	Second Coin
1.	P1	P2	11.	N1	P1	21.	D	P1
2.	P1	N1	12.	N1	P2	22.	D	P2
3.	P1	N2	13.	N1	N2	23.	D	N1
4.	P1	D	14.	N1	D	24.	D	N2
5.	P1	Q	15.	N1	Q	25.	D	Q
6.	P2	P1	16.	N2	P1	26.	Q	P1
7.	P2	N1	17.	N2	P2	27.	Q	P2
8.	P2	N2	18.	N2	N1	28.	Q	N1
9.	P2	D	19.	N2	D	29.	Q	N2
10.	P2	Q	20.	N2	Q	30.	Q	D

If the two coins are randomly sampled, find the probabilities of the following events:

A: At least one penny is selected.

B: The dime is selected.

C: Either the dime is selected or at least one penny is selected.

Solution   Since the sample space containing thirty samples is given in Table 4.9, and random sampling is being used, we can count the number of samples in each event of interest and find their probabilities.

If a sample contains P1, P2, or both, then at least one penny has been selected and event A has occurred. The samples in A are 1, 2, 3, 4, 5, 6, 7, 8, 9, 10, 11, 12, 16, 17, 21, 22, 26, and 27. There are eighteen samples in event A, and $\#A = 18$. We obtain

$$P(A) = \frac{\#A}{\#S} = \frac{18}{30} = \frac{3}{5}$$

If a sample contains D, then the dime has been selected and event B has occurred. The samples in B are 4, 9, 14, 19, 21, 22, 23, 24, 25, and 30. Since $\#B = 10$,

$$P(B) = \frac{\#B}{\#S} = \frac{10}{30} = \frac{1}{3}$$

If a sample contains P1, P2, or D (or any two of the three), then event C has occurred. The samples in event C are 1, 2, 3, 4, 5, 6, 7, 8, 9, 10, 11, 12, 14, 16, 17, 19, 21, 22, 23, 24, 25, 26, 27, and 30. Event C contains twenty-four samples, and

$$P(C) = \frac{\#C}{\#S} = \frac{24}{30} = \frac{4}{5}$$ ∎

In Example 4.13, you may have noticed that event C is a joining (union) of events A and B. Event C is the event that either event A or event B or both occurs. It is tempting to think that the probability of event C is the sum of the probabilities of events A and B, but this is not the case, as we can see from the results obtained in Example 4.13 ($\frac{24}{30} \neq \frac{18}{30} + \frac{10}{30}$). If you follow the five steps for finding the probability of an event as given in the box, you should obtain the correct probability. If you rely on your intuition or other seemingly reasonable methods, you may not get the correct probability.

EXAMPLE 4.14 A consumer has bought a package of five flashcubes. Unknown to the buyer, one of the flashcubes is defective (will not flash) and the other four are good. Two of the cubes are randomly chosen for use. Find the probability of each of the following events:

A: Both flashcubes are good.

B: One flashcube is good and one is defective.

C: The first flashcube is good or the second flashcube is good, or both are good.

Solution We will follow the usual steps for finding probabilities.

1. *Describe the experiment.* The experiment consists of choosing a random sample of $n = 2$ flashcubes from a total of $N = 5$ flashcubes.

2. *Describe the population.* The population is the collection of five flashcubes—one defective flashcube and four good flashcubes. We denote the defective flashcube as D and the four good cubes as G1, G2, G3, and G4.

3. *Calculate #S and list the samples, if feasible.* Since $n = 2$ and $N = 5$, $\#S = 5 \times 4 = 20$ samples in the sample space, as listed in Table 4.10.

4. *Count the samples in the events of interest.* The samples in each of the three events A, B, and C are listed below:

A: 6, 7, 8, 10, 11, 12, 14, 15, 16, 18, 19, and 20

B: 1, 2, 3, 4, 5, 9, 13, and 17

C: All twenty samples

Thus, $\#A = 12$, $\#B = 8$, $\#C = 20$.

Table 4.10

Sample space for Example 4.14

	First Flashcube	Second Flashcube		First Flashcube	Second Flashcube
1.	D	G1	11.	G2	G3
2.	D	G2	12.	G2	G4
3.	D	G3	13.	G3	D
4.	D	G4	14.	G3	G1
5.	G1	D	15.	G3	G2
6.	G1	G2	16.	G3	G4
7.	G1	G3	17.	G4	D
8.	G1	G4	18.	G4	G1
9.	G2	D	19.	G4	G2
10.	G2	G1	20.	G4	G3

5. *Find the probabilities of the events of interest.* Event A contains twelve samples, so

$$P(A) = \frac{\#A}{\#S} = \frac{12}{20} = \frac{3}{5}$$

Event B contains eight samples, so

$$P(B) = \frac{\#B}{\#S} = \frac{8}{20} = \frac{2}{5}$$

Event C contains twenty samples, so

$$P(C) = \frac{\#C}{\#S} = \frac{20}{20} = 1$$

Since $P(C) = 1$, event C will occur every time the experiment is performed. ∎

EXAMPLE 4.15   A supermarket has 500 apples for sale in its fresh fruit section. Although the apples appear to be identical, 50 of them are spoiled and the other 450 are good.

a. If you randomly choose one apple from the 500 for sale, is it unlikely that you choose a spoiled apple? Explain.

b. If you randomly choose two apples from the 500 for sale, is it unlikely that you choose two spoiled apples? Explain.

Solution   **a.** To assess how unlikely it is to choose a spoiled apple, we will calculate the probability of the event

A: A spoiled apple is chosen.

We will follow the usual steps to find $P(A)$.

1. *Describe the experiment.* The experiment consists of randomly choosing $n = 1$ apple from a total of $N = 500$ apples.
2. *Describe the population.* The population is the collection of 500 apples for sale. It is important to note that 50 are spoiled and 450 are good.
3. *Calculate #S and list the samples, if feasible.* Since $n = 1$ and $N = 500$, there are 500 samples in the sample space. Each sample contains one of the 500 apples. There are more samples than we care to list, but 50 of the samples contain a spoiled apple and 450 of the samples contain a good apple. Using S to denote spoiled and G to denote good, we could list the sample space as S1, S2, S3, . . . , S50, G1, G2, . . . , G450.
4. *Calculate #A.* All the samples that contain a spoiled apple are in event A. There are 50 samples, S1, S2, S3, . . . , S50, in event A, and $\#A = 50$.
5. *Find P(A).* The probability of event A is

$$P(A) = \frac{\#A}{\#S} = \frac{50}{500} = \frac{1}{10} = .10$$

If one apple were repeatedly chosen from a group of 500, of which 50 were spoiled, then about 10% of the time a spoiled apple would be chosen. This probability indicates that event A is not particularly unlikely. Although we might not expect to randomly choose a spoiled apple from the 500, we should not be surprised to see it happen, since it would occur 10% of the time.

b. Next, we want to calculate the probability of the event

> B: Two spoiled apples are chosen.

Since this event is associated with an experiment different from the one discussed in part **a**, we must go through all five steps again to find $P(B)$.

1. *Describe the experiment.* This experiment consists of randomly choosing $n = 2$ apples from a total of $N = 500$ apples.
2. *Describe the population.* The population is the collection of 500 apples for sale; 50 are spoiled and 450 are good.
3. *Calculate $\#S$ and list the samples, if feasible.* Since $n = 2$ and $N = 500$, $\#S = 500 \times 499 = 249{,}500$ samples in the sample space. Needless to say, we will not list them all, but typical samples are S1, S2; S1, S3; S1, S4; and so on.
4. *Calculate $\#B$.* Since we did not list all the samples, we will have to count the number of samples in event B by using a method similar to the one used to count all the samples. For a sample of two apples to contain two spoiled apples, both apples must be chosen from the 50 spoiled apples in the population. The number of different ways we may obtain samples of two spoiled apples from the 50 spoiled apples in the population is equivalent to the number of ways of sampling 2 objects from 50 objects. Thus, there are $50 \times 49 = 2{,}450$ samples that contain two spoiled apples, and $\#B = 2{,}450$.
5. *Find $P(B)$.* The event B contains 2,450 samples, and the sample space contains 249,500 samples. Since random sampling was used,

$$P(B) = \frac{\#B}{\#S} = \frac{2{,}450}{249{,}500} \approx .01$$

It is unlikely that you would choose two spoiled apples. This event would occur about 1% of the time if two apples were repeatedly chosen at random from a group of 500, of which 50 were spoiled. Although it is *possible* for the event B to occur, you would be very surprised if both apples you chose were spoiled. ∎

In step 4 of the solution to part **b** of Example 4.15, we counted the number of ways two spoiled apples could be sampled from the fifty spoiled apples in the population of 500 apples. We were interested in obtaining the entire sample from only a portion of the population. The method we used to count the number of ways to obtain such a sample can be stated in general. We offer the generalization as a helpful hint.

Helpful Hint

Consider an experiment that consists of randomly sampling n objects from a population of N objects. To count the number of ways that the n objects can be sampled from R objects (where R is less than N), multiply together n numbers: The first number is R, the second is $(R - 1)$, the third is $(R - 2)$, etc.

Recall the Meissen Monkey Band case introduced at the beginning of this chapter. The problem is similar to the one considered in Example 4.15, part **b**. In the example, we considered 500 apples, 50 of which were spoiled; in the Meissen Monkey Band case, the police had 31 figurines, 11 of which were accounted for. In Example 4.15, we were interested in the likelihood of randomly choosing two apples without getting a good one; the police were interested in the likelihood that the man could choose 19 figurines without including one of the 11 already accounted for. The police might have approached the problem as described in Example 4.16.

EXAMPLE 4.16   A set of 31 figurines contains 11 that are "marked." Suppose 19 figurines are chosen at random. What is the probability that no marked figurines are chosen?

Solution   **1.** *Describe the experiment.* The experiment consists of choosing 19 figurines from the set of 31.

2. *Describe the population.* The population is the set of 31 figurines, 11 of which are

3. *Calculate #S and list the samples, if feasible.* Since $n = 19$ and $N = 31$, the sample space has $31 \times 30 \times 29 \times \cdots \times 13$ samples. (#S is *very* large.)

4. *Count the samples in the event of interest.* The event E of interest consists of those samples that do not contain any of the 11 marked figurines—that is, all samples that are chosen from the 20 unmarked figurines. We count these samples as described in the previous Helpful Hint. The problem consists of sampling $n = 19$ objects from $R = 20$ objects. Thus, there are $20 \times 19 \times \cdots \times 2$ samples that contain only unmarked figurines. Multiplying these values would yield $\#E$.

5. *Find P(E).* Since random sampling was used,

$$P(E) = \frac{\#E}{\#S}$$

$$= \frac{20 \times 19 \times \cdots \times 2}{31 \times 30 \times \cdots \times 13} = \frac{12 \times 11 \times \cdots \times 2}{31 \times 30 \times \cdots \times 21} \approx .0000001$$

Therefore, E is quite a rare event. This was an important factor that helped the police make their inference in the Meissen Monkey Band case.   ∎

As we proceed through this text, experiments will become more complex, and your intuition will not always be able to provide an accurate assessment of the rarity of an event. This underscores the need for an understanding of probability theory to make inferences. Probability will provide a dependable procedure for evaluating how unlikely an event is (the rarity of an event), and this will ultimately enable us to measure the reliability of inferences.

Exercises (4.14–4.30)

Learning the Language

4.14   Define the following terms:
a. Random sampling   b. Event   c. Probability

4.15   What do $\#C$ and $P(C)$ represent?

Learning the Mechanics

4.16 In Exercise 4.4 we considered the experiment of drawing two numbers from a box containing slips of paper with the numbers 1, 2, 3, 4, 5, and 6 on them. The sample space is given in the table.

	First Draw	Second Draw		First Draw	Second Draw		First Draw	Second Draw
1.	1	2	11.	3	1	21.	5	1
2.	1	3	12.	3	2	22.	5	2
3.	1	4	13.	3	4	23.	5	3
4.	1	5	14.	3	5	24.	5	4
5.	1	6	15.	3	6	25.	5	6
6.	2	1	16.	4	1	26.	6	1
7.	2	3	17.	4	2	27.	6	2
8.	2	4	18.	4	3	28.	6	3
9.	2	5	19.	4	5	29.	6	4
10.	2	6	20.	4	6	30.	6	5

a. If the two numbers are randomly drawn from the six numbers in the box, assign an appropriate probability to each sample in the sample space.

b. List the samples in each of the following events:

A: Sum of the two numbers drawn is less than 8.

B: At least one of the two numbers drawn is larger than 3.

C: First number drawn is odd and second number drawn is even.

D: First number drawn is odd or second number drawn is even, or both.

c. Find the probability of each event in part b.

4.17 In Exercise 4.5 we considered the experiment of choosing three letters from A, B, C, and D. The sample space is given in the table.

	First Letter	Second Letter	Third Letter		First Letter	Second Letter	Third Letter
1.	A	B	C	13.	C	A	B
2.	A	B	D	14.	C	A	D
3.	A	C	B	15.	C	B	A
4.	A	C	D	16.	C	B	D
5.	A	D	B	17.	C	D	A
6.	A	D	C	18.	C	D	B
7.	B	A	C	19.	D	A	B
8.	B	A	D	20.	D	A	C
9.	B	C	A	21.	D	B	A
10.	B	C	D	22.	D	B	C
11.	B	D	A	23.	D	C	A
12.	B	D	C	24.	D	C	B

a. List the samples in each of the following events:

A: Letter A is chosen.

B: Letters chosen can be arranged to spell BAD.

C: Letter C is chosen first or second.

D: Letter C is chosen first and second.

b. If the letters are randomly chosen, find the probability of each event in part a.

4.18 In Exercise 4.6 we considered the experiment of choosing two marbles from a bag containing two red marbles (R1 and R2), two green marbles (G1 and G2), and one orange marble (O). The sample space is given in the table.

	First Marble	Second Marble		First Marble	Second Marble
1.	R1	R2	11.	G1	G2
2.	R1	G1	12.	G1	O
3.	R1	G2	13.	G2	R1
4.	R1	O	14.	G2	R2
5.	R2	R1	15.	G2	G1
6.	R2	G1	16.	G2	O
7.	R2	G2	17.	O	R1
8.	R2	O	18.	O	R2
9.	G1	R1	19.	O	G1
10.	G1	R2	20.	O	G2

a. List the samples in each of the following events:

A: Both marbles chosen are the same color.

B: Orange marble is not chosen.

C: Orange marble is chosen first.

D: Orange marble is chosen first or second.

b. Find the probability of each event in part a. Assume the marbles are randomly chosen.

4.19 Two numbers are randomly chosen from the numbers 1, 2, 3, 4, 5, 6, 7, 8, and 9. Find the probability of each of the following events:

A: Both numbers chosen are odd.

B: Both numbers chosen are even.

C: Number 1 is not chosen.

4.20 Two letters are randomly chosen from the letters A, B, C, D, and E. Find the probability of each of the following events:

A: Letter A is chosen.

B: First letter chosen appears in the alphabet before second letter chosen.

C: Letter B or letter D (or both) is chosen.

    D: First letter chosen is E.

    E: Second letter chosen is E.

Applying the Concepts

4.21 In Exercise 4.7 we considered the experiment of a customer picking the brand he liked best and the brand he liked second best of the three top-selling brands of razor blades (G, S, and W). The sample space is given in the table.

	Liked Best	Second Best		Liked Best	Second Best
1.	G	S	4.	S	W
2.	G	W	5.	W	S
3.	S	G	6.	W	G

    a. If the customer randomly chooses the two brands to rank first and second, what probability would you assign to each sample? (Note that a random selection would make sense if the customer actually likes the three brands equally well, but must make a first and second choice.)

    b. If the customer randomly chooses the two brands to rank first and second, what is the probability that brand S is ranked as second best?

4.22 In Exercise 4.9 we considered the experiment in which a student is going to buy two nonfiction books from the top five on the list of best sellers. If we denote the best seller as 1, the second best seller as 2, etc., the sample space may be written as shown in the table.

	First Book	Second Book		First Book	Second Book
1.	1	2	11.	3	4
2.	1	3	12.	3	5
3.	1	4	13.	4	1
4.	1	5	14.	4	2
5.	2	1	15.	4	3
6.	2	3	16.	4	5
7.	2	4	17.	5	1
8.	2	5	18.	5	2
9.	3	1	19.	5	3
10.	3	2	20.	5	4

    a. List the samples in each of the following events:

        A: Student buys the two best-selling books.

        B: Student buys at least one of the two best-selling books.

        C: Student buys exactly one of the two best-selling books.

    b. If the student randomly chooses the two books, find the probability of each event in part *a*.

4.23 A buyer for a large metropolitan department store must choose two firms from the four available to supply the store's fall line of men's slacks. The buyer has not dealt with any of the four firms before and considers their products to be equally attractive. Unknown to the buyer, two of the four firms are having serious financial problems that will result in their not being able to deliver the fall line of slacks on time. The other two firms are in good financial condition and will deliver the fall line of slacks on time. If the buyer randomly selects two firms from the four available, find the probability of each of the following events:

   A: Neither of the firms the buyer chooses will deliver the fall line of slacks on time.

   B: Exactly one of the chosen firms will deliver its fall line of slacks on time.

   C: At least one of the chosen firms will deliver its fall line of slacks on time.

4.24 One game that is very popular in many American casinos is roulette. Roulette is played by spinning a ball on a circular wheel that has been divided into thirty-eight arcs of equal length; these are numbered 00, 0, 1, 2, . . . , 35, 36. The number of the arc on which the ball comes to rest is the outcome of one play of the game. The numbers are also colored in the following manner:

   Red: 1, 3, 5, 7, 9, 12, 14, 16, 18, 19, 21, 23, 25, 27, 30, 32, 34, 36

   Black: 2, 4, 6, 8, 10, 11, 13, 15, 17,.20, 22, 24, 26, 28, 29, 31, 33, 35

   Green: 00, 0

Players may place bets on the table in a variety of ways, including bets on odd, even, red, black, high, low, etc. If the wheel is spun once, find the probability of each of the events given below. Carefully state any assumptions you made in finding these probabilities.

   A: Outcome is a black number.

   B: Outcome is a red number.

   C: Outcome is a low number (1–18).

   D: Outcome is 00 or 0.

4.25 A social psychologist wants to determine the effects of competition on performance. Each subject in the study is allowed to choose three opponents from a group of nine competitors in a particular game. Suppose that, of this group of nine competitors, four are of equal or greater ability than a particular subject and the remaining five are of lesser ability. If the subject in question chooses the opponents at random, find the probability of each of the following events:

   A: Subject chooses three opponents of lesser ability.

   B: Subject chooses three opponents of equal or greater ability.

4.26 Defective alternators have been mistakenly installed in three of the last six truck engines to emerge from an assembly line. It is not known which engines contain the defective alternators. If three of the six engines are randomly selected, find the probability of each of the following events:

   A: All three engines contain a defective alternator.

   B: None of the three engines contains a defective alternator.

4.27 The baseballs used at the major league level are individually hand-stitched and, hence, are not uniformly made. Most pitchers who throw curve balls prefer baseballs with high stitches due to the firmer grip they provide. From a box of twenty new baseballs, an umpire chooses five at random to start the game. If this particular box contains thirteen high-stitched baseballs, find the probability that none of the five balls chosen to start the game are preferred by a curve ball pitcher.

4.28 The accompanying table lists the principal worldwide producers of coal and their estimated amounts of annual production.

Nation	Millions of Tons
United States	556
Soviet Union	529
People's Republic of China	452
United Kingdom	162
Poland	160
West Germany	122

Source: Douglas M. Considine (ed.), *Energy Technology Handbook* (New York: McGraw-Hill, 1977), pp. 1–27.

a. If two of these six countries are randomly selected to join a panel studying worldwide energy problems, what is the probability that both the United States and the Soviet Union are selected?

b. What is the probability that neither the United States nor the Soviet Union is selected?

4.29 The Nobel Prize has been acknowledged as a prestigious award for human achievement. The table lists the countries having the most Nobel laureates in the first 79 years of the prize's history. If a random sample of three of these countries is selected, what is the probability that each of the countries selected has more than twenty Nobel laureates?

Country	Total Number of Nobel Laureates
United States	153
United Kingdom	74
Germany (East and West)	55
France	44
Sweden	24
Switzerland	14
Soviet Union	14
Netherlands	11
Denmark	11

Source: George T. Kurian, *The Book of World Rankings* (New York: Facts on File, Inc., 1979), p. 373.

4.30   In Exercise 4.13 we discussed the chances of having a federal income tax return audited. The accompanying table describes the 1984 returns of Schedule C filers who earned over $100,000.

	# of Returns	# of Audits	% Audited
Filers of Schedule C with Incomes of $100,000+	1,004,000	54,179	5.40

Source: Joseph A. Harb, "How to Survive a Tax Audit," *Consumers' Research*, Vol. 70, No. 3, March 1987, p. 22.

a.   If one of these returns is randomly selected, what is the probability it is audited? How is this probability related to the figure of 5.40% audited shown in the table?

b.   If one of these returns is randomly selected, what is the probability it is not audited?

c.   If two of these returns are randomly selected, what is the probability they are both audited?

d.   If two of these returns are randomly selected, what is the probability neither of them is audited?

4.3

PROBABILITY AND STATISTICAL INFERENCE

We have found probabilities of events when random samples are taken from populations, and we have interpreted these probabilities. This knowledge of probability can now be used to aid us in making inferences about a population based upon an observed sample. We will demonstrate the role of probability in making inferences with the following example.

EXAMPLE 4.17   A manufacturer markets ball-point pens in boxes of 1,000 pens each and claims that 85% of the pens in each box will write the first time they are used. A consumer group doubts that 85% of the pens really write the first time and suspects the percentage is lower. The consumer group is going to select a random sample of pens from a box the manufacturer marketed to test the manufacturer's claim.

a.   Suppose the consumer group randomly samples one pen from a box of 1,000 pens, and the pen does not write the first time it is used. Would you be willing to conclude that fewer than 85% of the pens write the first time? In other words, would you agree with the consumer group's suspicion?

b.   Suppose the consumer group randomly samples two pens from a box of 1,000 pens, and neither pen writes the first time. Would you agree with the consumer group's suspicion and reject the manufacturer's claim?

Solution   To answer these questions, we will use our knowledge of probability. We will calculate probabilities under the assumption that the manufacturer is correct, and then decide whether this assumption is appropriate. In answering both parts **a** and **b** we will follow the five steps for finding the probability of an event.

a. One pen is to be randomly chosen, and we are interested in the probability of the event

   A: Pen does not write the first time it is used.

   1. *Describe the experiment.* The experiment consists of randomly selecting $n = 1$ pen from a population of $N = 1,000$ pens.
   2. *Describe the population.* The population consists of 1,000 pens. We do not truly know how many of these pens will write the first time, but we will assume the manufacturer is correct and 85% of them will write the first time. Thus, we will assume that 850 (85% of 1,000) of the pens will write the first time they are used, and the remaining 150 will not write the first time.
   3. *Calculate #**S** and list the samples, if feasible.* Since $n = 1$ and $N = 1,000$, there are 1,000 samples in the sample space. We will not list each sample, but from the description of the population given in step 2, we know that 850 samples contain pens that will write the first time and 150 contain pens that will not. If we let D represent a defective pen and G represent a good one, then the sample space is D1, D2, . . . , D150, G1, G2, . . . , G850.
   4. *Calculate #A.* Event A contains the 150 samples that have pens that will not write the first time.
   5. *Find P(A).*

$$P(A) = \frac{\#A}{\#S} = \frac{150}{1,000} = .15$$

If the manufacturer's claim is correct, we would randomly choose a pen that does not write about 15% of the time in repeated performances of this experiment. This is not a particularly rare (unlikely) event. It is quite possible that we would choose one of the 150 pens that would not write even if the manufacturer's claim is true. Thus, there is not much support for the consumer group's suspicion, and we would not reject the manufacturer's claim.

Notice that we have *not* concluded that the manufacturer is correct. If the actual percentage of pens that write the first time is less than 85%, then the probability of choosing one that does not write is even larger than .15. In the most extreme case, if none of the pens write the first time, we would be certain to choose a pen that does not write the first time. Thus, the sample evidence neither refutes nor substantiates the manufacturer's claim.

b. We are now interested in the probability of the event

   B: Neither of two pens chosen will write the first time.

   1. *Describe the experiment.* The experiment consists of randomly sampling $n = 2$ pens from $N = 1,000$ pens.
   2. *Describe the population.* The population is the collection of 1,000 pens. Again, we assume that 850 pens will write the first time they are used and 150 pens will not.

3. *Calculate* #***S*** *and list the samples, if feasible.* Since $n = 2$ and $N = 1,000$, #***S*** $=$ $1,000 \times 999 = 999,000$ samples in the sample space. Typical samples are D1, D2; D1, D3; and so on.

4. *Calculate* #*B.* Using the same reasoning we employed in Example 4.15 (Section 4.2), we see that the samples in event *B* are those for which 2 pens have been chosen from the 150 that do not write the first time. Thus, #*B* $= 150 \times$ $149 = 22,350$ samples.

5. *Find P(B).*

$$P(B) = \frac{\#B}{\#S} = \frac{22,350}{999,000} \approx .02$$

If 85% of the pens write the first time, the probability of randomly choosing two pens that do not write the first time is only about .02. This event is unlikely to occur if the manufacturer's claim is correct. There are two possible interpretations we could make if two defective pens are selected:

1. We could conclude that the manufacturer's claim is correct and that a rare event—a very improbable sample—was observed. As a matter of fact, the event is so rare that it would occur only about 2 out of 100 times if two pens were selected from 1,000 over and over again a very large number of times.

2. Or, we could conclude that the percentage of pens that write the first time is smaller than 85%. This would support the consumer group's suspicions, since if fewer than 85% of the pens write the first time, the probability that two defective pens are selected increases. That is, the sample result becomes more likely to occur.

Which interpretation would you choose? Using the rare event approach to making inferences, we would choose the second interpretation. An event with a probability of only .02 is very unusual, so the sample result provides strong evidence that the manufacturer's claim about the population of pens is incorrect and that the consumer group's suspicions are justified. ∎

In Chapter 3 we discussed rare events in terms of means, standard deviations, and *z*-scores. Now, in Example 4.17, we have illustrated how this concept of rare events relates to probability. Measuring the probability of a sample is a more accurate measure of the rarity of an event than the measures used in Chapter 3. We know that if a sampled measurement is more than 2 standard deviations from the mean, it represents a rare event, but we cannot tell exactly how rare. That is, we cannot specify the exact probability of such an event. The Tchebysheff result guaranteed that the probability was at most .25; and our rule of thumb claimed that for practical experiments it was smaller. But we need to be more precise. The probability of an event provides a quantitative measure of its likelihood that we can determine precisely.

In Chapter 3 we defined *rare event* and gave some helpful hints for applying the rare event approach using means and standard deviations. We will now redefine *rare event* and rephrase these hints so that they may be used with probabilities.

Definition 4.5

An event that occurs with probability .05 or less is considered to be a **rare event.**

Helpful Hints

Follow these steps when using the rare event approach to making inferences:

1. Hypothesize the nature of the population of interest. That is, describe what items you are assuming the population contains.
2. Obtain a random sample from the population of interest, and calculate the probability of observing such a sample. (The event whose probability is to be calculated will be specified.)
3. If the calculated probability is .05 or smaller, then infer that the population of interest is somehow different from the hypothesized population. If the calculated probability is larger than .05, there is insufficient evidence to infer that the population of interest is different from the hypothesized population.

We have arbitrarily chosen .05 as the point at which we will declare an event to be rare. An experimenter may choose any value desired to represent a rare event, but the probability must be small (near 0). The value of .05 is one often chosen by experimenters when making inferences.

Before making inferences about populations, we must assume we know the nature of the population in order to calculate the probability of any event. If it is found that a rare event has occurred, under these assumptions, we will infer that there is something wrong with the assumptions and that the nature of the population is not what we assumed. Of course, there is a chance that our inference is incorrect. The nearer the calculated probability is to 0, the rarer the event, and the more confidence we have in our inference. Only if we observe a sample that could not possibly come from the (assumed) population, can we be 100% sure something is wrong with the assumed nature of the population.

A Caution

A few words of caution are in order before we leave this introduction to probability.

First, *not all experiments produce samples that are equally likely.* In this chapter we have only discussed the experiment of randomly sampling n items from a population of N items, and for such experiments, the samples have the same probability of occurrence. There are many other types of experiments and methods of sampling, and you should realize that not all of them can be analyzed using the methodology discussed in this chapter.

Second, *when using the rare event approach to make inferences, the choice of the event whose probability is to be evaluated and interpreted is extremely important.* The process of making inferences requires a theoretical basis and considerable experience. We are giving a simplistic view of this process in order to develop an understanding of the type of reasoning that is used when making inferences. If an inference is based on finding the probability of an event to be rare, but the event is not the one that should have been chosen, the inference is of little value. We will always specify the proper event to consider in any problem. Remember that there are theoretical reasons for choosing the appropriate event. As we proceed through the text, the reasons for the choice of events will become clearer.

Exercises (4.31–4.39)

Learning the Language

4.31 State the modified definition of a rare event.

Learning the Mechanics

4.32 A box contains ten marbles. Someone claims that the box contains three red marbles and seven blue marbles.
 a. Describe the items that the population is hypothesized to contain.
 b. What is the probability that a random sample of three items chosen from the population described in part a would contain three red marbles?
 c. What would you infer if a random sample of three marbles was entirely red? Why?

4.33 A bag contains eight coins. It is hypothesized that the coins are six pennies and two dimes.
 a. Describe the items that are assumed to be in the population.
 b. A random sample of two coins is taken from the bag. Find the probability of each of the following events if the population is as described in part a:

  P: Both coins are pennies.
  D: Both coins are dimes.

 c. If both randomly sampled coins were pennies, would you infer that the coins in the bag are other than hypothesized? Why?
 d. If both randomly sampled coins were dimes, would you infer that the coins in the bag are other than hypothesized? Why?

4.34 You are given a deck of 52 cards, which is supposed to contain four aces, four kings, four queens, etc. You shuffle the deck thoroughly and randomly draw four cards. If all four cards drawn are aces, what do you infer? Why? [*Hint:* What is the probability of observing four aces if the deck contains the cards it is supposed to contain?]

Applying the Concepts

4.35 The safety supervisor at a large manufacturing plant believes that half the accidents that occurred last year were due to mechanical failures of machines. The plant manager believes that fewer than half of last year's accidents were due to mechanical failures. An inspection of the plant's records indicates that there were 170 accidents last year.

a. Describe the population if the safety supervisor's belief is correct.

b. Five of last year's 170 accidents are randomly selected to be investigated. If the population is as described in part *a*, what is the probability that the investigation indicates none of the five accidents was due to mechanical failure?

c. If the investigation indicates that none of the five accidents was due to mechanical failure, would this support the plant manager's belief? Explain.

4.36 The accountant at a bank claims that at most 3% of the bank's records of accounts are in error. An auditing firm suspects the percentage of records in error is larger than 3%. A random sample of two account records is selected from the 10,000 accounts at the bank.

a. Describe the population, assuming that 3% of the bank's records are in error.

b. Find the probability that both sampled records are in error if the population is as described in part *a*.

c. If both sampled records are in error, would you infer that the auditing firm's suspicions are correct? Why?

d. Find the probability that neither of the sampled records is in error if the population is as described in part *a*.

e. If neither sampled account is in error, would you infer that the accountant's claim is incorrect? Why?

4.37 A professional football team must release eight players from its squad before the season begins in order to meet league limits on the total number of players. The coaches have narrowed their choice to ten players, eight of whom are black and two of whom are white. The ten players are of relatively equal value to the team, and the choice of which players to release is essentially a random selection.

a. If the players to be released are randomly selected, what is the probability that the eight black players are released?

b. What inferences might be made if the team chooses to release the eight black players? Why?

4.38 Political action committees (PACs) raise millions of dollars and campaign actively for both Republican and Democratic political candidates. Shown in the table at the top of page 176 are the percentages of Democrats, by geographic region, who donated $200 or more to the Democratic party or to any PAC.

Suppose a committee of 100 Democrats is formed that purports to replicate the regional percentages shown in the table. To verify this, a random sample of three committee members is selected and their home regions are noted.

a. If the committee makeup replicates the percentages given in the table, what is the probability that all three people sampled are from the Northeast?

b. If all three people sampled are from the Northeast, would you infer that the committee's makeup does not match the percentages given? Why?

Region	Democratic Percentage
Northeast	50
Midwest	14
South	15
West	21

Source: J. C. Green & J. L. Guth, "The Party Irregulars," *Psychology Today*, Vol. 18, No. 10, October 1984, p. 48. Reprinted with permission from *Psychology Today* Magazine. Copyright © 1984 (APA).

4.39 Refer to Exercise 4.38. The percentages of Republicans who donated $200 or more to their party or to a PAC are shown in the table for the same four regions of the country.

Region	Republican Percentage
Northeast	19
Midwest	17
South	28
West	36

Source: J. C. Green & J. L. Guth, "The Party Irregulars," *Psychology Today*, Vol. 18, No. 10, October 1984, p. 48. Reprinted with permission from *Psychology Today* Magazine. Copyright © 1984 (APA).

Suppose a committee of 100 Republicans is formed that claims to contain members from each region in the same percentages as those given in the table. To check the constituency of this committee, a random sample of three committee members is selected.

a. If the committee has the same percentages as claimed, what is the probability that all three people sampled are from the Northeast?

b. If all three people sampled are from the Northeast, would you infer that the committee's makeup does not match the given percentages? Why?

c. Compare your answers to those for Exercise 4.38 and discuss their differences.

4.4

COMPOUND EVENTS AND COMPLEMENTS (OPTIONAL)

Events may often be combined to form a new event, and such events are called **compound events.** Two ways in which compound events can be formed are given by the following definitions:

Definition 4.6

▬▬▬▬▬▬▬

The **union** of two events A and B is the event that occurs if either A or B, or both, occur on a single performance of the experiment. We will denote the union of the events A and B by $A \cup B$.

Definition 4.7

▬▬▬▬▬▬▬

The **intersection** of two events A and B is the event that occurs if both A and B occur on a single performance of the experiment. We will denote the intersection of A and B by $A \cap B$.

Venn diagrams are used to depict these compound events, as shown in Figure 4.1.

Figure 4.1
Venn diagram showing compound events $A \cap B$ and $A \cup B$

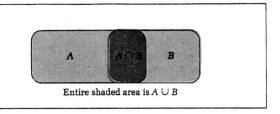

Entire shaded area is $A \cup B$

EXAMPLE 4.18   Consider the experiment of selecting one number at random from the numbers 1, 2, 3, 4, 5, and 6. (This is equivalent to tossing a fair die.) The sample space is the set of six possible outcomes $\{1, 2, 3, 4, 5, 6\}$. We define the following events:

   E: Observe an even number.

   F: Observe a number less than or equal to 3.

   a. Describe the event $E \cup F$.
   b. Describe the event $E \cap F$.
   c. Calculate $P(E \cup F)$ and $P(E \cap F)$.

Solution   **a.** The union of E and F, $E \cup F$, is the event that occurs if we observe either an even number or a number less than or equal to 3 (or both) when one number is selected from the numbers 1, 2, 3, 4, 5, and 6. The outcomes in the union of E and F are

   $$E \cup F = \{1, 2, 3, 4, 6\}$$

   See Figure 4.2 (page 178).

   b. The intersection of E and F, $E \cap F$, is the event that occurs if we observe *both* an even number and a number less than or equal to 3 when one number is selected

Figure 4.2
Union and intersection for
Example 4.18

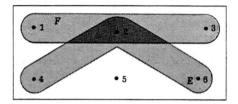

from the numbers 1, 2, 3, 4, 5, and 6. The only outcome in the intersection of E and F is

$$E \cap F = \{2\}$$

See Figure 4.2.

c. Recall that the probability of an event is the sum of the probabilities of the outcomes in the event. Since there are six possible outcomes for this experiment, each with a probability of $\frac{1}{6}$, we obtain

$$P(E \cup F) = P(1) + P(2) + P(3) + P(4) + P(6)$$
$$= \frac{1}{6} + \frac{1}{6} + \frac{1}{6} + \frac{1}{6} + \frac{1}{6} = \frac{5}{6}$$

and

$$P(E \cap F) = P(2) = \frac{1}{6} \qquad ■$$

Another way to define a new event with respect to an existing event is to consider the **complement** of an event.

Definition 4.8

The **complement** of an event A is the event that consists of all samples that are not in the event A—that is, the event that A does not occur. We will denote the complement of A by A^c.

The complement of A is depicted in a Venn diagram in Figure 4.3.

Figure 4.3
Event A and its complement, A^c

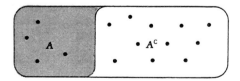

EXAMPLE 4.19   In Example 4.18 we considered the sample space $\{1, 2, 3, 4, 5, 6\}$ and the events

E: Observe an even number.

F: Observe a number less than or equal to 3.

a. Find E^c.

b. Find $[E \cup F]^c$.

c. Find $P(E^c)$ and $P([E \cup F]^c)$.

Solution   **a.** Since the event E is the observation of an even number, the event E^c is the observation of an odd number. Thus,

$$E^c = \{1, 3, 5\}$$

b. In Example 4.18 we found that

$$E \cup F = \{1, 2, 3, 4, 6\}$$

Thus, the complement of $E \cup F$ is

$$[E \cup F]^c = \{5\}$$

c. Since each sample has a probability of $\frac{1}{6}$,

$$P(E^c) = \frac{3}{6}$$

and

$$P([E \cup F]^c) = \frac{1}{6}$$ ∎

The relationships among unions, intersections, and complements provide the rules given in the box for calculating probabilities.

Rules for Calculating Probabilities

Additive Rule

The probability of the union of any two events A and B is the sum of the probabilities of events A and B minus the probability of the intersection of events A and B. That is,

$$P(A \cup B) = P(A) + P(B) - P(A \cap B)$$

Complement Rule

The probability of the complement of any event A is 1 minus the probability of event A. That is,

$$P(A^c) = 1 - P(A)$$

Notice that in the additive rule we must subtract $P(A \cap B)$ to obtain the correct value for $P(A \cup B)$.* This is because when summing $P(A)$ and $P(B)$, the probabilities of the samples in $A \cap B$ are added twice—once in $P(A)$ and once in $P(B)$. To get the complement rule, we use the fact that all the possible samples in a sample space are in either A or A^c ($A \cup A^c$ equals the sample space). Also, A and A^c are mutually exclusive, so that $P(A) + P(A^c) = 1$. The complement rule results from subtracting $P(A)$ from each side of this equation.

EXAMPLE 4.20   Hospital records show that 12% of all patients are admitted for surgical treatment, 16% are admitted for obstetrics, and 2% receive both obstetrics and surgical treatment.

a. What is the probability that a randomly selected patient is admitted for surgery, obstetrics, or both?

b. What is the probability that a randomly selected patient is admitted for something other than surgical treatment?

Solution   First, consider the following events:

S: Patient receives surgical treatment.

O: Patient receives obstetrical treatment.

From the statement of the example, we have

$P(S) = .12$ and $P(O) = .16$

and the probability of the intersection of S and O is

$P(S \cap O) = .02$

Now we can answer the questions.

a. We are asked to find the probability of the union of the events S and O. Using the additive rule of probability, we find

$$P(S \cup O) = P(S) + P(O) - P(S \cap O)$$
$$= .12 + .16 - .02 = .26$$

Thus, 26% of all patients admitted to the hospital receive either surgical treatment, obstetrical treatment, or both.

b. We are asked to find the probability of the complement of the event S. Using the complement rule of probability, we find

$$P(S^c) = 1 - P(S)$$
$$= 1 - .12 = .88$$

This means that 88% of all patients admitted to the hospital receive something other than surgical treatment.   ■

* If two events A and B do not intersect, they are said to be **mutually exclusive.** In this case, the intersection contains no samples, that is, $P(A \cap B) = 0$, and $P(A \cup B) = P(A) + P(B)$.

Exercises (4.40–4.47)

Learning the Language

4.40 Define the following items:
a. Complement of the event A
b. Intersection of the events A and B
c. Union of the events A and B

4.41 State the meaning of $A \cup B$, $A \cap B$, and A^c.

4.42 State the additive rule and the complement rule of probability.

Learning the Mechanics

4.43 A bag contains two red marbles and two green marbles. Two marbles are randomly chosen from the bag. The events E and F are defined as follows:

E: At least one red marble is drawn.

F: Exactly one green marble is drawn.

a. Identify the samples in the events E, F, $E \cup F$, $E \cap F$, and F^c.
b. Find $P(E)$, $P(F)$, $P(E \cup F)$, $P(E \cap F)$, and $P(F^c)$ by summing the probabilities of the appropriate samples.
c. Find $P(E \cup F)$ and $P(F^c)$ by using the probability rules.

4.44 Two numbers are chosen at random from the numbers 1, 2, 3, 4, and 5. The events E and F are defined as follows:

E: Sum of the two numbers is larger than 6.

F: Both numbers are even.

a. Identify the samples in the events E, F, $E \cup F$, $E \cap F$, and E^c.
b. Find $P(E)$, $P(F)$, $P(E \cup F)$, $P(E \cap F)$, and $P(E^c)$ by summing the probabilities of the appropriate samples.
c. Find $P(E \cup F)$ and $P(E^c)$ by using the probability rules.

Applying the Concepts

4.45 Refer to Exercise 4.23. A buyer for a large metropolitan department store must choose two firms from the four available to supply the stores's fall line of men's slacks. The buyer has not dealt with any of the four firms before and considers their products equally attractive. Unknown to the buyer, two of the four firms are having serious financial problems that will result in their not being able to deliver the fall line of slacks on time. The four firms may be identified as G1 and G2 (firms in good financial condition) and P1 and P2 (firms in poor financial condition). The possible samples identify the pairs of firms selected. If the buyer randomly selects the two firms from among the four, then the samples and their probabilities for this buying experiment are as listed in the table at the top of page 182.

	Sample	Probability			Sample	Probability
1.	G1, G2	$\frac{1}{12}$		7.	P1, G1	$\frac{1}{12}$
2.	G1, P1	$\frac{1}{12}$		8.	P1, G2	$\frac{1}{12}$
3.	G1, P2	$\frac{1}{12}$		9.	P1, P2	$\frac{1}{12}$
4.	G2, G1	$\frac{1}{12}$		10.	P2, G1	$\frac{1}{12}$
5.	G2, P1	$\frac{1}{12}$		11.	P2, G2	$\frac{1}{12}$
6.	G2, P2	$\frac{1}{12}$		12.	P2, P1	$\frac{1}{12}$

We define the following events:

E: At least one firm that is in good financial condition is selected.

F: The firm labeled P1 is selected.

a. Define the event $E \cap F$ as a specific collection of samples.
b. Define the event $E \cup F$ as a specific collection of samples.
c. Define the event E^c as a specific collection of samples.
d. Find $P(E)$, $P(F)$, $P(E \cap F)$, $P(E \cup F)$, and $P(E^c)$ by summing the probabilities of the appropriate samples.
e. Find the probabilities of $E \cup F$ and E^c by using the rules of probability.

4.46   A firm's accounting department claims that the firm will make a profit in the first quarter of the year in one of the ranges listed in the table, with probability as noted.

Profit Range	Probability
Under $75,000	.10
$75,000–99,999	.15
100,000–124,999	.25
125,000–149,999	.35
150,000–174,999	.10
175,000 or over	.05

Consider the following events:

A: Firm makes $99,999 or less.

B: Firm makes over $149,999.

C: Firm makes between $100,000 and $149,999.

Find $P(A \cap B)$, $P(B \cap C)$, $P(C^c)$, $P(A \cup C)$, and $P(B \cup C)$.

4.47   Refer to Exercise 4.24. One game that is very popular in many American casinos is roulette. Roulette is played by spinning a ball on a circular wheel that has been divided into thirty-eight arcs of equal length; these are numbered 00, 0, 1, 2, . . . , 35, 36. The number of the arc on which the ball comes to rest is the outcome of one play of the game. The numbers are also colored in the following manner:

Red: 1, 3, 5, 7, 9, 12, 14, 16, 18, 19, 21, 23, 25, 27, 30, 32, 34, 36

Black: 2, 4, 6, 8, 10, 11, 13, 15, 17, 20, 22, 24, 26, 28, 29, 31, 33, 35

Green: 00, 0

Players may place bets on the table in a variety of ways, including bets on odd, even, red, black, high, low, etc. Consider the following events (00 and 0 are not considered odd or even):

A: Outcome is an odd number.

B: Outcome is a black number.

C: Outcome is a low number (1–18).

Find $P(A)$, $P(B)$, $P(A \cap B)$, $P(A \cup B)$, and $P(C)$.

4.5

CONDITIONAL PROBABILITY AND INDEPENDENCE (OPTIONAL)

The probabilities of events that we have been discussing give the relative frequency of the occurrence of each event when the experiment is repeated a very large number of times. These probabilities are called **unconditional probabilities,** because no special conditions are assumed other than those that define the experiment.

Sometimes, we may want to alter the probability of an event when we have additional knowledge that might affect its outcome. Then, this probability is called the **conditional probability of the event.** For example, we have shown that the probability of observing an even number (event A) when drawing one number at random from the numbers 1, 2, 3, 4, 5, and 6 is $\frac{1}{2}$. However, suppose you are given the information that on a particular drawing the result was a number less than or equal to 3 (event B). Would you still think that the probability of observing an even number on that drawing is equal to $\frac{1}{2}$? If you make the assumption that B has occurred, then the sample space is reduced from six samples to three samples (namely, those contained in event B), and the reduced sample space contains the sampes $\{1, 2, 3\}$. Since the only even number of the three numbers in the reduced sample space of event B is the number 2, and since the number is randomly drawn, we conclude that the probability that event A occurs, given that event B has occurred, is $\frac{1}{3}$.

We will use the notation $P(A|B)$ to represent the probability of event A given that event B has occurred. For the simple example of drawing a number, we have

$$P(A|B) = \frac{1}{3}$$

In general, to calculate the probability of event A given that event B has occurred, we proceed as follows: We divide the probability of the part of event A that falls within the reduced sample space of event B, namely $P(A \cap B)$, by the total probability of the reduced sample space, namely $P(B)$. Thus, for the example of drawing a number with events A and B as defined above, we find

$$P(A|B) = \frac{P(A \cap B)}{P(B)} = \frac{P(2)}{P(1) + P(2) + P(3)} = \frac{\frac{1}{6}}{\frac{3}{6}} = \frac{1}{3}$$

The general formula for $P(A|B)$ is given in the next box.

> To find the **conditional probability that event A occurs given that event B occurs,** divide the probability that *both* A and B occur by the probability that B occurs, that is,
>
> $$P(A\,|\,B) = \frac{P(A \cap B)}{P(B)}$$
>
> where we assume that $P(B) \neq 0$.

EXAMPLE 4.21   Many medical researchers have conducted experiments to examine the relationship between cigarette smoking and cancer. Let A represent the event that an individual smokes, and let C represent the event that an individual develops cancer. Thus, $A \cap C$ is the event that an individual smokes and develops cancer, $A \cap C^c$ is the event that an individual smokes and does not develop cancer, etc. Assume that the probabilities associated with the four events are as shown in Table 4.11 for a certain section of the United States. How can these events and probabilities be used to examine the relationship between smoking and cancer?

Table 4.11

Event	Probability
$A \cap C$	.15
$A \cap C^c$	.25
$A^c \cap C$	.10
$A^c \cap C^c$	.50

Solution   One method of determining whether these probabilities indicate that smoking and cancer are related is to compare the conditional probability that an individual develops cancer given that he or she smokes with the conditional probability that an individual develops cancer given that he or she does not smoke. That is, we will compare $P(C\,|\,A)$ with $P(C\,|\,A^c)$.

The calculations are as follows:

$$P(C\,|\,A) = \frac{P(A \cap C)}{P(A)}$$

where the event A, a person smokes, is the union of the two mutually exclusive events, $A \cap C$ (a person smokes and develops cancer) and $A \cap C^c$ (a person smokes and does not develop cancer). Remembering that the probability of the union of two mutually exclusive events is the sum of the probabilities of those events, we obtain

$$P(A) = P(A \cap C) + P(A \cap C^c)$$
$$= .15 + .25 = .40$$

Then,

$$P(C\,|\,A) = \frac{P(A \cap C)}{P(A)} = \frac{.15}{.40} = .375$$

The conditional probability that a nonsmoking individual develops cancer is calculated in a similar manner:

$$P(C|A^c) = \frac{P(A^c \cap C)}{P(A^c)} = \frac{.10}{.60} = .167$$

where

$$P(A^c) = P(A^c \cap C) + P(A^c \cap C^c) = .10 + .50 = .60$$

or

$$P(A^c) = 1 - P(A) = 1 - .40 = .60$$

(either way we get the same answer).

Comparing the two conditional probabilities, we see that the probability that a smoker develops cancer, .375, is more than twice the probability that a nonsmoker develops cancer, .167. This does not prove that smoking *causes* cancer, but it does suggest a very pronounced link between smoking and cancer.   ■

EXAMPLE 4.22   The investigation of consumer product complaints by the Federal Trade Commission (FTC) has generated much interest by manufacturers in the quality of their products. A manufacturer of food processors conducted an analysis of a large number of consumer complaints and found that they fell into the six categories shown in Table 4.12. If a consumer complaint is received, what is the probability that the cause of the complaint was product appearance given that the complaint originated during the guarantee period?

Table 4.12

Distribution of product complaints

	Reason for Complaint			Totals
	Electrical	Mechanical	Appearance	
During Guarantee Period	18%	13%	32%	63%
After Guarantee Period	12%	22%	3%	37%
Totals	30%	35%	35%	100%

Solution   Let A represent the event that the cause of a particular complaint was product appearance, and let B represent the event that the complaint occurred during the guarantee period. Checking Table 4.12, you can see that 18% + 13% + 32% = 63% of the complaints occur during the guarantee period. Hence, $P(B) = .63$.

Again, from Table 4.12, the percentage of complaints that were caused by appearance and occurred during the guarantee period (the event $A \cap B$) is 32%. Therefore, $P(A \cap B) = .32$.

Using these probability values, we can calculate the conditional probability $P(A|B)$ that the cause of a complaint is appearance given that the complaint occurred during the guarantee time:

$$P(A|B) = \frac{P(A \cap B)}{P(B)} = \frac{.32}{.63} = .51$$

Consequently, the manufacturer can see that slightly more than half of the complaints that occurred prior to guarantee expiration were due to appearance (scratches, dents, or other imperfections on the surface of the food processor).   ■

If we multiply both sides of the formula for the conditional probability,

$$P(A|B) = \frac{P(A \cap B)}{P(B)}$$

by $P(B)$, we obtain the **multiplicative rule of probability.**

Multiplicative Rule of Probability

$$P(A \cap B) = P(A|B)P(B)$$

Since

$$P(B|A) = \frac{P(A \cap B)}{P(A)}$$

we also have

$$P(A \cap B) = P(B|A)P(A)$$

EXAMPLE 4.23   An agriculturalist, who is interested in planting wheat next year, is concerned with the following events:

    A: The production of wheat will be profitable.

    B: A serious drought will occur.

Based on available information, the agriculturalist believes that the probability is .01 that production of wheat will be profitable *assuming* a serious drought will occur in the same year, and that the probability is .05 that a serious drought will occur. That is,

    $P(A|B) = .01$ and $P(B) = .05$

Based on the information provided, what is the probability that a serious drought will occur *and* that a profit will be made? That is, find $P(A \cap B)$.

Solution   Using the multiplicative rule of probability, we obtain

$$P(A \cap B) = P(A|B)P(B)$$
$$= (.01)(.05) = .0005$$

The probability that a serious drought occurs *and* the production of wheat is profitable is only .0005. As we might intuitively expect, this event is very, very rare. ∎

An interesting point to consider when discussing conditional probabilities is whether the probability of an event A changes when we are given that an event B has occurred. In the experiment we described at the beginning of this section, where event A is defined as drawing an even number and event B is defined as drawing a number less than or equal to 3, the unconditional probability of A was found to be $P(A) = \frac{1}{2}$. However, the conditional probability was calculated as $P(A|B) = \frac{1}{3}$. In this case, knowing that B has occurred changes the probability that A will occur. This is not always the case, and if the probability does not change, we say that the two events A and B are **independent.**

Definition 4.9

Events A and B are **independent** if the occurrence of event B does not alter the probability of event A. That is, events A and B are independent if

$$P(A|B) = P(A)$$

When events A and B are independent, it will also be true that

$$P(B|A) = P(B)$$

Events that are not independent are said to be **dependent.**

EXAMPLE 4.24   Refer to the consumer product complaint study in Example 4.22. The percentages of complaints of various types during and after the guarantee period are shown in Table 4.12. As in Example 4.22, we define the following events:

     A: Cause of complaint is product appearance.

     B: Complaint occurred during the guarantee period.

Are A and B independent events?

Solution   Events A and B are independent if $P(A|B) = P(A)$. We calculated $P(A|B)$ in Example 4.22 to be .51, and from Table 4.12, we can see that

$$P(A) = .32 + .03 = .35$$

Therefore, $P(A|B)$ is not equal to $P(A)$, and A and B are not independent events. ∎

We will make two points about independent events. The first is that you should not trust your intuition when trying to decide whether two events are independent. In general, the way to check for independence is to calculate $P(A)$ and $P(A|B)$ to determine whether the values are equal.

The second point is that the probability of the intersection of independent events is very easy to calculate. Referring to the multiplicative rule for calculating the probability of an intersection, we find

$$P(A \cap B) = P(A|B)P(B)$$

Thus, since $P(A|B) = P(A)$ when A and B are independent, we have the following useful rule:

Multiplicative Rule for Independent Events

If events A and B are independent, the probability of the intersection of A and B equals the product of the probabilities of A and B. That is,

$$P(A \cap B) = P(A)P(B)$$

EXAMPLE 4.25   Suppose that one number is randomly selected from the numbers 1, 2, 3, 4, 5, and 6. Its value is noted, and the number is then reunited with the other five. The experiment is repeated, and a number is again randomly selected from the numbers 1, 2, 3, 4, 5, and 6. Assuming the two performances of the experiment are independent, what is the probability that both numbers chosen are 1's?

Solution   We define the events A and B as follows:

A: First number chosen is 1.

B: Second number chosen is 1.

The events A and B are independent, and we want to find the probability of $A \cap B$. The probabilities of the individual events A and B are

$$P(A) = \frac{1}{6} \quad \text{and} \quad P(B) = \frac{1}{6}$$

Using the multiplicative rule for independent events, we obtain

$$P(A \cap B) = P(A)P(B) = \left(\frac{1}{6}\right)\left(\frac{1}{6}\right) = \frac{1}{36}$$

Thus, the probability that both numbers are 1's is only $\frac{1}{36}$.   ■

Exercises (4.48–4.62)

Learning the Language

4.48   Define the following terms:
a. Conditional probability
b. Independence of two events A and B

4.49 State the multiplicative rule of probability if:
  a. The events are dependent.
  b. The events are independent.

4.50 What does the notation $P(A|B)$ mean?

Learning the Mechanics

4.51 A box contains two white, two red, and two blue poker chips. Two chips are randomly chosen and their colors are noted. Consider the following events:

  A: Both chips are the same color.

  B: Both chips are red.

  C: At least one chip is red.

Find $P(B|A)$, $P(B|A^c)$, $P(B|C)$, $P(A|C)$, and $P(C|A^c)$.

4.52 Two events, A and B, are such that $P(A) = .2$, $P(B) = .3$, and $P(A \cap B) = 0$.
  a. Find $P(A|B)$.
  b. Find $P(B|A)$.
  c. Are A and B independent? Why?

4.53 For two events, A and B, $P(A) = \frac{1}{2}$ and $P(B) = \frac{1}{3}$.
  a. If A and B are independent, find $P(A|B)$, $P(B|A)$, and $P(A \cap B)$.
  b. If A and B are dependent, and $P(A|B) = \frac{3}{5}$, find $P(A \cap B)$ and $P(B|A)$.

Applying the Concepts

4.54 A soap manufacturer has decided to market two new brands. An analysis of current market conditions and a review of the firm's past successes and failures with new brands have led the manufacturer to believe that the events and the probabilities of their occurrence in this marketing experiment are as given in the table (where, for example, SF means the first new brand succeeds and the second new brand fails).

Event	Probability
SS	.16
SF	.24
FS	.24
FF	.36

We define the following events:

  E: Both new brands succeed.

  F: At least one new brand succeeds.

  a. Find $P(E)$, $P(F)$, and $P(E \cap F)$.
  b. Find $P(E|F)$ and $P(F|E)$.

4.55 Each of a random sample of fifty people was asked to name his or her favorite soft drink. The responses are shown below:

Pepsi-Cola 18   Sprite   4
Coca-Cola 16   Nehi Orange 1
Mr. Pibb   4   Dr Pepper   1
Seven-Up   6

Suppose a person is selected at random from the survey. Let A be the event that the person preferred a soft drink bottled by the Coca-Cola Company (Coca-Cola, Mr. Pibb, or Sprite). Let B be the event that the person did *not* choose a cola (either Pepsi-Cola or Coca-Cola). Find the following:

a. $P(A)$ c. $P(A \cup B)$ e. $P(A \cap B^c)$
b. $P(B)$ d. $P(A \cap B)$ f. $P(A|B)$
g. Are A and B independent events? Why?

4.56 There are several methods of typing, or classifying, human blood. The most common procedure types blood into the general classifications of A, B, O, or AB. A method that is less well known is to examine phosphoglucomutase (PGM) and classify the blood into one of three main categories, 1-1, 2-1, or 2-2. Suppose that the percentages of people in PGM categories in a certain geographical region of the United States are those shown in the table. A person is to be chosen at random from this region.

		1-1	2-1	2-2
Race	White	50.4%	35.1%	4.5%
	Black	6.7%	2.9%	0.4%

a. What is the probability that a black person is chosen?
b. Given that a black is chosen, what is the probability that he or she is PGM type 1-1?
c. Given that a white is chosen, what is the probability that he or she is PGM type 1-1?

4.57 The table of percentages describes the 1,000 apartment units in a large suburban apartment complex. The manager of the complex is considering installing new carpets in all the apartments. Before doing so, he wants to wear-test a particular brand of carpeting for 6 months in one of the 1,000 apartments. He plans to choose one apartment at random from the 1,000 and install a test carpet.

| | | **Apartment Size** | | |
		One-bedroom	Two-bedroom	Three-bedroom
Location in Building	First floor	8%	30%	10%
	Second floor	22%	20%	10%

a. What is the probability that he will choose a first-floor, two-bedroom apartment?
b. Given that he chooses a second-floor apartment, what is the probability that it has three bedrooms?

c. Given that he chooses a one-bedroom apartment, what is the probability that the apartment is on the second floor?

d. Given that the apartment selected is on the first floor, what is the probability that it has two or three bedrooms?

4.58 A local YMCA has a membership of 1,000 people and operates facilities that include both a running track and an indoor swimming pool. Before setting up the new schedule of operating hours for the two facilities, the manager would like to know how many members regularly use each facility. A survey of the membership indicates that 65% use the running track, 45% use the swimming pool, and 5% use neither.

a. If one member is chosen at random, what is the probability that the member uses both the track and the pool?

b. If a member is chosen at random, what is the probability that the member uses only the track?

4.59 The probability that a certain electronics component fails when first used is .10. If it does not fail immediately, the probability that it lasts for 1 year is .99. What is the probability that a new component will last 1 year?

4.60 Consider an experiment that consists of drawing 1 card from a standard 52 card playing deck and define the following events:

A: Draw a black card.

B: Draw a face card (J, Q, K).

C: Draw a heart or spade.

a. Are A and B independent? Explain.

b. Are B and C independent? Explain.

4.61 A microwave oven manufacturer claims that only 10% of the ovens it makes will need repair in the first year. Suppose three recent customers are independently chosen.

a. If the manufacturer is correct, what is the probability that at least two of the three ovens will need repair in the first year?

b. If at least two of the three customers' ovens need repair in the first year, what inference may be made about the manufacturer's claim?

4.62 Psychologists tend to believe that aggressiveness and order of birth are related. To test this belief, a psychologist chose 500 elementary school students at random and administered to each a test designed to measure aggressiveness. Each student was classified in one of four categories. The percentages of students falling in the four categories are shown in the table.

	Firstborn	Not Firstborn
Aggressive	15%	15%
Not Aggressive	25%	45%

a. If one student is chosen at random from the 500, what is the probability that the student is a firstborn?

b. What is the probability that the student is aggressive?

c. What is the probability that the student is aggressive, given that the student was firstborn?

d. If we define the events *A* and *B* as follows:

   A: Student chosen is aggressive.

   B: Student chosen is firstborn.

are *A* and *B* independent events? Explain.

Chapter Summary

An **experiment** is any process that produces observations or measurements. The **sample space** is the collection of all the possible outcomes of an experiment. An **event** is a subset of the sample space.

When assigning probabilities to outcomes in a sample space, these rules must hold:

1. The probability of any outcome is between 0 and 1.

2. The sum of the probabilities of all the outcomes must equal 1.

The **probability of an event** is the sum of the probabilities of all the outcomes in the event.

The experiments of immediate interest to us consist of drawing a sample from a population. The sample space is the set of all possible samples, and the outcomes are samples. We restrict ourselves to sampling, in order, from a finite population. If all samples of a fixed size have the same probability of occurring, the experiment is called **random sampling.**

For random sampling, the probability of a sample is

$$\frac{1}{\text{Number of samples in the sample space}} = \frac{1}{\#S}$$

The probability of an event *E* is

$$P(E) = \frac{\text{Number of samples in } E}{\text{Number of samples in the sample space}} = \frac{\#E}{\#S}$$

To assist in counting the total number of samples or the number of samples in an event, we rely on two rules:

1. If a first thing can be done in *p* ways and a second thing can be done in *q* ways, then the two things can be done in order in *p* × *q* ways.

2. The number of ways of sampling *n* objects in order from a population of *N* objects is the product of *n* numbers: $N(N - 1)(N - 2) \cdots$.

The definition of a rare event, previously taken to be a measurement more than 2 standard deviations from the mean, is modified so that a **rare event** is now defined more specifically as an event that occurs with probability .05 or less.

The general approach to making inferences using rare events can now be described in these terms:

1. Hypothesize the nature of the population.
2. Obtain random samples from the population and calculate the probability of the appropriate event.
3. If the probability is .05 or less, infer that the population is somehow different from what was hypothesized.

Optional Topics

If A and B are events, the **union of A and B,** $A \cup B$, is the event that A or B, or both, occur. The **intersection of A and B,** $A \cap B$, is the event that both A and B occur. The probabilities of these compound events are related by the rule

$$P(A \cup B) = P(A) + P(B) - P(A \cap B)$$

The **complement** of an event A, A^c, is the event that A does not occur; for any event A,

$$P(A^c) = 1 - P(A)$$

The **conditional probability** of an event A, given that event B occurs, $P(A|B)$, is the probability of A when the sample space is reduced to outcomes in B; in symbols,

$$P(A|B) = \frac{P(A \cap B)}{P(B)}$$

If two events A and B are mutually exclusive, then

$$P(A \cup B) = P(A) + P(B)$$

Events A and B are **independent** if

$$P(A|B) = P(A) \quad \text{or, equivalently} \quad P(B|A) = P(B)$$

If A and B are independent, then

$$P(A \cap B) = P(A)P(B)$$

SUPPLEMENTARY EXERCISES (4.63–4.90)

Learning the Mechanics

4.63 Calculate $\#S$ for each of the following experiments in which n objects are sampled from N objects:

a. $n = 2$, $N = 6$ c. $n = 3$, $N = 15$
b. $n = 4$, $N = 9$ d. $n = 5$, $N = 20$

4.64   A jar contains three red jelly beans, two orange jelly beans, and one black jelly bean. Two jelly beans are randomly selected from the jar.
a.   Describe this experiment.
b.   Describe the population.
c.   Calculate $\#\boldsymbol{S}$.
d.   List all the samples in the sample space.
e.   List the samples in each of the following events:

   A: Both selected jelly beans are red.

   B: The black jelly bean is selected.

   C: At least one orange jelly bean is selected.

   D: Two black jelly beans are selected.

f.   Find the probability of each of the events given in part e.

4.65   Three letters are randomly chosen from the twenty-six letters in the alphabet. Find the probability of each of the following events:

   A: Letters chosen are A, B, and C.

   B: Letters chosen come from the first half of the alphabet.

   C: Letter Q is not chosen.

4.66   A box is supposed to contain fifteen green balls and five white balls. You suspect that this is not the case. You randomly sample three balls from the box and observe their colors.
a.   Find the probability of each of the following events (assume the box contains fifteen green balls and five white balls):

   G: Three chosen balls are green.

   W: Three chosen balls are white.

b.   If event G occurs, would you infer that the box does not contain what it is supposed to? Why?
c.   If event W occurs, would you infer that the box does not contain what it is supposed to? Why?

Applying the Concepts

4.67   A car rental agency has six vans available for rent. The agency does not know it, but one van has a faulty water pump, one van has a leak in the fuel line, and the other four are mechanically sound. A softball team rents two of the vans to go to a tournament. Assume their two rental vans are randomly selected.
a.   What is the probability both rented vans are mechanically sound?
b.   What is the probability the van with the faulty water pump is rented?
c.   What is the probability at least one van that is not mechanically sound is rented?

4.68   Five people are given a new brand of frozen pizza to try at home. The manufacturer of the pizza then randomly chooses two of these people and records their opinions about the pizza. Unknown to the manufacturer, three of the people thought the pizza was very good, and the other two thought it was only fair. Find the probability of each of the following events:

    A: Both people chosen thought the pizza was very good.

    B: At least one of the people chosen thought the pizza was very good.

    C: Both people chosen thought the pizza was only fair.

4.69 A manufacturer claims that 5% of the finished items coming off an assembly line each day are defective. A total of 5,000 items are produced one day, and three are randomly selected and inspected.

    a. If the manufacturer's claim is correct, what is the probability all three randomly selected items are defective?

    b. If all three randomly selected items are defective, would you infer the manufacturer's claim is incorrect? Why?

    c. If the manufacturer's claim is correct, what is the probability none of the three randomly selected items are defective?

    d. If none of the three randomly selected items are defective, would you infer that the manufacturer's claim is incorrect? Why?

4.70 The breakdown of 10,000 workers in a particular town according to their political affiliation and type of job held is given in the table. A worker is selected at random from the 10,000 workers in the town, and the worker's political affiliation and type of job are noted.

		Political Affiliation		
		Republican	Democrat	Independent
Type of Job	White-collar	1,200	1,200	600
	Blue-collar	2,300	4,300	400

    a. How many samples are possible for this experiment?

    b. What is the set of all samples called?

    c. Find the probability of each of the following events:

        A: Worker is a white-collar worker.

        B: Worker is a Republican.

        C: Worker is a Democrat.

        D: Worker is a white-collar worker and a Democrat.

4.71 Six people apply for two identical positions in a company. Four are minority applicants and the rest are nonminority. If all the applicants are equally qualified and the choice is essentially a random selection of two applicants from the six available, find the probability of each of the following events:

    A: Both people selected are nonminority.

    B: Both people selected are minority.

    C: At least one minority applicant is selected.

4.72 A fast-food restaurant chain with 700 outlets in the United States describes the geographic locations of its restaurants and their financial conditions over the past year with the accompanying table. One restaurant is randomly chosen from the 700 outlets. Using the description provided by the table, find the probability of each of the events defined below the table.

		Region			
		NE	SE	SW	NW
Financial Condition	Lost money	5	6	3	0
	Broke even	5	7	6	3
	Made money	270	162	131	102

A: Restaurant is located in the SE region.

B: Restaurant made money.

C: Restaurant is located in the NW region and lost money.

D: Restaurant is located in the NE or SE region.

4.73 Refer to Exercise 4.72. Suppose the fast-food restaurant chain is for sale and a potential buyer wants to check to be sure the financial conditions of the restaurants are accurately described in the table given in Exercise 4.72. The potential buyer randomly selects two restaurants and has an independent auditing firm provide a financial statement for last year.

a. Using the description provided by the table in Exercise 4.72, what is the probability both sampled restaurants lost money or broke even?

b. If both selected restaurants lost money or broke even, would you infer that the information in the table is not accurate? Why?

c. Considering your answers to parts a and b, should the potential buyer be wary of purchasing the fast-food chain? Explain.

4.74 A county welfare agency employs thirty welfare workers who interview prospective food stamp recipients. Periodically, the supervisor selects, at random, the forms completed by two workers and audits them for illegal deductions. Unknown to the supervisor, six of the workers have regularly been giving illegal deductions to applicants.

a. What is the probability that the first worker chosen has been giving illegal deductions?

b. What is the probability that neither of the two workers chosen has been giving illegal deductions?

c. What is the probability that both of the workers chosen have been giving illegal deductions?

4.75 Suppose you are to form a basketball team (five players) from eight available athletes. If the five players are randomly chosen, what is the probability that the tallest player is not chosen?

4.76 A mother claims her child can recognize the first ten letters of the alphabet. To test the mother's claim, you thoroughly mix cards marked with the twenty-six letters in the alphabet and ask the child to pick out first the letter A, then the letter B, and so on, through the letter J.

   a. If the child cannot recognize the letters and is essentially randomly picking out the letters you ask for, what is the probability that the child correctly identifies all ten letters?

   b. If all ten letters are correctly identified, would you infer that the child can indeed recognize these letters? Why?

4.77 A manufacturer of 35-mm. cameras knows that a shipment of thirty cameras sent to a large discount store contains six defective cameras. The manufacturer also knows that the store will choose two of the cameras at random, test them, and accept the shipment if neither is defective.

   a. What is the probability that the first camera chosen by the store will be defective?

   b. What is the probability that the shipment will be accepted?

4.78 Of the eight families in a particular neighborhood, six have incomes over $20,000 and two have incomes under $20,000. If four of these eight families are randomly selected to participate in a taste test for a newly developed type of instant breakfast food, what is the probability that all the chosen families have incomes over $20,000?

4.79 By mistake, a manufacturer of tape recorders includes three defective recorders in a shipment of ten going out to a small retailer. The retailer has decided to accept the shipment of recorders only if none are found to be defective when a random sample of the tape recorders is inspected.

   a. What is the probability the shipment will be accepted if the retailer inspects five of the recorders?

   b. What is the probability the shipment will be accepted if the retailer inspects six of the recorders?

4.80 Suppose you are purchasing cases of wine (twelve bottles per case) and that periodically you select a test case to determine the adequacy of the sealing process. To do this, you randomly select and test three bottles in the case. If a case contains one bottle of spoiled wine, what is the probability that it will not appear in your sample?

4.81 A nursery advertises that it has ten oak trees for sale. Unknown to a buyer, three of the trees have already been infected with a disease and will die within a year. If the buyer purchases two randomly selected trees, what is the probability that both trees will be healthy?

4.82 Five individuals apply for two vacancies in the shipping department of your plant. Two of the five people have superior credentials. You have been instructed to randomly choose two of the five applicants to fill the open positions. What is the probability that you choose the two with superior credentials?

4.83 If you are purchasing small lots of a manufactured product and it is very costly to test a single item, it may be desirable to test a sample of items from the lot rather than every item in the lot. For example, suppose that each lot contains ten items. You decide to sample four items per lot and reject the lot if you observe one or more defectives.

a. If the lot contains one defective item, what is the probability that you will accept the lot?

b. What is the probability that you will accept the lot if it contains four defective items?

4.84 The accompanying table lists the ten countries with the highest incidences of drug-related offenses. (The numbers are calculated per 100,000 inhabitants.)

Country	Drug-Related Offenses
Hong Kong	463.52
Finland	431.35
Sweden	296.90
United States	281.00
Bahamas	230.80
Jamaica	173.05
Trinidad and Tobago	113.83
New Zealand	77.40
Netherlands Antilles	65.77
Australia	64.10

Source: George T. Kurian, *The Book of World Rankings* (New York: Facts on File, Inc., 1979), p. 343.

a. Suppose a council of five countries is to be chosen from these ten to discuss means of reducing drug-related offenses. How many samples of five countries could be selected?

b. If the five countries are randomly sampled, what is the probability that every country selected has more than 100 drug-related offenses (per 100,000 inhabitants)?

c. If the five countries are randomly sampled, what is the probability that Hong Kong is not selected?

4.85 The June 1984 issue of *Psychology Today* reported that approximately 15% of blacks 18 to 34 years of age were attending colleges or universities in 1981. In 1964, this figure stood at only 5%.* Suppose a community has 1,000 black residents between the ages of 18 and 34 in 1981. To compare the percentage of this group who attend college with the national average of 15%, a random sample of five people is selected.

a. If 15% of this community's blacks aged 18 to 34 attend colleges or universities, what is the probability that all five people sampled are such students?

b. If all five people selected attend a college or university, what would you infer about this community? Why?

Supplementary Exercises for Optional Sections

4.86 In college basketball games, a player may be afforded the opportunity to shoot two consecutive foul shots (free throws).

* *Source:* Jeff Meer, "Civil Rights Indicators," *Psychology Today*, Vol. 18, No. 6, June 1984, p. 49. Reprinted with permission from *Psychology Today* Magazine. Copyright © 1984 (APA).

a. Suppose a particular player who scores on 80% of his foul shots has been awarded two free throws. If the two throws are considered independent, what is the probability that the player scores on both shots? Exactly one? Neither shot?

b. Suppose a particular player who scores on 80% of his first attempted foul shots has been awarded two free throws, and the outcome of the second shot is dependent on the outcome of the first shot. In fact, if this player makes the first shot, he makes 90% of the second shots, and if he misses the first shot, he makes 70% of the second shots. In this case, what is the probability that the player scores on both shots? Exactly one? Neither shot?

c. In parts a and b, we considered two ways of *modeling* the probability that a basketball player scores on two consecutive foul shots. Which model do you believe is a more realistic attempt to explain the outcome of shooting foul shots? That is, do you think two consecutive foul shots are independent or dependent? Explain.

4.87 Two companies, A and B, package and market a chemical substance and claim .15 of the total weight of the substance is sodium. However, a careful survey of 4,000 packages (half from each company) indicated the actual proportion varied around .15, with the results given in the table. Suppose a package is chosen at random from the 4,000 packages, and define events A–F as listed below the table.

		Proportion of Sodium			
		Less than .100	.100–.149	.150–.199	.200 and over
Brand	A	25%	10%	10%	5%
	B	5%	5%	10%	30%

A: Package chosen was brand A.

B: Package chosen was brand B.

C: Package chosen contained less than .100 sodium.

D: Package chosen contained between .100 and .149 sodium.

E: Package chosen contained between .150 and .199 sodium.

F: Package chosen contained at least .200 sodium.

Describe the characteristics of a package portrayed by the following events:

a. $A \cup B$ c. $A \cap D$ e. $(A \cap C) \cup (A \cap D)$

b. $B \cup F$ d. $E \cap B$

4.88 Refer to Exercise 4.87. Find the probabilities of the following events:

a. A, B, C, D, E, F c. $B \cap C$ e. $A \cap B$

b. $A \cup B$ d. $A \cap F$ f. $C \cup D$

4.89 A survey of the faculty of a large university yielded the breakdown given in the table (page 200) according to sex and marital status. Suppose a faculty member is chosen at random, and define events A–F as listed below the table.

		Marital Status			
		Married	Single	Divorced	Widowed
Sex	Male	60%	12%	8%	5%
	Female	10%	3%	2%	0%

A: Faculty member chosen is female.

B: Faculty member chosen is male.

C: Faculty member chosen is married.

D: Faculty member chosen is single.

E: Faculty member chosen is widowed.

F: Faculty member chosen is divorced.

Find the probability of each of the following events:

a. A, B, C, D, E, F d. $C \cup D$

b. $A \cup B$ e. $E \cap F$

c. $B \cap F$ f. $F \cup B$

4.90 The probability that a microcomputer salesperson sells a computer to a prospective customer on the first visit to the customer is .4. If the salesperson fails to make the sale on the first visit, the probability that the sale will be made on the second visit is .70. The salesperson never visits a prospective customer more than twice. What is the probability that the salesperson will make a sale to a particular customer?

CHAPTER 4 QUIZ

1. What is the role of probability in making inferences?

2. A population contains $N = 12$ items and a random sample of $n = 4$ items is selected from the population.

a. How many samples are in the sample space of this experiment?

b. What is the probability of observing each sample?

c. If an event contains 24 samples, what is the probability of that event?

3. Define the following terms:

a. Event   b. Random sampling

4. A person is asked to pick his first and second favorite television shows from a list of six prime-time shows. Two of the six shows are telecast by each of the major networks: two by ABC, two by CBS, and two by NBC. If the viewer picks his two favorite shows at random, find the probability of each of the following events:

A: He picks the two ABC shows.

B: He picks shows from two different networks.

5. A deck of cards used for poker contains four aces, four kings, four queens, etc.
- *a.* If four cards are randomly chosen from a deck used for poker, what is the probability all four are aces?
- *b.* If you were playing poker and the first four cards the dealer dealt to himself were aces, would you think the cards were being randomly dealt? Why?

CHAPTERS 1–4 CUMULATIVE QUIZ

1. We have measured how rare an event is in two different ways. One way is to measure the rarity in terms of distance (measured in standard deviations) from the mean, and the other way is to measure it in terms of the probability of the event. What advantage is there to using probability to measure the rarity of an event?

2. The following numbers represent a sample of four measurements: 0, 6, 1, 1
- *a.* Calculate the sample mean.
- *b.* Calculate the sample variance.
- *c.* Find the z-score associated with the value 6.

3. To perform a certain experiment, a behavioral psychology major requires fifty white mice that weigh at least 4 ounces. A supplier of white mice has fifty mice available, and it is known that their weights have a mean of 5 ounces and a standard deviation of 0.25 ounce. Will these white mice satisfy the weight requirements for the experiment? Explain.

4. The coach of a cross-country running team claims that 90% of the runners on the team run 10 miles or more each day that they practice. This coach supervises twenty runners, and you decide to select two runners randomly and observe how far they run in their next practice.
- *a.* Describe the experiment you are to conduct.
- *b.* Describe the population, assuming the coach's claim is correct.
- *c.* If the population is as described in part *b*, what is the probability both runners you select run less than 10 miles at the next practice?
- *d.* If both runners you select run less than 10 miles, would you doubt the coach's claim? Why?

On Your Own

Obtain a standard deck of 52 playing cards. An experiment will consist of drawing one card at random from the deck of cards and recording which card was observed. This will be simulated by shuffling the deck thoroughly and observing the top card. Consider the following two events:

- *A*: Card observed is a heart.
- *B*: Card observed is an ace, a king, a queen, or a jack.

a. Find $P(A)$ and $P(B)$.

b. Conduct the experiment ten times and record the observed card each time. Be sure to return the observed card to the deck each time and thoroughly shuffle the deck before making the next draw. After ten cards have been observed, calculate the proportion of observations that satisfy event A and event B. Compare the observed proportions with the true probabilities calculated in part a.

c. Conduct the experiment forty more times to obtain a total of fifty observed cards. Now, calculate the proportions of observations that satisfy event A and event B. Compare these proportions with those found in part b and the true probabilities found in part a.

d. Conduct the experiment fifty more times to obtain a total of 100 observations. Compare the observed proportions for the 100 trials with those found previously. What comments do you have concerning the different proportions found in parts b, c, and d as compared to the true probabilities found in part a? How do you think the observed proportions and true probabilities would compare if the experiment were conducted 1,000 times? 1,000,000 times?

Reference

Stephens, L. J. "Probability: An Important Tool in Police Science," *Journal of Police Science and Administration*, 1980, *8*(3), 353–354.

Chapter 5

INTRODUCTION TO SAMPLING DISTRIBUTIONS

THE BLIND TASTE TEST

Introduction

A soft drink company recently conducted an extensive advertising campaign that was based on challenging people to determine whether they preferred the company's brand or brand X in a blind taste test. The accompanying table presents the results of the taste tests conducted in Texas. The results of an experiment are often summarized by a number such as the percentage of people who prefer brand X, the mean grade-point average of a sample of students, or the variability in the closing price of a stock. In this chapter we will learn how to describe the results of such experiments and how to evaluate the corresponding probabilities.

	Percentage of People
Prefer company's brand	63%
Prefer brand X	35%
Undecided	2%

5.1

STATISTICS AND SAMPLING DISTRIBUTIONS

In Chapter 3 we discussed using numbers to describe samples, and we indicated the usefulness of numerical descriptive measures for making inferences. In Chapter 4 we calculated probabilities for random samples and discussed the role of probability in making inferences. In this chapter, we will continue to discuss the notions of numerical descriptive measures, probability, and making inferences. Before proceeding further, we will recall an important term first introduced in Chapter 1.

Definition 5.1

A **statistic** is a number calculated from the data in a sample. For each sample in a sample space, the value of a statistic is a uniquely determined number.

In Chapter 1, before we had introduced the term *sample*, we defined a *statistic* to be a number calculated from a collection of data. We are now specifying that the collected data form a sample. The statistics of greatest interest to us are numbers that are used to describe samples. The sample mean, $\bar{x}$, and the sample standard deviation, s, are two important statistics. Numerical descriptive measures may also be calculated for populations; they are then called **parameters.**

Definition 5.2

A **parameter** is a number calculated from the data in a population.

Examples of parameters are the population mean, μ, the population variance, σ^2, and the population standard deviation, σ.

As you will see, inferences are often made about population parameters using sample statistics. For example, if it is desired to estimate μ, the mean of a population, we could obtain a random sample from that population and calculate the sample mean, $\bar{x}$. The calculated value of $\bar{x}$ could then be used to estimate μ. Similarly, we could use the value of s^2, which is calculated from a sample, to estimate a population variance, σ^2. It is very important that you remember that statistics are sample quantities and parameters are population quantities.

Since we will be considering certain statistics and parameters over and over again in examples and exercises, it is convenient to denote them by symbols. Some of the symbols in the next box were given in previous chapters and are repeated to refresh your memory.

Symbols for Statistics

$\bar{x}$: Sample mean
s^2: Sample variance
s: Sample standard deviation
M: Sample median
R: Sample range

Symbols for Parameters

μ: Population mean
σ^2: Population variance
σ: Population standard deviation

In the following example we will calculate the probability of observing each possible value of a statistic that could be observed for a particular experiment.

EXAMPLE 5.1   A random sample of two measurements is drawn from a population containing the measurements 0, 2, 4, 6, and 8. The sample space for this experiment is given in Table 5.1.

Table 5.1

Sample space for Example 5.1

	First Number	Second Number		First Number	Second Number
1.	0	2	11.	4	6
2.	0	4	12.	4	8
3.	0	6	13.	6	0
4.	0	8	14.	6	2
5.	2	0	15.	6	4
6.	2	4	16.	6	8
7.	2	6	17.	8	0
8.	2	8	18.	8	2
9.	4	0	19.	8	4
10.	4	2	20.	8	6

a. Calculate the value of the sample mean, $\bar{x}$, for each sample in the sample space.
b. Find the probability of observing each different value of $\bar{x}$ found in part **a.**

Solution   **a.** For sample 1,

$$\bar{x} = \frac{0 + 2}{2} = 1$$

For sample 2,

$$\bar{x} = \frac{0 + 4}{2} = 2$$

and so on. The value of $\bar{x}$ for each sample is given in Table 5.2.

Table 5.2

Sample	1	2	3	4	5	6	7	8	9	10
Value of $\bar{x}$	1	2	3	4	1	3	4	5	2	3

Sample	11	12	13	14	15	16	17	18	19	20
Value of $\bar{x}$	5	6	3	4	5.	7	4	5	6	7

b. The different possible values of $\bar{x}$ are 1, 2, 3, 4, 5, 6, and 7. We want to find the probability of observing each of these values. The result $\bar{x} = 1$ occurs for samples 1 and 5, and is an event (as defined in Chapter 4) for this experiment. Recall that for random sampling, the probability of any event is

$$\frac{\text{Number of samples in the event}}{\text{Number of samples in the sample space}} = \frac{\#E}{\#S}$$

Therefore, the probability that $\bar{x} = 1$ is $\frac{2}{20} = \frac{1}{10}$. The probabilities for the other values of $\bar{x}$ (other events) are found in the same way, and are given in the table below. The possible values of $\bar{x}$ are shown in the first row, and the corresponding probabilities, denoted by $P(\bar{x})$, are in the second row.

$\bar{x}$	1	2	3	4	5	6	7
$P(\bar{x})$	$\frac{1}{10}$	$\frac{1}{10}$	$\frac{2}{10}$	$\frac{2}{10}$	$\frac{2}{10}$	$\frac{1}{10}$	$\frac{1}{10}$

■

If random samples were repeatedly drawn from the population of Example 5.1, we would observe a sample mean of 1 about $\frac{1}{10}$ of the time, a sample mean of 2 about $\frac{1}{10}$ of the time, etc. The table giving the values of $\bar{x}$ and their probabilities is called a **sampling distribution.**

In this chapter we will focus our attention on statistics with values that can be listed.* The sampling distribution of such a statistic may be defined as in Definition 5.3. In the next chapter we will generalize this definition for statistics with values that cannot be listed.

* Another term for a statistic with values that can be listed is **discrete random variable.**

Definition 5.3

The **sampling distribution** of a statistic is a rule that specifies the possible values that the statistic may assume and gives the probability of observing each of these values.

We will usually give the sampling distribution of a statistic in the form of a table that lists the values of the statistic and the corresponding probabilities. A sampling distribution may also be given by a graph or a formula.* Whatever its form, a sampling distribution must satisfy the two properties listed in the box.

Properties of Sampling Distributions

1. The probability of observing any particular value of a statistic must be between 0 and 1.
2. The sum of the probabilities of all the values of a statistic must be equal to 1.

These properties are very similar to the rules of probabilities for sample spaces given in Chapter 4. This is because a statistic assumes a unique value for each sample in a sample space.

EXAMPLE 5.2   A random sample of $n = 3$ observations is taken from a population containing the measurements 2, 6, 8, and 20.

a. Find the sampling distribution of the sample median, M.
b. Find the sampling distribution of the sample range, R.

Solution   The first step to finding a sampling distribution is to find the sample space, and we will do this using the steps given earlier in Chapter 4.

1. *Describe the experiment.* The experiment consists of randomly selecting $n = 3$ measurements from $N = 4$ measurements.
2. *Describe the population.* The population contains the four numbers 2, 6, 8, and 20.
3. *Count the samples and list them, if feasible.* Since $n = 3$ and $N = 4$, there are $4 \times 3 \times 2 = 24$ samples in the sample space. They are listed in Table 5.3 (page 208). We can now find the sampling distributions of M and R.
   a. The sampling distribution of M gives each value that M may assume and the probability of each value. The value of M for each sample is given in Table 5.4.

* When the term *discrete random variable* is used, a table, graph, or formula giving its values and probabilities is often called a **probability distribution** rather than a *sampling distribution.*

Table 5.3

Sample space for
Example 5.2

	First Number	Second Number	Third Number		First Number	Second Number	Third Number
1.	2	6	8	13.	8	2	6
2.	2	6	20	14.	8	2	20
3.	2	8	6	15.	8	6	2
4.	2	8	20	16.	8	6	20
5.	2	20	6	17.	8	20	2
6.	2	20	8	18.	8	20	6
7.	6	2	8	19.	20	2	6
8.	6	2	20	20.	20	2	8
9.	6	8	2	21.	20	6	2
10.	6	8	20	22.	20	6	8
11.	6	20	2	23.	20	8	2
12.	6	20	8	24.	20	8	6

Table 5.4

Sample	M	Sample	M	Sample	M	Sample	M
1.	6	7.	6	13.	6	19.	6
2.	6	8.	6	14.	8	20.	8
3.	6	9.	6	15.	6	21.	6
4.	8	10.	8	16.	8	22.	8
5.	6	11.	6	17.	8	23.	8
6.	8	12.	8	18.	8	24.	8

Remember that the median of each sample is the middle number when the measurements are arranged in increasing order.

Twelve samples result in a value of $M = 6$ and twelve result in a value of $M = 8$. Since random sampling is used, each value of M has a probability of $12/24 = 1/2$. The sampling distribution of M is given below:

M	6	8
$P(M)$	$1/2$	$1/2$

b. The range, R, for each sample is shown in Table 5.5. Recall that the range is the difference between the largest and smallest measurements.

Table 5.5

Sample	R	Sample	R	Sample	R	Sample	R
1.	6	7.	6	13.	6	19.	18
2.	18	8.	18	14.	18	20.	18
3.	6	9.	6	15.	6	21.	18
4.	18	10.	14	16.	14	22.	14
5.	18	11.	18	17.	18	23.	18
6.	18	12.	14	18.	14	24.	14

By counting the number of samples associated with each value of R and dividing these counts by 24 (the number of samples), the sampling distribution of R shown below was found:

R	6	14	18
$P(R)$	$6/24 = 1/4$	$6/24 = 1/4$	$12/24 = 1/2$

∎

Notice that for both sampling distributions in Example 5.2 the probabilities are proper fractions, and the sum of the probabilities for each sampling distribution equals 1.

After the sampling distribution of a statistic has been found, it is often of interest to find the probability that one or more values of the statistic would occur. The notation given in the box will be useful when expressing such probabilities. The use of this notation is demonstrated in the next example.

Notation for Probabilities of Statistics

Let X denote a statistic of interest, and let a be any number. Then:

1. $P(X = a)$ denotes the probability X is equal to a.
2. $P(X < a)$ denotes the probability X is less than a.
3. $P(X \leq a)$ denotes the probability X is less than or equal to a.
4. $P(X > a)$ denotes the probability X is greater than a.
5. $P(X \geq a)$ denotes the probability X is greater than or equal to a.

EXAMPLE 5.3   An experiment has a total of 5,040 random samples in its sample space. The possible values of $\bar{x}$ and the number of samples producing each of these values are given below:

Value of $\bar{x}$	5	6	7	8	9	10	11	12	13
Number of Samples	72	288	744	1,080	1,152	960	504	216	24

a. Give the sampling distribution of $\bar{x}$.
b. Find $P(\bar{x} = 9)$.
c. Find $P(\bar{x} \leq 8)$.
d. Find $P(\bar{x} < 9)$.
e. Find $P(\bar{x} \geq 11)$.
f. Find $P(\bar{x} > 12)$.

Solution   a. Since random sampling was used, we may find the probability for each value of $\bar{x}$ by dividing the number of samples producing each value by the total number of samples, 5,040. The sampling distribution is shown below:

$\bar{x}$	5	6	7	8	9	10	11	12	13
$P(\bar{x})$	$\dfrac{72}{5,040}$	$\dfrac{288}{5,040}$	$\dfrac{744}{5,040}$	$\dfrac{1,080}{5,040}$	$\dfrac{1,152}{5,040}$	$\dfrac{960}{5,040}$	$\dfrac{504}{5,040}$	$\dfrac{216}{5,040}$	$\dfrac{24}{5,040}$

b. The probability $\bar{x} = 9$ may be found directly from the sampling distribution.

$$P(\bar{x} = 9) = \frac{1,152}{5,040} = \frac{8}{35}$$

c. The notation $\bar{x} \le 8$ means $\bar{x}$ is less than or equal to 8. The values of $\bar{x}$ less than or equal to 8 are 5, 6, 7, and 8. Thus, we are interested in the event that $\bar{x} = 5$ or $\bar{x} = 6$ or $\bar{x} = 7$ or $\bar{x} = 8$. From the information given in the example, we know that there are 72 samples for which $\bar{x} = 5$; 288 samples for which $\bar{x} = 6$; 744 samples for which $\bar{x} = 7$; and 1,080 samples for which $\bar{x} = 8$. Thus, there are a total of $72 + 288 + 744 + 1,080 = 2,184$ samples for which $\bar{x} \le 8$. Dividing by 5,040, we obtain

$$P(\bar{x} \le 8) = \frac{2,184}{5,040} = \frac{13}{30}$$

We used the definition for finding the probability of an event for random sampling to calculate $P(\bar{x} \le 8)$. You may have noticed that it is also true that

$$P(\bar{x} \le 8) = P(\bar{x} = 5) + P(\bar{x} = 6) + P(\bar{x} = 7) + P(\bar{x} = 8)$$

$$= \frac{72}{5,040} + \frac{288}{5,040} + \frac{744}{5,040} + \frac{1,080}{5,040} = \frac{2,184}{5,040} = \frac{13}{30}$$

In other words, we could have obtained $P(\bar{x} \le 8)$ by adding the individual probabilities for the values 5, 6, 7, and 8. The two methods are essentially identical, and from now on we will add individual probabilities of a statistic to obtain the probability of a collection of values of the statistic.

d. $P(\bar{x} < 9) = P(\bar{x} = 5) + P(\bar{x} = 6) + P(\bar{x} = 7) + P(\bar{x} = 8)$

$$= \frac{72}{5,040} + \frac{288}{5,040} + \frac{744}{5,040} + \frac{1,080}{5,040} = \frac{2,184}{5,040} = \frac{13}{30}$$

Notice that the values less than 9 are 5, 6, 7, and 8, and that these are the same values less than or equal to 8. Thus, $P(\bar{x} < 9) = P(\bar{x} \le 8)$ for this experiment.

e. $P(\bar{x} \ge 11) = P(\bar{x} = 11) + P(\bar{x} = 12) + P(\bar{x} = 13)$

$$= \frac{504}{5,040} + \frac{216}{5,040} + \frac{24}{5,040} = \frac{744}{5,040} = \frac{31}{210}$$

f. There is only one value greater than 12, namely 13, so

$$P(\bar{x} > 12) = P(\bar{x} = 13) = \frac{24}{5,040} = \frac{1}{210}$$

■

The method used to calculate the probability that a statistic assumes any one of several possible values is summarized by the following rule:

The Additive Rule for Sampling Distributions

The probability that any one of two or more values of a statistic occurs is equal to the sum of their individual probabilities.

We will conclude this section with an example.

EXAMPLE 5.4   A real estate agent has five homes available to show to a potential buyer. Unknown to the realtor, the buyer would be willing to buy three of the homes, but would refuse to buy the other two. The real estate agent randomly selects two of the five homes to show to the potential buyer this afternoon. Let X be the number of these two homes that the buyer would be willing to buy.*

a. Find the sampling distribution of X. (Remember, we assume that the realtor randomly selects the homes to be shown. This assumption would be reasonable if the realtor has no idea which homes the potential buyer prefers.)
b. Find $P(X = 0)$. Explain what this means.
c. Find $P(X \geq 1)$. Explain what this means.

Solution   a. Let the three homes the person would buy be denoted by B1, B2, and B3. Let the other two homes be denoted by N1 and N2. Since the experiment consists of randomly selecting two of these homes, the number of samples in the sample space is $5 \times 4 = 20$, and these are listed in Table 5.6.

Table 5.6

Sample space for
Example 5.4

	First Home	Second Home		First Home	Second Home
1.	B1	B2	11.	B3	N1
2.	B1	B3	12.	B3	N2
3.	B1	N1	13.	N1	B1
4.	B1	N2	14.	N1	B2
5.	B2	B1	15.	N1	B3
6.	B2	B3	16.	N1	N2
7.	B2	N1	17.	N2	B1
8.	B2	N2	18.	N2	B2
9.	B3	B1	19.	N2	B3
10.	B3	B2	20.	N2	N1

* If the statistic of interest in an experiment is not a familiar one (such as the mean, $\bar{x}$, or the median, M), it is often denoted by the letter X.

Each of these samples has a probability of $\frac{1}{20}$, and we obtain the following sampling distribution for X, the number of the two homes shown that the buyer would be willing to buy:

X	0	1	2
$P(X)$	$\frac{2}{20}$	$\frac{12}{20}$	$\frac{6}{20}$

b. From the sampling distribution of X, we obtain

$$P(X = 0) = \frac{2}{20} = \frac{1}{10}$$

This means that the probability the buyer is shown no houses that he or she is willing to buy is $\frac{1}{10}$. If this were to happen, the buyer would be somewhat unlucky since it is a rather small probability.

c. $P(X \geq 1) = P(X = 1) + P(X = 2)$

$$= \frac{12}{20} + \frac{6}{20} = \frac{18}{20} = \frac{9}{10}$$

The probability the buyer would be shown at least one house that he or she would buy is $\frac{9}{10}$. This probability is quite large, and it indicates that this event is very likely to occur. ∎

COMPUTER
EXAMPLE 5.1

Minitab may be used to obtain a printout of the sampling distribution of a statistic X. Consider the sampling distribution given in part **a** of the solution to Example 5.4. After the values of X and the corresponding probabilities are used as input, the following display is generated using Minitab:

```
ROW    X    P(X)

 1     0    0.1
 2     1    0.6
 3     2    0.3
```

Minitab numbers each row that is printed, and since there are three X values (0, 1, and 2) the first column has the "ROW" numbers 1, 2, and 3. The second column gives the values of X, and the last column gives the probability of each value of X. When using probabilities with Minitab, you must express them in decimal form rather than as fractions. Thus, the original probabilities of $\frac{2}{20}$, $\frac{12}{20}$, and $\frac{6}{20}$ were converted to their decimal equivalents of 0.1, 0.6, and 0.3, respectively.

We can use Minitab to provide a nice, neat display of a sampling distribution. If we desired, we could cut off the "ROW" designation and draw in some lines to obtain the result given below:

```
X  |  P(X)
---+------
0  |  0.1
1  |  0.6
2  |  0.3
```

∎

Exercises (5.1–5.17)

Learning the Language

5.1 Define the following terms:
  a. Statistic   b. Parameter   c. Sampling distribution

5.2 What do the following symbols denote?
  a. $\bar{x}$ c. s e. R g. σ^2
  b. s^2 d. M f. μ h. σ

Learning the Mechanics

5.3 Explain why each of the following tables does not represent a sampling distribution:

  a.

X	0	1	4	10
P(X)	.1	.5	.3	.2

  b.

X	1	2	3	4
P(X)	.4	−.2	.5	.3

5.4 A random sample of two measurements is selected from a population containing the measurements 0, 2, 6, and 8. The sampling distribution of the sample mean, $\bar{x}$, is partially given below:

$\bar{x}$	1	3	4	5	7
$P(\bar{x})$	$\frac{1}{6}$	$\frac{1}{6}$	$\frac{2}{6}$	$\frac{1}{6}$	

  a. What is $P(\bar{x} = 7)$?
  b. Find $P(\bar{x} \le 4)$.
  c. Find $P(\bar{x} > 1)$.

5.5 Consider the following sampling distribution of a statistic, X:

X	−4	0	1	3
P(X)	.1	.3	.4	.2

  a. What are the possible values of X?
  b. What value of X is most probable?
  c. What is $P(X \ge 0)$?
  d. What is $P(X = 2)$?
  e. What is $P(X < 2)$?

5.6 A random sample of three measurements is selected from a population containing the values 0, 1, 2, and 10. The sample space for this experiment is given in the table at the top of page 214.

	First Number	Second Number	Third Number		First Number	Second Number	Third Number
1.	0	1	2	13.	2	0	1
2.	0	1	10	14.	2	0	10
3.	0	2	1	15.	2	1	0
4.	0	2	10	16.	2	1	10
5.	0	10	1	17.	2	10	0
6.	0	10	2	18.	2	10	1
7.	1	0	2	19.	10	0	1
8.	1	0	10	20.	10	0	2
9.	1	2	0	21.	10	1	0
10.	1	2	10	22.	10	1	2
11.	1	10	0	23.	10	2	0
12.	1	10	2	24.	10	2	1

Use this sample space to answer each of the following:

a. Find the sampling distribution of the sample median, M.

b. Let the largest measurement in each sample be denoted by L. Find the sampling distribution of L.

c. Find the sampling distribution of the sample variance, s^2.

5.7 A random sample of two measurements is selected from the population containing the values 1, 2, 3, 6, and 7.

a. Find the sampling distribution of $\bar{x}$.

b. Find the sampling distribution of R.

c. Let X be the first measurement selected in each sample. Find the sampling distribution of X.

Applying the Concepts

5.8 A first-year college student chooses two courses from among seven courses that all students must take to graduate. The student has no preferences, so these courses are randomly selected. Unknown to the student, two of the seven courses are being taught by professors who have been honored as "Teacher of the Year." Let X denote the number of courses the student chooses that are taught by a professor who has been "Teacher of the Year." The sampling distribution of X is given below. (You may derive this sampling distribution yourself if you want some practice.)

X	0	1	2
$P(X)$	$20/42 \approx .476$	$20/42 \approx .476$	$2/42 \approx .048$

a. Find $P(X = 2)$ and interpret this value.

b. Find $P(X > 0)$ and interpret this value.

5.9 Suppose that you have studied very hard for a history test and a sociology test you must take tomorrow. You feel very confident that you will pass both tests, that there is very little chance you will fail one test and pass the other, and that there is virtually no chance you will fail both tests. Let X be the number of tests you pass.

    a. Give an example of a sampling distribution for X that reflects your feelings.

    b. Suppose that you had not studied for the tests. Would the sampling distribution given in part a be appropriate? Explain.

5.10 Each of the following is a possible sampling distribution for the market demand for the number of new four-bedroom homes a building contractor will build next summer:

Sampling Distribution I

X = **Number of homes**	0	1	2	3	4	5	6	7
P(X)	.05	.13	.18	.25	.20	.11	.06	.02

Sampling Distribution II

X = **Number of homes**	2	3	4	5	6	7	8	9	10
P(X)	.03	.05	.06	.13	.16	.23	.15	.12	.07

    a. Which sampling distribution is more appropriate if the general rate of new homes being built is rather slow? Why?

    b. For each sampling distribution, find the probability that the contractor will be requested to build more than five homes.

    c. For each sampling distribution, find the probability that the contractor will be requested to build at most three homes.

    d. Give an example of a parameter that could be of interest to the building contractor.

5.11 Refer to Exercise 4.23 (Section 4.2). A buyer for a large metropolitan department store must choose two firms from the four available to supply the store's fall line of men's slacks. The buyer has not dealt with any of the four firms before and considers their products to be equally attractive. Unknown to the buyer, two of the four firms are having serious financial problems that will result in their not being able to deliver the fall line of slacks on time. The other two firms are in good financial condition and will deliver the fall line of slacks on time. The buyer randomly selects two firms from the four available. Let X be the number of firms selected that will deliver their slacks on time.

    a. Find the sampling distribution of X.

    b. Find $P(X = 0)$.

    c. Find $P(X = 2)$.

5.12 Refer to Exercise 4.26 (Section 4.2). Defective alternators have been mistakenly installed in three of the last six truck engines to emerge from an assembly line. It is not known which engines contain the defective alternators. Three of the six engines are randomly selected and the alternators are tested. Let X be the number of defective alternators that are found.

    a. Find the sampling distribution of X.

    b. Find the probability all three defective alternators are found.

    c. Find the probability at least one defective alternator is found.

5.13   Refer to Exercise 4.27 (Section 4.2). The baseballs used at the major league level are individually hand-stitched and, hence, are not uniformly made. Most pitchers who throw curve balls prefer baseballs with high stitches due to the firmer grip they provide. From a box of twenty new baseballs, an umpire chooses five at random to start the game. If this particular box contains thirteen high-stitched baseballs, then the sampling distribution of X, the number of balls chosen that would be preferred by a curve ball pitcher, is given below:

X	0	1	2	3	4	5
$P(X)$	$\dfrac{21}{15{,}504}$	$\dfrac{455}{15{,}504}$	$\dfrac{2{,}730}{15{,}504}$	$\dfrac{6{,}006}{15{,}504}$	$\dfrac{5{,}005}{15{,}504}$	$\dfrac{1{,}287}{15{,}504}$

  a. If a curve ball pitcher is pitching, and no baseballs with high stitches are selected, would you consider the pitcher to have been unlucky to observe this selection? Explain.
  b. What is the most probable number of high-stitched baseballs to be selected?
  c. Find $P(X \geq 3)$.

5.14   Refer to Exercise 4.11 (Section 4.1). Among weekly work schedules, the 40-hour work week is often considered to be the norm. In some countries, however, the average work week exceeds this. Shown in the accompanying table is the average number of working hours per week in manufacturing for seventeen countries with long work weeks.

Country	Average Hours Per Week	Country	Average Hours Per Week
Cyprus	45.0	Peru	48.1
Fiji	45.1	Thailand	48.1
Burma	45.6	Singapore	48.4
Gibraltar	45.6	Guyana	48.8
Mexico	45.6	Brunei	50.7
South Africa	46.1	Ecuador	51.0
Panama	47.1	South Korea	52.5
Syria	47.1	Egypt	57.0
Guatemala	47.2		

Source: George T. Kurian, *The Book of World Rankings* (New York: Facts on File, Inc., 1979), p. 214.

A random sample of five of these seventeen countries is selected in order to study their working habits more thoroughly. The sampling distribution of X, the number of countries chosen whose average work weeks exceed 50 hours, is given in the next table.

X	0	1	2	3	4
P(X)	$\dfrac{154{,}440}{742{,}560}$	$\dfrac{343{,}200}{742{,}560}$	$\dfrac{205{,}920}{742{,}560}$	$\dfrac{37{,}440}{742{,}560}$	$\dfrac{1{,}560}{742{,}560}$

a. What is the probability that at least one country whose work week exceeds 50 hours is selected?

b. Is it likely that all four countries whose work weeks exceed 50 hours are selected? Explain.

5.15 Refer to Exercise 4.12 (Section 4.1). At Po Folks, a family restaurant, the Po Plate is a meal consisting of four vegetables served with cornbread or biscuits. The list of vegetables from which to choose, as provided in the menu, is reproduced below.*

Turnip greens	Baked po-taters	Sliced tomaters
Corn on the cob	Fried po-taters	Cottage cheese
Green beans	Po-tater salad	Red beans & rice
Fried okra	Coleslaw	Rice & gravy
New po-taters & carrots		Seafood gumbo

A customer cannot decide which four vegetables to order and randomly selects four from the menu. The sampling distribution of X, the number of times potatoes are ordered (in one form or another), is given in the next table.

X	0	1	2	3	4
P(X)	$\dfrac{5{,}040}{24{,}024}$	$\dfrac{11{,}520}{24{,}024}$	$\dfrac{6{,}480}{24{,}024}$	$\dfrac{960}{24{,}024}$	$\dfrac{24}{24{,}024}$

a. What is the probability that one or two of the vegetables ordered are potatoes?

b. What is the probability that all of the vegetables ordered are potatoes?

*5.16 Refer to Exercise 5.14. We considered the experiment of randomly sampling five of seventeen countries to examine their working habits. The sampling distribution of X, the number of countries chosen whose average work weeks exceed 50 hours, is repeated for your convenience.

X	0	1	2	3	4
P(X)	$\dfrac{154{,}440}{742{,}560}$	$\dfrac{343{,}200}{742{,}560}$	$\dfrac{205{,}920}{742{,}560}$	$\dfrac{37{,}440}{742{,}560}$	$\dfrac{1{,}560}{742{,}560}$

a. Give the values of P(X) as decimals rounded to the nearest thousandth.

b. Use a computer package to produce a sampling distribution of X.

c. Sum the probabilities given in the sampling distribution of part b. Explain why this sum is not equal to exactly 1.

* Po Folks, Inc., 1983.

*5.17　Refer to Exercise 5.15. A random selection of four vegetables was taken from a menu. The sampling distribution of X, the number of times potatoes are ordered, is repeated for your convenience.

X	0	1	2	3	4
P(X)	$\frac{5,040}{24,024}$	$\frac{11,520}{24,024}$	$\frac{6,480}{24,024}$	$\frac{960}{24,024}$	$\frac{24}{24,024}$

　　a.　Convert the probabilities shown to decimals, rounded to the nearest thousandth.
　　b.　Use a computer package to produce a sampling distribution of X.
　　c.　Do the probabilities used in part b sum to 1? Explain why the sum should equal exactly 1, and why it does not.

5.2

THE MEAN OF A SAMPLING DISTRIBUTION

Let X denote the statistic of interest in an experiment, and suppose the table below represents the sampling distribution of X:

X	0	1	2
P(X)	$\frac{1}{4}$	$\frac{1}{2}$	$\frac{1}{4}$

Suppose the experiment producing this statistic was repeated a very large number of times, and the value of X was recorded each time the experiment was conducted. For example, if the experiment were conducted 1,000,000 times, we might observe the value zero 253,172 times, the value one 499,215 times, and the value two 247,613 times. A table summarizing the relative frequencies of these values (as discussed in Chapter 2) is shown below:

X	0	1	2
Relative Frequency	$\frac{253,172}{1,000,000} \approx .25$	$\frac{499,215}{1,000,000} \approx .50$	$\frac{247,613}{1,000,000} \approx .25$

As you can see, the relative frequency table and the sampling distribution of X would be very nearly identical. Consequently, we may view the sampling distribution of X as a relative frequency distribution for a population of data produced by repeating an experiment over and over again, a very large number of times.

　　A graph of the sampling distribution of X is shown in Figure 5.1. In this figure, the height of a rectangle located over a value of X is equal to its probability. This is the same procedure we used to draw bar graphs in Chapter 2, except that the height of a rectangle was equal to the relative frequency for that value of X. Thus, a sampling distribution describes a population of X values by giving each value in the population

Figure 5.1
A graph of the sampling distribution of X

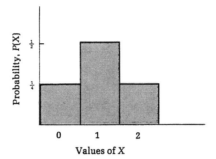

(the possible values of X) and the probability of each value of X (the relative frequency with which each value occurs). This population of data (the sampling distribution) possesses a mean, a variance, and a standard deviation. In this section we will discuss the **mean of a sampling distribution,** which is calculated using the formula given in Definition 5.4.

Definition 5.4

The **mean of the sampling distribution of a statistic X** is equal to the sum of each of the values of X multiplied by its probability. Denoting the mean of the sampling distribution of X by μ_X, we can write this in symbols by

$$\mu_X = \sum[X \cdot P(X)]$$

Since more than one statistic may be defined for an experiment, we must identify the mean of a sampling distribution by placing the statistic of interest as a subscript to the symbol for a population mean, μ. Thus, the mean of the sampling distribution of $\bar{x}$ is denoted by $\mu_{\bar{x}}$, the mean of the sampling distribution of s is denoted by μ_s, etc. Since each value of the statistic is multiplied by its probability, the mean of a sampling distribution is actually a weighted average of the values the statistic may assume.*

EXAMPLE 5.5   Find the mean of the sampling distribution of the statistic X that was introduced at the beginning of this section.

Solution   To calculate the mean of the sampling distribution, μ_X, we need to multiply each value of X by its probability. This is most easily done by setting up a new table that lists .these products next to the sampling distribution, as shown in Table 5.7 (page 220).

* When the terms *random variable* and *probability distribution* are used, the mean of the sampling distribution is often called the **expected value of X.**

Table 5.7

X	P(X)	X · P(X)
0	$\frac{1}{4}$	$0 \cdot \frac{1}{4} = 0$
1	$\frac{1}{2}$	$1 \cdot \frac{1}{2} = \frac{1}{2}$
2	$\frac{1}{4}$	$2 \cdot \frac{1}{4} = \frac{1}{2}$

The value of μ_X is the sum of the products in the third column of Table 5.7:

$$\mu_X = \sum[X \cdot P(X)] = 0 + \frac{1}{2} + \frac{1}{2} = \frac{2}{2} = 1$$

Figure 5.2 shows the graph of the sampling distribution of X and the location of its mean, μ_X.

Figure 5.2
Sampling distribution and μ_X
for Example 5.5

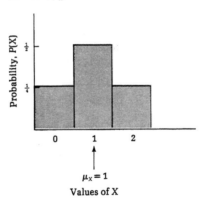

Since the sampling distribution of X in Example 5.5 is symmetric about the value 1, the fact that the mean is $\mu_X = 1$ is intuitively appealing. The mean of a sampling distribution will have the same meaning as the sample means calculated in Chapter 3. As you can see, the mean locates the center of the sampling distribution.

EXAMPLE 5.6   The following is the sampling distribution of a statistic X:

X	0	1	2
P(X)	$\frac{1}{10}$	$\frac{6}{10}$	$\frac{3}{10}$

a. Calculate the value of μ_X, the mean of this sampling distribution.
b. Graph the sampling distribution and locate μ_X on the graph.

Solution   **a.** To find the value of μ_X, we first calculate the product $X \cdot P(X)$ for each value of X. These products are given in Table 5.8. Summing the values of $X \cdot P(X)$, we obtain

$$\mu_X = \sum[X \cdot P(X)] = 0 + \frac{6}{10} + \frac{6}{10} = \frac{12}{10} = 1.2$$

Table 5.8

X	P(X)	X · P(X)
0	$\frac{1}{10}$	$0 \cdot \frac{1}{10} = 0$
1	$\frac{6}{10}$	$1 \cdot \frac{6}{10} = \frac{6}{10}$
2	$\frac{3}{10}$	$2 \cdot \frac{3}{10} = \frac{6}{10}$

b. A graph of the sampling distribution of the statistic X is shown in Figure 5.3.

Figure 5.3
A graph of the sampling distribution for Example 5.6

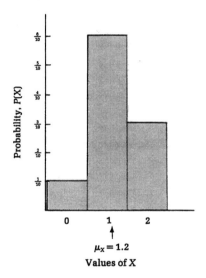

Since the sampling distribution in Example 5.6 is not symmetric (the probability of observing a 2 is larger than the probability of observing a 0), the mean, μ_X, is slightly larger than 1. Remember that when a set of data is not symmetric, the mean does not fall exactly in the center of the data. The relative occurrence of either larger or smaller values in the population shifts the value of the mean to above or below the center, respectively.

EXAMPLE 5.7   In Example 5.1 we found the sampling distribution of the sample mean, $\bar{x}$, associated with a random sample of two measurements taken from the population 0, 2, 4, 6, and 8. The sampling distribution of $\bar{x}$ is given below:

$\bar{x}$	1	2	3	4	5	6	7
$P(\bar{x})$	$\frac{1}{10}$	$\frac{1}{10}$	$\frac{2}{10}$	$\frac{2}{10}$	$\frac{2}{10}$	$\frac{1}{10}$	$\frac{1}{10}$

a. Discuss the information provided by the sampling distribution of $\bar{x}$.
b. Calculate $\mu_{\bar{x}}$, the mean of the sampling distribution of $\bar{x}$.
c. Graph the sampling distribution of $\bar{x}$ and locate $\mu_{\bar{x}}$ on the graph.

Solution   **a.** The sampling distribution describes a population of sample means such that $\frac{1}{10}$ of the numbers in the population are ones, $\frac{1}{10}$ are twos, $\frac{2}{10}$ are threes, $\frac{2}{10}$ are fours, $\frac{2}{10}$ are fives, $\frac{1}{10}$ are sixes, and $\frac{1}{10}$ are sevens. Notice that there are two distinct populations being discussed here. There is the *original population* of the five measurements 0, 2, 4, 6, and 8. Then there is the *population of sample means* described by the sampling distribution of $\bar{x}$. The population of $\bar{x}$'s contains an infinite number of measurements (each value is a 1, 2, 3, 4, 5, 6, or 7) and could be generated by randomly sampling two measurements from the original population an infinite number of times and calculating the mean of each of these samples.

b. To find the value of $\mu_{\bar{x}}$, the values of the products $\bar{x} \cdot P(\bar{x})$ are calculated and listed in Table 5.9. Summing the values of $\bar{x} \cdot P(\bar{x})$, we obtain

$$\mu_{\bar{x}} = \sum[\bar{x} \cdot P(\bar{x})] = \frac{1}{10} + \frac{2}{10} + \frac{6}{10} + \frac{8}{10} + \frac{10}{10} + \frac{6}{10} + \frac{7}{10}$$

$$= \frac{40}{10} = 4$$

The mean of the sampling distribution of $\bar{x}$ is 4.

Table 5.9

$\bar{x}$	$P(\bar{x})$	$\bar{x} \cdot P(\bar{x})$
1	$\frac{1}{10}$	$\frac{1}{10}$
2	$\frac{1}{10}$	$\frac{2}{10}$
3	$\frac{2}{10}$	$\frac{6}{10}$
4	$\frac{2}{10}$	$\frac{8}{10}$
5	$\frac{2}{10}$	$\frac{10}{10}$
6	$\frac{1}{10}$	$\frac{6}{10}$
7	$\frac{1}{10}$	$\frac{7}{10}$

c. Figure 5.4 shows a graph of the sampling distribution of $\bar{x}$, and $\mu_{\bar{x}} = 4$ is located on the graph.

Figure 5.4
Sampling distribution of $\bar{x}$
for Example 5.7

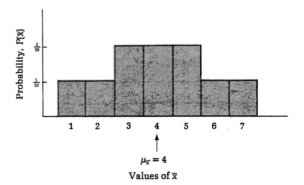

We have found $\mu_{\bar{x}}$, the mean of the population of sample means. This helps locate the center of the distribution of $\bar{x}$'s. As shown in Figure 5.4, the sample mean may assume the values 1, 2, 3, 4, 5, 6, and 7. The mean of this sampling distribution is 4.

■

EXAMPLE 5.8   Suppose you buy a $20,000 whole life insurance policy at an annual premium of $140. Actuarial tables show that the probability of death during the next year for a person of your age, sex, health, etc., is .001. The sampling distribution of the amount of money the insurance company could make for a policy of this type is shown in the table below. The amount of money that could be made is denoted as a gain, G.

Gain, G	$140	−$19,860
Probability, P(G)	.999	.001

a. Discuss the information provided by the sampling distribution.
b. Calculate the mean of the sampling distribution and interpret this value.

Solution   **a.** Examining the sampling distribution, we can see that the insurance company could gain $140, the annual premium you pay for this policy. The company will make this amount of money if you do not die next year, and the probability of this is .999. The insurance company could also have a negative gain (a loss) of −$19,860. This value represents the net loss the company would incur ($140 − $20,000 = −$19,860) if you die next year. The probability of this event is .001. In other words, if the insurance company were to insure many, many people with the same age, sex, health, etc., characteristics, approximately 999 times out of 1,000 the insurance company would gain the annual premium of $140. Approximately 1 time out of 1,000 the insurance company would have to pay a beneficiary $20,000 and would have a net loss of $19,860.

b. To find the mean gain, μ_G, the values of $G \cdot P(G)$ are calculated and listed in Table 5.10. Summing the values of $G \cdot P(G)$, we obtain

$$\mu_G = \sum[G \cdot P(G)] = \$139.86 + (-\$19.86) = \$120.00$$

The mean gain for the insurance company is $120. If the insurance company were to sell a very large number of these $20,000 whole life insurance policies to people similar to you, it would net (on the average) $120 per sale in the next year.

Table 5.10

G	P(G)	G · P(G)
$140	.999	$139.86
−$19,860	.001	−$19.86

A sketch of the sampling distribution of the insurance company's gain is shown in Figure 5.5 (page 224) and the location of the mean, $\mu_G = \$120$, is given. Since the relatively small positive gain of $140 occurs with such a high probability (.999) and the large negative gain of −$19,860 occurs with such a small probability (.001), the mean gain of $120 is heavily weighted toward the $140 value.

Figure 5.5
Sketch of sampling
distribution of G for
Example 5.8

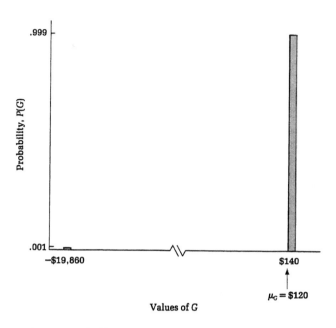

As we noted in Chapter 3, the mean (a measure of central tendency) provides only a partial description of any set of data. To describe a set of data fully, we need a measure of variability. In the next section we will discuss the variance and the standard deviation of a sampling distribution.

COMPUTER   Minitab is particularly useful for calculating the mean of a sampling distribution when
EXAMPLE 5.2   a statistic, X, assumes many different values. This sampling distribution was produced
               using Minitab:

```
ROW      X     P(X)

 1       2    0.01
 2       3    0.13
 3       5    0.07
 4       7    0.24
 5       9    0.23
 6      10    0.25
 7      12    0.07
```

The information in this distribution was used by Minitab to calculate the mean of the sampling distribution. The value was found to be

```
MEAN
7.85
```

Minitab was also used to produce the accompanying graph of the sampling distribution of X. Although the values of X range from 2 to 12, the largest probabilities
are associated with the values of 7, 9, and 10. This explains why the mean value of
7.85 is somewhat above 7, the midpoint between 2 and 12.

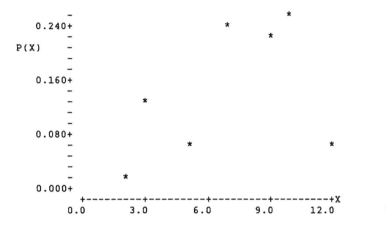

Exercises (5.18–5.29)

Learning the Language

5.18   Define the term *mean of a sampling distribution.*

5.19   What is denoted by μ_x?

Learning the Mechanics

5.20   Consider the following sampling distribution:

X	0	10	20	30
P(X)	.5	.2	.1	.2

a.   Find μ_x.
b.   Sketch a graph of the sampling distribution of X, and show the location of μ_x.

5.21   Consider the following sampling distribution:

X	−1	0	2	5
P(X)	.2	.4	.1	.3

a.   Find μ_x.
b.   Sketch a graph of the sampling distribution of X, and show the location of μ_x.

5.22 Consider the following sampling distributions:

(1)

X	0	5	10
P(X)	.1	.7	.2

(2)

X	0	5	10
P(X)	.7	.2	.1

(3)

X	0	1	10
P(X)	.1	.2	.7

a. Using your intuition, which distribution do you think has the smallest mean? The largest mean?

b. Calculate the mean for each sampling distribution and compare these values to your answers in part a.

Applying the Concepts

5.23 A company's marketing and accounting departments have determined that if the company markets its newly developed line of party favors, the following sampling distribution will describe the contribution of the new line to the firm's profit during a 6-month period:

X = Profit contribution	−$5,000*	$10,000	$30,000
P(X)	.3	.4	.3

* A negative contribution is a loss.

The company has decided it should market the new line of party favors if the mean contribution to profit for a 6-month period is over $10,000. Based on the sampling distribution, will the company market the new line?

5.24 On a particular busy holiday weekend, a national airline has many requests for standby flights at half the usual one-way air fare. Past experience has shown that these passengers have only about a 1 in 5 chance of getting on a standby flight. When they fail to get on a flight as a standby, their only other choice is to fly first class on the next flight out. Suppose that the usual one-way air fare to a certain city is $70 and the cost of flying first class is $90. Let X be the cost of a trip for a person flying standby. The sampling distribution of X is then given by the following:

X	$35	$90
P(X)	$\frac{1}{5}$	$\frac{4}{5}$

a. Calculate the average cost of a trip for a person flying standby.

b. In light of your answer to part a, should a passenger choose to fly as a standby on this trip? Explain.

5.25 Odds makers try to predict which football teams will win and by how many points (the *spread*). If the odds makers do this accurately, the probability that a person picks the winning team (adjusted by the spread) in any ball game is approximately $\frac{1}{2}$. Suppose you can get $6 for every $1 you risk if you pick the winners in three ball games (adjusted by the spread). Thus, for every $1 bet, you will either lose $1 or gain $5. If the odds makers are predicting the spread of each game correctly, the probability you win $5 is $\frac{1}{8}$ and the probability you lose $1 is $\frac{7}{8}$.

a. Find the sampling distribution of X if X is the amount of money you win with each bet of $1. (A loss is expressed as a negative win.)

b. Find μ_X.

c. If the odds makers are predicting the spread of each game correctly, would you make the $1 bet to try and pick the winners of three ball games? Explain.

5.26 A rehabilitation officer at a county jail questioned the inmates to determine how many previous convictions, X, each had prior to the one for which he or she was currently serving. The relative frequencies corresponding to X are given in the following table:

X	0	1	2	3	4
Relative Frequency	.16	.53	.20	.08	.03

If we can regard the relative frequencies as the approximate values for $P(X)$, find the mean number of previous convictions for an inmate.

5.27 Many of the smaller nations win few Olympic medals. At the 1976 Summer Olympics in Montreal, twenty-five of the countries winning at least one medal won less than seven medals. Let X be the number of medals won at the 1976 summer games by one country selected randomly from these twenty-five countries. The sampling distribution of X is given below.

X	1	2	3	4	5	6
$P(X)$	.32	.32	.04	.08	.12	.12

Find the mean number of medals won by these countries.

5.28 A buyer for a large department store wants to project the required inventory for men's ten-speed bicycles next spring. Experience has shown that demand (X) has approximately the sampling distribution shown in the table.

Demand, X = Number of bicycles	40	60	80	100	120	140	160	180
$P(X)$	.01	.06	.16	.24	.23	.15	.10	.05

a. Find μ_X.

b. What information does μ_X give you about the sampling distribution?

c. Sketch a graph of the sampling distribution and locate μ_X on the graph.

*d. Repeat parts a and c using a computer package.

5.29   The following quotation is from *The Book of World Rankings:*

> The most widely available and uniformly compiled statistics relating to cancer are mortality rates. Such data cover about 36% of the world population; in the case of developed countries the data are virtually complete, but in the case of developing countries the coverage is below 5%.*

There were thirty-nine countries reporting deaths from cancer as X, the percentage of all cancer-caused deaths for persons aged 55 and over. The sampling distribution of X is given in the accompanying table. (Due to the large number of values of X, the sampling distribution is given on two lines.)

X	88	87	86	85	84	83	82	81	80
$P(X)$	$2/39$	$2/39$	$2/39$	$4/39$	$4/39$	$2/39$	$2/39$	$2/39$	$4/39$

X	79	78	77	75	74	69	65	62	59	39
$P(X)$	$1/39$	$2/39$	$4/39$	$2/39$	$1/39$	$1/39$	$1/39$	$1/39$	$1/39$	$1/39$

a. Find the mean of X.
b. Explain the meaning of the value calculated in part a.
c. Convert the probabilities to decimals, retaining at least four decimal places. (If too few decimal places are retained, the value of the mean may be incorrect due to multiplying rounding errors.)
*d. Use a computer package to find the mean value of X.

5.3

THE VARIABILITY OF A SAMPLING DISTRIBUTION

In Section 5.2 we discussed the mean of a sampling distribution and indicated that the mean helps locate the center of a sampling distribution. To describe the dispersion, or spread, of a sampling distribution, we will calculate its **variance** and **standard deviation.** The variance of a sampling distribution is defined in the box.

Definition 5.5

The **variance of the sampling distribution of a statistic** X is denoted by σ_X^2, and is found by the following formula:

$$\sigma_X^2 = \sum [(X - \mu_X)^2 \cdot P(X)]$$

---

* *Source:*   George T. Kurian, *The Book of World Rankings* (New York: Facts on File, Inc., 1979), p. 297.

Using notation consistent with that used for the mean of a sampling distribution, we will denote the variance of a sampling distribution by writing the statistic of interest as a subscript to the symbol for a population variance, σ^2. For example, the variance of the sampling distribution of $\bar{x}$ is denoted by $\sigma_{\bar{x}}^2$. The variance of a sampling distribution is a weighted average [weighted by the probability $P(X)$] of the squared deviations of each value of X from the mean, μ_X. [Recall that the sample variance, s^2, was based on the squared deviations of sample measurements from their mean, $(X - \bar{x})^2$.]

If the value of μ_X is not an integer, calculating σ_X^2 using Definition 5.5 is quite difficult. As we did with the sample variance, s^2, we present a calculating formula for finding σ_X^2.

Calculating Formula for σ_X^2

$$\sigma_X^2 = \sum [X^2 \cdot P(X)] - \mu_X^2$$

To describe a sampling distribution, we will use the square root of the variance, which is the **standard deviation.**

Definition 5.6

The **standard deviation of the sampling distribution of a statistic X** is the positive square root of the variance of the sampling distribution. In symbols, with σ_X denoting the standard deviation of the sampling distribution,

$$\sigma_X = \sqrt{\sigma_X^2}$$

EXAMPLE 5.9   In Example 5.5 (Section 5.2) we found that the sampling distribution shown below has a mean of $\mu_X = 1$:

X	0	1	2
P(X)	$\frac{1}{4}$	$\frac{1}{2}$	$\frac{1}{4}$

a. Calculate σ_X^2, the variance of the sampling distribution of X.
b. Calculate σ_X, the standard deviation of the sampling distribution of X.
c. Locate the values $(\mu_X - 2\sigma_X)$ and $(\mu_X + 2\sigma_X)$ on a graph of the sampling distribution.

Solution   **a.** To calculate σ_X^2, we need to find the value of $\sum X^2 \cdot P(X)$. This is most easily done by constructing a table similar to Table 5.11. The value of $\sum X^2 \cdot P(X)$ is the sum of the values in the last column of the table:

$$\sum X^2 \cdot P(X) = 0 + \frac{1}{2} + 1 = 1\frac{1}{2}$$

The variance of the sampling distribution of X is

$$\sigma_X^2 = \sum [X^2 \cdot P(X)] - \mu_X^2 = 1\frac{1}{2} - (1)^2 = \frac{1}{2}$$

Table 5.11

Calculation of $\sum X^2 \cdot P(X)$

X	P(X)	X²	X² · P(X)
0	¼	0	$0 \cdot \frac{1}{4} = 0$
1	½	1	$1 \cdot \frac{1}{2} = \frac{1}{2}$
2	¼	4	$4 \cdot \frac{1}{4} = 1$

b. The value of σ_X is the positive square root of σ_X^2:

$$\sigma_X = \sqrt{\sigma_X^2} = \sqrt{\frac{1}{2}} \approx .7$$

The standard deviation of the sampling distribution of X is approximately .7.

c. We calculate

$$\mu_X - 2\sigma_X = 1 - 2(.7) = 1 - 1.4 = -.4$$

and

$$\mu_X + 2\sigma_X = 1 + 2(.7) = 1 + 1.4 = 2.4$$

These values, along with the value $\mu_X = 1$, are shown in Figure 5.6.

Figure 5.6
Sampling distribution for X
of Example 5.9

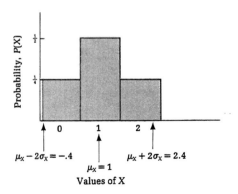

Since $-.4$ is less than 0 and 2.4 is more than 2, all the values of X for the sampling distribution are within 2 standard deviations of the mean. Recall that the rule of thumb given in Section 3.4 stated that for most sets of data resulting from real-life experiments, most of the measurements will fall within 2 standard deviations of the mean. Since there are only three values (0, 1, and 2) that X can assume, and there is little variability in the sampling distribution, all the measurements fall within 2 standard deviations of the mean. ■

EXAMPLE 5.10   The mean and standard deviation of the sampling distribution shown below are $\mu_X = 6.3$ and $\sigma_X = 1.6$,* respectively:

X	1	2	3	4	5	6	7	8
$P(X)$	.01	.02	.04	.07	.11	.20	.30	.25

a. Calculate $(\mu_X - 2\sigma_X)$ and $(\mu_X + 2\sigma_X)$, and find the probability that the statistic X falls between these values.

b. Calculate $(\mu_X - 3\sigma_X)$ and $(\mu_X + 3\sigma_X)$, and find the probability that the statistic X falls between these values.

Solution   **a.** $\mu_X - 2\sigma_X = 6.3 - 2(1.6) = 6.3 - 3.2 = 3.1$

$\mu_X + 2\sigma_X = 6.3 + 2(1.6) = 6.3 + 3.2 = 9.5$

A graph of the sampling distribution of X is given in Figure 5.7, and the values $(\mu_X - 2\sigma_X)$, μ_X, and $(\mu_X + 2\sigma_X)$ are shown. The values of X between 3.1 and 9.5

Figure 5.7
Sampling distribution for
Example 5.10

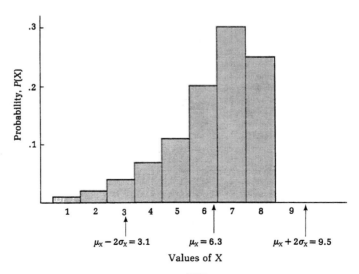

* The exact value of σ_X^2 is 2.53. Then $\sigma_X = \sqrt{2.53} \approx 1.6$.

are 4, 5, 6, 7, and 8. Summing the probabilities of these values as given by the sampling distribution, we obtain

$$.07 + .11 + .20 + .30 + .25 = .93$$

The probability that X falls within 2 standard deviations of the mean is .93. In other words, 93% of the values in the population described by the sampling distribution of X are within 2 standard deviations of the mean. This is certainly most of the measurements and agrees with the rule of thumb discussed earlier.

b. $\mu_X - 3\sigma_X = 6.3 - 3(1.6) = 6.3 - 4.8 = 1.5$

$\mu_X + 3\sigma_X = 6.3 + 3(1.6) = 6.3 + 4.8 = 11.1$

These values and the value of μ_X are shown in Figure 5.8. The only value of X that is not between 1.5 and 11.1 is the value 1. Thus, the probability that X is between 1.5 and 11.1 is

$$.02 + .04 + .07 + .11 + .20 + .30 + .25 = .99$$

Figure 5.8
Sampling distribution for
Example 5.10

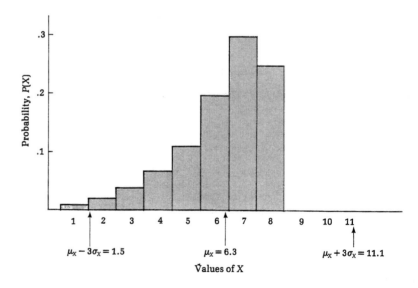

According to the rule of thumb, all or almost all measurements should be within 3 standard deviations of the mean. For this sampling distribution, 99% of the values of X are within 3 standard deviations of the mean, and this is certainly almost all the values. ∎

EXAMPLE 5.11   A company that specializes in data analysis tests all its applicants for employment by having them solve three short problems that are indicative of the type of work they will be required to perform. An applicant is given a score from 0 to 10 for each problem, and the mean score (rounded to the nearest integer) is recorded for each applicant. From the performances of previous applicants, the sampling distribution of mean scores has been found to be (approximately) as shown in the table below:

Mean Score $\bar{x}$	0	1	2	3	4	5	6	7	8	9	10
Probability $P(\bar{x})$	.001	.005	.010	.045	.060	.100	.150	.350	.200	.070	.009

a. Calculate the probability that an applicant has a mean score of less than 7.
b. Find the mean and standard deviation of the sampling distribution of $\bar{x}$.

Solution   **a.** We want to find $P(\bar{x} < 7)$. The values of $\bar{x}$ less than 7 are 0, 1, 2, 3, 4, 5, and 6. Thus,

$$P(\bar{x} < 7) = P(\bar{x} = 0) + P(\bar{x} = 1) + \cdots + P(\bar{x} = 6)$$
$$= .001 + .005 + .010 + .045 + .060 + .100 + .150$$
$$= .371$$

Approximately 37% of all applicants for employment with this company have a mean score of less than 7 for the three sample problems.

b. The mean and variance of the sampling distribution are most easily found by constructing a table similar to Table 5.12. Summing the values of $\bar{x} \cdot P(\bar{x})$, we obtain the mean,

$$\mu_{\bar{x}} = \sum[\bar{x} \cdot P(\bar{x})] = .000 + .005 + .020 + \cdots + .090 = 6.570$$

Table 5.12

$\bar{x}$	$P(\bar{x})$	$\bar{x} \cdot P(\bar{x})$	$\bar{x}^2$	$\bar{x}^2 \cdot P(\bar{x})$
0	.001	.000	0	0.000
1	.005	.005	1	.005
2	.010	.020	4	.040
3	.045	.135	9	.405
4	.060	.240	16	.960
5	.100	.500	25	2.500
6	.150	.900	36	5.400
7	.350	2.450	49	17.150
8	.200	1.600	64	12.800
9	.070	.630	81	5.670
10	.009	.090	100	.900
		$\mu_{\bar{x}} = \overline{6.570}$		$\sum \bar{x}^2 \cdot P(\bar{x}) = \overline{45.830}$

Using the calculating formula for $\sigma_{\bar{x}}^2$ and the values from Table 5.12, we obtain

$$\sigma_{\bar{x}}^2 = \sum[\bar{x}^2 \cdot P(\bar{x})] - \mu_{\bar{x}}^2 = 45.830 - (6.570)^2 = 2.6651$$

Then we can calculate the standard deviation of the sampling distribution:

$$\sigma_{\bar{x}} = \sqrt{2.6651} \approx 1.63$$

The mean of the sampling distribution is 6.57. This is the mean of the mean problem scores for all previous applicants for employment at the data analysis company. In other words, if many, many applicants are each given three short problems to solve, and the mean score for these problems is recorded for each applicant, the applicants would have an average (mean) score of approximately 6.57.

The sampling distribution of $\bar{x}$ is shown in Figure 5.9, and the values of $\mu_{\bar{x}}$, $(\mu_{\bar{x}} - 2\sigma_{\bar{x}})$, and $(\mu_{\bar{x}} + 2\sigma_{\bar{x}})$ are indicated. The values of $\bar{x}$ that are not within 2 standard deviations of the mean are 0, 1, 2, 3, and 10. Adding the probabilities of these values, we see that the probability that $\bar{x}$ is more than 2 standard deviations from the mean is $(.001 + .005 + .010 + .045 + .009) = .070$. Therefore, as usual, most of the measurements $(1 - .070 = .930 = 93\%)$ are within 2 standard deviations of the mean.

Figure 5.9
The sampling distribution of $\bar{x}$ for Example 5.11

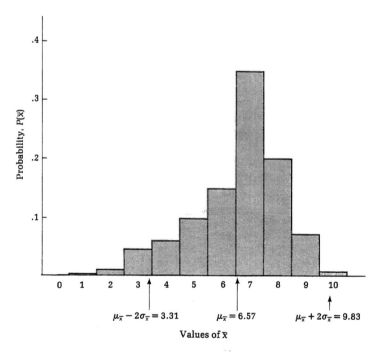

COMPUTER
EXAMPLE 5.3   Consider the sampling distribution given in Example 5.10:

X	1	2	3	4	5	6	7	8
P(X)	.01	.02	.04	.07	.11	.20	.30	.25

We stated that $\mu_X = 6.3$ and $\sigma_X = 1.6$. We will now calculate these values using Minitab. First, the sampling distribution is displayed by Minitab:

```
ROW    X    P(X)

 1     1    0.01
 2     2    0.02
 3     3    0.04
 4     4    0.07
 5     5    0.11
 6     6    0.20
 7     7    0.30
 8     8    0.25
```

Using this distribution, Minitab calculated the values of the mean and standard deviation to be

```
ROW    MEAN   STAN DEV

 1      6.3   1.59060
```

This value of the mean is exactly equal to the previously given value of $\mu_X = 6.3$, while this value of the standard deviation (1.59060) rounds to the previous value of $\sigma_X = 1.6$.

Finally, we used Minitab to construct a graph of this sampling distribution. The result provides essentially the same information as the graph shown in Figure 5.7:

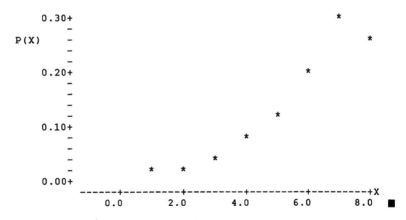

Learning the Language

5.30 Define the following terms:
a. Variance of a sampling distribution
b. Standard deviation of a sampling distribution

5.31 What do the symbols σ_X^2 and σ_X denote?

Learning the Mechanics

5.32 In Exercise 5.20 you were asked to calculate the mean of the sampling distribution given below. The value of the mean is $\mu_X = 10$.

X	0	10	20	30
P(X)	.5	.2	.1	.2

a. Find σ_X^2.   b. Find σ_X.

5.33 Consider the following sampling distribution:

X	1	2	3	5	10
P(X)	.1	.3	.2	.1	.3

a. Find the mean of the sampling distribution.
b. Find the variance.
c. Find the standard deviation.

5.34 Consider the following sampling distribution:

X	−4	−3	−2	−1	0	1	2	3	4
P(X)	.05	.07	.10	.15	.26	.15	.10	.07	.05

a. Calculate μ_X, σ_X^2, and σ_X.
b. Graph $P(X)$. Locate μ_X, $(\mu_X - 2\sigma_X)$, and $(\mu_X + 2\sigma_X)$ on the graph.
c. What is the probability that X is in the interval $(\mu_X - 2\sigma_X)$ to $(\mu_X + 2\sigma_X)$?

5.35 Consider the following sampling distributions:

(1)

X	0	1	2
P(X)	.3	.4	.3

(2)

X	0	1	2
P(X)	.1	.8	.1

a. Use your intuition to find the mean for each distribution. How did you arrive at your choice?
b. Which distribution appears to be more variable? Why?
c. Calculate μ_X and σ_X^2 for each distribution. Compare these answers to your answers in parts a and b.

Applying the Concepts

5.36 A hospital research laboratory purchases rats from a distributor for use in experiments. The distributor sells four different strains of rats, each at a different price, and fills requests with one of the four strains, depending on availability. The laboratory would like to estimate how much it will have to spend on rats in the next year. The table lists the four strains, the price per fifty rats for each strain, and the probability of the purchase of each strain.

Strain	X = Price per 50 rats	P(X)
A	$50.00	$1/10$
B	$75.00	$2/5$
C	$87.50	$3/10$
D	$100.00	$1/5$

a. What is the mean price paid for a shipment of fifty rats?
b. What is the variance of the price?
c. Graph $P(X)$. Locate μ_X and the interval $(\mu_X - 2\sigma_X)$ to $(\mu_X + 2\sigma_X)$ on the graph. What proportion of the time does X fall in this interval?

5.37 A maximum security prison reports that X, the number of escape attempts by prisoners per month, has the following sampling distribution:

X	0	1	2	3	4	5
P(X)	.22	.33	.25	.13	.05	.02

a. Find the mean number of escape attempts per month.
b. Find the variance of the number of escape attempts per month.
c. Find the standard deviation.
d. Graph the sampling distribution and locate μ_X, $(\mu_X - 2\sigma_X)$, and $(\mu_X + 2\sigma_X)$ on the graph.

5.38 The number of patients admitted per day to a hospital has a mean of 35 and a standard deviation of 10. If, on a given day, there are sixty beds available for new patients, do you think the hospital will have enough beds to accommodate the new patients admitted that day? Explain. [*Hint:* How many standard deviations is 60 above the mean of 35?]

5.39 The sampling distribution of X, the number of people who arrive at a cashier's counter in a bank per minute, is given below:

X	0	1	2	3	4
P(X)	.36	.38	.18	.06	.02

Verify that most of the values of X lie within 2 standard deviations of the mean.

5.40　Refer to Exercise 5.27. The sampling distribution of X, the number of medals won at the 1976 Summer Olympic Games in Montreal by countries winning six or fewer medals, is shown in the accompanying table. The mean number of medals won by these countries is $\mu_X = 2.72$.

X	1	2	3	4	5	6
P(X)	.32	.32	.04	.08	.12	.12

a. Find the standard deviation of the number of medals won by these countries.
b. What is the probability that X is within 2 standard deviations of the mean?

5.41　Refer to Exercise 5.28. A buyer for a large department store wants to project the required inventory for men's ten-speed bicycles next spring. Experience has shown that demand (X) has approximately the sampling distribution shown in the table. The mean demand is $\mu_X = 114.4$.

Demand, X = Number of bicycles	40	60	80	100	120	140	160	180
P(X)	.01	.06	.16	.24	.23	.15	.10	.05

a. Calculate σ_X^2.
b. Calculate σ_X.
c. Graph $P(X)$ and locate μ_X and the values $(\mu_X - 2\sigma_X)$ and $(\mu_X + 2\sigma_X)$ on the graph.
d. Using your answer from part c, how many bicycles would you order for next spring? Explain the reasoning behind your decision.
*e. Repeat parts b and c using a computer package.

5.42　Refer to Exercise 5.29. The sampling distribution of X, the percentage of deaths from cancer among all deaths of persons aged 55 and over in thirty-nine countries reporting such statistics, is shown below.* The mean percentage of deaths due to cancer for these countries is $\mu_X = 78.74$.

X	88	87	86	85	84	83	82	81	80
P(X)	$\frac{2}{39}$	$\frac{2}{39}$	$\frac{2}{39}$	$\frac{4}{39}$	$\frac{4}{39}$	$\frac{2}{39}$	$\frac{2}{39}$	$\frac{2}{39}$	$\frac{4}{39}$

X	79	78	77	75	74	69	65	62	59	39
P(X)	$\frac{1}{39}$	$\frac{2}{39}$	$\frac{4}{39}$	$\frac{2}{39}$	$\frac{1}{39}$	$\frac{1}{39}$	$\frac{1}{39}$	$\frac{1}{39}$	$\frac{1}{39}$	$\frac{1}{39}$

a. Calculate the standard deviation of the percentage of deaths due to cancer.
b. How many standard deviations is $X = 39$ below the mean?
c. What is the probability that X is more than 2 standard deviations from the mean?
*d. Repeat part a using a computer package.

* Source: George T. Kurian, The Book of World Rankings (New York: Facts on File, Inc., 1979). p. 297.

5.4

THE BINOMIAL EXPERIMENT

Many experiments produce qualitative data that may be classified into two distinct categories. Examples of such categories are Yes–No, Pass–Fail, Male–Female, Defective–Good, Success–Failure, and Right–Wrong.

Suppose we were to toss a coin five times and record the upward face each time. Each observation would be either a head or a tail, and we would obtain a total of five such observations. We may view these five observations as a sample obtained from an infinite population of heads and tails that could be generated by tossing the coin forever and recording the upward face after each toss. If the coin were balanced, we would consider this population to contain half heads and half tails. If the coin were not balanced, the population would contain some other fractions of heads and tails.

Many practical experiments have characteristics similar to those of the coin tossing experiment. In the box we specify the characteristics that an experiment must satisfy in order to be considered a **binomial experiment.**

Characteristics of a Binomial Experiment

1. The population contains an infinite number of observations, each of which may be classified into exactly one of two categories. The observations in one category will be denoted by S (for Success) and those in the other will be denoted by F (for Failure).
2. The fraction of successes in the population is denoted by π (the Greek letter pi). The fraction of failures is then $(1 - \pi)$.
3. The experiment consists of sampling n observations from the population and recording each sampled observation as S or F. The experiment must be performed in such a way that the following properties hold:
   a. The probability any one sampled observation is S is π, and the probability any one sampled observation is F is $(1 - \pi)$.
   b. The probability any sampled observation is S is not affected by what the other sampled observations happen to be.
4. We are interested in the statistic

   p = Fraction of successes observed in sample

The statistic p is called the **binomial statistic.** An essentially equivalent statistic that could be studied is the number of successes observed in the sample. We have chosen to use the fraction of successes in the sample, p, since inferences are usually made about the fraction of successes in the population, π, and it is natural to compare the sample fraction to the population fraction.

The classic example of a binomial experiment is the coin tossing experiment introduced earlier. In Example 5.12 we will discuss this experiment in more detail to help clarify the characteristics of a binomial experiment.

EXAMPLE 5.12   A coin is flipped in the air, coming to rest on a tabletop, and it is noted whether the upward face is a head or a tail. This is repeated until a total of ten observations are obtained. Let p equal the fraction of heads in the ten observations. Discuss the characteristics of a binomial experiment in reference to this experiment.

Solution   We will examine each of the characteristics and state them in terms of this experiment.

1. The population contains an infinite number of heads and tails. We may think of the population as the set of heads and tails that would be generated if the coin were tossed forever. Since we are interested in the fraction of heads observed, a head will be denoted by S and a tail by F.
2. The fraction of S's (heads) in the population is π. The value of π could be any value from 0 to 1, depending on the nature of the coin being flipped. If the coin were perfectly balanced, π would equal $\frac{1}{2}$. The fraction of F's (tails) in the population is $(1 - \pi)$.
3. The experiment consists of sampling ten observations from the population. This is accomplished by flipping the coin ten times. Each sampled observation is recorded as S (if a head appears) or F (if a tail appears). We may make the following assumptions, based upon the way the experiment is performed:
   a. The probability that any one sampled observation is S is π, and the probability of F is $(1 - \pi)$. For example, if the coin is balanced, $\pi = \frac{1}{2}$, and the probability that the first flip results in S (head) is $\frac{1}{2}$, the probability that the second flip results in S (head) is $\frac{1}{2}$, etc. Of course, the probability that any observation is F (tail) is also $\frac{1}{2}$ since $1 - \pi = \frac{1}{2}$.
   b. The probability that any sampled observation is S is not affected by what the other sampled observations happen to be. For example, suppose the first five tosses of the coin result in S's. If the coin is balanced, what is the probability the next toss is S? The answer is still $\frac{1}{2}$. The coin has no memory. It does not "realize" that the first five tosses resulted in S's. For a binomial experiment, the probability that any sampled observation is a success is always π, *regardless of the outcomes of the other sampled observations.*
4. We are interested in the statistic

   p = Fraction of successes (heads) observed in sample

   Since there are ten observations in the sample, the possible values of p are 0, $\frac{1}{10}$, $\frac{2}{10}$, $\frac{3}{10}$, $\frac{4}{10}$, $\frac{5}{10}$, $\frac{6}{10}$, $\frac{7}{10}$, $\frac{8}{10}$, $\frac{9}{10}$, and 1. ∎

For practical experiments the characteristics of a binomial experiment are rarely satisfied exactly. One reason is that populations of interest do not usually contain an infinite number of observations, although the number of observations may be very large. Under the conditions given in the next box, we will consider an experiment to be binomial even though the population is finite.

Conditions for an Experiment to Be Considered Binomial if the Population of Interest Is Finite

1. The population contains a large number, N, of observations, of which the fraction π are successes (denoted by S) and the fraction $(1 - \pi)$ are failures (denoted by F).
2. The experiment consists of randomly sampling n observations from the population and recording whether each sampled observation is S or F.
3. The sample size, n, must be small in comparison to both $N \cdot \pi$ and $N \cdot (1 - \pi)$.*
4. We are interested in the statistic

    p = Fraction of successes in sample

There are many experiments that appear to be binomial at first glance. You must be very careful to make certain that the characteristics and conditions listed in the boxes are satisfied before you conclude that an experiment results in a binomial statistic.

EXAMPLE 5.13   For each of the following examples, decide whether we may consider p to be a binomial statistic.

   a. There are 400,000 school-age children in a certain city. A rare neurological disorder can be expected in .0001 of these children. A random sample of twenty children is selected from this city to participate in a physical fitness program. Let p be the fraction of selected children who have the neurological disorder.

   b. Refer to the introductory example which considered the blind taste test. Suppose the soft drink company conducts a taste-preference survey of 100 randomly chosen consumers from Texas who state their preference for the company's brand or brand X. Suppose that 60% of all consumers prefer the company's brand. Let p be the fraction of the 100 randomly chosen consumers who prefer the company's brand.

Solution   **a.** We will check the conditions for a binomial experiment one by one until one is not met or all are met.

   1. The population contains a large number, 400,000, of observations, of which .0001 are successes (a child does have the neurological disorder) and .999 are

---

* The third condition guarantees that the probability of any sampled observation being S is practically unaffected by what the other sampled observations happen to be. As a general rule, we want n to be no more than 5% of $N \cdot \pi$ and $N \cdot (1 - \pi)$.

failures (a child does not have the neurological disorder). Note that a success is defined by the statistic of interest—it does not have to represent something "good."

2. The experiment consists of randomly sampling twenty children from the population and noting whether the child does (S) or does not (F) have the disorder.

3. The sample size is $n = 20$, and $N \cdot \pi = 400,000 \cdot (.0001) = 40$. Now there is a problem. The sample size is not small compared to $N \cdot \pi$.

This cannot be considered to be a binomial experiment.

b. Again, we will check the conditions for a binomial experiment one by one.

1. Although the number of consumers in the population is not stated, it is some very large number. A success is that a consumer prefers the soft drink company's brand, and $\pi = .60$. A failure is that a consumer prefers brand X, and $1 - \pi = .40$.

2. The experiment consists of randomly choosing 100 consumers and recording whether each prefers the company brand (S) or brand X (F).

3. Since N is very large, both $N \cdot \pi$ and $N \cdot (1 - \pi)$ will be large. For example, if $N = 1,000,000$, then $N \cdot \pi = (1,000,000)(.60) = 600,000$ and $N \cdot (1 - \pi) = 400,000$. Certainly, $n = 100$ is small compared to 600,000 and 400,000.

4. We are interested in the statistic

$p = $ Fraction of successes (consumers who prefer company brand) in sample

We may consider this example to be a binomial experiment. ∎

Now that we can identify a binomial statistic, we will examine its sampling distribution. Finding the probabilities associated with a binomial statistic is a much more complex task than finding the probabilities for statistics discussed previously. Fortunately, the probabilities have been tabulated for many binomial experiments, and we have provided the sampling distributions for some binomial statistics in Appendix Table I. A partial reproduction is shown in Table 5.13.

To use Table I we must know the sample size, n, and the fraction of successes in the population, π. In Table 5.13 we have indicated how to find the sampling distribution of the binomial statistic p when $n = 4$ and $\pi = .80$ by shading the value $\pi = .80$ at the top of the table and $n = 4$ in the leftmost column. The sampling distribution is also shaded, and the result is shown in Table 5.14.

The possible values of p, shaded in Table 5.13, are located in the second column, next to the value $n = 4$. The probabilities of observing these values are obtained from the body of Table 5.13 by proceeding down the column labeled $\pi = .80$ until you are opposite $n = 4$. These probabilities are also shaded in the table.

Notice that, in our example, the probabilities of the sampling distribution of p do not sum to exactly 1:

$.002 + .026 + .154 + .410 + .410 = 1.002$

This is because the values have been rounded to three decimal places. Also, if you examine Table 5.13, you will see entries (probabilities) in the body of the table that read ".00+." This means that these probabilities are slightly larger than 0, but are equal to .000 when rounded to three decimal places.

Table 5.13

A partial reproduction
of Appendix Table I,
Binomial Sampling
Distributions

n	p	.10	.20	.30	.40	π .50	.60	.70	.80	.90	p
2	0	.810	.640	.490	.360	.250	.160	.090	.040	.010	0
	$\frac{1}{2}$	.180	.320	.420	.480	.500	.480	.420	.320	.180	$\frac{1}{2}$
	1	.010	.040	.090	.160	.250	.360	.490	.640	.810	1
3	0	.729	.512	.343	.216	.125	.064	.027	.008	.001	0
	$\frac{1}{3}$	.243	.384	.441	.432	.375	.288	.189	.096	.027	$\frac{1}{3}$
	$\frac{2}{3}$	.027	.096	.189	.288	.375	.432	.441	.384	.243	$\frac{2}{3}$
	1	.001	.008	.027	.064	.125	.216	.343	.512	.729	1
4	0	.656	.410	.240	.130	.062	.026	.008	.002	.00+	0
	$\frac{1}{4}$	.292	.410	.412	.346	.250	.154	.076	.026	.004	$\frac{1}{4}$
	$\frac{2}{4}$	.049	.154	.265	.346	.375	.346	.265	.154	.049	$\frac{2}{4}$
	$\frac{3}{4}$	.004	.026	.076	.154	.250	.346	.412	.410	.292	$\frac{3}{4}$
	1	.00+	.002	.008	.026	.062	.130	.240	.410	.656	1
5	0	.590	.328	.168	.078	.031	.010	.002	.00+	.00+	0
	$\frac{1}{5}$	.328	.410	.360	.259	.156	.077	.028	.006	.00+	$\frac{1}{5}$
	$\frac{2}{5}$	.073	.205	.309	.346	.312	.230	.132	.051	.008	$\frac{2}{5}$
	$\frac{3}{5}$	.008	.051	.132	.230	.312	.346	.309	.205	.073	$\frac{3}{5}$
	$\frac{4}{5}$	.00+	.006	.028	.077	.156	.259	.360	.410	.328	$\frac{4}{5}$
	1	.00+	.00+	.002	.010	.031	.078	.168	.328	.590	1
6	0	.531	.262	.118	.047	.016	.004	.001	.00+	.00+	0
	$\frac{1}{6}$	.354	.393	.303	.187	.094	.037	.010	.002	.00+	$\frac{1}{6}$
	$\frac{2}{6}$	.098	.246	.324	.311	.234	.138	.060	.015	.001	$\frac{2}{6}$
	$\frac{3}{6}$	.015	.082	.185	.276	.312	.276	.185	.082	.015	$\frac{3}{6}$
	$\frac{4}{6}$	.001	.015	.060	.138	.234	.311	.324	.246	.098	$\frac{4}{6}$
	$\frac{5}{6}$	.00+	.002	.010	.037	.094	.187	.303	.393	.354	$\frac{5}{6}$
	1	.00+	.00+	.001	.004	.016	.047	.118	.262	.531	1

Table 5.14

Sampling distribution of p
when $n = 4$ and $\pi = .80$

p	0	$\frac{1}{4}$	$\frac{2}{4}$	$\frac{3}{4}$	1
P(p)	.002	.026	.154	.410	.410

Like the sampling distribution of any statistic, the sampling distribution of the binomial statistic p may be used to find the probability that a particular value of the statistic will occur, or the probability that any one of two or more values will occur. We present these uses in the next example.

EXAMPLE 5.14   The Heart Association claims that only 10% of the adults over 30 years of age in the United States can pass the minimum fitness requirements established by the president's Physical Fitness Commission. Suppose six adults are randomly selected from all the adults in the United States and each is given the fitness test. Let p equal the fraction of the six who pass the minimum fitness requirements. Assuming that the 10% figure stated by the Heart Association is correct, answer the following questions:

a. What is the sampling distribution of p?

b. What is the probability that none of the six pass the minimum requirements?

c. What is the probability that more than two of the six adults pass the minimum requirements?

d. Would you be surprised if more than two of the six sampled adults actually do pass the minimum requirements? Explain.

Solution   The experiment consists of randomly sampling $n = 6$ adults from a very large population (all people in the United States over 30 years old) that contains 10% S's (person passes the minimum requirements) and 90% F's (person does not pass the requirements). Since the population is so large, $n = 6$ is small compared to both $N \cdot \pi = N(.10)$ and $N \cdot (1 - \pi) = N(.90)$. Finally,

p = Fraction of S's in sample

is the statistic of interest.

Thus, all the conditions are satisfied and we may consider p a binomial statistic. We will use the portion of Appendix Table I shown in Table 5.13 to answer the questions of interest.

a. Consulting Table 5.13 and using the values $n = 6$, $\pi = .10$, we obtain the sampling distribution shown below:

p	0	$\frac{1}{6}$	$\frac{2}{6}$	$\frac{3}{6}$	$\frac{4}{6}$	$\frac{5}{6}$	1
$P(p)$	.531	.354	.098	.015	.001	.00+	.00+

b. If none of the people selected pass the test, then $p = 0$. From the sampling distribution in part a, we get $P(0) = .531$. That is, the probability none of the six pass the minimum requirements is .531. If the Heart Association's figure of 10% is accurate, over half the time six people are randomly selected from among those over 30, none would pass the minimum requirements.

c. We want to find $P(p > \frac{2}{6})$. Since $p = \frac{3}{6}$ and $p = \frac{4}{6}$ are the only values of p that have probabilities appreciably greater than 0, we obtain

$$P\left(p > \frac{2}{6}\right) = P\left(p = \frac{3}{6}\right) + P\left(p = \frac{4}{6}\right) = .015 + .001 = .016$$

(When summing probabilities from Table I, we may always consider probabilities of .00+ to be 0.)

Thus, more than two of the six people selected would pass the minimum requirements only .016 of the time.

d. It certainly would be surprising if more than two of the six people selected passed the minimum requirements. This rare event would occur with an extremely small probability of .016. Such an event may raise doubts about the Heart Association's claim.   ■

The mean, variance, and standard deviation of the sampling distribution of the binomial statistic p could be calculated using the definitions of Sections 5.2 and 5.3, but we can obtain these quantities more easily by using the formulas in the box. *Remember, these formulas apply only to binomial statistics.*

Mean, Variance, and Standard Deviation for a Binomial Statistic

$$\text{Mean:} \quad \mu_p = \pi$$

$$\text{Variance:} \quad \sigma_p^2 = \frac{\pi \cdot (1 - \pi)}{n}$$

$$\text{Standard deviation:} \quad \sigma_p = \sqrt{\frac{\pi \cdot (1 - \pi)}{n}}$$

where

n = Number of observations in the sample

π = Fraction of successes in the population

EXAMPLE 5.15   Let p be a binomial statistic with $n = 500$ and $\pi = .3$.

a. Find the mean, μ_p.
b. Find the variance, σ_p^2.
c. Find the standard deviation, σ_p.

Solution   **a.** Using the formula in the box, we compute the mean:

$$\mu_p = \pi = .3$$

If we repeated this binomial experiment ($n = 500$ and $\pi = .3$) a very large number of times, the mean of the fractions of successes observed would be approximately .3. The fraction of successes in most samples would be smaller or larger than .3, but their average value would be .3.

b. The variance is

$$\sigma_p^2 = \frac{\pi \cdot (1 - \pi)}{n} = \frac{.3(1 - .3)}{500} = \frac{.3(.7)}{500} = .00042$$

c. The standard deviation is

$$\sigma_p = \sqrt{\frac{\pi \cdot (1 - \pi)}{n}} = \sqrt{.00042} \approx .02$$

Most of the times we perform this binomial experiment we would expect the fraction of successes in the sample to be within 2 standard deviations of the mean. Thus, since

$$\mu_p - 2\sigma_p = .3 - 2(.02) = .3 - .04 = .26$$
$$\mu_p + 2\sigma_p = .3 + 2(.02) = .3 + .04 = .34$$

most of the sample fractions would be between .26 and .34.   ∎

EXAMPLE 5.16   A poll of twenty voters is randomly selected from the list of all registered voters in a large city. Let p be the fraction of voters out of the twenty polled who favor a particular candidate for mayor. Suppose we assume that 60% of all registered voters favor this candidate.

a. Find the mean and standard deviation of the sampling distribution of p.
b. Find the probability that p will fall within 2 standard deviations of its mean.
c. Interpret the answer to part b.

Solution   a. Since the sample of twenty voters was randomly selected from a large number of registered voters, we may consider p to be a binomial statistic. It is known that the fraction of all registered voters favoring the candidate is $\pi = .60$. Therefore, the mean and standard deviation of p are

$$\mu_p = \pi = .60$$
$$\sigma_p = \sqrt{\frac{\pi \cdot (1 - \pi)}{n}} = \sqrt{\frac{(.60) \cdot (.40)}{20}} = \sqrt{.012} \approx .11$$

b. Two standard deviations above and below the mean are, respectively:

$$\mu_p + 2\sigma_p = .60 + 2(.11) = .60 + .22 = .82$$
$$\mu_p - 2\sigma_p = .60 - 2(.11) = .60 - .22 = .38$$

Since $7/20 = .35$, $8/20 = .40$, $16/20 = .80$, and $17/20 = .85$, the values of p between .38 and .82 are $8/20$, $9/20$, $10/20$, $11/20$, $12/20$, $13/20$, $14/20$, $15/20$, and $16/20$. Using Appendix Table I, for $n = 20$ and $\pi = .60$, we find the probabilities

$$P\left(\frac{8}{20}\right) + P\left(\frac{9}{20}\right) + P\left(\frac{10}{20}\right) + P\left(\frac{11}{20}\right) + P\left(\frac{12}{20}\right) + P\left(\frac{13}{20}\right) + P\left(\frac{14}{20}\right) + P\left(\frac{15}{20}\right) + P\left(\frac{16}{20}\right)$$

$$= .035 + .071 + .117 + .160 + .180 + .166 + .124 + .075 + .035$$

$$= .963$$

The probability that p falls within 2 standard deviations of its mean (that is, between .38 and .82) is .963.

c. The rule of thumb tells us that for most sets of data most of the measurements are within 2 standard deviations of their mean. This is certainly true for the sampling distribution of this binomial statistic. If twenty voters were repeatedly polled, approximately 96% of the time the sample fraction of voters favoring the candidate would be between .38 and .82.   ∎

A natural question arising from Example 5.16 is: Does the poll support the assumption that 60% of the registered voters favor the candidate? Answering such questions— that is, making inferences from binomial experiments—is discussed in the next section. First, we offer a computer illustration of binomial sampling distributions.

COMPUTER
EXAMPLE 5.4

A very nice feature of Minitab is that it can generate the sampling distributions of binomial statistics. For example, consider a binomial statistic p with $n = 8$ and $\pi = .4$. The sampling distribution of p, as generated by Minitab, follows.

```
ROW      p       P(p)

  1    0.000   0.016796
  2    0.125   0.089580
  3    0.250   0.209019
  4    0.375   0.278692
  5    0.500   0.232243
  6    0.625   0.123863
  7    0.750   0.041288
  8    0.875   0.007864
  9    1.000   0.000655
```

Comparing this sampling distribution with the one given in Appendix Table I for $n = 8$ and $\pi = .4$, we find some minor differences. Minitab gives the possible values of p in decimal form (0.000, 0.125, 0.250, etc.) rather than as fractions (0, $\frac{1}{8}$, $\frac{2}{8}$, etc.). The probabilities are printed with six decimal places by Minitab. The probabilities given in Appendix Table I may be found by rounding the six-place decimals to three-place decimals.

As with any statistic, Minitab is able to produce a graph of a binomial sampling distribution. A computer-generated graph of the binomial sampling distribution with $n = 8$ and $\pi = .4$ is shown here.

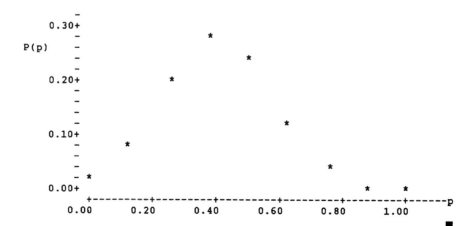

COMPUTER
EXAMPLE 5.5

In the last computer example we saw that Minitab may be used to replicate, essentially, the sampling distributions provided in Appendix Table I. In Table I, the only values of π used are .10, .20, . . . , .90. A real advantage of Minitab is that it can generate the sampling distribution of a binomial statistic for any value of π between 0 and 1. For example, suppose we are interested in a binomial experiment with a probability of success $\pi = .37$ and a sample size of $n = 5$. We could not find the probabilities of interest using Appendix Table I, but Minitab is capable of providing the desired information. The computer-generated binomial sampling distribution with $n = 5$ and $\pi = .37$ follows.

ROW	p	P(p)
1	0.0	0.099244
2	0.2	0.291430
3	0.4	0.342314
4	0.6	0.201042
5	0.8	0.059036
6	1.0	0.006934

■

Exercises (5.43–5.58)

Learning the Language

5.43   Define the following terms:
     a.   Binomial experiment   *b.*   Binomial statistic

5.44   What do the following symbols denote?
     a. p *b.* π *c.* μ_p *d.* σ_p^2 *e.* σ_p

Learning the Mechanics

5.45   A random sample of n observations is selected from a finite population of N observations. The fraction of successes in the population is π, and the fraction of failures is $(1 - \pi)$. The statistic of interest is p, the fraction of successes in the sample. For the values of N, n, and π given in each of the following, state whether p may be considered to be a binomial statistic:
     a. $N = 100,000$; $n = 50$; $\pi = .5$
     b. $N = 100,000$; $n = 50$; $\pi = .9995$
     c. $N = 100$; $n = 40$; $\pi = .5$
     d. $N = 100$; $n = 10$; $\pi = .2$
     e. $N = 1,000,000$; $n = 100$; $\pi = .99$

5.46   If p is a binomial statistic, calculate μ_p, σ_p^2, and σ_p for each of the following:
     a. $n = 20$; $\pi = .2$ *d.* $n = 100$; $\pi = .2$
     b. $n = 20$; $\pi = .8$ *e.* $n = 50$; $\pi = .2$
     c. $n = 100$; $\pi = .5$ *f.* $n = 500$; $\pi = .01$

Using the Tables

5.47   If p is a binomial statistic, use Appendix Table I to find the following probabilities:
     a. $P(p = \frac{2}{10})$ for $n = 10$, $\pi = .1$
     b. $P(p \leq \frac{5}{15})$ for $n = 15$, $\pi = .4$

c. $P(p > \frac{1}{5})$ for $n = 5$, $\pi = .5$

d. $P(p < \frac{10}{25})$ for $n = 25$, $\pi = .8$

e. $P(p \geq \frac{10}{15})$ for $n = 15$, $\pi = .9$

f. $P(p = \frac{2}{20})$ for $n = 20$, $\pi = .6$

5.48 a. If p is a binomial statistic with $n = 5$ and $\pi = .1$, give the sampling distribution of p and draw a graph of the distribution.

b. If p is a binomial statistic with $n = 5$ and $\pi = .5$, give the sampling distribution of p and draw a graph of the distribution.

c. If p is a binomial statistic with $n = 5$ and $\pi = .9$, give the sampling distribution of p and draw a graph of the distribution.

5.49 Suppose p is a binomial statistic with $n = 10$ and $\pi = .5$.

a. Find μ_p, σ_p^2, and σ_p.

b. Using Appendix Table I, graph the sampling distribution of p.

c. Locate the interval $(\mu_p - 2\sigma_p)$ to $(\mu_p + 2\sigma_p)$ on the graph. What is the probability that p is in this interval?

5.50 If p is a binomial statistic with $n = 25$ and $\pi = .9$:

a. Find μ_p, σ_p^2, and σ_p.

b. Graph the sampling distribution of p.

c. Locate the interval $(\mu_p - 2\sigma_p)$ to $(\mu_p + 2\sigma_p)$ on the graph. What is the probability that p is in this interval?

Applying the Concepts

5.51 A large university has determined from past records that the probability a student who registers for fall classes will have his or her schedule rejected (due to overfilled classrooms, clerical error, etc.) is .2. A total of 25,000 students registered this past fall.

a. Suppose that a random sample of 20 students is selected from the total of 25,000 and p is the fraction of these students who have their schedules rejected. Is p a binomial statistic? Explain.

b. Suppose that you sample the results of the first 100 students who register next fall and record p, the fraction of registrations that are rejected. Is p a binomial statistic? Explain.

c. If you identified p as a binomial statistic in either part a or part b, find μ_p, σ_p^2, and σ_p.

5.52 Suppose that 60% of all people who could be jurors in a large Florida city are in favor of capital punishment. A jury of twelve is to be randomly selected from among all the prospective jurors in this city.

a. What is the mean of the fraction of selected jurors who favor capital punishment?

b. What is the probability that none of the twelve jurors selected favors capital punishment?

c. If the jury were really selected at random, would you be surprised if none of the jurors favored capital punishment? Explain.

5.53 Over the years, a physician has found that one out of every ten diabetics receiving insulin develops antibodies against the hormone, thus requiring a more costly form of medication.

    a. Find the probability that none of the next five diabetic patients the physician treats will develop antibodies against insulin.

    b. Find the probability that at least one of the next five will develop antibodies.

    c. What assumptions are needed for solving this problem?

5.54 A problem of great concern to a manufacturer is the cost of repair and replacement required under a product's guarantee agreement. Assume it is known that 10% of all electronic pocket calculators purchased are returned for repair while their guarantee is still in effect. If a firm purchased twenty-five pocket calculators for its salespeople, what is the probability that at least five of these twenty-five calculators will need repair while the guarantee is still in effect?

5.55 A professor of literature decides to give a twenty-question true–false quiz on a reading assignment. She decides to pass anyone who gets at least fifteen of the answers correct. If someone randomly guesses at each of the twenty answers, what is the probability that person will pass the quiz?

5.56 Since the late 1970's, video display terminals (VDTs) have become widely used in homes and offices. One newspaper article addressing the possibility that VDTs cause eye problems among users stated:

> "It isn't known that VDTs cause eye problems," says Dr. Michael Smith, chief of motivation and stress research at the National Institute of Occupational Safety and Health (NIOSH). "What is known is that VDTs can aggravate existing problems."
>
> That may explain why VDT users have experienced migraine and weeks-long headaches, temporary loss of focal ability, burning and swollen eyes, and have been required to wear bifocals, trifocals and stronger lenses while using a VDT.*

According to the article, about 60% of all VDT users experience some type of eye problems. Suppose eight people who use VDTs are randomly sampled.

    a. What is the probability that none of the eight experiences eye problems?

    b. What is the probability that more than half of them experience eye problems?

5.57 Animal experiments are often conducted to determine whether certain chemicals are linked to cancer. Suppose that when a particular chemical is ingested in large doses by experimental rats, 24% of the rats contract thyroid tumors.

    a. If each rat in a group of 200 rats ingests a large dose of the chemical, what is the mean of the fraction of rats that would be free of thyroid tumors?

    b. Within what limits would we expect the fraction of rats free of thyroid tumors to fall? [*Hint:* Use the rule of thumb to assist in establishing the limits.]

   *c. A random sample of 15 rats ingests a large dose of the chemical. Use a computer package to find the sampling distribution of p, the fraction of the 15 rats that would be free of thyroid tumors.

   *d. Based on the sampling distribution found in part *c*, what is the probability that at least 9 of the 15 rats would be free of thyroid tumors?

5.58 In 1981, approximately 15% of all blacks 18 to 34 years old attended colleges or universities. Let p be the fraction of people attending a college or university in 1981 in a random sample of 500 blacks from this age group.

* *Source:* Virginia Mansfield, "Evidence Remains Incomplete, Blurred in Case of VDTs vs. Eyes of Mankind," *The Cincinnati Enquirer*, February 25, 1985, p. C-1.

a. Find μ_p, σ_p^2, and σ_p.

b. *Calculate the interval* $(\mu_p - 3\sigma_p)$ *to* $(\mu_p + 3\sigma_p)$. What does the rule of thumb state concerning the probability that p will be within this interval?

*c. A random sample of 20 blacks is obtained from this age group. Use a computer package to generate the sampling distribution of p, the fraction of the 20 people attending a college or university in 1981.

*d. Using your answer to part c, find the probability that the fraction attending a college or university would be $^3/_{20}$ or less.

5.5

USING THE BINOMIAL EXPERIMENT TO MAKE INFERENCES

Since binomial experiments are often used to describe real-world situations, many inferences involve the use of binomial statistics and their sampling distributions. We will continue our discussion of the rare event approach to making inferences with the following examples of binomial experiments.

EXAMPLE 5.17   A recent report stated that only 20% of all college graduates find work in the field of their undergraduate major. An employment agency believes the report's figure of 20% is too low and hopes that a poll of some recent college graduates will support their belief. A random sample of fifteen recent college graduates from across the country is selected, and the fraction, p, of those among the fifteen who found work in the field of their undergraduate major is recorded.

a. Assuming the figure of 20% stated in the report is correct, what is the probability that p, the fraction of college graduates polled who found work in their undergraduate major field, is larger than $^6/_{15}$?

b. If you observed more than six of the fifteen polled college graduates finding work in their college major, how would you interpret this result?

Solution   **a.** An examination of this experiment indicates that p may be considered a binomial statistic. Consequently, we may use Appendix Table I to find the sampling distribution of p. Since fifteen recent graduates are being polled, $n = 15$, and if the report's figure of 20% is correct, $\pi = .20$. Reading from the portion of Table I corresponding to $n = 15$ and $\pi = .20$, we find the probability that p is larger than $^6/_{15}$:

$$P\left(\frac{7}{15}\right) + P\left(\frac{8}{15}\right) + P\left(\frac{9}{15}\right) + \cdots + P\left(\frac{15}{15}\right) = .014 + .003 + .001 = .018$$

Note that $P(^{10}/_{15})$ through $P(^{15}/_{15})$ have values of .00+, so they may be ignored. Thus, if the value of 20% stated in the report is correct, the probability that p will be larger than $^6/_{15}$ is only .018. This is a rare event.

b. Since we do not expect to observe rare events, if the sample were to produce *any* value of p larger than $^6/_{15}$, we would infer that the report's figure of 20% is too low. That is, we would infer that the percentage of graduates who find work in their major field is actually larger than 20%. ∎

EXAMPLE 5.18   Refer to Example 5.17. Suppose that 5,000 college graduates had been randomly selected instead of fifteen. Let p equal the fraction of these graduates who find jobs in
the field of their college majors.

a. Calculate the mean and standard deviation of p if $\pi = .20$ (that is, assuming the
report's 20% value is accurate).
b. If $\pi = .20$, how many standard deviations does $p = .22$ lie above the mean?
c. If 1,100 of the college graduates in a random sample of 5,000 found work in the
field of their college majors (that is, $p = {}^{1,100}/_{5,000} = .22$), how would you interpret
the result?
d. If 1,050 of the college graduates in a random sample of 5,000 found work in the
field of their college majors (that is, $p = .21$), how would you interpret the result?

Solution   **a.** The statistic p is considered to be a binomial statistic with $n = 5,000$ and $\pi = .20$.
Thus,

$$\mu_p = \pi = .20$$

$$\sigma_p = \sqrt{\frac{\pi \cdot (1 - \pi)}{n}} = \sqrt{\frac{(.20)(.80)}{5,000}} \approx .006$$

The mean value of p that we would observe in repeated samplings is .20, and the
standard deviation is .006. This is a small standard deviation, and the sample fraction would not vary greatly from one sample of 5,000 graduates to another. (This is
due to the large sample size.)

b. Recall that a z-score counts the number of standard deviations between a measurement and the mean. The z-score for $p = .22$ is

$$z = \frac{p - \mu_p}{\sigma_p} = \frac{.22 - .20}{.006} \approx 3.33$$

If $\pi = .20$, then $p = .22$ is approximately 3.33 standard deviations above the mean.

c. If the observed fraction of the 5,000 sampled college graduates who found work
in the field of their undergraduate major is $p = .22$ or higher, we would infer that
the reported figure of 20% is incorrect. We should virtually never see a sample
fraction more than 3 standard deviations from the mean. If the true fraction of all
college graduates who find work in their major field is more than 20% (nearer 22%),
then the observed value of $p = .22$ would be more likely, so we infer that the
reported value of 20% is too low.

d. To interpret a value of $p = .21$, we will calculate the corresponding z-score. Assuming that $\pi = .20$, this z-score is

$$z = \frac{p - \mu_p}{\sigma_p} = \frac{.21 - .20}{.006} \approx 1.67$$

Since this z-score means that p is less than 2 standard deviations from the mean,
we do not consider it to be a rare event. There is not enough evidence to infer that
the reported figure of 20% is incorrect. ∎

In Example 5.17, where $n = 15$, we used Appendix Table I to find the probability of an event, and this probability helped us make an inference. In Example 5.18, where $n = 5,000$, we could not use Table I, so instead, we used a z-score to make an inference. Probability is the preferred method of judging the rarity of an event, but if the probability cannot be calculated and a z-score can, the z-score is a useful way to gauge the rarity of an event. In the next chapter we will show how to calculate more exact probabilities associated with z-scores.

EXAMPLE 5.19   In Example 5.13 (Section 5.4), we indicated that a blind taste test conducted by a soft drink company could be considered to be a binomial experiment. Suppose that a random sample of 100 consumers yields 63 who prefer the company's brand. Based on this sample, would you be willing to infer that more than 50% of the population prefer the company's brand?

Solution   This is a binomial experiment with $n = 100$ and $p =$ fraction of consumers who prefer company's brand. The fraction of the population that prefers the company's brand is denoted by π, and we want to see if the sample indicates that π is greater than .50 (50%). Since $n = 100$, we cannot calculate an exact probability to measure the rarity of the observed sample, and thus, we will rely on a z-score. To calculate the z-score, we will assume that $\pi = .50$, and we will determine whether the sample is contradictory to this assumption.

If $\pi = .50$, then

$$\mu_p = \pi = .50$$

$$\sigma_p = \sqrt{\frac{\pi \cdot (1 - \pi)}{n}} = \sqrt{\frac{.50(.50)}{100}} = .05$$

We observed 63 out of 100 consumers preferring the company's brand, so $p = {}^{63}\!/_{100} = .63$. The z-score associated with this value of p is

$$z = \frac{p - \mu_p}{\sigma_p} = \frac{.63 - .50}{.05} = \frac{.13}{.05} = 2.60$$

Thus, a value of $p = .63$ is 2.6 standard deviations above the mean if $\pi = .50$. This indicates that $p = .63$ is a rare event ($z > 2$) if 50% of the population prefer the company's brand, so we infer that more than 50% of the population prefer it. Note that we may infer that *more* than 50% of the population prefer the company's brand because $p = .63$ is 2.6 standard deviations *above* the mean if $\pi = .50$. If the sample fraction were 2.6 standard deviations *below* the mean ($z = -2.6$), it would indicate that *less* than 50% of the population prefer the company's brand.   ∎

Exercises (5.59–5.68)

Learning the Mechanics

5.59   Let p be a binomial statistic with $n = 20$. The value of π is not known, but someone claims the value is $\pi = .8$.

a. If $\pi = .8$, find $P(p \leq {}^{11}\!/_{20})$.

b. Considering your answer to part a, would you doubt that $\pi = .8$ if a random sample with $n = 20$ produces a value of $p \leq {}^{11}\!/_{20}$? Explain.

5.60 Let p be a binomial statistic with $n = 25$. The value of π is not known, but someone claims the value is $\pi = .6$.

a. If $\pi = .6$, find $P(p > {}^{20}\!/_{25})$.

b. If a random sample with $n = 25$ results in a value of $p > {}^{20}\!/_{25}$, do you doubt the claim that $\pi = .6$? Explain.

c. If $\pi = .6$, find $P(p \geq {}^{18}\!/_{25})$.

d. If a random sample with $n = 25$ results in a value of $p = {}^{18}\!/_{25}$, what is your interpretation? Why?

5.61 Let p be a binomial statistic with $n = 200$ and $\pi = .35$.

a. Calculate μ_p, σ_p^2, and σ_p.

b. Calculate a z-score for $p = .43$. Is $p = .43$ a rare event?

c. Calculate a z-score for $p = .30$. Is $p = .30$ a rare event?

d. Calculate a z-score for $p = .25$. Is $p = .25$ a rare event?

Applying the Concepts

5.62 An accountant believes that 10% of a large company's invoices contain errors. To check this theory the accountant randomly samples twenty-five invoices and finds that seven contain errors.

a. If the accountant's theory is correct, what is the probability that of the twenty-five invoices sampled, seven or more contain errors?

b. Since the sample had seven invoices with errors, what is your interpretation?

5.63 Suppose you are a purchasing officer for a large company. You have purchased ten million electrical switches and have been guaranteed by the supplier that the shipment will contain no more than 0.1% defectives. To check the shipment, you randomly sample 500 switches, test them, and find that four are defective. If the switches are as represented, calculate μ_p and σ_p for this sample of 500. Based on this evidence, do you think the supplier has complied with the guarantee? Explain. [*Hint:* Use μ_p and σ_p for $\pi = .001$ to see if a value of p as large as ${}^{4}\!/_{500}$ is probable.]

5.64 An experiment is to be conducted to determine whether an acclaimed psychic has extrasensory perception (ESP). Five different cards are shuffled and one is chosen at random. The psychic will then try to identify which card was drawn without seeing it. The experiment is to be repeated twenty times and p, the fraction of correct decisions, is computed. (Assume the twenty trials are independent.)

a. If the psychic is guessing—that is, if the psychic does not possess ESP—what is the value of π, the probability of a correct decision on each trial?

b. If the psychic is guessing, what is the mean of the fraction of correct decisions in twenty trials?

c. If the psychic is guessing, what is the probability that the fraction of correct decisions in twenty trials is ${}^{6}\!/_{20}$ or more?

d. Suppose that the psychic makes six correct decisions in twenty trials. Is there evidence to indicate that the psychic is *not* guessing and actually has ESP or is somehow cheating? Explain.

5.65 A newly synthesized drug is intended to reduce blood pressure. Twenty randomly selected hypertensive patients receive the new drug. Suppose eighteen or more of the patients' blood pressures drop.

a. Suppose the probability that a hypertensive patient's blood pressure drops is .5 *without treatment*. Then, what is the probability of observing eighteen or more blood pressure drops in a random sample of twenty treated patients if the drug is, in fact, ineffective in reducing blood pressure?

b. Considering the probability found in part a, do you think you have observed a rare event, or would you conclude that the drug was effective in reducing hypertension?

5.66 After a costly study, a market analyst claims that 12% of all consumers in a particular sales region prefer a certain noncarbonated beverage. To check the validity of this figure, you decide to conduct a survey in the region. You randomly sample $n = 400$ consumers and find that $p = {}^{31}/_{400}$ prefer the beverage.

a. Compute μ_p, σ_p^2, and σ_p, assuming the market analyst's claim is correct.

b. If the market analyst's claim is correct, is it likely that a random sample of 400 would produce a value of $p \leq {}^{31}/_{400}$? Explain.

c. Do the results of your survey disagree with the 12% figure given by the market analyst? Explain.

5.67 The following report is from *The Cincinnati Enquirer:*

> Despite scattered violence and a boycott call by the opposition, Pakistanis voted in large numbers Monday for a national legislature. It was the first election in the eight years since President Mohammed Zia ul-Haq seized power.
>
> Zia, who had hundreds of opposition leaders arrested before the vote, says the election is the first step toward an end to martial law and return to democratic government.
>
> The president had said he hoped 40% of the nation's 35 million voters would take part, and as the polls closed it appeared he would get his wish.*

A random sample of the nation's 35 million voters could be taken to investigate whether the 40% figure the president hoped for had been reached. Suppose that twenty-five of these potential voters are randomly selected, and p, the fraction who voted in the election, is calculated.

a. If the 40% figure is correct, what is the probability that p is less than or equal to $^8/_{25}$?

b. Considering the probability calculated in part a, if the sample produced $p = {}^8/_{25}$, would you conclude that the 40% figure is incorrect?

5.68 Refer to Exercise 5.67. A random sample of twenty-five voters is not very large. Suppose that ten times as many potential voters are randomly selected, and p, the fraction who voted in the election, is calculated.

---

* "Voter Turnout High in Pakistani Election," *The Cincinnati Enquirer*, February 26, 1985, p. A-9.

a. If the 40% figure is correct, is it likely that p is less than or equal to $^{80}\!/_{250}$?

b. Considering your answer to part a, if $p = {}^{80}\!/_{250}$ were observed, would you conclude that the 40% figure is incorrect?

c. Note that in this exercise and the previous one, the observed sample fractions, $^{80}\!/_{250}$ and $^{8}\!/_{25}$, are equal. Compare your answers for part b of both exercises. Discuss the difference in your answers.

Chapter Summary

A **statistic** is a number calculated from data in a sample. For each sample in a sample space, the value of the statistic is uniquely determined. A **parameter** is a number calculated from the data in an entire population. Statistics and parameters of particular interest to us are:

$\bar{x}$: Sample mean

s^2: Sample variance

s: Sample standard deviation

μ: Population mean

σ^2: Population variance

σ: Population standard deviation

M: Sample median

R: Sample range

X: Arbitrary statistic of interest

Every statistic has a **sampling distribution,** which is a rule that gives the probability of each value of the statistic. These probabilities must be numbers between 0 and 1, and the sum of the probabilities of all values of the statistic must equal 1.

The mean, μ_X, variance, σ_X^2, and standard deviation, σ_X, of the sampling distribution of any statistic, X, are given by

$$\mu_X = \sum[X \cdot P(X)] \qquad \sigma_X^2 = \sum[X^2 \cdot P(X)] - \mu_X^2 \qquad \sigma_X = \sqrt{\sigma_X^2}$$

These numbers provide descriptive measures of the distribution of X in the same sense that they would for a distribution of data as described in Chapter 3. Here, the data are the values of X.

A **binomial experiment** consists of sampling the observations from an infinite population for which the observations can be classified as either success (S) or failure (F). The fraction of successes in the population is denoted by π, and the fraction of failures is denoted by $(1 - \pi)$. For any sampled observation, the probability of success must be π, and it must not be affected by the outcome of other sampled observations. We are interested in the **binomial statistic,**

$p =$ Fraction of successes observed in sample

If a population has the characteristics required for a binomial experiment, except that it has a finite number N of observations, we can treat the experiment of sampling n

observations as a binomial experiment if n is small compared to both $N \cdot \pi$ and $N \cdot (1 - \pi)$. The sampling distribution of p is given in Appendix Table I for various values of n and π. The mean, variance, and standard deviation for p are given by

$$\mu_p = \pi \qquad \sigma_p^2 = \frac{\pi \cdot (1 - \pi)}{n} \qquad \sigma_p = \sqrt{\sigma_p^2}$$

If we assume a value for π in a binomial experiment and the experiment yields a value of p that is a rare event, then we can infer that our assumption about π was incorrect. The best measure of the rarity of the event of interest is the probability of the event, and we consider probabilities of .05 or less to represent rare events. If we cannot calculate the probability of interest, we will rely on a z-score to measure the rarity of an event. We consider z-scores less than -2 or greater than $+2$ to be rare events.

SUPPLEMENTARY EXERCISES (5.69–5.91)

Learning the Mechanics

5.69 Consider the following sampling distribution:

X	10	12	18	20
$P(X)$	.2	.2	.1	.5

    a. Calculate μ_X, σ_X^2, and σ_X.
    b. What is the probability that $X < 15$?
    c. Calculate $(\mu_X - 2\sigma_X)$ and $(\mu_X + 2\sigma_X)$.
    d. What is the probability that X is in the interval $(\mu_X - 2\sigma_X)$ to $(\mu_X + 2\sigma_X)$?

5.70 A random sample of two observations is taken from a population containing the measurements 1, 5, 9, and 11.
    a. Find the sample space for this experiment.
    b. Find the sampling distribution of $\bar{x}$.
    c. Find the sampling distribution of R.
    d. Calculate $\mu_{\bar{x}}$, $\sigma_{\bar{x}}^2$, and $\sigma_{\bar{x}}$.
    e. Calculate μ_R, σ_R^2, and σ_R.

5.71 A random sample of seventy-five marbles is selected from a jar containing 3,000 red marbles and 7,000 blue marbles. Let p equal the fraction of red marbles observed in the sample.
    a. Explain why p may be considered a binomial statistic.
    b. Calculate μ_p, σ_p^2, and σ_p.
    c. Find the z-score for $p = .43$. Is $p = .43$ a rare event? Explain.

Using the Tables

5.72 If p is a binomial statistic with $n = 20$ and $\pi = .6$, find:

a. $P(p = {}^{14}\!/_{20})$ e. $P({}^2\!/_{20} < p < {}^{17}\!/_{20})$

b. $P(p \leq {}^{10}\!/_{20})$ f. μ_p

c. $P(p > {}^{10}\!/_{20})$ g. σ_p^2

d. $P({}^8\!/_{20} \leq p \leq {}^{17}\!/_{20})$ h. σ_p

i. The probability that p is not in the interval $(\mu_p - 2\sigma_p)$ to $(\mu_p + 2\sigma_p)$

Applying the Concepts

5.73 An important function in any business is long-range planning. For example, additions to a firm's physical plant cannot be achieved overnight—construction must be planned several years in advance. A printing company is planning now for the warehouse space it will need in 5 years. The company anticipates a substantial growth in sales, but it cannot be certain exactly how many square feet of storage space it will need in 5 years. The company decides to estimate its needs using the following sampling distribution:

X = Square feet of storage space needed	10,000	15,000	20,000	25,000*	30,000	35,000
P(X)	.05	.15	.35	.25	.15	.05

What is μ_X, the mean number of square feet of storage space the printing company will need in 5 years?

5.74 An employee of a firm has an option to invest $1,000 in the company's bonds. At the end of 1 year, the company will buy back the bonds at a price determined by its profits for the year. From records of past years, the company predicts it will buy the bonds back at the following prices with the associated probabilities (X = Price paid for bonds):

X	$0	$500	$1,000	$1,500	$2,000
P(X)	.01	.22	.30	.22	.25

a. What is the probability the employee will receive $1,000 or less for the investment?

b. What is the mean price paid for the bonds?

c. What is the employee's mean profit?

d. Find σ_X^2 and σ_X for this sampling distribution.

5.75 Due to pollution, it is thought that 10% of the fish found in a particular river contain a level of mercury that is harmful to humans. To test this theory, environmentalists sample ten fish from the river and analyze each for the presence of a dangerous amount of mercury.

a. If the 10% contamination figure is correct, what is the probability that none of the ten fish contain a dangerous level of the substance?

b. What is the probability that no more than two of the fish contain a dangerous level of the substance?

5.76 Many minor operations at a hospital can properly be performed during the same day the patient is admitted. A hospital serving a large metropolitan area has found that in the past, 20% of newly admitted patients needing an operation are scheduled for same-day surgery. Suppose that ten patients are randomly selected from those admitted to the hospital for surgery over the past year. If p, the fraction in the sample of ten who receive same-day surgery, possesses a binomial sampling distribution:
a. What is the probability that exactly five of these ten patients have same-day surgery?
b. What is the probability that at most one of the ten has same-day surgery?
c. If the ten patients are selected from the admissions on a single given day, will p possess the characteristics of a binomial statistic? Explain.

5.77 The head librarian of a large library claims that 60% of the books in the library have been published since 1960.
a. If twenty books are chosen at random from the library, what is the probability that at least fifteen were published since 1960, assuming the claim is true?
b. What is the probability that fewer than nine books chosen were published after 1960 if the librarian's claim is true?

5.78 Suppose it is known that 10% of all radios produced by a manufacturer have defective tuning mechanisms. Your store receives a large shipment of the radios from which you choose ten radios to inspect. You have decided not to accept the shipment if you discover one or more defective radios. Before inspecting the ten radios, what is the probability that you will not accept the shipment?

5.79 The state highway patrol has determined that one out of every five calls for help originating from roadside call boxes is a hoax. Five calls for help have been received and five tow trucks are dispatched.
a. What is the probability that none of the calls was a hoax?
b. What is the probability that only three of the callers really needed assistance?
c. What assumptions do you have to make in order to solve this problem?

5.80 The owner of construction company A makes bids on jobs so that if awarded the job, the company will make a $10,000 profit. The owner of construction company B makes bids on jobs so that if awarded the job, the company will make a $15,000 profit. Each company describes the sampling distribution of X, the number of jobs the company is awarded per year, as follows:

Company A

X	2	3	4	5	6
P(X)	.05	.15	.20	.35	.25

Company B

X	2	3	4	5	6
P(X)	.15	.30	.30	.20	.05

    a. Find the mean number of jobs each company will be awarded in a year.

    b. What is the mean profit for each company?

    c. Find the variances and the standard deviations of the sampling distributions given above for each company.

    d. Graph $P(X)$ for both companies A and B. For each company, what proportion of the time will X fall in the interval $(\mu_X - 2\sigma_X)$ to $(\mu_X + 2\sigma_X)$?

5.81 A large cigarette manufacturer has determined that the probability a new brand of cigarettes will obtain a large enough market share to make production profitable is .3.

    a. If over the next 3 years this manufacturer introduces one new brand a year, what is the probability that at least one new brand will obtain a sufficient market share to make its production profitable?

    b. What is the probability that all three new brands will obtain a sufficient market share?

    c. What assumptions do you have to make in order to solve this problem?

5.82 If it is known that 10% of the finished products coming off an assembly line are defective, what is the probability that exactly one of the next four products coming off the line is defective? What assumptions do you have to make to solve this problem using the methodology of this chapter?

5.83 In recent years, the use of the telephone as a data collection instrument for public opinion polls has been steadily increasing. However, one of the major factors bearing on the extent to which the telephone will become an acceptable data collection tool in the future is the *refusal rate*—that is, the percentage of the eligible subjects actually contacted who refuse to take part in the poll. Suppose that past records indicate a refusal rate of 20% in a large city. A poll of twenty-five city residents is to be taken and p is the fraction of residents contacted by telephone who refuse to take part in the poll.

    a. Find the mean and variance of p.

    b. Find $P(p \leq 5/25)$.

    c. Find $P(p > 10/25)$.

5.84 The effectiveness of insecticides is often measured by the dose necessary to kill a certain percentage of insects. Suppose that a certain dose of a new insecticide is supposed to kill 80% of the exposed insects. To test the claim, twenty-five insects are put in contact with the insecticide.

    a. If the insecticide really kills 80% of the exposed insects, what is the probability that fewer than fifteen die?

    b. If you observed such a result, what would you conclude about the new insecticide? Explain your logic.

5.85 A physical fitness specialist claims that the probability is greater than .50 that an average adult male can improve his physical condition by spending 5 minutes per day on a certain exercise program. To test the claim, fifteen randomly selected adult males follow the program for a specified amount of time. Maximal oxygen uptake is measured before and after the program for each male and serves as the criterion for assessing physical condition. If the program is not really beneficial (assume the probability of improvement is .50), what is the probability that eleven or more of the fifteen men have improved maximal oxygen uptake? Suppose eleven or more showed increased maximal oxygen uptake. Assuming the specialist's exercise program is ineffective, would you regard $p \geq {}^{11}\!/_{15}$ as a rare event or would you conclude that, in actuality, the probability of improvement exceeds $\pi = .50$ and that the program is effective?

5.86   The following extract appeared in *The Cincinnati Enquirer* on February 25, 1985:

> The unnerving, high-pitched whir of the dentist's drill may soon become a sound of the past, says the head of a dental products company that has introduced a system to dissolve cavities painlessly.
>
> The chemical solution disintegrates cavities as the dentist gently scrapes the decay, said Roderick L. Mackenzie, chairman of Princeton Dental Products Inc. of New Brunswick, N.J. . . .
>
> Mackenzie said about half of all decayed teeth can be chemically treated while some teeth could be treated through a combination of chemical and mechanical techniques. The solution is not harmful to the tooth's healthy enamel, he said.*

A random sample of twenty patients, each having one cavity, is selected; assume that one-half of all decayed teeth can be treated with the chemical.

a.   What is the probability that 15 or more of the sampled patients can be treated with the chemical?

b.   What is the probability that fewer than half of the sampled patients can be treated with the chemical?

5.87   About 50% of all people who quit smoking will start smoking again within half a year.[†] A random sample of 100 ex-smokers is monitored for 6 months, and p, the fraction of people who are still not smoking after 6 months, is calculated.

a.   Find μ_p and σ_p.

b.   What are the smallest and largest fractions of the sampled people that could be expected to be still not smoking after 6 months?

*5.88   The table lists the ten American amusement facilities that attracted the most people in 1985.

Theme Park	1985 Attendance (in millions)
Walt Disney World	22.4
Disneyland	12.0
Knott's Berry Farm	3.5
Kings Island	3.0
Busch Gardens	2.9
Great Adventure	2.9
Six Flags Magic Mountain	2.6
Opryland USA	2.5
Six Flags over Texas	2.5
Great America	2.4

Source: Table, "America's Fun Spots," *Business and Society Review,* Vol. 58, Summer 1986, p. 88.

* "Chemical Melts Decay, Replacing Dental Drill," *The Cincinnati Enquirer*, February 25, 1985, p. D-3.

[†] *Source:*   Dave Beasley, "Quitting Time," *The Cincinnati Enquirer*, February 25, 1985, p. D-1.

A family randomly selects three of these attractions to visit next summer. Let X equal the number of places selected that had an attendance of 3.0 million or more in 1985. The sampling distribution of X is given in the table.

X	0	1	2	3
$P(X)$	.167	.500	.300	.033

Use a computer package to do the following:
a. Obtain a printout of this sampling distribution.
b. Obtain a graph of the distribution of X.
c. Calculate μ_X and σ_X.
d. Find the probability that the value of X is between $\mu_X - 2\sigma_X$ and $\mu_X + 2\sigma_X$.

*5.89   There is much competition in the personal computer (PC) software market. The fifteen most successful producers of PC software are listed in the table.

Company	% Market Share
Lotus	20.2
Ashton-Tate	10.4
IBM	8.9
Apple	6.2
Microsoft	5.6
MicroPro	4.5
Software Publishing	3.2
MultiMate*	2.6
Satellite Software	2.4
Autodesk	2.1
Computer Associates	1.5
Microrim	1.3
Innovative Software	1.2
Monogram	1.1
Open Systems	1.0

Source: Table, "Top Dogs in PC Software," *Business and Society Review,* Vol. 58, Summer 1986, p. 94.

A random sample of 6 of the 15 companies is selected for possible purchase of their software. Let X equal the number of the five top-selling companies selected. The sampling distribution of X follows:

X	0	1	2	3	4	5
$P(X)$	.042	.252	.420	.240	.045	.002

Use a computer package to do the following:
a. Obtain a printout of the sampling distribution of X.

    b. Calculate the values of μ_X and σ_X.

    c. Graph the sampling distribution of X. Mark the values of μ_X, $\mu_X - 2\sigma_X$, and $\mu_X + 2\sigma_X$ on the graph.

*5.90 Let p be a binomial statistic with $n = 12$ and $\pi = .25$.

    a. Use a computer package to obtain the sampling distribution of p.

    b. What are the values of μ_p and σ_p?

    c. Find $P(p \leq \frac{3}{12})$.

*5.91 Let p be a binomial statistic with $n = 17$ and $\pi = .66$.

    a. Use a computer package to obtain the sampling distribution of p.

    b. Find the values of μ_p and σ_p.

    c. Find $P(\mu_p - 2\sigma_p \leq p \leq \mu_p + 2\sigma_p)$.

CHAPTER 5 QUIZ

1. Define the following terms:

    a. Parameter   *b.* Statistic

2. Ninety percent of the students taking an introductory sociology course receive a passing grade. Fifteen students who are taking this course are randomly selected.

    a. What is the probability all fifteen students selected pass the course?

    b. What is the probability more than three of the fifteen selected students fail the course?

    c. What is the average fraction of fifteen randomly selected students that pass the course?

    d. What is the standard deviation of the fraction of fifteen randomly selected students that pass the course?

3. A box contains two pennies, a dime, and a quarter. Two coins are randomly selected, and X, the total value of the two coins, is noted. Find the sampling distribution of X.

4. Tickets in a raffle cost $100 each. The amount of money a person could make with the purchase of one ticket is denoted by the statistic X. The sampling distribution of X is shown below:

X	$400	$100	−$100
P(X)	.1	.2	.7

    a. What is $P(X > 0)$? Explain the meaning of this probability.

    b. Calculate μ_X and σ_X^2.

    c. Explain the meaning of the value of μ_X.

5. A physician thinks that 40% of his patients have illnesses that are psychosomatic in origin.

a. If the physician randomly samples twenty patients, what is the probability that three or fewer of the twenty patients have illnesses of psychosomatic origin?

b. Considering your answer to part *a*, what, if anything, would you infer if three of the twenty patients have psychosomatic illnesses?

CHAPTERS 1–5 CUMULATIVE QUIZ

1. Teachers at a large high school are concerned about the number of students who smoke cigarettes. Two hundred students are selected from the 1,637 students who attend the high school. Each selected student indicates whether he or she smokes cigarettes.
   a. Identify the population of interest.
   b. Identify the sample.
   c. Define a parameter that might be of interest to the teachers.

2. a. Explain what the term *inference* means.
   b. What should accompany any inference that is made?

3. Two numbers are randomly selected from the population containing the numbers 0, 1, 3, and 6.
   a. Give the sample space for this experiment.
   b. What is the probability at least one odd number is selected?
   c. Let *X* equal the sum of the two selected numbers. Give the sampling distribution of *X*.

4. Refer to Question 3. Using the sampling distribution found in part *c*, calculate μ_X, σ_X^2, and σ_X. Draw a graph of the sampling distribution of *X* and locate μ_X, $(\mu_X - 2\sigma_X)$, and $(\mu_X + 2\sigma_X)$ on the graph. What proportion of the time will the value of *X* be within 2 standard deviations of its mean?

5. A garment manufacturer claims that 99% of the garments shipped from the factory are free of defects. A random sample of 10,000 garments that are shipped is selected and *p*, the fraction of these garments with defects, is calculated.
   a. Calculate μ_p, σ_p^2, and σ_p.
   b. If the sample of 10,000 garments yields a value of $p = .02$, would you suspect the manufacturer's claim of 99% is false? [*Hint:* Calculate the z-score for $p = .02$, assuming the manufacturer's claim is valid.]

On Your Own

Consider the following statistics:

1. The fraction of people, *p*, who recover from a certain disease, out of a sample of five patients
2. The fraction of voters, *p*, in a sample of five who favor a method of tax reform
3. The fraction of hits, *p*, a baseball player gets in five official times at bat

In each case, p is a binomial statistic (or approximately so) with $n = 5$ trials. Assume that in each case the probability of success is .3. (What is a success in each of the examples?) Then the sampling distribution for p is the same for each of the three examples. To obtain a relative frequency histogram for p, conduct the following experiment: Place ten poker chips (or pennies, or marbles, or any ten *identical* items) in a bowl and mark three of the ten "success"—the remaining seven will represent failure. Randomly select a chip from the ten, observing whether it was a success or failure. Then return the chip to the bowl, and randomly select a second chip from the ten available chips and record this outcome. Repeat this process until a total of five trials has been conducted. Count the fraction, p, of successes observed in the five trials. Repeat the entire process 100 times to obtain 100 observed values of p.

a. Use the 100 values of p obtained from the simulation to construct a relative frequency histogram for p. Note that this histogram is an approximation to the sampling distribution of p.

b. Find the exact sampling distribution of p for $n = 5$ and $\pi = .3$ and compare these values with the approximations found in part **a.**

c. If you were to repeat the simulation an extremely large number of times (say 100,000), how do you think the relative frequency histogram and true sampling distribution would compare?

THE CENTRAL LIMIT THEOREM AND THE NORMAL DISTRIBUTION

AVERAGE LIFETIME OF LIGHT BULBS

Many manufacturers of consumer products use averages to describe important characteristics of their products. A copy of the front of a carton containing Sylvania light bulbs is shown in Figure 6.1 (page 268). At the bottom of the carton, it says "Avg. Life 1500 Hrs./Avg. Lumens 1585." Sylvania is telling the consumer that the lifetime of the population of all the bulbs of this kind has an average of 1,500 hours and that the average lumens given out for the population of all bulbs is 1,585.* If you buy (sample) n of these light bulbs and measure the lifetime of each bulb, what characteristics will the sampling distribution of the sample mean have? In Chapter 5 we partially answered this question when we introduced sampling distributions and examined the sampling distributions of statistics for relatively small samples. In this chapter we will expand our discussion of sampling distributions to include large samples. As you will see, we can state the characteristics of the sampling distribution of the sample mean based upon a large random sample from any population with any mean and standard deviation.

* A **lumen** is a unit of measure of light.

Figure 6.1
The front of a carton of
Sylvania light bulbs

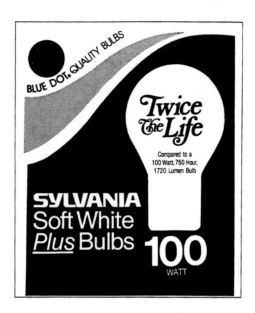

6.1

▬▬▬▬▬▬▬▬   THE CENTRAL LIMIT THEOREM

For the sampling distributions discussed in Chapter 5, the possible values of each statistic and the corresponding probabilities could be listed. Unfortunately, this is not always possible. For example, if the mean age of a sample of 500 Americans were to be observed, the number of possible values of this statistic (the sample mean) would be virtually unlimited. The statistics of interest for many experiments—particularly experiments with large samples—have too many values to list, and we cannot use the methods presented in Chapter 5 to find their sampling distributions. An intuitive introduction to the discussion of sampling distributions of statistics calculated from large samples is given in the following example.

EXAMPLE 6.1   The population of annual salaries of all employees at a large university has a mean $\mu = \$15,000$ and a standard deviation $\sigma = \$8,000$. A graph of this population is shown in Figure 6.2.

a. Discuss the graph of this population.
b. If a random sample of $n = 100$ salaries were observed and the sample mean, $\bar{x}$, calculated, where would you expect the value of $\bar{x}$ to be relative to $\mu = \$15,000$?
c. What would you expect the value of $\bar{x}$ to be if a second random sample of $n = 100$ salaries were chosen from the original population?

Figure 6.2
Graph of population of
university salaries

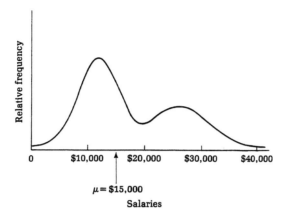

d. Suppose we consider every possible value of $\bar{x}$ that could be observed. What would the sampling distribution of $\bar{x}$ look like? How would the mean and standard deviation of this sampling distribution compare to the mean and standard deviation of the population of university salaries?

Solution a. The graph, which represents the distribution of the population of salaries, has two distinct peaks. This suggests that there may be two populations mixed together. One possible explanation is that the concentration of smaller salaries between $4,000 and $26,000 corresponds to the nonprofessional employees of the university, while the concentration of higher salaries between $20,000 and $40,000 corresponds to the professional employees. The graph is also highly skewed to the right. This can be explained by the fact that certain individuals, such as the president, the football coach, outstanding tenured faculty members, etc., would receive much higher salaries than the other employees.

b. We would expect the sample mean of 100 salaries to be near the population mean of $15,000, but we would not expect it to be exactly equal to $15,000. Although it is possible that a random sample could result in the lowest 100 salaries or the highest 100 salaries, we would be surprised if this happened. Thus, we would be surprised if $\bar{x}$ were a great distance from $\mu = \$15,000$.

c. The same comments made in part **b** could be made for any random sample, but we would expect the values of the sample mean to vary from sample to sample. The value of the sample mean for the second random sample should also be near $15,000, but the value of $\bar{x}$ for the second sample would probably be different from the value of $\bar{x}$ computed for the first sample.

d. Considering the answers to parts **b** and **c**, we would expect most of the possible values of $\bar{x}$ to be near $\mu = \$15,000$. The farther we get from $15,000, the less likely we are to see a value of $\bar{x}$. Also, we would expect the values of $\bar{x}$ to be approximately symmetric about $\mu = \$15,000$. Some would be larger and some would be smaller, but it is reasonable to expect that the average value of all possible values of $\bar{x}$ would be $15,000. In other words, $\bar{x}$ has a sampling distribution that centers about $\mu = \$15{,}000$, with the relative frequency diminishing as we move away from μ. We would also expect the values of $\bar{x}$ to be less variable than the salaries in the original population. We would expect fewer extreme values of $\bar{x}$ than we would of individual salaries, and most of the values of $\bar{x}$ will lie much closer to $\mu = \$15{,}000$ than the original salary measurements. The standard deviation of the sampling distribution of $\bar{x}$ will be smaller than the original population standard deviation of $\sigma = \$8{,}000$. The graph in Figure 6.3 shows a sampling distribution of $\bar{x}$ that reflects all these characteristics.

Figure 6.3
Sampling distribution of $\bar{x}$ for a sample of $n = 100$ university salaries

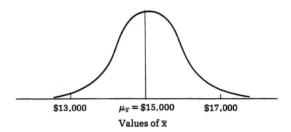

A sampling distribution with the shape of the graph shown in Figure 6.3 is sometimes described as **bell-shaped** or **mound-shaped.** We will refer to such a distribution as a **normal distribution.***

Many experiments generate statistics with sampling distributions that are very well approximated by a normal distribution. A few examples of statistics that could be approximately normally distributed are:

1. Errors made in measuring a person's blood pressure
2. Yearly rainfall at a certain weather station
3. Weights of loads of produce shipped to a supermarket

The normal distribution of such a statistic, X, is perfectly symmetric about its mean, μ_X, and its spread is determined by the value of its standard deviation, σ_X. In fact, there is a normal curve corresponding to every pair of values of μ_X and σ_X. Three of these normal curves are shown in Figure 6.4. You will see in the following sections that we will use the values of μ_X and σ_X to find probabilities associated with normal distributions. In this section we will examine the conditions that must be satisfied for $\bar{x}$ to have a sampling distribution that is approximately normal.

In Example 6.1 we presented an intuitive discussion of the sampling distribution of $\bar{x}$. It can be shown, in general, that the sampling distribution of $\bar{x}$ has the properties given in the box.

---

* The normal distribution curve is defined by the function

$$f(x) = \frac{1}{\sigma\sqrt{2\pi}}\, e^{-\frac{1}{2}[(x-\mu)/\sigma]^2} \qquad -\infty < x < \infty$$

For our purposes, it is not necessary to use this definition.

Figure 6.4
Several normal distributions, with different means and standard deviations

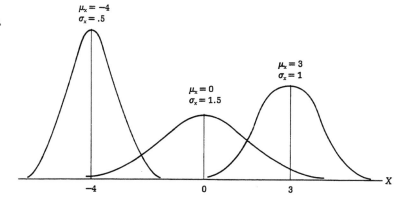

$\mu_x = -4$
$\sigma_x = .5$

$\mu_x = 3$
$\sigma_x = 1$

$\mu_x = 0$
$\sigma_x = 1.5$

Properties of the Sampling Distribution of $\bar{x}$

The sampling distribution of $\bar{x}$ will have the following properties when a random sample of n measurements is selected from *any* population with a mean μ and a standard deviation σ:

1. The mean of the sampling distribution of $\bar{x}$ equals the mean of the population from which the sample is taken. In symbols,

$$\mu_{\bar{x}} = \mu$$

2. The standard deviation of the sampling distribution of $\bar{x}$ equals the standard deviation of the sampled population divided by the square root of the sample size. That is,

$$\sigma_{\bar{x}} = \frac{\sigma}{\sqrt{n}}$$

3. The sampling distribution of $\bar{x}$ is approximately normal for a large sample size, n.

You can see from these properties that our intuitive presentation in Example 6.1 is in agreement with the properties given in the box. Property 1 assures us that the mean of the sampling distribution is equal to the mean of the sampled population. Recall that the mean of the population of salaries is $\mu = \$15,000$. So, $\mu_{\bar{x}} = \$15,000$ also.

Property 2 tells us how to calculate the standard deviation of the sampling distribution of $\bar{x}$. In Example 6.1 we were given the values of $\sigma = \$8,000$ and $n = 100$.

Substituting them into the formula for $\sigma_{\bar{x}}$, we get

$$\sigma_{\bar{x}} = \frac{\sigma}{\sqrt{n}} = \frac{\$8,000}{\sqrt{100}} = \$800$$

The value of the standard deviation of the sampling distribution of $\bar{x}$ is only one-tenth the value of the original population standard deviation.

Finally, property 3 assures us that the shape of the sampling distribution of $\bar{x}$ is approximately normal for a sufficiently large sample size. The justification for property 3 is contained in one of the most important theoretical results in statistics—the **Central Limit Theorem.**

The Central Limit Theorem

The sample mean, $\bar{x}$, of a large random sample from a population with mean, μ, and standard deviation, σ, possesses a sampling distribution that is approximately normal, *regardless of the distribution of the population from which the sample is obtained.* The larger the sample size, the better will be the normal approximation to the sampling distribution of $\bar{x}$.

The graphs in Figure 6.5 provide a visual representation of what the Central Limit Theorem is saying about the sampling distribution of $\bar{x}$. The four graphs at the left of the four rows represent the distributions of four different populations. The second graph in each row shows the sampling distribution of $\bar{x}$ based on a random sample of $n = 2$ measurements selected from the corresponding original population. The third graph in each row shows the sampling distribution of $\bar{x}$ for $n = 5$, and the last graph in each row shows the sampling distribution of $\bar{x}$ for $n = 30$. Notice that as the sample size is increased, the sampling distribution of $\bar{x}$ becomes more mound-shaped and less variable for each of the four original populations. For a sample size of $n = 30$, the four sampling distributions of $\bar{x}$ are very similar, even though the four original populations are quite different. Each of the sampling distributions for $n = 30$ appears to be normal.

EXAMPLE 6.2   A standardized test that is given nationally has a mean score of $\mu = 500$ and a standard deviation of $\sigma = 100$. A random sample of n test scores is selected from all the scores available.

a. Find $\mu_{\bar{x}}$ and $\sigma_{\bar{x}}$ if $n = 9$.
b. Find $\mu_{\bar{x}}$ and $\sigma_{\bar{x}}$ if $n = 400$.
c. Describe the shapes of the sampling distributions of $\bar{x}$ for parts **a** and **b**.
d. What is the largest and smallest value you would expect $\bar{x}$ to equal if a random sample of 400 test scores was selected?

Figure 6.5
Sampling distributions of $\bar{x}$
for different populations and
different sample sizes

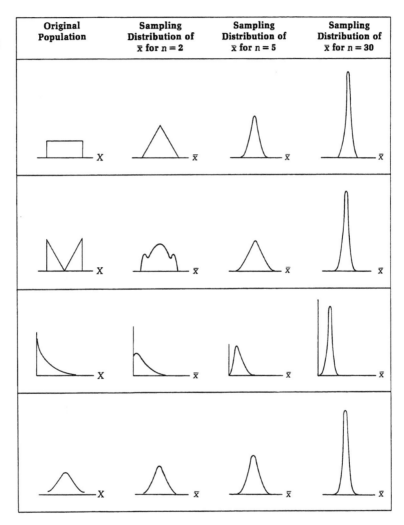

Solution The population from which the random sample is being taken is the collection of scores for all people who have taken the standardized test. We will refer to this population as the **original population** or the **sampled population.** This population has a mean of $\mu = 500$ and a standard deviation of $\sigma = 100$. Now that we have clarified what we know about the original population, we will proceed to answer part **a.**

a. The mean of the sampling distribution of $\bar{x}$ is equal to the mean of the sampled population. Since $\mu = 500$, we also have $\mu_{\bar{x}} = 500$. The standard deviation of the sampling distribution of $\bar{x}$ is the population standard deviation divided by $\sqrt{n}$.

Substituting the values of $\sigma = 100$ and $n = 9$ into the formula, we obtain

$$\sigma_{\bar{x}} = \frac{\sigma}{\sqrt{n}} = \frac{100}{\sqrt{9}} \approx 33.33$$

b. The mean of the sampling distribution is unaffected by sample size, so $\mu_{\bar{x}} = \mu = 500$. The standard deviation of the sampling distribution *does* depend on the sample size:

$$\sigma_{\bar{x}} = \frac{\sigma}{\sqrt{n}} = \frac{100}{\sqrt{400}} = 5$$

c. For part **a**, we cannot state the shape of the sampling distribution of $\bar{x}$. The Central Limit Theorem applies only when the sample size is large, and a sample size of $n = 9$ cannot be considered large.

　　For part **b**, we can state that the sampling distribution of $\bar{x}$ will have an approximately normal distribution. The sample size $n = 400$ is certainly large, and the Central Limit Theorem applies. The normal approximation should be very good since $n = 400$ is such a large sample size.

d. The rule of thumb states that for most sets of data most of the measurements will be within 2 standard deviations of the mean, and all or almost all of the measurements will be within 3 standard deviations of the mean. To be safe, we will choose $(\mu_{\bar{x}} - 3\sigma_{\bar{x}})$ as the smallest value we would expect $\bar{x}$ to be and $(\mu_{\bar{x}} + 3\sigma_{\bar{x}})$ as the largest value. Calculating these values, we obtain:

　　Smallest value expected $= \mu_{\bar{x}} - 3\sigma_{\bar{x}} = 500 - 3(5) = 485$
　　Largest value expected $= \mu_{\bar{x}} + 3\sigma_{\bar{x}} = 500 + 3(5) = 515$

If random samples of 400 test scores were repeatedly selected, very few, if any, of the sample means would be smaller than 485 or larger than 515. Figure 6.6 shows the sampling distribution of $\bar{x}$ and the location of the values 485 and 515.

Figure 6.6
Sampling distribution of $\bar{x}$
for $n = 400$

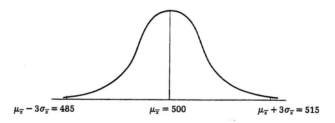

$\mu_{\bar{x}} - 3\sigma_{\bar{x}} = 485$ 　　　　$\mu_{\bar{x}} = 500$ 　　　　$\mu_{\bar{x}} + 3\sigma_{\bar{x}} = 515$ 　　　■

　　For any statistic, there will always be an original population from which a sample must be taken to calculate the statistic. This population will have a mean, μ, and a standard deviation, σ. There will also be a sampling distribution of the statistic, and it will have a mean and standard deviation. To help make a distinction between the standard deviation of the original population and the standard deviation of a sampling distribution, the latter is often called the **standard error of a statistic.**

Definition 6.1

The standard deviation of the sampling distribution of a statistic is called the **standard error of the statistic.**

When discussing the sampling distribution of the sample mean, $\bar{x}$, three sets of data must be distinguished. These are the original population, a sample of measurements (selected from the original population), and the sampling distribution of the sample mean. The terminology and notation used to discuss these three sets of data are very similar and can be confusing. To help reduce this possible confusion, a summary of the notation and terminology is presented in the boxes here and at the top of the next page.

Summary of Terminology and Notation

Original Population

The set of data from which the sample is taken

    Shape: The original population could have any shape

    Mean: μ

    Standard deviation: σ

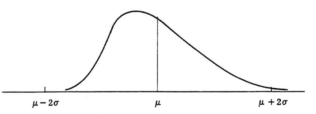

$\mu - 2\sigma$ μ $\mu + 2\sigma$

Large Random Sample

A sample of n measurements obtained from the original population

    Shape: Similar to the original population

    Mean: $\bar{x}$

    Standard deviation: s

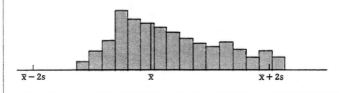

$\bar{x} - 2s$ $\bar{x}$ $x + 2s$

Sampling Distribution of $\bar{x}$

This set of data contains each possible value of the sample mean, $\bar{x}$, corresponding to all possible samples of n measurements.

Shape: Approximately normal (for large random samples)

Mean: $\mu_{\bar{x}} = \mu$

Standard error: $\sigma_{\bar{x}} = \dfrac{\sigma}{\sqrt{n}}$

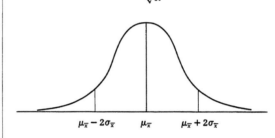

$$\mu_{\bar{x}} - 2\sigma_{\bar{x}} \qquad \mu_{\bar{x}} \qquad \mu_{\bar{x}} + 2\sigma_{\bar{x}}$$

The original population is the population from which a sample is taken. The distribution of data in the original population could have any shape, and to demonstrate that it does not have to be normally distributed, we have drawn a graph that is skewed to the right. The mean and standard deviation of the original population are denoted by μ and σ, respectively.

To apply the Central Limit Theorem, the sample must be large and random. A histogram of the sampled data would be similar to the graph of the original population, as shown above. (The degree of similarity would depend on the particular sample selected.) Statistics such as the sample mean, $\bar{x}$, and the sample standard deviation, s, could be calculated if the actual data in a sample were available.* The value of $\bar{x}$ could be used to estimate μ, and the value of s could be used to estimate σ.

If we repeatedly obtained large random samples of n measurements from the original population and calculated the sample mean, $\bar{x}$, for each sample, we would generate the sampling distribution of $\bar{x}$. The Central Limit Theorem assures us that the sampling distribution will be approximately normally distributed. The mean of the sampling distribution of $\bar{x}$ is denoted by $\mu_{\bar{x}}$ and is equal to μ, the mean of the original population. The standard error (standard deviation) of $\bar{x}$ is denoted by $\sigma_{\bar{x}}$ and equals the standard deviation of the original population divided by the square root of the sample size.

When applying the Central Limit Theorem, it is essential that you clearly identify the original population, the sample, the sampling distribution of the sample mean,

---

* In Chapter 3 we discussed the calculation of $\bar{x}$ and s. The formulas for these quantities are

$$\bar{x} = \frac{\sum x}{n} \quad \text{and} \quad s = \sqrt{\frac{n \sum x^2 - (\sum x)^2}{n(n-1)}}$$

and the notation associated with these sets of data. You should refer to the summary on pages 275–276 whenever you need to refresh your memory about these details. Before concluding with a computer illustration, we offer the following comments:

1. We will assume that the sampling distribution of $\bar{x}$ has an approximately normal distribution if a random sample of size $n = 30$ or more is obtained. In other words, we will consider a sample of thirty or more measurements to be a sufficiently large sample for the Central Limit Theorem to apply. The graphs in Figure 6.5 provide some justification for the choice of $n = 30$, and empirical evidence also supports this choice. Only if the original population is extremely skewed would a sample size larger than thirty be necessary to guarantee the approximate normality of the sampling distribution of $\bar{x}$.

2. If the original population has a normal distribution, the sampling distribution of $\bar{x}$ is exactly normally distributed. This is true for any sample size.

3. Throughout this section we have assumed (and in the remainder of the text we will assume) that the sampled population either is infinite or contains an extremely large number of measurements. If a finite population contains N measurements and a random sample of n measurements is selected, the standard error of $\bar{x}$ will depend on the sample size, n, *and* the population size, N. For a finite population, $\sigma/\sqrt{n}$ must be multiplied by a "correction factor" that takes into account the relationship between n and N. Therefore, the standard error of $\bar{x}$ for a finite population is given by

$$\sigma_{\bar{x}} = \left(\frac{\sigma}{\sqrt{n}}\right)\sqrt{\frac{N-n}{N-1}}$$

If N is extremely large relative to n, then the value of the term $\sqrt{(N-n)/(N-1)}$ will be very near 1, and $\sigma_{\bar{x}} \approx \sigma/\sqrt{n}$. In this text we will always assume this is the case.

COMPUTER EXAMPLE 6.1

Figure 6.6 in Example 6.2 is a graph of the sampling distribution of $\bar{x}$. It is a normal distribution with a mean of 500 and a standard deviation of 5. With the aid of Minitab, a computer-generated portrayal of this same graph was obtained, as shown in the accompanying printout.

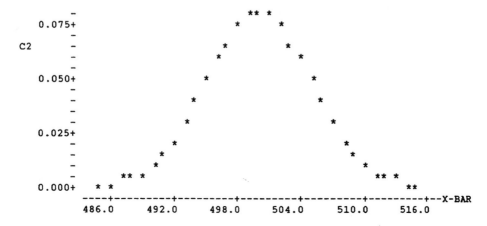

Comparing this graph to the normal curve shown in Figure 6.6, we see that they are very similar. Although the computer-generated graph is not a neat, smooth curve, it still gives a reasonable representation of the shape of a normal curve.   ■

Exercises (6.1–6.14)

Learning the Language

6.1 Define the term *standard error of a statistic*.

6.2 State the Central Limit Theorem.

6.3 If a random sample of n measurements is selected from a population with mean μ, and standard deviation σ, give the properties of the sampling distribution of $\bar{x}$.

6.4 What do the following symbols denote?

a. n c. σ e. $\bar{x}$ g. s^2 i. $\sigma_{\bar{x}}$

b. μ d. σ^2 f. s h. $\mu_{\bar{x}}$ j. $\dfrac{\sigma}{\sqrt{n}}$

Learning the Mechanics

6.5 Suppose that a random sample of n measurements is selected from a population with mean $\mu = 50$ and variance $\sigma^2 = 70$. For each of the following values of n, give the mean and standard error of the sampling distribution of the sample mean, $\bar{x}$.

a. $n = 10$ d. $n = 70$
b. $n = 25$ e. $n = 1{,}000$
c. $n = 100$ f. $n = 400$

6.6 Suppose that a random sample of $n = 100$ measurements is selected from a population with mean, μ, and standard deviation, σ. For each of the following values of μ and σ, give the values of $\mu_{\bar{x}}$ and $\sigma_{\bar{x}}$.

a. $\mu = 10$, $\sigma = 20$ c. $\mu = 50$, $\sigma = 300$
b. $\mu = 20$, $\sigma = 10$ d. $\mu = 30$, $\sigma = 200$

6.7 Refer to Exercise 6.6. Sketch a graph of the sampling distribution of $\bar{x}$ for parts a and c. Locate $\mu_{\bar{x}}$, $(\mu_{\bar{x}} - 3\sigma_{\bar{x}})$, and $(\mu_{\bar{x}} + 3\sigma_{\bar{x}})$ on these graphs.

6.8 Will the sampling distribution of $\bar{x}$ always be approximately normally distributed? Explain.

6.9 A random sample of $n = 800$ observations is selected from a population with $\mu = 100$ and $\sigma = 10$.

a. What are the largest and smallest values of $\bar{x}$ that you would expect to see?
b. How far, at the most, would you expect $\bar{x}$ to deviate from μ?
c. Did you have to know μ to answer part b? Explain.

Applying the Concepts

6.10 A can manufacturing company reports that the mean number of breakdowns per 8-hour shift on its machine-operated assembly line is 1.5. The standard deviation of the number of breakdowns is 1.1. A random sample of fifteen 8-hour shifts is selected, and the number of breakdowns on each shift is recorded. The sample mean, $\bar{x}$, is then calculated.

a. What is the original population in this exercise? What are the mean and standard deviation of this population?

b. What are the mean and standard error of the sampling distribution of $\bar{x}$?

c. Will the sampling distribution of $\bar{x}$ be approximately normal? Explain.

6.11 The Environmental Protection Agency (EPA) issues standards on air and water pollution that vitally affect the safety of consumers and the operations of industry. For example, the EPA states that manufacturers of vinyl chloride and similar compounds must limit the amount of these chemicals in plant air emissions to 10 parts per billion. One plant that manufactures vinyl chloride has a mean emission of 6 parts per billion and a standard deviation of 2.2 parts per billion. A random sample of air emissions from this plant for 40 working days is observed.

a. What is the shape of the sampling distribution of the sample mean air emission for a sample of 40 days? Why?

b. What are the values of $\mu_{\bar{x}}$ and $\sigma_{\bar{x}}$?

c. Is it likely that the mean emission for a sample of 40 days would exceed the EPA limit of 10 parts per billion? Explain.

6.12 To meet individual energy needs, different people require different amounts of nutrients. A graphic representation of the possible distribution of the level of requirement is shown in the accompanying figure.*

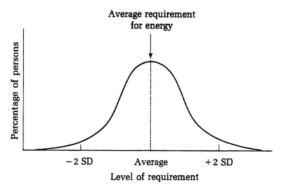

a. What is the shape of this distribution?

b. What can be said about the percentage of people having an energy requirement within 2 standard deviations of the mean? (In the graph the points 2 standard deviations from the mean (average) are labeled −2 SD and +2 SD.)

6.13 The average length of time a customer must wait for a clerk's assistance in a certain department store is 3.3 minutes, and the standard deviation is 2.6 minutes. A random

* *Source:* Adapted from George H. Beaton, "Uses and Limits of the Use of the Recommended Dietary Allowances for Evaluating Dietary Intake Data," *American Journal of Clinical Nutrition,* Vol. 41, January 1985, p. 156.

sample of 100 customers is observed in this department store. The mean length of time these customers must wait for a clerk for help is calculated.

a. What are the values of μ and σ? What population do these parameters describe?

b. What is the shape of the sampling distribution of the mean length of time the 100 customers must wait for help? Why?

c. Calculate $\mu_{\bar{x}}$ and $\sigma_{\bar{x}}$.

d. Graph the sampling distribution of $\bar{x}$ and locate the values $\mu_{\bar{x}}$, $(\mu_{\bar{x}} - 3\sigma_{\bar{x}})$, and $(\mu_{\bar{x}} + 3\sigma_{\bar{x}})$ on this graph.

*e. Repeat part d using a computer package.

6.14 The World Health Organization's recommended daily minimum for calories is 2,600 per individual. The average number of calories ingested per capita per day for countries worldwide, however, is approximately 2,460, and the standard deviation is approximately 300.

a. What is the shape of the sampling distribution of the sample mean number of calories per capita per day in a random sample of thirty countries?

b. What are the values of the mean and standard error of this sampling distribution?

*c. Use a computer package to obtain a graph of the sampling distribution described in parts a and b.

6.2

CALCULATING PROBABILITIES FOR THE SAMPLE MEAN

In Section 6.1 we stated that the sampling distribution of $\bar{x}$ is approximately normal when a large random sample is taken from any population. Our goal is to use the sampling distribution of $\bar{x}$ to make inferences about the population from which the sample is obtained. To do this, we must be able to use the sampling distribution of $\bar{x}$ to find the probabilities of events of interest. Since the values of $\bar{x}$ are too numerous to list when the sampling distribution is approximately normal, we cannot use the methods presented in Chapter 5 to find probabilities for a normal distribution. Probabilities are related to the normal curve by the following definition:

Definition 6.2

If a statistic X has a normal distribution and one value of X is randomly selected from the distribution, then the **probability that X is between any two numbers a and b** is the corresponding area under the normal curve between a and b.

Understanding the rationale for this definition is not essential for applying it to calculate probabilities.

In Figure 6.7, the area under the normal curve between two values, a and b, is shaded. If the sampling distribution of $\bar{x}$ is normal, as shown in Figure 6.7, then the

Figure 6.7
Probability as an area

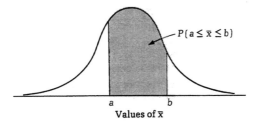

$P(a \leq \bar{x} \leq b)$

a   *b*

Values of $\bar{x}$

total area under the curve is 1, and the probability that a random sample would produce a value of $\bar{x}$ between *a* and *b* is equal to the area that is shaded.

But how do we find these areas? The actual calculation of such areas requires sophisticated mathematical methods. Fortunately, these areas have been tabulated and are presented in Appendix Table II. Although an infinitely large number of normal curves (one for each pair of values for the mean and standard error of $\bar{x}$) may be used to describe the various sampling distributions of $\bar{x}$, Table II was set up so that it may be used with any normal distribution. This was accomplished by constructing the table of areas in terms of z-scores. In symbolic form, the z-score corresponding to the sample mean, $\bar{x}$, is

$$z = \frac{\bar{x} - \mu_{\bar{x}}}{\sigma_{\bar{x}}}$$

A partial reproduction of Table II is shown in Table 6.1. Notice that z-scores are given in the left-hand column and top row of the table. The first decimal place of a

Table 6.1

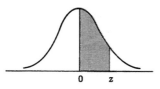

0   *z*

A partial reproduction of Appendix Table II, Normal Curve Areas

z	.00	.01	.02	.03	.04	.05	.06	.07	.08	.09
1.0	.341	.344	.346	.348	.351	.353	.355	.358	.360	.362
1.1	.364	.367	.369	.371	.373	.375	.377	.379	.381	.383
1.2	.385	.387	.389	.391	.393	.394	.396	.398	.400	.401
1.3	.403	.405	.407	.408	.410	.411	.413	.415	.416	.418
1.4	.419	.421	.422	.424	.425	.426	.428	.429	.431	.432
1.5	.433	.434	.436	.437	.438	.439	.441	.442	.443	.444
1.6	.445	.446	.447	.448	.449	.451	.452	.453	.454	.454
1.7	.455	.456	.457	.458	.459	.460	.461	.462	.462	.463
1.8	.464	.465	.466	.466	.467	.468	.469	.469	.470	.471
1.9	.471	.472	.473	.473	.474	.474	.475	.476	.476	.477
2.0	.477	.478	.478	.479	.479	.480	.480	.481	.481	.482

z-score may be found in the left-hand column, and the second decimal place may be found at the top of the table. For example, the first and second decimal places of a z-score of 1.37 are shaded in Table 6.1. The entry in the body of the table corresponding to a z-score of 1.37 is .415, and this value is also shaded in Table 6.1.

This entry, .415, is the area under the normal curve between the mean and a value 1.37 standard deviations above the mean. (Recall that a z-score measures the distance between a measurement and its mean in units of standard deviations.) Figure 6.8 shows the sampling distribution of a statistic, $\bar{x}$, that is approximately normal, with mean $\mu_{\bar{x}}$ and standard deviation $\sigma_{\bar{x}}$. The area under this curve between $\mu_{\bar{x}}$ and $(\mu_{\bar{x}} + 1.37\sigma_{\bar{x}})$ is .415. If the experiment producing this sampling distribution were performed, the probability that the value of $\bar{x}$ would be between $\mu_{\bar{x}}$ and $(\mu_{\bar{x}} + 1.37\sigma_{\bar{x}})$ is .415.

Figure 6.8
Normal distribution of $\bar{x}$

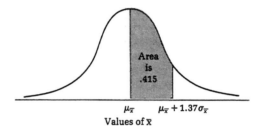

We have illustrated that Appendix Table II may be used to find the probability that $\bar{x}$ is between the mean and a value larger than the mean when the sampling distribution of $\bar{x}$ is approximately normal. This is the *only* type of probability that may be found directly from Table II. To find other areas under the normal curve, the general properties of the normal distribution listed in the box will be useful. By applying these properties, we will be able to make more extensive use of Table II.

Properties of the Normal Distribution

1. The total area under the normal curve is 1.
2. The distribution is perfectly symmetric about the mean. Half of the area under the normal curve lies above the mean, and half lies below.
3. There is no area defined above a single point (number), so the probability associated with a particular number is equal to 0.

EXAMPLE 6.3   A random sample of $n = 300$ measurements is selected from a population with mean $\mu = 50$ and standard deviation $\sigma = 20$. Find each of the following probabilities, where $\bar{x}$ is the mean of the sample of 300 measurements:

a. $P(50 < \bar{x} < 52)$
b. $P(50 \le \bar{x} \le 52)$
c. $P(48 < \bar{x} < 50)$

Solution   To find any probabilities involving $\bar{x}$, we must first specify its sampling distribution. To accomplish this, we begin by summarizing the information we have about the original population.

Original Population

  Shape: Unknown

  Mean: $\mu = 50$

  Standard deviation: $\sigma = 20$

Since the experiment is to take a random sample of $n = 300$ measurements from this population, we can state the properties of the sampling distribution of $\bar{x}$ as discussed in Section 6.1.

Sampling Distribution of $\bar{x}$

  Shape: Approximately normal

  Mean: $\mu_{\bar{x}} = \mu = 50$

  Standard error: $\sigma_{\bar{x}} = \dfrac{\sigma}{\sqrt{n}} = \dfrac{20}{\sqrt{300}} \approx 1.15$

We may now proceed to use Table II to find the probabilities of interest.

a. We want to find the probability that $\bar{x}$ falls between the mean $\mu_{\bar{x}} = 50$ and 52, a value larger than the mean. This area, which is shaded in Figure 6.9, is exactly the type of area we can find in Table II. Calculating the z-score for $\bar{x} = 52$, we obtain

$$z = \frac{\bar{x} - \mu_{\bar{x}}}{\sigma_{\bar{x}}} = \frac{52 - 50}{1.15} \approx 1.74$$

Figure 6.9
Shaded area corresponding
to $P(50 < \bar{x} < 52)$

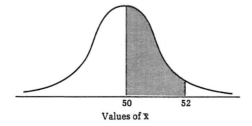

50   52
Values of $\bar{x}$

In the partial reproduction of Table II shown in Table 6.1, we find that the area corresponding to a z-score of 1.74 is .459. That is,

$$P(50 < \bar{x} < 52) = .459$$

b. The probability $P(50 \leq \bar{x} \leq 52)$ is identical to $P(50 < \bar{x} < 52)$. The probability does not change when we include the two values 50 and 52, because the area above each of these points is 0. Thus, we obtain the same answer as in part **a**:

$$P(50 \leq \bar{x} \leq 52) = P(50 < \bar{x} < 52) = .459$$

c. We want to find the probability that $\bar{x}$ falls between the mean $\mu_{\bar{x}} = 50$ and 48, a value smaller than the mean. In Figure 6.10, you can see that 48 is the same distance below $\mu_{\bar{x}} = 50$ as 52 is above it. The z-score corresponding to 48 is

$$z = \frac{\bar{x} - \mu_{\bar{x}}}{\sigma_{\bar{x}}} = \frac{48 - 50}{1.15} \approx -1.74$$

Figure 6.10
Distribution of $\bar{x}$ for
$\mu_{\bar{x}} = 50$ and $\sigma_{\bar{x}} = 1.15$

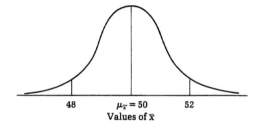

48          $\mu_{\bar{x}} = 50$ 52
Values of $\bar{x}$

This indicates that 48 is 1.74 standard deviations *below* $\mu_{\bar{x}} = 50$, which is the same distance 52 is *above* $\mu_{\bar{x}} = 50$. Since the normal distribution is symmetric,

$$P(48 < \bar{x} < 50) = P(50 < \bar{x} < 52) = .459$$ ∎

The method used to find the probability in Example 6.3c may be used in general.

Helpful Hint

Since the normal distribution is symmetric, if you want to find an area between the mean and a value smaller than the mean, find the answer by ignoring the minus sign of the z-score. That is, find the tabulated area corresponding to the positive value of the z-score.

EXAMPLE 6.4   Suppose the sampling distribution of $\bar{x}$ is approximately normal, with $\mu_{\bar{x}} = 1.2$ and $\sigma_{\bar{x}} = .4$.

a. Find $P(.96 < \bar{x} < 1.76)$.

b. Find $P(1.8 < \bar{x} < 2.04)$.

Solution **a.** A sketch of the desired area is shown in Figure 6.11. We cannot find this area directly from Table II. However, we can use the table to find the area between .96 and 1.2 and the area between 1.2 and 1.76. The z-score for .96 is

$$z = \frac{\bar{x} - \mu_{\bar{x}}}{\sigma_{\bar{x}}} = \frac{.96 - 1.2}{.4} = -.60$$

Figure 6.11

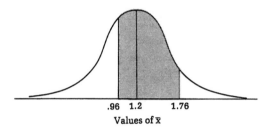

Values of $\bar{x}$

The z-score for 1.76 is

$$z = \frac{\bar{x} - \mu_{\bar{x}}}{\sigma_{\bar{x}}} = \frac{1.76 - 1.2}{.4} = 1.40$$

The areas corresponding to these z-scores are obtained from Table II and are shown in Figure 6.12.

Figure 6.12

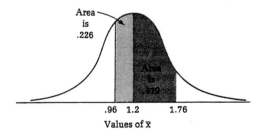

Area is .226

Area is .419

Values of $\bar{x}$

The total area between .96 and 1.76 is the sum of the two areas shown in Figure 6.12. Thus,

Total area = .226 + .419 = .645

Or,

$$P(.96 < \bar{x} < 1.76) = .645$$

b. A sketch of the desired area is shown in Figure 6.13. Again, we cannot find this area directly. However, we can find the area between 1.2 and 1.8 and the area between 1.2 and 2.04. The z-score for 1.8 is

$$z = \frac{\bar{x} - \mu_{\bar{x}}}{\sigma_{\bar{x}}} = \frac{1.8 - 1.2}{.4} = 1.50$$

The z-score for 2.04 is

$$z = \frac{\bar{x} - \mu_{\bar{x}}}{\sigma_{\bar{x}}} = \frac{2.04 - 1.2}{.4} = 2.10$$

Figure 6.13

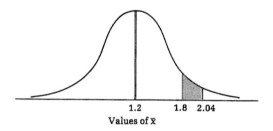

1.2   1.8 2.04
Values of $\bar{x}$

The areas corresponding to these z-scores are obtained from Table II and are shown in Figure 6.14. We are looking for the area *between* 1.8 and 2.04, so we obtain this area by subtraction:

$$.482 - .433 = .049$$

Or,

$$P(1.8 < \bar{x} < 2.04) = .049$$

Figure 6.14

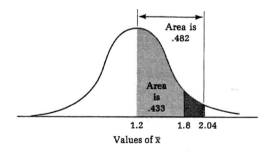

Area is
.482

Area
is
.433

1.2   1.8 2.04
Values of $\bar{x}$

Example 6.4 leads to the generalizations given in the box.

Helpful Hints

1. To find the area under a normal curve between a value *below the mean* and a value *above the mean*, use Table II to find the areas between the mean and each of these two values. The desired area is then found by adding these two areas.
2. To find the area under a normal curve between two values that are *on the same side of the mean*, find the areas between the mean and each of these values using Table II. Then, subtract the smaller area from the larger to find the desired area.

EXAMPLE 6.5   Suppose $\bar{x}$ has a sampling distribution that is approximately normal, with mean $\mu_{\bar{x}} = 20$ and standard error $\sigma_{\bar{x}} = 5$.

    a. Find $P(\bar{x} > 8.5)$.
    b. Find $P(\bar{x} < 8.5)$.
    c. Find $P(\bar{x} > 36.35)$.

Solution   **a.** A sketch of the desired area is shown in Figure 6.15. From the figure, we can see that the shaded area is composed of an area below the mean and all the area above the mean. To get the area of interest, we should add these two areas together. The z-score for $\bar{x} = 8.5$ is

$$z = \frac{\bar{x} - \mu_{\bar{x}}}{\sigma_{\bar{x}}} = \frac{8.5 - 20}{5} = -2.30$$

Figure 6.15

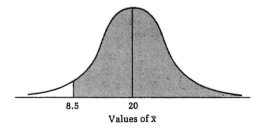

8.5   20
Values of $\bar{x}$

Using Table II, we find that the area corresponding to a z-score of −2.30 is .489. The area above the mean is equal to .500. These areas are shown in Figure 6.16 (page 288). Adding the areas together, we obtain

$$P(\bar{x} > 8.5) = .489 + .500 = .989$$

Figure 6.16

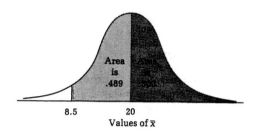

8.5   20
Values of $\bar{x}$

b. The area below 8.5 is shown in Figure 6.17. Since the total area under the curve is 1 and in part **a** we found that the area above 8.5 is .989, we can obtain the desired answer simply by subtracting:

$$P(\bar{x} < 8.5) = 1 - .989 = .011$$

Figure 6.17

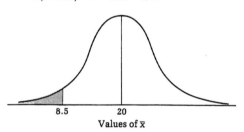

8.5   20
Values of $\bar{x}$

We could also find this answer by subtracting the area between 8.5 and 20 from .500, the total area to the left of the mean of 20. Of course, we would arrive at the same answer:

$$P(\bar{x} < 8.5) = .500 - .489 = .011$$

c. The area above 36.35 is shown in Figure 6.18. The z-score for $\bar{x} = 36.35$ is

$$z = \frac{\bar{x} - \mu_{\bar{x}}}{\sigma_{\bar{x}}} = \frac{36.35 - 20}{5} = \frac{16.35}{5} = 3.27$$

Figure 6.18

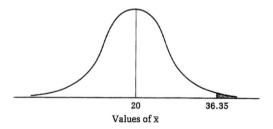

20   36.35
Values of $\bar{x}$

From Table II, we find that the area corresponding to a z-score of 3.27 is .499. Since this is the area between 20 and 36.35, we find the area (probability) of interest by subtracting:

$$P(\bar{x} > 36.35) = .500 - .499 = .001$$

Some general guidelines for using the normal distribution are given in the box.

Helpful Hints

When using the normal distribution to approximate the sampling distribution of $\bar{x}$ and to calculate probabilities for $\bar{x}$, the steps below should always be followed.

1. Specify the values of $\mu_{\bar{x}}$ and $\sigma_{\bar{x}}$.
2. Sketch a normal curve and shade the area of interest.
3. Calculate appropriate z-scores using the formula

$$z = \frac{\bar{x} - \mu_{\bar{x}}}{\sigma_{\bar{x}}}$$

4. Use Appendix Table II to find the area under the normal curve corresponding to each calculated z-score.
5. Refer to the sketch (step 2), and give the probability of interest as an area directly from Table II or by adding or subtracting appropriate areas.

EXAMPLE 6.6   The scores for an IQ test have a mean of $\mu = 100$ and a standard deviation of $\sigma = 15$. If the IQ test were given to a random sample of 100 people, what is the probability that these 100 people would have a mean test score larger than 103?

Solution   Since the sample size is $n = 100$, the sampling distribution of the sample mean, $\bar{x}$, will be approximately normal. The population of all possible IQ test scores from which the sample is selected has a mean $\mu = 100$ and a standard deviation $\sigma = 15$. Thus, the mean and standard error of the sampling distribution of $\bar{x}$ are $\mu_{\bar{x}} = 100$ and

$$\sigma_{\bar{x}} = \frac{\sigma}{\sqrt{n}} = \frac{15}{\sqrt{100}} = 1.5$$

We want to find the probability that $\bar{x}$ is greater than 103, that is, $P(\bar{x} > 103)$. The normal curve and the shaded area of interest are shown in Figure 6.19.

Figure 6.19

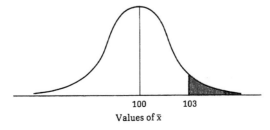

100   103
Values of $\bar{x}$

Converting $\bar{x} = 103$ to a z-score, we obtain

$$z = \frac{\bar{x} - \mu_{\bar{x}}}{\sigma_{\bar{x}}} = \frac{103 - 100}{1.5} = 2.00$$

From Figure 6.19 we can see that in order to find the area of interest, we should subtract the area corresponding to $z = 2.00$ from .500 (the area above the mean). From
Table II, the area corresponding to $z = 2.00$ is .477. So,

$$P(\bar{x} > 103) = .500 - .477 = .023$$

Thus, only a little more than 2% of the time would a random sample of 100 people
have a mean score above 103 on this IQ test. Notice that the z-score for $\bar{x} = 103$ is
$z = 2.00$, which means that $\bar{x} = 103$ is 2 standard deviations above the mean $\mu_{\bar{x}} = 100$.
As with any distribution, only a few measurements are more than 2 standard deviations
above the mean of a normal distribution. More specifically, for a normal distribution,
only about 2% of the measurements fall more than 2 standard deviations above the
mean.   ■

EXAMPLE 6.7   In the introductory example to this chapter, we indicated that Sylvania advertises an
average lifetime of 1,500 hours for a certain type of light bulb. If the standard deviation of the lifetimes of these bulbs is 100 hours, what is the probability that a random
sample of forty bulbs has a mean lifetime of less than 1,490 hours?

Solution   The sample size $n = 40$ is rather large, so the sampling distribution of the sample
mean, $\bar{x}$, will be approximately normal. The population of all bulb lifetimes from which
the sample is selected has a mean of 1,500 hours and a standard deviation of 100
hours. The mean of the sampling distribution of $\bar{x}$ is $\mu_{\bar{x}} = 1,500$ and the standard
error is

$$\sigma_{\bar{x}} = \frac{\sigma}{\sqrt{n}} = \frac{100}{\sqrt{40}} \approx 15.81$$

The probability of interest is that $\bar{x}$ is less than 1,490, that is, $P(\bar{x} < 1,490)$. The shaded
area representing this probability is shown in Figure 6.20.
Calculating the z-score corresponding to $\bar{x} = 1,490$, we get

$$z = \frac{\bar{x} - \mu_{\bar{x}}}{\sigma_{\bar{x}}} = \frac{1,490 - 1,500}{15.81} = -.63$$

Figure 6.20

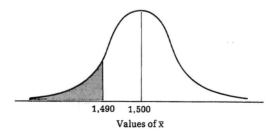

1,490   1,500
Values of $\bar{x}$

The area corresponding to $z = .63$ (the area between 1,490 and 1,500) is found in Table II to be .236. To get the probability of interest, we subtract this probability from .500:

$$P(\bar{x} < 1{,}490) = .500 - .236 = .264$$

Thus, a little over 25% of the time a random sample of forty of these light bulbs would have a mean lifetime less than 1,490 hours. We should not be surprised to see such a result. ■

COMPUTER EXAMPLE 6.2

In Example 6.7 we considered a sampling distribution of $\bar{x}$ that was approximately normally distributed. The values of the mean and standard error were found to be $\mu_{\bar{x}} = 1{,}500$ and $\sigma_{\bar{x}} \approx 15.81$. We then found that $P(\bar{x} < 1{,}490) = .264$. We will now use Minitab to recalculate this probability.

After inputting the values of $\mu_{\bar{x}} = 1{,}500$, $\sigma_{\bar{x}} = 15.81$, and $\bar{x} = 1{,}490$, we obtained the following information on the computer printout:

```
1.49E+03     0.2635
```

Minitab first reports the value of $\bar{x}$ (1,490 equals 1.49E + 03 in scientific notation). Also given is the probability (to four decimal places) of interest, namely, $P(\bar{x} < 1{,}490) =$ .2635. In Example 6.7 we found the probability to be .264, since Table II gives probabilities to only three decimal places.

It should be noted that Minitab will only (directly) provide the probability that a normally distributed sample mean is less than (less than or equal to) some value. Thus, for any value, say c, we can use Minitab to obtain $P(\bar{x} < c)$. This area under a normal curve is shown in Figure 6.21. In the next computer example we will use Minitab to find a probability that is not of the form $P(\bar{x} < c)$.

Figure 6.21
Shaded area showing probability provided by Minitab

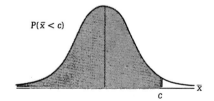

$P(\bar{x} < c)$

■

COMPUTER EXAMPLE 6.3

Refer to Example 6.6. The sampling distribution of $\bar{x}$ was approximately normal with $\mu_{\bar{x}} = 100$ and $\sigma_{\bar{x}} = 1.5$. We found that $P(\bar{x} > 103) = .023$. We cannot find this probability directly using Minitab. We can, however, find $P(\bar{x} \leq 103)$. Using the values of $\mu_{\bar{x}}$ and $\sigma_{\bar{x}}$ for this example, we obtained

```
103.0000     0.9772
```

Thus, $P(\bar{x} \leq 103) = .9772$. Since the total area under the normal curve is 1,

$$P(\bar{x} > 103) = 1 - P(\bar{x} \leq 103) = 1 - .9772 = .0228$$
■

Exercises (6.15–6.29)

Learning the Language

6.15 Define the term *probability* for a statistic with a normal sampling distribution.

6.16 What is denoted by the following formula?

$$z = \frac{\bar{x} - \mu_{\bar{x}}}{\sigma_{\bar{x}}}$$

Learning the Mechanics

6.17 A random sample of $n = 90$ measurements is taken from a population with $\mu = 15$ and $\sigma = 6$. Calculate the z-score for each of the following values of the sample mean, $\bar{x}$:

a. $\bar{x} = 15.0$ d. $\bar{x} = 14.5$
b. $\bar{x} = 17.3$ e. $\bar{x} = 15.5$
c. $\bar{x} = 16.1$ f. $\bar{x} = 13.0$

6.18 A random sample of $n = 400$ measurements is selected from a population with $\mu = 35.6$ and $\sigma^2 = 140.3$. Calculate the z-score for each of the following values of the sample mean, $\bar{x}$:

a. $\bar{x} = 34.4$ d. $\bar{x} = 35.0$
b. $\bar{x} = 40.1$ e. $\bar{x} = 34.1$
c. $\bar{x} = 33.2$ f. $\bar{x} = 37.3$

Using the Tables

6.19 If $\bar{x}$ is normally distributed with $\mu_{\bar{x}} = 60$ and $\sigma_{\bar{x}} = 10$, find the shaded area for each of the following:

a.

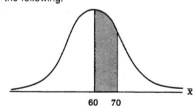

60   70

b.

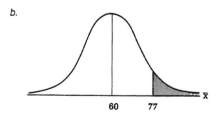

60   77

c.

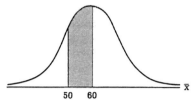

d.

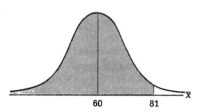

e.

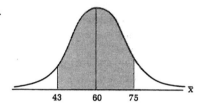

f.

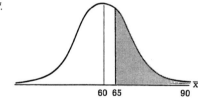

6.20   If $\bar{x}$ is normally distributed with $\mu_{\bar{x}} = 15$ and $\sigma_{\bar{x}} = 2$, find each of the following:

a. $P(\bar{x} > 15)$ d. $P(8.1 \leq \bar{x} \leq 14.3)$

b. $P(12 \leq \bar{x} \leq 18.2)$ e. $P(\bar{x} \leq 17.9)$

c. $P(\bar{x} < 10.6)$ f. $P(16.0 \leq \bar{x} \leq 20.7)$

6.21   A random sample of $n = 100$ observations is selected from a population with $\mu = 40$ and $\sigma = 20$. Find the following probabilities:

a. $P(\bar{x} \geq 40)$ c. $P(\bar{x} \leq 34.2)$

b. $P(37.6 \leq \bar{x} \leq 41.3)$ d. $P(\bar{x} \geq 44.1)$

6.22   A random sample of $n = 70$ observations is selected from a population with $\mu = 10$ and $\sigma = 3$. Find the following probabilities:

a. $P(\bar{x} \leq 11)$ c. $P(9.8 \leq \bar{x} \leq 10.3)$

b. $P(\bar{x} \leq 9.5)$ d. $P(\bar{x} \geq 9.2)$

Applying the Concepts

6.23   The distribution of the number of violent crimes per day in a city possesses a mean equal to 1.3 and a standard deviation equal to 1.7. A random sample of 50 days is observed, and the mean number of crimes for this sample, $\bar{x}$, is calculated.

a.   Give the mean and the standard error of the sampling distribution of $\bar{x}$.

b.   Will the sampling distribution be approximately normal? Explain.

c.   Find the probability that the sample mean, $\bar{x}$, is less than 1.

d.   Find the probability that $\bar{x}$ is greater than 1.9.

6.24   A random sample of fifty 6-ounce cans of tomato juice is drawn from all the cans of tomato juice a manufacturer produces in one week, and the contents of each sampled can is carefully measured. Prior experience has shown that the distribution of the contents has a mean of 6 ounces and a standard deviation of .06 ounce.

a.   What is the probability that the mean contents of the fifty sampled cans would be less than 5.97 ounces?

b.   What is the probability that the mean contents of the fifty sampled cans would be between 5.98 ounces and 6.02 ounces?

c.   How would the sampling distribution of the sample mean change if the sample size were increased from fifty to 100?

6.25   The amount of dissolved oxygen in rivers and streams depends on the water temperature and on the amounts of decaying organic matter from natural processes or human disturbances that are present in the water. The Council on Environmental Quality (CEQ) considers a dissolved oxygen content of less than 5 milligrams per liter of water to be unlikely to support aquatic life and, consequently, to be undesirable. Suppose that an industrial plant discharges its waste into a river and that the daily downstream dissolved oxygen content measurements are distributed with a mean equal to 5.3 milligrams per liter and a standard deviation of 1.6. A random sample of 40 days is obtained, and the downstream dissolved oxygen content for each day is noted.

a.   What is the probability that the sample mean dissolved oxygen content is less than 5 milligrams per liter of water?

b.   Would a sample size of 80 days increase or decrease the probability found in part a? Explain.

c.   Do you think an ecologist would be satisfied that the *mean* dissolved oxygen content is greater than 5 milligrams per liter? Explain why or why not.

6.26   The scores on a test designed to measure elementary school teachers' attitudes toward handicapped students are distributed with a mean score equal to 67 and a standard deviation equal to 10.8. A random sample of sixty-eight elementary school teachers takes this test.

   a. What is the probability the sixty-eight teachers have a mean test score above 70?

   b. What is the probability the sixty-eight teachers have a mean test score below 65?

   c. What is the probability the sixty-eight teachers have a mean test score between 63 and 71?

6.27 The accompanying figure is a representation of the distribution of the nutrient requirements for a particular class of individuals.* The recommended intake for nutrients is chosen to be the point 2 standard deviations (+2 SD) above the average requirement. Explain why between 2% and 3% of the individuals whose requirements this graph describes would require more than the recommended intake of nutrients.

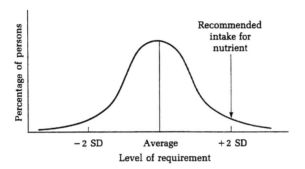

6.28 The amount of coffee dispensed by a vending machine has an average of 7 ounces per cup and a standard deviation of .24 ounce.

   a. What is the probability a random sample of fifty cups of coffee purchased from this machine would have a mean fill of less than 6.95 ounces?

   b. What is the probability a random sample of 100 cups of coffee purchased from this machine would have a mean fill of less than 6.95 ounces?

  c. Use a computer package to answer parts *a* and *b*.

6.29 Refer to Exercise 6.14. The average number of calories ingested per capita per day for countries worldwide is approximately 2,460 and the standard deviation is approximately 300. In Exercise 6.14 you calculated the mean and standard error of the sampling distribution of the sample mean number of calories per capita per day for a random sample of thirty countries. Use this information to answer the following questions.

   a. What is the probability that the sample mean is above 2,600, the minimal number of calories recommended per individual per day?

   b. What is the probability that the sample mean is less than 2,350?

  c. Use a computer package to answer parts *a* and *b*.

* *Source:* Adapted from George H. Beaton, "Uses and Limits of the Use of the Recommended Dietary Allowances for Evaluating Dietary Intake Data," *American Journal of Clinical Nutrition,* Vol. 41, January 1985, p. 156.

6.3

NORMAL APPROXIMATION FOR THE BINOMIAL STATISTIC

In Section 5.4 we presented the binomial experiment and the sampling distribution of the binomial statistic, p. Sampling distributions of p were tabulated (Appendix Table I) for relatively small sample sizes, that is, values of $n \leq 25$. If the sample size for a binomial experiment is large, it is impractical to give the sampling distribution of p in the form of a table. Fortunately, in many instances when the sample size is large, the sampling distribution of p may be approximated by a normal distribution. The conditions that must be met for p to have an approximately normal distribution and the properties of this distribution are given in the box.

Approximating a Binomial Sampling Distribution with a Normal Distribution

Let p be a binomial statistic based on a sample of n observations selected from a population with a fraction of successes, π, and a fraction of failures, $(1 - \pi)$. The sampling distribution of p will have the following properties:

1. The mean of the sampling distribution of p is

$$\mu_p = \pi$$

2. The standard error of the sampling distribution of p is

$$\sigma_p = \sqrt{\frac{\pi \cdot (1 - \pi)}{n}}$$

3. The sampling distribution of p is approximately normal if

$$\mu_p - 3\sigma_p > 0 \qquad \text{and} \qquad \mu_p + 3\sigma_p < 1$$

The values of μ_p and σ_p are the same as those given in Chapter 5. The statistic p has values only between 0 and 1, so we must be sure an approximately normal distribution can be "squeezed" between 0 and 1. Since virtually all the area for a normal distribution is within 3 standard deviations of the mean, the approximation should be appropriate whenever $\mu_p - 3\sigma_p > 0$ and $\mu_p + 3\sigma_p < 1$.

For example, consider a binomial experiment with $n = 100$ and $\pi = .3$. Then,

$$\mu_p = \pi = .3$$

$$\sigma_p = \sqrt{\frac{\pi \cdot (1 - \pi)}{n}} = \sqrt{\frac{(.3)(.7)}{100}} \approx .046$$

$$\mu_p - 3\sigma_p = .3 - 3(.046) = .162$$

$$\mu_p + 3\sigma_p = .3 + 3(.046) = .438$$

Since $\mu_p - 3\sigma_p > 0$ and $\mu_p + 3\sigma_p < 1$, the normal distribution may be used to approximate the sampling distribution of p. This is shown graphically in Figure 6.22.

Figure 6.22
The normal approximation of the sampling distribution of p ($n = 100$, $\pi = .3$)

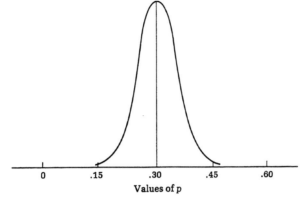

Values of p

On the other hand, consider a binomial experiment with $n = 100$ and $\pi = .01$. Now,

$$\mu_p = \pi = .01$$

$$\sigma_p = \sqrt{\frac{\pi \cdot (1 - \pi)}{n}} = \sqrt{\frac{(.01)(.99)}{100}} \approx .01$$

$$\mu_p - 3\sigma_p = .01 - 3(.01) = -.02$$

$$\mu_p + 3\sigma_p = .01 + 3(.01) = .04$$

Since $\mu_p - 3\sigma_p < 0$ for this experiment, we should not approximate the sampling distribution of p with a normal distribution. The fact that the normal curve cannot be "squeezed" between 0 and 1 is shown in Figure 6.23.

Figure 6.23
The sampling distribution of p ($n = 100$, $\pi = .01$) cannot be approximated by a normal distribution

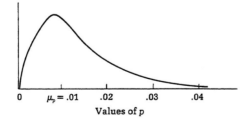

Values of p

The following example will demonstrate that even for a relatively small value of n, the normal distribution may provide an adequate approximation for a binomial statistic.

EXAMPLE 6.8   Consider a binomial experiment with $n = 20$, $\pi = .5$, and $p =$ fraction of successes in sample of $n = 20$.

a. Find $P(\%_{20} \leq p \leq {}^{14}\!/_{20})$, using the sampling distribution of p provided in Appendix Table I.

b. Show that the sampling distribution of p may be approximated by a normal distribution.

c. Using the normal distribution, approximate $P(\%_{20} \leq p \leq {}^{14}\!/_{20})$.

Solution   **a.** From the sampling distribution for $n = 20$ and $\pi = .5$ in Table I, we obtain the probabilities of interest:

$$P\left(\frac{6}{20} \leq p \leq \frac{14}{20}\right) = P\left(p = \frac{6}{20}\right) + P\left(p = \frac{7}{20}\right) + \cdots + P\left(p = \frac{14}{20}\right)$$

$$= .037 + .074 + \cdots + .037 = .958$$

b. To show that the sampling distribution of p may be approximated by a normal distribution, we need the values of μ_p and σ_p. They are

$$\mu_p = \pi = .5$$

$$\sigma_p = \sqrt{\frac{\pi \cdot (1 - \pi)}{n}} = \sqrt{\frac{(.5)(.5)}{20}} \approx .11$$

Since $\mu_p - 3\sigma_p = .5 - 3(.11) = .17$ is larger than 0 and $\mu_p + 3\sigma_p = .5 + 3(.11) = .83$ is less than 1, the sampling distribution of p may be approximated by a normal distribution.

c. To approximate the value of $P(\%_{20} \leq p \leq {}^{14}\!/_{20})$, we can use essentially the same method we used to find probabilities for $\bar{x}$ with the normal distribution. First, we sketch the approximate normal distribution of p and shade the area of interest, as shown in Figure 6.24. Next, we find the z-scores associated with $p = .3$ and $p = .7$. To find these z-scores, we must use the mean μ_p and standard error σ_p in the formula for z:

$$z = \frac{p - \mu_p}{\sigma_p}$$

Figure 6.24

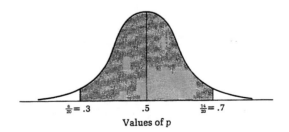

$\frac{6}{20} = .3$ $.5$ $\frac{14}{20} = .7$

Values of p

From part **b**, $\mu_p = .5$ and $\sigma_p = .11$. The z-score for $p = .3$ is therefore

$$z = \frac{p - \mu_p}{\sigma_p} = \frac{.3 - .5}{.11} \approx -1.82$$

The z-score for $p = .7$ is

$$z = \frac{p - \mu_p}{\sigma_p} = \frac{.7 - .5}{.11} \approx 1.82$$

As you may have noticed in Figure 6.24, .3 and .7 are the same distance from the mean, but .3 is 1.82 standard deviations below the mean while .7 is 1.82 standard deviations above it. From Appendix Table II, the area associated with a z-score of 1.82 is .466. Thus, using the normal approximation, we obtain

$$P\left(\frac{6}{20} \leq p \leq \frac{14}{20}\right) \approx .466 + .466 = .932$$

You can see that this approximate probability is close to the exact probability of .958 calculated in part **a**. ■

The purpose of Example 6.8 was to demonstrate how well the normal approximation works when it is properly used—even for the relatively small sample size of $n = 20$. However, since values up to $n = 25$ are given in Appendix Table I, the normal approximation should be used to calculate probabilities only when the value of n is greater than 25. For large values of n, the normal approximation will provide very nearly the exact probability that would be provided by the binomial sampling distribution.

Helpful Hint

If p is a binomial statistic for a sample of n observations selected from a population with a fraction π of successes, the z-score for a fraction p is

$$z = \frac{p - \mu_p}{\sigma_p} = \frac{p - \pi}{\sqrt{\dfrac{\pi \cdot (1 - \pi)}{n}}}$$

EXAMPLE 6.9   An important part of many electronic devices is a solid-state circuit. Since these circuits are mass-produced by machines, some form of quality control must be practiced. The manufacturing process must be monitored to make certain the fraction of defective items produced is kept at an acceptable level.

One method of dealing with this problem is **lot acceptance sampling,** in which a sample of the items produced is selected, and each item in the sample is carefully tested. Then, based on the fraction of defectives in the sample, the lot of items is accepted (inferring there is a small fraction of defectives in the entire lot) or rejected (inferring there is an unacceptable fraction of defectives in the entire lot). For example, suppose a manufacturer of solid-state circuits for television sets chooses 200 stamped circuits from the day's production and determines p, the fraction of defective circuits in the sample. The manufacturer is willing to accept up to 6% defectives in the entire day's production.

a. Describe the sampling distribution of p, assuming the fraction of defective circuits in the day's production is .06.

b. Use the normal approximation to find the probability that the fraction of defectives observed in a sample of 200 circuits is equal to or larger than .1. Assume that the sampling distribution of p is as described in part **a**.

Solution **a.** The statistic p, the fraction of defective circuits in the sample, is a binomial statistic. The sample size is $n = 200$ and the fraction of successes (defective circuits) in the population is assumed to be $\pi = .06$. Thus,

$$\mu_p = \pi = .06$$

$$\sigma_p = \sqrt{\frac{\pi \cdot (1 - \pi)}{n}} = \sqrt{\frac{(.06)(.94)}{200}} \approx .017$$

Note that $\mu_p - 3\sigma_p = .06 - 3(.017) = .009$ is larger than 0 and $\mu_p + 3\sigma_p = .06 + 3(.017) = .111$ is smaller than 1. The sampling distribution of p may be approximated by a normal distribution.

b. The probability of interest, $P(p \geq .1)$, is shown by the shaded area in Figure 6.25. The z-score corresponding to a value of $p = .1$ is

$$z = \frac{p - \mu_p}{\sigma_p} = \frac{.1 - .06}{.017} = 2.35$$

Figure 6.25

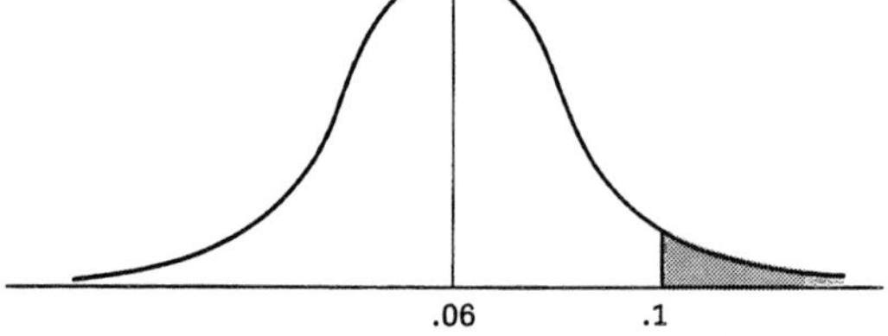

.06   .1

Values of p

Referring to Appendix Table II, we find that the area corresponding to $z = 2.35$ is .491. From Figure 6.25, we see that to find the probability of interest, .491 should be subtracted from .500. Therefore,

$$P(p \geq .1) = .500 - .491 = .009$$

The probability is extremely small (.009) that a fraction of .1 or more of a sample of 200 circuits will be found defective if, in fact, the true fraction of defectives in the population is .06. The manufacturer should be very concerned about the fraction of defective circuits in a day's production if the fraction of defectives observed in the sample is .1 or larger. ∎

COMPUTER EXAMPLE 6.4   In Example 6.9 we considered a binomial statistic p with $n = 200$ and $\pi = .06$. After calculating $\mu_p = .06$ and $\sigma_p \approx .017$, we found that $P(p \geq .1) = .009$. Since p is approximately normally distributed, we can use Minitab to find this probability. Using the values of $\mu_p = .06$, $\sigma_p = .017$, and $p = .1$, we obtain

$$0.1000 \qquad 0.9907$$

Thus, $P(p < .1) = .9907$. To find the probability of interest, we must subtract this probability from 1. That is,

$$P(p \geq .1) = 1 - P(p < .1) = 1 - .9907 = .0093 \approx .009 \qquad ∎$$

COMPUTER EXAMPLE 6.5   In Computer Example 6.4, we used Minitab to approximate a binomial probability with the normal distribution. In general, the sampling distribution of a binomial statistic is approximately normal if $\mu_p - 3\sigma_p > 0$ and $\mu_p + 3\sigma_p < 1$. With the aid of Minitab, we are able to provide visual support for the normal approximation to the binomial.

The accompanying figure is a computer-generated graph of the probabilities for a binomial statistic p with $n = 200$ and $\pi = .06$. To construct the graph, Minitab was first used to find the probabilities associated with the values of p equal to 0, $\frac{1}{200} = .005$, $\frac{2}{200} = .010$, $\frac{3}{200} = .015$, $\frac{4}{200} = .020$, $\frac{5}{200} = .025, \ldots, \frac{25}{200} = .125$. These values of p and their corresponding probabilities, $P(p)$, were used by Minitab to produce this graph.

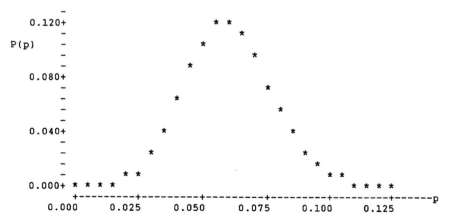

The computer-generated graph certainly appears to be mound-shaped and approximately normally distributed. From the graph we can see that $P(p = .125)$ is very near zero. Although the possible values of p could theoretically be as large as 1, the probabilities associated with values larger than $p = .125$ are essentially 0. ∎

Exercises (6.30–6.46)

Learning the Language

6.30 Let p be a binomial statistic based on a sample of n observations selected from a population with a fraction of successes, π. Give the properties of the sampling distribution of p.

6.31 What do the following symbols denote?

a. p c. μ_p e. $\sqrt{\dfrac{\pi \cdot (1 - \pi)}{n}}$

b. π d. σ_p f. $z = \dfrac{p - \mu_p}{\sigma_p}$

Learning the Mechanics

6.32 Let p be a binomial statistic based on a sample of $n = 80$ observations selected from a population with a fraction of successes $\pi = .70$. Calculate the z-score for each of the following values of p:

a. $p = .60$ c. $p = .80$ e. $p = .72$
b. $p = .70$ d. $p = .83$ f. $p = .55$

6.33 Let p be a binomial statistic based on a sample of $n = 123$ observations selected from a population with a fraction of successes $\pi = .4$. Calculate the z-score for each of the following values of p:

a. $p = .80$ c. $p = .47$ e. $p = .33$
b. $p = .24$ d. $p = .51$ f. $p = .38$

6.34 Let p be a binomial statistic based on a sample of n observations selected from a population with a fraction of successes, π. For each of the following determine whether the normal distribution may be used to approximate the sampling distribution of p:

a. $n = 300$; $\pi = .001$ d. $n = 726$; $\pi = .01$
b. $n = 30,000$; $\pi = .001$ e. $n = 181$; $\pi = .1$
c. $n = 30$; $\pi = .6$ f. $n = 607$; $\pi = .88$

6.35 Let p be a binomial statistic based on a sample of n observations selected from a population with a fraction of successes, π. For each of the following determine whether the normal distribution may be used to approximate the sampling distribution of p:

a. $n = 100,000$; $\pi = .995$ d. $n = 100$; $\pi = .008$
b. $n = 10,000$; $\pi = .995$ e. $n = 100$; $\pi = .500$
c. $n = 1,000$; $\pi = .995$ f. $n = 100$; $\pi = .992$

Using the Tables

6.36 If p is approximately normally distributed with $\mu_p = .600$ and $\sigma_p = .010$, find the shaded area for each of the following:

a.

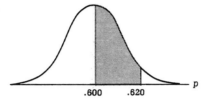

b.

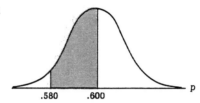

c.

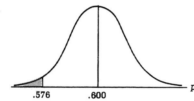

d.

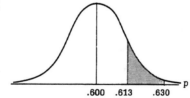

e.

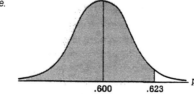

f.

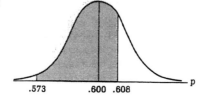

6.37 If p is approximately normally distributed with $\mu_p = .45$ and $\sigma_p = .06$, approximate each of the following:

    a. $P(.40 \leq p \leq .55)$ *d.* $P(.20 \leq p \leq .50)$

    b. $P(p \leq .30)$ *e.* $P(.33 \leq p \leq .41)$

    c. $P(p \leq .60)$ *f.* $P(p \geq .58)$

6.38 A random sample of $n = 200$ observations is selected from a population with a fraction of successes $\pi = .75$. Demonstrate that p, the fraction of successes in the sample, is approximately normally distributed, and approximate the following probabilities:

    a. $P(p \leq .79)$ *c.* $P(.73 \leq p \leq .84)$

    b. $P(p \leq .70)$ *d.* $P(p \geq .68)$

6.39 A random sample of $n = 150$ observations is selected from a population with a fraction of successes $\pi = .38$. Demonstrate that p, the fraction of successes in the sample, is approximately normally distributed, and approximate the following probabilities:

    a. $P(p \geq .38)$ *c.* $P(p \leq .29)$

    b. $P(.31 \leq p \leq .46)$ *d.* $P(p \geq .50)$

Applying the Concepts

6.40 Eighty percent of all adults in a large city favor an increased emphasis on the basics of education—reading, writing, and arithmetic. If 150 adults are randomly selected from among the total number of adults in the city:

    a. What is the approximate probability that the fraction of the adults surveyed who favor an increased emphasis on the basics of education will be at least .73?

    b. What is the approximate probability that the fraction of the adults surveyed who favor an increased emphasis on the basics of education will be .84 or more?

6.41 It is against the law to discriminate against job applicants because of race, religion, sex, or age. Forty percent of the individuals who apply for an accountant's position in a large corporation are over 45 years of age. If the company decides to choose fifty of a very large number of applicants for closer credential screening, claiming that the selection will be random and not age-biased, what is the approximate probability that the fraction of those chosen who are over 45 years of age will be at most .30?

6.42 A recent study involving attrition rates at a major university has shown that 43% of all incoming first-year students do not graduate within 4 years of entrance.

    a. If 200 first-year students are randomly sampled this year, and their progress through college is followed, what is the approximate probability that at least half will graduate within the next 4 years?

    b. What is the approximate probability that the fraction of first-year students graduating within 4 years will be between .50 and .60?

6.43 A credit card company claims that 80% of all clothing purchases in excess of $10 are made with credit cards. If this claim is valid and if p is the fraction of clothing purchases made with credit cards in a random sample of 100 clothing purchases in excess of $10, approximate the following:

    a. $P(p \leq .73)$ *b.* $P(.75 \leq p \leq .85)$

6.44   The Department of Labor is interested in the fraction of the employed work force in the United States who feel in danger of losing their jobs during the next year. A random sample of ninety members of the work force is taken and p, the fraction of people who feel in danger of losing their jobs during the next year, is observed. If 30% of the entire work force feel insecure about their jobs, what is the approximate probability that at most fifteen of the ninety members sampled feel insecure about their jobs?

6.45   Some 60% of all Americans exercise regularly. A random sample of 1,000 Americans is selected. Use a computer package to find the probability that at least 63% of those sampled exercise regularly.

6.46   Since delivering America's first test-tube baby in December 1981, the Eastern Virginia Medical School in Norfolk has helped "make" more than 100 other infants; it is recognized as one of the nation's premier *in vitro* programs. One-fourth of all women who attempt fertilization in the Norfolk laboratory become pregnant. This pregnancy rate is among the highest of any such program and approximates the percentages for healthy, fertile couples.[†]

   a.   Suppose 75 women attempt fertilization in the Norfolk laboratory. Is p, the fraction of women who become pregnant, approximately normally distributed? Explain.

   *b.   Use a computer package to find the approximate probability that 30 or more of the 75 women become pregnant.

6.4

USING LARGE-SAMPLE APPROXIMATIONS TO MAKE INFERENCES

By applying the Central Limit Theorem, we can use the mean of a large random sample to make an inference about a population mean, μ. Or, using a large sample, we can approximate the sampling distribution of a binomial statistic p with a normal distribution to make an inference about the fraction of successes in the population, π. In this section we will present some examples in which we make inferences using the rare event approach. We will formalize these inferential methods in the next chapter.

EXAMPLE 6.10   Many college and university professors have been accused of "grade inflation" over the past several years. This means they assign a higher grade to a student now than they would have given to a student of the same caliber in the past. A dean at a particular university wants to test the grade inflation theory. From past records, the dean finds that the mean grade-point average (GPA) of all graduates 10 years ago was 2.86. The standard deviation of the distribution of grade-point averages is approximately equal to .4. A random sample of seventy-five graduates from this year's class has a mean GPA of 3.02.

---

* *Source:* "Exercise" survey, *The Gallup Report*, No. 226, July 1984, p. 11.
† *Source:* Michael Gold, "The Baby Makers," *Science 85*, Vol. 6, No. 3, April 1985, p. 26.

a. If no grade inflation has occurred and the mean GPA of this year's entire graduating class is really 2.86, what is the probability that the sample of seventy-five graduates would have a mean GPA greater than or equal to 3.02?

b. If you observed a sample mean GPA of 3.02, how would you interpret it?

Solution **a.** The population the sample is taken from is the collection of grade-point averages of all students graduating this year. We are assuming this population has a mean $\mu = 2.86$ and a standard deviation $\sigma = .4$. Since $n = 75$ is a large sample size and the sample is random, the Central Limit Theorem assures us that the sampling distribution of $\bar{x}$ is approximately normal. The mean and standard error of the sampling distribution of $\bar{x}$ are

$$\mu_{\bar{x}} = \mu = 2.86$$

$$\sigma_{\bar{x}} = \frac{\sigma}{\sqrt{n}} = \frac{.4}{\sqrt{75}} \approx .05$$

Sketching the normal distribution and shading the area corresponding to $P(\bar{x} \geq 3.02)$, we obtain Figure 6.26. The z-score corresponding to $\bar{x} = 3.02$ is

$$z = \frac{\bar{x} - \mu_{\bar{x}}}{\sigma_{\bar{x}}} = \frac{3.02 - 2.86}{.05} = 3.20$$

Figure 6.26

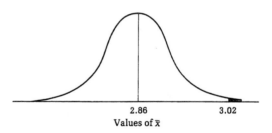

2.86   3.02
Values of $\bar{x}$

Consulting Appendix Table II, we find the area corresponding to $z = 3.20$ to be .499. Subtracting this area from .500 to obtain the area of interest, we arrive at the answer:

$$P(\bar{x} \geq 3.02) = .500 - .499 = .001$$

If the mean GPA for this year's graduating class were 2.86 and the standard deviation were .4, the same as the graduating class of 10 years ago, the probability that seventy-five randomly chosen graduates would have a mean GPA greater than or equal to 3.02 is only .001. There is very little chance this would happen.

b. Using the rare event approach to making inferences, there is overwhelming evidence that the mean GPA of this year's graduating class is higher than the mean GPA of the class of 10 years ago. We have observed a sample mean 3.20 standard deviations above $\mu_{\bar{x}} = 2.86$, and this would occur only .001 of the time. This indicates that grade inflation may very well have occurred.

However, this conclusion is valid only if the two graduating classes being compared were equal academically—that is, only if their academic abilities were equivalent. Before we conclude that grade inflation is the real explanation for the discrepancy, we should satisfy ourselves that the two classes really deserved to receive the same grades. ■

EXAMPLE 6.11 A candidate for political office claims he will receive 55% of the votes in the upcoming election. As a supporter of another candidate, you suspect that the claim of 55% is too high, and that the candidate making the claim will actually receive a smaller percentage of the votes. Hoping to provide support for your suspicions and enough evidence to refute the candidate's claim, you randomly sample eighty voters and find p, the fraction of voters who will vote for the candidate who claims to expect 55% of the votes.

  a. Assume the candidate's claim of 55% is correct. Discuss the sampling distribution of p.

  b. Suppose the random sample results in a value of $p = .4875$. Would you conclude the candidate's claim is wrong?

Solution **a.** The statistic p has a binomial sampling distribution. If the candidate's claim is correct, $\pi = .55$. Since $n = 80$, we obtain

$$\mu_p = \pi = .55$$

$$\sigma_p = \sqrt{\frac{\pi \cdot (1 - \pi)}{n}} = \sqrt{\frac{(.55)(.45)}{80}} \approx .056$$

To see if we can approximate the sampling distribution of p with a normal distribution, we calculate the following:

$$\mu_p - 3\sigma_p = .55 - 3(.056) = .382$$

$$\mu_p + 3\sigma_p = .55 + 3(.056) = .718$$

Since .382 is larger than 0 and .718 is smaller than 1, the sampling distribution of p is approximately normal.

  b. If the fraction of the sample who will vote for the candidate is .4875, this does not seem to agree with the candidate's claim that 55% will vote for him. If his claim is correct, he will easily win the election, but in the sample he did not receive support from even 50% of the voters. However, we must not rely on our intuition in order to make an inference in this situation. Instead, we should calculate how rare an event we have observed. That is, we should find $P(p \le .4875)$ if π really is .55. The area of interest is shown in Figure 6.27 (page 308). The z-score corresponding to $p = .4875$ is

$$z = \frac{p - \mu_p}{\sigma_p} = \frac{.4875 - .55}{.056} \approx -1.12$$

From Appendix Table II, the area corresponding to $z = 1.12$ is .369. Thus, the area of interest is $.500 - .369 = .131$. That is,

$$P(p \le .4875) = .131$$

Figure 6.27

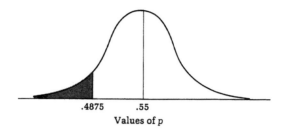

.4875   .55
Values of p

We now see that if 55% of the voters really will vote for the candidate, the probability is .131 that in a sample of eighty, the fraction of voters who will vote for him is less than or equal to .4875. In other words, if the candidate's claim is correct, we would observe a sample fraction equal to or smaller than the one we observed about 13% of the time that a random sample of eighty voters is selected. This is not a very rare event. The sample does not provide sufficient evidence to refute the candidate's claim. The sample evidence does not support the candidate's claim, but it is not strong enough to refute it. The sample has not substantiated your suspicions.   ■

EXAMPLE 6.12   A sociologist wants to estimate the mean number of television viewing hours per week for children in a certain city. A random sample of 500 children yields a mean of 38.7 hours of television viewing time per week. From other similar studies conducted by the sociologist, the standard deviation is known to be $\sigma = 12.4$ hours.

a. Describe the sampling distribution of the sample mean, $\bar{x}$.
b. What are the smallest and largest values you believe μ, the mean number of television viewing hours per week for all children in the city, might be?

Solution   **a.** The original population is the collection of the number of television viewing hours per week for every child in the city of interest. The value of μ, the mean of this population, is not given, but the standard deviation is $\sigma = 12.4$. Since a random sample of $n = 500$ is observed, the sampling distribution of $\bar{x}$ is approximately normal. Because the value of μ is unknown, the mean of the sampling distribution is also unknown. However, whatever the value of μ, we know that $\mu_{\bar{x}} = \mu$. The standard error of the sampling distribution is

$$\sigma_{\bar{x}} = \frac{\sigma}{\sqrt{n}} = \frac{12.4}{\sqrt{500}} \approx .55$$

b. Since we would not expect $\bar{x}$ to deviate from $\mu_{\bar{x}} = \mu$ by more than 3 standard deviations, we make the following estimates for the smallest and largest values μ might be:

Smallest value of μ: $\bar{x} - 3\sigma_{\bar{x}} = 38.7 - 3(.55) = 37.05$
Largest value of μ: $\bar{x} + 3\sigma_{\bar{x}} = 38.7 + 3(.55) = 40.35$

We are very confident that the value of μ, the mean number of television viewing hours per week for all children in this city, is somewhere in the interval 37.05 to 40.35. The only way this could be in error is if the observed value of $\bar{x} = 38.7$ were more than 3 standard deviations from the true value of $\mu_{\bar{x}} = \mu$. The probability that a random sample would produce a value of $\bar{x}$ more than 3 standard deviations from the mean is very, very small. The fact that this event is so unlikely explains why we have so much confidence in our inference about μ. ∎

We will conclude this section with some comments concerning rare events. In Chapter 5 we indicated that the best measure of the rarity of an event is the probability. A less precise measure of the rarity is the z-score. We want to emphasize that these statements also apply to normal distributions. For example, suppose we are interested in the event that a normally distributed statistic is more than 1.7 standard deviations above its mean. As shown in Figure 6.28, this probability (the area found in Appendix Table II) is .045, which is less than .05. Thus, the probability indicates that a value of the statistic more than 1.7 standard deviations above the mean is a rare event. Yet, this would not be viewed as a rare event if we considered only that the z-score of 1.7 is not more than 2 standard deviations from the mean. These should not be considered contradictory statements. The z-score is a rough measure of the rarity of an event, and for a normal distribution, events less than 2 standard deviations from the mean may be considered rare in terms of their probability. Remember, *the actual probability of an event is a better measure of its rarity than the z-score alone.*

Figure 6.28

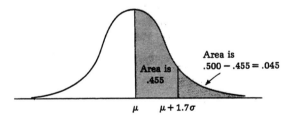

We are very confident... Area is .455   Area is .500 − .455 = .045

μ $\mu + 1.7\sigma$

Exercises (6.47–6.57)

Learning the Mechanics

6.47 It is claimed that the mean of a population is $\mu = 50$. Someone suspects that the mean is actually larger, and randomly samples $n = 60$ measurements from the population. Assuming the population standard deviation is $\sigma = 15$, which of the following values of the sample mean would indicate that μ is larger than 50? [*Hint:* Using a value of $\mu = 50$, calculate the probability of observing a sample mean as large as or larger than the stated value of $\bar{x}$ in each part.]

  a. $\bar{x} = 51.3$ b. $\bar{x} = 54.8$ c. $\bar{x} = 53.6$ d. $\bar{x} = 52.1$

6.48 It is claimed that the mean of a population is $\mu = 6.8$. Someone suspects that the mean is actually smaller, and randomly samples $n = 800$ measurements from the population. Assuming the population standard deviation is $\sigma = 1.9$, which of the following values of the sample mean would indicate that μ is smaller than 6.8? [*Hint:* Using a value of $\mu = 6.8$, calculate the probability of observing a sample mean as small as or smaller than the stated value of $\bar{x}$ in each part.]

    a. $\bar{x} = 6.65$ *b.* $\bar{x} = 6.73$ *c.* $\bar{x} = 6.50$ *d.* $\bar{x} = 7.00$

6.49 It is claimed that $\pi = .3$ is the fraction of successes in a population. Someone suspects that the value of π is actually smaller, and randomly samples $n = 200$ observations from the population. Which of the following values of the sample fraction would indicate that π is smaller than .3? [*Hint:* Using a value of $\pi = .3$, calculate the approximate probability of observing a sample fraction as small as or smaller than the given value of p in each part.]

    a. $p = .26$ *b.* $p = .22$ *c.* $p = .28$ *d.* $p = .34$

6.50 It is claimed that $\pi = .4$ is the fraction of successes in a population. Someone suspects that the value of π is actually larger, and randomly samples $n = 95$ observations from the population. Which of the following values of the sample fraction would indicate that π is larger than .4? [*Hint:* Using a value of $\pi = .4$, calculate the approximate probability of observing a sample fraction as large as or larger than the given value of p in each part.]

    a. $p = .54$ *b.* $p = .44$ *c.* $p = .48$ *d.* $p = .51$

Applying the Concepts

6.51 Due to a recent drought, conditions were favorable for the growth of a mold that produces the cancer-causing substance called *aflatoxin*. Researchers estimated that the mold affected 45% of the corn crop, thus making the corn unfit as feed for livestock.

    a. If a random sample of 500 ears of corn is taken, what is the approximate probability that .4 or fewer of the ears are affected, assuming the estimate of 45% is correct?

    b. Suppose a sample of 500 ears was actually taken and the fraction of ears that were affected was less than .4. What, if anything, would you conclude?

6.52 An educational researcher has developed an IQ test which she claims is not biased against black children. It is known that white children's scores on the test have a mean equal to 100 and a standard deviation equal to 15. In order to test the claim of non-bias, the IQ test is administered to a random sample of 200 black children.

    a. Assuming the researcher's claim is true, completely describe the sampling distribution of the mean IQ score for a sample of 200 black children.

    b. Assuming the researcher's claim is true, what is the probability that the sample has a mean IQ score less than 97?

    c. Suppose the sample mean is actually 96.5. How would you interpret this value of $\bar{x}$ in view of the researcher's claim? Explain.

    d. Suppose the sample mean is actually 98.5. How would you interpret this value of $\bar{x}$ in view of the researcher's claim? Explain.

6.53 Last year, a company initiated a program to compensate its employees for unused sick days, paying each employee a bonus of one-half the usual wage earned for each unused sick day. The question that naturally arises is "Did this policy motivate employees to use fewer allotted sick days?" *Before* last year, the number of sick days used by employees had an average of 7 days and a standard deviation of 2 days per year.

    a. Assuming these parameters did not change last year, find the probability that the sample mean number of sick days used by 100 employees chosen at random was less than or equal to 6.4 last year.

    b. Suppose the sample mean for the 100 employees was, in fact, 6.4. How would you interpret this result?

6.54 An advertising agency was hired to introduce a new product. After its campaign, it claimed that 30% of all consumers were familiar with the product. To check the claim, the manufacturer of the product surveyed 2,000 consumers. Of this number, 527 consumers had learned about the product through sources attributable to the campaign.

    a. What is the approximate probability that 527 or fewer of the 2,000 consumers would have learned about the product if the campaign was really 30% effective?

    b. Considering the answer to part *a*, do you think 30% of all consumers are familiar with the product? Explain.

6.55 A manufacturer of pencils has 700 pencils randomly chosen from each day's production and inspected for defects (chips, cracks, etc.). The manufacturer is willing to tolerate a fraction of up to .1 defectives in the production process. If the process fraction of defectives is more than .1, the process is considered out of control.

    a. Assuming that .1 of all pencils produced are defective, what is the approximate probability of observing eighty or more defectives in a day's sample of 700 pencils?

    b. If eighty defective pencils are observed in a day's sample, are you willing to infer that the process is out of control? Explain.

6.56 Studies by the National Institutes of Health confirm that dangerously high levels of blood cholesterol are linked to eggs in the diet:

> Each American consumes an average of 265 eggs a year. Thousands of servings of bacon and eggs, souffle, quiche, omelets and eggs Benedict are whipped up from the 68 billion eggs produced every year. All of these eggs also equal more than 17 trillion milligrams of cholesterol, which at high levels in the blood stream has recently been found to be a *direct cause* of heart disease.*

Senior citizens might be more concerned about the amount of cholesterol they ingest, and may thus tend to eat fewer eggs than the general populace. To see if this is correct, we could randomly sample 85 senior citizens and calculate $\bar{x}$, the mean number of eggs they eat per year.

    a. If senior citizens actually eat an average of 265 eggs annually, what is the probability that $\bar{x}$ is less than or equal to 240? Assume that the standard deviation of the number of eggs each American eats per year is 150.

* *Source:* Linda Villarosa, "Eggs: The Not-So-Sunny Side," *The Runner*, Vol. 7, No. 6, March 1985, p. 9. Copyright © 1985. CBS Magazines.

b. If a sample of 85 senior citizens yielded a value of $\bar{x} = 240$, would you be willing to conclude that senior citizens tend to eat fewer eggs? Why?

6.57 Approximately 72% of all U.S. citizens are registered to vote.* Suppose a city has conducted extensive voter registration drives for many years. A random sample of 300 potential voters is selected from this city, and p, the fraction who are registered to vote, is recorded. If 233 of the 300 people sampled are registered to vote, would you infer that this city has more than 72% of its citizens registered? [*Hint:* Calculate the approximate probability that $p \geq {}^{233}\!/_{300}$ when $\pi = .72$.]

Chapter Summary

A **normal distribution** has a bell-shaped curve:

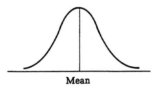

Mean

The curve is symmetric about the mean. The probability of a value occurring between two numbers is the area under the curve between those numbers. Appendix Table II gives the probabilities associated with a normal distribution in terms of z-scores.

The standard deviation of the sampling distribution of a statistic is also called the **standard error** of the statistic.

When a large random sample of n measurements is selected from a very large or infinite population with mean μ and standard deviation σ, then the sample mean, $\bar{x}$, has a distribution that is approximately normal with mean $\mu_{\bar{x}} = \mu$ and standard error $\sigma_{\bar{x}} = \sigma/\sqrt{n}$. This is the **Central Limit Theorem.** For our purposes, $n \geq 30$ will be sufficient to use the normal distribution to approximate the distribution of the statistic $\bar{x}$.

A binomial statistic p with mean $\mu_p = \pi$ and standard error $\sigma_p = \sqrt{\pi \cdot (1 - \pi)/n}$ will have a sampling distribution that is approximately normal if both $(\mu_p - 3\sigma_p)$ and $(\mu_p + 3\sigma_p)$ are between 0 and 1.

SUPPLEMENTARY EXERCISES (6.58–6.75)

Learning the Mechanics

6.58 Let p be a binomial statistic based on a sample of $n = 900$ observations selected from a population with a fraction of successes $\pi = .65$. Calculate the z-score for each of the following values of p:

* *Source:* "Voter Registration" survey, *The Gallup Report,* No. 224, May 1984, p. 10.

a. $p = .68$ d. $p = .62$
b. $p = .60$ e. $p = .69$
c. $p = .65$ f. $p = .66$

6.59 A random sample of 428 measurements is selected from a population with $\mu = 58$ and $\sigma^2 = 274$. Calculate the z-score for each of the following values of the sample mean, $\bar{x}$:

a. $\bar{x} = 55.2$ d. $\bar{x} = 60.1$
b. $\bar{x} = 58.2$ e. $\bar{x} = 59.1$
c. $\bar{x} = 50.9$ f. $\bar{x} = 56.5$

Using the Tables

6.60 A random sample of $n = 75$ observations is selected from a population with $\mu = 120$ and $\sigma^2 = 410$.
a. Find $\mu_{\bar{x}}$ and $\sigma_{\bar{x}}$.
b. What is the shape of the sampling distribution of $\bar{x}$?
c. Find $P(\bar{x} \le 118)$.
d. Find $P(115 \le \bar{x} \le 123)$.
e. Find $P(\bar{x} \le 124.6)$.
f. Find $P(\bar{x} \ge 122.7)$.

6.61 A random sample of $n = 500$ observations is selected from a population with a fraction of successes $\pi = .8$. Demonstrate that p, the fraction of successes in the sample, is approximately normally distributed, and approximate the following probabilities:
a. $P(p \ge .76)$ c. $P(p \ge .83)$
b. $P(.75 \le p \le .81)$ d. $P(.8 \le p \le .82)$

Applying the Concepts

6.62 An admissions officer for a law school indicates that 35% of the applicants meet all ten requirements and 95% meet at least eight of the ten requirements.
a. If a random sample of 300 applicants is taken, what is the approximate probability that 210 or fewer of the 300 will fail to meet all ten requirements?
b. What is the approximate probability that 280 or more of the 300 will meet at least eight of the ten requirements?

6.63 Assume that 16% of the black population in the United States suffers from sickle-cell anemia. If 1,000 black people are sampled at random, what is the approximate probability that:
a. The fraction who have the disease is .175 or more?
b. The fraction who have the disease is .140 or less?
c. The fraction of people in the sample who have the disease is between .130 and .180?

6.64 Over the last month, a large supermarket chain received many consumer complaints about the quantity of chips in 9-ounce bags of a particular brand of potato chips.

Suspecting that the complaints were merely the result of the potato chips settling to the bottom of the bags during shipping, but wanting to be able to assure its customers they were getting their money's worth, the managers of the chain decided to examine the next shipment of chips received by their largest store. Thirty-five 9-ounce bags were randomly selected from the shipment, their contents weighed, and the sample mean weight computed. The chain's management decided that if the sample mean was less than 8.95 ounces, the shipment would be refused and a complaint registered with the potato chip company. Assume the distribution of weights of the contents of all the potato chip bags in question has a mean of 8.9 ounces and a standard deviation of .13 ounce. What is the probability that the supermarket chain's investigation will lead to refusal of the shipment?

6.65 Electric power plants that use water for cooling their condensers sometimes discharge heated water into rivers, lakes, or oceans. It is known that water heated above certain temperatures has a detrimental effect on the plant and animal life in the water. Suppose it is known that the increased temperature of the heated water discharged by a certain power plant on any given day has a distribution with a mean of 5°C and a standard deviation of .5°C.

a. For fifty randomly selected days, what is the probability that the average increase in temperature of the discharged water is greater than 5.0°C?

b. Less than 4.8°C?

6.66 A loan officer in a large bank has been assigned to screen sixty loan applications during the next week. Her past record indicates that she turns down 20% of the applicants.

a. What is the approximate probability that forty-one or more of the sixty applications will be approved?

b. What is the approximate probability that between forty-five and fifty of the sixty applications will be approved?

6.67 This past year, an elementary school began using a new method to teach arithmetic to first graders. A standardized test, administered at the end of the year, was used to measure the effectiveness of the new method. The distribution of past scores on the standardized test produced a mean of 75 and a standard deviation of 10.

a. If the new method is no different from the old method, what is the probability that a random sample of thirty-six students who were taught by the new method would have a mean score on the standardized test greater than 79?

b. If a random sample of thirty-six students who were taught by the new method had a mean score of 81 on the standardized test, would you conclude that the new method produces different results from the old method? Explain.

6.68 Contrary to our intuition, very reliable decisions concerning the proportion of a large group of consumers favoring a particular product or a particular social issue can be based upon relatively small samples. For example, suppose the target population of consumers contains 50,000,000 people and we want to decide whether the proportion of consumers, π, in the population that favor some product (or issue) is larger than some value, say .2. Suppose you randomly select a sample as small as 1,600 from the

50,000,000 and you observe the fraction, p, of consumers in the sample who favor the new product.

a. Assuming that $\pi = .2$, find the mean and standard deviation of p.

b. Suppose that 400 (or 25%) of the sample of 1,600 consumers favor the new product. Why might this sample result lead you to conclude that π (the proportion of consumers favoring the product in the population of 50,000,000) is larger than .2? [*Hint:* Calculate the approximate probability that p is greater than or equal to .25 when $\pi = .2$.]

6.69 Golf balls that do not meet a manufacturer's shape specifications are referred to as being "out of round" and may be sold as rejects. Assume that 5% of the balls produced by a particular machine are out of round. What is the approximate probability that a random sample of 200 balls produced by the machine contains 8% or more balls that are out of round?

6.70 The distribution of the number of characters printed per second by a particular kind of line printer at a computer terminal has the following parameters: $\mu = 45$ characters per second, $\sigma^2 = 4$.

a. Describe the sampling distribution of the mean number of characters printed per second for random samples of 60 seconds.

b. Find the probability that the sample mean for a random sample of 60 seconds will be between 44.5 and 45.3 characters per second.

c. Find the probability that the sample mean will be less than 44 characters per second.

6.71 The distribution of the number of loaves of bread sold per day by a large grocery store over the past 5 years has a mean of 250 and a variance of 2,025.

a. Describe the sampling distribution of the mean number of loaves of bread sold per 50 randomly selected shopping days.

b. Give the probability that the mean number of loaves sold per 50 shopping days is between 240 and 250.

c. Give the probability that the mean is greater than 265 loaves.

6.72 The distribution of the number of barrels of oil produced by a particular oil well each day for the past 3 years has a mean of 400 and a variance of 5,625.

a. Describe the sampling distribution of the mean number of barrels produced per day for samples of 40 production days drawn from the past 3 years.

b. What is the probability that the sample mean will be greater than 425? Less than 400?

6.73 After a culture of bacteria is subjected to a particular drug, the length of time until all bacteria in the culture die has a mean of 30 minutes and a standard deviation of 5 minutes. A new drug has been developed which, it is hoped, will reduce the time until the bacteria die. A random sample of forty cultures of the bacteria is subjected to the new drug, and $\bar{x}$, the mean length of time until all bacteria in each culture die, is calculated.

a. If the new drug actually performs the same as the original drug, what is the probability that $\bar{x}$ is less than or equal to 29 minutes?

b. If $\bar{x}$ is observed to equal 29 minutes, would you conclude that the effectiveness of the new drug is different from that of the original drug? Explain.

c. Suppose a random sample of 140 cultures had been used instead of only forty cultures. Rework parts a and b using $n = 140$ instead of $n = 40$.

*6.74 According to *The Gallup Report* for July 1984, approximately 18% of all Americans jog.* If a random sample of 2,000 Americans is surveyed, use a computer package to approximate the probability that 400 or more of the 2,000 jog.

*6.75 Refer to Exercise 6.74. In the same issue of *The Gallup Report*, the average distance joggers jog was reported to be 2.5 miles.[†] Assume that the standard deviation of the distances jogged is 1.5 miles. A random sample of forty joggers is selected from all joggers in a city, and $\bar{x}$, the mean distance they jog, is calculated.

a. If the joggers in this city jog an average of 2.5 miles and the standard deviation is 1.5 miles, use a computer package to find the probability that $\bar{x}$ is 3 miles or more.

b. If the value of $\bar{x}$ equals 3 miles, what would you infer about the distances jogged by the joggers in this city? Explain.

CHAPTER 6 QUIZ

1. The percentage of fat in the bodies of American men has a mean of 15% and a standard deviation of 2%. A random sample of 100 American men is selected, and the percentage of fat in the body of each man is calculated. Completely describe the sampling distribution of $\bar{x}$, the mean percentage of fat in the bodies of the 100 sampled men.

2. Refer to Question 1. What is the probability that the mean percentage of fat in the bodies of the 100 sampled men would be lower than 14.5%?

3. If on a spin of a roulette wheel, a gambler bets that the observed number will be black, the gambler has a probability of $^{18}/_{38}$ of winning. If the gambler bets the observed number will be black on 1,000 spins of a roulette wheel, what is the approximate probability the gambler wins at least half of the bets?

4. A large company instituted an extensive safety campaign. Before the campaign, 12% of the company's employees had at least one accident per year. A year after the campaign, a random sample of 270 employees is taken, and p, the fraction of these employees that had at least one accident in the last year, is calculated.

a. If the percentage of employees having at least one accident per year is the same after the campaign as before, what is the approximate probability that $p \leq {}^{25}/_{270}$?

* *Source:* "Jogging" survey, *The Gallup Report*, No. 226, July 1984, p. 10.
† *Source:* "Distances Run" survey, *The Gallup Report*, No 226, July 1984, p. 10.

b. If p is observed to be $^{25}\!/_{270}$, would you conclude that the percentage of employees having at least one accident per year has decreased? Explain.

CHAPTERS 1–6 CUMULATIVE QUIZ

1. Two cards are randomly selected from six cards. The six cards consist of one 2, two 3's, one 4, and two 6's.
   a. Give the sample space for this experiment.
   b. What is the probability neither of the two cards selected is a 3?
   c. Let X be the value of the second card selected. Give the sampling distribution of X.

2. A sample of five grades on an introductory political science test is as follows:
   94, 76, 80, 32, 88
   a. Calculate the mean and median for this sample.
   b. Calculate the variance and standard deviation for this sample.

3. Students in a self-paced course may take each unit test more than once, if they so desire. The maximum number of times a student may take any one test is five. Let X be the number of times a student takes a unit test. The sampling distribution of X is given below:

X	1	2	3	4	5
$P(X)$	.43	.32	.11	.04	.10

   a. Calculate μ_X, σ_X^2, and σ_X.
   b. Graph the sampling distribution of X and locate the values μ_X, $(\mu_X - \sigma_X)$, and $(\mu_X + \sigma_X)$ on the graph.
   c. What is the probability that X is in the interval $(\mu_X - \sigma_X)$ to $(\mu_X + \sigma_X)$?

4. The length of time it takes boys in a secondary school to run a mile has a distribution with a mean of 450 seconds and a standard deviation of 40 seconds. One boy is timed at 350 seconds.
   a. Find the z-score for a time of 350 seconds.
   b. Do you think there are many boys in secondary school who can run a mile faster than the boy who ran it in 350 seconds? Explain.

5. A network television department that sells commercial time to advertisers claims the number of homes reached per afternoon by a certain daytime soap opera has a mean of 4.5 million homes and a standard deviation of 5 million.
   a. If the network's claim is correct, what is the probability that a random sample of 40 days would have a mean of 4.3 million or fewer homes reached by the soap opera?
   b. If a sample of 40 days had a mean of 4.3 million homes reached by the soap opera, would you conclude that the network's claim is incorrect? Explain.

On Your Own

For large values of n the computational effort involved in working with the binomial sampling distribution is considerable. Fortunately, the normal distribution provides a good approximation to the binomial distribution in many instances. This exercise was designed to demonstrate how well the normal distribution approximates the binomial distribution.

a. Let the statistic, p, have a binomial sampling distribution with $n = 10$ and $\pi = .5$. Using the binomial sampling distribution (Table I), find the exact probability that p takes on a value in each of the following intervals: $(\mu_p - \sigma_p)$ to $(\mu_p + \sigma_p)$, $(\mu_p - 2\sigma_p)$ to $(\mu_p + 2\sigma_p)$, and $(\mu_p - 3\sigma_p)$ to $(\mu_p + 3\sigma_p)$.

b. Find the probabilities requested in part **a** using a normal approximation to the given binomial statistic.

c. Compare each of the three probabilities found by the binomial sampling distribution to the corresponding probabilities found by the normal approximation.

d. Let p have a binomial sampling distribution with $n = 20$ and $\pi = .5$. Repeat parts **a**, **b**, and **c**. Notice that the probability estimates provided by the normal approximation are more accurate for $n = 20$ than for $n = 10$.

e. Let p have a binomial sampling distribution with $n = 10$ and $\pi = .1$. Repeat parts **a**, **b**, and **c**. Notice that the probability estimates provided by the normal approximation are poor in this case. Explain why this occurs.

Chapter 7

INFERENCES ABOUT ONE POPULATION

A GALLUP REPORT

Introduction

A report of a 1981 Gallup poll is shown in Figure 7.1* (page 320). The poll was conducted to investigate the opinions of Americans about the possibility of voting on major issues facing the nation, as well as for candidates. The report concludes that voter turnout would improve substantially if the voters were allowed to vote on major issues.

This conclusion is an inference based on a sample of Americans. In the next to last paragraph of the report we see that 1,553 adults were sampled, and the percentages given in the report were calculated from this sample. The last paragraph provides an indication of how accurately these percentages estimate the corresponding percentages for all adult Americans. That is, the last paragraph provides a statement of reliability. In this chapter we will present statistical methods for making inferences such as this from sample data. We will also discuss terms such as *95% confidence* and *error attributable to sampling*, which appear in the last paragraph.

* George Gallup, "Major Issues Would Draw More Voters," *The Cincinnati Enquirer*, May 29, 1981, p. B4.

Major Issues Would Draw More Voters

BY GEORGE GALLUP

PRINCETON, N.J.—If Americans could vote on major *issues* facing the nation, as well as *candidates*, voter turn-out in national elections—now the worst of any major democracy in the world—could improve substantially.

This conclusion is based on the results of a recent Gallup Poll in which national samples of those who voted in the 1980 presidential election and those who failed to vote were questioned about the related proposals that have been offered as ways of increasing turnout in national elections.

When non-voters were asked if they would be more likely or less likely to vote in national elections if they could vote on important national issues as well as candidates, as many as half (48%) said they would be more likely to do so.

The implications of these findings are far-reaching: If half the non-voters who say they would be more likely to vote actually did so turnout would increase to almost 80%, a level equal to that found in other major democratic nations.

Turnout in the 1980 presidential election was the lowest in 32 years. In fact, only 54% of those eligible took the trouble to vote in November.

As the survey results indicate, measures which would provide the public with avenues for direct expression of their views might well enhance their participation in candidate elections. This may have particular application in election years similar to 1980, when many prospective voters were less than enthusiastic about the major-party nominees.

FURTHER EVIDENCE of the appeal of voting on issues as well as for candidates is seen in the results of a question on a constitutional amendment, described by its sponsors as the Voter Initiative Amendment. According to this amendment, a national referendum would be held on any issue when 3% of all voters in the previous presidential election signed petitions demanding such a vote.

This amendment is favored by a 2-to-1 margin among all persons interviewed, with voters and non-voters in last fall's election holding similar views.

A recent report by the Committee for the Study of the American Electorate indicated that most conventional proposals for increasing voter turnout have had little effect. Participation was not notably higher in states that used postcard registration or allowed voters to register at the polls on Election Day.

The current survey results lend support to the view of proponents of initiative and referendum that offering citizens an opportunity to vote on major issues of the day would encourage greater participation by giving them a greater voice in national affairs.

Here is the first question asked:

Would you be more likely or less likely to vote in national elections if you could vote on important national issues as well as on candidates, or wouldn't it make any difference?

Here are the results, based on the responses of non-voters in the 1980 presidential election. Of particular interest are the views of young adults, whose voting record has consistently been poor.

VOTE ON ISSUES
(Based on non-voters)

	More likely	Less likely	No difference	No opinion
NATIONAL	48%	2%	43%	7%
Under 30 years	56%	3%	36%	5%
30–49 years	43%	2%	46%	9%
50 and older	41%	1%	50%	8%

A second question, asked of the total sample, was this:

The U.S. Senate will consider a proposal that would require a national vote—that is, a referendum—on any issue when 3% of all voters who voted in the most recent presidential election sign petitions asking for such a nation-wide vote. How do you feel about this plan—do you favor or oppose such a plan?

As in the case of the earlier question, younger adults and independents are among the most enthusiastic supporters of the initiative. Both these groups have below average records of election participation.

Here are the results by key population groups:

VOTER INITIATIVE REFERENDUM

	Favor	Oppose	No opinion
NATIONAL	52%	23%	25%
Voters in 1980	53%	25%	22%
Non-voters	50%	19%	31%
Under 30 years...........	59%	19%	22%
30–49 years	51%	25%	24%
50 and older	48%	23%	29%
Republicans	51%	26%	23%
Democrats	49%	20%	31%
Independents	58%	23%	19%

The latest results are based on in-person interviews with 1,553 adults, 18 and over, conducted in more than 300 scientifically-selected localities across the nation during the period April 10–13.

For results based on a sample of this size, one can say with 95% confidence that the error attributable to sampling and other random effects could be three percentage points in either direction.

Figure 7.1

Source: George Gallup, "Major Issues Would Draw More Voters." *The Cincinnati Enquirer*, May 29, 1981, p. B4.

7.1

THE ELEMENTS OF A TEST OF A HYPOTHESIS

In Chapter 6 we used the sampling distribution of the sample mean, $\bar{x}$, and the rare event approach to make informal inferences about a population mean, μ. Similarly, we used the sampling distribution of the sample proportion, p, to make inferences about a population proportion, π. In this chapter, we present formal structures for making inferences about population parameters such as μ and π. We will begin by introducing a **test of hypothesis.**

Definition 7.1

A **hypothesis** is a conjecture about the nature of a population.

Hypotheses are usually phrased in terms of population parameters. The following are some examples of hypotheses:

1. $\mu = 6$ (*A population mean equals* 6.)
2. $\mu > 6$ (*A population mean is greater than* 6.)
3. $\sigma^2 = 30$ (*A population variance equals* 30.)
4. $\pi \neq .4$ (*A population proportion is not equal to* .4.)
5. $\mu < 100$ (*A population mean is less than* 100.)

Definition 7.2

A **test of hypothesis** is a statistical procedure used to make a decision about the value of a population parameter.

In the remainder of this text, we will find that there are certain elements common to all tests of hypotheses. These elements are introduced and discussed briefly in this section. They will be discussed in more detail in the next section, where examples of tests of hypotheses will be presented. To help you see the connection between the informal tests, which we were informally conducting using the rare event approach, and the formal structure, which we will use from now on, we will refer to Example 6.11 (Section 6.4). In Example 6.11, we considered the following situation:

A candidate for political office claims he will receive 55% of the votes in the upcoming election. As a supporter of another candidate, you suspect that the claim of 55% is too high, and that the candidate making the claim will actually receive a smaller percentage of the votes. Hoping to provide support for your suspicions and enough evidence to refute the candidate's claim, you randomly sample eighty voters and find p, the fraction of voters who will vote for the candidate who claims to expect 55% of the votes.

Examining this example, we can find two hypotheses that are of interest:

Hypothesis 1:   The candidate claims he will receive 55% of the votes ($\pi = .55$).

Hypothesis 2:   The supporter of another candidate suspects the claim of 55% is too high ($\pi < .55$).

In a test of hypothesis, the hypotheses of interest are given special names. The first hypothesis is called the **null hypothesis** and is defined in the following box.

Definition 7.3

The **null hypothesis,** denoted by H_0, specifies the value of a population parameter. The experiment is conducted to determine whether this specified value is unreasonable.

In the experiment we are considering, the supporter of a second candidate is the one conducting the experiment. Therefore, the null hypothesis of interest, the one that the experimenter suspects is unreasonable, is

H_0: $\pi = .55$

(that is, the first candidate will receive 55% of the vote).

The second element of a test of hypothesis is the **alternative hypothesis.**

Definition 7.4

The **alternative hypothesis,** denoted by H_a, gives an opposing conjecture to that given in the null hypothesis. The experiment is conducted to determine whether the alternative hypothesis is supported.

In the example being considered, the supporter of a second candidate suspects the 55% figure is too high. This is the hypothesis to be supported, and thus the alternative hypothesis is

H_a: $\pi < .55$

(that is, the first candidate will receive less than 55% of the vote).

Consider the two hypotheses we have given:

H_0: $\pi = .55$ H_a: $\pi < .55$

You may have noticed that these hypotheses are not all-inclusive, because the values of $\pi > .55$ are not included under either hypothesis. To make the hypotheses all-inclusive, we could write the null hypothesis as H_0: $\pi \geq .55$, and if you prefer this method of expressing H_0, you may use it. However, in this text we will always express the null hypothesis as a strict equality, and thus specify a unique value for the parameter in the null hypothesis. We prefer this method because the specified value is used when conducting the test.

The alternative hypothesis is often called the **research hypothesis,** because this hypothesis expresses the theory that the experimenter, or researcher, believes to be true.

The next element in a test of hypothesis is a **test statistic.**

Definition 7.5

A **test statistic** is a quantity calculated from the sample that is used when making a decision about the hypotheses of interest.

After a sample of eighty voters was obtained and p, the proportion of the voters in favor of the first candidate, was found, we calculated a z-score using

$$z = \frac{p - .55}{\sigma_p}$$

This z-score compared the observed value of the sample proportion p to the value $\pi = .55$ specified in the null hypothesis. This could serve as a test statistic in this example.

To interpret the value of the test statistic, it is necessary to introduce the fourth element of a test of hypothesis, the **rejection region.**

Definition 7.6

A **rejection region** specifies the values of the test statistic for which the null hypothesis is rejected (and for which the alternative hypothesis is accepted).

The rejection region identifies the values of the test statistic that support the alternative hypothesis and would be improbable (rare) if the null hypothesis were true. Since we do not expect to observe rare events (improbable values of the test statistic), we will reject the null hypothesis when the sample produces such a value. For the political candidate example, any time the sample proportion, p, was less than .55, this would provide some support for the alternative hypothesis ($\pi < .55$). We would decide that there is enough support to conclude that π is less than .55 if the observed value of p represents a rare event. Since a value of p more than 2 standard errors below .55 would certainly be rare, and this would result in a z-score less than -2, a possible rejection region is

Reject H_0 if $z < -2$

In the remainder of this text, we will test hypotheses about many different parameters based on different sampling methods. However, every test of hypothesis will include the four elements we have introduced: the null hypothesis (H_0), the alternative hypothesis (H_a), the test statistic, and the rejection region. The goal of any test of hypothesis is to decide which hypothesis—the null or the alternative—should be accepted. Since any decision will be based on the partial information about a population contained in a sample, there will always be a possibility of an incorrect decision. Table 7.1 summarizes four possible situations that can arise in a test of hypothesis.

Table 7.1

Possible decisions and consequences for a test of hypothesis

Possible Decisions	True State of the Population	
	H_0 Is True	H_a Is True
Reject H_0 (Accept H_a)	Type I error	Correct decision
Accept H_0	Correct decision	Type II error

The two right-hand columns of the table indicate the two possible states of the population; that is, either H_0 is true or H_a is true. The two rows give the two possible decisions; either reject H_0 (and accept H_a) or accept H_0. The body of the table contains the possible consequences of making these decisions.

If the null hypothesis is rejected and, in fact, the null hypothesis is true, an error has been made. This is called a **Type I error.** A **Type II error** would occur if the null hypothesis were accepted and, in fact, the alternative hypothesis is true. A correct decision would be made if the alternative hypothesis is true and the null hypothesis is rejected, or if the null hypothesis is true and it is accepted.

For the voting example, recall that the hypotheses of interest are H_0: $\pi = .55$ and H_a: $\pi < .55$. For this situation a Type I error would be made if we concluded that the candidate was going to get less than 55% of the vote, but the candidate actually will get 55% of the vote. A Type II error would be made if we concluded that the candidate was going to get 55% of the vote, but the candidate actually will get less than 55% of the vote.

Since we can never eliminate the possibility of making a Type I error or a Type II error when using samples to make inferences, we will consider the probabilities of making these errors. The probability of a Type I error is denoted by the Greek letter alpha, α. That is,

$\alpha = P(\text{Type I error})$

$= P(\text{Rejecting } H_0 \text{ if } H_0 \text{ is true})$

The probability of a Type II error is denoted by the Greek letter beta, β. That is,

$\beta = P(\text{Type II error})$

$= P(\text{Accepting } H_0 \text{ if } H_0 \text{ is false})$

We would like both α and β to be near 0, but this is generally not possible. Since the experimenter wants to conclude that H_a is true (reject H_0), we will be most concerned that α is a small probability, such as .01 or .05. In other words, we will want to make sure that if H_0 is true, it will be very rare that we will reject H_0. The experimenter is free to choose the value of α, that is, to determine how rare an observed event must be in order to reject H_0.

Determining the value of β for the tests of hypotheses we will present is difficult, so we will not attempt this. As you will see, this will not create any problems for us. We will discuss β and accepting H_0 with the aid of specific examples in the next section.

Nonstatistical "tests of hypotheses" are often conducted in the everyday world. A courtroom trial provides a good example of such a test, as shown by Example 7.1.

EXAMPLE 7.1   The prosecuting attorney in a trial attempts to show that the defendant is guilty. The trial can be thought of as a test of hypothesis, since a decision is to be made as to whether the defendant is guilty or innocent.

  a. State the null and alternative hypotheses of interest to the prosecuting attorney.
  b. Define the Type I error and Type II error for this situation.
  c. Discuss α and β for this situation.

Solution   **a.** Since the prosecuting attorney wants to show that the defendant is guilty, this specifies the alternative hypothesis, and the hypotheses are:

      H_0:   Defendant is innocent

      H_a: Defendant is guilty

  b. In general, a Type I error is rejecting H_0 if H_0 is true. For this example, a Type I error would occur if the defendant were found guilty if, in fact, the defendant is innocent. A Type II error occurs if H_0 is accepted and H_0 is false. Thus, a Type II error would be pronouncing the defendant innocent if, in fact, the defendant is guilty.

  c. Both the errors defined in part **b** are very serious, and the judicial system is designed so that innocent people should not be convicted and guilty people should not go free. Unfortunately, the system is not perfect, and errors are made. Since we would not want an innocent person to receive a jail sentence, or any type of conviction, we want to be sure a Type I error is rarely made. In other words, α, the probability of a Type I error, should be very small. We would also like β, the probability of finding a defendant innocent if, in fact, he or she is guilty, to be very small. However, since such an effort is made to ensure that α is small, β cannot be controlled quite as well. It is not uncommon to hear of a trial where an "obviously guilty" person is found "innocent." The person is often freed due to a technicality. These technicalities are needed to help ensure that innocent people are set free, but they also allow some guilty people to go free. Most people believe it is worse to convict one innocent person than occasionally to free a guilty person. Thus, α is kept extremely small at the cost of β being somewhat larger.   ■

We will find exactly the same kind of phenomenon as illustrated in Example 7.1 in hypothesis testing: By keeping α small we avoid accepting our research (alternative) hypothesis if the null hypothesis is true. To do otherwise would invite the criticism that we have biased our research in order to prove our own hypothesis. The sacrifice we make to keep α small is that the chance of accepting the null hypothesis if our research hypothesis is true (β) may be greater than we prefer. In short, we require substantial evidence before accepting our research hypothesis; in effect, the null hypothesis is assumed true until proven otherwise.

Exercises (7.1–7.5)

Learning the Language

7.1   Define the following terms:
- a.   Hypothesis
- b.   Test statistic
- c.   Test of hypothesis
- d.   Rejection region
- e.   Null hypothesis
- f.   Type I error
- g.   Alternative hypothesis
- h.   Type II error

7.2   What do the following symbols denote?
   a. H_0 b. H_a c. α d. β

Applying the Concepts

7.3   Each person who is going to fly on a major airline is electronically searched for dangerous weapons. Each of these searches can be viewed as a nonstatistical test of the following hypotheses:

H_0:   Person is carrying weapons

H_a: Person is carrying no weapons

- a.   Which of the hypotheses is the research hypothesis? The null hypothesis? The alternative hypothesis?
- b.   Define a Type I error and a Type II error in terms of this example.
- c.   Define α and β in terms of this example.
- d.   Which error do you believe is more serious, a Type I error or a Type II error?
- e.   Considering your answer to part d, what should the relative values of α and β be? (Which, if either, should be smaller?)

7.4   A car manufacturer is considering the production of a new model. Based on the results of many tests of the model, cost analyses, market analyses, etc., the manufacturer will decide between the following hypotheses:

H_0:   Model will not be profitable

H_a: Model will be profitable

If the manufacturer accepts H_0, the new model will not be produced. On the other hand, if the manufacturer accepts H_a, the new model will be put into production.

a.   Define a Type I error and a Type II error in terms of this example.

b.   Discuss the consequences of making each of the errors discussed in part a.

c.   Discuss what you believe the relative values of α and β should be. (Should one be smaller than the other?)

7.5   A dental laboratory can check for mercury vapor contamination by mechanically monitoring the air quality within the laboratory. Based on the results of the monitoring, a decision will be made concerning the following hypotheses:

H_0:   Laboratory is contaminated by mercury vapor

H_a: Laboratory is not contaminated

a.   Discuss the consequences of rejecting H_0 (and accepting H_a) if H_0 is really true.

b.   Discuss the consequences of accepting H_0 if H_0 is really false.

c.   Define α and β in terms of this example.

d.   Should either α or β (or both) be near 0? Why?

7.2

A LARGE-SAMPLE TEST OF HYPOTHESIS ABOUT A POPULATION MEAN, μ

We are now ready to present some statistical procedures that are commonly used to make inferences. These procedures will be based upon sampling distributions such as the normal distribution. To ensure that the procedures are valid (that is, to ensure that a sampling distribution has the required properties), we must be careful that certain conditions are met. As we introduce each statistical procedure, we will state when that procedure is valid.

In this section we will present a test of hypothesis about a population mean, μ. Since the test will use the information contained in a large random sample, we will refer to it as a **large-sample test of hypothesis about μ.**

When to Use a Large-Sample Test of Hypothesis About μ

A large-sample test of hypothesis about the mean, μ, can be used validly only if:

1. A random sample is taken from the population.

2. The sample size is large.

These are essentially the same conditions as those given in the Central Limit Theorem in Chapter 6.* When these conditions are satisfied, we are assured that the sampling distribution of the sample mean is approximately normal. Note that the population of measurements from which the sample is selected may have any distribution. Remember, we will consider a sample size of $n \geq 30$ measurements to be a large sample.

The large-sample test of hypothesis about μ is given in the accompanying box. The testing procedure is stated in terms of the elements of a test of hypothesis introduced in Section 7.1.

Large-Sample Test of Hypothesis About μ

H_0: $\mu = \mu_0$

H_a: $\mu > \mu_0$ or H_a: $\mu < \mu_0$ or H_a: $\mu \neq \mu_0$

Test statistic: $z = \dfrac{\bar{x} - \mu_0}{s/\sqrt{n}}$

Rejection region: Use Appendix Table III.

The value of μ specified in the null hypothesis is denoted by μ_0 (read "mu nought"). The symbol μ_0 represents a number which the experimenter would determine. For example, the experimenter might wish to test H_0: $\mu = 12$, and μ_0 would be equal to 12. There are three possible alternative hypotheses given in the box. In any particular problem, only one of the alternatives would be chosen. The appropriate alternative is the one that reflects the relationship between μ and μ_0 the experimenter hypothesizes to be true.

The test statistic is the same for all three possible alternatives, and is based on the z-score given in Section 6.2:

$$z = \frac{\bar{x} - \mu_{\bar{x}}}{\sigma_{\bar{x}}} = \frac{\bar{x} - \mu_{\bar{x}}}{\sigma/\sqrt{n}} \approx \frac{\bar{x} - \mu_{\bar{x}}}{s/\sqrt{n}}$$

Since the standard deviation of the population, σ, will not be known in practical problems, the sample standard deviation, s, is substituted for σ in the test statistic. Using s to estimate σ in the z-score is valid when a large random sample has been obtained. If H_0 is true, then $\mu_{\bar{x}}$ equals μ_0, and the test statistic, z, has a normal sampling distribution. When the test statistic is calculated for an observed value of $\bar{x}$, it measures the distance between $\bar{x}$ and μ_0 in units of (approximate) standard errors.

The rejection region is found by using Appendix Table III. The rationale for this table and its use will be demonstrated in the examples.

---

* To be more precise, the population from which the sample is taken must also have a finite mean and standard deviation. This will always hold for the inference problems we will discuss and for practical problems in general.

EXAMPLE 7.2   Building specifications in a certain city require that residential sewer pipe have a minimum mean breaking strength of 2,400 pounds per foot. A contractor has been having problems with pipe bought from a particular manufacturer, and the contractor believes that this manufacturer's pipe does not meet the minimum standard for mean breaking strength of 2,400 pounds per foot. In an attempt to substantiate this belief, the contractor tests a random sample of fifty-five sections of the manufacturer's sewer pipe and obtains the following statistics on breaking strength:

$\bar{x} = 2{,}340$ pounds per foot $s = 200$ pounds per foot

Is there sufficient evidence to conclude that the contractor is correct? Use $\alpha = .10$.

Solution   The population of interest is the set of breaking strengths of all sewer pipes the manufacturer produces. Since a large sample ($n = 55$) is randomly selected from this population, it is valid to use a large-sample test of hypothesis.

To give the elements of the test as specified in the box, we must first select the appropriate null and alternative hypotheses. It is usually easier to identify the alternative hypothesis first. The contractor (the experimenter) wants to see if the mean breaking strength is less than 2,400 pounds per foot, that is, $\mu < 2{,}400$. This specifies the alternative hypothesis, and the corresponding null hypothesis is $\mu = 2{,}400$. The elements of the test are

H_0: $\mu = 2{,}400$ H_a: $\mu < 2{,}400$

Test statistic: $z = \dfrac{\bar{x} - 2{,}400}{s/\sqrt{n}}$

Rejection region: Reject H_0 if $z < -1.28$.

Before we use the sample data to calculate the test statistic and interpret the result, we will discuss in more detail how to determine the rejection region. The rejection region was obtained from Appendix Table III, which is reproduced here, in Table 7.2, for convenience. To use this table, you must know the value of α and the type of alternative hypothesis ($<$, $>$, or $\neq$). For this example, $\alpha = .10$ and the alternative hypothesis is

Table 7.2

A reproduction of Appendix Table III, z-Values for Rejection Regions for Large-Sample Tests of Hypotheses

Value of α	Type of Alternative Hypothesis, H_a		
	$>$	$<$	$\neq$
$\alpha = .10$	Reject H_0 if $z > 1.28$	Reject H_0 if $z < -1.28$	Reject H_0 if $z > 1.65$ or $z < -1.65$
$\alpha = .05$	Reject H_0 if $z > 1.65$	Reject H_0 if $z < -1.65$	Reject H_0 if $z > 1.96$ or $z < -1.96$
$\alpha = .01$	Reject H_0 if $z > 2.33$	Reject H_0 if $z < -2.33$	Reject H_0 if $z > 2.58$ or $z < -2.58$

$\mu < 2,400$. Thus, the rejection region is given in the top row of the table, under the symbol $<$, as indicated by the shading in Table 7.2.

The rejection region specifies the calculated values of the test statistic that are considered to be rare events if the null hypothesis is true. Thus, it indicates when the null hypothesis should be rejected. Since the alternative hypothesis is $\mu < 2,400$, only values of $\bar{x}$ less than 2,400 would provide evidence that the alternative hypothesis is true. Thus, only if $\bar{x}$ is considered to be a rare event that supports the alternative hypothesis, will the null hypothesis be rejected. Note that *rejection of the null hypothesis is equivalent to acceptance of the alternative hypothesis.*

Table 7.2 tells us to reject the null hypothesis (at $\alpha = .10$) when $z < -1.28$. The reason for this is shown in Figure 7.2. If the null hypothesis is true, and, in fact, μ does equal 2,400, then we want the probability of rejecting H_0 (the probability of a Type I error) to be .10. This probability is indicated in Figure 7.2. The value 1.28 is found by consulting Appendix Table II. The area under the curve between 0 and -1.28 is .400, so the area under the curve to the left of -1.28 is $.500 - .400 = .100$ (or .10). Thus, if the null hypothesis is true, we would observe a computed value of $z < -1.28$ only 10% of the time.

Figure 7.2

Normal curve area

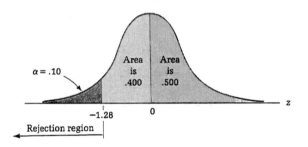

The other values in Table 7.2 were obtained from Appendix Table II in the same way. Normal curves were drawn and the area under the curve corresponding to α was indicated for each situation. The three situations ($<$, $>$, and $\neq$) are shown in Figure 7.3. The z-values corresponding to the areas shown in the figure for the three values of α (.10, .05, and .01) can be obtained from Table II. These are the only values of α we will use in this text, so we will always refer to Table 7.2 (or Appendix Table III) to obtain a rejection region. If you need a rejection region for a different value of α, you will have to obtain it by using Appendix Table II, as described above.

We will now complete the test of hypothesis by finding the calculated value of the test statistic and interpreting the result.

From the statement of the example we obtain $\bar{x} = 2,340$, $s = 200$, and $n = 55$. The calculated value of the test statistic is

$$z = \frac{2,340 - 2,400}{200/\sqrt{55}} = \frac{-60}{26.97} = -2.22$$

The value of $z = -2.22$ means that $\bar{x} = 2,340$ is (approximately) 2.22 standard errors below $\mu_0 = 2,400$. Examining the rejection region, we see that H_0 is rejected if

Figure 7.3
Areas used to find rejection
regions

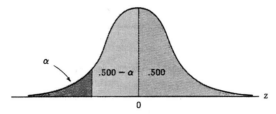

a. Form of H_a: $<$

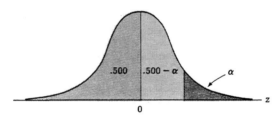

b. Form of H_a: $>$

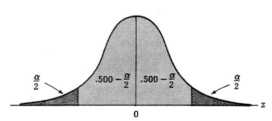

c. Form of H_a: $\neq$

$z < -1.28$. Since -2.22 is less than -1.28, we will reject H_0 and make the following inference: There is sufficient evidence to conclude that the mean breaking strength of the manufacturer's sewer pipe is less than 2,400 pounds per foot. The contractor is justified in believing that the manufacturer's pipe does not meet the minimum specifications. Since the test was conducted with $\alpha = .10$, we are confident our inference is correct.   ■

The last sentence in Example 7.2 is a statement of reliability for the inference that was made. When the null hypothesis is rejected in a test of hypothesis, the inference should always be accompanied by a statement of reliability. The following are three popular ways of phrasing such a statement:

1. We are confident the inference is correct since the test was conducted with $\alpha = .10$.
2. We are confident the inference is correct since the test was conducted at the .10 **level of significance.**
3. We are 90% confident the inference is correct.

The first two statements are essentially identical. The probability of a Type I error, α, is also called the **significance level** (or **level of significance**). The third statement is phrased in terms of $(1 - \alpha)100\%$ confidence. All three are equivalent and you may use whichever you prefer.

Remember, alternative hypotheses are always expressed as inequalities and, in this text, null hypotheses are always expressed as equalities. We need to specify one value of μ in H_0 to be used in the test statistic, because we need to determine the values that contradict the null hypothesis and support the alternative hypothesis from the sampling distribution of the test statistic. Any value of the test statistic that leads to rejection of, say, H_0: $\mu = 2,400$ in favor of H_a: $\mu < 2,400$, would also lead to rejection of any value of μ greater than 2,400. For this reason, the null hypothesis is given as the equality rather than the more inclusive inequality $\mu \geq 2,400$.

Before considering another example of a test of hypothesis, review the basic steps given in the accompanying box for choosing the appropriate null and alternative hypotheses.

Helpful Hints

When choosing the null and alternative hypotheses, take the following steps:

1. The experiment is conducted to see if there is support for some hypothesis. This will be the alternative hypothesis, expressed as an inequality.

    Example: H_a: $\mu < 2,400$

2. State the null hypothesis as an equality. The parameter being tested is set equal to the number specified in H_a.

    Example: H_0: $\mu = 2,400$

EXAMPLE 7.3   A research psychologist plans to administer a test designed to measure self-confidence to a random sample of fifty professional athletes. The psychologist theorizes that professional athletes tend to be more self-confident than others. Since the nationwide average score on the test is known to be 72, the theory may be partially validated if it can be shown that the mean score for all professional athletes, μ, exceeds 72.

Suppose the sample mean and standard deviation of the fifty scores are

$$\bar{x} = 74.1 \qquad s = 13.3$$

Do these data support the research hypothesis of the psychologist? Use $\alpha = .05$.

Solution   Since a large sample (fifty test scores) is randomly selected from a population of test scores, it is valid to use a large-sample test of hypothesis. The research hypothesis of the psychologist is that the mean score for all professional athletes exceeds 72, that is, $\mu > 72$. The corresponding null hypothesis is $\mu = 72$. The elements of the test are

$$H_0: \quad \mu = 72 \qquad H_a: \quad \mu > 72$$

Test statistic: $\quad z = \dfrac{\bar{x} - 72}{s/\sqrt{n}}$

Rejection region: Reject H_0 if $z > 1.65$.

The rejection region was obtained from Appendix Table III, and is shown in Figure 7.4. Since $\alpha = .05$ and the alternative hypothesis is $\mu > 72$, we will reject H_0 only if $\bar{x}$ is more than 1.65 standard errors *above* $\mu_0 = 72$. In other words, H_0 will be rejected only if $z > 1.65$.

Figure 7.4
Rejection region for
Example 7.3

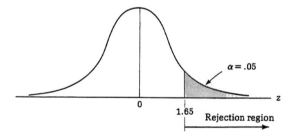

Since $\bar{x} = 74.1$ and $s = 13.3$, the calculated value of the test statistic is

$$z = \frac{74.1 - 72}{13.3/\sqrt{50}} = 1.12$$

The rejection region tells us that z must be larger than 1.65 in order to reject H_0. Since $z = 1.12$ is not larger than 1.65, we will not reject H_0. What interpretation should we make? There are two remaining possibilities:

1. Accept the null hypothesis.
2. Make no decision about the hypotheses.

If we were to accept the null hypothesis, we would run the risk of making a Type II error (accepting H_0 if H_0 is false). The probability of the Type II error, β, would provide a measure of reliability if we were to accept the null hypothesis. Although we will not actually calculate any values of β, we can make some intuitive comments about β.

In Figure 7.5 (page 334) we have drawn two graphs to show the areas corresponding to α and β. If H_0 is true, then the sampling distribution of the test statistic z would be as shown in Figure 7.5a, and the value of α would equal the shaded area under the curve above the rejection region.

If H_a is true, then the sampling distribution of the test statistic would be as shown in Figure 7.5b. In this case, the mean of the sampling distribution is larger than that of Figure 7.5a, and the entire distribution is shifted to the right. The value of β, the

Figure 7.5
Graphs showing α and β

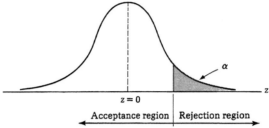

a. Sampling distribution if H_0 is true ($\mu = 72$).
In this case, $\bar{x}$ has mean 72 and
$z = (\bar{x} - 72)/\sigma_{\bar{x}}$ will have mean 0.

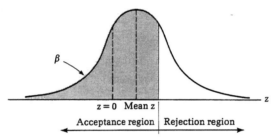

b. Sampling distribution if H_a is true ($\mu > 72$).
In this case, $\bar{x}$ has mean greater than 72 and
$z = (\bar{x} - 72)/\sigma_{\bar{x}}$ will have mean greater than 0.

probability of accepting H_0 if H_a is true, would equal the shaded area under the curve corresponding to the acceptance region.

If H_a is true, but the value of the parameter is nearly equal to the value specified in H_0, the two graphs shown in Figure 7.5 will be nearly identical, and the value of β will be slightly less than $(1 - \alpha)$. Since we always make sure that α is small (near 0), β could be rather large (near 1). Because we do not control the value of β, and the chance of making an error could be very large if we were to accept the null hypothesis, we will never make this decision. If we cannot reject the null hypothesis, we will refuse to make any decision at all.

Thus, the interpretation for this example is that there is insufficient evidence to conclude that the mean score for all professional athletes is higher than the nationwide average of 72. ∎

You can now see that when statistical procedures are used, inferences are made only when we are confident the inference is correct. If the data do not provide sufficient evidence for us to believe the inference is reliable, we will say the evidence is insufficient and refuse to make an inference.

The following helpful hints will aid you in interpreting tests of hypotheses:

Helpful Hints

A test of hypothesis leads to one of two interpretations:

1. If the calculated value of the test statistic is in the rejection region, the null hypothesis is rejected. The meaning of accepting the alternative hypothesis is explained, and a statement of reliability is given.
2. If the calculated value of the test statistic is not in the rejection region, the null hypothesis is neither accepted nor rejected. It is stated that there is insufficient evidence to make a decision.

Since we will be conducting many tests of hypotheses in the remainder of this text, we will always follow the same format. The steps we will follow are listed in the next box, and we suggest that you follow the same steps when working exercises.

Steps to Follow When Testing a Hypothesis

1. Identify the population and parameter of interest.
2. Give the null and alternative hypotheses.
3. Give the test statistic, and state the conditions required for the validity of the test procedure being used.
4. Give the rejection region for the desired value of α.
5. Calculate the value of the test statistic.
6. Interpret the results.

Examples 7.2 and 7.3 illustrated the $<$ and $>$ forms of H_a as shown in Figure 7.3**a** and **b**. These forms of the test are referred to as **one-tailed tests.** The remaining form for H_a, namely $\neq$, as shown in Figure 7.3**c**, is referred to as a **two-tailed test.** The next example illustrates this form.

EXAMPLE 7.4   A nutritionist believes that 12-ounce boxes of breakfast cereal should contain an average of 1.2 ounces of bran. However, the nutritionist suspects that a popular cereal has a different mean bran content, so he carefully analyzes the contents of a random sample of sixty 12-ounce boxes of the cereal to determine their bran content. The following statistics are calculated:

$\bar{x} = 1.170$ ounces of bran $s = .111$ ounce of bran

Do the data provide sufficient evidence to conclude that the mean content of bran differs from 1.2 ounces? Test at the .05 level of significance.

Solution   We will conduct the test following the steps given in the box. The population of interest is the bran contents of all the boxes of this cereal. The parameter of interest is μ, the mean bran content of all the boxes.

The nutritionist wants to see if the cereal's mean content of bran, μ, *differs* from 1.2 ounces, and the elements of the test are as follows:

$$H_0: \quad \mu = 1.2 \qquad H_a: \quad \mu \neq 1.2$$

$$\text{Test statistic:} \quad z = \frac{\bar{x} - 1.2}{s/\sqrt{n}}$$

Since a large ($n = 60$) random sample is taken from the population of interest, it is valid to conduct a large-sample test of hypothesis.

Rejection region: Reject H_0 if $z > 1.96$ or $z < -1.96$.

The rejection region, which was obtained from Appendix Table III, is shown in Figure 7.6. Since the alternative is $\mu \neq 1.2$, we will reject H_0 if $\bar{x}$ is more than 1.96 standard errors *above* 1.2 or more than 1.96 standard errors *below* 1.2. Notice that the value of α is split, with half for each part of the rejection region.

Figure 7.6
Rejection region for
Example 7.4

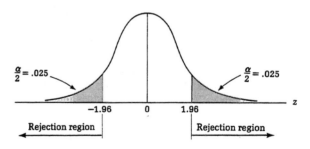

The calculated value of the test statistic is

$$z = \frac{1.170 - 1.2}{.111/\sqrt{60}} = \frac{-.030}{.014} = -2.09$$

Since the value $z = -2.09$ is in the lower end of the rejection region, we conclude that the cereal's mean content of bran differs from 1.2 ounces. It appears that the mean content of bran is less than 1.2 ounces (since $\bar{x} = 1.170$ is less than 1.2, and z is in the lower end of the rejection region). We have 95% confidence that this inference is correct. ∎

COMPUTER
EXAMPLE 7.1

In Example 7.4 we were interested in μ, the mean bran content of 12-ounce boxes of a breakfast cereal. To test the hypotheses of interest, a random sample of sixty bran contents was obtained. The data used to conduct the test are given below:

1.09	1.23	1.29	1.11	1.33	1.13	1.10	1.20	1.11	1.30	1.12
1.25	1.09	1.16	1.12	1.08	1.08	1.08	1.14	1.16	1.20	1.11
1.20	1.14	1.10	1.11	1.08	1.08	1.18	1.24	1.11	1.08	1.19
1.08	1.21	1.37	1.32	1.67	1.08	1.08	1.14	1.37	1.14	1.09
1.08	1.08	1.09	1.15	1.31	1.14	1.21	1.14	1.29	1.40	1.08
1.08	1.33	1.10	1.10	1.11						

Recall that the elements of the test are

H_0: $\mu = 1.2$ H_a: $\mu \neq 1.2$

Test statistic: $z = \dfrac{\bar{x} - 1.2}{s/\sqrt{n}}$

Rejection region: Reject H_0 if $z > 1.96$ or $z < -1.96$.

When the summary statistics, $\bar{x}$ and s, are given for a set of data, it is easy to calculate the value of the test statistic using a calculator. Given the actual sample data, it is a good idea to use a computer package to aid in the calculations. Using the sixty bran measurements, Minitab yielded the following printout:

```
TEST OF MU = 1.2000 VS MU N.E. 1.2000
THE ASSUMED SIGMA = 0.111

            N      MEAN    STDEV    SE MEAN       Z    P VALUE
C6         60    1.1705   0.1111    0.0143    -2.06      0.040
```

The first line indicates that we are testing the null hypothesis $\mu = 1.2$ versus the alternative hypothesis $\mu \neq 1.2$ ($\neq$ is abbreviated "N.E."). To obtain this printout, we must give Minitab a value of the standard deviation. Since the value of the population standard deviation, σ, is not known, we substitute the value of the sample standard deviation, s. We used Minitab to find the value of s for the given sample, and then used that to conduct the test. On the printout, Minitab reminds us of this by printing "THE ASSUMED SIGMA = 0.111." Other information that is provided about this sample includes $n = 60$, $\bar{x} = 1.1705$, $s = 0.1111$, $s/\sqrt{n} = 0.0143$ (this is the standard error of the mean), and $z = -2.06$. We had calculated a value of the test statistic of $z = -2.09$. The value given by Minitab is slightly different because we rounded the values of $\bar{x}$ and s to three decimal places, whereas Minitab used more decimal places. Once the calculated value of z is obtained, we would interpret the results in the same way. In this case, we are 95% confident that the mean bran content is less than 1.2 ounces.

We have not yet said anything about the p-value of 0.040 that is given on the printout. We will discuss p-values in the next section. For the moment, ignore this value when examining Minitab printouts. ∎

Exercises (7.6–7.20)

Using the Tables

7.6   A large-sample test of hypothesis is conducted to test the following:

$$H_0: \quad \mu = 53 \qquad H_a: \quad \mu \neq 53$$

   a.   Give the rejection region if $\alpha = .01$.
   b.   Give the rejection region if $\alpha = .10$.
   c.   Give the rejection region if $\alpha = .05$.

7.7   A large-sample test of the null hypothesis $H_0: \mu = 120$ is conducted with $\alpha = .05$. Give the rejection region for each of the following alternative hypotheses:
   a. $H_a: \mu > 120$ b. $H_a: \mu \neq 120$ c. $H_a: \mu < 120$

7.8   A large-sample test of hypothesis is conducted to test the following:

$$H_0: \quad \mu = 10 \qquad H_a: \quad \mu > 10$$

   a.   Give the rejection region if $\alpha = .10$.
   b.   Give the rejection region if $\alpha = .05$.
   c.   Explain the difference in your answers to parts a and b.

Learning the Mechanics

7.9   A random sample of n observations is taken from a population with mean μ. Complete a test of hypothesis for each of the following situations:
   a. $H_0: \mu = 10$; $H_a: \mu \neq 10$; $n = 70$; $\bar{x} = 11.2$; $s = 7.3$; $\alpha = .05$
   b. $H_0: \mu = 76$; $H_a: \mu > 76$; $n = 50$; $\bar{x} = 80$; $s^2 = 413$; $\alpha = .10$
   c. $H_0: \mu = .2$; $H_a: \mu < .2$; $n = 40$; $\bar{x} = .19$; $s = .02$; $\alpha = .01$

7.10   A random sample of n observations is selected from a population with mean μ and variance σ^2. Complete a test of hypothesis for each of the following situations:
   a. $H_0: \mu = 5,000$; $H_a: \mu > 5,000$; $n = 200$; $\bar{x} = 6,000$; $s = 9,000$; $\alpha = .05$
   b. $H_0: \mu = 5,000$; $H_a: \mu > 5,000$; $n = 200$; $\bar{x} = 6,000$; $s = 5,000$; $\alpha = .01$
   c. $H_0: \mu = 12.7$; $H_a: \mu \neq 12.7$; $n = 150$; $\bar{x} = 10.3$; $s^2 = 210.6$; $\alpha = .05$

7.11   A random sample of n observations is selected from a population with mean μ and variance σ^2. Complete a test of hypothesis for each of the following situations:
   a. $H_0: \mu = 0$; $H_a: \mu \neq 0$; $n = 175$; $\bar{x} = 6.5$; $s^2 = 496.3$; $\alpha = .01$
   b. $H_0: \mu = 65$; $H_a: \mu < 65$; $n = 86$; $\bar{x} = 63.8$; $s = 10.3$; $\alpha = .10$
   c. $H_0: \mu = 65$; $H_a: \mu < 65$; $n = 186$; $\bar{x} = 63.8$; $s = 10.3$; $\alpha = .10$

7.12   A random sample of forty-nine observations produced the following sums:

$$\sum x = 50.3 \qquad \sum x^2 = 68$$

   a.   Test the null hypothesis that $\mu = 1.18$ against the alternative hypothesis that $\mu < 1.18$. Use $\alpha = .05$.
   b.   Test the null hypothesis that $\mu = 1.18$ against the alternative hypothesis that $\mu \neq 1.18$. Use $\alpha = .05$.

7.13   In a test of hypothesis, who or what determines the size of the rejection region?

7.14 If you test a hypothesis and reject the null hypothesis in favor of your research hypothesis, does your test prove that the research hypothesis is correct? Explain.

Applying the Concepts

7.15 An automobile manufacturer believes that the mean mileage per gallon of one of its new models exceeds the mean EPA (Environmental Protection Agency) rating of 43 miles per gallon. To gain evidence to support its beliefs, the manufacturer randomly selects forty of the cars and records the miles per gallon for each car over a 100-mile course. The mean and standard deviation for the sample of forty cars are $\bar{x} = 43.6$ and $s = 1.3$ miles per gallon.

a. Since the manufacturer wants to show that the mean miles per gallon for the cars exceeds 43, what should you choose for your alternative and null hypotheses?

b. Do the data provide sufficient evidence to support the manufacturer's belief? Use $\alpha = .05$.

7.16 A pain reliever currently being used in a hospital is known to bring relief to patients in a mean time of 3.5 minutes. To compare a new pain reliever with the one currently in use, the new drug is administered to a random sample of fifty patients. The mean time to relief for the sample of patients is 2.8 minutes and the variance is 1.3. Do the data provide sufficient evidence to conclude that the new drug was effective in reducing the mean time until a patient receives relief from pain? Test using $\alpha = .10$.

7.17 A machine is set to produce bolts with a mean length of 1 inch. Bolts that are too long or too short do not meet the customer's specifications and must be rejected. To avoid producing too many rejects, the bolts produced by the machine are sampled from time to time and tested as a check to see whether the machine is still operating properly, that is, producing bolts with a mean length of 1 inch. Suppose fifty bolts have been sampled, and $\bar{x} = 1.02$ inches and $s = .04$ inch. At the $\alpha = .01$ significance level, does the sample evidence indicate that the machine is producing bolts with a mean not equal to 1 inch; that is, is the production process out of control?

7.18 To measure how people feel about certain groups, some sociologists have employed a "feeling thermometer." They describe this instrument as follows:

I have here a card on which there is something that looks like a thermometer. We call it a feeling thermometer because it measures your feelings toward groups. Here's how it works. If you don't know too much about a group or don't feel particularly warm or cold toward them, then you should place them in the middle, at the fifty degree mark. If you have a warm feeling toward a group, or feel favorably toward it, you would give it a score somewhere between fifty and one hundred degrees, depending on how warm your feeling is toward the group. On the other hand, if you don't feel very favorably toward some of these groups—if there are some you don't care for too much—then you would place them somewhere between zero and fifty degrees.*

* P. E. Converse, J. D. Dotson, W. J. Hoag, and W. H. McGee III, *American Social Attitudes Data Sourcebook, 1947–1978* (Cambridge, Mass.: Harvard University Press, 1980), p. 38. Reprinted by permission. Copyright © 1980 by The President and Fellows of Harvard College.

In 1964 Americans gave big business a mean feeling thermometer score of approximately 60.2. In 1966 a sample of 1,272 Americans were asked to use the feeling thermometer to express their attitudes toward big business. The mean score for this sample was 60.1 and the standard deviation was 19.9. If the sample was randomly selected, is there sufficient evidence to conclude that in 1966 the mean feeling thermometer score for all Americans toward big business was less than 60.2? Test at the $\alpha = .10$ level of significance.

7.19   Refer to Exercise 7.18. In 1976 a sample of 1,769 Americans were asked to express their attitudes toward big business. Using the feeling thermometer, they produced a mean score of 48.4. The standard deviation of these scores was 21.0. If the sample was randomly selected, is there sufficient evidence to conclude that in 1976 the mean score associated with all Americans was less than 50? Use $\alpha = .01$.

*7.20 The state association of home builders claims that the mean cost of a new house in Florida is $122,000. One realtor, however, believes this figure is too high. A random sample of thirty new houses sold in Florida this year is obtained. The prices (in dollars) are given below:

114,479	123,823	124,461	114,979	118,479	109,009
115,487	124,117	113,508	128,806	130,815	118,342
108,371	106,709	113,017	121,274	137,561	97,690
132,602	48,637	119,434	131,334	94,925	123,664
115,876	117,429	114,688	113,291	120,258	126,562

a. What is the research hypothesis that the realtor wants to support? Identify the alternative and null hypotheses.
b. Use a computer package to see if the sample tends to support the realtor's belief. Test at the .01 significance level.

7.3

OBSERVED SIGNIFICANCE LEVELS: p-VALUES

In the procedure for conducting a test of hypothesis as described in Section 7.2, the rejection region and, correspondingly, the value of α are stated before calculating the value of the test statistic and making an interpretation. This is the procedure we will follow when conducting tests of hypotheses in this text, although when the results of statistical tests are reported in journals, a slightly different format is often used. Rather than specifying a value of α and giving an appropriate rejection region, journal articles report the extent to which the calculated value of the test statistic disagrees with the null hypothesis. A measure of this disagreement is called the **observed significance level** or **p-value** for the test.

Definition 7.7

The **observed significance level,** or **p-value,** for a statistical test of hypothesis is the probability (assuming the null hypothesis were true) of observing a value of the test statistic that supports the alternative hypothesis at least as well as the value calculated from the sample data.

In previous chapters we used the rare event approach to make inferences. Recall that this involved obtaining a sample from a population and calculating how rare an event the observed sample was in relation to some claim made about the population. Loosely speaking, the observed significance level of a test is just a measure of how rare an event the calculated value of the test statistic is compared to the null hypothesis. The p-value for a test of hypothesis will be calculated in the following example.

EXAMPLE 7.5   Refer to Example 7.2. A contractor wants to determine whether the mean breaking strength of the sewer pipe produced by a manufacturer is less than 2,400 pounds per foot. A random sample of fifty-five sections of the manufacturer's sewer pipe yields the following statistics on breaking strength:

$\bar{x} = 2{,}340$ pounds per foot $s = 200$ pounds per foot

The null hypothesis, alternative hypothesis, and test statistic are:

$H_0: \quad \mu = 2{,}400 \qquad H_a: \quad \mu < 2{,}400$

$\text{Test statistic:} \quad z = \dfrac{\bar{x} - 2{,}400}{s/\sqrt{n}}$

Find the observed significance level (p-value) for this test.

Solution   The calculated value of the test statistic is

$$z = \frac{2{,}340 - 2{,}400}{200/\sqrt{55}} = -2.22$$

Since the alternative hypothesis is that $\mu < 2{,}400$, the calculated value, $z = -2.22$, supports this hypothesis, and any value smaller than -2.22 would support the alternative hypothesis even better. Therefore, the observed significance level for this test is

$p\text{-value} = P(z \leq -2.22)$

To find this probability, we use Table II, which contains normal curve areas. The area of interest is shown in Figure 7.7 (page 342), and from Table II we obtain

$p\text{-value} = P(z \leq -2.22) = .5 - .487 = .013$

Figure 7.7
p-value for Example 7.5

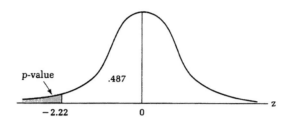

Since p-value = .013, if the null hypothesis were true (that is, if $\mu = 2{,}400$), it is quite unlikely that a random sample would produce a z-value of -2.22 or less. If the results of this test were reported, a statement similar to the following could be made: "The sample data provide support for the hypothesis that the mean breaking strength of this manufacturer's pipe is less than 2,400 pounds. The observed significance level is .013."

It would be up to each person who reads such a report to decide if there is enough evidence (support) to conclude that the mean strength is indeed less than 2,400 pounds. How one would make such a decision, and how this process is related to the testing procedure given in Section 7.2, will be discussed shortly. First, we will calculate another p-value.　■

EXAMPLE 7.6　Refer to Example 7.4. A nutritionist believes that 12-ounce boxes of a popular cereal have a mean bran content that differs from 1.2 ounces. A random sample of sixty 12-ounce boxes of this cereal produces the following statistics:

$$\bar{x} = 1.170 \text{ ounces of bran} \qquad s = .111 \text{ ounce of bran}$$

Find the p-value for this test of hypothesis.

Solution　The first three elements of the test are:

$$H_0: \quad \mu = 1.2 \qquad H_a: \quad \mu \neq 1.2$$

$$\text{Test statistic:} \quad z = \frac{\bar{x} - 1.2}{s/\sqrt{n}}$$

The calculated value of the test statistic is

$$z = \frac{1.170 - 1.2}{.111/\sqrt{60}} = -2.09$$

Any z-value less than -2.09 would give better support to the alternative hypothesis that $\mu \neq 1.2$. Since this is a two-tailed test, however, a z-value of *positive* 2.09 would support the alternative hypothesis as well as the calculated value of $z = -2.09$. Also, any values greater than 2.09 would provide greater support. Thus, the observed significance level for this test, as shown in Figure 7.8, is found to be:

$$p\text{-value} = P(z \leq -2.09 \text{ or } z > 2.09)$$
$$= (.5 - .482) + (.5 - .482) = .018 + .018 = .036$$

Figure 7.8
p-value for Example 7.6

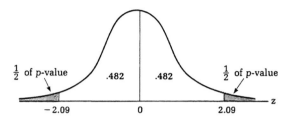

For this example, the data support the alternative hypothesis at an observed signifi-cance level of .036. ∎

As indicated by the last two examples, exactly how a p-value is calculated for a large-sample test of hypothesis about μ depends on the alternative hypothesis. The general way to find a p-value for these tests is given in the box.

Calculating p-Values for Large-Sample Tests About μ

For the test of interest, give H_0, H_a, and the test statistic. Using the sam-ple data, calculate the value of the test statistic. Depending on the form of the alternative hypothesis, calculate the p-value as follows:

1. H_a: $\mu > \mu_0$:

p-value $= P(z >$ calculated z-value$)$

2. H_a: $\mu < \mu_0$:

p-value $= P(z <$ calculated z-value$)$

3. H_a: $\mu \neq \mu_0$:

p-value $= P(z <$ negative calculated z-value$)$
$+ P(z >$ positive calculated z-value$)$

We have defined what a p-value is and have shown how to calculate p-values. We will now discuss how to interpret a p-value. That is, we will indicate how an individ-ual can use a p-value to make a decision when testing a hypothesis.

Consider a large-sample test of hypothesis with H_0: $\mu = 75$, H_a: $\mu > 75$, and $\alpha = .05$. In this situation we would reject H_0 if $z > 1.65$ (see Table III of Appendix). What is the p-value for any value of the test statistic ($z > 1.65$) that would lead us to reject H_0? To answer this, consider Figure 7.9 (page 344).

If the calculated value of z is greater than 1.65, then p-value $= P(z >$ calculated value) will be a smaller area than the one corresponding to $\alpha = .05$. That is, the p-value will be less than $\alpha = .05$ whenever the calculated z-value is in the rejection region.

Figure 7.9
Rejection region for
$H_a: \mu > 75; \alpha = .05$

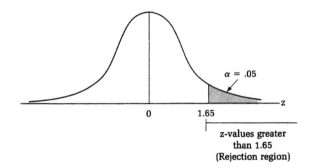

$\alpha = .05$

0   1.65

z

z-values greater
than 1.65
(Rejection region)

In other words, saying we would reject H_0 if $z > 1.65$ is the same as saying we would reject H_0 if p-value $< .05$. This type of reasoning may be applied to any test of hypothesis. The use of a p-value to interpret test results is summarized in the box.

Using p-Values to Make Decisions

1. Specify the value of α that is desired.

2. If the observed significance level (p-value) of the test is less than the value of α, then reject the null hypothesis.

3. If the p-value is not less than the value of α, then there is insufficient evidence to make a decision.

EXAMPLE 7.7   Refer to Example 7.3. A psychologist theorizes that professional athletes tend to be more self-confident than others. The theory may be partially validated if professional athletes who take a test designed to measure self-confidence have a mean score exceeding 72, the nationwide average on this test. The sample mean and standard deviation of scores for a random sample of fifty professional athletes are

$\bar{x} = 74.1$ $s = 13.3$

a. Calculate the p-value for this test.

b. In Example 7.3, the test was conducted with $\alpha = .05$. Considering the p-value calculated in part **a**, should H_0 be rejected with $\alpha = .05$?

Solution   **a.** The first three elements of the test are

$$H_0: \quad \mu = 72 \qquad H_a: \quad \mu > 72$$

$$\text{Test statistic:} \quad z = \frac{\bar{x} - 72}{s/\sqrt{n}}$$

The calculated value of the test statistic is

$$z = \frac{74.1 - 72}{13.3/\sqrt{50}} = 1.12$$

The p-value (see Figure 7.10) for this test is

$$p = P(z > 1.12) = .5 - .369 = .131$$

Figure 7.10
p-value for Example 7.7

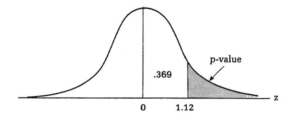

b. Comparing the p-value to α, we see that p-value = .131 is not less than $\alpha = .05$. Thus, H_0 should not be rejected. There is insufficient evidence to conclude that the mean score for all professional athletes is higher than the nationwide average of 72. This is, of course, the same interpretation as the one given in Example 7.3.

∎

There are several important reasons for understanding p-values. First, using p-values to report test results enables readers to select the value of α that they would be willing to tolerate for the test being considered. A reader must be careful, however, when doing this. Suppose one believes that $\alpha = .05$ is appropriate for a particular test, and the reported observed significance level is p-value = .08. One must be sure not to reject H_0 in this case. It is tempting to say that even though $p = .08$ is not less than $\alpha = .05$, "it is close enough." If one were to proceed along these lines, the chance of a Type I error would no longer be as small as desired, and there would be an increased risk of incorrectly rejecting H_0. It is best to select carefully the value of α that reflects the seriousness of making a Type I error for a particular test of hypothesis, and to adhere strictly to the value selected.

Second, we have stated that p-values are often used to report the results of tests in journal articles. There is another important use of p-values. If a computer is used to help analyze data for a test of hypothesis (as in Computer Example 7.2 to follow), the information (output) provided by the computer will usually include a p-value. Both of these uses of p-values are essentially the same. The computer output may be thought of as a report of the results of a test of hypothesis. The computer will not select a value of α, nor interpret the test results, but will simply report a p-value. It is the experimenter who must choose an appropriate value for α, compare the computed p-value to the chosen α, and interpret the results.

Finally, the main reason we have discussed p-values is so that you may interpret any p-values that you see in journal articles or on computer printouts. In these situations, you will not have to calculate the p-values. Therefore, we will not discuss how to calculate p-values for tests of hypotheses presented in later sections of this text. You should, however, be able to use reported p-values to interpret test results.

COMPUTER EXAMPLE 7.2 In Example 7.6 we found the p-value for a test of hypothesis concerning the mean bran content of 12-ounce boxes of a popular cereal. In Computer Example 7.1, we showed the results of using Minitab to test the hypothesis of interest. The Minitab printout is repeated here.

```
TEST OF MU = 1.2000 VS MU N.E. 1.2000
THE ASSUMED SIGMA = 0.111

            N     MEAN    STDEV   SE MEAN        Z   P VALUE
    C6     60   1.1705   0.1111    0.0143    -2.06     0.040
```

When Minitab is used to conduct a large-sample test of hypothesis, the p-value is part of the information provided on the printout. In this case, Minitab calculated the p-value to be 0.040. This is slightly larger than 0.036, the value we obtained in Example 7.6. The two slightly different, but essentially equivalent, values are due to the fact that Minitab used more decimal places when finding the value of the test statistic.

When the test was performed in Example 7.6, the level of significance was set at $\alpha = .05$. Since the p-value $= .036$ is less than .05, we should reject the null hypothesis that $\mu = 1.2$. This agrees with the decision reached in Example 7.6.

Most popular computer packages report p-values for any test of hypothesis that is conducted. This is an attractive feature of such packages and is an aid in the interpretation of test results. ∎

Exercises (7.21–7.29)

Learning the Mechanics

7.21 A random sample of n observations is taken from a population with mean μ. Calculate the observed significance level for each of the following situations.
a. H_0: $\mu = 16$; H_a: $\mu > 16$; $n = 50$; $\bar{x} = 17.3$; $s^2 = 33$
b. H_0: $\mu = 16$; H_a: $\mu \neq 16$; $n = 100$; $\bar{x} = 18.9$; $s = 12.6$
c. H_0: $\mu = 25$; H_a: $\mu < 25$; $n = 120$; $\bar{x} = 24.4$; $s^2 = 6.4$

7.22 Refer to Exercise 7.21. For which parts of the exercise would H_0 be rejected if:
a. $\alpha = .01$?
b. $\alpha = .05$?
c. $\alpha = .10$?

7.23 It is desired to test H_0: $\mu = 70$ against H_a: $\mu > 70$. Which of the following situations, based on random samples of n observations, provides more support for H_a? (Calculate a p-value in each case.)

a. $n = 500$; $\bar{x} = 73.2$; $s = 60$
b. $n = 100$; $\bar{x} = 70.9$; $s^2 = 20$

Applying the Concepts

7.24 A new drug is advertised to reduce a patient's blood pressure an average of more than 6 units after the first week of medication. An experiment was conducted in which forty patients were treated with the drug for 1 week. The mean and standard deviation for this sample of blood pressure reductions were 8.7 and 6.8, respectively. The experimenter plans to report that the data support the advertised claim. What is the observed significance level associated with testing H_0: $\mu = 6$ against H_a: $\mu > 6$?

7.25 A small component recently developed for use in personal computers may not last as long as larger components, whose mean longevity is 3,000 hours. A random sample of thirty-five of the small components provided the following longevity statistics:

$$\bar{x} = 2{,}950 \text{ hours} \qquad s = 100 \text{ hours}$$

On the basis of this sample, it is desired to see if there is evidence to conclude that the mean longevity of the small component is less than that of the larger component.
a. Calculate the p-value to measure the amount of support for the hypothesis that the mean longevity of the small component is less than that of the larger component.
b. Using a value of $\alpha = .01$, would you conclude that the small component does not last as long, on the average, as the larger component? Explain.

7.26 The contents label on the back of a 16-ounce box of sugar-coated cereal states that the mean sugar content of all boxes of this brand is 3.0 ounces. A consumer group conducts an experiment to see if the mean sugar content differs from 3.0 ounces. It randomly samples fifty 16-ounce boxes of cereal of this brand and measures the sugar content of each. The results are summarized below:

$$\bar{x} = 3.06 \text{ ounces} \qquad s = 0.2 \text{ ounce}$$

a. Keeping in mind that the consumer group is interested in showing that the mean sugar content differs from 3.0 ounces, calculate the observed significance level for this situation.
b. What interpretation would you make in this situation? Explain.

7.27 New York's Nutrition Education and Training (NET) program was developed to improve children's dietary choices by supplementing nutritional education in the classroom as well as by changing the lunchroom fare. One researcher, Ardyth Harris Gillespie, monitored public elementary schools in upstate New York in order to evaluate changes in snacking when schools adopted the NET program.[*] The snacks eaten in and out of the home were evaluated for children in grades 3 through 6. Based on what the children ate and the frequency of consumption, Gillespie computed a "snack score" for each child. The higher the snack score, the more nutritious the snacks.

[*] A. H. Gillespie, "Evaluation of Nutrition Education and Training Mini-Grant Programs," *Journal of Nutrition Education*, Vol. 16, No. 1, March 1984, pp. 8–12.

a. The at-home snacking habits were evaluated for 333 students who were involved in the NET program. Snack scores were obtained for each student prior to the program and again afterward. Subtracting the earlier score from the later score gave the change in each student's snack score; these were the data that were analyzed. The sample mean change in snack scores was greater than 0, and the increase was stated to have an observed significance level of less than .05. Interpret these findings.

b. Snacking habits away from home were similarly evaluated for 325 students. Again, the mean difference between snack scores was greater than 0. The observed significance level was less than .05. Interpret this result.

*7.28 According to advertisements, a new fertilizer will increase soybean production by an average of more than 5 bushels per acre. Thirty-five randomly selected soybean farmers were persuaded by the manufacturer to try the new product. Each farmer used the new fertilizer on a 10-acre plot of soybeans and put a previously used fertilizer on a similar 10-acre plot of soybeans. The increase in production, measured in bushels per acre, for each farm is shown below. The manufacturer of the new fertilizer would like to report that the results of this experiment support the claim that the mean production is increased by more than 5 bushels per acre.

5.96	6.09	6.32	2.39	5.35	3.92	4.13
5.52	4.54	4.15	5.31	5.67	3.62	4.85
4.66	4.16	3.73	4.45	6.55	3.60	5.26
6.79	6.30	5.83	7.07	4.59	6.15	5.86
4.36	6.73	4.19	6.28	3.83	6.63	4.35

a. Use a computer package to find the p-value associated with this experiment.

b. Based on your answer to part a, do you think the manufacturer should state that there is support for the advertised claim? Explain.

*7.29 Refer to Exercise 7.20. A realtor believes that the state association of home builders' stated mean cost for a new house in Florida is too high. The prices (in dollars) for a sample of thirty new homes sold in Florida this year are shown below:

114,479	123,823	124,461	114,979	118,479	109,009
115,487	124,117	113,508	128,806	130,815	118,342
108,371	106,709	113,017	121,274	137,561	97,690
132,602	48,637	119,434	131,334	94,925	123,664
115,876	117,429	114,688	113,291	120,258	126,562

The hypotheses of interest are

H_0: $\mu = 122{,}000$ H_a: $\mu < 122{,}000$

The test was conducted with $\alpha = .01$.

a. Use a computer package to find the p-value for this test.

b. What interpretation would you make in this case? Explain.

7.4

A LARGE-SAMPLE CONFIDENCE INTERVAL FOR A POPULATION MEAN, μ

In Section 7.2 we discussed making a decision about the value of a population mean using a test of hypothesis. In this section we will discuss another type of inference. Rather than make a decision about the value of a population mean, we will *estimate* its value. To estimate a population parameter we will form a **confidence interval.**

The confidence interval we will present in this section will use the information contained in a large random sample, and we will refer to it as a **large-sample confidence interval for μ.** The conditions that must be met to assure the validity of this confidence interval are given in the box.

When to Form a Large-Sample Confidence Interval for μ

To assure the validity of a large-sample confidence interval for the mean, μ, of a population, the following conditions must be met:

1. A random sample must be taken from the population.
2. The sample size must be large.

These are the same conditions that must be met to perform a valid large-sample test of hypothesis. In both situations we must be sure that the sampling distribution of the sample mean is approximately normal. The formula for finding a large-sample confidence interval is given in the box.

Large-Sample Confidence Interval for μ

The large-sample confidence interval for μ is expressed as

$$\bar{x} - z\left(\frac{s}{\sqrt{n}}\right) \quad \text{to} \quad \bar{x} + z\left(\frac{s}{\sqrt{n}}\right)$$

The value of z is obtained from Appendix Table IV.

The confidence interval could also be expressed as

$$\bar{x} \pm z\left(\frac{s}{\sqrt{n}}\right)$$

We prefer the form given in the box and will use this style throughout the text. The other form may, of course, be used if preferred.

We will demonstrate how to find confidence intervals and discuss the interpretation of the results in Example 7.8.

EXAMPLE 7.8   A large hospital wants to estimate the average length of time previous patients have remained in the hospital. To accomplish this objective, a random sample of 100 patients' records is obtained from all the previous patients' records. From the data the following statistics are calculated:

$$\bar{x} = 7.84 \text{ days} \qquad s^2 = 88.85$$

Find a large-sample confidence interval for the mean length of time all previous patients remained in the hospital. Form the interval so that you will have 95% confidence in the result.

Solution   The population of interest is the collection of the lengths of time all previous patients have remained in the hospital. The parameter of interest is μ, the mean length of time patients have stayed in the hospital. We want to form a confidence interval for this parameter. The confidence interval is given by

$$\bar{x} - z\left(\frac{s}{\sqrt{n}}\right) \qquad \text{to} \qquad \bar{x} + z\left(\frac{s}{\sqrt{n}}\right)$$

Since a large ($n = 100$) random sample has been selected from the population of interest, it is valid to form a large-sample confidence interval for the mean of this population.

From the statement of the example, we observe that $\bar{x} = 7.84$, $s^2 = 88.85$, $s = \sqrt{88.85}$, and $n = 100$. Thus, to find the desired interval all we need is the appropriate z-value from Appendix Table IV. A reproduction of Table IV is given in Table 7.3, where we can see that to have 95% confidence, the value of z must be equal to 1.96. (We will explain why this value is appropriate after calculating the interval.)

Table 7.3

A reproduction of Appendix Table IV, z-Scores for Large-Sample Confidence Intervals

	Confidence Desired		
	90%	95%	99%
Value of z	1.65	1.96	2.58

Substituting into the formula, we calculate the interval:

$$7.84 - 1.96\left(\frac{\sqrt{88.85}}{\sqrt{100}}\right) \qquad \text{to} \qquad 7.84 + 1.96\left(\frac{\sqrt{88.85}}{\sqrt{100}}\right)$$

$$7.84 - 1.85 \qquad \text{to} \qquad 7.84 + 1.85$$

$$5.99 \qquad \text{to} \qquad 9.69$$

Thus, we estimate the mean length of stay in the hospital for all the previous patients to fall in the interval from 5.99 days to 9.69 days. As a statement of reliability we will say that we are 95% confident this interval actually contains the true mean length of stay. ■

Now, how was the value of z found in Table 7.3 calculated? In Figure 7.11 we have drawn a normal curve representing the sampling distribution of $\bar{x}$ for a large random sample. The value of μ and points 1.96 standard errors from μ are shown in the figure. (Recall that $\mu_{\bar{x}}$, the mean of the sampling distribution of $\bar{x}$, equals μ, the mean of the original population.)

Figure 7.11
Sampling distribution of $\bar{x}$

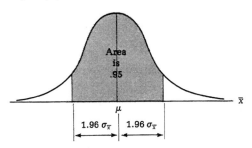

In Appendix Table II, we find that the area under the normal curve between the boundaries 1.96 standard errors above and below μ is equal to .475 + .475 = .95. If a sample yields a value of $\bar{x}$ that is within these boundaries, then the interval $(\bar{x} - 1.96\sigma_{\bar{x}})$ to $(\bar{x} + 1.96\sigma_{\bar{x}})$ will contain μ. If $\bar{x}$ falls more than 1.96 standard errors from μ, then the interval will not contain μ. This is the basis for the confidence we expressed in the interval found in Example 7.8.

We cannot be *certain* that the interval we calculated in Example 7.8 (5.99 to 9.69) actually contains the true value of μ, but we are 95% confident that it does. This confidence is derived from the knowledge that if we were to repeatedly select random samples of 100 measurements from this population and form confidence intervals based on each of the many samples selected, then approximately 95% of the intervals would contain μ. We do not know (unless we look at all the patients' records and calculate μ) whether the interval we formed is one of the 95% that contain μ or one of the 5% that do not, but the odds favor the interval's containing μ.

To change the degree of confidence we have in an interval, all we need to change is the z-value. In Table 7.3, we have provided the correct z-values for 90%, 95%, and 99% confidence. If you need any other level of confidence, you will have to calculate the appropriate z-value using Appendix Table II. The correct z-value will be the one for which the area between $-z$ and $+z$ under the standard normal curve equals the confidence desired. You may want to use Table II to verify that the z-values given in Table 7.3 satisfy this requirement.

As with tests of hypotheses, we will follow a standard format when estimating parameters with confidence intervals. We have listed the steps we will follow in the box on page 352, and we encourage you to follow the same steps when forming confidence intervals.

Steps to Follow When Forming a Confidence Interval

1. Identify the population and parameter of interest.
2. Give the procedure (formula) to be used, and state the conditions required for the validity of this procedure.
3. Calculate the interval.
4. Interpret the results.

We will now present another example of estimating a population mean with a confidence interval.

EXAMPLE 7.9   Unoccupied seats on flights cause the airlines to lose revenue. Suppose a large airline wants to estimate its average number of unoccupied seats per flight over the past year. To accomplish this, the records of 225 flights are randomly selected from the files and the number of unoccupied seats is noted for each of the sampled flights. The sample mean and standard deviation are

$$\bar{x} = 11.6 \text{ seats} \qquad s = 4.1 \text{ seats}$$

Estimate μ, the mean number of unoccupied seats per flight during the past year, using a 99% confidence interval.

Solution   The population of interest is the number of unoccupied seats per flight for each flight over the past year, and we want to estimate μ, the mean of this population. The general form for a large-sample confidence interval for μ is

$$\bar{x} - z\left(\frac{s}{\sqrt{n}}\right) \qquad \text{to} \qquad \bar{x} + z\left(\frac{s}{\sqrt{n}}\right)$$

Since a large random sample has been selected from the population of interest, it is valid to form a large-sample confidence interval for the mean of this population. From the statement of the example, we obtain $\bar{x} = 11.6$, $s = 4.1$, and $n = 225$. From Table 7.3, we find that the value of z for 99% confidence is 2.58. The desired interval is

$$11.6 - 2.58\left(\frac{4.1}{\sqrt{225}}\right) \qquad \text{to} \qquad 11.6 + 2.58\left(\frac{4.1}{\sqrt{225}}\right)$$

$$11.6 - .71 \qquad \text{to} \qquad 11.6 + .71$$

$$10.89 \qquad \text{to} \qquad 12.31$$

With 99% confidence we estimate the mean number of unoccupied seats per flight to be between 10.89 and 12.31 during the past year. We emphasize that the confidence is derived from the procedure used. If we were to repeatedly apply this procedure to different random samples, approximately 99% of the interval formed would contain the true value of μ. ∎

When you are choosing whether to form a confidence interval or to test a hypothesis in a particular problem, consider the objective of the experiment. When testing hypotheses, the main objective is to make a decision about the value of a parameter. When forming a confidence interval, the main objective is to estimate the value of a parameter. It is important to realize that whether you conduct a test of hypothesis or form a confidence interval in any given situation depends on the objective of the experimenter—decision-making or estimation.*

In the remainder of this text, we will be forming confidence intervals for many different parameters. The interpretations of all such intervals will be similar to the interpretations of the confidence intervals given in this section. The following box summarizes how to interpret a confidence interval for a population mean.

Interpretation of a Confidence Interval for a Population Mean

When we form a 95% confidence interval for μ, we express our confidence in the interval with a statement such as, "we can be 95% confident that μ lies between the lower and upper bounds of the confidence interval." In a particular application we express the meaning of μ in words and substitute the appropriate numerical values for the bounds.

The statement actually reflects our confidence in the process rather than in one particular interval formed. We know that if different large random samples were obtained from the same population, then repeated applications of the same procedure would result in different lower and upper bounds. Furthermore, we know that 95% of the resulting intervals will contain μ, and 5% will not contain μ.

Usually there is no way to determine whether any particular interval is one of the 95% that contain μ or one of the 5% that do not. However, since the procedure results in a "correct" interval 95% of the time, we say we are 95% confident that any one interval contains μ. This provides a measure of reliability for confidence intervals.

Similar statements could be made for 90% or 99% confidence intervals.

Figure 7.12 (page 354) shows what would typically happen if a number of samples were drawn from the same population and a confidence interval for μ were calculated

---

* If one were careful, confidence intervals could be used to make decisions in hypothesis-testing situations. The difficulty is that the level of significance, α, can be easily misinterpreted if this were done. For this reason, in this text we will use confidence intervals only for estimating purposes.

Figure 7.12
Repeatedly forming
confidence intervals for μ

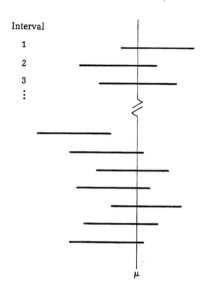

for each sample. The true value of μ is indicated by a vertical line in the figure. Different confidence intervals, resulting from different random samples, are shown as horizontal line segments. Note that the value of μ is constant, but since $\bar{x}$ varies from sample to sample, the confidence intervals vary from sample to sample. Most of the confidence intervals would contain μ, but some of them would not contain μ. If a 99% confidence interval were calculated for each sample, then in the long run, 99% of the confidence intervals that were formed would contain μ.

COMPUTER
EXAMPLE 7.3

In Example 7.8 we formed a confidence interval for the mean length of time all previous patients remained in a hospital. The interval was calculated using the values of $\bar{x}$ and s from a random sample of 100 patients' records. The 100 observations follow:

4	1	6	13	7	23	3	7	11	12	7	3	13
1	3	3	2	12	13	14	10	1	10	31	7	3
1	11	16	2	13	8	5	6	3	1	3	4	8
1	1	4	16	2	4	17	1	3	1	7	4	14
2	38	12	5	1	17	1	30	14	2	6	3	3
7	6	39	2	3	12	11	8	5	1	63	4	1
1	4	4	2	2	1	1	8	4	2	15	1	9
11	1	22	10	3	1	2	3	5				

Minitab used these data to calculate a 95% confidence interval for the mean of interest. The printout is reproduced here:

```
THE ASSUMED SIGMA =9.43
```

	N	MEAN	STDEV	SE MEAN	95.0 PERCENT C.I.
C4	100	7.840	9.426	0.943	(5.990, 9.690)

As with large-sample tests of hypotheses, Minitab needs to be given a value for the standard deviation. The printed information "THE ASSUMED SIGMA = 9.43" indicates that we used the value of the sample standard deviation $s = 9.43 \approx \sqrt{88.85} = \sqrt{s^2}$. The resulting 95% confidence interval is 5.99 to 9.69. This is identical to the interval found in Example 7.8, and the interpretation of the interval is the same. We are 95% confident that the mean length of stay in the hospital for all previous patients is between 5.99 and 9.69 days. ∎

Exercises (7.30–7.45)

Learning the Language

7.30 Define the term *confidence interval*.

Using the Tables

7.31
 a. What z-value is used to form a large-sample 95% confidence interval?
 b. What z-value is used to form a large-sample 90% confidence interval?
 c. What z-value is used to form a large-sample 99% confidence interval?

Learning the Mechanics

7.32 A random sample of sixty observations produced a mean $\bar{x} = 75$ and a standard deviation $s = 12$.
 a. Find a 90% confidence interval for the population mean, μ.
 b. Find a 95% confidence interval for μ.
 c. Find a 99% confidence interval for μ.
 d. What is the effect on the width of a confidence interval as n is held fixed and the confidence level is increased?

7.33 A random sample of n measurements was selected from a population with mean μ and variance σ^2. Calculate a 95% confidence interval for each of the following combinations of n, $\bar{x}$, and s^2:
 a. $n = 50$, $\bar{x} = 40$, $s^2 = 30$
 b. $n = 200$, $\bar{x} = 40$, $s^2 = 30$
 c. $n = 100$, $\bar{x} = 50$, $s^2 = 20$
 d. $n = 100$, $\bar{x} = 50$, $s^2 = 40$

7.34 A random sample of fifty observations from a population produced the following summary statistics:

$$\sum x = 390 \qquad \sum x^2 = 7{,}212$$

 a. Find a 99% confidence interval for μ.
 b. Find a 90% confidence interval for μ.

7.35 Explain what is meant by the statement, "We are 95% confident that an interval estimate contains μ."

7.36 Is a large-sample confidence interval valid if the population from which the sample is taken is not normally distributed? Explain.

7.37 The mean and standard deviation of a random sample of n measurements are equal to 13.6 and 4.1, respectively.

    a. Find a 95% confidence interval for μ if $n = 100$.

    b. Find a 95% confidence interval for μ if $n = 400$.

    c. Find the widths of the confidence intervals found in parts *a* and *b*. What is the effect on the width of a confidence interval of quadrupling the sample size, while holding the confidence level fixed?

Applying the Concepts

7.38 A fact long known but little understood is that twins, in their early years, tend to have lower IQ's and pick up language more slowly than nontwins. Recently, psychologists have found that the slower intellectual growth of most twins may be caused by benign parental neglect. Suppose that it is desired to estimate the mean attention time given to twins per week by their parents. A random sample of forty-six sets of $2\frac{1}{2}$-year-old twin boys is taken, and at the end of 1 week, the attention time given to each pair is recorded. The results are as follows:

    $\bar{x} = 22$ hours $\qquad s = 16$ hours

Using the data, find a 90% confidence interval for the mean attention time given to all $2\frac{1}{2}$-year-old twin boys by their parents.

7.39 Suppose a large labor union wants to estimate the mean number of hours per month a union member is absent from work. The union decides to randomly sample 320 of its members and monitor their working time for 1 month. At the end of the month, the total number of hours absent from work is recorded for each employee. If the mean and standard deviation of the sample are $\bar{x} = 9.6$ hours and $s = 6$ hours, find a 95% confidence interval for the true mean number of hours absent per month per employee.

7.40 Automotive engineers are continually improving their products. Suppose a new type of brake light has been developed by General Motors. As part of a product safety evaluation program, General Motors' engineers want to estimate the mean driver response time to the new brake light. (Response time is the length of time from the point that the brake is applied until the driver in the following car takes some corrective action.) Fifty drivers are selected at random, and the response time (in seconds) for each driver is recorded, yielding the following results: $\bar{x} = .72$, $s^2 = .022$. Estimate the mean driver response time to the new brake light using a 99% confidence interval.

7.41 A sociologist wants to estimate the average number of television viewing hours per American family per week. A random sample of 400 families yields a mean of 32.6 hours and a variance of 97.6. Estimate the mean viewing time with a 90% confidence interval.

7.42 How useful do farmers find meteorological information? Researchers addressing this question surveyed Texas farmers in the counties of Bailey, Floyd, Lubbock, Lynn, Parmer, Terry, Bell, McLennan, and Williamson. A variety of information about the producers in these counties was obtained by a mail survey, some of which is summarized in the accompanying table.

	Item	Number of Farmers Surveyed	Mean	Standard Deviation
General information about producers	Years of operation	244	26	16
	Acres in operation	251	892	1,120

Source: Adapted from K. C. Vining, C. A. Pope III, and W. A. Dugas, Jr., "Usefulness of Weather Information to Texas Agricultural Producers," *Bulletin of the American Meteorological Society,* Vol. 65, No. 12, December 1984, Table 1, p. 1317.

a. Find a 95% confidence interval for the mean number of years of operation for all farms in these counties.

b. Find a 95% confidence interval for the mean number of acres in operation for all farms in these counties.

7.43 Refer to Exercise 7.42. The farmers surveyed were also asked to rank various types of agricultural weather information that they might use in making decisions. The rankings were on a scale of 1 to 5, where 1 meant the information was unimportant and 5 indicated great importance. The results for two of the types of weather information are summarized in the table.

	Item	Number of Farmers Surveyed	Mean	Standard Deviation
Rankings for types of weather information	Animal stress reports	172	2.4	1.5
	Freeze warnings	205	4.0	1.3

Source: Adapted from K. C. Vining, C. A. Pope III, and W. A. Dugas, Jr., "Usefulness of Weather Information to Texas Agricultural Producers," *Bulletin of the American Meteorological Society,* Vol. 65, No. 12, December 1984, Table 2 p. 1317.

a. Find a 99% confidence interval for the mean ranking of the importance of animal stress reports for all farmers in these counties.

b. Find a 99% confidence interval for the mean ranking of the importance of freeze warnings for all farmers in these counties.

*7.44 As an aid in the establishment of personnel requirements, the director of a hospital wants to estimate the mean number of people admitted to the emergency room during a 24-hour period. The director randomly samples fifty-five different 24-hour periods and determines the number of admissions for each. The sample data are shown here. With the aid of a computer package, estimate the mean number of admissions per 24-hour period with a 95% confidence interval.

21	19	19	24	13	21	5	20	28	16	22
26	23	16	20	21	22	30	17	26	9	18
14	20	23	21	18	16	11	15	25	11	32
20	19	27	20	20	22	27	17	18	20	4
14	23	26	15	17	19	23	26	20	24	15

*7.45 Refer to Exercise 7.28. A manufacturer wanted to advertise that a new fertilizer increased soybean production by an average of more than 5 bushels per acre. A random sample of thirty-five soybean farmers used the new fertilizer and reported the increases in production. The thirty-five increases in bushels per acre are shown.

5.96	6.09	6.32	2.39	5.35	3.92	4.13
5.52	4.54	4.15	5.31	5.67	3.62	4.85
4.66	4.16	3.73	4.45	6.55	3.60	5.26
6.79	6.30	5.83	7.07	4.59	6.15	5.86
4.36	6.73	4.19	6.28	3.83	6.63	4.35

a. Using a computer package, find a 99% confidence interval for the mean increase in production.

b. Considering your answer to part a, what claim do you think the manufacturer is entitled to advertise regarding increased soybean production? Explain.

7.5

SMALL-SAMPLE INFERENCES ABOUT A POPULATION MEAN, μ

In Sections 7.2 and 7.4 we used large random samples to test hypotheses and form confidence intervals for population means. Unfortunately, it is not always possible to obtain a large sample, and inferences must sometimes be made using small samples. In this section we will test hypotheses and form confidence intervals using small samples.

In 1908, W. S. Gosset, a scientist who worked for a Guinness brewery, published an article discussing the **t distribution.** Under the pen name of Student, Gosset discussed the sampling distribution of the quantity

$$t = \frac{\bar{x} - \mu}{s/\sqrt{n}}$$

The main result of Gosset's work is that if small random samples are taken from a normal distribution with mean μ, then the t statistic will have a sampling distribution similar to that of the z statistic: mound-shaped, symmetric, and with mean 0. The main difference between the sampling distributions of t and z is that the t distribution is more variable than the z distribution. This is because s will vary quite a bit from sample to sample for small samples, while s remains fairly constant for large samples. Thus, the t statistic contains two random quantities ($\bar{x}$ and s), while z essentially contains one ($\bar{x}$).

The variability in the sampling distribution of t depends on the sample size, n. A convenient way of expressing this dependence is to say that the t statistic has **($n - 1$) degrees of freedom.** [Recall that the quantity ($n - 1$) is the divisor that appears in the formula for s^2. This number plays a key role in the sampling distribution of s^2 and will appear in discussions of other statistics in later chapters.] The smaller the number of

Figure 7.13
Standard normal
z distribution and
t distribution with 4 df

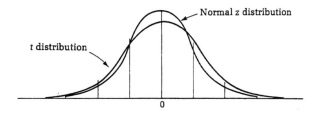

degrees of freedom associated with the t statistic, the more variable its sampling distribution will be.

In Figure 7.13 we have drawn both the sampling distribution of z and the sampling distribution of t with 4 degrees of freedom (df). You can see that the t distribution is similar to, but more variable than, the z distribution. When we conduct tests of hypotheses or form confidence intervals using small samples, we will need tables of **t-values** that reflect the greater variability of the t statistic. These tables are provided in the Appendix and will be discussed in the examples.

Exactly the same conditions must be met either to test a hypothesis about μ using a small sample or to form a confidence interval for μ using a small sample. The conditions that must be met to make the use of the t statistic valid for small-sample inferences are given in the box.

**When to Use the t Statistic to Make Inferences
About a Population Mean, μ**

To assure the validity of a small-sample test of hypothesis about a population mean, μ, or to form a confidence interval for μ, the following conditions must be met:

1. A random sample must be taken from the population.
2. The population must be approximately normally distributed. Visually, this means that a graph of the population of measurements would resemble the one below.

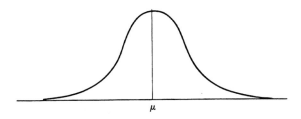

In this section we will make inferences about populations that are approximately normally distributed. Strictly speaking, to obtain an *exact t* distribution, random samples would need to be taken from a population that is *exactly* normally distributed. However, it has been shown that the *t* distribution is rather insensitive to moderate departures from normality. As long as a population of interest is mound-shaped (approximately normal), the use of the *t* statistic produces valid results.

The methods for performing a test of hypothesis or finding a confidence interval using small samples are very similar to the ones used with large samples. The main difference is that different tables will be used to find *t*-values rather than *z*-values. The small-sample test of hypothesis about μ and confidence interval for μ are given in the accompanying boxes. We will demonstrate their use with examples.

Small-Sample Test of Hypothesis About μ

H_0: $\mu = \mu_0$

H_a: $\mu > \mu_0$ or H_a: $\mu < \mu_0$ or H_a: $\mu \neq \mu_0$

Test statistic: $t = \dfrac{\bar{x} - \mu_0}{s/\sqrt{n}}$

Rejection region: See Appendix Table V, using $(n - 1)$ df.

Small-Sample Confidence Interval for μ

The small-sample confidence interval for μ is expressed as

$$\bar{x} - t\left(\frac{s}{\sqrt{n}}\right) \quad \text{to} \quad \bar{x} + t\left(\frac{s}{\sqrt{n}}\right)$$

The value of t is obtained from Appendix Table VI, using $(n - 1)$ df.

EXAMPLE 7.10   When food prices began their rapid increase in the early 1970's, some of the major television networks began periodically purchasing a grocery basket full of food at supermarkets around the country. They always bought the same items at each store so they could compare food prices. Suppose you want to estimate the mean price for a grocery basket of food in a specific geographical region of the country. You purchase the specified items at a random sample of twenty supermarkets in the region. The mean and standard deviation of the costs at the twenty supermarkets are

$\bar{x} = \$66.84$ $s = \$5.63$

The distribution of the costs of the grocery basket at all supermarkets is approximately normally distributed. Form a 95% confidence interval for the mean cost, μ, of a grocery basket of food for this region.

Solution   We want to form a confidence interval for a population mean using a small sample ($n = 20$). To do this, we will follow the same steps we used with large samples.

The population of interest is the collection of costs for the grocery basket of food at all the supermarkets in the region. The parameter of interest is μ, the true mean cost of the grocery basket of food in this region. It is stated in the example that this population is approximately normally distributed and that the sample was randomly selected from this population. Thus, it is valid to form a small-sample confidence interval for μ. The confidence interval is

$$\bar{x} - t\left(\frac{s}{\sqrt{n}}\right) \quad \text{to} \quad \bar{x} + t\left(\frac{s}{\sqrt{n}}\right)$$

From the example statement, we obtain $\bar{x} = 66.84$, $s = 5.63$, and $n = 20$. To calculate the interval, we need to find the t-value from Appendix Table VI. A reproduction of part of Table VI is given in Table 7.4. To use this table, we need to know the desired level of confidence and the degrees of freedom. We are told to use 95% confidence; since $n = 20$, the degrees of freedom are $n - 1 = 20 - 1 = 19$ df. In Table 7.4, we find that the t-value is 2.09, as indicated by the shading.

Table 7.4

A partial reproduction of Appendix Table VI, t-Values for Confidence Intervals

df	Confidence Level		
	90%	95%	99%
15	1.75	2.13	2.95
16	1.75	2.12	2.92
17	1.74	2.11	2.90
18	1.73	2.10	2.88
19	1.73	2.09	2.86
20	1.72	2.09	2.85
21	1.72	2.08	2.83
22	1.72	2.07	2.82
23	1.71	2.07	2.81
24	1.71	2.06	2.80
25	1.71	2.06	2.79
26	1.71	2.06	2.78
27	1.70	2.05	2.77
28	1.70	2.05	2.76
z	1.65	1.96	2.58

We calculate the interval to be

$$66.84 - 2.09\left(\frac{5.63}{\sqrt{20}}\right) \quad \text{to} \quad 66.84 + 2.09\left(\frac{5.63}{\sqrt{20}}\right)$$

$$66.84 - 2.63 \quad \text{to} \quad 66.84 + 2.63$$

$$64.21 \quad \text{to} \quad 69.47$$

We are 95% confident that the interval from $64.21 to $69.47 contains μ, the true mean cost of the grocery basket of food in this region. ■

If you examine Table 7.4, you will notice that as the degrees of freedom increase, the t-values decrease. This is because as the degrees of freedom increase, the t distribution becomes less variable and looks more and more like the z distribution. As a matter of fact, the last row of the table contains the z-values for large-sample confidence intervals. This is because for sample sizes of 30 or more (degrees of freedom of 29 or more), the t-values and z-values are nearly equal. This partially explains why we chose $n = 30$ as the cutoff for large samples.

EXAMPLE 7.11   Suppose that medical researchers have devised a test that quantitatively measures a person's risk of heart disease. The test scores for the general public are approximately normally distributed with a mean of 50, and the higher the test score, the higher the risk of heart disease. An advocate of long-distance running hypothesizes that marathon runners should have a lower risk of heart disease than the general public. A random sample of seventeen marathon runners is selected, and they are tested for their risk of heart disease. Their test scores have a mean of 43.7 and a standard deviation of 4.2. Is there sufficient evidence to conclude that marathon runners have a lower mean score than the general public on the test measuring the risk of heart disease? Use $\alpha = .01$.

Solution   We want to test a hypothesis about a population mean, μ, and we will follow the steps for testing a hypothesis given in Section 7.2.

The population of interest is the set of scores of all marathon runners on the test designed to measure the risk of heart disease, and the parameter of interest is μ, the mean of this population.

The general public has a mean test score of 50, and the running advocate wants to see if there is a lower risk of heart disease in marathon runners, that is, it is desired to show that $\mu < 50$, where μ is the mean test score for marathon runners. The elements of the test are

$$H_0: \quad \mu = 50 \qquad H_a: \quad \mu < 50$$

$$\text{Test statistic:} \quad t = \frac{\bar{x} - 50}{s/\sqrt{n}}$$

It is stated in the example that the test scores for the general public are approximately normally distributed, and if marathon runners have the same risk of heart disease (if the null hypothesis is true), then the population of interest is approximately normal. Since a random sample has been taken from this population, we may conduct a small-sample test of hypothesis.

Rejection region: Reject H_0 if $t < -2.58$.

Appendix Table V was used to find the rejection region. A reproduction of part of Table V is given in Table 7.5. To use this table, you need to know three things: the value of α, whether the test is one-tailed or two-tailed, and the degrees of

Table 7.5

A partial reproduction of Appendix Table V, t-Values for Rejection Regions for Small-Sample Tests of Hypotheses

df	$\alpha = .10$ One-tailed test	$\alpha = .10$ Two-tailed test	$\alpha = .05$ One-tailed test	$\alpha = .05$ Two-tailed test	$\alpha = .01$ One-tailed test	$\alpha = .01$ Two-tailed test
15	1.34	1.75	1.75	2.13	2.60	2.95
16	1.34	1.75	1.75	2.12	2.58	2.92
17	1.33	1.74	1.74	2.11	2.57	2.90
18	1.33	1.73	1.73	2.10	2.55	2.88
19	1.33	1.73	1.73	2.09	2.54	2.86
20	1.33	1.72	1.72	2.09	2.53	2.85
21	1.32	1.72	1.72	2.08	2.52	2.83
22	1.32	1.72	1.72	2.07	2.51	2.82
23	1.32	1.71	1.71	2.07	2.50	2.81
24	1.32	1.71	1.71	2.06	2.49	2.80
25	1.32	1.71	1.71	2.06	2.49	2.79
26	1.32	1.71	1.71	2.06	2.48	2.78
27	1.31	1.70	1.70	2.05	2.47	2.77
28	1.31	1.70	1.70	2.05	2.47	2.76
z	1.28	1.65	1.65	1.96	2.33	2.58

freedom (df). For this example, $\alpha = .01$, and it is a one-tailed test since the alternative hypothesis is $\mu < 50$. The sample size is $n = 17$, so there are $n - 1 = 16$ df. The shaded entry in Table 7.5 is the t-value that corresponds to 16 df, $\alpha = .01$, and a one-tailed test; the t-value is 2.58. Because the alternative hypothesis is $\mu < 50$, we reject H_0 only if the calculated value of the test statistic (t) is less than the negative t-value. (If the alternative hypothesis were $\mu > 50$, we would reject H_0 if $t > 2.58$.) The rejection region is shown in Figure 7.14.

Using the values given in the example, $\bar{x} = 43.7$, $s = 4.2$, and $n = 17$, we calculate the test statistic to be

$$t = \frac{43.7 - 50}{4.2/\sqrt{17}} = \frac{-6.3}{1.02} = -6.18$$

Figure 7.14
The rejection region for Example 7.11

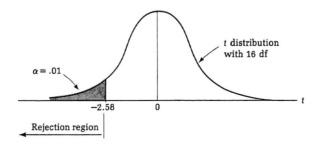

The value $t = -6.18$ is certainly less than the boundary value -2.58 given for the rejection region, so we will reject H_0. With 99% confidence, we can say that marathon runners have a lower mean score than the general public on the test to measure the risk of heart disease. If the test provides an accurate assessment of the risk of heart disease, we can feel very confident that marathon runners have a lower risk than the general public.   ∎

Since Appendix Table V provides only t-values and does not specifically state the rejection region (as Table III does for large-sample tests), some helpful hints for determining the rejection region are given in the box.

Helpful Hints

When using Appendix Table V to find a rejection region, proceed as follows:

1. Determine the value of α, the number of degrees of freedom (df), and whether the test is one-tailed or two-tailed.
2. Find the t-value in Table V that corresponds to the three components specified in step 1.
3. Using the t-value from Table V, give the rejection region as follows:
   a. For H_a: $\mu < \mu_0$, reject H_0 if $t < -$(Table value)
   b. For H_a: $\mu > \mu_0$, reject H_0 if $t > $ (Table value)
   c. For H_a: $\mu \neq \mu_0$, reject H_0 if $t < -$(Table value)
    or $t > $ (Table value)

EXAMPLE 7.12   A serious problem facing strawberry growers is the control of nematodes. These organisms compete with the plants for nutrients in the soil, thereby reducing yield. To help alleviate this problem, fields are usually fumigated. The fumigants currently in use result in an average yield of 3 pounds of marketable fruit per test plot. Recently, a new fumigant has been developed, and it is being tested to compare it with the standard fumigants. The new fumigant is applied to eight randomly selected test plots which yield an average of 3.3 pounds of marketable fruit and a standard deviation of .6 pound. Is there sufficient evidence to conclude that the new fumigant results in a different mean yield of marketable fruit? Assume the yields of marketable fruit are approximately normally distributed, and conduct the test at the .10 level of significance.

Solution   The population of interest is the set of yields of marketable fruit per test plot when using the new fumigant. The parameter of interest is μ, the mean yield of marketable fruit grown with the new fumigant. The elements of the test are

$$H_0: \ \mu = 3 \qquad H_a: \ \mu \neq 3$$

$$\text{Test statistic:} \quad t = \frac{\bar{x} - 3}{s/\sqrt{n}}$$

A small random sample has been selected from the population of interest, which is approximately normally distributed. Thus, we may conduct a small-sample test of hypothesis.

Rejection region: Reject H_0 if $t < -1.89$ or $t > 1.89$.

The rejection region was found using Table V with $\alpha = .10$, df $= 7$, and a two-tailed test.

Using the information given in the example, we obtain

$$t = \frac{3.3 - 3}{.6/\sqrt{8}} = \frac{.3}{.21} = 1.43$$

The calculated value of the test statistic is *not* in the rejection region, so we cannot reject H_0. There is insufficient evidence to conclude that the new fumigant results in a different mean yield of fruit per test plot than the fumigants currently in use.   ■

In the examples used in this section we have been careful to state that the populations of interest are approximately normally distributed. While many real-life populations do have approximately normal distributions, you should keep in mind that many do not. When actually making a small-sample inference, the population of interest must be mound-shaped so that the t statistic will produce credible results.* For cases in which you suspect that the distribution is distinctly nonnormal, you should use a *nonparametric* method of analysis (the topic of Chapter 12) if a small sample must be used. Another possibility—if it is economically feasible—is to take a large sample from the population of interest, because no assumptions must be made about the population of interest in order to make large-sample inferences about the mean.

COMPUTER
EXAMPLE 7.4

In this final computer example for Chapter 7, we will demonstrate several features of Minitab.

Refer to Example 7.11, concerning the risk of heart disease for marathon runners. A random sample of seventeen marathon runners was selected and tested to measure their risk of heart disease. The test scores for the general public are approximately normally distributed with a mean of 50, and the question was whether marathon runners have a lower mean score. The elements of the test were

H_0: $\mu = 50$ H_a: $\mu < 50$

Test statistic: $t = \dfrac{\bar{x} - 50}{s/\sqrt{n}}$

Rejection region: Reject H_0 if $t < -2.58$ ($\alpha = .01$).

The sample of seventeen marathon runners' test scores is shown:

```
38.2   40.7   48.4   48.8   46.6   43.8   35.6   41.3   43.7   43.2   44.9
45.1   43.9   53.3   39.4   43.1   42.9
```

* There are statistical procedures to check the assumption of normality, but they are beyond the scope of this text. In practical problems, an experimenter often knows the approximate shape of the distribution of the population of interest, and this knowledge may be used in choosing the appropriate statistical procedure.

Using these data, Minitab was employed to conduct the test of hypothesis of interest. The computer printout follows.

```
TEST OF MU = 50.00 VS MU N.E. 50.00

           N      MEAN    STDEV    SE MEAN        T    P VALUE
C5        17     43.70     4.20       1.02    -6.18     0.0000
```

The information provided by this printout is very similar to that produced when Minitab performs a large-sample test of hypothesis. The hypotheses of interest are stated and sample summary statistics follow. The value of the test statistic is shown to be $t = -6.18$, the same value obtained in Example 7.11. Since this value is less than -2.58, we conclude with 99% confidence that marathon runners have a lower mean score than the general public on the test to measure the risk of heart disease.

Note that the p-value equals 0.0000. In other words, to four decimal places the observed significance level is zero. With this sample we could have rejected the null hypothesis of interest with virtually any value of α—even with $\alpha = .0001$.

For the same sample of seventeen test scores, we will now display the Minitab printout for a 99% confidence interval:

```
           N      MEAN    STDEV    SE MEAN    99.0 PERCENT C.I.
C5        17     43.70     4.20       1.02    (   40.72,   46.68)
```

This looks just like the printout that would be obtained if we used Minitab to form a confidence interval using a large sample. We are alerted to the fact that this is a small-sample interval since the sample size $n = 17$ is shown. From the calculated interval we make the following interpretation: We are 99% confident that marathon runners have a mean test score between 40.72 and 46.68. Note that the largest value we think the mean could be for marathon runners, 46.68, is well below 50, the mean test score for the general public.

Finally, let us use Minitab to construct a box and whisker plot of the sample data.

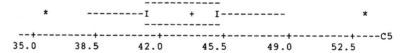

We have stated that the population of test scores for the general public is normally distributed. The box and whisker plot gives some visual indication that the population of test scores for marathon runners is mound-shaped. Examining the graph, we see that the data are quite symmetric. Most of the data fall near the middle, and a few measurements lie at either end.

A word of caution must be given. A graph of the sample data cannot prove whether the population is normally distributed. If a graph of a sample were highly skewed in either direction, however, one would be dubious about using the small-sample inference techniques discussed in this section. Ideally, an experimenter should have some knowledge about the shape of the population distribution. ■

Exercises (7.46–7.64)

Learning the Language

7.46 Define the following terms:
a. t statistic   b. Degrees of freedom

7.47 What do the symbols t and df denote?

Using the Tables

7.48 A random sample of n observations is selected from a normal population to test the null hypothesis that $\mu = 10$. Give the rejection region for each of the following combinations of H_a, α, and n:
a. H_a: $\mu \neq 10$; $\alpha = .05$; $n = 20$
b. H_a: $\mu > 10$; $\alpha = .01$; $n = 25$
c. H_a: $\mu > 10$; $\alpha = .10$; $n = 10$
d. H_a: $\mu < 10$; $\alpha = .01$; $n = 13$
e. H_a: $\mu \neq 10$; $\alpha = .10$; $n = 17$
f. H_a: $\mu < 10$; $\alpha = .05$; $n = 8$

7.49 A random sample of n observations is selected from a normal population with mean μ. For each of the following combinations of sample size and confidence level, give the value of t needed to form a confidence interval for μ:
a. $n = 15$;   confidence level of 90%
b. $n = 6$;   confidence level of 99%
c. $n = 16$;   confidence level of 99%
d. $n = 28$;   confidence level of 95%
e. $n = 28$;   confidence level of 90%

Learning the Mechanics

7.50 A random sample of five measurements from a normally distributed population yielded $\bar{x} = 4.8$ and $s = 1.5$.
a. Test the null hypothesis that the mean of the population is 6 against the alternative hypothesis, $\mu < 6$. Use $\alpha = .10$.
b. Test the null hypothesis that the mean of the population is 6 against the alternative hypothesis, $\mu \neq 6$. Use $\alpha = .10$.
c. Form a 95% confidence interval for μ.
d. Form a 90% confidence interval for μ.

7.51 A random sample of six measurements from a normally distributed population produced the following observations: 4, 6, 2, 7, 6, 5
a. Calculate $\bar{x}$ and s.
b. Test the null hypothesis that $\mu = 2.4$ against the alternative hypothesis that $\mu > 2.4$. Use $\alpha = .01$.

c. Test the null hypothesis that $u = 2.4$ against the alternative hypothesis that $\mu \neq 2.4$. Use $\alpha = .01$.

d. Test the null hypothesis that $\mu = 2.4$ against the alternative hypothesis that $\mu \neq 2.4$. Use $\alpha = .05$.

e. Find a 95% confidence interval for μ.

f. Find a 99% confidence interval for μ.

Applying the Concepts

We will not always state that the population of interest is normally distributed, but for the exercises below, this condition is at least approximately met. Be sure to state the conditions that must be met to make valid inferences for each exercise.

7.52   Pulse rate is an important measure of the fitness of a person's cardiovascular system. The mean pulse rate for American adult males is approximately 72 heart beats per minute. A random sample of twenty-one American adult males who jog at least 15 miles per week had a mean pulse rate of 52.6 beats per minute, and a variance equal to 10.4. Find a 95% confidence interval for the mean pulse rate of all American adult males who jog at least 15 miles per week.

7.53   A consumer protection group is concerned that a catsup manufacturer is filling its 20-ounce family-size containers with less than 20 ounces of catsup. The group purchases ten family-size bottles of this catsup, weighs the contents, and finds that the mean weight is equal to 19.86 ounces and the standard deviation is equal to .22 ounce.

a. Do the data provide sufficient evidence for the consumer group to conclude that the mean fill per family-size bottle is less than 20 ounces? Test using $\alpha = .05$.

b. If the test in part a were conducted on a periodic basis by the company's quality control department, would the consumer protection group be more concerned about a Type I error or a Type II error? The probability of making this type of error is called the **consumer's risk.**

c. The catsup company is also interested in the mean amount of catsup per bottle. It does not want to overfill the bottles. For the test conducted in part a, is the company concerned about a Type I error or a Type II error? The probability of making this type of error is called the **producer's risk.**

7.54   Refer to Exercise 7.53. Find a 90% confidence interval for the mean number of ounces of catsup being dispensed. Does the interpretation of this interval agree with the interpretation of the test conducted in part a of Exercise 7.53?

7.55   The application of adrenaline is the prevailing treatment to reduce eye pressure in glaucoma patients. Theoretically, a new synthetic drug will cause the same mean drop in pressure (5.5 units) without the side effects caused by adrenaline. The new drug is given to five glaucoma patients and the reductions in pressure for the patients are 4.0, 3.8, 5.7, 5.3, and 4.6 units.

a. Look at the data. Based on your intuition, do you think that the mean reduction in pressure for the new drug differs from the mean reduction produced by adrenaline?

b. Use a statistical test to answer the question in part *a*. Do the data provide sufficient evidence to indicate that the mean reduction in pressure due to the new drug is different from that produced by adrenaline? Test using $\alpha = .05$.

7.56 One way of determining whether red pine trees are growing properly is to measure the diameter of the main stem of the tree at the age of 4 years. The main-stem growth for a random sample of seventeen 4-year-old red pine seedlings produced a mean and standard deviation equal to 11.3 and 3.1 inches, respectively. Find a 99% confidence interval for the mean main-stem growth of a population of 4-year-old red pine trees.

7.57 A cigarette manufacturer advertises that its new low-tar cigarette "contains on average no more than 4 milligrams of tar." You have been asked to test the claim using the following sample information: $n = 25$, $\bar{x} = 4.10$ milligrams, $s = .14$ milligram. Does the sample information disagree with the manufacturer's claim? Test using $\alpha = .01$.

7.58 A company purchases large quantities of naphtha in 50 gallon drums. Because the purchases are ongoing, small shortages in the drums can represent a sizable loss to the company. Suppose the company samples the contents of twenty drums, measures the naphtha in each, and calculates $\bar{x} = 49.70$ gallons and $s = .32$ gallon. Do the sample statistics provide sufficient evidence to indicate that the mean fill per 50 gallon drum is less than 50 gallons? Use $\alpha = .10$.

7.59 A psychologist was interested in knowing whether male heroin addicts' assessments of self-worth differ from those of the general male population. On a test designed to measure assessment of self-worth, the mean score for males from the general population is 48.6. A random sample of twenty-five scores achieved by heroin addicts yielded a mean of 44.1 and a variance of 38.44. Do the data indicate a difference in assessment of self-worth between male heroin addicts and the general male population? Test using $\alpha = .01$.

7.60 A problem that occurs in certain types of mining operations is the release of mildly radioactive byproducts. These byproducts may be discharged into our freshwater supply during processing of ores. The EPA has issued a regulation that sets the maximum level for naturally occurring radiation in drinking water at 5 picocuries per liter of water. A random sample of twenty-four water specimens from a city's water supply produced the sample statistics $\bar{x} = 4.61$ picocuries per liter and $s = .87$ picocurie per liter.
a. Do these data provide sufficient evidence to indicate that the mean level of radiation is safe (below the EPA's maximum setting)? Test using $\alpha = .01$.
b. Why should you want to use a small value of α for the test in part *a*?

7.61 In a study of body composition, size, and physiological responses, a sample of thirteen female high-school gymnasts were tested. One characteristic that the researchers measured was the percentage of body fat. The mean percentage of body fat was found to be 13.09 and the standard deviation was 5.12.* Find a 90% confidence interval for the mean percentage of body fat for all female high-school gymnasts.

* *Source:* R. J. Moffatt, B. Surina, B. Golden, and N. Ayres, "Body Composition and Physiological Characteristics of Female High-School Gymnasts," *Research Quarterly for Exercise & Sport*, Vol. 55, No. 1, March 1984, p. 81.

7.62 Various methods can be used to estimate the percentage of body fat in male athletes. A recent study included a sample of nineteen male gymnasts on college teams. Their mean percentage of body fat was 6.5, and the standard deviation was 2.4.*

a. Find a 90% confidence interval for the mean percentage of body fat for all male college gymnasts.

b. Compare the confidence interval found in part a to the one found in Exercise 7.61. Intuitively, what do these two intervals seem to indicate? [*Note:* We will discuss proper statistical procedures for comparing two means in Chapter 8.]

*7.63 Four fossils of humerus bones were unearthed at an archeological site in East Africa. The length-to-width ratios of the bones were 6, 9, 10, and 10. Humerus bones from the same species of animal tend to have approximately the same length-to-width ratio. Suppose it is known that species A has a mean ratio of 8.5. It can be assumed that the four unearthed bones are all from the same unknown species. If an archeologist believes the bones to be from an animal of species A, do the data provide sufficient evidence to contradict the archeologist's theory?

a. Use a computer package to test the hypothesis of interest at the .10 level of significance.

b. What is the *p*-value associated with this test?

*7.64 Suppose you want to estimate the mean percentage of gain in per-share value for growth-type mutual funds over a specific 2-year period. Ten mutual funds are randomly selected from the population of all the commonly listed funds. The percentage gain figures are shown below (negative gains indicate losses):

12.1	−3.7	7.6	6.8	−2.3
4.6	8.4	18.1	9.2	30

a. Use a computer package to find a 90% confidence interval for μ, the mean percentage gain for the population of funds.

b. Use a computer package to find a 99% confidence interval for μ.

7.6

LARGE-SAMPLE INFERENCES ABOUT A POPULATION PROPORTION, π

In Section 5.4 we introduced the binomial experiment and the binomial statistic, *p*. Recall that *p* is the fraction of successes observed in a random sample taken from a population that contains a fraction of successes, π. In Section 6.3 we indicated that when the sample size is large, the sampling distribution of *p* is approximately normal.

* *Source:* W. E. Sinning, D. G. Dolny, K. D. Little, L. N. Cunningham, A. Racaniello, S. F. Siconolfi, and J. L. Sholes, "Validity of 'Generalized' Equations for Body Composition Analysis in Male Athletes," *Medicine and Science in Sports and Exercise*, Vol. 17, No. 1, February 1985, p. 125.

In this section we will test hypotheses about π and form confidence intervals for π using large samples. These methods will be based on the normal distribution and will be very similar to the large-sample methods used to make inferences about μ.

The conditions that must be met to make large-sample inferences about π are given in the box. These conditions guarantee that the sampling distribution of p is approximately normal and are the same as the conditions that were given in Section 6.3.

When to Make Large-Sample Inferences About a Population Proportion, π

To make a valid test of hypothesis about a population proportion, π, or to form a confidence interval for π, the following conditions must be met:

1. A random sample must be taken from the population.
2. The sample must be large enough so that both of the following inequalities hold:

$$\pi - 3\sqrt{\frac{\pi \cdot (1 - \pi)}{n}} > 0 \quad \text{and} \quad \pi + 3\sqrt{\frac{\pi \cdot (1 - \pi)}{n}} < 1$$

Recall that π is the mean of the sampling distribution of p and that $\sqrt{\pi \cdot (1 - \pi)/n}$ is the standard error of the sampling distribution. The second condition in the box assures us that an approximately normal distribution can be fit between 0 and 1, the range of the values of p.

The large-sample test of hypothesis about π and the confidence interval for π are given in the following boxes. Their use will be demonstrated by examples.

Large-Sample Test of Hypothesis About π

H_0: $\pi = \pi_0$

H_a: $\pi > \pi_0$ or H_a: $\pi < \pi_0$ or H_a: $\pi \neq \pi_0$

Test statistic: $z = \dfrac{p - \pi_0}{\sqrt{\dfrac{\pi_0 \cdot (1 - \pi_0)}{n}}}$

Rejection region: See Appendix Table III.

Notice that the denominator of the test statistic, which is the standard error of p, uses the value of π_0 for π [and $(1 - \pi_0)$ for $(1 - \pi)$]. This is because the test statistic is computed assuming H_0 is true. We will reject this hypothesis only if the sample evidence convinces us to do so. Also, notice that the same table is used to find the rejection region as is used for large-sample tests of hypotheses about μ. This is possible because the test statistic is approximately normally distributed in both cases.

Large-Sample Confidence Interval for π

The large-sample confidence interval for π is expressed as

$$p - z \sqrt{\frac{p \cdot (1 - p)}{n}} \quad \text{to} \quad p + z \sqrt{\frac{p \cdot (1 - p)}{n}}$$

The value of z is obtained from Appendix Table IV.

Since the objective of forming a confidence interval is to estimate π, we will not know its value. Thus, we cannot calculate $\sqrt{\pi \cdot (1 - \pi)/n}$, the true standard error of p. Instead, we replace π with p, the sample estimate of π. This is valid for large samples. Table IV is the same table we used when forming large-sample confidence intervals for μ. Again, this is possible because we are relying on the normal sampling distributions of the estimates $\bar{x}$ and p to form a large-sample confidence interval for μ or π.

EXAMPLE 7.13   The reputations (and hence, the sales) of many businesses can be severely damaged by shipments of manufactured items that contain an unusually large percentage of defectives. For example, a manufacturer of flashbulbs does not want to make shipments of bulbs if more than 3% are defective. A random sample of 300 bulbs is selected from a very large shipment, each is tested, and fifteen are found to be defective. Does this provide sufficient evidence at the .05 level of significance to indicate that this shipment of flashbulbs should not be sent?

Solution   The population of interest is the large shipment of flashbulbs, and each bulb is either defective or nondefective. We are interested in π, the true fraction of defectives in the population. We want to know if the shipment should not be sent, that is, if the percentage of defectives is larger than 3%. This indicates that H_a is $\pi > .03$ and H_0 is $\pi = .03$.
   The elements of this test are

H_0: $\pi = .03$ H_a: $\pi > .03$

Test statistic: $z = \dfrac{p - .03}{\sqrt{(.03)(.97)/n}}$

A random sample of 300 bulbs has been selected from this population, and we want to determine whether this sample is large enough to guarantee that p is approximately normal. First, we must be sure that $\pi - 3\sqrt{\pi \cdot (1 - \pi)/n} > 0$ and $\pi + 3\sqrt{\pi \cdot (1 - \pi)/n} < 1$. However, to check this, we need a value for π, and the value we will use is $\pi_0 = .03$, the value given in H_0 and the one used in the test statistic. Calculating the quantities of interest, we obtain

$$.03 - 3\sqrt{\frac{(.03)(.97)}{300}} = .03 - .0295 = .0005 > 0$$

$$.03 + 3\sqrt{\frac{(.03)(.97)}{300}} = .03 + .0295 = .0595 < 1$$

This indicates that it is valid to conduct the large-sample test about π. Notice that .0005 is just barely greater than 0, which indicates that the sample size of 300 is just large enough to validate our use of this test.

Rejection region: Reject H_0 if $z > 1.65$.

The rejection region was obtained from Table III using $\alpha = .05$.

Before we calculate the test statistic, we must find the value of p, the fraction of defectives (successes) in the sample. Since there were fifteen defectives in the sample of 300 bulbs, $p = {}^{15}\!/_{300} = .05$. Calculating the test statistic, we obtain

$$z = \frac{.05 - .03}{\sqrt{(.03)(.97)/300}} = \frac{.02}{.0098} = 2.04$$

Since 2.04 is greater than 1.65, we will reject the null hypothesis. With 95% confidence, the fraction of defective bulbs in the entire shipment is larger than 3%. The shipment should not be sent. ∎

EXAMPLE 7.14   An experiment was conducted to investigate the possibility that a drug suppresses the effects of shock in cats. A sample of twenty-eight cats was induced into a state of shock, and then given the drug. Eighteen of these cats showed signs that the shock symptoms had been suppressed. Form a 99% confidence interval for π, the proportion of all cats for which the drug would be effective in suppressing the effects of shock.

Solution   The population of interest is the collection of observations (effects of shock suppressed or not) that would result if all cats were in shock and given the drug. The parameter of interest is π, the proportion of all cats for which the drug would be effective in suppressing the effects of shock. The large-sample confidence interval is

$$p - z\sqrt{\frac{p \cdot (1 - p)}{n}} \qquad \text{to} \qquad p + z\sqrt{\frac{p \cdot (1 - p)}{n}}$$

The sample size of $n = 28$ does not appear very large, and we must be sure that

$$\pi - 3\sqrt{\pi \cdot (1 - \pi)/n} > 0 \qquad \text{and} \qquad \pi + 3\sqrt{\pi \cdot (1 - \pi)/n} < 1$$

However, we do not know the value of π, since that is what we want to estimate, so we will substitute p for π in order to check these conditions. Although we would like to use the true value of π when checking to see if p is approximately normally distributed, substituting p will be an adequate check when π is unknown.

From the example statement, we know that eighteen out of twenty-eight cats showed signs of suppressed shock symptoms, so $p = {}^{18}\!/_{28} \approx .64$. Performing further calculations, we obtain

$$.64 - 3\sqrt{\frac{(.64)(.36)}{28}} = .64 - .27 = .37 > 0$$

$$.64 + 3\sqrt{\frac{(.64)(.36)}{28}} = .64 + .27 = .91 < 1$$

Thus, even though the sample size is only $n = 28$, we can feel comfortable in using a normal approximation to the sampling distribution of p.

Since 99% confidence is desired, we obtain $z = 2.58$ from Appendix Table IV. Substituting $z = 2.58$, $p = .64$, and $n = 28$ into the above formula, we get

$$.64 - 2.58\sqrt{\frac{(.64)(.36)}{28}} \quad \text{to} \quad .64 + 2.58\sqrt{\frac{(.64)(.36)}{28}}$$

$$.64 - .23 \quad \text{to} \quad .64 + .23$$

$$.41 \quad \text{to} \quad .87$$

With 99% confidence we estimate that the true value of π is between .41 and .87. For somewhere between 41% and 87% of all cats the drug would be effective in suppressing the effects of shock. It certainly seems that the drug is useful for treating cats in shock.   ∎

Extreme caution must be used if you want to make inferences about the effectiveness of the drug discussed in Example 7.14 for treating other animals (humans, for example) in shock. We have sampled only from the population of cats, and it is about this population only that we can validly make inferences. It seems reasonable to suspect that the drug would have a similar effect on animals that are biologically similar to cats. However, to find out whether this is the case, the drug should be tested on the other animals (that is, on samples taken from the other populations of interest), so that inferences can be based on these results.

One final comment should be made about Example 7.14. The interval we found is very wide—from .41 to .87. We do not have a very precise idea of what the value of π really is. The main reason for this is that the sample size ($n = 28$) is relatively small. If we were to take a larger sample, the width of the confidence interval would decrease. In the next section we will discuss how to find the appropriate sample size to obtain a confidence interval of a desired width.

Exercises (7.65–7.79)

Learning the Mechanics

7.65 A random sample of sixty-four observations is selected from a binomial population with proportion of successes π. The computed value of p is .45.
    a. Construct a 95% confidence interval for π.
    b. Construct a 90% confidence interval for π.
    c. Test H_0: $\pi = .5$ against H_a: $\pi \neq .5$. Use $\alpha = .05$.
    d. Test H_0: $\pi = .5$ against H_a: $\pi < .5$. Use $\alpha = .05$.

7.66 A random sample of 100 observations is selected from a binomial population with probability of success π. The computed value of p is .9.
    a. Test H_0: $\pi = .8$ against H_a: $\pi > .8$. Use $\alpha = .01$.
    b. Test H_0: $\pi = .8$ against H_a: $\pi > .8$. Use $\alpha = .10$.
    c. Test H_0: $\pi = .85$ against H_a: $\pi \neq .85$. Use $\alpha = .05$.
    d. Form a 95% confidence interval for π.
    e. Form a 99% confidence interval for π.

7.67 A random sample of ninety observations is selected from a binomial population and results in a value of $p = .65$. It is desired to test the null hypothesis that the population parameter, π, is equal to .70 against the alternative hypothesis that $\pi < .70$.
    a. Noting that $p = .65$, what does your intuition tell you? Do you think the sample value of $p = .65$ provides sufficient evidence to conclude that $\pi < .70$?
    b. Use the large-sample z test to test H_0: $\pi = .70$ against H_a: $\pi < .70$. Use $\alpha = .10$. How do the test results compare with your intuitive decision from part *a*?

Applying the Concepts

7.68 There is much debate about whether an employer should be allowed to force an employee to retire at age 65. A survey was conducted to estimate π, the fraction of workers retiring at age 65 who would have preferred to stay on the job. In a random sample of 300 Americans over 65 years of age, it was found that twenty-one would have preferred to stay on the job. Find a 95% confidence interval for π.

7.69 The United States Commission on Crime wants to estimate the fraction of crimes related to firearms in an area that has one of the highest crime rates in the country. The commission randomly selects 600 files of recently committed crimes in the area and finds 380 in which a firearm reportedly was used. Find a 99% confidence interval for π, the true fraction of crimes in which some type of firearm reportedly was used.

7.70 A method currently used by doctors to screen women for possible breast cancer fails to detect cancer in 15% of the women who actually have the disease. A new method has been developed which researchers hope will be able to detect cancer more accurately. A random sample of seventy women known to have breast cancer was screened using the new method. Of these, the new method failed to detect cancer in six women. Do the data provide sufficient evidence to indicate that the new screening method is better than the one currently in use? Test using $\alpha = .05$.

7.71 A recent report stated that only 20% of all college graduates find work in the field of their undergraduate major. A random sample of 400 graduates from across the country found 100 working in the field of their undergraduate major. Does the sample provide evidence at the $\alpha = .05$ level of significance to indicate that the percentage given in the report was too low?

7.72 A tire manufacturer interested in estimating the proportion of defective automobile tires it produces tested a random sample of 490 tires and found twenty-seven to be defective. Find a 95% confidence interval for π, the true fraction of tires produced by the firm that are defective.

7.73 A company is interested in how well their new computer billing operation is working, so a company statistician randomly samples 400 bills that are ready for mailing and checks them for errors. Twenty-four are found to contain at least one error. Find a 90% confidence interval for π, the true proportion of bills that contain at least one error.

7.74 A producer of frozen orange juice claims that 20% or more of all orange juice drinkers prefer its product. To test the validity of this claim, a competitor randomly samples 200 orange juice drinkers and finds that only thirty-three prefer the producer's brand. Does the sample evidence refute the producer's claim? Test at the $\alpha = .10$ level of significance.

7.75 Last year, a local television station determined that 70% of the people who watch news at 11:00 P.M. watch its station. The station's management believes that the current audience share may have changed. In an attempt to determine whether this is so, the station questioned a random sample of eighty local viewers and found that sixty watched its news show. Does the sample evidence support the management's belief? Test at the $\alpha = .10$ level of significance.

7.76 Shoplifting is a constant problem for retailers. In the past, a large department store found that one in twelve people entering its store engaged in shoplifting. To reduce the incidence of shoplifting, the store recently hired more security guards and introduced other methods of surveillance. After instituting this new security program, the store randomly selected 750 shoppers and observed their behavior. Forty were found to be shoplifting. Is there sufficient evidence to conclude that the new security measures have reduced the incidence of shoplifting? Test using $\alpha = .01$.

7.77 The jobless figures compiled by the United States Bureau of the Census showed that in early 1982 16.4% of the labor force of a large industrial city was unemployed. However, some critics believe that these figures underestimate actual unemployment, since so-called "discouraged workers," who have given up hope of finding a job, are not counted by the bureau. Suppose that in a random sample of 1,000 members of the labor force, 186 are found to be unemployed (some discouraged workers are included). Estimate the unemployment rate when discouraged workers are included. Use a 99% confidence interval.

7.78 A newspaper article investigating the art and science of lie detection reported humbling results:

People are surprisingly inept at detecting lies, new research shows. . . .

Dozens of studies have found that people's accuracy at detecting lies usually exceeds chance by very little. While guessing alone would give a rate of 50% accuracy, in the recent studies the best rate of accuracy for any group has never exceeded 60%, and is most often near chance.

This is true even for those in professions where lie detection is at a premium.[*]

Suppose that a police detective is asked to interview fifty-five potential witnesses of crimes and state if they are lying or telling the truth. If the detective correctly assesses the veracity of 60% of the potential witnesses, would you conclude that this detective can correctly tell more than 50% of the time whether someone is lying? Use $\alpha = .05$.

7.79 Would you generally favor or generally oppose "an agreement between the United States and the Soviet Union for an immediate, verifiable freeze on testing and production of nuclear weapons"? When 1,590 Americans were surveyed on this proposal between September 28 and October 1, 1984, 78% favored a verifiable freeze on testing and production of nuclear weapons.[†] Find a 95% confidence interval for the fraction of all Americans who favor such a nuclear freeze.

7.7

SELECTING THE SAMPLE SIZE

When planning to sample from a population, an important question to answer is, "How large should the sample be?" If the sample is too small, the experimenter will not obtain enough information about the population. On the other hand, we do not want to sample any more observations than necessary, since an expenditure of resources—time and money—is required to obtain sample observations. In this section we will discuss how to find the appropriate sample size to form a confidence interval for a population mean, μ, or a population proportion, π.

The quantity $z(s/\sqrt{n})$ provides a measure of the amount of information provided by a confidence interval for μ. This quantity is added to and subtracted from the sample mean, $\bar{x}$, when forming a confidence interval for μ, and it is equal to half the width of the confidence interval. Since $z(s/\sqrt{n})$ is the farthest we believe μ could be from $\bar{x}$, it is called **the bound on the error of estimation.** To determine what sample size to use to form a confidence interval for μ, it is customary to specify the bound on the error of estimation, B, to specify the desired confidence (to determine z), and to solve the following equation for n:

$$B = z\frac{s}{\sqrt{n}}$$

[*] Daniel Goleman, "Not Quite The Truth," *The Cincinnati Enquirer*, March 8, 1985, p. D-1. Copyright © 1985 by The New York Times Company. Reprinted by permission.
[†] *Source:* "Nuclear Freeze" survey, *The Gallup Report*, No. 229, October 1984, p. 4.

Similarly, to find the sample size required for estimating π to within a certain bound, B, we would solve the equation

$$B = z\sqrt{\frac{p \cdot (1 - p)}{n}}$$

The algebraic solutions to these equations are given in the boxes.

Sample Size for Estimating μ

To estimate μ to within a bound, B, use the following formula to find the sample size, n:

$$n = \frac{z^2 s^2}{B^2}$$

Using a specified confidence, the value of z is found in Appendix Table IV.

Sample Size for Estimating π

To estimate π to within a bound, B, use the following formula to find the sample size, n:

$$n = \frac{z^2 p \cdot (1 - p)}{B^2}$$

Using a specified confidence, the value of z is found in Appendix Table IV.

EXAMPLE 7.15   In Example 7.9 we formed a 99% confidence interval for μ, the mean number of un-occupied seats per flight during the past year for a large airline. The resulting interval was

$$\bar{x} - z\left(\frac{s}{\sqrt{n}}\right) \quad \text{to} \quad \bar{x} + z\left(\frac{s}{\sqrt{n}}\right)$$

$$11.6 - 2.58\left(\frac{4.1}{\sqrt{225}}\right) \quad \text{to} \quad 11.6 + 2.58\left(\frac{4.1}{\sqrt{225}}\right)$$

$$11.6 - .71 \quad \text{to} \quad 11.6 + .71$$

$$10.89 \quad \text{to} \quad 12.31$$

The bound on the error of estimation for this interval is .71. How large a sample would need to be taken to reduce the bound to .50?

Solution   To find the sample size we need to calculate

$$n = \frac{z^2 s^2}{B^2}$$

Since 99% confidence is being used, $z = 2.58$. Also, $s = 4.1$ and $B = .50$. Thus, we obtain

$$n = \frac{(2.58)^2 (4.1)^2}{(.50)^2} = 447.58$$

To estimate μ to within .50 with 99% confidence, we need to randomly sample about 448 flights. Notice that this sample size is much larger than the original sample size of $n = 225$. This is the price that must be paid to obtain a smaller bound on the error of estimation.   ■

The sample size equations for estimating μ and π that are given in the boxes contain the sample quantities s^2 and p, respectively. Usually, these values would be obtained from previous samples, but a previous sample does not always exist. If a value for s^2 or p is not available, use the suggested approximations given in the box.

Helpful Hints
━━━━━━━━━━━━━━━

1. If a value of s^2 is not available when finding a sample size for μ, s can be approximated by one-fourth the range. Thus, substitute $(\text{Range}/4)^2$ for s^2.
2. If a value of p is not available when finding a sample size for π, substitute the value of .5 for p.

In the first case, we presume that enough information is generally known about the population to at least approximate the range. In the second case, by substituting .5 for p, we will be sure to get within the desired bound, since the largest value that $p \cdot (1 - p)$ can be is .25, which occurs when $p = .5$. The use of $p = .5$ will result in a computed sample size that is at least as large as required to obtain the desired confidence and bound.

EXAMPLE 7.16   Refer to the introductory example at the beginning of this chapter. A Gallup poll was taken to estimate the fraction of voters that favored the idea of a voter initiative referendum. How many voters must be sampled in order to estimate this fraction to within .03 with 95% confidence? Did the Gallup organization take a sufficiently large sample to support its final claim?

Solution   It is desired to find a sample size to estimate π, and we must calculate

$$n = \frac{z^2 p \cdot (1 - p)}{B^2}$$

The bound is $B = .03$, and since 95% confidence is desired, the z-value is 1.96. A value of p has not been given, so we will use $p = .5$. Substituting into the equation, we obtain

$$n = \frac{(1.96)^2 (.5)(.5)}{(.03)^2} = 1{,}067$$

Approximately 1,067 voters should be sampled to estimate the fraction who favor a voter initiative referendum to within .03 with 95% confidence. Since the Gallup organization sampled 1,553, this provides support for their claim that "with 95% confidence . . . the error attributable to sampling and other random effects could be three percentage points in either direction."   ■

The calculated value of $n = 1{,}067$ in Example 7.16 gives the appropriate sample size to obtain the desired bound using a *random sample*. The surveys conducted by Gallup and other pollsters are sometimes *not* random, and methods other than the ones given in this text must be used to evaluate the sample results. When we state that the sample of 1,553 provides support for their claim, we are presuming that the Gallup poll is based on a random sample.

Exercises (7.80–7.94)

Learning the Mechanics

7.80 Find the sample size needed to estimate μ to satisfy each of the following:
a. Bound is 3; $s = 40$; 95% confidence desired
b. Bound is 1; $s = 40$; 95% confidence desired
c. Bound is 3; $s = 80$; 99% confidence desired
d. Bound is 6; $s = 80$; 90% confidence desired

7.81 Find the sample size needed to estimate π to satisfy each of the following:
a. Bound is .01; p is .9; 99% confidence desired
b. Bound is .05; p is .9; 99% confidence desired
c. Bound is .02; p is .6; 95% confidence desired
d. Bound is .03; p is .4; 90% confidence desired

7.82 If you want to estimate a population mean to within .1 with 95% confidence, and from a previous sample you have calculated s^2 to be 1.5, how many observations must be included in a sample to meet the desired bound?

7.83 Find the sample size needed to estimate a binomial proportion, π, to within .02 with 90% confidence if:
a. A previous sample produced $p = .7$.
b. There is no previous estimate of π, and you want to be sure to be within the desired bound.

7.84 You want to estimate a population mean to within .2 with 90% confidence. However, a previous sample is not available to provide an estimate of σ^2. The measurements in the population range between 39 and 42. Find the sample size that will produce the desired bound.

Applying the Concepts

7.85 A psychologist who works for the National Football League wants to estimate μ, the mean number of days spent away from home by professional football players during a year. Past records indicate that the distribution of the number of days spent away from home has a standard deviation of 10 days.

    a. How many professional football players should be included in the sample if the researcher wants to be 90% confident that the estimate is within 2 days of the true value of μ?

    b. To reduce the sample size required, would the psychologist have to increase or decrease the desired confidence?

7.86 Before a bill to increase federal price supports for farmers comes before the United States Congress, a representative would like to know how nonfarmers feel about the issue. Approximately how many nonfarmers should the representative survey in order to estimate the true proportion favoring this bill to within .05 with 99% confidence?

7.87 The owner of a large turkey farm knows from previous years that the largest profit is made by selling turkeys when their average weight is 15 pounds. Approximately how many turkeys must be sampled in order to estimate their true mean weight to within .5 pound with 95% confidence? Assume prior knowledge indicates a standard deviation equal to about 2 pounds.

7.88 To estimate the mean age (in months) at which an American Indian child learns to walk, how many Indian children must be sampled if the researcher desires an estimate to be within 1 month of the true mean with 99% confidence? Assume the researcher knows that the age of walking of these children ranges from 8 to 26 months.

7.89 If you want to estimate the proportion of operating automobiles that are equipped with air pollution controls, approximately how large a sample would be required to estimate π to within .02 with 90% confidence?

7.90 Suppose a department store wants to estimate μ, the average age of the customers who shop in its contemporary apparel department, correct to within 2 years with 95% confidence. Approximately how large a sample would be required? The management guesses that the ages of these customers range from 15 to 45.

7.91 The EPA standard for the amount of suspended solids that can be discharged into rivers and streams is a maximum of 60 milligrams per liter daily, with a maximum monthly average of 30 milligrams per liter. Suppose you want to test a randomly selected sample of n water specimens in order to estimate the mean daily rate of pollution produced by a mining operation. If you want your estimate correct to within 1 milligram with 90% confidence, how many water specimens would you have to include in your sample? Assume prior knowledge indicates that pollution readings in water samples taken during a given day have a standard deviation equal to 5 milligrams.

7.92   Suppose you are a retailer and you want to estimate the proportion of your customers who are shoplifters. You decide to select a random sample of shoppers and check closely to determine whether they steal any merchandise while in the store. Suppose that experience suggests the percentage of shoplifters is near 5%. How many customers should you include in your sample if you want to estimate the proportion of shoplifters in your store correct to within .02 with 95% confidence?

7.93   In "A Pollsters' Sampler," Gerald M. Goldhaber examines inaccuracy in public opinion polls. One key is, of course, the sample size. Goldhaber discusses the relationship between the size of the sample and the sampling error associated with estimating a population proportion. The accompanying table summarizes this relationship.

Sample Size	Sampling Error
2,400	±2.0%
1,536	±2.5
1,067	±3.0
784	±3.5
600	±4.0
474	±4.5
384	±5.0
267	±6.0
196	±7.0
150	±8.0
119	±9.0
96	±10.0
45	±15.0

Source: Gerald M. Goldhaber, "A Pollsters' Sampler," *Public Opinion*, Vol. 7, No. 3, June/July 1984, p. 48.

To see how this table was derived, answer the following questions.

a. How large a random sample is needed to estimate a population proportion to within .02 with 95% confidence? Compare your answer to the sample size of 2,400 associated with a sampling error of ±2.0%.

b. How large a random sample is needed to estimate a population proportion to within .05 with 95% confidence? Compare your answer to the sample size of 384 associated with a sampling error of ±5.0%.

c. Verify that the other sample sizes in the table yield the stated sampling errors.

7.94   Refer to Exercise 7.42. A sample of 224 farms from certain Texas counties produced 26 as the mean number of years in operation and 16 as the standard deviation.* How many farms would need to be sampled to estimate the mean number of years in operation to within 1.75 with 95% confidence?

* *Source:* K. C. Vining, C. A. Pope III, and W. A. Dugas, Jr., "Usefulness of Weather Information to Texas Agricultural Producers," *Bulletin of the American Meteorological Society*, Vol. 65, No. 12, December 1984, p. 1317.

7.8

INFERENCES ABOUT A POPULATION VARIANCE, σ^2 (OPTIONAL)

Many practical problems involve inferences about a population mean or proportion. It is also sometimes of interest to make an inference about a population variance, σ^2. Intuitively, it seems that we should in some way use the sample variance, s^2, to make inferences about σ^2. As a matter of fact, inferences will use the quantity

$$\frac{(n-1)s^2}{\sigma^2}$$

which has a sampling distribution called a **chi-square (χ^2) distribution** when the population from which the sample is taken is normally distributed. Several chi-square distributions are shown in Figure 7.15. As with the t distribution, the shape of the chi-square distribution depends on its degrees of freedom. Looking at Figure 7.15, you can also see that the chi-square distribution is not symmetric. It is bounded below by 0, and is skewed to the right.

Figure 7.15
Several chi-square distributions

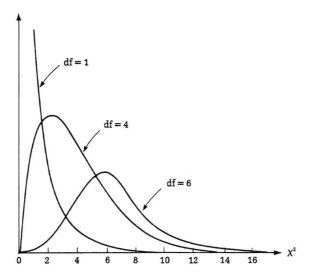

Chi-square tables, which will be used when testing hypotheses about σ^2 or forming confidence intervals about σ^2, are given in the Appendix. These tables will be introduced in the examples.

The conditions that must be met to test a hypothesis about σ^2 or form a confidence interval for σ^2 are given in the box at the top of page 384.

When to Use the Chi-Square Statistic to Make Inferences About a Population Variance, σ^2

To make a valid test of hypothesis about a population variance, σ^2, or to form a confidence interval for σ^2, the following conditions must be met:

1. A random sample must be taken from the population.
2. The population must be approximately normally distributed.

The test of hypothesis about σ^2 and confidence interval for σ^2 are given in the next two boxes. We will demonstrate these procedures with examples.

Test of Hypothesis About σ^2

H_0: $\quad \sigma^2 = \sigma_0^2$

H_a: $\quad \sigma^2 > \sigma_0^2 \qquad$ or $\qquad H_a$: $\quad \sigma^2 < \sigma_0^2 \qquad$ or $\qquad H_a$: $\quad \sigma^2 \neq \sigma_0^2$

Test statistic: $\quad \chi^2 = \dfrac{(n-1)s^2}{\sigma_0^2}$

Rejection region: Use Appendix Table VII, with $(n-1)$ df.

Confidence Interval for σ^2

The confidence interval for σ^2 is expressed as

$$\frac{(n-1)s^2}{\text{Larger table value}} \quad \text{to} \quad \frac{(n-1)s^2}{\text{Smaller table value}}$$

The table values are obtained from Appendix Table VIII, using $(n-1)$ df.

EXAMPLE 7.17   A quality control supervisor in a soup cannery knows that the exact amount of soup each can contains varies due to uncontrollable factors. Although the mean amount of soup placed in the cans is important, the variation in this amount is also important. If σ^2, the variance of the amount of soup put in the cans, is too large, some cans will be overfilled while others will be underfilled. The manager of the cannery would like the standard deviation of the amount of soup put in the cans to be less than .03 ounce. To determine whether the canning process is meeting this specification, the supervisor randomly selects ten cans, measures their contents, and finds the sample standard deviation to be .018 ounce. Do the data provide sufficient evidence to indicate that the variability of the amount of soup put in the cans meets the manager's specifications? Assume that the distribution of the amount of soup put in all cans is approximately normal, and conduct the test at the .05 level of significance.

Solution   The population of interest is the amount of soup put in all cans, and the parameter of interest is σ^2, the variance of this population.

The manager of the cannery wants the standard deviation, σ, to be less than .03 ounce. This means that σ^2 should be less than $(.03)^2 = .0009$. This specifies the alternative hypothesis, and the elements of the test of hypothesis are

H_0: $\sigma^2 = .0009$ H_a: $\sigma^2 < .0009$

Test statistic: $\chi^2 = \dfrac{(n-1)s^2}{.0009}$

The population is approximately normally distributed, and a random sample has been obtained from the population. It is valid to proceed with a test of hypothesis about σ^2.

Rejection region: Reject H_0 if $\chi^2 < 3.33$.

The rejection region was obtained from Appendix Table VII. Since this is a one-tailed test, part A of the table was used. Table VIIA is partially reproduced in Table 7.6. Since $n = 10$, there are $n - 1 = 9$ df. Using $\alpha = .05$, and the column for $<$, we find the value 3.33, as indicated by the shading in Table 7.6.

Table 7.6

A partial reproduction of Appendix Table VIIA, χ^2-Values for Rejection Regions for One-Tailed Tests

| df | $\alpha = .10$ | | $\alpha = .05$ | | $\alpha = .01$ | |
	One-tailed test, $>$	One-tailed test, $<$	One-tailed test, $>$	One-tailed test, $<$	One-tailed test, $>$	One-tailed test, $<$
1	2.71	0.0158	3.84	0.00393	6.64	0.000157
2	4.61	0.211	6.00	0.103	9.21	0.0201
3	6.25	0.584	7.82	0.352	11.4	0.115
4	7.78	1.0636	9.50	0.711	13.3	0.297
5	9.24	1.61	11.1	1.15	15.1	0.554
6	10.6	2.20	12.6	1.64	16.8	0.872
7	12.0	2.83	14.1	2.17	18.5	1.24
8	13.4	3.49	15.5	2.73	20.1	1.65
9	14.7	4.17	17.0	3.33	21.7	2.09
10	16.0	4.87	18.3	3.94	23.2	2.56

In the example statement we are given that $s = .018$, so $s^2 = (.018)^2$. Calculating the value of the test statistic, we obtain

$$\chi^2 = \frac{(10 - 1)(.018)^2}{.0009} = 3.24$$

Since 3.24 is less than 3.33, we conclude that the variance of the amount of soup put in the cans is less than .0009, or equivalently, the standard deviation is less than .03 ounce. We are 95% confident that the canning process meets the manager's specifications.   ■

If the alternative hypothesis in Example 7.17 were $H_a: \sigma^2 > \sigma_0^2$, we would reject H_0 if $\chi^2 > 17.0$ (from Table VIIA). If the alternative hypothesis were $H_a: \sigma^2 \neq \sigma_0^2$, we would reject H_0 if $\chi^2 < 2.70$ or $\chi^2 > 19.0$. These values were obtained from Appendix Table VIIB, which is partially reproduced in Table 7.7.

Table 7.7

A partial reproduction
of Appendix Table VIIB,
χ^2-Values for Rejection
Regions for Two-Tailed Tests

df	$\alpha = .10$ Smaller value	$\alpha = .10$ Larger value	$\alpha = .05$ Smaller value	$\alpha = .05$ Larger value	$\alpha = .01$ Smaller value	$\alpha = .01$ Larger value
1	0.00393	3.84	0.000982	5.02	0.0000393	7.88
2	0.103	6.00	0.0506	7.38	0.0100	10.6
3	0.352	7.82	0.216	9.35	0.0717	12.9
4	0.711	9.50	0.484	11.1	0.207	14.9
5	1.15	11.1	0.831	12.8	0.412	16.8
6	1.64	12.6	1.24	14.5	0.676	18.6
7	2.17	14.1	1.69	16.0	0.990	20.3
8	2.73	15.5	2.18	17.5	1.34	22.0
9	3.33	17.0	2.70	19.0	1.73	23.6
10	3.94	18.3	3.25	20.5	2.16	25.2

EXAMPLE 7.18   Test scores are often used to discriminate among individuals applying for the same job, students applying to graduate school, attorneys seeking admission to the bar, etc. Suppose an employment agency uses a 500-point examination to help in determining which job applicants are best qualified for certain positions. The variability in these test scores should be considered when evaluating the test results. For example, if all applicants should somehow score exactly the same score on the test (no variability among the scores), the test would be of no value in deciding which applicants should be employed. A large amount of variability among the test scores would be desirable in order to differentiate the relative merits of the applicants. To evaluate the variability of the test scores, the employment agency randomly selects 100 test scores and calculates $s^2 = 127$. Use this information to form a 95% confidence interval for σ^2, the variability for *all* test scores. Assume the population of all test scores is mound-shaped.

Solution The population of interest is the collection of all test scores. The parameter of interest is σ^2, the variance of all test scores. We want to form a confidence interval for σ^2. The general form of the confidence interval is

$$\frac{(n-1)s^2}{\text{Larger table value}} \quad \text{to} \quad \frac{(n-1)s^2}{\text{Smaller table value}}$$

Since this population is mound-shaped (approximately normal) and a random sample of test scores has been obtained, it is valid to form a confidence interval for σ^2.

We are told that $s^2 = 127$ and $n = 100$, so all we need to find are the table values required in the denominators. A reproduction of part of Appendix Table VIII is shown in Table 7.8. Since $n = 100$, we should use $n - 1 = 99$ df. Examining the table, however, we see that there is no row corresponding to 99 df, so we will use the closest value, which is 100 df. Since 95% confidence is desired, the two table values are 74.2 and 130.0, as indicated by the shading in Table 7.8. Calculating the interval, we obtain

$$\frac{(100-1)(127)}{130.0} \quad \text{to} \quad \frac{(100-1)(127)}{74.2}$$

$$96.72 \quad \text{to} \quad 169.45$$

At 95% confidence, the variance of the test scores for all applicants is between 96.72 and 169.45.

Table 7.8

A partial reproduction of Appendix Table VIII, χ^2-Values for Confidence Intervals for σ^2

df	90% confidence Smaller value	90% confidence Larger value	95% confidence Smaller value	95% confidence Larger value	99% confidence Smaller value	99% confidence Larger value
25	14.6	37.7	13.1	40.7	10.5	46.9
26	15.4	38.9	13.8	41.9	11.2	48.3
27	16.2	40.1	14.6	43.2	11.8	49.7
28	16.9	41.3	15.3	44.5	12.5	51.0
29	17.7	42.6	16.1	45.7	13.1	52.3
30	18.5	43.8	16.8	47.0	13.8	53.7
40	26.5	55.8	24.4	59.3	20.7	66.8
50	34.8	67.5	32.4	71.4	28.0	79.5
60	43.2	79.1	40.5	83.3	35.5	92.0
70	51.8	90.5	48.8	95.0	43.3	104.0
80	60.4	102.0	57.2	107.0	51.2	116.0
90	69.1	113.0	65.7	118.0	59.2	128.0
100	77.9	124.0	74.2	130.0	67.3	140.0

To interpret the confidence interval found in Example 7.18, it is better to consider the standard deviation. Since we believe σ^2 to be between 96.72 and 169.45, this means that the standard deviation should be between approximately 10 and 13 ($\sqrt{96.72} \approx 10$ and $\sqrt{169.45} \approx 13$). Since almost all the test scores would lie within 3 standard deviations of the mean, the employment agency can evaluate whether the amount of variability in the test scores is sufficient for its purposes. If a standard deviation of 10 (or 13) is deemed to be too small to allow the agency to discriminate among the applicants, the structure of the test should be changed.

Exercises (7.95–7.105)

Learning the Language

7.95 Define the term *chi-square statistic*.

7.96 What does the symbol χ^2 denote?

Using the Tables

7.97 A random sample of n observations is selected from a normal population to test the null hypothesis that $\sigma^2 = 25$. Give the rejection region for each of the following combinations of H_a, α, and n.

     a. H_a: $\sigma^2 \neq 25$; $\alpha = .05$; $n = 20$
     b. H_a: $\sigma^2 > 25$; $\alpha = .01$; $n = 25$
     c. H_a: $\sigma^2 > 25$; $\alpha = .10$; $n = 10$
     d. H_a: $\sigma^2 < 25$; $\alpha = .01$; $n = 13$
     e. H_a: $\sigma^2 \neq 25$; $\alpha = .10$; $n = 7$
     f. H_a: $\sigma^2 < 25$; $\alpha = .05$; $n = 10$

7.98 A random sample of fifteen observations is selected from a normal population with variance σ^2. Give the smaller χ^2-value and the larger χ^2-value that would be used to form a confidence interval for σ^2 for each of the following levels of confidence:
     a.   95%   *b.*   90%   *c.*   99%

Learning the Mechanics

7.99 A random sample of five measurements gave $\bar{x} = 3.1$ and $s^2 = 4.41$.

     a.   What conditions must be met in order to test a hypothesis about (or estimate) σ^2?
     b.   Suppose the conditions in part *a* are satisfied. Test the null hypothesis that $\sigma^2 = 1$ against the alternative hypothesis that $\sigma^2 > 1$. Use $\alpha = .05$.
     c.   Test the null hypothesis that $\sigma^2 = 1$ against the alternative hypothesis that $\sigma^2 \neq 1$. Use $\alpha = .05$.
     d.   Find a 90% confidence interval for σ^2.

7.100 Suppose that in Exercise 7.99, $n = 100$, $\bar{x} = 3.1$, and $s^2 = 4.41$.

     a.   Test the null hypothesis, H_0: $\sigma^2 = 1$, against the alternative hypothesis, H_a: $\sigma^2 > 1$.
     b.   Compare your test results with those of Exercise 7.99. Explain the similarity or the difference in the results of the two tests.
     c.   Find a 90% confidence interval for σ^2. Compare this confidence interval with the confidence interval obtained in Exercise 7.99 and note the effect of an increase in sample size on the width of the interval.

Applying the Concepts

In the exercises below, the populations of interest are (approximately) normally distributed.

7.101 A marine biologist wants to use male angelfish for experimental purposes because she thinks that the variability in weights among male angelfish is small. The biologist randomly samples sixteen male angelfish and finds that their mean weight is 4.1 pounds and the variance is 3. Find a 95% confidence interval for the variability in weights of all male angelfish.

7.102 Refer to Exercise 7.101. It is suggested that the marine biologist use parrotfish instead of male angelfish in the experiment. Since these are more difficult to obtain, the biologist decides to use parrotfish only if there is evidence that the variance of their weights is less than 4. A random sample of ten parrotfish produces a mean of 4.3 pounds and a variance of 2. Is there sufficient evidence for the biologist to claim that the variability in weights among parrotfish is small enough to justify their use in the experiment? Test at $\alpha = .05$.

7.103 A gunlike apparatus has been devised to replace the needle in administering vaccines. The apparatus, which is connected to a large supply of the vaccine, can be set to inject different amounts of the serum, but the variance in the amount of serum injected into an individual must not be greater than .06 to ensure proper inoculation. A random sample of twenty-five injections resulted in a variance of .135. Do the data provide sufficient evidence to indicate that the gun is not working properly? Use $\alpha = .10$.

7.104 It is essential in the manufacture of machinery to utilize parts that conform to specifications. In the past, the diameters of ball-bearings produced by a certain manufacturer had a variance of .00156. To cut costs, the manufacturer instituted a less expensive production method. The variance of the diameters of 100 randomly sampled bearings produced by the new process was .00211. Do the data provide sufficient evidence to indicate that the diameters of ball-bearings produced by the new process are more variable than those produced by the old process? Use $\alpha = .01$.

7.105 To perform an experiment, a chemist has to use a substance that contains 50% sodium nitrate. The chemist suspects that a particular batch of the substance has not been mixed thoroughly, thus causing the amount of sodium nitrate to vary from one portion of the batch to another. The results of twenty randomly selected 10-milliliter samples yield a sample standard deviation equal to .05 milliliter. Estimate the true variance of the amount of sodium nitrate in 10-milliliter samples selected from the batch. Use a 95% confidence interval.

Chapter Summary

Two methods for making inferences about population parameters are the **test of hypothesis,** to make a decision about a specific value for the parameter, and the **confidence interval,** to estimate the value of the parameter.

A test of hypothesis begins with a hypothesized value for the parameter and attempts to reject this value based on sample information. Four elements make up the test.

1. **Null hypothesis, H_0:**   Specifies an assumed value for the parameter that an experiment may show is unreasonable
2. **Alternative (research) hypothesis, H_a:**   Gives the opposing conjecture to that given in the null hypothesis; an experiment may support this hypothesis
3. **Test statistic:**   Is calculated and used to make a decision about the null hypothesis
4. **Rejection region:**   Specifies the values of the test statistic for which we can reject the null hypothesis with the confidence we require

To reject a true null hypothesis is to make a **Type I error.** We want the probability, α, of this sort of error to be small enough so that we can be reasonably confident that it will not occur; the values .10, .05, and .01 are commonly used for α, depending on the corresponding degree of confidence or reliability desired. A **Type II error** consists of accepting a false null hypothesis. We avoid making this kind of error by inferring only that the null hypothesis is false, or else that the test is inconclusive.

The ***p*-value,** or **observed significance level,** for a statistical test of hypothesis is the probability of observing a value of the test statistic that supports the alternative hypothesis at least as well as the value obtained from the sample data. The null hypothesis can be rejected if the *p*-value is less than α.

To form a confidence interval for a population parameter, we estimate the value of the parameter using a sample statistic and then infer that the population parameter must lie in the interval from the calculated value minus a bound, B, to the calculated value plus B. The number B is called the **bound on the error of estimation;** its value depends on sample statistics and the degree of confidence required. Typically, we want to be able to say that we are sufficiently confident that the parameter is within B of the calculated value.

In this chapter, tests of hypotheses and confidence intervals have been applied to four situations.

 I. *Situation:*   Any population, large random sample ($n \geq 30$)

 Parameter: Population mean μ

 Test of hypothesis:

   H_0: $\mu = \mu_0$, where μ_0 is the hypothesized value

   H_a: Choose one of $\mu < \mu_0$, $\mu > \mu_0$,   or $\mu \neq \mu_0$

   Test statistic: $z = \dfrac{\bar{x} - \mu_0}{s/\sqrt{n}}$

   where s is the sample standard deviation used as an approximation to the population standard deviation

   Rejection region: Use Table III.

Confidence interval:

Obtain z from Table IV for the confidence required.

Calculate $\bar{x}$, s, and $B = z(s/\sqrt{n})$.

Interval: $\bar{x} - B$ to $\bar{x} + B$

II. *Situation:* Population with an approximately normal distribution, small random sample ($n < 30$)

Parameter: Population mean μ

Test of hypothesis:

H_0: $\mu = \mu_0$, where μ_0 is the hypothesized value

H_a: $\mu < \mu_0$, $\mu > \mu_0$, or $\mu \neq \mu_0$

Test statistic: Student t distribution with $(n - 1)$ df,

$$t = \frac{\bar{x} - \mu_0}{s/\sqrt{n}}$$

Rejection region: Use Table V.

Confidence interval:

Obtain t from Table VI, using $(n - 1)$ df, for the confidence required.

Calculate $\bar{x}$, s, and $B = t(s/\sqrt{n})$.

Interval: $\bar{x} - B$ to $\bar{x} + B$

III. *Situation:* Binomial experiment, random sample with n sufficiently large that both

$$\pi - 3\sqrt{\frac{\pi \cdot (1 - \pi)}{n}} \quad \text{and} \quad \pi + 3\sqrt{\frac{\pi \cdot (1 - \pi)}{n}}$$

are between 0 and 1. Use either assumed or estimated values of π to test this condition.

Parameter: Population proportion π

Test of hypothesis:

H_0: $\pi = \pi_0$, where π_0 is the hypothesized value

H_a: $\pi < \pi_0$, $\pi > \pi_0$, or $\pi \neq \pi_0$

Test statistic: $z = \dfrac{p - \pi_0}{\sqrt{\dfrac{\pi_0 \cdot (1 - \pi_0)}{n}}}$

Rejection region: Use Table III.

Confidence interval:

Obtain z from Table IV for the confidence required.

Calculate p and $B = z\sqrt{\dfrac{p \cdot (1 - p)}{n}}$.

Interval: $p - B$ to $p + B$

IV. (*Optional*) *Situation:* Population with an approximately normal distribution, random sample of size n

Parameter: Population variance σ^2

Test of hypothesis:

$H_0:$ $\sigma^2 = \sigma_0^2$,   where σ_0^2 is the hypothesized value

$H_a:$ $\sigma^2 < \sigma_0^2$, $\sigma^2 > \sigma_0^2$,   or $\sigma^2 \neq \sigma_0^2$

Test statistic: Chi-square with $(n - 1)$ df,

$$\chi^2 = \frac{(n - 1)s^2}{\sigma_0^2}$$

Rejection region: Use Table VII.

Confidence interval:

Obtain two values, a larger value and a smaller value, from Table VIII, for the confidence required.

Calculate s, $\dfrac{(n - 1)s^2}{\text{Smaller table value}}$, and $\dfrac{(n - 1)s^2}{\text{Larger table value}}$.

Interval: $\dfrac{(n - 1)s^2}{\text{Larger table value}}$ to $\dfrac{(n - 1)s^2}{\text{Smaller table value}}$

In Situations I and III, it may be necessary to determine a sample size n large enough to make the confidence interval as narrow as we want, that is, to obtain a value of B as small as we want. The equations can be solved for n:

I. $n = \dfrac{z^2 s^2}{B^2}$ **III.** $n = \dfrac{z^2 p \cdot (1 - p)}{B^2}$

The value for z is obtained from Table IV; s^2 may be approximated by one-fourth the range, and p may be approximated by .5 if necessary.

SUPPLEMENTARY EXERCISES (7.106–7.142)

In each exercise, carefully state the condition(s) required for the validity of the statistical procedure used.

Using the Tables

7.106   A random sample of 200 observations is selected from a population with mean μ and variance σ^2.

a.   Give the rejection region for testing $H_0: \mu = 35$ versus $H_a: \mu < 35$, with $\alpha = .05$.

b.   Give the z-value for forming a 99% confidence interval for μ.

    c. Give the z-value for forming a 90% confidence interval for μ.

    d. Give the rejection region for testing H_0: $\mu = 20$ versus H_a: $\mu \neq 20$, with $\alpha = .01$.

7.107 A random sample of twenty-five measurements is taken from a normally distributed population with mean μ and standard deviation σ.

    a. (*Optional*) Give the rejection region for testing H_0: $\sigma^2 = 30$ versus H_a: $\sigma^2 < 30$, with $\alpha = .05$.

    b. Give the t-value for forming a 90% confidence interval for μ.

    c. Give the rejection region for testing H_0: $\mu = 75$ versus H_a: $\mu < 75$, with $\alpha = .01$.

    d. (*Optional*) Give the rejection region for testing H_0: $\sigma^2 = 30$ versus H_a: $\sigma^2 \neq 30$, with $\alpha = .10$.

    e. (*Optional*) Give the smaller and larger chi-square values for forming a 95% confidence interval for σ^2.

    f. Give the rejection region for testing H_0: $\mu = 75$ versus H_a: $\mu > 75$, with $\alpha = .05$.

    g. Give the t-value for forming a 99% confidence interval for μ.

Learning the Mechanics

7.108 A random sample of twenty observations selected from a normal population produced $\bar{x} = 72.6$ and $s = 19.4$.

    a. Form a 90% confidence interval for the population mean.

    b. Test H_0: $\mu = 80$ against H_a: $\mu < 80$. Use $\alpha = .05$.

    c. Test H_0: $\mu = 80$ against H_a: $\mu \neq 80$. Use $\alpha = .01$.

    d. Form a 99% confidence interval for μ.

    e. How large a sample would be required to estimate μ to within 3.0 with 95% confidence?

7.109 A random sample of $n = 200$ yields $p = .29$.

    a. Test H_0: $\pi = .35$ against H_a: $\pi < .35$. Use $\alpha = .05$.

    b. Test H_0: $\pi = .35$ against H_a: $\pi \neq .35$. Use $\alpha = .05$.

    c. Form a 95% confidence interval for π.

    d. Form a 99% confidence interval for π.

    e. How large a sample would be required to estimate π to within .05 with 99% confidence?

7.110 A random sample of 175 measurements has a mean $\bar{x} = 8.2$ and a standard deviation $s = .79$.

    a. Form a 95% confidence interval for μ.

    b. Test H_0: $\mu = 8.3$ against H_a: $\mu \neq 8.3$. Use $\alpha = .05$.

    c. Test H_0: $\mu = 8.4$ against H_a: $\mu \neq 8.4$. Use $\alpha = .05$.

7.111 (*Optional*) A random sample of forty-one observations from a normal population possesses a mean $\bar{x} = 88$ and a standard deviation $s = 6.9$.

    a. Form a 90% confidence interval for σ^2.

    b. Form a 99% confidence interval for σ^2.

    c. Test H_0: $\sigma^2 = 30$ against H_a: $\sigma^2 > 30$. Use $\alpha = .05$.

    d. Test H_0: $\sigma^2 = 30$ against H_a: $\sigma^2 \neq 30$. Use $\alpha = .05$.

Applying the Concepts

7.112 Failure to meet payments on student loans guaranteed by the government has been a major problem for both banks and the United States government. Approximately 50% of all students loans guaranteed by the government are in default. A random sample of 350 loans to college students in one region of the United States indicates that 147 loans are in default. Do the data indicate that the proportion of loans in default in this area of the country differs from the proportion of all loans in the United States that are in default? Use $\alpha = .01$.

7.113 In order to be effective, the mean lifetime of a particular mechanical component used in a space craft must be larger than 1,100 hours. Due to the prohibitive cost of the components, only three can be tested under simulated space conditions. The lifetimes (in hours) of the three components were recorded with the following summary results: $\bar{x} = 1{,}173.6$ and $s = 36.3$. Do the data provide sufficient evidence to conclude that the component will be effective? Use $\alpha = .01$.

7.114 The mean score on a Peace Corps application test, based on many tests conducted over a long period of time, is 80. Ten prospective applicants have taken a course designed to improve their scores on the test. After completing the course, the applicants had test scores with mean equal to 86.1 and standard deviation equal to 12.4.
 a. Do the data provide sufficient evidence to conclude that students taking the course will have a higher mean score than those who do not? Test using $\alpha = .05$.
 b. The observed significance level of this test is approximately p-value = .08. Explain what this means.

7.115 A sporting goods manufacturer who produces both white and yellow tennis balls claims that more than 75% of all tennis balls sold are yellow. A marketability study of the purchases of white and yellow tennis balls at a number of stores showed that of 470 cans of balls sold, 410 were yellow and 60 were white. Is there sufficient evidence to support the manufacturer's claim? Test using $\alpha = .01$.

7.116 A discount store claims that its steel-belted radial tires last at least as long as those of a major tire company, X. The following experiment was performed to test this claim. On each of forty cars, one discount tire and one company X tire were mounted on the rear axle. After each car was driven 8,000 miles, the tires were inspected for wear. Suppose the tires of company X show less wear on thirty-two of the cars. Is there sufficient evidence to conclude that the discount store's claim is incorrect? Test using $\alpha = .10$. [*Hint:* Let π equal the proportion of discount store tires that last as long as those of company X. Then conduct an appropriate hypothesis test concerning π.]

7.117 During past harvests, a farmer has averaged 68.2 bushels of corn per acre. After using a new fertilizer, the farmer notes the yield of corn for four randomly selected fields of equal size. The mean yield is 72.4 bushels per acre and the standard deviation is 2.2 bushels.
 a. If these data truly represent a random sample of corn yields that the farmer might expect (now and in the future) when using the new fertilizer, do they suggest a

difference in the mean yield of corn per acre for the new fertilizer as compared to the mean yield of past years? Test using $\alpha = .05$.

b. Note that the four yield measurements were selected from within the same year. Are these measurements a random sample selected from the population of interest to the farmer? If not, what information do the data provide the farmer?

7.118 The EPA sets a limit of 5 parts per million on PCB (a dangerous substance) in water. A major manufacturing firm producing PCB for electrical insulation discharges small amounts from the plant. The company management, attempting to control the PCB in its discharge, has given instructions to halt production if the mean amount of PCB in the effluent exceeds 3 parts per million. A random sampling of fifty water specimens produced the following statistics: $\bar{x} = 3.1$ parts per million and $s = .5$ part per million.

a. Do these statistics provide sufficient evidence to halt the production process? Use $\alpha = .01$.

b. If you were the plant manager, would you want to use a large or a small value for α for the test in part a? Explain.

7.119 If the rejection of the null hypothesis of a particular test would cause your firm to go out of business, would you want α to be small or large? Explain.

7.120 A company is interested in estimating μ, the mean number of days of sick leave taken by all its employees. The firm's statistician selects at random 100 personnel files and notes the number of sick days taken by each employee. The following sample statistics are computed: $\bar{x} = 12.2$ days and $s = 8.3$ days.

a. Estimate μ using a 90% confidence interval.

b. How many personnel files would the statistician have to select in order to estimate μ to within 1 day with 99% confidence?

7.121 A large mail-order company has placed an order for 5,000 electric can openers with a supplier on the condition that no more than 2% of the can openers will be defective. To check the shipment, the company tests a random sample of 400 of the can openers, and finds eleven defective.

a. Does this provide sufficient evidence to indicate that the proportion of defective can openers in the shipment exceeds .02? Test using $\alpha = .05$.

b. Suppose the company wants to estimate the proportion of defective can openers in the shipment correct to within .04 with 95% confidence. Approximately how large a sample would be required?

7.122 A university is considering a change in the way students pay for their education. Presently, the students pay $16 per credit hour. The university is contemplating charging each student a set fee of $240 per quarter, regardless of how many credit hours are taken. To see if this would be economically feasible, the university would like to know how many credit hours, on the average, each student takes per quarter. A random sample of 250 students yields a mean of 14.1 credit hours and a standard deviation of 2.3.

a. Estimate the mean number of credit hours per student per quarter using a 99% confidence interval.

b. Use this confidence interval to obtain an interval estimate for the mean tuition fee per student per quarter under the current fee system. Interpret the result.

7.123 In checking the reliability of a bank's records, auditing firms sometimes ask a sample of the bank's customers to confirm the accuracy of their savings account balances as reported by the bank. Suppose an auditing firm is interested in estimating the proportion of a bank's savings accounts on which the bank and the customer disagree on the balance. Of 200 savings account customers questioned by the auditors, fifteen said their balance disagreed with that reported by the bank.

a. Use a 95% confidence interval to estimate the actual proportion of the bank's savings accounts on which the bank and the customer disagree on the balance.

b. The bank claims that the true proportion of accounts on which there is disagreement is at most .05. You, as an auditor, doubt this claim. Does the sample evidence support your suspicion? Test at the .10 significance level.

7.124 The Chamber of Commerce of a small seaside resort would like to know the mean number of hours of labor required to clear litter from its public beach on weekends. A random sample of the labor expended on each of fifteen randomly selected Sunday mornings produced a mean of 3.6 hours and a standard deviation of .6 hour. Estimate the mean amount of labor required per Sunday morning to clear the beach of litter. Use a 95% confidence interval.

7.125 The strength of a pesticide dosage is often measured by the proportion of pests that dosage will kill. To determine this proportion for a particular dosage of rat poison, 250 rats are fed the dosage of poison and 215 die. Use a 90% confidence interval to estimate the true proportion of rats that will succumb to the dosage.

7.126 A meteorologist wants to estimate the mean amount of snowfall per year in Spokane, Washington. A random sampling of the recorded snowfalls for 20 years produces a sample mean equal to 54 inches and a variance of 92.

a. Estimate the true mean annual amount of snowfall in Spokane using a 99% confidence interval.

b. If you were purchasing snow-removal equipment for a city, what numerical descriptive measure of the distribution of depth of snowfall would be of most interest to you? Would it be the mean?

7.127 A leading cigarette manufacturer claims its cigarettes contain an average of less than 16 milligrams of tar. To check this claim, a random sample of cigarettes will be chosen and the mean amount of tar per cigarette will be estimated. Assuming previous information indicates that the amount of tar per cigarette has a standard deviation of 2.5 milligrams, how many cigarettes should be sampled to estimate the true mean to within .45 milligram with 95% confidence?

7.128 A university dean is interested in determining the proportion of students who receive some sort of financial aid. Rather than examine the records for all students, the dean randomly selects 200 students and finds that 118 are receiving financial aid. Use a 95% confidence interval to estimate the true proportion of students who receive aid.

7.129 Before approval is given for the use of a new insecticide, the United States Department of Agriculture (USDA) requires that several tests be performed to see how the substance will affect wildlife. In particular, the USDA would like to know the proportion of starlings that will die after being exposed to the insecticide. A random sample of eighty starlings is caught and fed food treated with the insecticide. Within 10 days, ten starlings die. Use a 99% confidence interval to estimate the true proportion of starlings that will be killed by the substance.

7.130 A recent poll of 200 college-age women from across the country indicated that 170 approved of women seeking professional careers. Find a 95% confidence interval for the true proportion of college-age women who approve of women seeking professional careers.

7.131 Officials at a high school claim that at least 85% of the students who have graduated from the school have received a college degree or are enrolled in a college program. A random sample of sixty former students indicates that forty-seven have received or are working toward a college degree.

    a. Do the data contradict the high school official's claim? Use $\alpha = .10$.

    b. The observed significance level for this test is p-value = .074. Interpret this value.

7.132 Many people think that a national lobby's successful fight against gun control legislation is reflecting the will of a minority of Americans. A random sample of 4,000 citizens yielded 2,250 who are in favor of gun control legislation. Use a 99% confidence interval to estimate the true proportion of Americans who favor gun control legislation.

7.133 In the past, a chemical company produced 880 pounds of a certain type of plastic per day. Now, using a newly developed and cheaper process, the mean daily yield of plastic for the first 50 days of production was 871 pounds and the standard deviation was 21 pounds.

    a. Do the data provide sufficient evidence to indicate that the mean daily yield for the new process is less than for the old procedure? (Test using $\alpha = .01$.)

    b. Was the sample randomly selected? How might this affect the test results?

7.134 Under the headline "'Ms.' Title Bothers Newspaper Editors," *The Cincinnati Enquirer* ran the following article:

> "Ms." is rapidly losing respect among editors, another indication that the influence of the women's movement on language is waning, says the author of a language survey conducted by Indiana University.
>
> Richard L. Tobin said the sixth annual English usage survey taken by the university's School of Journalism found the acceptability of the courtesy title favored by feminists had been cut in half over the last three years. The survey was released Feb. 28.
>
> Just 28.4% of the 150 newspaper editors and 50 magazine editors responding to the new survey approved of "Ms.," compared with 57.3% in 1982, said Tobin, an adjunct professor of journalism.
>
> Tobin said the latest results, combined with last year's overwhelming 70.4% against "chairperson" in place of the once-standard "chairman," led him to conclude that the influence of the women's movement on language is lessening.*

Assuming that the 200 editors were randomly sampled, find a 95% confidence interval for the fraction of all editors who approved of "Ms." at the time of the new survey.

---

* "'Ms.' Title Bothers Newspaper Editors," *The Cincinnati Enquirer*, March 4, 1985, p. A-2.

7.135 A survey of 3,528 Americans in 1971 asked, "What do you think is the ideal number of children for the average family?" The following statistics were obtained:*

$\bar{x} = 2.8$ children $s = 1.1$

  a. Find a 99% confidence interval for the 1971 mean response for all Americans.

  b. How many Americans would need to be sampled now in order to estimate the mean response to the same question to within .03 with 99% confidence? Assume that the variability of the data will remain about the same.

7.136 A 1984 Gallup poll was conducted to obtain responses to the following question:

> As you may know, the Administration's 1985 budget projects a deficit of about $180 billion—that is, the Federal government would spend about $180 billion more than it takes in. Do you feel that budget deficits of this size are or are not likely to cut the economic recovery short?

The responses of the 820 adults surveyed are summarized below:

Yes	No	No opinion
65%	21%	14%

  a. Find a 90% confidence interval for the fraction of all adults who would reply "Yes" to this question.

  b. How many people would need to be surveyed to estimate the fraction of all adults who would reply "Yes" to within .02 with 90% confidence?

7.137 We tend to think of America as a prosperous nation. Unfortunately, some families at times cannot afford to buy food. A 1984 survey of 1,562 adults, aged 18 and older, found that 20% of those polled had at times during the previous year lacked enough money to buy food for their families.‡ Find a 99% confidence interval for the true fraction of all Americans who at times lacked enough money to buy food for their families.

7.138 (*Optional*) A machine used to fill beer cans must operate so that the amount of beer actually dispensed varies very little. If too much beer is released, the cans will overflow, causing waste. If too little beer is released, the cans will not contain enough beer, causing complaints from customers. A random sample of the fills for twenty cans yielded a standard deviation of .09 ounce. Estimate the true variance of the fills using a 95% confidence interval.

7.139 (*Optional*) Ophthalmologists use a special instrument to measure intraocular pressure in glaucoma patients. The device now in general use is known to yield readings of this pressure with a variance of 10.3. The variance of five pressure readings on the same eye by a newly developed instrument is equal to 9.8. Does this provide sufficient

---

* *Source:* P. E. Converse, J. D. Dotson, W. J. Hoag, and W. H. McGee III, *American Social Attitudes Data Sourcebook, 1947–1978* (Cambridge, Mass.: Harvard University Press, 1980), p. 126. Reprinted by permission. Copyright © 1980 by The President and Fellows of Harvard College.
† "Effect of Deficit on Economic Recovery" survey, *The Gallup Report*, No. 223, April 1984, p. 12.
‡ *Source:* "Money for Food" survey, *The Gallup Report*, Nos. 220/221, January/February 1984, p. 23.

evidence to indicate that the new instrument is more reliable (yields pressure readings with a smaller variance) than the instrument currently in use? (Use $\alpha = .05$.)

7.140 (*Optional*) The variance in the diameters of screw-top lids must not be larger than .38 square millimeters to ensure that the lids will fit properly on glass jars. A random sample of fifteen lids yielded a sample variance of .61. Do the data provide sufficient evidence to conclude that the variance in the lid diameters is larger than .38? (Use $\alpha = .05$.)

*7.141 To help consumers assess the risks they are taking, the Food and Drug Administration (FDA) publishes information on the amount of nicotine found in all commercial brands of cigarettes. A new cigarette has recently been marketed, and the FDA randomly samples twenty-four cigarettes of this brand and measures their nicotine contents. The sample measurements (in milligrams) are:

27.4	26.7	26.2	23.6	24.6	23.8	26.5	27.3
28.6	28.9	24.6	25.5	26.9	24.4	25.6	26.0
25.4	27.9	29.2	22.7	25.2	24.6	25.2	27.7

a. Use a computer package to form a 90% confidence interval for μ, the true mean nicotine content per cigarette of the new brand.

b. Use a computer package to test the null hypothesis $\mu = 27$ versus the alternative hypothesis $\mu < 27$. Use $\alpha = .01$.

c. What is the p-value of the test conducted in part a?

*7.142 A health researcher wants to investigate the number of cavities per child for children under the age of 12 who live in a specified environment. The number of cavities per child for a random sample of thirty-five children under the age of 12 is given.

4	2	3	7	3	1	3	10	2
3	2	1	1	3	5	4	4	0
5	3	2	2	6	3	9	2	0
1	2	10	3	0	4	1	4	

a. It is desired to determine whether the mean number of cavities per child for all children in this environment is less than 4.1. Use a computer package to test the hypotheses of interest. Use $\alpha = .10$.

b What is the p-value of the test conducted in part a? If the value of $\alpha = .05$ (instead of $\alpha = .10$) had been used, would your interpretation be the same? Explain.

c. Use a computer package to form a 90% confidence interval for the true mean number of cavities per child in this region.

CHAPTER 7 QUIZ

1. An experimenter is testing the following hypotheses:

$$H_0: \quad \pi = .4 \qquad H_a: \quad \pi < .4$$

a. Define a Type I error and a Type II error for this test.

b. Define α and β for this test.

2. A psychologist employed by a medical clinic examines the records of a random sample of 1,000 of the clinic's patients and finds that 60% of their illnesses are of a psychosomatic nature. Form a 99% confidence interval for the proportion of all the clinic's patients with illnesses of a psychosomatic nature.

3. Refer to Question 2. How many patients' records must be sampled to estimate the proportion of all the clinic's patients with psychosomatic illnesses to within .03 with 99% confidence?

4. A rehabilitation officer wants to estimate the average number of convictions for juvenile delinquents from a certain city. A random sample of forty-five juvenile delinquents from this city had a mean of 1.8 convictions and a standard deviation of .4. Find a 95% confidence interval for the mean number of convictions for all juvenile delinquents in this city.

5. In a certain state, fields planted in corn yield an average of 67 bushels of corn per acre. The state university has developed a new variety of corn which is thought to produce a higher mean yield per acre. A random sample of seventy fields is planted with the new variety, and they yield an average of 70.1 bushels per acre and a standard deviation of 2.6 bushels. Do the data provide sufficient evidence at the .05 level of significance to indicate that the new variety of corn produces a higher mean yield per acre?

CHAPTERS 5–7 CUMULATIVE QUIZ

1. A gun club in a certain city claims that at most 15% of the city's residents are in favor of registering hand guns. A random sample of 500 city residents found 80 who favored the registration of hand guns.

   a. Do the data provide sufficient evidence to indicate that more than 15% of the city's residents are in favor of registering hand guns? Use $\alpha = .05$.

   b. Let π be the true fraction of all the city's residents who are in favor of registering hand guns. If $\pi = .15$, what is the probability that in a sample of 500 residents, the sample fraction favoring hand gun registration, p, is $^{80}/_{500}$ or larger? How does this answer relate to the test results in part a?

2. A population has a mean of $\mu = 1.5$ and a variance of $\sigma^2 = .45$. For a random sample of $n = 2$ measurements selected from this population, the sample mean, $\bar{x}$, has the following sampling distribution:

$\bar{x}$	0.0	0.5	1.0	1.5	2.0
$P(\bar{x})$	.01	.06	.21	.36	.36

   a. Calculate $\mu_{\bar{x}}$ and $\sigma_{\bar{x}}^2$. Compare these values to μ and σ^2.

   b. Fully describe the sampling distribution of $\bar{x}$ for a random sample of $n = 65$ measurements taken from the population with $\mu = 1.5$ and $\sigma^2 = .45$.

3. *a.* When is it valid to consider the sampling distribution of $\bar{x}$ to be approximately normal?

   b. When is it valid to consider the sampling distribution of p to be approximately normal?

4. The number of pounds of potatoes a supermarket sells per week has an approximately normal distribution with mean μ. To estimate the value of μ, the number of pounds of potatoes sold is recorded for a random sample of 10 weeks. The results of this sample yielded a mean of 1,075 pounds and a standard deviation of 154 pounds. Form a 95% confidence interval for μ.

5. Refer to Question 4. How large a sample needs to be taken to estimate μ to within 50 pounds with 95% confidence?

On Your Own

Choose a population pertinent to your major area of interest that has an unknown mean (or, if the population is binomial, that has an unknown probability of success). For example, a marketing major may be interested in the proportion of consumers who prefer a particular product. A sociology major may be interested in estimating the proportion of people in a certain socioeconomic group or the mean income of people living in a certain part of a city. A political science major may want to estimate the proportion of an electorate in favor of a certain candidate, amendment, or presidential policy. A nursing major might want to find the average length of time patients stay in the hospital or the average number of people treated daily in the emergency room. We could continue with examples, but the point should be clear—choose something of interest to you.

Define the parameter you want to estimate and conduct a **pilot study** to obtain an initial estimate of the parameter of interest, and, more importantly, an estimate of the variability associated with the estimator. A pilot study is a small experiment (perhaps twenty to thirty observations) used to gain some information about the population of interest. The purpose is to help plan more elaborate future experiments. Based upon the results of your pilot study, determine the sample size necessary to estimate the parameter to within a reasonable bound (of your choice) with a 95% confidence interval.

INFERENCES COMPARING TWO POPULATIONS

COMPARING CITY LIVING AND COUNTRY LIVING

Figure 8.1 (page 404) is a reproduction of part of a newspaper article* comparing the cost of living in the city to that in the country. This figure compares the prices of several items in Cincinnati, Ohio (the "city"), to those in Greensburg, Indiana (the "country"). These represent a sample of prices from all the items sold in Cincinnati, and a sample of prices from all the items sold in Greensburg. Many practical problems involve the comparison of two populations, and inferences are made by sampling from both populations. In this chapter we will present methods for comparing two population means, two population proportions, and two population variances.

* Carol Sanger, "Is It Really Cheaper in the Country?" *The Cincinnati Enquirer*, August 28, 1981, p. B1.

Figure 8.1
City versus country
Source: Carol Sanger,
"Is It Really Cheaper
in the Country?" *The*
Cincinnati Enquirer,
August 28, 1981, p. B1.
Reproduced by permission.

CHEAPER HERE:

ITEM	CINCINNATI	GREENSBURG
Band-Aid Sheer Strips	$1.59	$1.77
Flair pen	$ .79	$ .87
Frigidaire refrigerator	$799.95	$1,089.99
GE microwave oven	$369.99	$599
Kodak Colorburst camera	$21.50	$21.88
Magnavox color TV	$449	$499.95
Pampers diapers	$2.69	$3.24
Rain Dance car wax	$5.99	$8.09

CHEAPER THERE:

ITEM	CINCINNATI	GREENSBURG
Clairol Crazy Curl	$13.99	$13.27
Milk—1 gallon	$2.09	$1.89
Reynolds Wrap	$ .67	$ .63
Secret Roll-on deodorant	$1.59	$1.29
Tide	$2.29	$2.08
Samsonite carry-on	$59	$55
Wilson basketball	$49.95	$24.95
Wilson tennis racket	$49.95	$36.95

8.1

INDEPENDENT AND DEPENDENT SAMPLES

To make inferences about two populations, we must obtain two samples—one from each population. There are many methods by which the two samples could be obtained; we will discuss two of them in this text. These methods result in either **independent** or **dependent** samples.

Definition 8.1

If two samples are selected, one from each of two populations, then the two samples are **independent** if the selection of objects from one population is unrelated to the selection of objects from the other population.

Definition 8.2

If two samples are selected, one from each of two populations, then the two samples are **dependent** if for each object selected from one population an object is chosen from the other population to form a pair of similar objects.*

The following examples will help clarify what it means for samples to be independent or dependent.

EXAMPLE 8.1   A sociologist wants to compare the mean starting salaries for men and women who graduate from college this year. Two methods by which the samples could be obtained are described below.

Method I: From the population of all males graduating from college this year, a random sample of 100 graduates is chosen, and the starting salary of each sampled graduate is recorded. A second random sample of 100 female graduates is then selected from all those available this year. The starting salary of each sampled female is also recorded, and a comparison of the male and female salaries is then made.

Method II: Since the graduates' starting salaries may be related to factors such as college major and grade-point average, the male and female graduates could be matched to form similar pairs. Before actually selecting any samples, pairs of male and female graduates would be formed in such a way that each pair has the same major and approximately the same grade-point average. A random sample of 100 of these matched pairs would then be selected, and the starting salary of each male and female recorded. A comparison of the male and female salaries would then be made.

One of these methods describes independent samples and the other describes dependent samples. Classify each method as being independent or dependent.

Solution   **Method I:**   This method of sampling results in independent samples. The selection of the male graduates is completely unrelated to the selection of the female graduates. Two different random samples are taken, one from each population.

Method II: Since males and females were matched with respect to college major and grade-point average, the samples are dependent. For each male selected, a similar (in terms of major and grade-point average) female was selected.   ■

* There are other ways of sampling, in addition to pairing observations, that would result in dependent samples. In this text, however, all dependent samples will be paired samples.

You may find that, intuitively, the second method of sampling in Example 8.1 seems better. By pairing male and female graduates according to grade-point average and college major, we seem to be making a fairer comparison of salaries. If independent samples were selected, we might sample highly qualified males and poorly qualified females (or vice versa). This could not happen with dependent samples. We will compare independent and dependent sampling methods in more detail later in this chapter.

The key to recognizing two independent samples is to realize that they are always two *different* random samples.* By contrast, dependent samples always consist of matched, or paired, observations.

EXAMPLE 8.2   A new mathematics course for college students is being studied. It is desired to compare the students' level of mathematical achievement after 1 month in the course to their level at the outset. A random sample of twenty students is chosen at the beginning of the course and given a test to measure their level of mathematical achievement at this time. One month later, each of the twenty students is given a second test to measure the level of achievement, and a comparison of the two sets of test scores is made. Do the two sets of test scores represent independent or dependent samples?

Solution   The same twenty students who are given the first test are also given the test 1 month later. Thus, there are twenty pairs of test scores—one for each student—and the samples are dependent. In many experiments, pairs of observations are formed by taking two measurements for the same individual (or other experimental object). This is a common method of obtaining dependent samples.   ■

It is very important that you be able to recognize whether two samples are independent or dependent. Different statistical methods are used to analyze these two different types of samples. This will be demonstrated in the following sections.

Exercises (8.1–8.8)

Learning the Language

8.1   Define the following terms:
  a. Independent samples   b. Dependent samples

Applying the Concepts

8.2   An experiment is conducted at a large university to compare the grade-point averages of students who belong to a fraternity or sorority to students who do not belong to such organizations. A random sample of 50 students who belong to a fraternity or

---

* Although independent samples could be obtained by taking samples that are not random, we will discuss only random samples in this text.

sorority is selected, and a second random sample of 100 students is selected from among the students who are not members of a fraternity or sorority. The grade-point average of each student sampled is recorded. Do these two samples of grade-point averages represent independent or dependent samples?

8.3 A medical researcher wants to compare the incidence of heart disease for men in a certain city to that of women in the same city. A random sample of 500 men is selected from all males in the city, and 33 have a history of heart disease. A second random sample of 500 women is obtained, and 21 have a history of heart disease. Are these two samples independent or dependent?

8.4 An experiment is conducted to compare two methods of teaching a simple task to 3-year-old children. Six pairs of 3-year-old identical twins are used in the experiment. One child from each pair is randomly assigned to Method I and the other to Method II. The length of time it takes each child to learn the task is recorded. Do these two sets of times represent independent or dependent samples?

8.5 A psychologist wants to compare adult reaction times to two different stimuli, A and B. An experiment will be conducted in which adults will be given either stimulus A or stimulus B, and their reaction times will be recorded.
   a. Describe how you would obtain two independent samples of reaction times.
   b. Describe how you would obtain two dependent samples of reaction times.

8.6 In the introduction to this chapter, we presented a comparison of the cost of certain items in the city and in the country. Examine the data shown in Figure 8.1. Do the data represent independent or dependent samples?

8.7 Two researchers have studied the impact of an innovative work schedule on the employees of a public university library:

> Flexitime (also called flextime, flexible scheduling) refers to the practice of giving employees the opportunity to choose, usually within limits, the hours each workday which will be spent on the job. A basic premise of flexitime is that life also exists outside of the work environment and, in fact, is more satisfying when work and nonwork activities are as harmonious as possible.*

Library employees completed questionnaires prior to and 6 months after flexitime was implemented. To help evaluate the effectiveness of flexitime, the second set of responses was compared to the employees' original responses. Are the two sets of questionnaire responses independent or dependent samples?

8.8 The Japanese and American commercial rivalry has intrigued researchers concerned with the psychology of management and motivation:

> In recent years, the Japanese have made noticeable inroads into American markets, with such products as automobiles, color television, videotape recorders, computers, and calcu-

---

* C. S. Saunders and R. Saunders, "Effects of Flexitime on Sick Leave, Vacation Leave, Anxiety, Performance, and Satisfaction in a Library Setting," *The Library Quarterly*, Vol. 55, No. 1, January 1985, p. 71.

lators. At the same time, Americans have watched their own productivity reach an all-time low. Because of these developments, many have assumed that the Japanese must therefore be doing something right in managing their businesses, while Americans are doing something wrong. Thus there is a rush to discover the Japanese secrets and apply them in the United States. . . .

The message of . . . popular books and articles is often carried by management consultants who use their experiences in Japanese and American organizations as supporting evidence of their beliefs. Yet it is unusual to find the theories and points of view specifically tied to scientific measures of Japanese–American differences in management orientation, style, or other relevant characteristics.*

What is the best way to compare Japanese and American managers? One set of researchers obtained two samples of data for comparison:

The purpose of the present study was to directly measure some of the motivations and values of a sample of Japanese and American managers to see if they parallel the popular claims.

The Japanese data collection took place in the fall of 1981 within the context of two-day management seminars conducted by the senior author and Douglas W. Bray of AT&T in Tokyo and Osaka, Japan. Two groups of American managers employed in Bell System telephone companies had been administered the same set of questionnaires between 1976 and 1980; they constituted the American sample.†

Are these two samples independent or dependent?

8.2

LARGE-SAMPLE INFERENCES ABOUT $\mu_1 - \mu_2$, THE DIFFERENCE BETWEEN TWO POPULATION MEANS: INDEPENDENT SAMPLES

In this section we will discuss methods for comparing two population means using two large, independent samples. The procedures we will use to make inferences about two population means are modifications of the procedures we used to make inferences about one population mean. In this section we will utilize the z statistic for comparing two population means. In the next section we will employ the t statistic for making small-sample inferences about two population means.

The notation presented in the box will be used when making inferences about two population means. As you can see, the symbols are the same as those used for one sample, but each has a subscript to denote which population or sample is involved.

---

* A. Howard, K. Shudo, and M. Umeshima, "Motivation and Values Among Japanese and American Managers," *Personnel Psychology*, Vol. 36, No. 4, Winter 1983, pp. 883–884.
† Ibid., pp. 884–885,

Notation for Comparing Two Population Means

Population I   Population II

  Mean: μ_1 Mean: μ_2
  Variance: σ_1^2 Variance: σ_2^2

Sample I   Sample II

  Mean: $\bar{x}_1$ Mean: $\bar{x}_2$
  Variance: s_1^2 Variance: s_2^2
  Sample size: n_1 Sample size: n_2

To compare two population means, inferences are traditionally phrased in terms of the difference between the means, $\mu_1 - \mu_2$. When testing hypotheses about $\mu_1 - \mu_2$ or forming confidence intervals for $\mu_1 - \mu_2$, we will use the sampling distribution of $\bar{x}_1 - \bar{x}_2$, the difference in the sample means. The conditions that must be met to use large samples to make valid inferences about $\mu_1 - \mu_2$ are given in the box.

When to Make Large-Sample Inferences About $\mu_1 - \mu_2$,
the Difference Between Two Population Means

To make a valid test of hypothesis about $\mu_1 - \mu_2$, or to form a confidence interval for $\mu_1 - \mu_2$, the following conditions must be met:

1. Two random samples must be taken—one from each of the two populations of interest.
2. The two samples must be independent.
3. Both samples must be large, that is, $n_1 \geq 30$ and $n_2 \geq 30$.

As we have indicated above, to make inferences about $\mu_1 - \mu_2$, we will be interested in the statistic $\bar{x}_1 - \bar{x}_2$. The conditions given above guarantee that the sampling distribution of $\bar{x}_1 - \bar{x}_2$ will have the properties listed in the box at the top of page 410.

Since the sampling distribution of $\bar{x}_1 - \bar{x}_2$ is approximately normal, the large-sample procedures used to make inferences about $\mu_1 - \mu_2$ are very similar to those used to make large-sample inferences about μ. The large-sample test of hypothesis about $\mu_1 - \mu_2$ and the confidence interval for $\mu_1 - \mu_2$ are given in boxes on the next page.

Properties of the Sampling Distribution of $\bar{x}_1 - \bar{x}_2$

When the conditions stated in the previous box are satisfied, the following properties hold:

1. The sampling distribution of $\bar{x}_1 - \bar{x}_2$ is approximately normal.

2. The mean of the sampling distribution of $\bar{x}_1 - \bar{x}_2$ is $\mu_1 - \mu_2$.

3. The standard error of the sampling distribution of $\bar{x}_1 - \bar{x}_2$ is

$$\sigma_{(\bar{x}_1 - \bar{x}_2)} = \sqrt{\frac{\sigma_1^2}{n_1} + \frac{\sigma_2^2}{n_2}}$$

Large-Sample Test of Hypothesis About $\mu_1 - \mu_2$

H_0: $\mu_1 - \mu_2 = D_0$

H_a: $\mu_1 - \mu_2 > D_0$ or H_a: $\mu_1 - \mu_2 < D_0$

   or H_a: $\mu_1 - \mu_2 \neq D_0$

Test statistic: $z = \dfrac{(\bar{x}_1 - \bar{x}_2) - D_0}{\sqrt{\dfrac{s_1^2}{n_1} + \dfrac{s_2^2}{n_2}}}$

Rejection region: Use Appendix Table III.

The symbol D_0 denotes the hypothesized difference between μ_1 and μ_2. The value of D_0 will often be 0, but it could be any number. If $D_0 = 0$, then the null hypothesis is stating that there is no difference between μ_1 and μ_2; that is, $\mu_1 = \mu_2$. Any of the three usual alternative hypotheses may be selected.

Since the values of the population variances, σ_1^2 and σ_2^2, are generally not known in practical problems, s_1^2 and s_2^2 are substituted in the standard error of $\bar{x}_1 - \bar{x}_2$. As is usual for a large-sample test, the rejection region is obtained from Appendix Table III.

Large-Sample Confidence Interval for $\mu_1 - \mu_2$

The large-sample confidence interval for $\mu_1 - \mu_2$ is expressed as

$$(\bar{x}_1 - \bar{x}_2) - z \sqrt{\frac{s_1^2}{n_1} + \frac{s_2^2}{n_2}} \quad \text{to} \quad (\bar{x}_1 - \bar{x}_2) + z \sqrt{\frac{s_1^2}{n_1} + \frac{s_2^2}{n_2}}$$

The value of z is obtained from Appendix Table IV.

We will demonstrate these procedures with examples.

EXAMPLE 8.3   A restaurant has changed its menu so that most of the meals now offered cost less than the meals it used to offer. The new meals are also less expensive to prepare, and the owner of the restaurant wants to compare the mean net daily income obtained with the lower-priced meals to the previous mean net daily income obtained with the higher-priced meals. A random sample of fifty net daily incomes for the earlier period is selected, and a second random sample of thirty net daily incomes for the current period is selected. A summary of the results of the two samples is shown in Table 8.1.

Table 8.1

Higher-Priced Meals	Lower-Priced Meals
$n_1 = 50$	$n_2 = 30$
$\bar{x}_1 = \$185$	$\bar{x}_2 = \$166$
$s_1 = \$41$	$s_2 = \$53$

Do these samples provide sufficient evidence for the owner to conclude that the mean net daily income has changed since the menu was changed? Test using $\alpha = .05$.

Solution   We want to test a hypothesis about the difference between two population means. Population I is the collection of all net daily incomes associated with the higher-priced meals, and μ_1 is the mean of these measurements. Population II is the corresponding set of data for the lower-priced meals, with mean μ_2.

The owner wants to know whether the mean net daily income has changed (either increased or decreased), so the alternative hypothesis is $\mu_1 - \mu_2 \neq 0$. The elements of the test of hypothesis are

$$H_0: \quad \mu_1 - \mu_2 = 0 \qquad\qquad H_a: \quad \mu_1 - \mu_2 \neq 0$$

$$\text{Test statistic:} \quad z = \frac{(\bar{x}_1 - \bar{x}_2) - 0}{\sqrt{\dfrac{s_1^2}{n_1} + \dfrac{s_2^2}{n_2}}}$$

Both the samples are random and large ($n_1 = 50$ and $n_2 = 30$). It is implied that the two samples are independent, since two different random samples were selected. Thus, it is valid to conduct a large-sample test of hypothesis to compare μ_1 to μ_2.

Rejection region: Reject H_0 if $z > 1.96$ or $z < -1.96$.

The rejection region was found using Appendix Table III and is shown in Figure 8.2 on page 412.

We now calculate

$$z = \frac{185 - 166}{\sqrt{\dfrac{(41)^2}{50} + \dfrac{(53)^2}{30}}} = \frac{19}{11.28} = 1.68$$

[*Note:* From Table 8.1, $s_1 = 41$ and $s_2 = 53$; so $s_1^2 = (41)^2$ and $s_2^2 = (53)^2$.]

As you can see, the calculated value of the test statistic, $z = 1.68$, does not fall in the rejection region. We have not observed an event that is rare enough to warrant

Figure 8.2
Rejection region for
Example 8.3

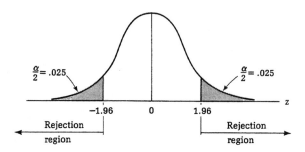

rejection of H_0. The samples do not provide sufficient evidence to conclude that the mean daily net income has changed since the menu was changed. ■

Since we could not conclude that μ_1 is different from μ_2 in Example 8.3, we might want to perform further analyses and learn more about what the value of the actual difference between μ_1 and μ_2 might be. This can be accomplished by forming a confidence interval for $\mu_1 - \mu_2$.

EXAMPLE 8.4   Find a 95% confidence interval for the difference in mean net daily incomes obtained with the higher-priced and lower-priced meals of Example 8.3.

Solution   In Example 8.3 we demonstrated the validity of conducting a large-sample test of hypothesis about $\mu_1 - \mu_2$. The same conditions that made the test of hypothesis valid also allow us to form a large-sample confidence interval. The confidence interval is given by

$$(\bar{x}_1 - \bar{x}_2) - z\sqrt{\frac{s_1^2}{n_1} + \frac{s_2^2}{n_2}} \quad \text{to} \quad (\bar{x}_1 - \bar{x}_2) + z\sqrt{\frac{s_1^2}{n_1} + \frac{s_2^2}{n_2}}$$

$$(185 - 166) - 1.96\sqrt{\frac{(41)^2}{50} + \frac{(53)^2}{30}} \quad \text{to} \quad (185 - 166) + 1.96\sqrt{\frac{(41)^2}{50} + \frac{(53)^2}{30}}$$

$$19 - 22.11 \quad \text{to} \quad 19 + 22.11$$

$$-3.11 \quad \text{to} \quad 41.11$$

Thus, we estimate that $\mu_1 - \mu_2$, the difference in mean net daily incomes, falls in the interval −$3.11 to $41.11. In other words, we estimate that μ_1, the mean net daily income for the higher-priced meals, could be larger than μ_2, the mean for the lower-priced meals, by as much as $41.11; or μ_1 could be smaller than μ_2 by as little as $3.11. Also, since the interval contains 0, it is possible that μ_1 is equal to μ_2. We have 95% confidence that the actual value of $\mu_1 - \mu_2$ is in the calculated interval. ■

The interpretation of the confidence interval in Example 8.4 agrees with the interpretation for the test of hypothesis in Example 8.3. In both instances, we could not determine whether μ_1 is larger than, smaller than, or equal to μ_2. When interpreting a confidence interval for $\mu_1 - \mu_2$, the interpretation will depend upon whether 0 is included in the calculated interval. To help you interpret confidence intervals, we provide the hints given in the box.

Helpful Hints

When interpreting a confidence interval for $\mu_1 - \mu_2$, there are three distinct situations to consider:

1. If the confidence interval contains 0, then μ_1 could be equal to μ_2, or μ_1 could be larger than μ_2, or μ_1 could be smaller than μ_2.
2. If the confidence interval contains only positive numbers, then we will infer (at the specified confidence level) that μ_1 is larger than μ_2 (or $\mu_1 - \mu_2 > 0$).
3. If the confidence interval contains only negative numbers, then we will infer (at the specified confidence level) that μ_1 is smaller than μ_2 (or $\mu_1 - \mu_2 < 0$).

EXAMPLE 8.5   An experiment was conducted to compare two methods (Method I and Method II) of teaching spelling to children. The objective of the experiment was to estimate the difference between μ_1, the mean score on a standardized spelling test for all children who are taught spelling by Method I, and μ_2, the mean score on the test for all children who are taught by Method II. A large random sample of children was taught spelling by Method I and a second large random sample was taught by Method II. At the conclusion of the school year, all the sampled children were given the standardized test, and these test scores were used to form a 99% confidence interval for $\mu_1 - \mu_2$.

How would you interpret the confidence interval if it was found to be from 4 to 10?

Solution   With 99% confidence we infer that the value of $\mu_1 - \mu_2$ is between 4 and 10. Since the interval contains only positive numbers, this indicates that μ_1 is larger than μ_2. We conclude that children who are taught by Method I have a higher mean score on the test than those taught by Method II. The true mean score for Method I is at least 4 points higher than the true mean score for Method II, and it may be 10 points higher.   ∎

Exercises (8.9–8.25)

Learning the Language

8.9   What do the following symbols denote?

a. μ_1 and μ_2 d. $\bar{x}_1$ and $\bar{x}_2$ g. n_1 and n_2

b. $\mu_1 - \mu_2$ e. $\bar{x}_1 - \bar{x}_2$ h. $\sigma_{(\bar{x}_1 - \bar{x}_2)}$

c. σ_1^2 and σ_2^2 f. s_1^2 and s_2^2 i. D_0

Using the Tables

8.10   Two large, independent random samples are selected from populations with means μ_1 and μ_2.

a. Give the appropriate z-value for forming a 95% confidence interval for $\mu_1 - \mu_2$.
b. Give the rejection region for testing H_0: $\mu_1 - \mu_2 = 0$ against H_a: $\mu_1 - \mu_2 \neq 0$ with $\alpha = .01$.
c. Give the rejection region for testing H_0: $\mu_1 - \mu_2 = 0$ against H_a: $\mu_1 - \mu_2 > 0$ with $\alpha = .05$.
d. Give the appropriate z-value for forming a 90% confidence interval for $\mu_1 - \mu_2$.
e. Give the rejection region for testing H_0: $\mu_1 - \mu_2 = 0$ against H_a: $\mu_1 - \mu_2 < 0$ with $\alpha = .01$.

Learning the Mechanics

8.11 You select two independent random samples, forty observations from population 1 and fifty from population 2. The sample means and variances are shown in the table.

Sample 1	Sample 2
$n_1 = 40$	$n_2 = 50$
$\bar{x}_1 = 3.7$	$\bar{x}_2 = 4.2$
$s_1^2 = .27$	$s_2^2 = .33$

a. Form a 95% confidence interval for $\mu_1 - \mu_2$, the difference in the means of populations 1 and 2.
b. Test the null hypothesis H_0: $\mu_1 - \mu_2 = 0$ against the alternative hypothesis H_a: $\mu_1 - \mu_2 \neq 0$. Use $\alpha = .05$.

8.12 Two independent random samples were selected from populations with means μ_1 and μ_2, respectively. The sample sizes, means, and standard deviations are shown in the table.

Sample 1	Sample 2
$n_1 = 75$	$n_2 = 75$
$\bar{x}_1 = 20.3$	$\bar{x}_2 = 24.4$
$s_1 = 6.1$	$s_2 = 8.9$

a. Form a 90% confidence interval for $\mu_1 - \mu_2$.
b. Form a 99% confidence interval for $\mu_1 - \mu_2$.
c. Test the null hypothesis H_0: $\mu_1 - \mu_2 = 0$ against the alternative hypothesis H_a: $\mu_1 - \mu_2 < 0$. Use $\alpha = .01$.

8.13 Two random and independent samples produced the results shown in the table.

Sample 1	Sample 2
$n_1 = 110$	$n_2 = 160$
$\bar{x}_1 = 6.8$	$\bar{x}_2 = 4.7$
$s_1^2 = 93.2$	$s_2^2 = 117.6$

    a.   Test H_0: $\mu_1 - \mu_2 = 0$ against H_a: $\mu_1 - \mu_2 > 0$. Use $\alpha = .10$.

    b.   Test H_0: $\mu_1 - \mu_2 = 0$ against H_a: $\mu_1 - \mu_2 \neq 0$. Use $\alpha = .05$.

    c.   Form a 95% confidence interval for $\mu_1 - \mu_2$.

8.14   Are the test and confidence interval procedures given in this section valid if the sampled populations are not normally distributed? Explain.

Applying the Concepts

8.15   An experiment has been conducted at a university to compare the mean number of study hours expended per week by student athletes with the mean number of hours expended by nonathletes. A random sample of 55 athletes produced a mean equal to 20.6 hours of study per week and a standard deviation equal to 5.3 hours. A second random sample of 200 nonathletes produced a mean equal to 23.5 hours per week and a standard deviation equal to 4.1 hours.

    a.   Describe the two populations involved in the comparison.

    b.   Do the samples provide sufficient evidence to conclude that there is a difference in the mean number of hours of study per week between athletes and non-athletes? Test using $\alpha = .01$.

    c.   Construct a 99% confidence interval for $\mu_1 - \mu_2$.

    d.   Would a 95% confidence interval for $\mu_1 - \mu_2$ be narrower or wider than the one you found in part *c*? Why?

8.16   A dental laboratory has developed a new type of band for children who wear braces. The new bands are designed to be more comfortable, look better, and—it is hoped —realign teeth more rapidly than the standard braces now in use. An experiment is conducted to compare the mean wearing times necessary to correct a specific type of misalignment when the standard braces are used and when the new bands are used. One hundred children are randomly assigned the two types of treatment— fifty to each type. A summary of the data is given in the table.

Standard Braces	New Bands
$\bar{x}_1 = 410$ days	$\bar{x}_2 = 380$ days
$s_1 = 45$ days	$s_2 = 60$ days

    a.   Is there sufficient evidence to conclude that the new bands do not have to be worn as long as the standard braces, on the average? Use $\alpha = .01$.

    b.   Find a 95% confidence interval for the difference in mean wearing times for the two types of braces.

8.17   Suppose it is desired to compare two physical education training programs for pre-adolescent girls. A total of eighty girls are randomly selected, with forty assigned to each program. After 18 weeks in the program, each girl is given a fitness test that yields a score between 0 and 100. The means and variances of the scores for the two groups are shown in the table at the top of page 416. Calculate a 99% confidence interval for the true difference in mean fitness scores for girls trained using these two programs.

	n	$\bar{x}$	s^2
Program 1	40	78.7	201.6
Program 2	40	75.3	259.2

8.18 A distributor of soft drink vending machines knows from experience that the mean number of drinks a machine will sell per day varies according to the location of the machine. At a boys' club, two machines are placed in what the distributor believes to be two different optimal locations. The machines are observed for 30 days, and the number of drinks sold per day for each machine is recorded. The means and standard deviations of the number of drinks sold per day at the two locations are given in the table. Based on the data, can the distributor conclude that either location is better than the other? Test at the $\alpha = .05$ level of significance.

Machine at Location 1	Machine at Location 2
$\bar{x}_1 = 32.5$	$\bar{x}_2 = 28.5$
$s_1 = 6.0$	$s_2 = 5.5$

8.19 An experiment was conducted to compare the yield of two varieties of tomato, A and B. Forty plants of each variety were randomly selected and planted within the same field. The yields, recorded in kilograms of tomatoes produced for each plant, possessed means of 10.5 kilograms per plant for variety A and 9.3 kilograms per plant for variety B. The variances for samples A and B were 2.1 and 2.8, respectively. Do the data provide sufficient evidence to conclude that there is a difference between the mean weights of tomatoes produced per plant for the two varieties? Test using $\alpha = .05$.

8.20 It is often said that economic status is related to the commission of crimes. To test this theory, a sociologist selected a random sample of seventy people in a certain city who had no record of criminal conviction, and their annual incomes were noted. Similarly, a random sample of sixty criminals (each one a first-time offender) was selected, and the annual income (prior to arrest) was recorded for each. The annual incomes were recorded in thousands of dollars, and the means and variances for these data are shown in the table. Do the data provide sufficient evidence to indicate that the mean income of criminals prior to committing their first offense is lower than that for the noncriminal public? Test usíng $\alpha = .05$.

	$\bar{x}$	s^2
Criminals	13.3	24.2
Noncriminals	15.4	42.6

8.21 A large supermarket chain is interested in determining whether a difference exists between the mean shelf-life (in days) of brand S bread and brand H bread. Random samples of fifty freshly baked loaves of each brand were tested, with the results given in the table.

Brand S	Brand H
$\bar{x}_1 = 4.1$	$\bar{x}_2 = 5.2$
$s_1 = 1.2$	$s_2 = 1.4$

a. Is there sufficient evidence to conclude that a difference does exist between the mean shelf-lives of brand S and brand H? Test at the $\alpha = .05$ level.

b. Let μ_1 represent the mean shelf-life for brand S, and let μ_2 represent the mean shelf-life for brand H. Find a 90% confidence interval for $\mu_1 - \mu_2$.

8.22 Two manufacturers of corrugated fiberboard each claim that the strength of their product tests at more than 360 pounds per square inch, on the average. As a result of consumer complaints, a consumer products testing firm believes that firm A's product is stronger than firm B's. To test its belief, 100 fiberboards were chosen randomly from firm A's inventory and 100 were chosen from firm B's inventory. The strength (in pounds per square inch) of each fiberboard was determined, with the results shown in the accompanying table. Does the sample information support the belief of the consumer products testing firm? Test at the .05 significance level.

Firm A	Firm B
$\bar{x}_1 = 365$	$\bar{x}_2 = 352$
$s_1 = 23$	$s_2 = 41$

8.23 Give a practical example where the following hypotheses would be appropriate:

a. $H_0: \mu_1 - \mu_2 = 0$ and $H_a: \mu_1 - \mu_2 > 0$

b. $H_0: \mu_1 - \mu_2 = 0$ and $H_a: \mu_1 - \mu_2 < 0$

c. $H_0: \mu_1 - \mu_2 = 0$ and $H_a: \mu_1 - \mu_2 \neq 0$

8.24 The efficacy of having 6-year-old children tutor their peers in reading has been investigated by studying two groups during the school year. The treatment group was involved in peer tutoring, while the control group was not. At the end of the study all the children took a criterion-referenced test to measure reading skills, with the results summarized in the accompanying table.

Group	Sample Size	Mean	Standard Deviation
Treatment	49	19.98	4.75
Control	46	14.93	6.02

Source: D. G. Reay, G. Von Harrison, and C. Gottfredson, "The Effect on Pupil Reading Achievement of Teacher Compliance with Prescribed Methodology," *Research in Education*, No. 32, November 1984, p. 21.

a. Is there evidence to conclude that the mean score for the treatment group is larger than the mean score for the control group? Use $\alpha = .01$.
b. The researchers reported that the mean score of the treatment group exceeded that of the control group, and that the difference was significant at the .01 level. Does this agree with your answer to part a?

8.25 Refer to Exercise 8.8. To compare Japanese and American managers, tests were given that measured how strongly certain factors motivated them. Two factors of interest were advancement and money. Sample results of the motivational tests are shown in the table.

	American		Japanese	
	$\bar{x}$	s	$\bar{x}$	s
Advancement	16.75	4.75	23.92	3.20
Money	14.80	3.60	18.12	2.90
Sample size	$n_1 = 211$		$n_2 = 100$	

Source: A. Howard, K. Shudo, and M. Umeshima, "Motivation and Values Among Japanese and American Managers," *Personnel Psychology*, Vol. 36, No. 4, Winter 1983, p. 893.

a. Is there sufficient evidence to conclude that American managers' average motivation regarding advancement differs from that of the Japanese managers? Use $\alpha = .05$.
b. The article reported an observed significance level of p-value $< .001$ in testing for a difference in financial motivation between American and Japanese managers. Interpret this result.
c. Find a 99% confidence interval for the difference in the mean test scores for financial motivation of American versus Japanese managers.

8.3

SMALL-SAMPLE INFERENCES ABOUT $\mu_1 - \mu_2$, THE DIFFERENCE BETWEEN TWO POPULATION MEANS: INDEPENDENT SAMPLES

In this section we will present methods for making inferences about the difference between two population means using small samples. As in the case of a single mean, the inferences will be based on the t distribution, and the t distribution is appropriate only if certain conditions are met. These conditions are given in the next box.

Since the two populations must have mound-shaped distributions with equal variances, the distributions appear as shown in Figure 8.3. The two population distributions are identical in appearance except for the fact that the means, μ_1 and μ_2, may be different.

When to Use the _t_ Distribution to Make Inferences About $\mu_1 - \mu_2$, the Difference Between Two Population Means

To use the t distribution to make a valid test of hypothesis about $\mu_1 - \mu_2$, or to form a confidence interval for $\mu_1 - \mu_2$, the following conditions must be met:

1. Two random samples must be taken—one from each of the two populations of interest.
2. The two samples must be independent.
3. **Both of the populations must be (approximately) normally distributed.**
4. **The population variances must be equal.**

Figure 8.3
Two normally distributed populations with equal variances ($\mu_1 > \mu_2$)

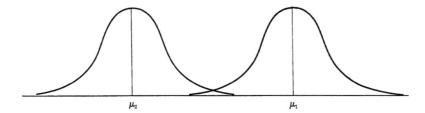

μ_2 μ_1

Since the two population variances are equal, we will say that they have a common variance, σ^2; that is, $\sigma_1^2 = \sigma_2^2 = \sigma^2$. Since the value of σ^2 will have to be estimated, we will use the information contained in both samples to construct a **pooled sample estimate** of σ^2. The pooled estimate of σ^2, denoted by s_p^2, is a function of the sample variances, s_1^2 and s_2^2, and the sample sizes, n_1 and n_2. The formula for s_p^2 is given in the box.

Formula for s_p^2, the Pooled Estimate of σ^2

$$s_p^2 = \frac{(n_1 - 1)s_1^2 + (n_2 - 1)s_2^2}{n_1 + n_2 - 2}$$

Recall that in Section 7.5 we defined the degrees of freedom for one sample as $(n - 1)$, that is, 1 less than the sample size. Since we are now pooling the information from two samples to estimate σ^2, the degrees of freedom associated with the pooled sample variance s_p^2 is equal to the sum of the degrees of freedom for the two samples. This is the denominator of the formula for s_p^2, $(n_1 - 1) + (n_2 - 1) = n_1 + n_2 - 2$.

Small-sample test statistics and confidence intervals for $\mu_1 - \mu_2$ will contain s_p^2, and when using the tables to find t-values, $(n_1 + n_2 - 2)$ degrees of freedom will be used. The small-sample test of hypothesis about $\mu_1 - \mu_2$ and confidence interval for $\mu_1 - \mu_2$ are given in the boxes.

Small-Sample Test of Hypothesis About $\mu_1 - \mu_2$

H_0: $\mu_1 - \mu_2 = D_0$

H_a: $\mu_1 - \mu_2 > D_0$ or H_a: $\mu_1 - \mu_2 < D_0$

   or H_a: $\mu_1 - \mu_2 \neq D_0$

Test statistic: $t = \dfrac{(\bar{x}_1 - \bar{x}_2) - D_0}{\sqrt{\dfrac{s_p^2}{n_1} + \dfrac{s_p^2}{n_2}}}$

Rejection region: Use Appendix Table V, with $(n_1 + n_2 - 2)$ df.

As with the large-sample test for $\mu_1 - \mu_2$, D_0 is the hypothesized difference between μ_1 and μ_2, and D_0 is often 0.

Small-Sample Confidence Interval for $\mu_1 - \mu_2$

The small-sample confidence interval for $\mu_1 - \mu_2$ is expressed as

$$(\bar{x}_1 - \bar{x}_2) - t\sqrt{\dfrac{s_p^2}{n_1} + \dfrac{s_p^2}{n_2}} \quad \text{to} \quad (\bar{x}_1 - \bar{x}_2) + t\sqrt{\dfrac{s_p^2}{n_1} + \dfrac{s_p^2}{n_2}}$$

The value of t is obtained from Appendix Table VI, using $(n_1 + n_2 - 2)$ df.

The following examples will demonstrate the use of these procedures.

EXAMPLE 8.6   A television network wanted to investigate whether major sports events or first-run movies attract more viewers. Thirteen evenings that had programs devoted to major sports events were randomly selected, and another random sample of fifteen evenings that had first-run movies was selected. The number of viewers (obtained from a television viewer rating firm) was recorded for each program. The populations of number of viewers per program for major sports events and for first-run movies are both approximately normally distributed with the same variability. The television network's samples produce the results shown in Table 8.2.

Do the data provide sufficient evidence to indicate a difference in the mean number of viewers for major sports events and first-run movies? Test at the .05 level of significance.

Table 8.2

Sports	Movie
$n_1 = 13$	$n_2 = 15$
$\bar{x}_1 = 6.8$ million	$\bar{x}_2 = 5.3$ million
$s_1 = 1.8$ million	$s_2 = 1.6$ million

Solution The populations of interest are the number of viewers per program for major sports events and the number of viewers per program for first-run movies. With μ_1 being the mean number of sports viewers per evening of sports programming and μ_2 being the mean number of movie viewers per evening, we want to detect a difference between μ_1 and μ_2—if a difference exists. Thus, the elements of the test of hypothesis are

$$H_0: \quad \mu_1 - \mu_2 = 0 \qquad H_a: \quad \mu_1 - \mu_2 \neq 0$$

$$\text{Test statistic:} \quad t = \frac{(\bar{x}_1 - \bar{x}_2) - 0}{\sqrt{\dfrac{s_p^2}{n_1} + \dfrac{s_p^2}{n_2}}}$$

The populations are approximately normally distributed with equal variances, and different random samples were selected from the two populations. This implies that the samples are also independent, and all the conditions are met for a small-sample test of hypothesis to be conducted.

Rejection region: Reject H_0 if $t < -2.06$ or $t > 2.06$.

The rejection region was found using Table V and $n_1 + n_2 - 2 = 13 + 15 - 2 = 26$ df. The rejection region is shown in Figure 8.4.

Figure 8.4
Rejection region for
Example 8.6

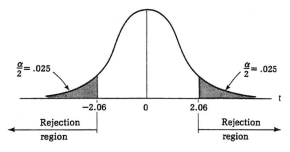

Before calculating the test statistic, we must first calculate s_p^2, the pooled estimate of σ^2:

$$\begin{aligned}
s_p^2 &= \frac{(n_1 - 1)s_1^2 + (n_2 - 1)s_2^2}{n_1 + n_2 - 2} \\
&= \frac{(13 - 1)(1.8)^2 + (15 - 1)(1.6)^2}{13 + 15 - 2} \\
&= \frac{74.72}{26} = 2.87
\end{aligned}$$

The calculated value of the test statistic is

$$t = \frac{(6.8 - 5.3) - 0}{\sqrt{\dfrac{2.87}{13} + \dfrac{2.87}{15}}} = \frac{1.5}{.64} = 2.34$$

Since the observed value of $t = 2.34$ falls in the rejection region, the samples provide sufficient evidence to indicate that the mean number of viewers differs for major sports events and first-run movies shown in the evening. Or, we can say that the test results are statistically significant at the $\alpha = .05$ level of significance. Because the calculated value of t was in the positive, or upper, tail of the t distribution, it appears that the mean number of viewers for sports events exceeds that for movies. ∎

EXAMPLE 8.7   An experiment is conducted to investigate the effect a drug has on the time to complete a task. Forty people are divided, at random, into two groups, with twenty people in each group. One group (the control group) is given a placebo, while the second group (the experimental group) is given the drug. Each person is then asked to perform the task, and the length of time it takes each person to complete the task is recorded. The populations of all times, both for the control group and for the experimental group, are approximately normally distributed, with equal variances. The times required by the control group have an average of 14.8 minutes and a variance of 3.9. For the experimental group, the average is 12.3 minutes and the variance is 4.3. Estimate the difference between μ_1, the mean length of time to complete the task for the control group, and μ_2, the mean length of time to complete the task for the experimental group, using a 95% confidence interval.

Solution   The populations of interest are the times to complete the task for all people receiving the placebo and all people receiving the drug. We will let μ_1 represent the mean time to complete the task for those receiving the placebo, while μ_2 represents the corresponding mean for those receiving the drug.

We want to find a small-sample confidence interval for $\mu_1 - \mu_2$, and the general form of the interval is

$$(\bar{x}_1 - \bar{x}_2) - t\sqrt{\frac{s_p^2}{n_1} + \frac{s_p^2}{n_2}} \quad \text{to} \quad (\bar{x}_1 - \bar{x}_2) + t\sqrt{\frac{s_p^2}{n_1} + \frac{s_p^2}{n_2}}$$

The populations are approximately normally distributed, and their variances are equal. Since two random and independent samples are taken from these populations, it is valid to form a small-sample confidence interval for $\mu_1 - \mu_2$. (In many experiments, to obtain two random and independent samples, a sample of subjects is selected and then randomly split into two groups.) The data are summarized in Table 8.3.

Table 8.3

	Control	Experimental
	$n_1 = 20$	$n_2 = 20$
	$\bar{x}_1 = 14.8$ minutes	$\bar{x}_2 = 12.3$ minutes
	$s_1^2 = 3.9$	$s_2^2 = 4.3$

We must first calculate s_p^2:

$$s_p^2 = \frac{(n_1 - 1)s_1^2 + (n_2 - 1)s_2^2}{n_1 + n_2 - 2}$$

$$= \frac{(20 - 1)(3.9) + (20 - 1)(4.3)}{20 + 20 - 2} = \frac{155.8}{38} = 4.1$$

To calculate the confidence interval, we need to find the t-value from Appendix Table VI. Since $n_1 + n_2 - 2 = 20 + 20 - 2 = 38$ df, we use the last row of the table (the z-value) to obtain $t = 1.96$. Notice that we cannot form a large-sample confidence interval for $\mu_1 - \mu_2$, because to do so *both* the samples must be large. In this case, both samples are small ($n_1 = 20$ and $n_2 = 20$), so we must find the pooled estimate of σ^2 and form a small-sample confidence interval. However, since the *total* number of degrees of freedom is over 28, we use 1.96 as the t-value.

We now calculate the confidence interval:

$$(14.8 - 12.3) - 1.96 \sqrt{\frac{4.1}{20} + \frac{4.1}{20}} \quad \text{to} \quad (14.8 - 12.3) + 1.96 \sqrt{\frac{4.1}{20} + \frac{4.1}{20}}$$

$$2.5 - 1.26 \quad \text{to} \quad 2.5 + 1.26$$

$$1.24 \quad \text{to} \quad 3.76$$

With 95% confidence, the value of $\mu_1 - \mu_2$ is between 1.24 and 3.76. We conclude that the value of μ_1 is larger than the value of μ_2. People using the placebo average somewhere between 1.24 and 3.76 minutes *more* to complete the task than people using the drug. The drug seems to reduce the mean time required to complete the task. ∎

As indicated in Example 8.7, unless both samples are large, a large-sample procedure cannot be used to compare μ_1 and μ_2. Therefore, if one sample (or both) is small ($n_1 < 30$ or $n_2 < 30$), we will use a small-sample method (when all other required conditions are met). If the number of degrees of freedom ($n_1 + n_2 - 2$) is greater than 28, the last row of Table V or Table VI must be used to find the appropriate t-value.

EXAMPLE 8.8   A new method of teaching reading to "slow learners" has been developed, and it is desired to compare this method to a standard method that is currently being used. The comparison will be based upon the results of a reading test given at the end of a 6-month learning period. The distributions of the test scores for both methods are believed to be approximately normal, and the variability of both distributions is the same. A random sample of eight slow learners is taught by the new method, and a second random sample of twelve slow learners is taught by the standard method. The two samples are independent. All twenty children are taught by qualified instructors under similar conditions for a 6-month period. The results of the reading test given at the end of this period are summarized in Table 8.4 (page 424).

Table 8.4

New Method	Standard Method
$n_1 = 8$	$n_2 = 12$
$\bar{x}_1 = 76.63$	$\bar{x}_2 = 72.42$
$s_1 = 4.84$	$s_2 = 5.73$

Do the data indicate that the mean test score for the new method is higher than the mean test score for the standard method? Use $\alpha = .10$.

Solution   The populations of interest are the test scores for all children taught by the new method and all children taught by the standard method.

Letting μ_1 equal the true mean test score for the new method and μ_2 equal the true mean test score for the standard method, we want to show that μ_1 is larger than μ_2 ($\mu_1 - \mu_2 > 0$). (Note that if we reversed the definitions of μ_1 and μ_2, we would want to show that $\mu_1 - \mu_2 < 0$.) The elements of the test of hypothesis are

$$H_0: \quad \mu_1 - \mu_2 = 0 \qquad H_a: \quad \mu_1 - \mu_2 > 0$$

$$\text{Test statistic:} \quad t = \frac{(\bar{x}_1 - \bar{x}_2) - 0}{\sqrt{\dfrac{s_p^2}{n_1} + \dfrac{s_p^2}{n_2}}}$$

The populations are approximately normally distributed with equal variances, and random samples were independently selected from the two populations. Thus, it is valid to conduct a small-sample test of hypothesis for the difference between two means.

Rejection region: Reject H_0 if $t > 1.33$.

Using 18 df, we referred to Appendix Table V to find the rejection region, which is shown in Figure 8.5.

Calculating s_p^2, we obtain

$$s_p^2 = \frac{(n_1 - 1)s_1^2 + (n_2 - 1)s_2^2}{n_1 + n_2 - 2}$$

$$= \frac{(8 - 1)(4.84)^2 + (12 - 1)(5.73)^2}{8 + 12 - 2}$$

$$= \frac{525.1411}{18} = 29.1745$$

Figure 8.5
Rejection region for
Example 8.8

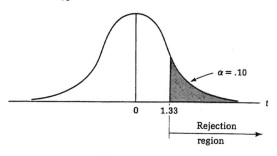

The calculated value of the test statistic is

$$t = \frac{(76.63 - 72.42) - 0}{\sqrt{\dfrac{29.1745}{8} + \dfrac{29.1745}{12}}} = 1.71$$

Since $t = 1.71$ is larger than 1.33, we will reject H_0. With 90% confidence, the true mean test score for the new method of teaching reading to slow learners is larger than the true mean score for the standard method. Although the test was conducted at a relatively high significance level ($\alpha = .10$) and the samples were relatively small, there is sufficient evidence to indicate that the new method may be superior to the one currently being used. However, before the standard method is discarded, it would probably be wise to conduct further experiments with larger samples and to conduct the test at a lower level of significance (say, $\alpha = .01$). ■

The two-sample t statistic is a useful tool for comparing two population means when the appropriate conditions are met. It has also been shown to retain its usefulness when the sampled populations are only approximately normally distributed. And, when the sample sizes are equal, the assumption of equal population variances can be relaxed. That is, when $n_1 = n_2$, σ_1^2 and σ_2^2 can be quite different and the test statistic will still possess, approximately, a t distribution. When the experimental situation does not satisfy the conditions that must be met, two large random samples could possibly be selected from the populations. If this is not feasible, nonparametric tests (which are presented in Chapter 12) can be used.

COMPUTER
EXAMPLE 8.1

Minitab may be used to compare two population means with the small-sample methods discussed in this section. To demonstrate such a result, we will refer to Example 8.8. A new and a standard method of teaching reading were compared using test scores for samples of eight and twelve children, respectively. In Example 8.8, we used summary statistics to conduct the analysis. Minitab uses the actual sample data, and the test scores for the two methods are shown here:

ROW	NEW	STANDARD
1	83	75
2	74	68
3	80	79
4	68	73
5	79	71
6	73	78
7	80	76
8	76	69
9		68
10		59
11		77
12		76

The elements of the test of interest are repeated:

$$H_0: \quad \mu_1 - \mu_2 = 0 \qquad H_a: \quad \mu_1 - \mu_2 > 0$$

$$\text{Test statistic:} \quad t = \frac{(\bar{x}_1 - \bar{x}_2) - 0}{\sqrt{\dfrac{s_p^2}{n_1} + \dfrac{s_p^2}{n_2}}}$$

Rejection region: Reject H_0 if $t > 1.33$ ($\alpha = .10$).

Using the two samples of data, Minitab produced the following printout:

```
TWOSAMPLE T FOR NEW VS STANDARD
             N      MEAN     STDEV    SE MEAN
NEW          8     76.63      4.84      1.71
STANDARD    12     72.42      5.73      1.65

95 PCT CI FOR MU NEW - MU STANDARD: (-0.9707, 9.387)
TTEST MU NEW = MU STANDARD (VS GT): T=1.71 P=0.052 DF=18.0
```

The first line of the printout informs us that we are comparing "NEW" versus "STANDARD" using the two-sample t procedure. The next part of the printout provides summary statistics about the samples. Notice that the means and standard deviations are identical to those given in Table 8.4.

When Minitab compares two means using a small-sample analysis, it automatically forms a confidence interval and tests a hypothesis. The next-to-last line of the printout is a 95% confidence interval for $\mu_1 - \mu_2$. We need not interpret this result because we are interested in testing a hypothesis about $\mu_1 - \mu_2$. The information we desire is given on the last line of the printout.

We are first told in abbreviated form what the null and alternative hypotheses are. Next is the value of the test statistic, "T = 1.71." This is the same value calculated in Example 8.8. Since this value is larger than 1.33, we would reject H_0 with 90% confidence that the mean test score for the new method exceeds the mean for the standard method.

Minitab reports the p-value for this test to be equal to 0.052. This also indicates that the null hypothesis should be rejected with $\alpha = .10$ because the p-value is smaller than α. Note, however, that if it were desired to have $\alpha = .05$, then the null hypothesis would not be rejected. The observed significance level is 0.052, and the null hypothesis should be rejected only when the observed significance level (p-value) is smaller than α.   ■

Exercises (8.26–8.37)

Learning the Language

8.26   What is denoted by s_p^2?

Using the Tables

8.27   Independent random samples are selected from two normally distributed populations with means μ_1 and μ_2 and equal variances. The sample sizes are $n_1 = 17$ and $n_2 = 10$.

a. Give the rejection region for testing $H_0: \mu_1 - \mu_2 = 0$ against $H_a: \mu_1 - \mu_2 > 0$. Use $\alpha = .01$.

b. Give the rejection region for testing $H_0: \mu_1 - \mu_2 = 0$ against $H_a: \mu_1 - \mu_2 \neq 0$. Use $\alpha = .10$.

c. Give the appropriate t-value for forming a 99% confidence interval for $\mu_1 - \mu_2$.

d. Give the rejection region for testing $H_0: \mu_1 - \mu_2 = 0$ against $H_a: \mu_1 - \mu_2 < 0$, with $\alpha = .05$.

e. Give the appropriate t-value for forming a 95% confidence interval for $\mu_1 - \mu_2$.

Learning the Mechanics

8.28 Independent random samples from two normally distributed populations with equal variances produced the results listed in the table.

Sample 1	Sample 2
2.1	3.4
3.6	2.8
1.4	4.1
3.0	3.9
2.9	
1.8	

a. Calculate $\bar{x}_1$, $\bar{x}_2$, s_1^2, and s_2^2.

b. Calculate s_p^2, the pooled estimate of σ^2.

c. Form a 90% confidence interval for $\mu_1 - \mu_2$.

d. Test $H_0: \mu_1 - \mu_2 = 0$ against $H_a: \mu_1 - \mu_2 < 0$. Use $\alpha = .05$.

e. Form a 99% confidence interval for $\mu_1 - \mu_2$.

f. Test $H_0: \mu_1 - \mu_2 = 0$ against $H_a: \mu_1 - \mu_2 \neq 0$. Use $\alpha = .05$.

8.29 Independent random samples selected from normal populations with equal variances produced the sample means and standard deviations shown in the table.

Sample 1	Sample 2
$n_1 = 15$	$n_2 = 10$
$\bar{x}_1 = 10.3$	$\bar{x}_2 = 7.5$
$s_1 = 3.5$	$s_2 = 5.4$

a. Test $H_0: \mu_1 - \mu_2 = 0$ versus $H_a: \mu_1 - \mu_2 \neq 0$. Use $\alpha = .05$.

b. Form a 95% confidence interval for $\mu_1 - \mu_2$.

c. Form a 90% confidence interval for $\mu_1 - \mu_2$.

d. Test $H_0: \mu_1 - \mu_2 = 0$ versus $H_a: \mu_1 - \mu_2 > 0$. Use $\alpha = .05$.

Applying the Concepts

In the following exercises, assume that it has been determined that the two populations of interest are normally distributed. State any additional conditions required for the validity of the statistical procedures used.

8.30 Some statistics students complain that pocket calculators give other students an unfair advantage during statistics examinations. To check this contention, forty-five students were randomly assigned to two groups—twenty-three to use calculators and twenty-two to perform calculations by hand. The students then took a statistics examination that required some arithmetic calculations. The means and variances of the test scores for the two groups are shown in the table. Do the data provide sufficient evidence to indicate that students taking this particular examination obtain higher scores when using a calculator? Test using $\alpha = .10$.

	n	$\bar{x}$	s^2
Calculators	23	80.7	49.5
No calculators	22	78.9	60.4

8.31 An experiment is conducted to determine the effect of a drug on the time to complete a task. Twenty people are divided at random into two groups, with ten in each group. One group is given a placebo, while the second group is administered an experimental drug thought to increase the time required to complete the task. The times required by the control group to complete the task have an average of 10.2 minutes and a variance of 4.1. For the experimental group, the times have an average of 13.1 minutes and a variance of 5.8. Test the null hypothesis that there is no difference between the mean times required to complete the task by subjects who receive the placebo and those who receive the drug, against the alternative hypothesis that the mean time associated with the drug is greater than that associated with the placebo. Use $\alpha = .10$.

8.32 To compare two methods of teaching reading, randomly selected groups of elementary school children were assigned to each of the two teaching methods for a 6-month period. The criterion for measuring achievement was a reading comprehension test. The results are shown in the table. Estimate the true difference between the mean scores on the comprehensive test for method 1 versus method 2. Use a 95% confidence interval to estimate the difference.

	Number of children per group	$\bar{x}$	s^2
Method 1	11	64	52
Method 2	14	69	71

8.33 A manufacturing company is interested in determining whether there is a significant difference between the average number of units produced per day by two machine operators. A random sample of ten daily outputs by each operator was selected from the past year. The summary data on the number of items produced per day are shown in the table.

Operator 1	Operator 2
$n_1 = 10$	$n_2 = 10$
$\bar{x}_1 = 35$	$\bar{x}_2 = 31$
$s_1^2 = 17.2$	$s_2^2 = 19.1$

a. Do the samples provide sufficient evidence at the .10 significance level to conclude that a difference does exist between the mean daily outputs of the two machine operators?

b. Find a 90% confidence interval for $\mu_1 - \mu_2$.

8.34 Refer to Exercises 7.61 and 7.62, where you estimated the mean percentages of body fat for all female high-school gymnasts and all male college gymnasts, respectively, using 90% confidence intervals. The relevant data are given in the table. Find a 90% confidence interval for the difference in the mean percentages of body fat for female high-school gymnasts versus male college gymnasts.

Female High-School Gymnasts	Male College Gymnasts
$n_1 = 13$	$n_2 = 19$
$\bar{x}_1 = 13.09$	$\bar{x}_2 = 6.50$
$s_1 = 5.12$	$s_2 = 2.40$

Sources: R. J. Moffatt, B. Surina, B. Golden, and N. Ayres, "Body Composition and Physiological Characteristics of Female High-School Gymnasts," *Research Quarterly for Exercise and Sport,* Vol. 55, No. 1, March 1984, p. 81; W. E. Sinning, D. G. Dolny, K. D. Little, L. N. Cunningham, A. Racaniello, S. F. Siconolfi, and J. L. Sholes, "Validity of Generalized' Equations for Body Composition Analysis in Male Athletes," *Medicine and Science in Sports and Exercise,* Vol. 17, No. 1, February 1985, p. 125.

8.35 In an analysis of physical characteristics of 3- and 4-year-old children, weights (in kilograms) were recorded for samples of 21 boys and 21 girls. The data are summarized in the accompanying table. Is there significant evidence to indicate a difference in mean weights between preschool boys and preschool girls? Test at the $\alpha = .10$ level of significance.

	Boys	Girls
Mean weight	17.2	16.4
Standard deviation	2.5	1.9

Source: S. J. Erbaugh, "The Relationship of Stability Performance and the Physical Growth Characteristics of Preschool Children," *Research Quarterly for Exercise and Sport,* Vol. 55, No. 1, March 1984, p. 12.

*8.36 Suppose you are the personnel manager of a company and you suspect a difference in the mean amount of work time lost due to sickness for two types of employees— those who work at night versus those who work during the day. In particular, you suspect that the mean time lost for the night shift exceeds the mean for the day shift. To check your theory, you randomly sample the records for ten employees for each shift category and record the number of days lost due to sickness within the past year. The data are shown in the table.

Night Shift, 1		Day Shift, 2	
21	2	13	18
10	19	5	17
14	6	16	3
33	4	0	24
7	12	7	1
$\bar{x}_1 = 12.8$		$\bar{x}_2 = 10.4$	
$\sum x_1^2 = 2{,}436$		$\sum x_2^2 = 1{,}698$	

a. Let μ_1 and μ_2 represent the mean number of days lost per year due to sickness for the night and day shifts, respectively. Use a computer package to test H_0: $\mu_1 - \mu_2 = 0$ against H_a: $\mu_1 - \mu_2 > 0$. Use $\alpha = .05$.

b. Use a computer package to form a 90% confidence interval for $\mu_1 - \mu_2$.

*8.37 Suppose your manufacturing plant purifies its liquid waste and discharges the water into a local river. An EPA inspector has collected water specimens of the discharge of your plant and water specimens in the river upstream from your plant. Each water specimen is divided into five parts, the bacteria count is read for each part, and the mean count for each specimen is reported. The table lists the average bacteria count readings for six specimens taken at each of the two locations.

Plant Discharge	Upstream
30.1	29.7
36.2	30.3
33.4	26.4
28.2	27.3
29.8	31.7
34.9	32.3

a. Why might the bacteria count readings tend to be approximately normally distributed?

b. Do the data provide sufficient evidence to indicate that the mean of the bacteria counts for the discharge exceeds the mean of the counts taken upstream? Use a computer package to test this hypothesis at the .10 level of significance.

c. What is the p-value for the test conducted in part b?

8.4

INFERENCES ABOUT $\mu_1 - \mu_2$, THE DIFFERENCE BETWEEN TWO POPULATION MEANS: DEPENDENT SAMPLES

In Sections 8.2 and 8.3 we presented methods that are appropriate for making inferences about the difference between two population means when the two samples are independent. However, many experiments are designed so that the samples are dependent, and the methods we previously presented are inappropriate. In this section we will discuss making inferences about the difference between two population means when the samples are dependent.

Recall that in Section 8.1 we stated that dependent samples result in pairs of observations. The sample data for paired observations might appear as shown in Table 8.5, which shows five pairs of observations. The first pair of numbers has a 10 in sample 1 and an 8 in sample 2; the second pair has a 6 in sample 1 and a 5 in sample 2; etc.

Table 8.5

Example of data for dependent samples

Pair	Sample 1	Sample 2
1	10	8
2	6	5
3	12	10
4	15	12
5	4	5

When making inferences using dependent samples, we will consider the differences between the pairs of measurements. In Table 8.6 we give these differences (denoted by D) in the last column of the table. The first difference is $10 - 8 = 2$; the second is $6 - 5 = 1$; etc. Notice that in order to calculate each difference, we always subtracted the measurement in sample 2 from the corresponding measurement in sample 1. The differences must always be taken in the same order, and in this text we will always subtract sample 2 from sample 1.

Table 8.6

Differences for dependent samples of Table 8.5

Pair	Sample 1	Sample 2	Difference, D
1	10	8	2
2	6	5	1
3	12	10	2
4	15	12	3
5	4	5	−1

For dependent samples, inferences about the difference between two population means will be based upon the differences in the paired observations. The resulting sample of differences is a single sample, so we can then use one-sample procedures (presented in Chapter 7) to make inferences. Since a relatively small number of pairs of observations is usually selected when dependent samples are used, we will present the small-sample procedures from Section 7.5.

The conditions that must be met to use the procedures discussed in this section are essentially the same as those given for use of the t statistic in Section 7.5. In the next box we restate all these conditions in terms of differences.

**When to Use the t Statistic to Make
Inferences Using Dependent Samples**

To make valid inferences about the difference between two population means using the differences between paired measurements (dependent samples), the following conditions must be met:

1. A random sample of differences must be selected from the population of all possible differences.
2. The population of differences must be (approximately) normally distributed.

We want to make an inference about the difference between two population means, $\mu_1 - \mu_2$. To accomplish this, we are considering a population of differences that has a mean which we will denote by μ_D. However, $\mu_D = \mu_1 - \mu_2$ because taking differences first and then averaging is arithmetically equivalent to averaging first and then taking differences. Thus, from now on we will discuss making inferences about μ_D, the mean of the population of differences. Keep in mind that this is the same as the difference between the two population means.

A test of hypothesis about μ_D and a confidence interval for μ_D are given in the accompanying boxes. The value of D_0, the hypothesized mean difference, will usually be 0. To emphasize that the sample of differences are the data of interest, we have added a subscript D to $\bar{x}$, s, and n. Thus, n_D is the number of sample differences (number of pairs of measurements), and $\bar{x}_D$ and s_D are calculated from the differences (denoted by D) using the following formulas:

$$\bar{x}_D = \frac{\sum D}{n_D} \qquad s_D = \sqrt{\frac{n_D \sum D^2 - (\sum D)^2}{n_D(n_D - 1)}}$$

Test of Hypothesis About μ_D

$H_0: \quad \mu_D = D_0$

$H_a: \quad \mu_D > D_0 \qquad$ or $\qquad H_a: \quad \mu_D < D_0 \qquad$ or $\qquad H_a: \quad \mu_D \neq D_0$

Test statistic: $\quad t = \dfrac{\bar{x}_D - D_0}{\dfrac{s_D}{\sqrt{n_D}}}$

Rejection region: Use Appendix Table V, with $(n_D - 1)$ df.

Confidence Interval for μ_D

The confidence interval for μ_D is expressed as

$$\bar{x}_D - t\left(\frac{s_D}{\sqrt{n_D}}\right) \qquad \text{to} \qquad \bar{x}_D + t\left(\frac{s_D}{\sqrt{n_D}}\right)$$

The value of t is found in Appendix Table VI, using $(n_D - 1)$ df.

If a large sample ($n_D \geq 30$) of differences is obtained, the large-sample procedures from Sections 7.2 and 7.4 would be used. This involves using z-values rather than t-values, and the condition that the population of differences be normally distributed is not necessary for the validity of the inferences.

EXAMPLE 8.9   In Example 8.8, we compared two methods of teaching reading to slow learners using two independent random samples. Suppose it is possible to measure the "reading IQ's" of the slow learners before they are subjected to a teaching method. Eight pairs of slow learners with similar reading IQ's are randomly selected, and one member of each pair is randomly assigned to the standard teaching method while the other is assigned to the new method. Do the data in Table 8.7 (page 434) support the hypothesis that the population mean reading test score for slow learners taught by the new method is greater than the mean reading test score for those taught by the standard method? Assume that the population of all differences is mound-shaped, and use $\alpha = .10$.

Solution   The population of interest is the set of all differences between pairs of test scores for the new and standard methods, and the parameter of interest is μ_D, the mean difference between all pairs of test scores. Since we want to show that the mean for the new

Table 8.7

Pair	New Method	Standard Method	Difference D	D^2
1	77	72	5	25
2	74	68	6	36
3	82	76	6	36
4	73	68	5	25
5	87	84	3	9
6	69	68	1	1
7	66	61	5	25
8	80	76	4	16
			$\sum D = 35$	$\sum D^2 = 173$

method is greater than the mean for the standard method, we want to show that $\mu_D > 0$. The elements of the test are

$$H_0: \quad \mu_D = 0 \qquad H_a: \quad \mu_D > 0$$

$$\text{Test statistic:} \quad t = \frac{\bar{x}_D - 0}{\dfrac{s_D}{\sqrt{n_D}}}$$

The population is assumed to be approximately normally distributed, and a random sample has been selected from this population. It is valid to use the t statistic to test a hypothesis about μ_D.

Rejection region: Reject H_0 if $t > 1.42$.

The rejection region was obtained from Appendix Table V, using $n_D - 1 = 8 - 1 = 7$ df. The rejection region is shown in Figure 8.6.

Figure 8.6
Rejection region for
Example 8.9

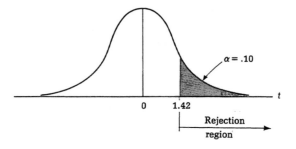

Before finding the calculated value of the test statistic, we must calculate $\bar{x}_D$ and s_D. In Table 8.7, we have given the differences, D, and the squared differences, D^2. We also found the sums of these quantities. Using this information, we can apply the formulas given earlier to find $\bar{x}_D$ and s_D:

$$\bar{x}_D = \frac{\sum D}{n_D} = \frac{35}{8} = 4.375$$

$$s_D = \sqrt{\frac{n_D \sum D^2 - (\sum D)^2}{n_D(n_D - 1)}} = \sqrt{\frac{8(173) - (35)^2}{8(7)}} = 1.69$$

Calculating the test statistic, we obtain

$$t = \frac{4.375 - 0}{\dfrac{1.69}{\sqrt{8}}} = \frac{4.375}{.60} = 7.29$$

Because this value of t falls in the rejection region, we conclude that the mean test score for slow learners taught by the new method exceeds the mean score for those taught by the standard method. We have 90% confidence in this conclusion (since $\alpha = .10$). ∎

The kind of experiment described in Example 8.9, in which observations are paired and the differences are analyzed, is often called a **paired difference experiment.** In many cases, a paired difference experiment can provide more information about the difference between population means than an independent samples experiment. The idea is to compare population means by examining the differences between measurements obtained within pairs of experimental units (objects, people, etc.) that were very similar prior to the experiment. Taking the differences removes sources of random variation. For example, when two children are taught to read by different methods, the observed difference in achievement may be due to a difference in the effectiveness of the two teaching methods *or* it may be due to differences in the initial reading levels and IQ's of the two children (random error). To reduce the effect of differences in the ability of the children on the observed differences in reading achievement, the two teaching methods are assigned to pairs of children who are likely to have similar intellectual potential (that is, children with nearly equal IQ's). The effect of this pairing is to remove the large source of variation that would be present if children with different abilities were randomly assigned to the two samples.

Some other examples for which the paired difference experiment would be appropriate are the following:

1. Suppose you want to estimate the difference $(\mu_1 - \mu_2)$ in mean price per gallon between two major brands of premium gasoline. If you choose two independent random samples of stations for each brand, the variability in price due to geographical location may be large. To eliminate this source of variability you could choose pairs of stations of similar size, one station for each brand, in close geographical proximity, and use the sample of differences between the prices of the brands to make an inference about $\mu_1 - \mu_2$.

2. Suppose a college placement center wants to estimate the difference $(\mu_1 - \mu_2)$ in mean starting salaries for men and women graduates who seek jobs through the

center. If the center independently samples men and women, the starting salaries may vary due to different college majors and differences in grade-point averages. To eliminate these sources of variability, the placement center could match male and female job seekers according to their majors and grade-point averages. Then the differences between the starting salaries of each pair in the sample could be used to make an inference about $\mu_1 - \mu_2$. This application will be discussed in Example 8.10.

3. Suppose we want to estimate the difference $(\mu_1 - \mu_2)$ in mean absorption rate into the bloodstream for two drugs that relieve pain. If we independently sample people, the absorption rates might vary due to factors such as age, weight, sex, blood pressure, etc. In fact, there are many possible sources of variability, and pairing individuals who are similar in all the possible sources would be quite difficult. However, it may be possible to obtain two measurements *on the same person*. First, we administer one of the two drugs and record the time until absorption. After a sufficient amount of time, the other drug is administered and a second measurement on absorption time is obtained. The differences between the measurements for each person in the sample could then be used to estimate $\mu_1 - \mu_2$. This procedure would be advisable only if the amount of time elapsed between drugs is sufficient to guarantee little or no carryover effect. Otherwise, it would be better to use different people matched as closely as possible on the factors thought to be most important.

EXAMPLE 8.10   To compare the starting salaries of male and female college graduates, an experiment is conducted in which matched pairs of male and female graduates are selected. Each pair is formed by choosing a male and a female with the same major and similar grade-point averages. A random sample of ten such pairs is selected, and the starting annual salary of each person is recorded. The results are shown in Table 8.8, and the differences between the pairs of salaries have been calculated. Use the sample data to form a 95% confidence interval for the difference between the mean salaries of males and females. Assume that the population of differences is approximately normal.

Table 8.8

Pair	Male	Female	Difference D
1	$14,300	$13,800	$ 500
2	16,500	16,600	−100
3	15,400	14,800	600
4	13,500	13,500	0
5	18,500	17,600	900
6	12,800	13,000	−200
7	14,500	14,200	300
8	16,200	15,100	1,100
9	13,400	13,200	200
10	14,200	13,500	700

Solution The population of interest—the population of differences—is assumed to be approximately normal, and a random sample has been chosen from this population. We may proceed to find a confidence interval for μ_D, the mean of the differences between male and female salaries.

The general form of the confidence interval is

$$\bar{x}_D - t\left(\frac{s_D}{\sqrt{n_D}}\right) \quad \text{to} \quad \bar{x}_D + t\left(\frac{s_D}{\sqrt{n_D}}\right)$$

First we calculate $\sum D$ and $\sum D^2$:

$$\sum D = 500 + (-100) + \cdots + 700 = 4,000$$
$$\sum D^2 = (500)^2 + (-100)^2 + \cdots + (700)^2 = 3,300,000$$

Then,

$$\bar{x}_D = \frac{\sum D}{n_D} = \frac{4,000}{10} = 400$$

$$s_D = \sqrt{\frac{n_D \sum D^2 - (\sum D)^2}{n_D(n_D - 1)}} = \sqrt{\frac{10(3,300,000) - (4,000)^2}{10(9)}}$$

$$= 434.61$$

From Appendix Table VI, using 9 df, we obtain $t = 2.26$. Substituting into the formula for the confidence interval, we find

$$400 - 2.26\left(\frac{434.61}{\sqrt{10}}\right) \quad \text{to} \quad 400 + 2.26\left(\frac{434.61}{\sqrt{10}}\right)$$

$$400 - 310.60 \quad \text{to} \quad 400 + 310.60$$

$$89.40 \quad \text{to} \quad 710.60$$

With 95% confidence, μ_D is between 89.40 and 710.60. We conclude that the mean annual starting salary for males exceeds the mean starting salary for females by at least $89.40 and perhaps by as much as $710.60. ∎

In Section 8.1 we indicated that using dependent samples may be better in some instances than using independent samples. We will now present a way to determine whether a paired difference experiment provides more or less information than an experiment with independent samples.

Consider the data for Example 8.10. For the moment, we will ignore the fact that the samples are dependent, and we will analyze the data as though the samples were independent. Using the results of Section 8.3, we will form a 95% confidence interval for the difference between the mean salary for males and the mean salary for females.

Using the two samples for males and females, we calculate the quantities given in Table 8.9. We also find that $s_p^2 = 2,670,111.11$.

Table 8.9

Males	Females
$n_1 = 10$	$n_2 = 10$
$\bar{x}_1 = \$14{,}930$	$\bar{x}_2 = \$14{,}530$
$s_1^2 = 3{,}009{,}000$	$s_2^2 = 2{,}331{,}222.22$

The confidence interval for independent samples is

$$(\bar{x}_1 - \bar{x}_2) - t\sqrt{\frac{s_p^2}{n_1} + \frac{s_p^2}{n_2}} \quad \text{to} \quad (\bar{x}_1 - \bar{x}_2) + t\sqrt{\frac{s_p^2}{n_1} + \frac{s_p^2}{n_2}}$$

From Appendix Table VI, we find that $t = 2.10$ for $n_1 + n_2 - 2 = 10 + 10 - 2 = 18$ df. The interval is calculated to be

$$(14{,}930 - 14{,}530) - 2.10\sqrt{\frac{2{,}670{,}111.11}{10} + \frac{2{,}670{,}111.11}{10}}$$

$$\text{to} \quad (14{,}930 - 14{,}530) + 2.10\sqrt{\frac{2{,}670{,}111.11}{10} + \frac{2{,}670{,}111.11}{10}}$$

$$400 - 1{,}534.61 \quad \text{to} \quad 400 + 1{,}534.61$$

$$-1{,}134.61 \quad \text{to} \quad 1{,}934.61$$

Analyzing the data as though the two samples were independent produces a confidence interval that is about *five times wider* than the confidence interval for the paired analysis. Even though the two samples are not independent, similar data might be obtained if independent samples were actually taken. The huge difference in the width of the two confidence intervals provides strong evidence that dependent samples provide more information than independent samples when comparing male and female starting salaries.

We must point out that **analyzing dependent samples as though they were independent is not valid for making inferences about $\mu_1 - \mu_2$.** Notice that we did not interpret the second interval that we calculated. The only reason for calculating the interval was to get an idea of how wide an interval might result from independent samples.

Note: The pairing of the observations is determined *before* the experiment is performed (that is, by the *design* of the experiment). A paired difference experiment is *never* obtained by pairing the sample observations after the measurements have been acquired. Such is the stuff of which statistical lies are made!

COMPUTER EXAMPLE 8.2

Refer to Example 8.10. Based on a random sample of ten matched pairs of male and female graduates, a 95% confidence interval for the difference between the mean salaries of males and females was found. We will now see how Minitab performs the same analysis.

The male and female salaries given in Table 8.8 were entered into the computer, and Minitab was used to calculate the differences between these pairs of salaries.

The data are shown below:

ROW	PAIR	MALE	FEMALE	DIFF.
1	1	14300	13800	500
2	2	16500	16600	-100
3	3	15400	14800	600
4	4	13500	13500	0
5	5	18500	17600	900
6	6	12800	13000	-200
7	7	14500	14200	300
8	8	16200	15100	1100
9	9	13400	13200	200
10	10	14200	13500	700

In Section 7.5, we used Minitab to help make inferences about the population mean for a small random sample. Since the analysis for a paired difference experiment uses the one sample of differences, the Minitab analyses for paired difference experiments are analogous to the analyses considered in Section 7.5. The accompanying printout gives a 95% confidence interval for the mean difference in salaries.

	N	MEAN	STDEV	SE MEAN	95.0 PERCENT C.I.
DIFF.	10	400	435	137	(89, 711)

As before, the Minitab result differs slightly from ours due to rounding. We found the interval of interest to be from 89.40 to 710.60, whereas Minitab reports it to be from 89 to 711. Regardless of the number of decimal places used, the interpretation is essentially the same. We may conclude with 95% confidence that the mean annual starting salary for males is between $89 and $711 more than the mean starting salary for females. ∎

Exercises (8.38–8.52)

In the following exercises, assume that it has been determined that the two populations of interest are normally distributed. State any additional conditions required for the validity of the statistical procedures used.

Learning the Language

8.38 What do the following symbols denote?
a. D b. μ_D c. $\bar{x}_D$ d. s_D e. n_D

Using the Tables

8.39 A random sample of twelve paired measurements is observed, and the difference between the first measurement in each pair and the corresponding second measurement is calculated. It is desired to make inferences about μ_D, the mean of the population of all possible differences.
a. Give the appropriate t-value for forming a 95% confidence interval for μ_D.
b. Give the appropriate t-value for forming a 90% confidence interval for μ_D.

   c. Give the rejection region for testing $H_0: \mu_D = 0$ against $H_a: \mu_D < 0$ using $\alpha = .01$.

   d. Give the rejection region for testing $H_0: \mu_D = 0$ against $H_a: \mu_D \neq 0$ using $\alpha = .05$.

Learning the Mechanics

8.40 Data for a random sample of six paired observations are shown in the table.

Pair	Sample from Population 1	Sample from Population 2
1	6	3
2	2	0
3	5	7
4	10	6
5	8	5
6	4	2

   a. Calculate the difference between each pair of observations by subtracting observation 2 from observation 1. Calculate $\bar{x}_D$ and s_D.

   b. If μ_1 and μ_2 are the means of populations 1 and 2, respectively, express μ_D in terms of μ_1 and μ_2.

   c. Form a 95% confidence interval for μ_D.

   d. Test the null hypothesis $H_0: \mu_D = 0$ versus the alternative hypothesis $H_a: \mu_D \neq 0$. Use $\alpha = .05$.

   e. Form a 90% confidence interval for μ_D.

   f. Test $H_0: \mu_D = 0$ against $H_a: \mu_D > 0$ using $\alpha = .05$.

8.41 The data for a random sample of ten paired observations are shown in the table.

Pair	Sample from Population 1	Sample from Population 2	Pair	Sample from Population 1	Sample from Population 2
1	50	55	6	48	54
2	55	61	7	50	53
3	47	50	8	47	52
4	63	68	9	65	65
5	51	54	10	50	56

   a. Do the data provide sufficient evidence to conclude that $\mu_D < 0$? Test using $\alpha = .05$.

   b. Form a 90% confidence interval for μ_D.

Applying the Concepts

8.42 A new weight-reducing technique, consisting of a liquid protein diet, is currently undergoing tests by the Food and Drug Administration (FDA) before its introduction into the market. A typical test performed by the FDA is the following: The weights of a random sample of five people are recorded before they are introduced to the liquid protein diet. The five individuals are then instructed to follow the liquid protein diet for 3 weeks. At the end of this period, their weights (in pounds) are again recorded. The results are listed in the table. Construct a 95% confidence interval for the true mean difference in weight before and after the diet is used.

Person	Weight Before Diet	Weight After Diet
1	150	143
2	195	190
3	188	185
4	197	191
5	204	200

8.43 A company is interested in hiring a new secretary. Several candidates are interviewed and the choice is narrowed to two possibilities. The final choice will be based on typing ability. Six letters are randomly selected from the company's files, and each candidate is required to type each one. The number of words typed per minute is recorded for each candidate–letter combination. The data are listed in the table.

Letter	Candidate 1	Candidate 2
1	62	59
2	60	60
3	65	61
4	58	57
5	59	55
6	64	60

a. Do the data provide sufficient evidence to indicate a difference in the mean number of words typed per minute by the two candidates? Test using $\alpha = .10$.

b. Find a 90% confidence interval for the difference in the mean typing rates.

8.44 A farmer was interested in determining which of two soil fumigants, A or B, is more effective in controlling the number of parasites in a particular crop. To compare the fumigants, four small fields were divided into two equal areas, then fumigant A was applied to one part and fumigant B to the other. Crop samples of equal size were taken from each of the eight plots and the number of parasites per square foot was counted. The data are shown in the table. Do the data provide sufficient evidence to indicate a difference in the mean levels of parasites for the two fumigants? Test at the .01 significance level.

Field	A	B
1	15	9
2	5	3
3	8	6
4	8	4

8.45 In the past, many bodily functions were thought to be beyond conscious control. However, recent experimentation suggests that it may be possible for a person to control certain body functions if that person is trained in a program of biofeedback exercises. An experiment is conducted to show that blood pressure levels can be consciously reduced in people trained in this program. The blood pressure measurements (in millimeters of mercury) listed in the table represent readings before and after the training of six subjects. Is there sufficient evidence to conclude that the training produces a reduction in mean blood pressure for people trained in this program? Use $\alpha = .05$.

Subject	Before	After
1	136.9	130.2
2	201.4	180.7
3	166.8	149.6
4	150.0	153.2
5	173.2	162.6
6	169.3	160.1

8.46 A manufacturer of automobile shock absorbers was interested in comparing the durability of its shocks with the durability of shocks produced by its biggest competitor. To make the comparison, six of the manufacturer's and six of the competitor's shocks were randomly selected and one of each brand was installed on the rear wheels of each of six cars. After the cars had been driven 20,000 miles, the strength of each shock absorber was measured, coded, and recorded. The results are given in the table.

Car Number	Manufacturer's	Competitor's
1	8.8	8.4
2	10.5	10.1
3	12.5	12.0
4	9.7	9.3
5	9.6	9.0
6	13.2	13.0

a. Do the data present sufficient evidence to conclude that there is a difference in the mean strength of the two types of shock absorbers after 20,000 miles of use? Let $\alpha = .05$.

b. Construct a 95% confidence interval for μ_D. Interpret this confidence interval.

8.47 Suppose the data in Exercise 8.46 are based on independent random samples.

a. Do the data provide sufficient evidence to indicate a difference between the mean strengths for the two types of shocks? Let $\alpha = .05$.

b. Construct a 95% confidence interval for $\mu_1 - \mu_2$.

c. Compare the confidence intervals you obtained in Exercise 8.46 and part b of this exercise. Which is wider? To what do you attribute the difference in width? Assuming in each case that the appropriate assumptions are satisfied, which interval provides you with more information about $\mu_1 - \mu_2$? Explain.

d. Are the results of an unpaired analysis valid when the data have been collected from a paired experiment?

8.48 A company has many plants throughout the world that produce the same product. The number of units produced at each of five randomly selected plants on a particular day was recorded. After the arrangement of the assembly line at each plant was modified, the sampling procedure was repeated. The results are given in the table.

Plant Number	Before	After
1	90	93
2	94	96
3	91	92
4	85	88
5	88	90

a. Are the two samples independent? Explain.

b. Do the data present sufficient evidence to conclude that there is a difference in the mean daily output of the company's plants before and after the modification of their assembly lines? Use $\alpha = .05$.

c. Construct a 90% confidence interval for μ_D.

8.49 American health workers agree that obesity in children is developmentally dangerous, but suitable weight control methods are a source of dispute:

In recent years the investigation and treatment of obesity among adolescents has received increasing attention. A number of factors seem to be contributing to this growing concern. The prevalence of obesity in adolescents has been estimated to be between 15% and 30%. . . . [and] obese adolescents are at higher risk for developing diabetes, hypertension, and cardiovascular disease in adulthood. More immediately, adolescent obesity has been found to have undesirable psychosocial implications, including lowered self-esteem, problems with family relationships, and discrimination by peers. . . .

In response to these concerns, a small body of research investigating the efficacy of treatment programs for adolescent weight loss, and a plethora of commercial weight loss summer camps have emerged. Unfortunately, the research findings do not appear to be reflected in the typical treatment programs available to most adolescents.*

In California, the Stanford Adolescent Weight Loss Camp was created to investigate and alleviate such obesity:

The goals were threefold: (1) to provide an intensive behavioral program to facilitate eating and exercise habit change, (2) to provide practice and supervision of eating and exercise habits during a portion of each camp day, and (3) to involve parents in the support of their

* M. A. Southam, B. G. Kirkley, A. Murchison, and R. I. Berkowitz, "A Summer Day Camp Approach to Adolescent Weight Loss," *Adolescence*, Vol. 19, No. 76, Winter 1984, pp. 855–856.

child's habit changes. Thus, the camp was structured so that the teenagers came to the Stanford University campus four mornings per week for either a four- or an eight-week session. Each daily session was composed of four periods: aerobic exercise, behavioral counseling, sports skills instruction, and an eating laboratory. Weigh-ins were held weekly. In addition, parent seminars were conducted one evening per week.*

Seven adolescents attending the eight-week session were weighed prior to and at the end of the program. The initial weight was subtracted from the final weight, producing a mean difference of -7.5 and a standard deviation of 4.0.

a. Is there sufficient evidence to conclude that adolescents who attend this camp experience a mean weight loss? Use $\alpha = .05$.

b. Find a 95% confidence interval for the mean difference in the weights of adolescents before and after this program.

8.50 Refer to Exercise 8.7. A sample of library employees was surveyed shortly before and 6 months after starting a flexitime work schedule. One item on the employee questionnaire was used to measure attitudes toward personnel productivity. It was concluded from comparison of the two sets of responses that no increase in personnel productivity resulted from flexible scheduling. The observed level of significance associated with this conclusion was given as $p \le .01$.[†] Interpret these results.

*8.51 An assertiveness training course has just been added to the services offered by a college counseling center. To measure its effectiveness, ten students are given a test at the beginning of the course and again at the end. A high score on the test implies high assertiveness. The test scores are shown in the table. Do the data provide sufficient evidence to conclude that the mean score on the test is higher after taking the course than before?

Student	Before	After	Student	Before	After
1	50	65	6	56	70
2	62	68	7	49	48
3	51	52	8	67	69
4	41	43	9	42	53
5	63	60	10	57	61

a. Use a computer package to test the desired hypothesis at the .05 level of significance.

b. What is the p-value for the test conducted in part a?

*8.52 The *National Survey for Professional, Administrative, Technical, and Clerical Pay* is conducted annually to determine whether federal pay scales are commensurate with private sector salaries. The government workers and private workers in the study are

* Ibid., p. 859.

† *Source:* C. S. Saunders and R. Saunders. "Effects of Flexitime on Sick Leave. Vacation Leave, Anxiety, Performance, and Satisfaction in a Library Setting," *The Library Quarterly*, Vol. 55, No. 1, January 1985, pp. 81, 85.

matched as closely as possible on job level and experience before the salaries are compared. Suppose the data in the accompanying table represent annual salaries for twelve pairs of individuals in the sample. Use a computer package to place a 99% confidence interval on the difference between the mean salaries of the private and government workers.

Pair	Private	Government	Pair	Private	Government
1	$16,250	$15,280	7	$20.540	$18,860
2	28,990	27,170	8	22,750	23,270
3	18,850	19,240	9	30.260	27,830
4	49,900	48,810	10	54,730	56,160
5	27,040	27,950	11	21,840	19,790
6	24,960	23,920	12	18,820	18,460

8.5

LARGE-SAMPLE INFERENCES ABOUT $\pi_1 - \pi_2$, THE DIFFERENCE BETWEEN TWO POPULATION PROPORTIONS: INDEPENDENT SAMPLES

In this section we will present methods for comparing two population proportions using two large, independent random samples. Some useful notation for making inferences about two population proportions is given in the box. As before, we add a subscript to each symbol to indicate which population, or sample, is being considered.

Notation for Comparing Two Population Proportions

Population I

   Proportion: π_1

Population II

   Proportion: π_2

Sample I

   Proportion: p_1
   Sample size: n_1

Sample II

   Proportion: p_2
   Sample size: n_2

Inferences about two population proportions are stated in terms of the difference between the proportions, $\pi_1 - \pi_2$. Inferences are based upon the sampling distribution of the corresponding statistic, the difference in the sample proportions, $p_1 - p_2$. The conditions that must be met to use large samples to make valid inferences about $\pi_1 - \pi_2$ are given in the next box (page 446).

**When to Make Large-Sample Inferences About $\pi_1 - \pi_2$,
the Difference Between Two Population Proportions**

To make a valid test of hypothesis about $\pi_1 - \pi_2$, or to form a confidence
interval for $\pi_1 - \pi_2$, the following conditions must be met:

1. Two random samples must be taken—one from each of the two
   populations of interest.
2. The two samples must be independent.
3. Both samples must be large ($n_1 \geq 30$ and $n_2 \geq 30$).

In order to make a large-sample inference about one population proportion, π, we
required the sample size, n, to be large enough so that both the following hold:

$$\pi - 3\sqrt{\frac{\pi \cdot (1 - \pi)}{n}} > 0 \quad \text{and} \quad \pi + 3\sqrt{\frac{\pi \cdot (1 - \pi)}{n}} < 1$$

A similar requirement could be given for making inferences about $\pi_1 - \pi_2$, and it would
involve π_1, π_2, n_1, and n_2. Rather than give this requirement, we point out that as
long as the value of $\pi_1 - \pi_2$ is not near 1 (or -1), sample sizes of $n_1 \geq 30$ and
$n_2 \geq 30$ will be sufficient to make large-sample inferences about $\pi_1 - \pi_2$. When
comparing two proportions, π_1 and π_2, their values are usually similar, and we may
simply check to be sure that $n_1 \geq 30$ and $n_2 \geq 30$. If the values of π_1 and π_2 are
extremely different, a formal statistical procedure is probably not required to compare
them.

When the conditions given in the box above are satisfied, the sampling distribution
of $p_1 - p_2$ has the following properties:

Properties of the Sampling Distribution of $p_1 - p_2$

When the conditions stated in the previous box are satisfied, the following properties hold:

1. The sampling distribution of $p_1 - p_2$ is approximately normal.
2. The mean of the sampling distribution of $p_1 - p_2$ is $\pi_1 - \pi_2$.
3. The standard error of the sampling distribution of $p_1 - p_2$ is

$$\sigma_{(p_1 - p_2)} = \sqrt{\frac{\pi_1 \cdot (1 - \pi_1)}{n_1} + \frac{\pi_2 \cdot (1 - \pi_2)}{n_2}}$$

Inferences about $\pi_1 - \pi_2$ are based on the sampling distribution of $p_1 - p_2$ described in the previous box. The large-sample test of hypothesis about $\pi_1 - \pi_2$ and
the confidence interval for $\pi_1 - \pi_2$ are given in the next two boxes.

Large-Sample Test of Hypothesis About $\pi_1 - \pi_2$

H_0: $\pi_1 - \pi_2 = D_0$

H_a: $\pi_1 - \pi_2 > D_0$ or H_a: $\pi_1 - \pi_2 < D_0$

or H_a: $\pi_1 - \pi_2 \neq D_0$

Test statistic: When $D_0 = 0$,

$$z = \frac{(p_1 - p_2) - 0}{\sqrt{\dfrac{p(1 - p)}{n_1} + \dfrac{p(1 - p)}{n_2}}}$$

and when $D_0 \neq 0$,

$$z = \frac{(p_1 - p_2) - D_0}{\sqrt{\dfrac{p_1(1 - p_1)}{n_1} + \dfrac{p_2(1 - p_2)}{n_2}}}$$

where

$$p = \frac{n_1 p_1 + n_2 p_2}{n_1 + n_2}$$

Rejection region: Use Appendix Table III.

Large-Sample Confidence Interval for $\pi_1 - \pi_2$

The large-sample confidence interval for $\pi_1 - \pi_2$ is expressed as

$$(p_1 - p_2) - z \sqrt{\frac{p_1(1 - p_1)}{n_1} + \frac{p_2(1 - p_2)}{n_2}}$$

to

$$(p_1 - p_2) + z \sqrt{\frac{p_1(1 - p_1)}{n_1} + \frac{p_2(1 - p_2)}{n_2}}$$

The value of z is obtained from Appendix Table IV.

As usual, D_0 is the hypothesized difference between the parameters π_1 and π_2. The form of the test statistic depends on the value of D_0. If $D_0 = 0$, then $\pi_1 - \pi_2 = 0$, or $\pi_1 = \pi_2 = \pi$, when H_0 is true. Thus, for the standard error of $p_1 - p_2$ in the test statistic, a single value, a **pooled estimate of π,** denoted by p, is substituted for π_1 and π_2. The pooled estimate of π uses the information from both samples to calculate one sample proportion, p.

When D_0 is any value other than 0, then $\pi_1 \neq \pi_2$ when H_0 is true. In this case, p_1 is substituted for π_1, and p_2 is substituted for π_2 in the standard error of $p_1 - p_2$.

EXAMPLE 8.11   Suppose a presidential candidate wants to compare the preferences of registered voters in the northeastern United States (NE) to those in the southeastern United States (SE). Such a comparison would help determine where to concentrate campaign efforts. The candidate hires a professional pollster to randomly choose 1,000 registered voters in the northeast and 1,000 registered voters in the southeast and interview them to determine their voting preferences. The objective is to use this sample information to form a confidence interval for $\pi_1 - \pi_2$, the difference between the proportions of all registered voters in the northeast and all registered voters in the southeast who plan to vote for this presidential candidate. The results of the sample are given in Table 8.10. Form the desired confidence interval using 95% confidence.

Table 8.10

NE	SE
$n_1 = 1,000$	$n_2 = 1,000$
$p_1 = .54$	$p_2 = .47$

Solution   The populations of interest are the voting preferences of all registered voters in the northeast and the southeast, with π_1 being the proportion of the northeast voters who plan to vote for a presidential candidate, and π_2 being the corresponding proportion for the southeast.

Two very large random samples were taken from these populations, and the values $p_1 = .54$ and $p_2 = .47$ indicate that the values of π_1 and π_2 are not very different. Thus, it is valid to form a large-sample confidence interval for $\pi_1 - \pi_2$

Using Appendix Table IV, we find the z-value for a 95% confidence interval to be $z = 1.96$. Calculating the confidence interval, we obtain

$$(p_1 - p_2) - z\sqrt{\frac{p_1(1-p_1)}{n_1} + \frac{p_2(1-p_2)}{n_2}} \quad \text{to} \quad (p_1 - p_2) + z\sqrt{\frac{p_1(1-p_1)}{n_1} + \frac{p_2(1-p_2)}{n_2}}$$

$$(.54 - .47) - 1.96\sqrt{\frac{(.54)(.46)}{1,000} + \frac{(.47)(.53)}{1,000}} \quad \text{to} \quad (.54 - .47) + 1.96\sqrt{\frac{(.54)(.46)}{1,000} + \frac{(.47)(.53)}{1,000}}$$

$$.07 - .044 \quad \text{to} \quad .07 + .044$$

$$.026 \quad \text{to} \quad .114$$

Thus, we estimate the value of $\pi_1 - \pi_2$ to be in the interval .026 to .114. We infer that there are between 2.6% and 11.4% more registered voters in the northeast than in the southeast who plan to vote for the presidential candidate. This indicates that the candidate might direct a stronger campaign effort to the southeast than to the northeast. We have 95% confidence that the interval formed contains $\pi_1 - \pi_2$. ■

EXAMPLE 8.12   Numerous antismoking campaigns have been sponsored by both federal and private agencies. The American Cancer Society is interested in the effects of these campaigns in one section of the country. In particular, the American Cancer Society would like to compare π_1, the proportion of adults living in this part of the country who smoked in 1980, to π_2, the proportion of adults living in this part of the country who smoked in 1988. Independent random samples of adults are obtained in each of these years, and the results are given in Table 8.11.

Do the data indicate that the fraction of smokers decreased over this 8-year period in this section of the country? Use $\alpha = .10$.

Table 8.11

1980	1988
Sample size: 500	Sample size: 800
Number of smokers: 153	Number of smokers: 222

Solution The populations of interest are the observations of whether people smoke for all people in this section of the country in 1980 and 1988. The parameters of interest, π_1 and π_2, were defined in the statement of the example.

We would like to show that the fraction of smokers has decreased, and this would mean that $\pi_1 - \pi_2 > 0$. The elements of the test are

$$H_0: \quad \pi_1 - \pi_2 = 0 \qquad H_a: \quad \pi_1 - \pi_2 > 0$$

$$\text{Test statistic:} \quad z = \frac{(p_1 - p_2) - 0}{\sqrt{\dfrac{p(1-p)}{n_1} + \dfrac{p(1-p)}{n_2}}}$$

Two independent random samples were selected from these populations, and the sample sizes are certainly large. We will proceed with a large-sample test of hypothesis about $\pi_1 - \pi_2$.

Rejection region: From Appendix Table III, reject H_0 if $z > 1.28$.

Before we calculate the test statistic, we must first find p_1, p_2, and p. In words, p_1 is the sample fraction of smokers in 1980, and p_2 is the sample fraction of smokers in 1988. Thus,

$$p_1 = \frac{153}{500} = .3060 \quad \text{and} \quad p_2 = \frac{222}{800} = .2775$$

Using the formula for p, we find

$$p = \frac{n_1 p_1 + n_2 p_2}{n_1 + n_2} = \frac{500(.3060) + 800(.2775)}{500 + 800} = \frac{153 + 222}{1{,}300}$$

$$= \frac{375}{1{,}300} = .2885$$

Notice that p, the pooled estimate of $\pi_1 = \pi_2 = \pi$, is actually the total number of smokers for both samples divided by the total sample size for both samples. When p is calculated, the two samples are essentially lumped together as one. This provides a reasonable estimate of $\pi = \pi_1 = \pi_2$.

Calculating the test statistic, we obtain

$$z = \frac{(.3060 - .2775) - 0}{\sqrt{\dfrac{(.2885)(.7115)}{500} + \dfrac{(.2885)(.7115)}{800}}} = \frac{.0285}{.0258} = 1.10$$

Since $z = 1.10$ is not greater than 1.28, we cannot reject H_0. There is insufficient evidence to conclude that the fraction of smokers decreased from 1980 to 1988 in this section of the country. ∎

Exercises (8.53–8.64)

Learning the Language

8.53 What do the following symbols denote?
a. π_1 and π_2
b. $\pi_1 - \pi_2$
c. p_1 and p_2
d. $p_1 - p_2$
e. $\sigma_{(p_1 - p_2)}$

Using the Tables

8.54 Two large, independent random samples are selected from two binomial populations with fractions of success π_1 and π_2.
a. Give the z-value used to form a 90% confidence interval for $\pi_1 - \pi_2$.
b. Give the rejection region for testing $H_0: \pi_1 - \pi_2 = 0$ against $H_a: \pi_1 - \pi_2 > 0$. Use $\alpha = .01$.
c. Give the z-value used to form a 99% confidence interval for $\pi_1 - \pi_2$.
d. Give the rejection region for testing $H_0: \pi_1 - \pi_2 = 0$ against $H_a: \pi_1 - \pi_2 \neq 0$, using $\alpha = .05$.
e. Give the rejection region for testing $H_0: \pi_1 - \pi_2 = 0$ against $H_a: \pi_1 - \pi_2 < 0$. Test at the .10 level of significance.

Learning the Mechanics

8.55 Independent random samples are selected from two binomial populations with fractions of success π_1 and π_2. Construct a 95% confidence interval for $\pi_1 - \pi_2$ for each of the following situations:
a. $n_1 = 100$, $p_1 = .72$, $n_2 = 100$, $p_2 = .61$
b. $n_1 = 130$, $p_1 = .16$, $n_2 = 210$, $p_2 = .25$
c. $n_1 = 70$, $p_1 = .53$, $n_2 = 60$, $p_2 = .48$

8.56 Independent random samples of size $n_1 = 400$ and $n_2 = 500$ were selected from two binomial populations. The sample from population 1 produced 200 successes, and the sample from population 2 produced 300 successes.

a. Test H_0: $\pi_1 - \pi_2 = 0$ against H_a: $\pi_1 - \pi_2 \neq 0$. Use $\alpha = .05$.
b. Test H_0: $\pi_1 - \pi_2 = .2$ against H_a: $\pi_1 - \pi_2 \neq .2$. Use $\alpha = .05$.
c. Form a 90% confidence interval for $\pi_1 - \pi_2$.
d. Form a 99% confidence interval for $\pi_1 - \pi_2$.
e. Test H_0: $\pi_1 - \pi_2 = 0$ against H_a: $\pi_1 - \pi_2 < 0$. Use $\alpha = .01$.

Applying the Concepts

8.57 Two surgical treatments are widely used for a specific type of cancer. To compare their success rates, random samples of patients who had undergone the two types of surgery were obtained, and those who showed no relapse after 1 year were counted. Do the data in the table give enough evidence to indicate that the two success rates differ within the 1-year period? Use $\alpha = .05$.

	Procedure A	Procedure B
n	100	100
Number of successes	78	87

8.58 A new insect spray, type A, is compared with another, type B, that is in wide use. Equal amounts are sprayed in two rooms of equal size, and then 200 insects are released in each room. One hour later, the dead insects are counted, with the results shown in the table.

	Spray A	Spray B
Number of insects	200	200
Number of dead insects	120	80

a. Are the data sufficient to indicate that spray A is a better insecticide than spray B? Test using $\alpha = .05$.
b. Find a 90% confidence interval for the difference between the proportions of insects killed by the two sprays.

8.59 When making purchases, people often receive incorrect change. Is it more likely that shortchanging will be reported, or excess change? To investigate, a psychologist had fifty randomly selected customers receive $1 too much change and fifty receive $1 too little change. Twenty-three of the first group reported the error, while forty-one of those who had been shortchanged complained. Estimate the true difference in the proportions of people who report the two types of error. Use a 99% confidence interval.

8.60 To market a new cigarette, a tobacco company employs two advertising agencies—one operating in the east and one in the west. After 6 months, random samples of smokers are taken from the two regions and asked about their cigarette preference.

The numbers favoring the new brand are shown in the table. Are the data sufficient to indicate a regional difference in the proportions preferring the new brand? Test using $\alpha = .05$.

	Sample Size	Number Preferring New Brand
East	500	12
West	450	15

8.61 Moving van lines are required by law to publish a Carrier Performance Report each year. One statistic called for is the percentage of shipments that incurred a $50 or larger claim for loss or damage. Suppose company A and company B each decide to estimate this figure by sampling their records, and they obtain the data given in the table.

	Company A	Company B
Number of shipments delivered	9,542	6,631
Number of shipments with claim of $50 or larger	1,653	501

a. Use a 95% confidence interval to estimate the true difference in the proportions of shipments made by company A and company B that result in claims of at least $50 for loss or damage.

b. Test the null hypothesis that no difference exists between the true percentages of shipments made by companies A and B that result in claims of at least $50, versus the alternative hypothesis that a difference does exist. Use $\alpha = .05$.

8.62 Suppose a firm switches its table salt container from a cylinder (costly) to a rectangular box (cheap). The results shown in the table are obtained from 1,000 households sampled nationwide before the switch and 1,000 sampled afterward.

	Before	After
Sample size	1,000	1,000
Number of households using firm's brand	475	305

a. Estimate the true difference in the percentage of households using the firm's salt before and after the repackaging. Use a 90% confidence interval.

b. The vice-president in charge of sales claims that the switch to the box has shrunk the company's market share. Does the sample evidence support this claim at the .05 significance level?

8.63 When asked in the fall of 1984 if they favor or oppose tax increases to reduce the federal budget deficit, independent samples of Republicans and Democrats responded as shown in the accompanying table. Find a 99% confidence interval for the true difference in the fractions of Republicans versus Democrats who in 1984 favored a tax increase to reduce the federal budget deficit.

	Favor	Oppose	No Opinion
Republicans	154	373	22
Democrats	243	325	24

Source: "Tax Increases to Reduce Deficit" survey, *The Gallup Report*, No. 229, October 1984, p. 6.

8.64 In autumn of 1984, 784 men and 806 women were surveyed to determine their opinions regarding prayer in public schools. The accompanying table gives the percentages for each sex who favor prayer in public schools, oppose it, or have no opinion. Do the data indicate that there is a difference in the percentages of men and women who favor prayer in public schools? Use $\alpha = .05$.

	Favor	Oppose	No Opinion
Male	66%	31%	3%
Female	71%	26%	3%

Source: "Prayer in Public Schools" survey, *The Gallup Report*, No. 229, October 1984, p. 7.

8.6

SELECTING THE SAMPLE SIZES

To discover what sample sizes are appropriate for estimating the difference between two parameters ($\mu_1 - \mu_2$ or $\pi_1 - \pi_2$) within a specified bound and with a specified degree of reliability, we will use a method similar to that employed in Section 7.7 for estimating one parameter. Namely, we will set the bound, B, equal to half the desired width of the confidence interval we seek. Then we will solve for the sample sizes.

The resulting equation for $\mu_1 - \mu_2$ is

$$B = z \sqrt{\frac{s_1^2}{n_1} + \frac{s_2^2}{n_2}}$$

The equation for $\pi_1 - \pi_2$ is

$$B = z \sqrt{\frac{p_1(1 - p_1)}{n_1} + \frac{p_2(1 - p_2)}{n_2}}$$

To solve either of these equations for the appropriate sample sizes, a specific relationship between n_1 and n_2 must be stated. (Otherwise, we would be trying to solve an equation with two unknowns, n_1 and n_2, so there would be no unique solution.) In this text we will always make the sample sizes equal; that is, $n_1 = n_2$. For this relationship, the algebraic solutions of the two equations given above appear in the boxes. If a different relationship between n_1 and n_2 were desired (say, $n_1 = 2n_2$), other algebraic solutions to the two equations would have to be found.

Sample Sizes for Estimating $\mu_1 - \mu_2$

To estimate $\mu_1 - \mu_2$ to within a bound B when the two samples are equal ($n_1 = n_2$), the appropriate sample sizes are

$$n_1 = n_2 = \frac{z^2(s_1^2 + s_2^2)}{B^2}$$

The value of z is obtained from Appendix Table IV, using a specified confidence.

Sample Sizes for Estimating $\pi_1 - \pi_2$

To estimate $\pi_1 - \pi_2$ to within a bound B when the two samples are equal ($n_1 = n_2$), the appropriate sample sizes are

$$n_1 = n_2 = \frac{z^2[p_1(1 - p_1) + p_2(1 - p_2)]}{B^2}$$

The value of z is obtained from Appendix Table IV, using a specified confidence.

The equations in the boxes contain the sample quantities s_1^2, s_2^2, p_1, and p_2. As for the single-sample problem, if s_1^2 or s_2^2 is not known from a previous sample, we divide the corresponding population range by 4 to estimate the standard deviation. If p_1 or p_2 is not known from previous sampling, we substitute .5 when finding the sample sizes.

EXAMPLE 8.13   New fertilizer compounds are designed to increase yields. A study is underway to determine how μ_1, the mean yield of wheat when a new fertilizer is used, compares to μ_2, the mean yield with a popular fertilizer. The difference in the mean yields per acre will be estimated to within .25 bushel with 99% confidence. The range for past wheat harvests is about 3 bushels per acre. Using equal sample sizes, decide how many 1-acre plots of wheat must be sampled for each fertilizer.

Solution   We want to find equal sample sizes, n_1 and n_2, for estimating $\mu_1 - \mu_2$. The bound is $B = .25$, and since 99% confidence is desired, the z-value is $z = 2.58$ (from Appendix Table IV). Since the range in yields of wheat per acre is 3, we may approximate the standard deviations by

$$s_1 = s_2 \approx \frac{\text{Range}}{4} = \frac{3}{4} = .75$$

Using the sample size formula for $\mu_1 - \mu_2$, we obtain

$$n_1 = n_2 = \frac{z^2(s_1^2 + s_2^2)}{B^2}$$

$$= \frac{(2.58)^2[(.75)^2 + (.75)^2]}{(.25)^2}$$

$$= 119.8152 \approx 120$$

Approximately 120 acres of wheat must be sampled for *each* fertilizer to estimate the mean difference in yields per acre to within .25 bushel with 99% confidence. This would require experimenting with a total of 240 acres—a rather costly proposition. The experimenter might now reconsider such an extensive survey. The sample size could be reduced by allowing a larger bound (say, $B = .50$ or $B = 1.0$) or by reducing the desired confidence to 95% or 90%. Knowing *before* the experiment is begun how much effort will be needed to obtain a certain amount of information is very helpful. If the determined sample sizes prove unsatisfactory, the experiment can be redesigned. ∎

EXAMPLE 8.14   A production supervisor suspects that two machines turn out defective items in different proportions. Previous sampling of their products indicates that π_1 is approximately .03 and π_2 is about .04. To estimate the difference in the proportions of defective items to within .005 with 95% confidence, how many items must the supervisor sample from each machine's output? (Assume that $n_1 = n_2$.)

Solution   We want to find the appropriate sample sizes to estimate $\pi_1 - \pi_2$. The bound is $B = .005$; we will use $p_1 = .03$ and $p_2 = .04$; and $z = 1.96$ (from Appendix Table IV). The sample size formula for $\pi_1 - \pi_2$ yields

$$n_1 = n_2 = \frac{z^2[p_1(1 - p_1) + p_2(1 - p_2)]}{B^2}$$

$$= \frac{(1.96)^2[(.03)(.97) + (.04)(.96)]}{(.005)^2}$$

$$= 10{,}372.32 \approx 10{,}372$$

You can see that this would be a very tedious, and perhaps costly, sampling procedure. If the supervisor insists on estimating $\pi_1 - \pi_2$ to within .005 with 95% confidence, over 10,000 items from each machine will have to be checked. The supervisor might want to consider another sampling scheme. ∎

Exercises (8.65–8.74)

Learning the Language

8.65 What does the symbol B denote?

Learning the Mechanics

8.66 Find the appropriate values of n_1 and n_2 (assume $n_1 = n_2$) needed to estimate $\mu_1 - \mu_2$ to within:

a. A bound on the error of estimation equal to 3 with 95% confidence; from a previous experiment it is known that $s_1 = 12$ and $s_2 = 15$

b. A bound on the error of estimation equal to 5 with 99% confidence; the range of each population is 40

c. A bound on the error of estimation equal to .4 with 90% confidence; assume that $s_1^2 = 3.4$ and $s_2^2 = 5.3$

8.67 Assuming that $n_1 = n_2$, find the appropriate sample sizes needed to estimate $\pi_1 - \pi_2$ for each of the following situations:

a. Bound = .01 with 99% confidence; assume that a previous experiment yields $p_1 = .3$ and $p_2 = .6$

b. Bound = .05 with 90% confidence; assume that approximate values of π_1 and π_2 are unknown

c. Bound = .05 with 90% confidence; assume that a previous experiment yields $p_1 = .1$ and $p_2 = .2$

Applying the Concepts

8.68 As new car prices continue to rise, buyers are tending to take loans for longer than 36 months, with 48 months being a popular alternative. Suppose you plan to survey potential buyers in your sales region to estimate the difference between the proportion of buyers in the over 40 age group who favor 48-month automobile loans and the proportion in the 40 and under age group who favor them. If you intend to select random samples of the same size from each group, how many potential automobile buyers should be included in your samples to estimate the difference in the proportions to within .05 with 95% confidence?

8.69 One reason high-school seniors are encouraged to attend college is that the job opportunities are much better for those with college degrees than for those without. A high-school counselor wants to estimate the difference in mean incomes per day between college graduates and high-school graduates who have not gone on to college. Suppose it is decided to compare the daily incomes of 30-year-olds, and the range of daily incomes for both groups is approximately $200 per day. How many people from each group should be sampled in order to estimate the true difference in mean daily incomes to within $10 per day with 90% confidence?

8.70 Rat damage creates a large financial loss in the production of sugar cane. One aspect of the problem that has been investigated by the United States Department of Agriculture concerns the optimal place to locate rat poison. To be more effective in reducing rat damage, should the poison be located in the middle of the field or on the outer perimeter? One way to answer this question is to determine where the most damage occurs. If damage is measured by the proportion of cane stalks that have been damaged by rats, how many stalks from each section of the field should be sampled in order to estimate the true difference between proportions of stalks damaged in the two sections to within .02 with 95% confidence?

8.71 Consider investing $1,000 in a stock, and suppose you are interested in the dividend rate on your $1,000 investment at the end of 5 years. Consider two types of stocks in particular: electrical utilities and oil companies. To conduct your study, you plan to randomly select n oil stocks and n electrical utility stocks. For each stock, you will check the records, calculate the number of shares of stock you could have purchased 5 years ago for $1,000, and then calculate the dividend rate (in percent) that the stock would be paying today on your $1,000 investment. Suppose you think the dividend rates will vary over a range of roughly 25%. How large should n be if you want to estimate the difference in the mean rates of dividend return correct to within 3% with 99% confidence?

8.72 A television manufacturer wants to compare the proportion of its best sets that need repair within 1 year to those of a competitor. If it is desired to estimate the difference in proportions to within .05 with 90% confidence, and if the manufacturer plans to sample equal numbers of sets for each brand, how many buyers of each brand must be sampled? Assume the proportion of sets that need repair will be about .2 for both brands.

8.73 Refer to Exercise 8.35. Samples of 21 boys and 21 girls were weighed. The data (in kilograms) are summarized in the table. How large would equal-sized samples need to be to estimate the difference in mean weights to within 1 kilogram with 95% confidence?

	Boys	**Girls**
Mean weight	17.2	16.4
Standard deviation	2.5	1.9

Source: S. J. Erbaugh, "The Relationship of Stability Performance and the Physical Growth Characteristics of Preschool Children," *Research Quarterly for Exercise and Sport*, Vol. 55, No. 1, March 1984, p. 12.

8.74 Refer to Exercise 8.63. The data at the top of page 458 reflect the opinions of Republicans and Democrats surveyed about raising taxes to reduce the federal budget deficit. How many Republicans and Democrats would need to be sampled to estimate the difference in the proportions favoring a tax increase to within .05 with 99% confidence? Assume that the true proportions are close to the 1984 sample proportions, and that the two sample sizes are to be equal.

	Favor	Oppose	No Opinion
Republicans	154	373	22
Democrats	243	325	24

Source: "Tax Increases to Reduce Deficit," survey, *The Gallup Report,* No. 229, October 1984, p. 6.

8.7

COMPARING TWO POPULATION VARIANCES: INDEPENDENT SAMPLES (OPTIONAL)

It is often of practical interest to use the techniques developed so far in this chapter to compare the means or proportions of two populations. But there are also times when it is desirable to compare two population variances. For example, when two devices are available for producing precision measurements (scales, calipers, thermometers, etc.), we might want to compare the variability of the measurements of the devices before deciding which one to buy. Or, when two standardized tests can be used to rate applicants, the variability of the scores for both tests should be taken into consideration before deciding which test to use.

For problems like these we need to develop a statistical procedure to compare population variances. The common statistical procedure for comparing two population variances, σ_1^2 and σ_2^2, makes an inference about the ratio σ_1^2/σ_2^2. In this section, we will test hypotheses about σ_1^2/σ_2^2 and form confidence intervals for σ_1^2/σ_2^2.

To make an inference about the ratio σ_1^2/σ_2^2, it seems reasonable to collect two samples and use the corresponding ratio of the sample variances, s_1^2/s_2^2. As a matter of fact, inferences can be based upon the sampling distribution of s_1^2/s_2^2, and we will call this the **F distribution.** The procedures we present in this chapter are valid when the conditions listed in the box are met.

When to Use the F Distribution to Make Inferences About $\dfrac{\sigma_1^2}{\sigma_2^2}$, the Ratio of Two Population Variances

To make valid inferences about σ_1^2/σ_2^2, the following conditions must be met:

1. Two random samples must be taken—one from each of the two populations of interest.
2. The two samples must be independent.
3. Both of the populations must be (approximately) normally distributed.

The distribution of $F = s_1^2/s_2^2$ depends on the degrees of freedom associated with s_1^2 and the degrees of freedom associated with s_2^2, which are $(n_1 - 1)$ and $(n_2 - 1)$, respectively. We say that there are **$(n_1 - 1)$ numerator degrees of freedom and $(n_2 - 1)$ denominator degrees of freedom,** and when discussing F distributions, we will always give the numerator degrees of freedom first and the denominator degrees of freedom second. An F distribution with 7 and 9 degrees of freedom is shown in Figure 8.7. As you can see, the distribution is skewed to the right, because s_1^2/s_2^2 cannot be less than 0 but can increase without bound.

Figure 8.7
An F distribution with 7 numerator degrees of freedom and 9 denominator degrees of freedom

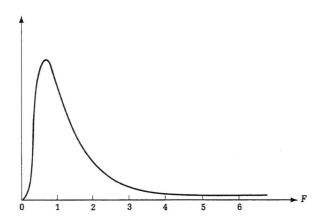

We need to be able to find F-values corresponding to the tail areas of the F distribution in order to make inferences about σ_1^2/σ_2^2. The upper-tail* F-values corresponding to areas of .10, .05, .025, and .01 are given in Appendix Tables IX, X, XI, and XII, respectively. Table X is partially reproduced in Table 8.12 (page 460). This table gives F-values that correspond to an area of .05 in the upper tail. The columns give the degrees of freedom $(n_1 - 1)$ for the numerator sample variance s_1^2, and the rows give the degrees of freedom $(n_2 - 1)$ for the denominator sample variance s_2^2. Thus, if the numerator degrees of freedom are 7 and the denominator degrees of freedom are 9, we look in the seventh column and ninth row to find the value $F = 3.29$. As shown in Figure 8.8, the tail area to the right of 3.29 for the F distribution with 7 and 9 df is .05.

To find a lower-tail area for an F distribution, we will make use of the fact that the **reciprocal of an upper-tail F-value for a specific area gives the lower-tail F-value (for the same area) with the numerator and denominator degrees of freedom reversed.** Thus, if we want a lower-tail area of .05 for an F distribution with 7 and 9 df, we would first reverse the degrees of freedom and use 9 in the numerator and 7 in the denominator. Looking in the table for 9 and 7 df, we find the value $F = 3.68$. Taking the reciprocal of

* For the F distribution, "upper-tail" refers to the right end of the curve, "lower-tail" to the left end.

Table 8.12

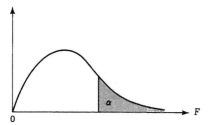

A partial reproduction of Appendix Table X, Percentage Points of the F Distribution, $\alpha = .05$

				Numerator Degrees of Freedom					
		2	3	4	5	6	7	8	9
	1	161.4	199.5	215.7	224.6	230.2	234.0	236.8	238.9	240.5
	2	18.51	19.00	19.16	19.25	19.30	19.33	19.35	19.37	19.38
	3	10.13	9.55	9.28	9.12	9.01	8.94	8.89	8.85	8.81
	4	7.71	6.94	6.59	6.39	6.26	6.16	6.09	6.04	6.00
	5	6.61	5.79	5.41	5.19	5.05	4.95	4.88	4.82	4.77
	6	5.99	5.14	4.76	4.53	4.39	4.28	4.21	4.15	4.10
	7	5.59	4.74	4.35	4.12	3.97	3.87	3.79	3.73	3.68
	8	5.32	4.46	4.07	3.84	3.69	3.58	3.50	3.44	3.39
	9	5.12	4.26	3.86	3.63	3.48	3.37	3.29	3.23	3.18
	10	4.96	4.10	3.71	3.48	3.33	3.22	3.14	3.07	3.02
	11	4.84	3.98	3.59	3.36	3.20	3.09	3.01	2.95	2.90
	12	4.75	3.89	3.49	3.26	3.11	3.00	2.91	2.85	2.80
	13	4.67	3.81	3.41	3.18	3.03	2.92	2.83	2.77	2.71
	14	4.60	3.74	3.34	3.11	2.96	2.85	2.76	2.70	2.65

(Denominator Degrees of Freedom along the rows 1–14)

Figure 8.8
An F distribution with
7 and 9 df

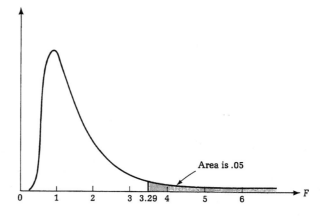

Area is .05

this number, we find the lower-tail *F*-value of interest:

$$F = \frac{1}{3.68} = .27$$

As shown in Figure 8.9, the tail area to the left of .27 for the *F* distribution with 7 and 9 degrees of freedom is .05.

Figure 8.9
An *F* distribution with 7 and 9 df

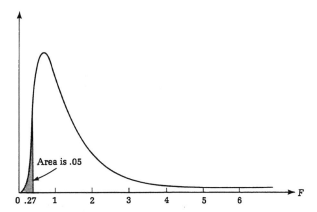

We will further demonstrate the use of the *F* tables in the examples. To find a lower-tail area, however, remember the following:

Helpful Hint

To find a lower-tail area for an *F* distribution with $(n_1 - 1)$ numerator degrees of freedom and $(n_2 - 1)$ denominator degrees of freedom, follow these steps:

1. Using the *F* table with an upper-tail area equal to the area of interest, look up the upper-tail *F*-value, but **reverse the degrees of freedom;** i.e., use $(n_2 - 1)$ numerator degrees of freedom, and $(n_1 - 1)$ denominator degrees of freedom.
2. The lower-tail *F*-value of interest is the reciprocal of the upper-tail *F*-value found in step 1.

The test of hypothesis about σ_1^2/σ_2^2 and the confidence interval for σ_1^2/σ_2^2 are given in the boxes on page 462.

Test of Hypothesis About $\dfrac{\sigma_1^2}{\sigma_2^2}$

H_0: $\dfrac{\sigma_1^2}{\sigma_2^2} = 1$

H_a: $\dfrac{\sigma_1^2}{\sigma_2^2} > 1$ or H_a: $\dfrac{\sigma_1^2}{\sigma_2^2} < 1$ or H_a: $\dfrac{\sigma_1^2}{\sigma_2^2} \neq 1$

Test statistic: $F = \dfrac{s_1^2}{s_2^2}$

Rejection region: See Appendix Tables IX, X, XI, and XII. For an upper-tail F-value, use $(n_1 - 1)$ numerator df and $(n_2 - 1)$ denominator df. For a lower-tail value, use the procedure described in the preceding box.

Confidence Interval for $\dfrac{\sigma_1^2}{\sigma_2^2}$

The $p \cdot 100\%$ confidence interval for σ_1^2/σ_2^2 is expressed as

$$\left(\frac{s_1^2}{s_2^2}\right)\left(\frac{1}{F_1}\right) \quad \text{to} \quad \left(\frac{s_1^2}{s_2^2}\right)(F_2)$$

where F_1 and F_2 are obtained from the F table for an upper-tail area of $(1 - p)/2$. The degrees of freedom for F_1 are $(n_1 - 1)$ and $(n_2 - 1)$, while the degrees of freedom for F_2 are reversed to $(n_2 - 1)$ and $(n_1 - 1)$.

Notice that the only null hypothesis we will test is that $\sigma_1^2/\sigma_2^2 = 1$, or equivalently, $\sigma_1^2 = \sigma_2^2$. In order for the test statistic to have an F distribution, the null hypothesis must be true and the two population variances must be equal.

If a 90% confidence interval is desired, then $p = .90$, and the appropriate F table is one for an upper-tail area of $(1 - p)/2 = (1 - .9)/2 = .10/2 = .05$. Similarly, if a 95% confidence interval is desired, the upper-tail area of interest is .025.

EXAMPLE 8.15   A scientist wants to compare the metabolic rates of white mice subjected to different drugs. The weights of the mice may affect their metabolic rates, and so the experimenter wants to obtain mice that are relatively homogeneous with regard to weight. Two different suppliers have mice available, and the experimenter would like to know if there is a difference in the variability of the weights of mice from the two suppliers. The scientist weighs a random sample of eighteen mice from supplier 1 and an independent random sample of thirteen mice from supplier 2. The resulting data are summarized in Table 8.13.

Table 8.13

Supplier 1	Supplier 2
$n_1 = 18$	$n_2 = 13$
$\bar{x}_1 = 4.21$ ounces	$\bar{x}_2 = 4.18$ ounces
$s_1^2 = .019$	$s_2^2 = .049$

Do the data provide sufficient evidence to indicate a difference in the variability of the weights of mice obtained from the two suppliers? Test at the .10 level of significance, and assume that the populations of weights are approximately normal.

Solution The populations of interest are the weights of all white mice available from the two suppliers. We want to compare σ_1^2, the population variance of weights of mice from supplier 1, to σ_2^2, the population variance of weights of mice from supplier 2.

We want to see if the data indicate that the population variances differ, that is, $\sigma_1^2/\sigma_2^2 \neq 1$. Using this as the alternative hypothesis, the elements of this test are

$$H_0: \quad \frac{\sigma_1^2}{\sigma_2^2} = 1 \qquad H_a: \quad \frac{\sigma_1^2}{\sigma_2^2} \neq 1$$

$$\text{Test statistic:} \quad F = \frac{s_1^2}{s_2^2}$$

Since the populations of interest are approximately normal, and independent random samples were selected from these two populations, we may proceed with a test of hypothesis about σ_1^2/σ_2^2.

Rejection region: Reject H_0 if $F < .42$ or $F > 2.62$.

The rejection region was obtained from Appendix Table X. Since a two-tailed test is being conducted, we will reject H_0 if F is either too large or too small. We want $\alpha = .10$, so the total area for the rejection region is .10, with an area of $\frac{1}{2}(.10) = .05$ being placed in both the upper tail and the lower tail, as shown in Figure 8.10. We have

Figure 8.10
Rejection region for
Example 8.15

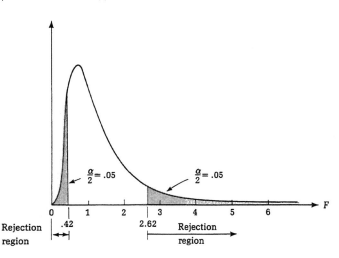

$n_1 - 1 = 17$ numerator df and $n_2 - 1 = 12$ denominator df. In Table X, the closest we can find to 17 numerator df is 15, so we use this as an approximation and obtain the upper-tail F-value of 2.62. The lower-tail F-value is found by using $n_2 - 1 = 12$ numerator df and $n_1 - 1 = 17$ denominator df. Finding the corresponding F-value in Table X, we get $F = 2.38$. The lower-tail value is the reciprocal of this number; that is, $1/2.38 = .42$. The rejection region is shown in Figure 8.10.

Calculating the test statistic, we obtain

$$F = \frac{s_1^2}{s_2^2} = \frac{.019}{.049} = .39$$

Since $F = .39$ is less than .42, we will reject H_0. With 90% confidence, there is sufficient evidence to indicate a difference in the variability of weights of mice obtained from the two suppliers. Since the calculated value of the test statistic was in the lower tail of the rejection region, it appears that the weights of mice obtained from supplier 1 tend to be more homogeneous (less variable) than those obtained from supplier 2. On the basis of this evidence, we would advise the experimenter to purchase mice from supplier 1.   ∎

Helpful Hints

When using the F tables in the Appendix to find a rejection region, take the following steps:

1. Determine the value of α, the numerator df $(n_1 - 1)$, the denominator df $(n_2 - 1)$, and the alternative hypothesis.
2. Choose the procedure appropriate to the alternative hypothesis.
   a. If H_a: $\sigma_1^2/\sigma_2^2 > 1$, use the F table for an upper-tail area equal to α. Find the table value using $(n_1 - 1)$ and $(n_2 - 1)$ degrees of freedom, and give the rejection region as

   $$\text{Reject } H_0 \text{ if } F > \text{Table value}$$

   b. If H_a: $\sigma_1^2/\sigma_2^2 < 1$, use the F table for an upper-tail area equal to α. Find the table value using $(n_2 - 1)$ and $(n_1 - 1)$ degrees of freedom (remember to reverse the df), and give the rejection region as

   $$\text{Reject } H_0 \text{ if } F < \frac{1}{\text{Table value}}$$

   c. If H_a: $\sigma_1^2/\sigma_2^2 \neq 1$, use the F table for an upper-tail area equal to $\alpha/2$ (remember to divide α in half). Find a first table value F_1 using $(n_1 - 1)$ and $(n_2 - 1)$ df and a second table value F_2 using $(n_2 - 1)$ and $(n_1 - 1)$ df. The rejection region is given as

   $$\text{Reject } H_0 \text{ if } F > F_1 \quad \text{or} \quad F < \frac{1}{F_2}$$

Since the F tables in the Appendix provide only F-values and do not explicitly state rejection regions, we provide the helpful hints given in the preceding box for finding rejection regions for tests of hypotheses about σ_1^2/σ_2^2.

EXAMPLE 8.16   An investor believes that although the price of stock 1 usually exceeds that of stock 2, stock 1 represents a riskier investment, where the risk of a given stock is measured by the variation in daily price changes. Suppose we obtain a random sample of twenty-five daily price changes for stock 1 and twenty-five for stock 2. The sample results are summarized in Table 8.14. Compare the risks associated with the two stocks by testing the null hypothesis that the variances of the price changes for the stocks are equal against the alternative that the price variance of stock 1 exceeds that of stock 2. Use $\alpha = .05$. Assume the populations of stock price changes are approximately normal.

Table 8.14

Stock 1	Stock 2
$n_1 = 25$	$n_2 = 25$
$\bar{x}_1 = .250$	$\bar{x}_2 = .125$
$s_1 = .76$	$s_2 = .46$

Solution   The populations of interest are the daily price changes for the two stocks. We will let σ_1^2 equal the population variance in prices for stock 1 and σ_2^2 equal the population variance for stock 2.

We want to show that σ_1^2 is larger than σ_2^2. The elements of the test are

$$H_0: \quad \frac{\sigma_1^2}{\sigma_2^2} = 1 \qquad H_a: \quad \frac{\sigma_1^2}{\sigma_2^2} > 1$$

$$\text{Test statistic:} \quad F = \frac{s_1^2}{s_2^2}$$

The populations are approximately normally distributed, and independent random samples have been obtained from the two populations. We will proceed with the test of hypothesis.

Rejection region: Reject H_0 if $F > 1.98$.

The rejection region was obtained from Appendix Table X, since $\alpha = .05$ and the test is one-tailed. Since the test is upper-tailed, the table value of 1.98 is found directly using $n_1 - 1 = 24$ and $n_2 - 1 = 24$ df. We calculate

$$F = \frac{s_1^2}{s_2^2} = \frac{(.76)^2}{(.46)^2} = 2.73$$

Since F exceeds the table value of 1.98, we conclude that the variance of the daily price changes for stock 1 exceeds that for stock 2. It appears that stock 1 is a riskier investment than stock 2. How much reliability can we place in this inference? Only one time in twenty (since $\alpha = .05$), on the average, would this statistical test lead us to conclude erroneously that σ_1^2 exceeds σ_2^2 if, in fact, they are equal. ∎

EXAMPLE 8.17 Refer to Example 8.15 and find a 90% confidence interval for the ratio of the variances of the weights of mice obtained from the two suppliers. For convenience, the data are repeated in Table 8.15.

Table 8.15

Supplier 1	Supplier 2
$n_1 = 18$	$n_2 = 13$
$\bar{x}_1 = 4.21$ ounces	$\bar{x}_2 = 4.18$ ounces
$s_1^2 = .019$	$s_2^2 = .049$

Solution The conditions that were satisfied to make the test of hypothesis valid also make a confidence interval valid.

Since 90% confidence is desired, $p = .90$, and $(1 - p)/2 = .05$. Thus, we will use Table X to find the two F-values—one with $n_1 - 1 = 17$ df and $n_2 - 1 = 12$ df, and the other with $n_2 - 1 = 12$ df and $n_1 - 1 = 17$ df. These values are $F_1 = 2.62$ (approximately) and $F_2 = 2.38$.

Using the formula for the confidence interval for σ_1^2/σ_2^2, we obtain

$$\left(\frac{s_1^2}{s_2^2}\right)\left(\frac{1}{F_1}\right) \quad \text{to} \quad \left(\frac{s_1^2}{s_2^2}\right)(F_2)$$

$$\left(\frac{.019}{.049}\right)\left(\frac{1}{2.62}\right) \quad \text{to} \quad \left(\frac{.019}{.049}\right)(2.38)$$

$$.15 \quad \text{to} \quad .92$$

According to this confidence interval, we estimate that σ_1^2, the variance in the weights of mice obtained from supplier 1, could be as small as .15 or as large as .92 times the size of σ_2^2, the variance in the weights of mice obtained from supplier 2. Since all the values in the confidence interval are less than 1, we may infer that $\sigma_1^2 < \sigma_2^2$. This is the same interpretation we made in Example 8.15. Again, we have 90% confidence in our inference. ∎

As a final application, consider the comparison of population variances as a check of the assumption that $\sigma_1^2 = \sigma_2^2$, which is needed for the two-sample t test. Rejection of the null hypothesis $\sigma_1^2 = \sigma_2^2$ would indicate that the assumption is invalid. [*Note:* Nonrejection of the null hypothesis *does not* imply that the assumption *is* valid.] We will illustrate with an example.

EXAMPLE 8.18 In Example 8.8 (Section 8.3) we used the two-sample t statistic to compare the mean test scores of two groups of slow learners who had been taught to read using two different methods. The data are repeated in Table 8.16 for convenience. The use of the t statistic was based on the assumption that the population variances of the test scores were equal for the two methods. Check this assumption using $\alpha = .10$.

Table 8.16

New Method	Standard Method
$n_1 = 8$	$n_2 = 12$
$\bar{x}_1 = 76.9$	$\bar{x}_2 = 72.7$
$s_1 = 4.85$	$s_2 = 6.35$

Solution   The conditions for performing a valid F test to compare σ_1^2 to σ_2^2 are met when performing a t test to compare μ_1 to μ_2. Thus, we will proceed with the test of interest. We want to determine whether the population variances differ, so the elements of the test are

$$H_0: \quad \frac{\sigma_1^2}{\sigma_2^2} = 1 \qquad H_a: \quad \frac{\sigma_1^2}{\sigma_2^2} \neq 1$$

Test statistic: $F = \dfrac{s_1^2}{s_2^2}$

Rejection region: Reject H_0 if $F > 3.01$ or $F < .27$.

The rejection region was found using Appendix Table X, since $\alpha/2 = .10/2 = .05$. The table value of 3.01 corresponds to $n_1 - 1 = 7$ df and $n_2 - 1 = 11$ df. The value .27 is the reciprocal of the approximate table value for $n_2 - 1 = 11$ df and $n_1 - 1 = 7$ df. Actually, we used 10 df and 7 df (since the table does not have a listing for 11 df in the numerator), which yielded a table value of 3.64, and $1/3.64 = .27$.

Calculating the test statistic, we obtain

$$F = \frac{(4.85)^2}{(6.35)^2} = .58$$

Since F is greater than .27 but less than 3.01, we do not reject the null hypothesis that the population variances of the reading test scores are equal. It is here that the temptation to misuse the F test is strongest. **We cannot conclude that the data justify the use of the t statistic.** This is equivalent to accepting H_0, and we have repeatedly warned against this conclusion, because the probability of a Type II error, β, is unknown. The α level of .10 protects us only against rejecting H_0 if it is true. ∎

In summary, the F test may prevent us from abusing the t procedure when we obtain a value of F that leads to a rejection of the assumption that $\sigma_1^2 = \sigma_2^2$. But, when the F statistic does not fall in the rejection region, we know little more about the validity of the assumption than before we conducted the test.

Exercises (8.75–8.83)

Learning the Language

8.75   Define the term F distribution.

Using the Tables

8.76   Independent random samples are selected from two normally distributed populations with variances σ_1^2 and σ_2^2. The sample sizes are $n_1 = 10$ and $n_2 = 11$.

a. Give the rejection region for testing $H_0: \sigma_1^2/\sigma_2^2 = 1$ against $H_a: \sigma_1^2/\sigma_2^2 > 1$. Use $\alpha = .05$.

b. Give the rejection region for testing $H_0: \sigma_1^2/\sigma_2^2 = 1$ against $H_a: \sigma_1^2/\sigma_2^2 \neq 1$, with $\alpha = .10$.

c. Give the rejection region for testing $H_0: \sigma_1^2/\sigma_2^2 = 1$ against $H_a: \sigma_1^2/\sigma_2^2 < 1$, with $\alpha = .01$.

d. Give the two F-values, F_1 and F_2, used to form a 90% confidence interval for σ_1^2/σ_2^2.

e. Repeat parts a–d with $n_1 = 16$ and $n_2 = 6$.

Learning the Mechanics

8.77 Independent random samples were selected from each of two normally distributed populations, $n_1 = 15$ from population 1 and $n_2 = 27$ from population 2. The means and variances for the two samples are given in the table.

Sample 1	Sample 2
$n_1 = 15$	$n_2 = 27$
$\bar{x}_1 = 31.7$	$\bar{x}_2 = 37.4$
$s_1^2 = 3.94$	$s_2^2 = 9.73$

a. Test $H_0: \sigma_1^2/\sigma_2^2 = 1$ against $H_a: \sigma_1^2/\sigma_2^2 \neq 1$. Use $\alpha = .05$.

b. Form a 95% confidence interval for σ_1^2/σ_2^2.

c. Test $H_0: \sigma_1^2/\sigma_2^2 = 1$ against $H_a: \sigma_1^2/\sigma_2^2 < 1$. Use $\alpha = .05$.

8.78 Independent random samples were selected from each of two normally distributed populations, $n_1 = 6$ from population 1 and $n_2 = 4$ from population 2. The data are shown in the table.

Sample 1	Sample 2
3.1	2.3
4.3	2.4
1.2	3.7
1.7	2.9
.6	
3.4	

a. Calculate s_1^2 and s_2^2.

b. Form a 90% confidence interval for σ_1^2/σ_2^2.

c. Form a 95% confidence interval for σ_1^2/σ_2^2.

d. Test $H_0: \sigma_1^2/\sigma_2^2 = 1$ against $H_a: \sigma_1^2/\sigma_2^2 > 1$. Use $\alpha = .10$.

e. Test $H_0: \sigma_1^2/\sigma_2^2 = 1$ against $H_a: \sigma_1^2/\sigma_2^2 \neq 1$. Use $\alpha = .10$.

Applying the Concepts

In the following exercises, assume that it has been determined that the two populations of interest are normally distributed. State any additional conditions required for the validity of the statistical procedures used.

8.79 A series of experiments has been conducted to compare the quantity of hemoglobin in the blood of men and women who are between the ages of 20 and 30 years. One phase of the study deals with a comparison of the variability in the hemoglobin measurements between the two groups. In random samples of twenty-five women and twenty men, all between the ages of 20 and 30 years, a researcher has recorded the amount of hemoglobin in the blood of each. (The quantity of hemoglobin is measured as a percentage of the total amount of blood.) The results are shown in the table. Form a 90% confidence interval for the ratio of the variance of the amount of hemoglobin in women's blood to the corresponding variance for men.

Women	Men
$n_1 = 25$	$n_2 = 20$
$\bar{x}_1 = 42.7$	$\bar{x}_2 = 41.8$
$s_1^2 = 18.3$	$s_2^2 = 8.5$

8.80 Refer to Exercise 8.33 (Section 8.3). A manufacturing company is interested in determining whether a significant difference exists between the mean number of units produced per day by two different machine operators. Two independent random samples yielded the data in the table.

Operator 1	Operator 2
$n_1 = 10$	$n_2 = 10$
$\bar{x}_1 = 35$ units	$\bar{x}_2 = 31$ units
$s_1^2 = 17.2$	$s_2^2 = 19.1$

a. In order to conduct the hypothesis test required in Exercise 8.33, you had to assume $\sigma_1^2 = \sigma_2^2$. Use an F test with $\alpha = .10$ to test H_0: $\sigma_1^2/\sigma_2^2 = 1$ against H_a: $\sigma_1^2/\sigma_2^2 \neq 1$.

b. Form a 90% confidence interval for σ_1^2/σ_2^2. Explain the significance of this result.

8.81 The goalie is generally regarded as the most important player on a hockey team. One measure of a goalie's ability is called the *goals against (GA) average*—that is, the average number of goals the goalie gives up per game. However, most National Hockey League coaches agree that consistency in performance is just as important as the GA average. A consistent goalie is one whose number of goals given up per game varies only slightly from game to game. Two goalies with similar GA averages are competing for the starting position on a hockey team. The coach will choose the starter on the basis of the better GA average only if there is no evidence of a difference in the consistency of the two goalies based on their performances in ten exhibition games (ten games per goalie). Otherwise, the more consistent goalie will win the starting position. The results of the exhibition games are given in the table. What decision does the coach make? (Test at the $\alpha = .05$ level of significance.)

	Goalie A	Goalie B
Number of games	10	10
$\bar{x}$ (GA average)	3.3	3.1
s^2	.68	2.77

8.82 Suppose a firm has been experimenting with two different arrangements of its assembly line. It has been determined that both arrangements yield approximately the same average number of finished units per day. To obtain an arrangement that produces greater process control, you suggest that the arrangement with the smaller variance in the number of finished units produced per day be permanently adopted. Two independent random samples yield the results shown in the table. Do the samples provide sufficient evidence at the .10 significance level to conclude that the variances of the two arrangements differ? If so, which arrangement would you choose? If not, what would you suggest the firm do?

Assembly Line 1	Assembly Line 2
$n_1 = 21$ days	$n_2 = 21$ days
$s_1^2 = 1,432$	$s_2^2 = 3,761$

8.83 The quality control department of a paper company measures the brightness (a measure of reflectance) of finished paper on a periodic basis throughout the day. Two instruments that are available to measure the paper specimens are subject to error, but they can be adjusted so that the mean readings for a control paper specimen are the same for both instruments. Suppose you are concerned about the precision of the two instruments and want to compare the variability in the readings of instrument 1 to those of instrument 2. Five brightness measurements were made on a single paper specimen using each of the two instruments. The data are shown in the table. Form a 95% confidence interval for the ratio of the variance of the measurements obtained by instrument 1 to the variance of the measurements obtained by instrument 2.

Instrument 1	Instrument 2
29	26
28	34
30	30
28	32
30	28

Chapter Summary

Inferences comparing two populations can be made based on samples from each. Methods similar to those described in Chapter 7 are applied to the differences between the means or proportions, or to the ratio of the variances. The type of method used depends on the kind of samples selected. Samples are **independent** if the selection of objects from one population is unrelated to the selection from the other; samples are **dependent** if objects are paired so that similar ones are selected from each population. A confidence interval applied to the difference between two parameters yields an inference about which is larger: If the interval contains only positive numbers, the first is larger; if it contains only negative numbers, the second is larger; if the interval contains 0, no conclusion can be drawn.

Dependent sampling on two populations where the difference of paired measurements is obtained is called a **paired difference experiment.** The mean, μ_D, of the paired differences can be treated by the methods of Chapter 7; see Situations I and II in the Chapter 7 Summary.

For independent sampling, four situations were considered in this chapter. They are analogous to the four situations considered in Chapter 7.

I. *Situation:* Any populations; large ($n_1 \geq 30$ and $n_2 \geq 30$), independent random samples from each

   Parameter: $\mu_1 - \mu_2$, the difference between the population means

   Test of hypothesis:

      H_0: $\mu_1 - \mu_2 = D_0$

      H_a: $\mu_1 - \mu_2 > D_0$,   or $\mu_1 - \mu_2 < D_0$,   or $\mu_1 - \mu_2 \neq D_0$

      Test statistic: $z = \dfrac{(\bar{x}_1 - \bar{x}_2) - D_0}{\sqrt{\dfrac{s_1^2}{n_1} + \dfrac{s_2^2}{n_2}}}$

   where s_1 and s_2 are the sample standard deviations used to approximate the population standard deviations

   Rejection region: Use Appendix Table III.

  Confidence interval:

   Obtain z from Appendix Table IV.

   Calculate $\bar{x}_1$, $\bar{x}_2$, s_1, s_2, and $B = z \sqrt{\dfrac{s_1^2}{n_1} + \dfrac{s_2^2}{n_2}}$

   Interval: $(\bar{x}_1 - \bar{x}_2) - B$ to $(\bar{x}_1 - \bar{x}_2) + B$

II. *Situation:* Populations with approximately normal distributions having the same variance, small ($n < 30$) random samples from each

   Parameter: $\mu_1 - \mu_2$

Test of hypothesis:

H_0: $\mu_1 - \mu_2 = D_0$

H_a: $\mu_1 - \mu_2 > D_0$, or $\mu_1 - \mu_2 < D_0$, or $\mu_1 - \mu_2 \neq D_0$

Test statistic: $t = \dfrac{(\bar{x}_1 - \bar{x}_2) - D_0}{\sqrt{\dfrac{s_p^2}{n_1} + \dfrac{s_p^2}{n_2}}}$

where s_p^2 is the **pooled sample estimate** of σ^2 and is equal to

$$\frac{(n_1 - 1)s_1^2 + (n_2 - 1)s_2^2}{n_1 + n_2 - 2}$$

Rejection region: Use Appendix Table V, with $(n_1 + n_2 - 2)$ df.

Confidence interval:

Obtain t from Appendix Table VI.

Calculate $\bar{x}_1$, $\bar{x}_2$, s_p, and $B = t\sqrt{\dfrac{s_p^2}{n_1} + \dfrac{s_p^2}{n_2}}$

Interval: $(\bar{x}_1 - \bar{x}_2) - B$ to $(\bar{x}_1 - \bar{x}_2) + B$

III. *Situation:* Binomial experiments, large ($n_1 \geq 30$ and $n_2 \geq 30$) random samples from each

Parameter: $\pi_1 - \pi_2$, the difference between the population proportions

Test of hypothesis:

H_0: $\pi_1 - \pi_2 = D_0$

H_a: $\pi_1 - \pi_2 > D_0$, or $\pi_1 - \pi_2 < D_0$, or $\pi_1 - \pi_2 \neq D_0$

Test statistic:

$$z = \frac{p_1 - p_2 - D_0}{\sqrt{\dfrac{p_1(1 - p_1)}{n_1} + \dfrac{p_2(1 - p_2)}{n_2}}} \qquad \text{for } D_0 \neq 0$$

$$z = \frac{p_1 - p_2}{\sqrt{\dfrac{p(1 - p)}{n_1} + \dfrac{p(1 - p)}{n_2}}} \qquad \text{for } D_0 = 0$$

where

$$p = \frac{n_1 p_1 + n_2 p_2}{n_1 + n_2}$$

is a **pooled estimate** for the common proportion

Rejection region: Use Appendix Table III.

Confidence interval:

Obtain z from Appendix Table IV.

Calculate p_1, p_2, and $B = z\sqrt{\dfrac{p_1(1 - p_1)}{n_1} + \dfrac{p_2(1 - p_2)}{n_2}}$

Interval: $(p_1 - p_2) - B$ to $(p_1 - p_2) + B$

IV. (*Optional*) *Situation:* Independent random samples from two populations with approximately normal distributions

Parameter: $\dfrac{\sigma_1^2}{\sigma_2^2}$, the ratio of the population variances

Test of hypothesis:

H_0: $\dfrac{\sigma_1^2}{\sigma_2^2} = 1$

H_a: $\dfrac{\sigma_1^2}{\sigma_2^2} > 1$, or $\dfrac{\sigma_1^2}{\sigma_2^2} < 1$, or $\dfrac{\sigma_1^2}{\sigma_2^2} \neq 1$

Test statistic: $F = \dfrac{s_1^2}{s_2^2}$

Rejection region: Use Appendix Tables IX, X, XI, and XII for upper-tail values with $(n_1 - 1)$ numerator df and $(n_2 - 1)$ denominator df. For lower-tail values, reverse the degrees of freedom and take the reciprocal of the table value. The use of these tables is summarized on page 464.

Confidence interval:

Obtain F_1 and F_2 from Appendix Tables IX, X, XI, and XII using $(n_1 - 1)$ and $(n_2 - 1)$ df for F_1 and $(n_2 - 1)$ and $(n_1 - 1)$ df for F_2.

Interval: $\left(\dfrac{s_1^2}{s_2^2}\right)\left(\dfrac{1}{F_1}\right)$ to $\left(\dfrac{s_1^2}{s_2^2}\right)(F_2)$

In the first and third situations listed above, it may be necessary to choose sample sizes large enough to make the confidence interval as narrow as we want. We will make the sample sizes equal, that is, $n_1 = n_2 = n$. For a given bound B, and z obtained from Appendix Table IV, the values for n are

I. $n = \dfrac{z^2(s_1^2 + s_2^2)}{B^2}$ **III.** $n = \dfrac{z^2[p_1(1 - p_1) + p_2(1 - p_2)]}{B^2}$

SUPPLEMENTARY EXERCISES (8.84–8.125)

For some statistical procedures to be valid, it is necessary to determine that the populations of interest are normally distributed. Assume that this has been done when required. State any other conditions required for the validity of the statistical procedures used.

Using the Tables

8.84 Two large, independent random samples are selected from populations with means μ_1 and μ_2.

a. Give the appropriate z-value for forming a 99% confidence interval for $\mu_1 - \mu_2$.

b. Give the rejection region for testing H_0: $\mu_1 - \mu_2 = 0$ against H_a: $\mu_1 - \mu_2 > 0$, using $\alpha = .10$.

8.85 Independent random samples are selected from two populations with means μ_1 and μ_2. The sample sizes are $n_1 = 8$ and $n_2 = 9$.

a. Give the rejection region for testing H_0: $\mu_1 - \mu_2 = 0$ against H_a: $\mu_1 - \mu_2 \neq 0$ at the .05 level of significance.

b. Give the appropriate t-value for forming a 90% confidence interval for $\mu_1 - \mu_2$.

c. (*Optional*) Give the appropriate F-values, F_1 and F_2, used to form a 95% confidence interval for σ_1^2/σ_2^2.

d. (*Optional*) Give the rejection region for testing H_0: $\sigma_1^2/\sigma_2^2 = 1$ against H_a: $\sigma_1^2/\sigma_2^2 < 1$ with $\alpha = .01$.

e. (*Optional*) Give the rejection region for testing H_0: $\sigma_1^2/\sigma_2^2 = 1$ against H_a: $\sigma_1^2/\sigma_2^2 \neq 1$ with $\alpha = .10$.

8.86 A random sample of seventeen paired measurements is selected, and the difference between the first and second measurement of each pair is calculated. The parameter of interest is μ_D, the mean of the population of all possible differences.

a. Give the appropriate t-value for forming a 95% confidence interval for μ_D.

b. Give the rejection region for testing H_0: $\mu_D = 0$ against H_a: $\mu_D < 0$ with $\alpha = .01$.

8.87 Two large, independent random samples are selected from binomial populations with fractions of success π_1 and π_2.

a. Give the rejection region for testing H_0: $\pi_1 - \pi_2 = 0$ against H_a: $\pi_1 - \pi_2 > 0$ with $\alpha = .05$.

b. Give the z-value for forming a 95% confidence interval for $\pi_1 - \pi_2$.

Learning the Mechanics

8.88 Independent random samples were selected from two normally distributed populations with means μ_1 and μ_2, respectively. The sample sizes, means, and variances are shown in the table.

Sample 1	Sample 2
$n_1 = 12$	$n_2 = 14$
$\bar{x}_1 = 17.8$	$\bar{x}_2 = 15.3$
$s_1^2 = 74.2$	$s_2^2 = 60.5$

a. Test the null hypothesis H_0: $\mu_1 - \mu_2 = 0$ versus the alternative hypothesis H_a: $\mu_1 - \mu_2 > 0$. Use $\alpha = .05$.

b. Form a 99% confidence interval for $\mu_1 - \mu_2$.

c. How large must n_1 and n_2 be if you want to estimate $\mu_1 - \mu_2$ to within two units with 99% confidence?

8.89 (*Optional*) Independent random samples were selected from two normally distributed populations with means μ_1 and μ_2 and variances σ_1^2 and σ_2^2. The sample sizes, means, and variances are shown in the table.

Sample 1	Sample 2
$n_1 = 20$	$n_2 = 15$
$\bar{x}_1 = 123$	$\bar{x}_2 = 116$
$s_1^2 = 31.3$	$s_2^2 = 120.1$

a. Form a 95% confidence interval for σ_1^2/σ_2^2.
b. Test H_0: $\sigma_1^2 = \sigma_2^2$ versus H_a: $\sigma_1^2 \neq \sigma_2^2$. Use $\alpha = .05$.
c. Would you be willing to use a t test to test the null hypothesis H_0: $\mu_1 - \mu_2 = 0$ versus the alternative H_a: $\mu_1 - \mu_2 \neq 0$? Why?

8.90 Independent random samples are taken from two populations. The results of these samples are shown in the table.

Sample 1	Sample 2
$n_1 = 135$	$n_2 = 148$
$\bar{x}_1 = 12.1$	$\bar{x}_2 = 8.3$
$s_1^2 = 2.1$	$s_2^2 = 3.0$

a. Form a 90% confidence interval for $\mu_1 - \mu_2$.
b. Test H_0: $\mu_1 - \mu_2 = 0$ versus H_a: $\mu_1 - \mu_2 \neq 0$. Use $\alpha = .01$.
c. What sample sizes would be required if you want to estimate $\mu_1 - \mu_2$ to within .2 with 90% confidence?

8.91 Independent random samples were selected from two binomial populations. The sizes and number of observed successes for each sample are shown in the table.

	Sample 1	Sample 2
Sample size	200	300
Number of successes	110	130

a. Test the null hypothesis H_0: $\pi_1 - \pi_2 = 0$ versus the alternative hypothesis H_a: $\pi_1 - \pi_2 > 0$. Use $\alpha = .01$.
b. Form a 95% confidence interval for $\pi_1 - \pi_2$.
c. What sample sizes would be required in order to estimate $\pi_1 - \pi_2$ to within .01 with 95% confidence?

8.92 A random sample of five pairs of observations were selected, one of each pair from a population with mean μ_1, the other from a population with mean μ_2. The data are shown in the table at the top of page 476.

Pair	First Value	Second Value
1	28	22
2	31	27
3	24	20
4	30	27
5	22	20

a. Test the null hypothesis H_0: $\mu_D = 0$ versus H_a: $\mu_D \neq 0$. Use $\alpha = .05$.

b. Form a 95% confidence interval for μ_D.

Applying the Concepts

8.93 To compare the rate of return an investor can expect on tax-free municipal bonds with the rate of return on taxable bonds, an investment advisory firm randomly samples ten bonds of each type and computes the annual rate of return over the past 3 years for each bond. The rate of return is then adjusted for taxes, assuming the investor is in a 30% tax bracket. The means and standard deviations for the adjusted returns are given in the table.

Tax-Free Bonds	Taxable Bonds
$\bar{x}_1 = 9.8\%$	$\bar{x}_2 = 9.3\%$
$s_1 = 1.1\%$	$s_2 = 1.0\%$

a. Test to determine whether there is a difference in the mean rates of return between tax-free and taxable bonds for investors in the 30% tax bracket. Use $\alpha = .05$.

b. (*Optional*) Test the assumption that the two population variances are equal. Use $\alpha = .10$.

c. (*Optional*) Form a 90% confidence interval for σ_1^2 / σ_2^2.

8.94 Management training programs are often instituted in order to teach supervisory skills and thereby increase productivity. Suppose a company psychologist administers a set of examinations to each of ten randomly chosen supervisors before such a training program begins and then administers similar examinations at the end of the program. The examinations are designed to measure supervisory skills, with higher scores indicating greater skill. The results of the tests are shown in the table. Test to determine whether the data indicate that the mean score on the examinations after the training program is greater than the mean score before training. Use $\alpha = .10$.

Supervisor	Before Training	After Training	Supervisor	Before Training	After Training
1	63	78	6	72	85
2	93	92	7	91	99
3	84	91	8	84	82
4	72	80	9	71	81
5	65	69	10	80	87

8.95 Lack of motivation is a problem of many students in inner-city schools. To cope with this problem, an experiment was conducted to determine whether motivation could be improved by allowing students greater choice in the structures of their curricula. Two schools with similar student populations were chosen and fifty students were randomly selected from each to participate in the experiment. School A permitted its fifty students to choose only the courses they wanted to take. School B permitted its students to choose their courses and also to choose when and from which instructors to take the courses. The measure of student motivation was the number of times each student was absent from or late for a class during a 20-day period. The means and variances for the two samples are shown in the table. Do the data provide sufficient evidence to indicate that students from school B were late or absent less often than those from school A? Use $\alpha = .10$.

School A	School B
$\bar{x}_A = 20.5$	$\bar{x}_B = 19.6$
$s_A^2 = 26.2$	$s_B^2 = 24.1$

8.96 Will premium gasoline provide an increase in the mileage per gallon obtained by your automobile in comparison with the mileage for standard gasoline? Is the higher price of premium gasoline worth the increase (assuming an increase exists)? To assist in answering these questions, a government agency randomly selected 100 automobiles from its fleet and divided them into two equal groups. Each car was driven until it consumed 100 gallons of gasoline. One group used regular gasoline, the other used premium At the conclusion of the experiment, the miles per gallon were calculated for each automobile. The means and standard deviations for the two samples are shown in the table.

	Number of Cars	Mean Number of Miles per Gallon	s
Regular	50	19.3	2.1
Premium	50	22.0	1.7

a. Estimate the difference in mean gasoline mileages between regular and premium gasolines using a 95% confidence interval.

b. Suppose the agency wants to estimate the difference in mean mileages to within .5 mile per gallon with 95% confidence. How many cars would the agency have to include in each sample?

8.97 Advertising companies often try to characterize the average user of a client's product so that their ads can be targeted at particular segments of the buying community. Suppose a new movie is about to be released and an advertising company wants to know whether to aim the advertising at people under or over 25 years old. Individuals from both groups will be given an advance showing and then polled to determine their feelings about the movie. How many individuals should be included from each age group if the advertising company wants to estimate the difference in the proportions of those who like the movie to within .05 with 90% confidence?

8.98 Two banks, bank 1 and bank 2, each independently sampled forty and fifty of their business accounts, respectively, and counted the number of the bank's services (loans, checking, savings, investment counseling, etc.) each sampled business was using. Both banks offer the same services. A summary of the data is given in the table. Do the samples yield sufficient evidence to conclude that the average number of services used by bank 1's business customers is significantly greater (at the $\alpha = .10$ level) than the average number of services used by bank 2's customers?

Bank 1	Bank 2
$n_1 = 40$	$n_2 = 50$
$\bar{x}_1 = 2.2$	$\bar{x}_2 = 1.8$
$s_1 = 1.15$	$s_2 = 1.10$

8.99 A consumer protection agency wants to compare the work of two electrical contractors in order to evaluate their safety records. The agency plans to inspect residences in which each of these contractors has done the wiring in order to estimate the difference in the proportions of residences that are electrically deficient. Suppose the proportions of residences with deficient work are expected to be about .10 for both contractors. How many homes wired by each contractor should be inspected in order to estimate the difference in proportions to within .05 with 90% confidence?

8.100 The intraocular pressure of glaucoma patients is often reduced by treatment with adrenaline. To compare a new synthetic drug with adrenaline, seven randomly chosen glaucoma patients were treated with both drugs, one eye with adrenaline and one with the synthetic drug. The reduction in pressure in each eye was then recorded, as shown in the table. Do the data provide sufficient evidence to indicate a difference in the mean reductions in eye pressure for the two drugs? Test using $\alpha = .10$.

Patient	Adrenaline	Synthetic
1	3.5	3.2
2	2.6	2.8
3	3.0	3.1
4	1.9	2.4
5	2.9	2.9
6	2.4	2.2
7	2.0	2.2

8.101 Suppose you have been offered similar jobs in two different locales. To help in deciding which job to accept, you would like to compare the cost of living in the two cities. One of your primary concerns is the cost of housing, so you obtain a copy of a newspaper from each locale and study the classified ads. One convenient method for getting a general idea of prices is to compute the prices on a per square foot basis. This is done by dividing the price of the house by the heated area (in square feet) of the house. Random samples of sixty-three advertisements in city A and seventy-eight in city B produce the results given in the table. Is there evidence that the mean housing price per square foot differs in the two locales? Test using $\alpha = .05$.

City A	City B
$\bar{x}_1 = \$23.40$ per square foot	$\bar{x}_2 = \$25.20$ per square foot
$s_1 = \$2.50$ per square foot	$s_2 = \$2.80$ per square foot

8.102 Refer to Exercise 8.101. You also want to compare food prices in the two cities. You develop a list of fifteen food items of various types and obtain prices from the newspaper ads for a supermarket chain with a store in each city. The results are shown in the table. Do the data provide sufficient evidence to indicate a difference between the mean food prices in the two cities? Use $\alpha = .01$.

Food Item	City A	City B	Food Item	City A	City B
1	$2.49	$2.55	9	$3.83	$3.75
2	0.35	0.37	10	2.93	3.11
3	5.12	5.05	11	1.03	1.25
4	1.33	1.52	12	2.13	2.05
5	0.78	0.85	13	6.25	6.25
6	3.03	2.98	14	2.14	2.30
7	0.25	0.35	15	1.98	1.97
8	4.16	4.29			

8.103 Some power plants are located near rivers or oceans so that the available water can be used for cooling the condensers in the plants. As part of an environmental impact study, suppose a power company wants to estimate the mean difference in water temperature between the discharge of its plant and the offshore waters. How many sample measurements must be taken at each site in order to estimate the true mean difference to within .2°C with 95% confidence? Assume the range in readings will be about 4°C at each site and the same number of readings is to be taken at each site.

8.104 The use of preservatives by food processors has become a controversial issue. Suppose two preservatives are extensively tested and determined to be safe for use in meats. A processor wants to compare the preservatives for their effects on retarding spoilage. Suppose a random sample of fifteen cuts of fresh meat are treated with preservative A and another random sample of fifteen are treated with preservative B. Then, the number of hours until spoilage begins is recorded for each of the thirty cuts of meat. The results are summarized in the table.

Preservative A	Preservative B
$\bar{x}_1 = 106.4$ hours	$\bar{x}_2 = 96.5$ hours
$s_1 = 10.3$ hours	$s_2 = 13.4$ hours

a. Is there evidence of a difference in mean times until spoilage begins between the two preservatives at the $\alpha = .05$ level?

b. Can you recommend another way to conduct this experiment that may provide more information?

c. (Optional) Form a 95% confidence interval for σ_1^2/σ_2^2.

8.105 A physiologist wanted to study the effects of birth control pills on exercise capacity. Five female subjects who had never taken the pills before were randomly selected, and their maximal oxygen uptake was measured (in milliliters per kilogram of body weight) during a treadmill session. The five subjects then took birth control pills for a specified length of time and their uptakes were measured again, as given in the table. Do the data provide sufficient evidence to indicate that the mean maximal oxygen uptake after taking the pills for a specified period of time is less than the mean maximal oxygen uptake before taking the pills? Use $\alpha = .01$.

Subject	Maximal Oxygen Uptake Before	After
1	35.0	29.5
2	36.5	33.5
3	36.0	32.0
4	39.0	36.5
5	37.5	35.0

8.106 An economist wants to investigate the difference in unemployment rates between an urban industrial community and a university community in the same state. The economist interviews a random sample of 525 potential members of the work force in the industrial community and a second random sample of 375 in the university area. Of these, forty-seven and twenty-two, respectively, are unemployed. Use a 95% confidence interval to estimate the difference in unemployment rates in the two communities.

8.107 A careful auditing is essential to all businesses, large and small. Suppose a firm wants to compare the performance of two auditors it employs. One measure of auditing performance is error rate, so the firm decides to randomly sample 200 pages from the work of each auditor and carefully examine each page for errors. Suppose the number of pages on which at least one error is found is seventeen for auditor A and twenty-five for auditor B. Test to determine whether the data indicate a difference in the true error rates for the two auditors. Use $\alpha = .01$.

8.108 Since tourism is the largest industry in the state of Florida, the economy of the state depends heavily on the number of tourists who visit Florida annually and on the mean amount of money tourists spend while they are in the state. Suppose a study is conducted during two consecutive years, say 1988 and 1989, to compare the mean expenditures of tourists in Florida. Random samples of 325 and 375 tourists (a family is treated as one tourist) are selected in 1988 and 1989, respectively, and the total expenditure in the state is recorded for each. The results are summarized in the table. Form a 90% confidence interval for the difference in mean expenditures per tourist in 1988 and 1989.

1988	1989
$n_1 = 325$	$n_2 = 375$
$\bar{x}_1 = \$676$	$\bar{x}_2 = \$853$
$s_1 = \$554$	$s_2 = \$715$

8.109 It is desired to compare two drugs that cause the pupil of the eye to dilate. Initially, five randomly chosen rabbits are to be used in studying the drugs. Each rabbit has one drug randomly assigned to the left eye and the other drug assigned to the right eye. The results are listed in the table. Do the data provide sufficient evidence to indicate a difference in the true mean lengths of time until the pupil returns to normal for the two drugs? Test at the $\alpha = .10$ level of significance.

Rabbit	Drug A	Drug B
1	52	51
2	50	61
3	49	53
4	63	68
5	58	60

8.110 A large shipment of produce contains Valencia and navel oranges. To determine whether there is a difference in percentages of nonmarketable fruit between the two varieties, random samples of 850 Valencia and 1,500 navel oranges were independently selected and the numbers of nonmarketable oranges were counted. Thirty Valencia and ninety navel oranges from these samples were nonmarketable. Do the data provide sufficient evidence to indicate a difference in percentages of Valencia and navel oranges that are nonmarketable? Test at the $\alpha = .05$ level of significance.

8.111 (*Optional*) When new instruments are developed to perform chemical analyses of products (food, medicine, etc.), they are usually evaluated with respect to two criteria: accuracy and precision. *Accuracy* refers to the ability of the instrument to identify correctly the nature of the components in the product and the quantities of each of these components. *Precision* refers to the consistency of the instrument in identifying the components. Thus, a large variability in the identification of a single sample of a product indicates a lack of precision. Suppose a pharmaceutical firm is considering two instruments designed to identify the components of certain drugs. As part of a comparison of precision, ten test-tube samples of a well-mixed batch of a drug are randomly selected and then five are analyzed by instrument A and five by instrument B. The data shown in the table at the top of page 482 are the percentages of the primary component of the drug measured by the instruments. Do the data provide evidence of a difference in the precision of the two instruments? Use $\alpha = .10$.

Instrument A	Instrument B
43	46
48	49
37	43
52	41
45	48

8.112 A large department store plans to renovate one of its floors, and this will result in increasing the floor space for one department. The management has narrowed the decision about which department to enlarge to two departments: men's clothing and sporting goods. The final decision will be based on mean sales—the department that has the greater mean will be enlarged. The sales data from the last 12 months are shown in the table.

Month	Men's Clothing	Sporting Goods	Month	Men's Clothing	Sporting Goods
1	$15,726	$17,533	7	$15,525	$16,774
2	11,243	10,895	8	15,799	16,223
3	22,325	19,449	9	16,449	16,135
4	23,494	21,500	10	16,993	17,834
5	12,676	18,925	11	19,832	18,429
6	13,492	21,426	12	32,434	34,565

a. Use the data to form a 95% confidence interval for the mean difference in monthly sales for the two departments.

b. On the basis of the confidence interval formed in part a, can you make a recommendation to the store management as to which department should be enlarged?

c. Are you certain that all the conditions required for the validity of the statistical procedures were met in this case? Was the sample of differences actually random?

8.113 A politician conducted a sample survey to evaluate a television advertising campaign. Both before and after the advertising campaign, random samples were taken from among the voters to determine voter preference in the coming election. The table shows the results of the surveys.

	Sample Size	Number Who Prefer the Politician
Before advertising campaign	200	85
After advertising campaign	300	139

a. Estimate the difference in the proportions of voters who favor the politician before and after the advertising campaign. Use a 95% confidence interval.

b. What size samples need to be taken in order to estimate the difference in proportions to within .04 with 95% confidence?

8.114 The following experiment was conducted to compare two coatings designed to improve the durability of the soles of jogging shoes: A $\frac{1}{8}$-inch thick layer of coating 1 was applied to one of a pair of shoes and an equally thick layer of coating 2 was applied to the other shoe. Ten randomly selected joggers were given pairs of treated shoes and were instructed to record the number of miles covered before the coating on each shoe wore through in any place. The results are given in the table.

Jogger	Coating 1	Coating 2	Jogger	Coating 1	Coating 2
1	892	985	6	853	875
2	904	953	7	780	895
3	775	775	8	695	725
4	435	510	9	825	858
5	946	895	10	750	812

a. Do the data provide sufficient evidence to indicate a difference in the mean number of miles of wear by the two coatings? Test using $\alpha = .05$.
b. Use a 95% confidence interval to estimate the true difference in mean number of miles of wear between the two coatings.
c. Why is the design chosen for this experiment preferable to independent random sampling?

8.115 Many college and university professors have been accused of "grade inflation" over the past several years. This means they assign higher grades now than in the past, even though students' work may be of the same caliber or less. If grade inflation has occurred, the mean grade-point average of today's students should exceed the mean of 10 years ago. To test the grade inflation theory at one university, a business professor randomly selects seventy-five business majors who are graduating with the present class and fifty who graduated 10 years ago. The grade-point average of each student is recorded, and the summary results are shown in the table. Test to determine whether the data support the hypothesis of grade inflation in the business school of this university. Use $\alpha = .05$. (Assume that the quality of the students' work is actually comparable for the two periods.)

10 Years Ago	Present
$n_1 = 50$	$n_2 = 75$
$\bar{x}_1 = 2.82$	$\bar{x}_2 = 3.04$
$s_1 = .43$	$s_2 = .38$

8.116 Smoke detectors are highly recommended safety devices for early fire detection in homes and businesses. It is extremely important that the devices work properly. Suppose that 100 brand A smoke detectors are tested and twelve fail to operate. When given the same test, fifteen out of ninety brand B detectors fail. Form a 90% confidence interval to estimate the difference in the fractions of defective smoke detectors produced by the two companies.

8.117 The federal government wants to know whether salary discrimination exists between men and women in the private sector. Suppose random samples of fifteen women and twenty-two men are drawn from the population of first-level managers in the private sector, and the annual salary of each is recorded. The information is summarized in the table.

Women	Men
$n_1 = 15$	$n_2 = 22$
$\bar{x}_1 = \$18,400$	$\bar{x}_2 = \$19,700$
$s_1 = \$2,300$	$s_2 = \$3,100$

a. Do the data provide sufficient evidence to indicate that the mean salary of male managers exceeds the mean salary of female managers? Use $\alpha = .10$.

b. (*Optional*) Conduct a test to determine whether the data indicate that the assumption of equal salary variances is false. Use $\alpha = .10$.

8.118 An automobile manufacturer wants to estimate the difference in the mean miles per gallon rating for two models of its cars. If the range of ratings is expected to be about 6 miles per gallon for each model, how many cars of each type must be tested in order to estimate the difference in means to within .5 mile per gallon with 95% confidence?

8.119 The state of Florida now requires all high-school students to pass a literacy test before receiving a high school diploma. A student who fails the test can enroll in a refresher course and retake the test at a later date. To evaluate the effectiveness of the refresher course, a random sample of eight students' test scores were compared before and after; the results are shown in the table. Do the data provide sufficient evidence to conclude that the mean score on the literacy test after the refresher course is greater than the mean score before taking the course? (Use $\alpha = .05$.)

Student	Before	After	Student	Before	After
1	45	49	5	57	53
2	52	50	6	55	61
3	63	70	7	60	62
4	68	71	8	59	67

8.120 (*Optional*) A drug currently used to reduce the heart rate of patients before surgery works very well in some people but has little effect on others. A new drug that is hoped to be more consistent has been developed. To test it, twenty-four randomly chosen dogs were divided into two equal groups. One group was injected with the new drug, and the other group was injected with the old drug. The reduction in heart rate for each dog was recorded, with the results summarized in the table. Do the data provide sufficient evidence to conclude that the variation in the heart rate reductions is less for the new drug than for the old drug? Use $\alpha = .05$.

Old Drug	New Drug
$n_1 = 12$	$n_2 = 12$
$s_1^2 = 14.3$	$s_2^2 = 8.2$

8.121 A football fan decides to place bets according to the predictions of one of two newspaper columnists. To decide which columnist to follow, both writers' predictions are randomly sampled over the preceding weeks and the number of correct predictions for each is noted. The results are shown in the table. Do the data provide sufficient evidence to indicate that one of the columnists is better at picking winners? Use $\alpha = .10$.

Columnist	Predictions Sampled	Correct Predictions
1	60	48
2	50	42

8.122 Pharmaceutical companies testing new drugs routinely employ statistical analyses to interpret the test results, as the following report illustrates:

> Heart attack victims had fewer second attacks after combined use of aspirin and a drug that also hinders blood clot formation, according to a study funded by a company that produces the drug.
>
> Dr. Christian R. Klimt, an epidemiologist at the Maryland Medical Research Institute, . . . presented the results of the study of combined use of aspirin and Persantine, made by the German drug firm Boehringer Ingelheim.
>
> The 1,563 patients in the study who received aspirin plus Persantine experienced 30% fewer heart attacks after one year and 24% fewer after two years than 1,565 patients who received a placebo in capsules identical to the Persantine capsules, Klimt said.
>
> University of Texas neurologist William Fields criticized the study, saying it failed to examine whether aspirin alone was responsible for the reduction in second heart attacks rather than the aspirin–Persantine combination.
>
> Klimt said two earlier studies found no significant reduction in second heart attacks among people who received aspirin alone, compared with a group that received a placebo.*

Notice that the article gives only the *difference* in the percentages of heart attack victims for the patients who received aspirin plus Persantine versus those who received a placebo. The percentages of heart attack victims for each group are not given. In parts a and b, hypothetical percentages for the two groups are given. In part c you are asked to compare the confidence intervals you calculate. In particular, does the choice of the individual percentages have a significant effect on the calculated intervals?

* "Drug Seen Aiding Heart Patients," *The Cincinnati Enquirer*, March 13, 1985, p. D-4.

    a. Suppose that 30% of the patients given aspirin plus Persantine suffered a second heart attack in the first year, compared to 60% of the patients given a placebo. Find a 95% confidence interval for the true difference in the percentages of second heart attacks for the two treatments.

    b. Suppose that 50% of the aspirin–Persantine patients suffered a second heart attack in the first year, compared to 80% of the placebo patients. Again, find a 95% confidence interval for the true difference in the percentages for the two treatments.

    c. Compare your answers to parts *a* and *b*.

8.123 Public service advertising has for years urged the use of seat belts. Independent random samples of American drivers polled in 1982 and 1984 were asked, "Thinking about the last time you got in the car, did you use a seat belt, or not?" Each sample contained approximately 1,500 responses, which are summarized in the table.

| | **Seat Belt Usage** | | |
	Yes, Did	No, Did Not	No Opinion
1984	25%	74%	1%
1982	17%	81%	2%

Source: "Seat Belt Usage" survey, *The Gallup Report*, No. 226, July 1984, p. 5.

    a. Estimate the difference in the percentages of all Americans who answered yes to the question in 1984 as compared to 1982. Use a 95% confidence interval.

    b. How large must equal size samples have been to estimate the true difference in percentages to within .02 with 95% confidence?

*8.124 In 1983 and 1984 the Hertz Corporation compared the cost of owning and operating a new compact car in cities throughout the United States. The costs per mile for twenty cities are given in the accompanying table. The study included such expenses as loan rates, insurance, taxes, maintenance, and gasoline.

City	1984	1983	City	1984	1983
Los Angeles	60.69¢	56.86¢	Minneapolis	47.58¢	44.99¢
San Francisco	58.16	54.61	San Diego	47.53	45.34
New York City	57.25	54.50	Detroit	46.47	43.44
Miami	53.47	50.03	Cleveland	45.80	43.21
Chicago	52.74	49.46	Pittsburgh	45.44	43.56
Denver	51.91	47.74	Milwaukee	45.42	44.23
St. Louis	51.57	47.53	Atlanta	45.10	41.59
Seattle	50.58	47.06	Dallas	45.08	42.27
Houston	49.31	44.09	Washington, D.C.	45.00	42.11
Boston	48.48	45.34	Cincinnati	44.65	42.68

Source: "The Cost of Driving a Mile" table, *The Kentucky Enquirer*, March 23, 1985, p. A-1.

    a. Do the data for the years 1983 and 1984 represent dependent or independent samples?

b. With the aid of a computer package, form a 95% confidence interval for the difference in mean cost per mile for all U.S. cities in 1984 versus 1983.

*8.125 An experiment was performed to evaluate the effects of two additives on gasoline consumption. Fifteen cars with varying engine sizes—and thus varying demands for gasoline—were available for the test. The experimenter randomly assigned eight cars to receive additive A and seven cars to receive additive B. Each car was driven 100 miles over the same course, and the gasoline consumption was recorded for each. The sample data are given in the table (measurements are in terms of miles per gallon).

Additive A	Additive B
23.6	19.9
20.6	16.4
17.5	17.6
12.4	17.3
19.7	20.9
15.0	25.0
16.8	22.6
16.0	

a. Use a computer package to find a 95% confidence interval for the true difference in the mean gasoline consumption with the two additives.
b. Do you think there is a better way to conduct this experiment? Explain.

CHAPTER 8 QUIZ

1. Explain the difference between independent samples and dependent samples.

2. Two basketball players engage in a foul-shooting contest to compare the fraction of shots achieved by each player. Each player takes 100 shots, and player A sinks 93 while player B makes 86. Is there sufficient evidence to indicate a difference in the fractions of foul shots made by each player? Test at the .05 level of significance.

3. An agricultural firm is interested in comparing the average yearly rainfall in two locations available for purchase. A random sample of seven yearly rainfalls is obtained for location 1 and a second random sample of seven yearly rainfalls is obtained for location 2. Using the data in the table, form a 90% confidence interval for $\mu_1 - \mu_2$, the difference in the mean yearly rainfall at location 1 and location 2.

Location 1	Location 2
$n_1 = 7$	$n_2 = 7$
$\bar{x}_1 = 123$ inches	$\bar{x}_2 = 112$ inches
$s_1 = 13$ inches	$s_2 = 19$ inches

4. Refer to Question 3. Suppose a random sample of 7 years is selected, and the yearly rainfall (in inches) at each location is recorded. The data are given in the table.

Year	Location 1	Location 2	Year	Location 1	Location 2
1	99	84	5	146	141
2	110	105	6	128	120
3	135	120	7	133	125
4	115	100			

a. Form a 90% confidence interval for the mean of the differences in yearly rainfall for location 1 versus location 2.

b. Compare your answer to part *a* with the answer to Question 3. For this experimental situation, do you think independent samples or dependent samples are better? Why?

5. An experiment was conducted to compare the eating habits of families whose mothers work outside the home with families whose mothers do not work. In particular, it was desired to compare the mean number of times per week that the families eat at fast-food restaurants. Independent random samples of forty families of each type were selected, and the number of weekly fast-food outings was recorded. The data are summarized in the table. Do the data indicate at the .05 level of significance that families whose mothers work outside the home eat at fast-food restaurants more often than families whose mothers stay at home?

Mothers Working Outside the Home	Mothers Not Working Outside the Home
$n_1 = 40$	$n_2 = 40$
$\bar{x}_1 = 1.7$	$\bar{x}_2 = .9$
$s_1 = 2.0$	$s_2 = 1.3$

CHAPTERS 6–8 CUMULATIVE QUIZ

1. Many schools now use volunteers to tutor children who are having difficulty learning. A school decided to experiment with this method of tutoring last year. A random sample of fifty fifth-graders were tutored by volunteers throughout the school year and then were given a standardized achievement test. Their mean score was 78. From past years, it is known that the mean score of fifth-graders on the test is 76 and the standard deviation is 8.

a. If the distribution of scores on the standardized test for students who receive tutoring is the same as the scores in the past, what is the probability that the fifty sampled children would have a mean score of 78 or higher on the test?

b. Considering your answer to part a, what, if anything, do you conclude?

2. Radio stations sometimes conduct prize giveaways to try to increase their share of the listening audience. Suppose a station manager calls 300 randomly selected households in a city and finds that sixty-five have members who regularly listen to the station. The station then conducts a 2-month promotional contest and follows it with a survey of 500 randomly chosen households. The survey shows that 154 households have members who regularly listen to the station.

   a. Use a 90% confidence interval to estimate the difference between the proportions of those who regularly listen to the station before and after the promotional contest.

   b. Place a 95% confidence interval on the proportion of those who listen to the station after the promotion is over.

3. Experiments are to be conducted to investigate the mean length of time it takes patients to obtain relief of pain with each of two new drugs. The standard deviations of the lengths of time to pain relief are approximately 3.5 minutes for both drugs.

   a. How many patients need to be sampled to estimate the mean length of time it takes the first pain reliever to act, to within .75 minute with 95% confidence?

   b. How many patients need to be sampled to estimate the difference in the mean lengths of time it takes the two drugs to act, to within .25 minute with 99% confidence? Do you think these would be reasonable sample sizes upon which to conduct the experiment?

4. A woman has placed a bird feeder in her backyard and is interested in determining the average number of birds per day that visit it. She randomly chooses 5 days and observes the number of birds visiting the feeder each day. These values are 23, 31, 18, 30, and 27.

   a. Calculate $\bar{x}$ and s.

   b. Form a 90% confidence interval for the mean number of birds visiting the feeder per day.

5. A new fertilizer has been developed that supposedly increases the yield of tomato plants. Before using this fertilizer on a general scale, a vegetable farmer wants to be sure it performs as advertised. The farmer takes six plots of land of equal size and uses the new fertilizer on half of each plot, while using the old brand on the other half of each plot. The yield of tomato plants in the sectioned plots is recorded (in pounds) as shown in the table. Is there sufficient evidence that the mean yield is greater with the new fertilizer than with the old? Test at the .05 significance level.

Plot	New Fertilizer	Old Fertilizer	Plot	New Fertilizer	Old Fertilizer
1	265	278	4	310	295
2	280	270	5	285	281
3	278	267	6	290	300

On Your Own

We have now discussed two methods of collecting data to compare two population means. In many experimental situations, a decision must be made either to collect two independent samples or to conduct a paired difference experiment. The importance of this decision cannot be overemphasized, since the amount of information obtained and the cost of the experiment are both directly related to the method of experimentation chosen.

Choose two populations (pertinent to your major area of study) that have unknown means and for which you could collect two independent samples as well as paired observations. Before conducting the experiment, state which method of sampling you think will provide more information, and why. Then, to compare the two methods, first perform the independent sampling procedure by collecting ten observations from each population (a total of twenty measurements), and then perform the paired difference experiment by collecting ten pairs of observations.

Construct two 95% confidence intervals, one for each experiment you conducted. Which method provided the narrower confidence interval and thus more information? Does this agree with your preliminary expectations?

Chapter 9

REGRESSION AND CORRELATION: A STRAIGHT-LINE RELATIONSHIP

ANALYZING CRIME RATES

Reporting and analyzing crime rates is an important function of many law enforcement agencies. An article by D. Heaukulani in the *Journal of Police Science and Administration*, titled "The Normal Distribution of Crime" (1975, pp. 312–318), illustrates a useful statistical tool for the analysis and prediction of crime rates. Using the data shown in Table 9.1 (page 492), the author demonstrates that the annual value of the United States' crime index is related to the size of the population, and that these two variables have a straight-line relationship.

In this chapter we will discuss what it means for two variables to have a *straight-line relationship* and how this relationship may be used to make predictions or estimations. We will also discuss what it means for two variables to be correlated.

Table 9.1

Year	Population	Actual Crime Index	Year	Population	Actual Crime Index
1963	188,531,000	2,259,081	1969	201,921,000	4,989,747
1964	191,334,000	2,604,426	1970	202,184,772	5,568,197
1965	193,818,000	2,780,015	1971	206,256,000	5,995,211
1966	195,857,000	3,243,400	1972	208,232,000	5,891,924
1967	197,864,000	3,802,273	1973	209,851,000	8,638,375
1968	199,861,000	4,466,573			

United States population and crime index

Source: Reprinted by permission of the *Journal of Police Science and Administration,* copyright 1975 by Northwestern University School of Law, Vol. 3, No. 3, pp. 312–318.

9.1

EXPLORATORY DATA ANALYSIS: THE SCATTERPLOT

In order to discuss the relationship between two variables, it is helpful to plot the sample data. For example, suppose an accountant at a brewery wants to study the relationship between the amount of money (measured in $1,000 units) spent monthly on advertising and the brewery's total monthly sales (also measured in $1,000 units). The accountant randomly samples seven months and records the advertising expenditures and total sales for each month. The results are shown in Table 9.2, with advertising expenditure denoted by X and total sales denoted by Y.

Table 9.2

Month	Advertising Expenditure X, Thousands of Dollars	Total Sales Y, Thousands of Dollars
1	0.9	100.7
2	0.7	80.6
3	0.8	93.8
4	1.0	91.1
5	0.5	52.2
6	0.3	31.0
7	0.7	94.5

Total sales versus advertising expenditure

The graph shown in Figure 9.1 was constructed based on the data from Table 9.2. The graph, called a **scatterplot,** is drawn on the **rectangular coordinate system.** Each of the plotted points may be identified by the corresponding values of X and Y, which are found on the horizontal and vertical axes, respectively. For example, in Figure 9.1 this relationship has been shown for the point (0.5, 52.2). The notation (0.5, 52.2) gives the **coordinates of the point** and indicates that $X = 0.5$ and $Y = 52.2$.

Figure 9.1
Scatterplot for data in
Table 9.2

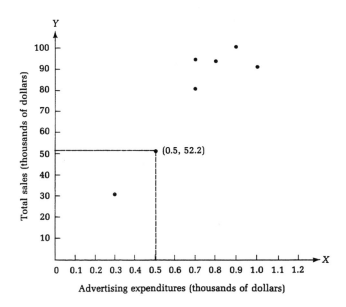

Advertising expenditures (thousands of dollars)

Examining the scatterplot, you can see that Y tends to increase as X increases. The scatterplot provides visual evidence to indicate that the brewery's total sales are related to advertising expenditure. As with other exploratory data analysis methods, drawing the scatterplot is a first step in studying such a relationship. Being encouraged that a relationship exists, we would conduct further analyses. We will discuss how to proceed with the analysis in the rest of this chapter.

EXAMPLE 9.1   The introduction to this chapter provided data comparing the annual value of the United States crime index to the size of the population. Denoting the crime index by Y and the population size by X, we repeat the data in Table 9.3.

Table 9.3

United States population
and crime index

Year	Population	Actual Crime Index	Year	Population	Actual Crime Index
1963	188,531,000	2,259,081	1969	201,921,000	4,989,747
1964	191,334,000	2,604,426	1970	202,184,772	5,568,197
1965	193,818,000	2,780,015	1971	206,256,000	5,995,211
1966	195,857,000	3,243,400	1972	208,232,000	5,891,924
1967	197,864,000	3,802,273	1973	209,851,000	8,638,375
1968	199,861,000	4,466,573			

Source: Reprinted by permission of the *Journal of Police Science and Administration*, copyright 1975 by Northwestern University School of Law, Vol. 3, No. 3, pp. 312–318.

a. Draw a scatterplot of the data.

b. Discuss the plot drawn in part **a**.

Solution   **a.** A scatterplot of the data in Table 9.3 is shown in Figure 9.2. The population values given on the X-axis are shown in millions. The crime index values given on the Y-axis are also given in millions. Note that due to the large values of X, the X-axis has been broken.

Figure 9.2
Scatterplot for population
and crime index data

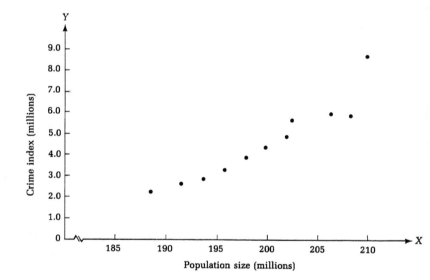

b. The scatterplot indicates that the annual value of the United States crime index is related to the size of the population. As the population increases in size, the crime index also tends to increase. ∎

COMPUTER
EXAMPLE 9.1

As you will see, the arithmetic calculations that will be required in this chapter are somewhat more complicated than those required in previous chapters. Because of this, Minitab will be very helpful throughout this chapter.

In this example we will use Minitab to construct a scatterplot for a set of data. The same data will be used in all the computer examples for the remainder of the chapter.

The accompanying table shows the number of hours worked per week and the first-semester grade-point average (GPA) for a sample of fifty freshmen who were enrolled in Calculus I at a large university. The fifty students were randomly selected from all Calculus I students who worked up to 20 hours per week at a part-time job.

With the aid of Minitab a scatterplot of these data, comparing GPA to the number of hours worked, was constructed. The scatterplot is also reproduced here.

Data for Computer
Example 9.1

ROW	WORKHRS	GPA	ROW	WORKHRS	GPA
1	18	2.94	26	12	2.90
2	15	2.80	27	10	2.70
3	16	2.85	28	8	2.93
4	19	2.64	29	14	2.39
5	17	2.59	30	20	2.11
6	20	2.86	31	0	2.97
7	0	3.16	32	0	3.23
8	14	2.81	33	17	2.56
9	16	2.78	34	18	2.83
10	12	2.74	35	12	2.81
11	8	3.13	36	16	2.33
12	15	3.04	37	14	2.83
13	18	2.76	38	20	2.86
14	12	3.06	39	12	2.78
15	17	2.48	40	20	2.39
16	16	2.81	41	19	2.48
17	8	2.82	42	0	2.86
18	0	3.00	43	16	3.12
19	15	2.64	44	19	2.60
20	20	2.83	45	16	2.69
21	16	2.54	46	16	2.81
22	20	2.76	47	12	3.00
23	0	3.01	48	0	2.63
24	16	2.64	49	8	2.67
25	19	2.62	50	12	2.86

```
           -
           -  *
   3.15+   *  *                   *                   *
           -                             *
 GPA       -  2                          *        *
           -  *                   *                          *
           -  *                                     *                   2
   2.80+   -                      *      2      2  * 3    *        *
           -                          *  *                      *   *
           -  *                   *                 * 2         *
           -                                             2      2
           -                                       *
   2.45+   -                                       *      *
           -                             *                        *
           -                                       *
           -
           -
   2.10+   -                                             *
           -
           +---------+---------+---------+---------+---------+------WORKHRS
          0.0       4.0       8.0      12.0      16.0      20.0
```

Scatterplot for Computer Example 9.1

   When plotting the data, Minitab placed an asterisk (*) at each coordinate of the
"GPA–WORKHRS" data. If two students worked the same number of hours and had
the same, or almost the same, GPA, then Minitab placed a "2" in the scatterplot. This
indicates that really there should be two asterisks at essentially the same point. For
example, looking at the graph we see a "2" among the points plotted for students
who worked zero hours per week. Examining the original data, we see that students

18 and 23 both worked zero hours and had GPAs of 3.00 and 3.01, respectively. Since Minitab could not print two asterisks this close together, the number "2" was printed. Similarly, the value "3" in the graph for those working 16 hours indicates that three of these GPAs were nearly equal (namely, those for students 9, 16, and 46).

Examining the graph, we see that there tends to be a slight decrease in GPA as the number of hours worked increases. This is best seen by comparing the GPA values at 0 hours to those at 20 hours.   ■

Exercises (9.1–9.6)

Applying the Concepts

9.1   When student seating in a large lecture course is by personal choice, is seat location related to a student's grade? In particular, do students who choose to sit forward tend to obtain better grades than those who choose seats in the rear? The grades of ten randomly selected statistics students are recorded in the table, along with their row number (rows numbered from front to back).

Grade, Y	93	68	89	98	66	86	73	55	80	71
Row, X	7	1	4	3	25	12	9	30	13	27

a. Draw a scatterplot of the data.

b. Does the scatterplot indicate that students sitting in the front rows tend to have higher grades than those sitting in the back?

9.2   The number of golf courses in the United States with at least eighteen holes and the United States divorce rate are given in the table for twelve different years.

Year	Number of 18-Hole and Larger Golf Courses in the U.S.	Number of Divorces in the U.S. (in millions)
1960	2,725	2.9
1965	3,769	3.5
1970	4,845	4.3
1975	6,282	6.5
1977	6,551	8.0
1978	6,699	8.6
1979	6,787	8.8
1980	6,856	9.9
1981	6,944	10.8
1982	7,059	11.5
1983	7,125	11.6
1984	7,230	12.3

Source: U.S. Bureau of the Census, Statistical Abstract of the United States: 1986 (Washington, D.C.: Government Printing Office, 1985), pp. 35, 230.

a. Draw a scatterplot of the data. Let Y be the divorce rate and X be the number of golf courses.

b. Interpret the graph drawn in part a.

9.3 Underinflating or overinflating tires can increase tire wear. A new tire was tested for wear at different pressures, with the results shown in the table.

Pressure X, Pounds per Square Inch	Mileage Y, Thousands
30	29
31	32
32	36
33	38
34	37
35	33
36	26

a. Draw a scatterplot for the data.

b. Discuss the graph drawn in part a.

9.4 Is the number of games won by a major league baseball team in a season related to the team's batting average? The accompanying table shows the number of games won and the batting averages for the fourteen teams in the American League for the 1986 season.

Team	Number of Games Won Y	Team Batting Average X
Cleveland	84	.284
New York	90	.271
Boston	95	.271
Toronto	86	.269
Texas	87	.267
Detroit	87	.263
Minnesota	71	.261
Baltimore	73	.258
California	92	.255
Milwaukee	77	.255
Seattle	67	.253
Kansas City	76	.252
Oakland	76	.252
Chicago	72	.247

Source: Official American League Averages 1986. New York: The American League of Professional Baseball Clubs, pp. 2, 24–27.

a. If you were to consider the relationship between the number of games won by a major league team and the team's batting average, would you expect that the number of games won would tend to increase or decrease as batting average increases? Explain.

b. Construct a scatterplot for the data. Does the pattern revealed by the scatterplot agree with your answer to part a?

*9.5 The manager of a clothing store wants to investigate the relationship between the number of sales clerks on duty and the value of the merchandise lost due to shoplifting. The dollar volume lost for each of thirty-four randomly selected weeks was recorded, along with the number of sales clerks working each week. The number of sales clerks was constant within a given week, but varied from week to week. The data are given in the accompanying table. (Since the data will also be used in subsequent exercises, you should save them in the computer, or on a disk, if possible.)

Week	Number of Sales Clerks X	Loss Y	Week	Number of Sales Clerks X	Loss Y
1	14	161	18	18	109
2	14	163	19	18	128
3	10	189	20	15	141
4	10	146	21	20	104
5	17	121	22	12	189
6	16	161	23	19	104
7	14	194	24	14	187
8	14	169	25	18	100
9	11	186	26	15	122
10	16	178	27	12	170
11	17	91	28	15	152
12	16	151	29	12	201
13	18	103	30	13	196
14	15	146	31	13	162
15	16	127	32	15	144
16	12	163	33	14	157
17	20	119	34	12	173

a. What do you think should be the relationship between the number of sales clerks and the loss due to shoplifting?

b. Use a computer package to obtain a scatterplot for the data.

c. Discuss the plot drawn in part b. Does it support your answer to part a?

*9.6 To investigate the relationship between yield of potatoes, Y, and level of fertilizer application, X, an experimenter divided a field into forty plots of equal size. Different amounts of fertilizer were applied to each of the first eight plots. The same amounts of fertilizer were applied to the next eight plots, and this was repeated until all the plots had been fertilized. The yield of potatoes (in pounds) and the fertilizer application (in pounds) were recorded for each plot. The data are shown in the table. (Since the data will also be used in subsequent exercises, you should save them in the computer, or on a disk, if possible.)

Plot	X	Y	Plot	X	Y
1	1.0	24.9	21	3.0	31.9
2	1.5	27.5	22	3.5	33.0
3	2.0	34.2	23	4.0	30.9
4	2.5	34.5	24	4.5	33.1
5	3.0	31.6	25	1.0	26.1
6	3.5	31.4	26	1.5	25.6
7	4.0	41.1	27	2.0	30.7
8	4.5	36.1	28	2.5	28.6
9	1.0	28.2	29	3.0	30.7
10	1.5	22.2	30	3.5	33.0
11	2.0	27.9	31	4.0	34.6
12	2.5	25.0	32	4.5	38.5
13	3.0	29.4	33	1.0	31.1
14	3.5	35.1	34	1.5	29.2
15	4.0	33.2	35	2.0	26.8
16	4.5	28.9	36	2.5	31.5
17	1.0	21.4	37	3.0	31.4
18	1.5	30.8	38	3.5	36.7
19	2.0	34.0	39	4.0	40.7
20	2.5	26.6	40	4.5	39.2

a. Draw a scatterplot of the data.
b. Does the fertilizer appear to be effective? Explain.

9.2

THE EQUATION OF A STRAIGHT LINE

As stated in the introduction to this chapter, we will discuss what it means for two variables to have a straight-line relationship. In this section we will review the correspondence between the equation of a straight line and the graph of a straight line.

In Figure 9.3 (page 500) we see the graph of a straight line. Any point on the line may be identified by the corresponding values of X and Y, which are found on the horizontal and vertical axes, respectively. For example, in Figure 9.3 it is indicated that the point with coordinates (2, 3) is on the line.

For each graph of a straight line there is an algebraic expression, called **the equation of a straight line,** that expresses the relationship between the two variables X and

Figure 9.3
The graph of a straight line

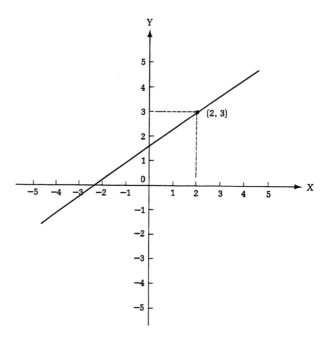

Y. For the equation to reflect the appropriate relationship, the coordinates of every point on the line must satisfy the equation, and vice versa.

EXAMPLE 9.2   **a.** Draw a graph of the equation $Y = 2X + 3$. (The graph will be a straight line.)
b. Choose an arbitrary point on the line, and show that the coordinates of the point satisfy the equation.

Solution   **a.** To draw the graph, we must determine the coordinates of at least two points that satisfy the equation. This is accomplished by substituting values of X into the equation and finding the corresponding values of Y. We choose to use the three values of $X = -2$, $X = 0$, and $X = 1$. Substituting into the equation $Y = 2X + 3$, we obtain

$$Y = 2(-2) + 3 = -4 + 3 = -1$$
$$Y = 2(0) + 3 = 0 + 3 = 3$$
$$Y = 2(1) + 3 = 2 + 3 = 5$$

The three coordinates we have found are summarized in the table,

X	-2	0	1
Y	-1	3	5

and the graph is shown in Figure 9.4. Since two points determine a straight line, three coordinates are not necessary, but it is a good idea always to find the coordinates of at least three points to provide a check of your work.

Figure 9.4
A graph of the equation
$Y = 2X + 3$

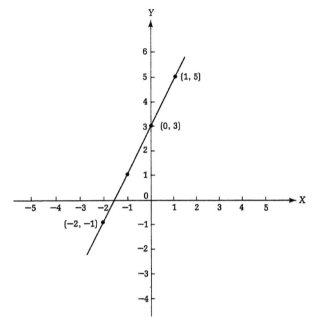

b. As shown in Figure 9.4, another point on the line has coordinates $(-1, 1)$. Substituting $X = -1$ into the equation, we obtain

$$Y = 2(-1) + 3 = 1$$

Thus, this point does satisfy the equation, and any other point you picked that is on the line would also.   ■

In Figure 9.5 (page 502) we have redrawn the graph of the equation $Y = 2X + 3$. We have indicated the point where the line crosses the Y-axis. This point is called the **Y-intercept,** and since any point on the Y-axis has an X-coordinate equal to 0, we say the Y-intercept is equal to 3.

In Figure 9.5 we have also shown that a 1-unit increase in X is associated with a 2-unit increase in Y. The amount of change in Y for a 1-unit increase in X is called the **slope** of the line. The slope of the line measures the rate of change in Y as compared to the increase in X. The slope of the line shown in Figure 9.5 is $\frac{2}{1} = 2$. If you examine the equation $Y = 2X + 3$, you can see that the coefficient of X is the slope, 2, and the constant that is added to $2X$ is the Y-intercept, 3. When the equation of a straight line

Figure 9.5
A graph of the equation
$Y = 2X + 3$, showing
general characteristics

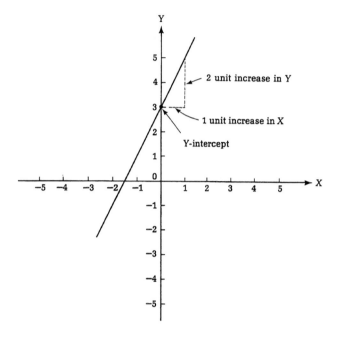

is written in this manner, it is called the **slope–intercept form** of a straight line. The variable Y is called the **dependent variable** and X is called the **independent variable.**

The Slope–Intercept Form of a Straight Line

$Y = mX + b$

where

Y = Dependent variable
m = Slope
X = Independent variable
b = Y-intercept

EXAMPLE 9.3   The relationship between temperatures expressed in degrees Fahrenheit, °F, and degrees Celsius, °C, is given by the equation

$$F = \left(\frac{9}{5}\right) C + 32$$

Let the dependent variable be F and the independent variable be C, and answer the following:

a. How do you know that this is an equation of a straight line?
b. What are the slope and the Y-intercept of this line?
c. Graph the line.

Solution **a.** Since the equation $F = (\%_5)C + 32$ is in the slope–intercept form of a straight line (with Y replaced by F, and X replaced by C), the graph of the equation is a straight line.

b. The coefficient of the dependent variable is the slope. Thus, the slope of this equation is $\%_5$. The constant 32 is the Y-intercept. This means that a temperature of 0°C is equal to a temperature of 32°F. Since the slope is $\%_5$, for every 1° increase in temperature on the Celsius scale, the temperature on the Fahrenheit scale increases by $\%_5$°.

c. A graph of the equation is given in Figure 9.6. The graph was drawn by plotting the points given in the table.

C	0	5	10
F	32	41	50

Figure 9.6
A graph of the equation
$F = (\%_5)C + 32$

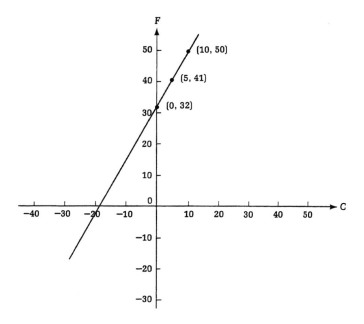

In statistics, the equation of a straight line is usually written in the following manner:

$Y = b + mX$

Thus, it is similar to the slope–intercept form, but on the right-hand side of the equation the Y-intercept is given first, and the X term (involving the slope) is given second.

EXAMPLE 9.4   Consider the following equation of a straight line:

$Y = 2 - X$

a. Graph this line.
b. What are the Y-intercept and the slope of this line?

Solution   **a.** The values of X and Y in the table satisfy the equation, and they were used to graph the line shown in Figure 9.7. Notice that this line slopes downward. As X *increases, Y decreases.*

X	2	0	4
Y	0	2	-2

b. The Y-intercept and slope may be obtained from either the equation of the line or the graph. Examining the equation, we see that the constant on the right-hand side is 2, so the Y-intercept equals 2. This is also evident from the graph. Since the coefficient of X is -1, the slope is -1. The *negative* slope means that for every unit increase in the value of X, the value of Y *decreases* by 1 unit. This can also be seen in the graph.

Figure 9.7
A graph of the equation
$Y = 2 - X$

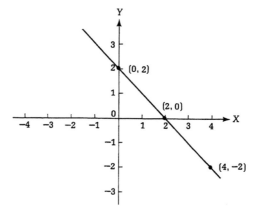

In the next box we summarize the information about the equation of a straight line provided in this section. In this text, an equation will always be written so that the dependent variable, Y, has a coefficient of 1 and appears on the left-hand side of the equation. If an equation were given to you in another form, you would have to rewrite it in this form before finding the Y-intercept and the slope.

The Equation of a Straight Line

$$Y = b + mX$$

where

Y = Dependent variable

b = Y-intercept (point where the line crosses the Y-axis)

m = Slope (amount Y increases or decreases for each unit increase in X)

X = Independent variable

The equation of a straight line is also called a **linear equation.** Thus, if two variables satisfy a linear, or straight-line, equation, we will say that the two variables are **linearly related.** A similar way to say this is that there is a **linear relationship** between Y and X.

Exercises (9.7–9.12)

Learning the Language

9.7 Define each of the following important terms, which were introduced in this section:
a. Slope of a line   b. Y-intercept of a line

9.8 Consider the equation $Y = 4 + 3X$. Identify the independent variable and the dependent variable.

Learning the Mechanics

9.9 Graph each of the following lines:
a. $Y = 3 + 2X$ c. $Y = -2 + 3X$ e. $Y = 4 - 3X$
b. $Y = 1 - X$ d. $Y = 5X$ f. $Y = -1 - 2X$

9.10 Give the slope and the Y-intercept for each of the lines graphed in Exercise 9.9.

Applying the Concepts

9.11 Suppose a book publisher has spent $20,000 to produce a new book and the producer will receive $10 for each copy sold. The publisher's profit, Y, on this book can then be given by the equation

$$Y = -20,000 + 10X$$

where X equals the number of books sold.
a. Graph the equation given above.
b. What are the slope and the Y-intercept for this equation?
c. Explain what the slope and the Y-intercept mean, in terms of this problem.

9.12 A man is on a strict diet and he loses 2 pounds every week. When the man began the diet, he weighed 250 pounds, and he plans to continue the diet for 1 year. The man's weight, Y, during that year may be expressed by the equation

$Y = 250 - 2X$

where X is the number of weeks he has been on the diet.
a. Graph the equation relating the man's weight to the number of weeks on the diet.
b. What are the slope and the Y-intercept of this equation?
c. Explain what the slope and the Y-intercept mean, in terms of this problem.

9.3

A PROBABILISTIC MODEL

In Section 9.2 we discussed the equation of a straight line. This type of equation describes an exact relationship between two variables X and Y, since for each value of X exactly one value of Y is determined. An equation that models an exact relationship between variables is called a **deterministic model.** The relationship between two variables cannot always be modeled exactly, and in this case we will use a **probabilistic model.**

For example, suppose you want to model the length of time it takes a person to respond to a stimulus as a function of the percentage of a certain drug in the bloodstream. If we believe that Y, the reaction time (in seconds), will be *exactly* .5 plus 1.5 times X, the percentage of the drug in the bloodstream, we would use the deterministic model

$Y = .5 + 1.5X$

Using this model (equation), we can always determine the exact value of Y when the value of X is known. There is no allowance for variability in the value of Y (at a given value of X) in this model.

Because the deterministic model does not allow for variability in the value of Y, it is not very realistic in this situation. If we repeatedly measured individuals' reaction times at one given amount of drug in the bloodstream, we would expect the reaction times to vary. There will surely be some variation in reaction times due to **random phenomena** that cannot be modeled (such as slight instrument measuring errors), or due to unincluded variables. It is more realistic to discard the deterministic model and use a probabilistic model that accounts for this random error.

Our probabilistic model will be made up of two parts. One part is the deterministic equation given above, and the other part is an **error term.** For example, a probabilistic model relating the response time, Y, to the percentage of drug would be given by

$Y = .5 + 1.5X + $ Error term

Figure 9.8**a** shows the reaction times, Y, for five values of X, the percentage of drug in the blood, using the deterministic model. All the responses must fall exactly on the line because a deterministic model leaves no room for error. Figure 9.7**b** shows a possible set of reaction times for the same values of X when using a probabilistic model. Note that the deterministic part of the model (the straight line itself) is the same. By including a random error term, the reaction times are allowed to vary about this line. Since reaction times would vary for a given value of X, the probabilistic model is a more reasonable model than a deterministic model.

Figure 9.8
Possible reaction times, Y, for five drug dosages, X

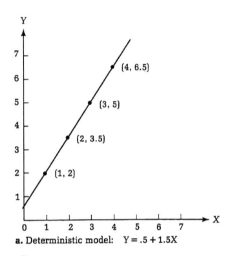

a. Deterministic model: $Y = .5 + 1.5X$

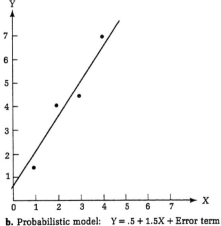

b. Probabilistic model: $Y = .5 + 1.5X + $ Error term

The general form of a probabilistic model for a linear relationship between two variables X and Y is given in the box on page 508. As you can see, Greek symbols, β_0 and β_1, are used to represent the Y-intercept and the slope of the model, respectively, since both are unknown parameters that must be estimated from experimental (sample) data. Estimation of these parameters is the subject of the next section.

A Probabilistic Model for a Linear Relationship

$$Y = \beta_0 + \beta_1 X + \varepsilon$$

where

$$Y = \text{Dependent variable}$$
$$X = \text{Independent variable}$$
$$\beta_0 \ (\text{beta zero}) = Y\text{-intercept of the line}$$
$$\beta_1 \ (\text{beta one}) = \text{Slope of the line}$$
$$\varepsilon \ (\text{epsilon}) = \text{Random error term}$$

It is helpful to follow these six steps when considering a probabilistic linear relationship between two variables:

Step 1: Define the elements in the model in terms of the problem of interest.

Step 2: Draw a scatterplot of the sample data.

Step 3: Use sample data to estimate β_0 and β_1.

Step 4: Specify the properties of the distribution of the random error term, and estimate any unknown parameters of this distribution.

Step 5: Statistically check the usefulness of the model.

Step 6: If the model proves useful, use it for prediction and estimation (If the model is not found to be useful, we would have to rethink the entire problem. Perhaps some other model is appropriate, or a larger sample should be examined, or the two variables are not related.)

EXAMPLE 9.5   In Section 9.1 we considered the relationship between the amount of money (measured in $1,000 units) that a brewery spends monthly on advertising and the total monthly sales (measured in $1,000 units). Write a linear model relating total sales, Y, to advertising expenditure, X, and define each term in the model. (This is step 1 of the six-step process given above.)

Solution   The probabilistic linear model is given by

$$Y = \beta_0 + \beta_1 X + \varepsilon$$

For this example, these terms are

Y = Brewery's total monthly sales, measured in $1,000 units

X = Brewery's total monthly advertising costs, measured in $1,000 units

β_0 = Y-intercept, the amount of total sales per month to be expected if no money were spent on advertising

β_1 = Slope of the line, the increase (or decrease) in total sales per month that is associated with a $1,000 (1-unit) increase in advertising

ε = Random error term; this term allows for variability in total sales for a fixed monthly advertising expenditure   ∎

In Section 9.1, a scatterplot of the data was constructed using a sample of seven data points (see Figure 9.1). This is step 2 of the modeling procedure outlined above, and steps 3–6 will be discussed in Sections 9.4–9.9.

Exercises (9.13–9.18)

Learning the Language

9.13 Define the following important terms, which were introduced in this section:
a. Deterministic model
b. Probabilistic model

9.14 State what each of the symbols β_0, β_1, and ε represent.

Applying the Concepts

9.15 When is it appropriate to use a deterministic model to describe the relationship between two variables? Give a practical example of two variables that you think have a deterministic relationship.

9.16 When is it appropriate to use a probabilistic model to describe the relationship between two variables? Give a practical example of two variables that you think should be modeled by a probabilistic relationship.

9.17 Refer to Exercise 9.5. We compared Y, the dollar loss due to shoplifting each week, to X, the number of sales clerks working each week. A probabilistic linear model relating these two variables is given by

$$Y = \beta_0 + \beta_1 X + \varepsilon$$

Define each term in this model.

9.18 Refer to Exercise 9.6. It was desired to investigate the relationship between the yield of potatoes, Y, and the amount of fertilizer, X, applied to plots of land. Define each term in the model

$$Y = \beta_0 + \beta_1 X + \varepsilon$$

9.4

FITTING THE MODEL: THE METHOD OF LEAST SQUARES

In this section we will discuss the third step in the consideration of a probabilistic linear relationship between two variables:

Step 3:   Use sample data to estimate β_0 and β_1.

Consider Example 9.5 (Section 9.3), where we discussed the relationship between a brewery's total monthly sales and monthly advertising expenditures. A scatterplot of the advertising expenditure and total sales from a random sample of seven months is shown in Figure 9.9. (The scatterplot was first shown in Figure 9.1.) The scatterplot indicates that it seems reasonable that X and Y are linearly related.

Figure 9.9
Scatterplot of monthly sales and advertising expenditures

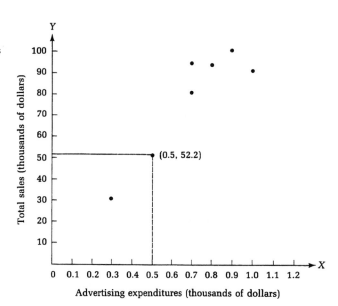

In Figure 9.10 we have redrawn the scatterplot from Figure 9.9 and added a straight line that reflects one possible relationship between Y and X. We have also indicated the deviation (vertical distance) between each point in the sample and the straight line. The sample data will be used to determine a unique equation of a straight line that best reflects the linear relationship between Y and X. This line will be found by using the **method of least squares.** The method of least squares chooses as the best line that line for which the sum of squared deviations between each sample point and the line is the smallest.

Figure 9.10
A straight line visually fit to
the data

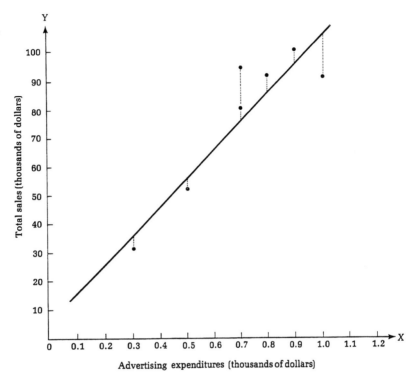

The line for which the sum of squared deviations is the smallest is called the **least squares line,** the **regression line,** or the **least squares prediction equation.** The least squares line is an estimate of $Y = \beta_0 + \beta_1 X$, the deterministic part of the probabilistic model, and the equation is written as

$$\hat{Y} = b_0 + b_1 X$$

where $\hat{Y}$ (read: Y-hat) indicates that this is an estimated quantity, and b_0 and b_1 are estimates of β_0 and β_1, respectively. The values of b_0 and b_1 that satisfy the least squares criterion are found by using the formulas given in the box on page 512. The summary statistics S_{XY} and S_X, used to calculate b_1, will also be used in later calculations. This notation was selected because S_{XY} involves sums of X and Y values, while S_X involves sums of X values only.

We will demonstrate the use of these formulas by finding the least squares prediction equation for the advertising expenditure–sales example. The preliminary calculations needed to find b_0 and b_1 are shown in Table 9.4. The values of X and Y were originally given in Table 9.2. Recall that X represents advertising expenditure and Y represents total sales. (Both are measured in thousands of dollars.)

Formulas for the Least Squares Estimates

$$b_1 = \frac{S_{XY}}{S_X} \qquad b_0 = \bar{Y} - b_1\bar{X}$$

where

$$\bar{Y} = \frac{\sum Y}{n} \qquad S_{XY} = n\sum (XY) - \left(\sum X\right)\left(\sum Y\right)$$

$$\bar{X} = \frac{\sum X}{n} \qquad S_X = n\sum X^2 - \left(\sum X\right)^2$$

$n =$ Number of pairs of X and Y values in the sample

Table 9.4

Preliminary calculations for the advertising–sales example

X	Y	XY	X^2
0.9	100.7	90.63	0.81
0.7	80.6	56.42	0.49
0.8	93.8	75.04	0.64
1.0	91.1	91.10	1.00
0.5	52.2	26.10	0.25
0.3	31.0	9.30	0.09
0.7	94.5	66.15	0.49
$\sum X = 4.9$	$\sum Y = 543.9$	$\sum (XY) = 414.74$	$\sum X^2 = 3.77$

Using the values from Table 9.4, we calculate S_{XY} and S_X.

$$S_{XY} = n\sum (XY) - \left(\sum X\right)\left(\sum Y\right) = 7(414.74) - (4.9)(543.9)$$
$$= 238.07$$

$$S_X = n\sum X^2 - \left(\sum X\right)^2 = 7(3.77) - (4.9)^2 = 2.38$$

Since the formula for b_0 contains b_1, we calculate b_1 first:

$$b_1 = \frac{S_{XY}}{S_X} = \frac{238.07}{2.38} = 100.029412 \approx 100.029$$

Then,

$$b_0 = \bar{Y} - b_1\bar{X} = \frac{\sum Y}{n} - b_1\left(\frac{\sum X}{n}\right)$$

$$= \frac{543.9}{7} - 100.029412\left(\frac{4.9}{7}\right) = 7.679412 \approx 7.679$$

(Notice that when calculating b_0, we retain six decimal places in the value used for b_1. This is done to minimize round-off errors.)

The least squares line is

$$\hat{Y} = b_0 + b_1X = 7.679 + 100.029X$$

The graph of this line is shown in Figure 9.11. The Y-intercept of the least squares line is 7.679, and it is an estimate of the monthly sales (in thousands of dollars) for the brewery if no money were spent on advertising. The slope of the least squares line is 100.029, and it is an estimate of the increase in monthly sales (in thousands of dollars) for each $1,000 increase in advertising expenditure.

Figure 9.11
The line $\hat{Y} = 7.679 + 100.029X$
fit to the data

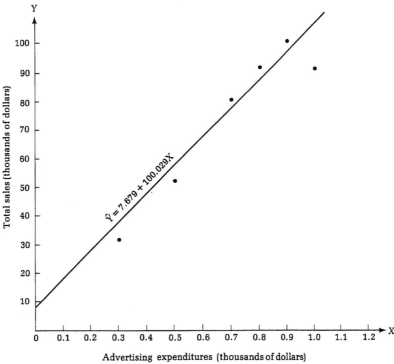

The predicted value of Y for a given value of X can be obtained by substituting the value of X into the least squares prediction equation. For example, when X = 0.6 (a $600 advertising expenditure), we predict Y to be

$$\hat{Y} = 7.679 + 100.029(0.6) = 67.696$$

Thus, we would predict monthly sales of $67,696 for a $600 advertising expenditure. We will provide a measure of reliability for this prediction in Section 9.9.

COMPUTER
EXAMPLE 9.2

In Computer Example 9.1, we used Minitab to construct a scatterplot of GPA versus number of hours worked per week for a random sample of fifty students. With GPA as the dependent variable and work hours as the independent variable, we used Minitab to find the least squares line for these data. The accompanying printout was obtained.

```
The regression equation is
GPA = 3.01 - 0.0178 WORKHRS

Predictor        Coef       Stdev      t-ratio
Constant       3.00718     0.06465       46.51
WORKHRS       -0.017795    0.004437      -4.01

s = 0.1962     R-sq = 25.1%     R-sq(adj) = 23.5%

Analysis of Variance

SOURCE         DF         SS          MS
Regression      1       0.61898     0.61898
Error          48       1.84747     0.03849
Total          49       2.46645

Unusual Observations
Obs.  WORKHRS      GPA      Fit  Stdev.Fit  Residual  St.Resid
 30     20.0     2.1100   2.6513   0.0411    -0.5413    -2.82R
 36     16.0     2.3300   2.7225   0.0305    -0.3925    -2.03R
 43     16.0     3.1200   2.7225   0.0305     0.3975     2.05R
 48      0.0     2.6300   3.0072   0.0647    -0.3772    -2.04R
```

In producing the least squares prediction equation, which is highlighted on the printout, Minitab also provided other information. We will examine other parts of this printout in the remaining sections of this chapter.* For now, we will concentrate on what Minitab has labeled "the regression equation," which we will often call the least squares prediction equation.

If we let $Y = $ GPA and $X = $ WORKHRS, the probabilistic linear model is written as

$$Y = \beta_0 + \beta_1 X + \varepsilon$$

From the highlighted portion of the printout we obtain the estimates of β_0 and β_1, respectively, to be

$$b_0 = 3.01 \quad \text{and} \quad b_1 = -0.0178$$

We would write the least squares line as

$$\hat{Y} = 3.01 - 0.0178X$$

Note that Minitab used "GPA" and "WORKHRS" to refer to the variables of interest. Also, it is presumed that you know that this is a prediction equation, so Minitab does

---

* This chapter will address only the material contained in the upper portion of the printout. For this reason, in succeeding examples we will not reproduce the parts headed "Analysis of Variance" and "Unusual Observations."

not print a "hat" over the dependent variable GPA. The scatterplot from Computer Example 9.1 is reproduced with a graph of the least squares line added.

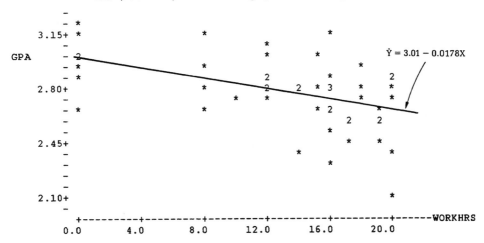

The Y-intercept of the least squares line is 3.01, and it is an estimate of the average GPA of students who do not hold a job. The slope of the least squares line, -0.0178, is an estimate of the decrease in GPA for each additional hour worked per week at a part-time job. ∎

Exercises (9.19–9.27)

Learning the Language

9.19 Describe the method of least squares.

9.20 What do the symbols Y, b_0, and b_1 mean?

Learning the Mechanics

9.21 The following quantities were calculated for a sample of $n = 10$ data points:

$$\sum X = 20 \qquad \sum Y = 100 \qquad \sum(XY) = 75 \qquad \sum X^2 = 50$$

a. Calculate the values of S_X and S_{XY}.
b. Calculate the value of b_1.
c. Calculate the value of b_0.
d. Give the least squares prediction equation.

9.22 Consider the six data points shown in the table:

X	1	2	3	4	5	6
Y	1	2	2	3	5	5

a. Calculate $\sum X$, $\sum X^2$, $\sum Y$, and $\sum(XY)$, and give the value of n.
b. Calculate S_x and S_{xy}.
c. Calculate the least squares estimates of β_0 and β_1.
d. Draw a scatterplot of the data and graph the least squares line. Does the line fit the data?

9.23 Consider the five data points shown in the table:

X	−2	−1	0	1	2
Y	4	3	3	1	−1

a. What are the least squares estimates of β_0 and β_1?
b. Plot the data points and graph the least squares line. Does the line pass through the data points?
c. Predict the value of Y for $X = 1$.
d. Predict the value of Y for $X = -1.5$.

Applying the Concepts

9.24 In recent years, physicians have used the so-called "dividing reflex" to reduce abnormally rapid heartbeats in humans by briefly submerging the patient's face in cold water. The reflex, triggered by cold temperatures, is an involuntary neural response that shuts off circulation to the skin, muscles, and internal organs, and diverts extra oxygen-carrying blood to the heart, lungs, and brain. A research physician conducted an experiment to investigate the effects of various cold water temperatures on the pulse rate of small children. The data for seven 6-year-olds appear in the table.

Child	Water Temperature X, °F	Pulse Rate Decrease Y, Beats/Minute
1	68	2
2	65	5
3	70	1
4	62	10
5	60	9
6	55	13
7	58	10

a. Find the least squares line for the data.
b. Construct a scatterplot for the data. Then graph the least squares line as a check on your calculations.
c. If the water temperature is 60°F, predict the drop in pulse rate for a 6-year-old child.

9.25 An appliance company is interested in relating the sales rate of 17-inch color television sets to the price per set. To do this, the company randomly selected 15 weeks in the past year and recorded the number of sets sold and the price at which sets were being sold during that week. The data are shown in the table.

Week	Number of Sets Sold Y	Price X, Dollars	Week	Number of Sets Sold Y	Price X, Dollars
1	55	350	9	20	400
2	54	360	10	45	340
3	25	385	11	50	350
4	18	400	12	35	335
5	51	370	13	30	330
6	20	390	14	30	325
7	45	375	15	53	365
8	19	390			

a. Find the least squares line relating Y to X.

b. Plot the data and graph the least squares line as a check on your calculations.

*9.26 Refer to Exercise 9.5. The dollar volume lost for each of thirty-four randomly selected weeks, along with the number of sales clerks working each week, were recorded. The data are repeated here for your convenience.

Week	Number of Sales Clerks X	Loss Y	Week	Number of Sales Clerks X	Loss Y
1	14	161	18	18	109
2	14	163	19	18	128
3	10	189	20	15	141
4	10	146	21	20	104
5	17	121	22	12	189
6	16	161	23	19	104
7	14	194	24	14	187
8	14	169	25	18	100
9	11	186	26	15	122
10	16	178	27	12	170
11	17	91	28	15	152
12	16	151	29	12	201
13	18	103	30	13	196
14	15	146	31	13	162
15	16	127	32	15	144
16	12	163	33	14	157
17	20	119	34	12	173

a. Use a computer package to find the least squares prediction equation.

b. Predict the dollar loss if 15 sales clerks were working.

*9.27   Refer to Exercise 9.6. The yield of potatoes and level of fertilizer applied to experimental plots of equal size were recorded. The data are repeated here.

Plot	X	Y	Plot	X	Y
1	1.0	24.9	21	3.0	31.9
2	1.5	27.5	22	3.5	33.0
3	2.0	34.2	23	4.0	30.9
4	2.5	34.5	24	4.5	33.1
5	3.0	31.6	25	1.0	26.1
6	3.5	31.4	26	1.5	25.6
7	4.0	41.1	27	2.0	30.7
8	4.5	36.1	28	2.5	28.6
9	1.0	28.2	29	3.0	30.7
10	1.5	22.2	30	3.5	33.0
11	2.0	27.9	31	4.0	34.6
12	2.5	25.0	32	4.5	38.5
13	3.0	29.4	33	1.0	31.1
14	3.5	35.1	34	1.5	29.2
15	4.0	33.2	35	2.0	26.8
16	4.5	28.9	36	2.5	31.5
17	1.0	21.4	37	3.0	31.4
18	1.5	30.8	38	3.5	36.7
19	2.0	34.0	39	4.0	40.7
20	2.5	26.6	40	4.5	39.2

a. Use a computer package to find the least squares estimates for β_0 and β_1.
b. According to your least squares line, approximately how many pounds of potatoes would you expect from a plot to which 3.75 pounds of fertilizer were applied?
c. Graph the least squares line on a scatterplot of the data. Does the line fit the data?

9.5

THE RANDOM ERROR COMPONENT

In Example 9.5 (Section 9.3) we defined the probabilistic model

$$Y = \beta_0 + \beta_1 X + \varepsilon$$

relating total monthly sales of a brewery, Y, to the monthly advertising expenditure, X. We also drew a scatterplot for a sample of seven months' data. In Section 9.4 we found the least squares prediction equation to be

$$\hat{Y} = b_0 + b_1 X = 7.679 + 100.029X$$

We will now discuss the fourth step to be followed when considering a probabilistic linear relationship between two variables:

Step 4: Specify the properties of the distribution of the random error term, and estimate any unknown parameters of this distribution.

In the remainder of this chapter we will assume that the distribution of the random error term, ε, has the following properties:

Properties of the Random Error Term, ε

Property 1: The average of the errors over an infinitely long series of experiments is 0 for each value of the independent variable X.

Property 2: The variance of the errors is equal to a constant, σ^2, for all values of X.

Property 3: The errors are normally distributed.

Property 4: The error associated with one value of Y has no effect on the errors associated with other values of Y. Thus, any two errors are independent.

The meanings of the first three properties are portrayed in Figure 9.12, which shows the distributions of errors for three different values of X, namely, X_1, X_2, and X_3. The errors in each case are distributed about the line $Y = \beta_0 + \beta_1 X$, the deterministic part of the probabilistic model. At each value of X the errors form a normal distribution (property 3). Some errors are positive (the observed values of Y are above the line),

Figure 9.12
The distribution of errors

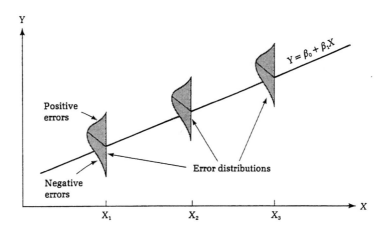

some errors are negative (the observed values of Y are below the line), and the average error is 0 (property 1). All the distributions have the same amount of spread, or variability, and thus there is a constant variance, σ^2 (property 2).

Property 4 states that any two errors must be independent. If the sample data are randomly chosen, this property will be satisfied. Thus, we will always be sure to obtain randomly sampled observations when finding a least squares line and using it to make inferences.

The only unknown parameter that appears in the distribution of the errors is the constant variance, σ^2. Since σ^2 is a measure of the variability of the errors about the line $Y = \beta_0 + \beta_1 X$, it makes sense to estimate σ^2 by using comparable errors in the sample data about the least squares line, $\hat{Y} = b_0 + b_1 X$. An estimate of σ^2 will use the quantity $\sum(Y - \hat{Y})^2$, which is the sum of the squared deviations of the observed values of Y from the least squares estimates, $\hat{Y}$. To obtain s^2, an estimate of σ^2, we must divide $\sum(Y - \hat{Y})^2$ by $(n - 2)$, the appropriate degrees of freedom associated with this quantity. Thus, σ^2 is estimated by

$$s^2 = \frac{\sum(Y - \hat{Y})^2}{n - 2}$$

It would be very difficult to calculate s^2 using this formula. Instead, we will use the equivalent formula, which is given in the box.

Formula for Calculating s^2

$$s^2 = \frac{S_Y - b_1 S_{XY}}{n(n - 2)}$$

where

$$S_Y = n \sum Y^2 - \left(\sum Y\right)^2$$

Warning: When calculating s^2, be certain to carry at least six significant figures for the value of b_1, in order to avoid a substantial round-off error in the calculated value of s^2.

For the advertising–sales example, we have already made some preliminary calculations in Section 9.4. These calculations are repeated in Table 9.5 for convenience. We have also calculated the value of $\sum Y^2$, since it is needed to calculate s^2.

The value of S_Y is calculated to be

$$S_Y = n \sum Y^2 - \left(\sum Y\right)^2 = 7(46{,}350.59) - (543.9)^2$$
$$= 28{,}626.92$$

Recall that the value of b_1 to six significant figures and the value of S_{XY} are

$$b_1 = 100.029412 \qquad S_{XY} = 238.07$$

Table 9.5

X	Y	XY	X^2	Y^2
0.9	100.7	90.63	0.81	10,140.49
0.7	80.6	56.42	0.49	6,496.36
0.8	93.8	75.04	0.64	8,798.44
1.0	91.1	91.10	1.00	8,299.21
0.5	52.2	26.10	0.25	2,724.84
0.3	31.0	9.30	0.09	961.00
0.7	94.5	66.15	0.49	8,930.25
$\sum X = \overline{4.9}$	$\sum Y = \overline{543.9}$	$\sum (XY) = \overline{414.74}$	$\sum X^2 = \overline{3.77}$	$\sum Y^2 = \overline{46,350.59}$

Calculations for the advertising–sales example

The value of s^2 is then calculated to be

$$s^2 = \frac{S_Y - b_1 S_{XY}}{n(n-2)} = \frac{28,626.92 - 100.029412(238.07)}{7(7-2)}$$

$$= \frac{4,812.917886}{35} = 137.511940 \approx 137.512$$

As you will subsequently see in Sections 9.6 and 9.9, s^2 will be used when making inferences.

The quantities $\sum X$, $\sum Y$, S_X, S_Y, and S_{XY} were used in finding b_0, b_1, and s^2. Some of these quantities will also be used again in later calculations. In any regression problem, it is a good idea to first calculate all these quantities, so that you will have them to use in further calculations. Also, remember to retain the value of b_1 to six significant figures.

COMPUTER EXAMPLE 9.3 In the last computer example (Section 9.4), we found the least squares prediction equation relating GPA to the number of hours worked. The printout is reproduced here.

```
The regression equation is
GPA = 3.01 - 0.0178 WORKHRS

Predictor        Coef        Stdev      t-ratio
Constant      3.00718      0.06465       46.51
WORKHRS      -0.017795     0.004437      -4.01

s = 0.1962      R-sq = 25.1%      R-sq(adj) = 23.5%
```

The value highlighted on the printout is s, the square root of s^2, the estimate of σ^2. To obtain the value of s^2 we simply square the value of s given in the printout. Thus, the estimated value of σ^2 is

$$s^2 = (0.1962)^2 = .03849$$

■

Exercises (9.28–9.37)

Learning the Language

9.28 State the four properties of the distribution of the random error term, ε.

9.29 What do the symbols s^2 and σ^2 mean?

Learning the Mechanics

9.30 The following quantities were calculated from a sample of $n = 15$ data points:

$$b_1 = 4.3 \qquad S_Y = 8{,}100 \qquad S_{XY} = 1{,}608$$

Calculate s^2, the estimate of σ^2.

9.31 The following six data points were given in Exercise 9.22:

X	1	2	3	4	5	6
Y	1	2	2	3	5	5

And the following quantities were calculated:

$$b_1 = .857143 \qquad \sum Y = 18 \qquad S_{XY} = 90$$

a. Calculate $\sum Y^2$ and S_Y.
b. Calculate s^2.

9.32 In Exercise 9.23 you were given these five data points:

X	−2	−1	0	1	2
Y	4	3	3	1	−1

And the following quantities were calculated:

$$b_1 = -1.2 \qquad \sum Y = 10 \qquad S_{XY} = -60$$

Calculate s^2 using these data.

9.33 Consider the following five data points:

X	0	1	2	3	4
Y	3	5	7	9	11

a. Calculate the values of b_0, b_1, and S_Y.
b. Calculate the value of s^2.
c. Plot the data points and graph the least squares line.
d. Does the value of s^2 found in part b describe the variability of the points about the line graphed in part c?

Applying the Concepts

9.34 An electronics dealer believes that there is a linear relationship between the number of hours of quadraphonic programming on a city's FM stations and sales of quadraphonic systems. Records for the dealer's sales during the last 6 months and the amount of quadraphonic programming for those months are given in the table.

Month	Average Amount of Quadraphonic Programming X, Hours	Number of Quadraphonic Systems Sold Y
1	33.6	7
2	36.3	10
3	38.7	13
4	36.6	11
5	39.0	14
6	38.4	18

a. Find the least squares prediction equation.
b. Plot the data and graph the least squares line.
c. Calculate the value of s^2.

9.35 A company keeps extensive records on its new salespeople on the premise that sales should increase with experience. A random sample of seven new salespeople produced the data on experience and sales shown in the table.

Months on Job X	Monthly Sales Y, Thousands of Dollars
2	2.4
4	7.0
8	11.3
12	15.0
1	0.8
5	3.7
9	12.0

a. Find the least squares estimates of β_0 and β_1.
b. Calculate the value of s^2.

*9.36 In Exercise 9.26 you used the computer to find the least squares line relating loss due to shoplifting and the number of sales clerks working. What is the value of s^2?

*9.37 In Exercise 9.27 you used the computer to find estimates of β_0 and β_1 for data representing the yield of potatoes per plot and amount of fertilizer applied. What is the value of s^2?

9.6

MAKING INFERENCES ABOUT THE SLOPE, β_1

Now that we have defined the probabilistic model, estimated β_0 and β_1, discussed the distribution of the error term ε, and estimated σ^2, we will proceed to the fifth step in the modeling procedure.

Step 5: Statistically check the usefulness of the model.

Suppose that two variables, X and Y, are not linearly related. What could be said about the values of β_0 and β_1 in the probabilistic model

$$Y = \beta_0 + \beta_1 X + \varepsilon$$

if X does not help to predict Y? This would mean that we would not expect the value of Y to increase or decrease as the value of X increases. In other words, β_1, the slope of the deterministic part of the model, would equal 0. A graph of a model with slope equal to 0 and Y-intercept equal to β_0 is shown in Figure 9.13.

Figure 9.13
A graph of the model
$Y = \beta_0 + \varepsilon$
$(\beta_1 = 0)$

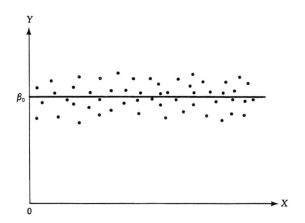

To determine whether X helps to predict Y in a linear manner, we must determine whether β_1 differs from 0. Therefore, if we want to show that X and Y are linearly related $(\beta_1 \neq 0)$, we need to test the hypotheses

$$H_0: \quad \beta_1 = 0 \quad \text{and} \quad H_a: \quad \beta_1 \neq 0$$

If the data provide support for the alternative hypothesis, we will conclude that X does help to predict Y using the linear model. Thus, this test will provide an indication of whether the model is useful.

The test of hypothesis about β_1 is given in the box. This procedure is valid when the distribution of ε has the four properties given in Section 9.5.

A Test of Hypothesis About β_1, the Slope of the Line $Y = \beta_0 + \beta_1 X$

H_0: $\beta_1 = 0$

H_a: $\beta_1 > 0$ or H_a: $\beta_1 < 0$ or H_a: $\beta_1 \neq 0$

Test statistic: $t = \dfrac{b_1 \sqrt{S_x}}{\sqrt{ns^2}}$

Rejection region: Use Appendix Table V, with $(n - 2)$ df.

Notice that 0 is the only value of the slope about which we will test hypotheses. The test statistic has a t distribution, and the rejection region is found by using Table V with $(n - 2)$ df, the degrees of freedom associated with s^2. All the quantities needed to calculate the test statistic were calculated in previous sections.

We may also form a confidence interval for the slope, β_1. This confidence interval is shown in the box.

Confidence Interval for the Slope, β_1

The confidence interval for β_1 is expressed as

$$b_1 - t\sqrt{\frac{ns^2}{S_x}} \quad \text{to} \quad b_1 + t\sqrt{\frac{ns^2}{S_x}}$$

The value of t is obtained from Appendix Table VI, using $(n - 2)$ df.

We will now perform a test of hypothesis to determine whether there is a linear relationship between monthly sales and advertising expenditure in the example discussed in previous sections. When we calculated $b_1 = 100.029$ in Section 9.4, we had to calculate $S_x = 2.38$. We have also already found $s^2 = 137.512$, and there were $n = 7$ pairs of observations. Using these data, we will test to determine whether there is sufficient evidence to indicate that monthly sales tend to increase as advertising expenditures increase. We will use $\alpha = .05$. The test of interest is:

H_0: $\beta_1 = 0$ H_a: $\beta_1 > 0$

Test statistic: $t = \dfrac{b_1 \sqrt{S_x}}{\sqrt{ns^2}}$

Rejection region: Reject H_0 if $t > 2.02$.

The rejection region was found using Appendix Table V, with $\alpha = .05$, the appropriate column for a one-tailed test, and $n - 2 = 5$ df.

Calculating the value of the test statistic, we obtain

$$t = \frac{b_1\sqrt{S_x}}{\sqrt{ns^2}} = \frac{100.029\sqrt{2.38}}{\sqrt{7(137.512)}} = 4.97$$

Since $t = 4.97 > 2.02$, we will reject H_0. We are 95% confident that monthly sales are linearly related to advertising expenditures, and that the slope is positive. In other words, the amount spent on advertising does help to predict sales, and an increase in advertising expenditure is associated with an increase in sales.

What interpretation will be made if a calculated t-value does not fall in the rejection region for a test about β_1? We know from previous discussions of the philosophy of hypothesis testing that we will *not* accept the null hypothesis when such a t-value is found. That is, we will not conclude that $\beta_1 = 0$. Data from additional sampling might provide sufficient evidence to conclude that $\beta_1 \neq 0$, or a more complex relationship (other than linear) may exist between Y and X.

Finally we will form a 90% confidence interval for β_1, the slope from the advertising–sales example. The general form of the confidence interval is

$$b_1 - t\sqrt{\frac{ns^2}{S_x}} \quad \text{to} \quad b_1 + t\sqrt{\frac{ns^2}{S_x}}$$

Using Appendix Table VI, we find the value of t to be 2.02. Substituting this value and the other quantities into the formula, we obtain

$$100.029 - 2.02\sqrt{\frac{7(137.512)}{2.38}} \quad \text{to} \quad 100.029 + 2.02\sqrt{\frac{7(137.512)}{2.38}}$$
$$100.029 - 40.624 \quad \text{to} \quad 100.029 + 40.624$$
$$59.405 \quad \text{to} \quad 140.653$$

We are 90% confident that the value of β_1 is between 59.405 and 140.653. This means that each $1,000 increase in advertising expenditure is associated with at least a $59,405 increase in monthly sales, and perhaps as much as a $140,653 increase in monthly sales. This is a rather large interval due to the small sample ($n = 7$) that was taken. We could obtain a narrower interval by collecting more information, that is, by increasing the sample size.

COMPUTER
EXAMPLE 9.4   Consider the data on GPA and hours worked we have been using. We would like to determine whether there is sufficient evidence to conclude that GPA tends to decrease in a linear manner as the number of hours worked per week increases. The test is to be conducted at the .01 level of significance. The elements of the test are:

H_0: $\beta_1 = 0$ H_a: $\beta_1 < 0$

Test statistic: $t = \dfrac{b_1\sqrt{S_x}}{\sqrt{ns^2}}$

Rejection region: Reject H_0 if $t < -2.33$.

The rejection region was found using Appendix Table V, with $\alpha = .01$, the appropriate column for a one-tailed test, and $n - 2 = 48$ df.

Minitab may be used to calculate the value of the test statistic. As a matter of fact, this information is provided on the same printout given in the previous computer example. The printout is reproduced here with the information of interest highlighted. The value of the test statistic (called a "t-ratio" by Minitab) associated with the variable "WORKHRS" is -4.01. Since this value is less than -2.33, we may conclude with 99% confidence that GPAs tend to decrease in a linear manner as the number of hours worked per week increases.

```
The regression equation is
GPA = 3.01 - 0.0178 WORKHRS
```

Predictor	Coef	Stdev	t-ratio
Constant	3.00718	0.06465	46.51
WORKHRS	-0.017795	0.004437	-4.01

```
s = 0.1962      R-sq = 25.1%      R-sq(adj) = 23.5%
```

Also given for "WORKHRS" are values for "Coef" and "Stdev." The "Coef" value of -0.017795 is the value of b_1 given in the regression equation, where it was rounded to -0.0178. This is sometimes referred to as the "coefficient of WORKHRS."

The confidence interval formula for β_1 is

$$b_1 - t\sqrt{\frac{ns^2}{S_x}} \quad \text{to} \quad b_1 + t\sqrt{\frac{ns^2}{S_x}}$$

The term $\sqrt{ns^2/S_x}$ is sometimes called the standard deviation of b_1. The computer value of $\sqrt{ns^2/S_x}$ is given by Minitab to be the "Stdev" value of 0.004437.

This information, along with a t-value, may be used to form a confidence interval for β_1. For example, to form a 99% confidence interval for β_1, from Appendix Table VI we get $t = 2.58$ and the interval is calculated to be

$$-.017795 - (2.58)(.004437) \quad \text{to} \quad -.017795 + (2.58)(.004437)$$
$$-.017795 - .0114475 \quad \text{to} \quad -.017795 + .0114475$$
$$-.0292425 \quad \text{to} \quad -.0063475$$

This means that we are 99% confident that each hour increase in the number of hours worked is associated with a decrease of between .006 and .029 in GPA. ∎

Exercises (9.38–9.48)

Learning the Mechanics

9.38 The following quantities are calculated for a random sample of $n = 20$ data points:

$$b_1 = .64 \qquad S_x = 12.8 \qquad s^2 = .37$$

a. Test H_0: $\beta_1 = 0$ against H_a: $\beta_1 \neq 0$. Use $\alpha = .05$.
b. Form a 95% confidence interval for β_1.
c. Test H_0: $\beta_1 = 0$ against H_a: $\beta_1 > 0$. Use $\alpha = .05$.

9.39 The following quantities are calculated for a random sample of $n = 8$ data points:

$$b_1 = -6.4 \qquad S_x = 21.6 \qquad s^2 = 12.5$$

a. Test $H_0: \beta_1 = 0$ against $H_a: \beta_1 < 0$. Use $\alpha = .01$.
b. Form a 90% confidence interval for β_1.

9.40 Consider the following six data points:

X	0	2	3	4	7	9
Y	7	4	5	4	2	1

a. Calculate b_0 and b_1.
b. Calculate s^2.
c. Form a 99% confidence interval for β_1.
d. Test $H_0: \beta_1 = 0$ against $H_a: \beta_1 \neq 0$. Use $\alpha = .01$.
e. Test $H_0: \beta_1 = 0$ against $H_a: \beta_1 < 0$. Use $\alpha = .10$.

9.41 Consider the following four data points:

X	10	15	20	25
Y	1.2	1.4	2.1	2.8

a. Is there sufficient evidence to conclude that X and Y are linearly related? Use $\alpha = .05$.
b. Form a 95% confidence interval for β_1.

Applying the Concepts

9.42 A breeder of thoroughbred horses thinks that there may be a linear relationship between the gestation period and the length of a horse's life. The information in the table was supplied to the breeder by the owners of several thoroughbred stables. [*Note:* Horses have the greatest known variation of gestation period of any species.]

Horse	Gestation Period X, Days	Lifetime Y, Years
1	416	24.0
2	279	25.5
3	298	20.0
4	307	21.5
5	356	22.0
6	403	23.5
7	265	21.0

a. Fit a least squares line to the data. Plot the data points and graph the least squares line as a check on your calculations.

b. Do the data provide sufficient evidence to support the horse breeder's hypothesis? Test using $\alpha = .05$.

c. Find a 90% confidence interval for β_1.

9.43 A group of eight children ranging from 10 to 12 years old was given a verbal test in order to study the relationship between the number of words used and the silence interval before response. The tester believes that a linear relationship exists between the two variables. Each subject was asked a series of questions, and the total number of words used in answering was recorded. The time (in seconds) before the subject responded to each question was also recorded. The data are given in the table.

Subject	Total Words Y	Total Silence Time X, Seconds	Subject	Total Words Y	Total Silence Time X, Seconds
1	61	23	5	91	17
2	70	37	6	63	21
3	42	38	7	71	42
4	52	25	8	55	16

a. Write a probabilistic linear model relating total words to total silence time, and use the least squares method to estimate the deterministic part of the model.

b. Does X contribute information for the prediction of Y? Test using $\alpha = .05$.

9.44 In an investigation into the possibility of dysfunction in depth perception in schizophrenics, subjects were asked to align images at different distances from a light source. Eleven schizophrenic patients took part in the experiment. The distance and perception score recorded for each subject are listed in the table. (Better depth perception is reflected by higher perception scores.)

Subject	Distance from Light X, Feet	Perception Score Y	Subject	Distance from Light X, Feet	Perception Score Y
1	6	22	7	12	13
2	10	16	8	8	17
3	4	25	9	4	26
4	5	26	10	8	22
5	10	18	11	12	19
6	6	18			

a. Construct a scatterplot for the data.

b. Use the least squares method to estimate the Y-intercept and the slope of a linear model.

c. Plot the least squares line on your scatterplot.

d. Test the adequacy of the model by testing the null hypothesis $H_0: \beta_1 = 0$ against the alternative hypothesis, $H_a: \beta_1 < 0$ (that the average perception score decreases as the distance of the subject from the light increases). Use $\alpha = .05$.

9.45 Buyers are often influenced by bulk advertising of a product. For example, suppose you have a product that sells for 25¢. If it is advertised at 2 for 50¢, 3 for 75¢, or 4 for $1, some people may think they are getting a bargain and buy it in these quantities. To test this theory, a store manager advertised an item for equal periods of time at five different bulk rates and observed the data listed in the table. Do the data provide sufficient evidence to indicate that sales increase as the number in the bulk increases? Test at the .01 level of significance.

Advertised Number in Bulk Sale X	Volume Sold Y
1	27
2	36
3	34
4	63
5	52

9.46 A large car rental agency sells its cars after using them for a year. Among the records kept for each car are mileage and maintenance costs for the year. To evaluate the performance of a particular car model in terms of maintenance costs, the agency wants to use a 95% confidence interval to estimate the mean increase in maintenance costs for each additional 1,000 miles driven. Assume the relationship between maintenance cost and miles driven is linear. Use the data in the table to accomplish the objective of the rental agency.

Car	Miles Driven X, Thousands	Maintenance Cost Y, Dollars
1	54	326
2	27	159
3	29	202
4	32	200
5	28	181
6	36	217

*9.47 In Exercise 9.26 you used the computer to find the least squares line relating shoplifting loss to number of sales clerks working per week.

a. Is there evidence at the .05 level of significance to conclude that the amount lost due to shoplifting decreases as the number of sales clerks increases?

b. Find a 95% confidence interval for β_1.

*9.48 In Exercise 9.27 you used the computer to find least squares estimates of β_0 and β_1. The data of interest were yields of potatoes and amounts of fertilizer applied to plots of land.

  a. Is there evidence that potato yield is linearly related to the amount of fertilizer applied? Use $\alpha = .01$.

  b. Find a 99% confidence interval for β_1.

9.7

THE PEARSON PRODUCT MOMENT COEFFICIENT OF CORRELATION

In this section and the next section, we will continue to discuss ways to evaluate the usefulness of the model. Once we are convinced that there is a linear relationship between two variables, Y and X (from the results of the test of hypothesis discussed in Section 9.6), it is helpful to have a measure of the **strength of the linear relationship.** In Figure 9.14, we have drawn graphs representing two possible linear relationships between two variables, X and Y. However, the strengths of the relationships are quite different.

Figure 9.14
Linear relationships between X and Y

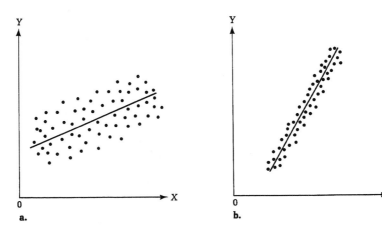

a.

b.

Which of the graphs in Figure 9.14 do you think indicates the stronger linear relationship between Y and X? We think you will agree it is the graph shown in part **b** of the figure. There is little variability about the line, and the slope of the line is quite steep. As the value of X increases, it is clearly associated with an increase in Y. The weaker relationship is shown in part **a** of Figure 9.14. Although Y tends to increase as

X increases, the relationship is not as strong as that shown in part **b**. This is because there is a gradual slope to the line, and there is a great deal of variability in the data.

If two variables have a strong linear relationship, they are often said to be "highly correlated." When people discuss the **correlation** between two variables, they are actually discussing the strength of the linear relationship. A numerical descriptive measure of the correlation between two variables is provided by the **Pearson product moment coefficient of correlation.**

Definition 9.1

The **Pearson product moment coefficient of correlation,** or simply the **coefficient of correlation,** is a measure of the strength of the linear relationship between two variables, X and Y. It is denoted by r, and is found by the formula

$$r = \frac{S_{XY}}{\sqrt{S_X \cdot S_Y}}$$

Note that the formula for the **correlation coefficient, r,** involves quantities that were calculated in previous sections. In fact, the numerator of r is the same as the numerator of b_1. Thus, r will be positive when the slope of the regression line is positive and negative when the slope of the regression line is negative. If $b_1 = 0$ (indicating X does not help to predict Y in a linear manner), then r will also be equal to 0. The value of r is scaleless and will always be between -1 and $+1$, regardless of the units of X and Y.

A value of r near or equal to 0 means that the sample shows a very weak linear relationship between X and Y. The closer r becomes to $+1$ or -1, the stronger the linear relationship between X and Y in the sample. As a matter of fact, if $r = +1$ or $r = -1$, all the sample points must fall exactly on the regression line. In Figure 9.15 we have drawn several scatterplots portraying relationships between values of r and sample data.

Using the data for the advertising–sales example, we will calculate the value of r. First, we list all the quantities we will need that were previously calculated:

$S_{XY} = 238.07$

$S_X = 2.38$

$S_Y = 28{,}626.92$

The value of r is

$$r = \frac{S_{XY}}{\sqrt{S_X \cdot S_Y}} = \frac{238.07}{\sqrt{(2.38)(28{,}626.92)}} = .912$$

Figure 9.15
Scatterplots and the values
of r

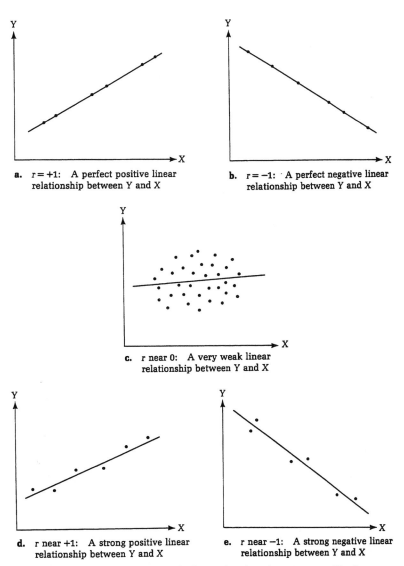

a. $r = +1$: A perfect positive linear
relationship between Y and X

b. $r = -1$: A perfect negative linear
relationship between Y and X

c. r near 0: A very weak linear
relationship between Y and X

d. r near +1: A strong positive linear
relationship between Y and X

e. r near −1: A strong negative linear
relationship between Y and X

The fact that r is positive and near 1 indicates that there is a strong positive linear relationship between X and Y in the sample. This can be seen by examining Figure 9.16
(page 534).

We must be careful not to jump to any unwarranted conclusions when interpreting
the coefficient of correlation. For instance, the brewery may be tempted to conclude
that the best thing it can do to increase sales is to spend a great deal of money on
advertising. That is, they may conclude that there is a **causal relationship** between

Figure 9.16
Scatterplot for
advertising–sales data

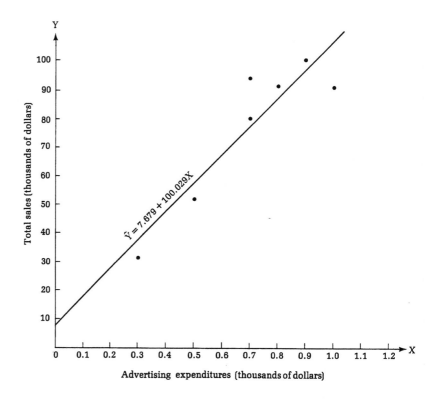

Advertising expenditures (thousands of dollars)

advertising expenditures and sales, with increased advertising expenditures causing increased sales. However, *high correlation does not imply causality.* In fact, many factors have probably contributed to the increase of both the advertising expenditures and the sales. The economy may have inflated, or perhaps more products are sold by the brewery. *We must be careful not to infer a causal relationship on the basis of high sample correlation.* The appropriate interpretation of a high correlation in the sample is simply that there is a strong linear relationship between X and Y in the sample data.

The correlation coefficient r measures the strength of the linear relationship between X-values and Y-values in the sample, and a similar linear coefficient of correlation exists for the population from which the data points were selected. The population correlation coefficient is denoted by the symbol ρ (rho). As you might expect, ρ is estimated by the corresponding sample statistic, r. Or, rather than estimating ρ, we might want to test the hypothesis $H_0: \rho = 0$ against $H_a: \rho \neq 0$—that is, test the hypothesis that X contributes no information for the prediction of Y using the straight-line model against the alternative that the two variables are at least linearly related.

However, we have already performed this identical test in Section 9.6 when we tested $H_0: \beta_1 = 0$ against $H_a: \beta_1 \neq 0$. When we tested the null hypothesis $H_0: \beta_1 = 0$

in the advertising–sales example, the data led to a rejection of the null hypothesis at the $\alpha = .05$ level of significance. This means that the null hypothesis $H_0: \rho = 0$ would also be rejected at the .05 level. The difference between the values of b_1, the slope of the regression line, and r, the correlation coefficient, is the scale of measurement. Therefore, some of the information these two quantities provide about the utility of the least squares model is redundant. We will use the slope to make inferences about the existence of a positive or negative linear relationship between two variables, and we will use r to judge the strength of the relationship.

9.8

THE COEFFICIENT OF DETERMINATION

In Section 9.7 we presented the correlation coefficient r. The meaning of the value of r is quite clear if r is near $+1$, -1, or 0. However, it is not as easy to interpret a value such as $r = .4$. Does such a value indicate a strong or weak linear relationship between two variables? To help answer this question, consider the **coefficient of determination.**

Definition 9.2

The **coefficient of determination** is the square of the coefficient of correlation. It is denoted by r^2, and its value is the fraction of the variability in the dependent variable Y that can be attributed to a linear relationship between Y and X.

Since r is always between -1 and $+1$, r^2 is always between 0 and 1. In Figure 9.17 we have drawn several graphs to demonstrate the relationships between values of r^2 and sample data. In part **a** of the figure, all the sample points fall on a straight line, and $r^2 = 1$. Since the values of Y do not vary from the least squares line, all (100%) of the variation in the Y-values is attributable to the linear relationship between Y and X. There is no unexplained variability in the values of Y.

Figure 9.17
Values of r^2 for different samples

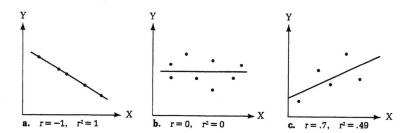

a. $r = -1$, $r^2 = 1$ b. $r = 0$, $r^2 = 0$ c. $r = .7$, $r^2 = .49$

REGRESSION AND CORRELATION: A STRAIGHT-LINE RELATIONSHIP

In part **b** of Figure 9.17, there is no linear relationship between Y and X for the sample data, and $r^2 = 0$. Thus, none of the variation in the values of Y is attributable to a linear relationship between Y and X. All of the variation in the Y-values is unexplained.

Finally, in part **c** of Figure 9.17, there is a linear trend, and there is variability about the regression line. For this set of data, $r^2 = .49$. This indicates that approximately half of the variability in the values of Y is attributable to the linear relationship between Y and X. Approximately half of the variation in Y is left unexplained. This unexplained variability is said to be due to the experimental error.

For the advertising–sales example, we calculated r to be .912. The coefficient of determination is then equal to

$$r^2 = (.912)^2 = .832$$

Thus, approximately 83% of the variation in monthly sales is explained by (attributable to) the linear relationship between sales and advertising. This leaves 17% of the variability due to error. Since most of the variability in sales is accounted for by the advertising expenditures, the linear relationship is quite strong.

The coefficient of determination, r^2, provides a measure of the strength of the linear relationship between two variables X and Y that is easier to interpret than the coefficient of correlation, r. For example, when people are asked to describe the strength of the linear relationship between two variables for which $r = .5$, they often answer that this represents a "medium-strength" relationship—one that is neither weak nor strong. However, consider that $r^2 = (.5)^2 = .25$. This means that only 25% of the variation in the Y-values is attributable to the linear relationship between Y and X. This leaves 75% of the variation due to error. We think you will agree that this indicates a weak relationship between Y and X.

In the introduction to this chapter, we stated that a correlation exists between the size of the United States population and the crime index. The data used in the study for the years 1963–1973 are repeated for convenience in Table 9.6. Using these data, the value of the correlation coefficient is found to be $r = .94$. The value of the coefficient of determination is $r^2 = (.94)^2 = .88$. This indicates a strong relationship between the United States population size and the crime index.

Table 9.6

United States population and crime index

Year	Population	Actual Crime Index	Year	Population	Actual Crime Index
1963	188,531,000	2,259,081	1969	201,921,000	4,989,747
1964	191,334,000	2,604,426	1970	202,184,772	5,568,197
1965	193,818,000	2,780,015	1971	206,256,000	5,995,211
1966	195,857,000	3,243,400	1972	208,232,000	5,891,924
1967	197,864,000	3,802,273	1973	209,851,000	8,638,375
1968	199,861,000	4,466,573			

Source: Reprinted by permission of the *Journal of Police Science and Administration*, copyright 1975 by Northwestern University School of Law, Vol. 3, No. 3, pp. 312–318.

COMPUTER
EXAMPLE 9.5

In the computer examples in this chapter we have been comparing the GPAs of fifty students to the number of hours worked per week. Minitab may be used to calculate both the coefficient of correlation and the coefficient of determination. The accompanying printout has been studied in previous examples. The value of the coefficient of determination, r^2, is highlighted.

```
The regression equation is
GPA = 3.01 - 0.0178 WORKHRS

Predictor      Coef        Stdev       t-ratio
Constant       3.00718     0.06465     46.51
WORKHRS        -0.017795   0.004437    -4.01

s = 0.1962     R-sq = 25.1%    R-sq(adj) = 23.5%
```

Note that Minitab gives the value of "R-sq" as a percentage. We would report the value of the coefficient of determination to be $r^2 = .251$. This means that 25.1% of the variability in the students' GPAs can be attributed to the linear relationship between GPAs and number of hours worked per week. This is not a very strong relationship. About 75% of the variability in the GPA values is still left unexplained and is due to error. This should not be surprising, since we would expect many other variables besides the number of hours worked to be related to GPA. Thus, it might be useful to have a model that includes other variables such as high school GPA, college major, athletic commitments, and number of credit hours being taken. We will not discuss models that include more than one variable in this text.*

Minitab was also used to calculate the value of the coefficient of correlation, r. The result is shown here:

```
Correlation of GPA and WORKHRS = -0.501
```

Note that the value of $r = -0.501$ is negative because GPAs tend to decrease as the number of hours worked per week increases. Also, $r^2 = (-0.501)^2 = .251$, the value of the coefficient of determination previously given. We remind you to be careful in using only the value of r in interpreting the strength of the linear relationship. The value of r^2 is more easily interpreted when considering the strength.

Finally, we should comment that in Computer Example 9.4, we found that there was overwhelming evidence to conclude that GPA and number of hours worked are linearly related. Recall that the value of the test statistic was $t = -4.01$. We now see, however, that the linear relationship is not a strong one. The value of the test statistic provides only the weight of the evidence that a linear relationship exists, whether weak or strong. Thus, for these data we see that there is strong evidence of some linear relationship between GPA and hours worked, but the value of r^2 indicates that it is a rather weak relationship.

* If you are interested in such models you might consult *Statistics*, 4th edition, by J. T. McClave and F. H. Dietrich II (San Francisco: Dellen Publishing Comapny, 1988).

The accompanying scatterplot, with a graph of the least squares line, helps to make this point. We can see the negative trend in the relationship between the two variables, but we can also see that there is a great deal of variability about the least squares line; thus, the relationship is weak.

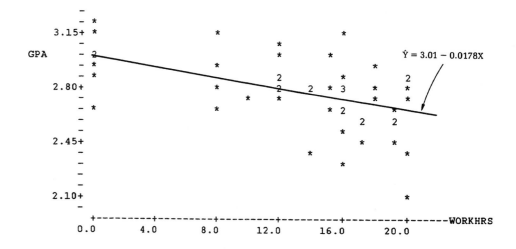

Exercises (9.49–9.60)

Learning the Language

9.49 Define the following important terms, which were introduced in the last two sections:

a. Coefficient of correlation

b. Coefficient of determination

9.50 What do the symbols r and r^2 represent?

Learning the Mechanics

9.51 The following quantities were calculated for a sample of $n = 9$ data points:

$$\sum X = 6 \quad \sum Y = 8 \quad \sum (XY) = 10 \quad \sum X^2 = 9 \quad \sum Y^2 = 12$$

Calculate the values of r and r^2.

9.52 Consider the six data points shown in the table, which were also given in Exercises 9.22 and 9.31:

X	1	2	3	4	5	6
Y	1	2	2	3	5	5

Calculate the values of r and r^2.

9.53 Consider the five data points shown in the table, which were also given in Exercises 9.23 and 9.31:

X	−2	−1	0	1	2
Y	4	3	3	1	−1

Calculate the values of r and r^2.

Applying the Concepts

9.54 Refer to Exercise 9.1. A scatterplot was drawn to compare students' grades with where they choose to sit in class. It was thought that students who sit near the front might obtain better grades than those who choose seats in the rear. The grades of ten randomly selected statistics students are recorded in the table, along with their row number (rows numbered from front to back).

Grade, Y	93	68	89	98	66	86	73	55	80	71
Row, X	7	1	4	3	25	12	9	30	13	27

a. Calculate r and r^2.

b. Do the data provide sufficient evidence to indicate that students sitting in the front of the classroom tend to receive better grades? Test H_0: $\beta_1 = 0$ against H_a: $\beta_1 < 0$. Use $\alpha = .05$.

9.55 The sample data listed in the accompanying table were first given in Exercise 9.2. Find the correlation coefficient and the coefficient of determination for the data and interpret your results.

Year	Number of 18-hole and Larger Golf Courses in the U.S.	Number of Divorces in the U.S. (in millions)
1960	2,725	2.9
1965	3,769	3.5
1970	4,845	4.3
1975	6,282	6.5
1977	6,551	8.0
1978	6,699	8.6
1979	6,787	8.8
1980	6,856	9.9
1981	6,944	10.8
1982	7,059	11.5
1983	7,125	11.6
1984	7,230	12.3

Source: U.S. Bureau of the Census, *Statistical Abstract of the United States: 1986* (Washington, D.C.: Government Printing Office, 1985), pp. 35, 230.

9.56   Is the maximal oxygen uptake, a measure often used by physiologists to indicate an individual's state of cardiovascular fitness, related to the performance of distance runners? Six long-distance runners submitted to treadmill tests to establish their maximal oxygen uptake. The results, along with each runner's best mile time (in seconds), are shown in the table.

Athlete	Maximal Oxygen Uptake X, Milliliters/Kilogram	Mile Time Y, Seconds
1	63.3	241.5
2	60.1	249.8
3	53.6	246.1
4	58.8	232.4
5	67.5	237.2
6	62.5	238.4

a. Is there sufficient evidence to indicate that as X increases, Y decreases? That is, test $H_0: \beta_1 = 0$ against $H_a: \beta_1 < 0$, using $\alpha = .05$.

b. Calculate r and r^2.

9.57   Is there a correlation between verbal scores and math scores on the Scholastic Aptitude Test (SAT)? The mean verbal score and the mean math score of college-bound high school seniors for eleven years are given in the table.

Year	Mean SAT Score Verbal	Math	Year	Mean SAT Score Verbal	Math
1967	466	492	1978	429	468
1970	460	488	1979	427	467
1974	444	480	1980	424	466
1975	434	472	1981	424	466
1976	431	472	1982	426	467
1977	429	470			

Source: U.S. Bureau of the Census, Statistical Abstract of the United States: 1984 (Washington, D.C.: Government Printing Office, 1983), p. 158.

a. Use the data to find the correlation coefficient.

b. Find the coefficient of determination and explain its meaning in terms of the problem.

9.58   It seems reasonable that the cost of a new house would be related to the cost of construction materials. The price index for new one-family houses sold and the producer price index for construction materials for nine years are given in the table. Calculate r and r^2, and interpret these values.

Year	Price Index for New One-Family Houses	Producer Price Index for Construction Materials
1973	67.5	138.5
1974	73.8	160.9
1975	81.7	174.0
1976	88.7	187.7
1978	114.5	228.3
1979	130.8	251.4
1980	145.2	266.4
1981	157.4	283.0
1982	161.5	288.0

Source: U.S. Bureau of the Census, *Statistical Abstract of the United States: 1984* (Washington, D.C.: Government Printing Office, 1983), p. 739.

*9.59   In Exercise 9.47 you found that there was a linear relationship between monetary losses due to shoplifting and the number of sales clerks. Use a computer package to calculate r and r^2 for this example. Interpret these values.

*9.60   In Exercise 9.48 you concluded that the yield of potatoes per plot is linearly related to the amount of fertilizer applied. Use a computer package to calculate r and r^2 for this example. Interpret these values.

9.9

USING THE LEAST SQUARES LINE FOR ESTIMATION AND PREDICTION

When we are satisfied that there is a linear relationship between two variables and that the relationship is fairly strong, we will put the least squares line to its intended use. We will use the least squares line to make inferences (estimates or predictions) about the values of Y for values of X of interest. This is the sixth and final step in considering a linear relationship between two variables.

Step 6:   If the model proves useful, use it for prediction and estimation. ·

The most common uses of a probabilistic model for making inferences can be divided into two categories. One use of the model is to estimate the mean, or average, value of the dependent variable Y for a specific value of X. For the advertising–sales example, the brewery might want to estimate the average sales for all months with an advertising expenditure of $600. In other words, they would like to estimate the average sales if, month after month after month, $600 were spent on advertising.

A second use of the model is to predict a particular value of Y for a specific value of X. For example, the brewery may want to predict the sales for next month if $600 is to be spent on advertising. In the first use of the model, we are attempting to estimate the mean value of Y for a very large number of performances of the experiment at the specific X-value. In the second use, we are trying to predict the result of a single performance of the experiment at the specific X-value. To help distinguish between these two uses of the model, we will always be careful to use the phrases "*estimate* the mean value of Y," and "*predict* the particular value of Y."

We will use the least squares prediction equation

$$\hat{Y} = b_0 + b_1 X$$

both to estimate the mean value of Y and to predict a particular value of Y for a specific value of X. For the advertising—sales example, we have found the least squares line to be

$$\hat{Y} = 7.679 + 100.029X$$

Substituting a value of $X = 0.6$ (a \$600 advertising expenditure) into the equation, we have obtained

$$\hat{Y} = 7.679 + 100.029(0.6) = 67.696$$

This value (in thousands of dollars) is the *estimated* mean sales for all months in which $X = 0.6$. The identical value is also the *predicted* sales in a single month in which $X = 0.6$. The estimated mean value and predicted individual value are identical, as shown in Figure 9.18.

Figure 9.18
Estimated mean value and predicted individual value of sales, Y, for $X = 0.6$

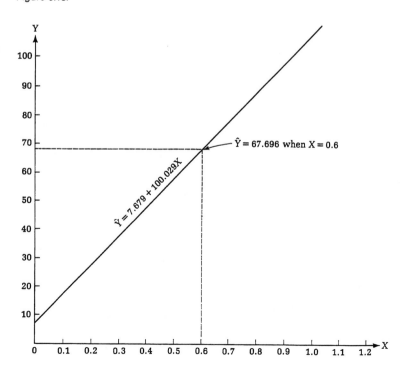

Although the estimated and predicted values are identical, there is a difference in these two uses of the model. The difference lies in the reliability with which we may make an estimate or a prediction. To measure the reliability, we will form a **confidence interval** for the mean value of Y and a **prediction interval** for the particular value of Y. The forms of these intervals are given in the next two boxes.

Confidence Interval for the Mean Value of Y at a Specific Value of X

A confidence interval for the mean value of Y at a specific value of X is expressed as

$$\hat{Y} - t \sqrt{s^2 \left[\frac{1}{n} + \frac{n(X - \bar{X})^2}{S_X} \right]}$$

to

$$\hat{Y} + t \sqrt{s^2 \left[\frac{1}{n} + \frac{n(X - \bar{X})^2}{S_X} \right]}$$

The value of t is found in Appendix Table VI, using $(n - 2)$ df.

Prediction Interval for a Particular Value of Y at a Specific Value of X

A prediction interval for a particular value of Y at a specific value of X is expressed as

$$\hat{Y} - t \sqrt{s^2 \left[1 + \frac{1}{n} + \frac{n(X - \bar{X})^2}{S_X} \right]}$$

to

$$\hat{Y} + t \sqrt{s^2 \left[1 + \frac{1}{n} + \frac{n(X - \bar{X})^2}{S_X} \right]}$$

The value of t is found in Appendix Table VI, using $(n - 2)$ df.

Notice that the confidence interval and prediction interval are very similar. As a matter of fact, the only difference is that the prediction interval has the value 1 added under the square root sign. This will cause a prediction interval at any specific value of X to be wider than the corresponding confidence interval. This should have some intuitive appeal, since we should be better able to estimate the average outcome of an experiment than to predict what will happen when an experiment is performed once.

It is valid to form a confidence interval or a prediction interval when the properties of the distribution of the random error term (given in Section 9.5) are satisfied.

EXAMPLE 9.6   For the advertising–sales example, find a 95% confidence interval for the mean sales per month when the advertising expenditure is $600.

Solution   For a $600 advertising expenditure, $X = 0.6$, and we have found that

$$\hat{Y} = 7.679 + 100.029(0.6) = 67.696$$

We have also calculated $s^2 = 137.512$, $\bar{X} = 0.7$, and $S_X = 2.38$, and the value of n is 7. The t-value for a 95% confidence interval is found to be 2.57 (from Appendix Table VI, using $n - 2 = 5$ df). Substituting these values into the formula for a confidence interval, we obtain

$$\hat{Y} - t\sqrt{s^2\left[\frac{1}{n} + \frac{n(X - \bar{X})^2}{S_X}\right]} \quad \text{to} \quad \hat{Y} + t\sqrt{s^2\left[\frac{1}{n} + \frac{n(X - \bar{X})^2}{S_X}\right]}$$

$$67.696 - 2.57\sqrt{137.512\left[\frac{1}{7} + \frac{7(0.6 - 0.7)^2}{2.38}\right]} \quad \text{to} \quad 67.696 + 2.57\sqrt{137.512\left[\frac{1}{7} + \frac{7(0.6 - 0.7)^2}{2.38}\right]}$$

$$67.696 - 12.509 \quad \text{to} \quad 67.696 + 12.509$$

$$55.187 \quad \text{to} \quad 80.205$$

We are 95% confident that the mean sales per month will be between $55,187 and $80,205 when the advertising expenditure is $600 per month.   ∎

EXAMPLE 9.7   For the advertising–sales example, form a 95% prediction interval for the sales in one month in which the advertising expenditure is $600.

Solution   The form of the prediction interval is

$$\hat{Y} - t\sqrt{s^2\left[1 + \frac{1}{n} + \frac{n(X - \bar{X})^2}{S_X}\right]} \quad \text{to} \quad \hat{Y} + t\sqrt{s^2\left[1 + \frac{1}{n} + \frac{n(X - \bar{X})^2}{S_X}\right]}$$

Substituting the values given in Example 9.6 for the quantities of interest, we obtain

$$67.696 - 2.57\sqrt{137.512\left[1 + \frac{1}{7} + \frac{7(0.6 - 0.7)^2}{2.38}\right]}$$

to

$$67.696 + 2.57\sqrt{137.512\left[1 + \frac{1}{7} + \frac{7(0.6 - 0.7)^2}{2.38}\right]}$$

$$67.696 - 32.630 \quad \text{to} \quad 67.696 + 32.630$$

$$35.066 \quad \text{to} \quad 100.326$$

We are 95% confident that the sales for one month will be between $35,066 and $100,326 when $600 is spent on advertising that month. Note how much wider the prediction interval is than the confidence interval is.   ∎

A comparison of the confidence interval and the prediction interval found in Examples 9.6 and 9.7 is given in Figure 9.19.

Figure 9.19
A 95% confidence interval and a 95% prediction interval for Y when X = 0.6

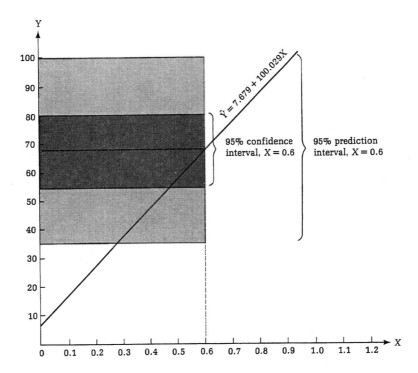

Caution

We must add a word of caution. Be careful not to use the least squares prediction equation to estimate or predict for values of X that fall outside the range of the values of X contained in the sample data. By conducting the test of hypothesis about the slope β_1 and examining r and r^2, we can determine whether the linear model is appropriate for the sampled data. Although the model might provide an excellent fit to the data over the range of X-values in the sample, the relationship between Y and X may be different outside these sampled values. For example, in the advertising-sales example the sampled advertising expenditures, X, ranged between 0.3 and 1.0. The least squares prediction equation is appropriate for values of X between these two figures. It should not be expected that sales will keep increasing at the same rate as advertising expenditures increase. The model may change outside the experimental region, as shown in Figure 9.20 (page 546). Thus, it would be appropriate to use the least squares prediction equation for values of X less than 0.3 or larger than 1.0. Failure to heed this warning may lead to serious errors of estimation or prediction.

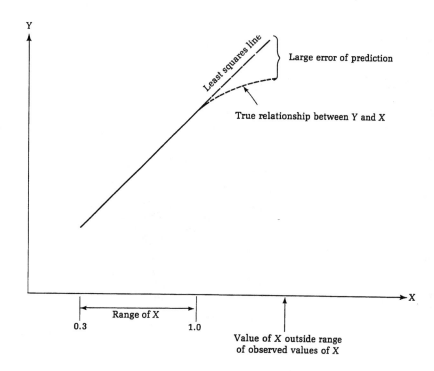

Figure 9.20
Change in the relationship between *Y* and *X* outside the sampled values of *X*

COMPUTER
EXAMPLE 9.6

We will now use the GPA and hours worked per week example to form a confidence interval and a prediction interval. Recall that we concluded that GPA and number of hours worked are linearly related, but the relationship is fairly weak. The value of $r^2 = .251$ is rather low. Although ideally we would like the relationship to be stronger, we can still use the least squares line for estimation and prediction for purposes of demonstration.

With one command Minitab will find both a confidence interval and a prediction interval for any value of the independent variable, which in this case is the number of hours worked per week. For example, let's consider making inferences using a value of 7 hours of work per week. With this value Minitab produced the following printout. The 95% confidence interval and 95% prediction interval are highlighted.

```
   Fit    Stdev.Fit        95% C.I.              95% P.I.
2.8826      0.0389     ( 2.8043, 2.9609)    ( 2.4804, 3.2849)
```

The "95% C.I." (confidence interval) may be interpreted as follows: We are 95% confident that all students who work 7 hours per week have a mean GPA between 2.80 and 2.96.

The "95% P.I." (prediction interval) has the following meaning: We are 95% confident that the GPA for one student who works 7 hours per week is between 2.48 and 3.28.

As always, the prediction interval is wider than the confidence interval. Also, since in the sample the number of hours worked per week ranged between 0 and 20, we should be dubious about using values larger than 20 to make estimates or predictions. ∎

Exercises (9.61–9.70)

Learning the Language

9.61 What is the difference between estimation and prediction?

9.62 What is the difference between a confidence interval and a prediction interval?

Learning the Mechanics

9.63 The data from Exercise 9.23 are again repeated below:

X	−2	−1	0	1	2
Y	4	3	3	1	−1

In previous exercises, the following quantities were calculated:

$$b_0 = 2 \quad b_1 = -1.2 \quad s^2 = .533 \quad S_x = 50 \quad \bar{X} = 0$$

a. Find a 90% confidence interval for the mean value of Y when $X = -1$.
b. Find a 90% prediction interval for Y when $X = -1$.
c. Find a 95% prediction interval for Y when $X = 0$.
d. Find a 95% confidence interval for the mean value of Y when $X = 2$.

9.64 The following quantities were computed for $n = 22$ data points:

$$\sum X = 44 \quad \sum Y = 66 \quad \sum X^2 = 113 \quad \sum Y^2 = 215 \quad \sum (XY) = 152$$

a. Find the least squares prediction equation.
b. Calculate s^2.
c. Find a 95% confidence interval for the mean value of Y when $X = 1$.
d. Find a 95% prediction interval for Y when $X = 1.5$.
e. Find a 90% prediction interval for Y when $X = 0$.
f. Find a 90% confidence interval for the mean value of Y when $X = 2$.

Applying the Concepts

9.65 Certain dosages of a new drug developed to reduce a smoker's reliance on tobacco may reduce a person's pulse rate to dangerously low levels. To investigate the drug's effect on pulse rate, different dosages of the drug were administered to six randomly selected patients, and 30 minutes later the decrease in each patient's pulse rate was recorded. The results are given in the table at the top of page 548.

Patient	Dosage X, Cubic Centimeters	Decrease in Pulse Rate Y, Beats/Minute
1	2.0	15
2	1.5	9
3	3.0	18
4	2.5	16
5	4.0	23
6	3.0	20

a. Is there evidence of a linear relationship between change in pulse rate and drug dosage? Test at $\alpha = .10$.

b. Find a 99% confidence interval for the mean decrease in pulse rate corresponding to a dosage of 3.5 cubic centimeters.

c. Find a 99% prediction interval for the decrease in pulse rate corresponding to a dosage of 3.5 cubic centimeters.

9.66 In planning for an orientation meeting with new accounting majors, the chairperson of the Accounting Department wants to emphasize the importance of doing well in the major courses in order to get a better-paying job after graduation. To support this point, the chairperson plans to show that there is a strong positive correlation between starting salaries for recent accounting graduates and their grade-point averages in the major courses. Records of seven recent graduates are randomly selected and the data obtained are listed in the table.

Grade-Point Average in Major Courses X	Starting Salary Y, Thousands of Dollars
2.58	11.5
3.27	13.8
3.85	14.5
3.50	14.2
3.33	13.5
2.89	11.6
2.23	10.6

a. Find the least squares prediction equation.

b. Plot the data and graph the line as a check on your calculations.

c. Find the values of r and r^2, and interpret them.

d. Find a 95% prediction interval for a graduate whose grade-point average is 3.2.

e. What is the mean starting salary for graduates with grade-point averages equal to 3.0? Use a 95% confidence interval.

9.67 Does a linear relationship exist between the Consumer Price Index (CPI) and the Dow Jones Industrial Average (DJA)? A random sample of 10 months selected from the past several years produced the following corresponding DJA and CPI data:

DJA, Y	660	638	639	597	702	650	579	570	725	738
CPI, X	13.2	14.2	13.7	15.1	12.6	13.8	15.7	16.0	11.3	10.4

a. Find the least squares line relating the DJA, Y, to the CPI, X.

b. Do the data provide sufficient evidence to indicate that X contributes information for the prediction of Y? Test using $\alpha = .05$.

c. Suppose you are interested in the value of the DJA when the CPI is at 15.0. Should you calculate a 95% prediction interval for a particular value of the DJA, or a 95% confidence interval for the mean value of the DJA? Explain the difference.

d. Calculate both intervals considered in part c when the CPI is 15.0.

9.68 Will the national 55-mile-per-hour highway speed limit provide a substantial savings in fuel? To investigate the relationship between automobile gasoline consumption and driving speed, a small economy car was driven twice over a stretch of interstate freeway at each of six speeds. The miles per gallon measurements for each of the twelve trips are given in the table.

Miles per Hour	Miles per Gallon
50	34.8, 33.6
55	34.6, 34.1
60	32.8, 31.9
65	32.6, 30.0
70	31.6, 31.8
75	30.9, 31.7

a. Fit a least squares line to the data.

b. Is there sufficient evidence to conclude that there is a linear relationship between gasoline consumption and speed? Test using $\alpha = .05$.

c. Construct a 95% confidence interval for the mean miles per gallon when the speed is 72 miles per hour.

d. Construct a 95% prediction interval for miles per gallon when the speed is 58 miles per hour.

*9.69 In Exercise 9.26 you used the computer to find the least squares line relating shoplifting losses to the number of sales clerks working per week.

a. Use this result and the computer to construct a 90% prediction interval for the value of goods that would be lost in one week if 12 sales clerks were working.

b. Construct a 95% confidence interval for the average loss for weeks when 15 sales clerks work.

c. Would you be willing to predict the losses in a week when 30 sales clerks work? Explain.

*9.70 In Exercise 9.27 you used the computer to find the least squares line relating potato yield to the amount of fertilizer applied.

a. Use this result and the computer to construct a 99% confidence interval for the mean potato yield for plots of land that received 2 pounds of fertilizer.

b. Find a 95% prediction interval for the potato yield for one plot of land with 4 pounds of fertilizer.

c. For amounts of fertilizer between 1 and 4.5 pounds, we have seen that potato yield tends to increase. What will happen to potato yields sooner or later as the amount of fertilizer is increased beyond 4.5 pounds? How is this related to using the least squares line for estimation and prediction?

9.10

■ AN EXAMPLE

In the previous sections we have presented the basic elements needed to use a linear model for estimation or prediction. In this final section we will apply what we have discussed to a practical example. In doing so, we will follow the six steps originally listed in Section 9.3.

Suppose a fire insurance company wants to relate the amount of fire damage to houses to the distance between a house and the nearest fire station. In particular, the insurance company wants to investigate this relationship for brick (or stone) houses that are valued at $60,000. (If a relationship is found for these houses, the insurance company will develop a more complex model to include other types of houses.) A random sample of fifteen fires at houses fitting the description is selected. The amount of damage, Y, and the distance, X, to the nearest fire station are recorded for each fire. The results are given in Table 9.7.

Table 9.7

Fire damage data

Distance to Fire Station X, Miles	Fire Damage Y, Thousands of Dollars	Distance to Fire Station X, Miles	Fire Damage Y, Thousands of Dollars
3.4	26.2	2.6	19.6
1.8	17.8	4.3	31.3
4.6	31.3	2.1	24.0
2.3	23.1	1.1	17.3
3.1	27.5	6.1	43.2
5.5	36.0	4.8	36.4
0.7	14.1	3.8	26.1
3.0	22.3		

Step 1: First, we will define each term in the probabilistic linear model:

$$Y = \beta_0 + \beta_1 X + \varepsilon$$

Y = Amount of fire damage to a house, in thousands of dollars

X = Mileage between a house and the nearest fire station

β_0 = Amount of fire damage to a house that is 0 miles from (right next to) a fire station

β_1 = Change (increase) in fire damage for each mile added to the distance between a house and the nearest fire station

ε = Random error term

Step 2: Draw a scatterplot of the data. A scatterplot of the data in Table 9.7 is shown in Figure 9.21. Examining the figure, we can see that the fire damage tends to increase as the distance to the fire station increases. The graph indicates that these two variables are linearly related, and we will proceed with the analysis expecting to verify this relationship.

Figure 9.21
Scatterplot for the fire damage data

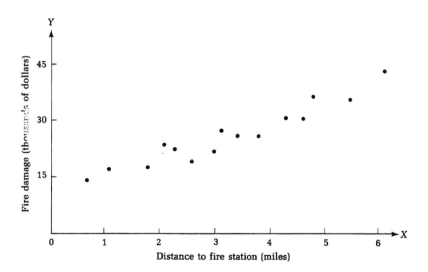

Step 3: Use the sample data to estimate β_0 and β_1. Before calculating b_0 and b_1, the estimates of β_0 and β_1, we make the preliminary calculations listed in Table 9.8 (page 552). We also calculate the following quantities:

$$S_X = n\sum X^2 - \left(\sum X\right)^2 = 15(196.16) - (49.2)^2 = 521.76$$
$$S_Y = n\sum Y^2 - \left(\sum Y\right)^2 = 15(11,376.48) - (396.2)^2 = 13,672.76$$
$$S_{XY} = n\sum(XY) - \left(\sum X\right)\left(\sum Y\right) = 15(1,470.65) - (49.2)(396.2) = 2,566.71$$

$$\bar{X} = \frac{\sum X}{n} = \frac{49.2}{15} = 3.28$$

$$\bar{Y} = \frac{\sum Y}{n} = \frac{396.2}{15} = 26.413$$

Table 9.8

X	Y	XY	X^2	Y^2
3.4	26.2	89.08	11.56	686.44
1.8	17.8	32.04	3.24	316.84
4.6	31.3	143.98	21.16	979.69
2.3	23.1	53.13	5.29	533.61
3.1	27.5	85.25	9.61	756.25
5.5	36.0	198.00	30.25	1,296.00
0.7	14.1	9.87	.49	198.81
3.0	22.3	66.90	9.00	497.29
2.6	19.6	50.96	6.76	384.16
4.3	31.3	134.59	18.49	979.69
2.1	24.0	50.40	4.41	576.00
1.1	17.3	19.03	1.21	299.29
6.1	43.2	263.52	37.21	1,866.24
4.8	36.4	174.72	23.04	1,324.96
3.8	26.1	99.18	14.44	681.21
$\sum X = 49.2$	$\sum Y = 396.2$	$\sum(XY) = 1,470.65$	$\sum X^2 = 196.16$	$\sum Y^2 = 11,376.48$

The least squares estimates of β_0 and β_1 are

$$b_1 = \frac{S_{XY}}{S_X} = \frac{2,566.71}{521.76} = 4.919331 \approx 4.919$$

$$b_0 = \bar{Y} - b_1\bar{X} = 26.413 - 4.919331(3.28) = 10.277594 \approx 10.278$$

The least squares prediction equation is

$$\hat{Y} = 10.278 + 4.919X$$

The scatterplot of the data has been duplicated in Figure 9.22, and a graph of the prediction equation has been added. The value of 10.278 is the estimated fire damage (in thousands of dollars) to a house that is right next to the fire station ($X = 0$). The value of 4.919 is the estimated increase in fire damage for each mile added to the distance between a house and the nearest fire station.

Step 4: We will now specify the properties of the distribution of the random error component, ε, and estimate the variance, σ^2. The properties of the distribution are that the errors are normally distributed with mean 0 and constant variance σ^2. Also, the errors are independent. Although these properties may not be exactly satisfied (they rarely are for any practical problem), we will assume that they are approximately met for this example.

To estimate σ^2, we calculate s^2 to be

$$s^2 = \frac{S_Y - b_1 S_{XY}}{n(n-2)} = \frac{13,672.76 - (4.919331)(2,566.71)}{15(13)}$$

$$= \frac{1,046.263929}{15(13)} = 5.365456 \approx 5.365$$

Figure 9.22
Least squares prediction
equation for the fire
damage data

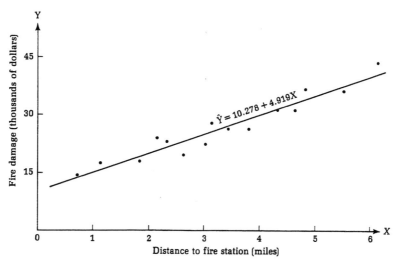

Step 5: Now we must make sure that the least squares equation is useful for prediction and estimation. First, we will test the null hypothesis that the slope is 0, that
is, that there is no linear relationship between Y and X, against the alternative hypothesis that fire damage increases as the distance from the fire station increases.
The elements of the test are

$$H_0: \quad \beta_1 = 0 \qquad H_a: \quad \beta_1 > 0$$

$$\text{Test statistic:} \quad t = \frac{b_1\sqrt{S_x}}{\sqrt{ns^2}}$$

Rejection region (using $\alpha = .05$): Reject H_0 if $t > 1.77$.

The rejection region was found for a one-tailed test using Appendix Table V, with
$n - 2 = 13$ df.
   Calculating the test statistic, we obtain

$$t = \frac{4.919\sqrt{521.76}}{\sqrt{15(5.376)}} = 12.51$$

This large t-value leaves little doubt that the linear model is useful for describing
the relationship between distance to the nearest fire station and the amount of
fire damage to $60,000 brick houses.
   To further examine the usefulness of the model, we will calculate the coefficient
of correlation, r, and the coefficient of determination, r^2:

$$r = \frac{S_{XY}}{\sqrt{S_X \cdot S_Y}}$$

$$= \frac{2,566.71}{\sqrt{(521.76)(13,672.76)}} = .961$$

This high correlation (near 1) indicates a strong linear relationship between fire damage and distance to the fire station. The value of the coefficient of determination is

$$r^2 = (.961)^2 = .924$$

This means that approximately 92% of the variability in the sampled fire damages (Y-values) is explained by the distance to the fire station, X. Only 8% of the variability in the Y-values is left to error. All signs point to a very strong linear relationship between Y and X. The model should be very useful for making predictions and estimations.

Step 6: We are now ready to use the least squares prediction equation. Suppose the insurance company wants to estimate the mean amount of fire damage that would occur if $60,000 brick houses located 3.5 miles from the nearest fire station caught fire. The insurance company would be most interested in *estimating the mean*, since it would hope to insure many houses. The estimated value of the fire damage for $X = 3.5$ is

$$\hat{Y} = 10.278 + 4.919(3.5) = 27.495$$

For a 95% confidence interval, the appropriate t-value is 2.16 (from Appendix Table VI, with $n - 2 = 13$ df). Substitution yields the interval

$$\hat{Y} - t\sqrt{s^2\left[\frac{1}{n} + \frac{n(X - \bar{X})^2}{S_x}\right]} \quad \text{to} \quad \hat{Y} + t\sqrt{s^2\left[\frac{1}{n} + \frac{n(X - \bar{X})^2}{S_x}\right]}$$

$$27.495 - 2.16\sqrt{5.376\left[\frac{1}{15} + \frac{15(3.5 - 3.28)^2}{521.76}\right]} \quad \text{to} \quad 27.495 + 2.16\sqrt{5.376\left[\frac{1}{15} + \frac{15(3.5 - 3.28)^2}{521.76}\right]}$$

$$27.495 - 1.294 \quad \text{to} \quad 27.495 + 1.294$$

$$26.201 \quad \text{to} \quad 28.789$$

The insurance company may be 95% confident that the mean amount of fire damage will be between $26,201 and $28,789 for $60,000 brick houses that are 3.5 miles from the nearest fire station.

An individual home owner would be more concerned with *predicting* the fire damage to one house. Suppose you own a $60,000 brick house that is 3.5 miles from the nearest fire station. The 95% prediction interval for the amount of fire damage to this house is found by

$$\hat{Y} - t\sqrt{s^2\left[1 + \frac{1}{n} + \frac{n(X - \bar{X})^2}{S_x}\right]} \quad \text{to} \quad \hat{Y} + t\sqrt{s^2\left[1 + \frac{1}{n} + \frac{n(X - \bar{X})^2}{S_x}\right]}$$

$$27.495 - 2.16\sqrt{5.376\left[1 + \frac{1}{15} + \frac{15(3.5 - 3.28)^2}{521.76}\right]}$$

to

$$27.495 + 2.16\sqrt{5.376\left[1 + \frac{1}{15} + \frac{15(3.5 - 3.28)^2}{521.76}\right]}$$

$$27.495 - 5.173 \quad \text{to} \quad 27.495 + 5.173$$

$$22.322 \quad \text{to} \quad 32.668$$

We can be 95% confident that the fire damage to an individual house 3.5 miles from the nearest fire station would be between $22,322 and $32,668.

One final caution: Since all the sample values of X fall between 0.7 and 6.1, we would not use this prediction equation for values of X that are not within this range. Remember that it is dangerous to use the model outside the region in which the sample data fall.

Chapter Summary

Graphing corresponding values of two random variables, X and Y, yields a **scatterplot**. A **linear relationship** between the two variables is one that can be expressed by an equation of the form $Y = b + mX$. Here, Y is the **dependent variable** and X is the **independent variable.** The graph of this equation in rectangular coordinates is a straight line; the line has **slope** (the change in Y for each unit increase in X) equal to m and **Y-intercept** (the point where the line crosses the Y-axis) equal to b. This linear relationship is **deterministic** in the sense that any value for X exactly determines the corresponding Y-value.

Adding a random error term, ε, to the deterministic linear relationship, we obtain a more realistic model, called a **probabilistic model,** for the relationship between two variables:

$$Y = \beta_0 + \beta_1 X + \varepsilon$$

Sample data are used to estimate β_0 and β_1. The **method of least squares** yields a line that minimizes the squared differences between observed values and those on the line. This line is called the **least squares line,** or **regression line,** and is expressed by the equation $\hat{Y} = b_0 + b_1 X$, where b_0 and b_1 are the estimates for β_0 and β_1:

$$b_1 = \frac{S_{XY}}{S_X} \qquad b_0 = \bar{Y} - b_1 \bar{X}$$

where

$$S_{XY} = n \sum (XY) - \left(\sum X \right) \left(\sum Y \right)$$
$$S_X = n \sum X^2 - \left(\sum X \right)^2$$

The distribution of the random error term ε must be specified. We consider only the case where for each value of X the error term is independent of errors at other values and has a normal distribution with mean 0 and a common variance σ^2. The variance can be estimated by

$$s^2 = \frac{S_Y - b_1 S_{XY}}{n(n-2)}$$

To determine whether X helps to predict Y in a linear manner, test the following hypothesis about β_1:

H_0: $\beta_1 = 0$
H_a: $\beta_1 > 0$, or $\beta_1 < 0$, or $\beta_1 \neq 0$

Test statistic: $t = \dfrac{b_1\sqrt{S_x}}{\sqrt{ns^2}}$

Rejection region: Use Appendix Table V, with $(n - 2)$ df.

The following procedure is used to find a confidence interval for β_1:

> Obtain t from Appendix Table VI, using $(n - 2)$ df.
>
> Calculate $B = t\sqrt{\dfrac{ns^2}{S_x}}$
>
> Interval: $b_1 - B$ to $b_1 + B$

The strength of the linear relationship between Y and X is measured by the **coefficient of correlation,**

$$r = \dfrac{S_{XY}}{\sqrt{S_X \cdot S_Y}}$$

The closer this coefficient is to $+1$ or -1, the stronger the linear relationship. A high correlation indicates only a relationship and does not imply cause and effect. The **coefficient of determination,** r^2, indicates the fraction of the variability in Y that can be attributed to the linear relationship between Y and X, as opposed to the error effect.

The regression line can be used to estimate the mean value of Y for a given X within the range of sample data. The reliability of this estimate is measured by a **confidence interval for the mean:**

> Obtain t from Appendix Table VI, using $(n - 2)$ df.
>
> Calculate $\hat{Y}$ and $B = t\sqrt{s^2\left[\dfrac{1}{n} + \dfrac{n(X - \bar{X})^2}{S_X}\right]}$
>
> Interval: $\hat{Y} - B$ to $\hat{Y} + B$

The regression line can also be used to predict a specific value of Y using a **prediction interval.** This wider interval is similar to the confidence interval and is given by

$$\hat{Y} - t\sqrt{s^2\left[1 + \frac{1}{n} + \frac{n(X - \bar{X})^2}{S_X}\right]} \quad \text{to} \quad \hat{Y} + t\sqrt{s^2\left[1 + \frac{1}{n} + \frac{n(X - \bar{X})^2}{S_X}\right]}$$

SUPPLEMENTARY EXERCISES (9.71–9.85)

Learning the Mechanics

9.71   The following quantities were computed for a sample of $n = 20$ data points:

$\sum X = 26$ $\sum X^2 = 83.8$ $\sum Y = 540$

$\sum Y^2 = 14{,}605$ $\sum(XY) = 672$

a. Find the least squares prediction equation.
b. Calculate s^2.
c. Is there sufficient evidence to conclude that X and Y are linearly related? Use $\alpha = .10$.
d. Calculate the values of r and r^2.
e. Find a 95% prediction interval for Y when $X = 1.8$.
f. Find a 95% confidence interval for the mean value of Y when $X = 1.8$.

9.72 Consider the following ten data points:

X	3	5	6	4	3	7	6	5	4	7
Y	4	3	2	1	2	3	3	5	4	2

a. Calculate the values of r and r^2.
b. Plot the data. Do the values you calculated in part a appear to be reasonable?
c. Is there sufficient evidence to indicate that X and Y are linearly related? Use $\alpha = .10$.

Applying the Concepts

9.73 A study was conducted to determine whether the final grade in an introductory sociology course was related to a student's performance on a verbal ability test administered before college entrance. The verbal test scores and final grades for a random sample of ten students are shown in the table.

Student	Verbal Ability Test Score X	Final Introductory Sociology Grade Y
1	39	65
2	43	78
3	21	52
4	64	82
5	57	92
6	47	89
7	28	73
8	75	98
9	34	56
10	52	75

a. Find the least squares line.
b. Plot the data points and graph the least squares line.
c. Do the data provide sufficient evidence to indicate that a positive correlation exists between verbal score and final grade? Use $\alpha = .01$.
d. Find a 95% confidence interval for the slope β_1.
e. Predict a student's final grade in the course when the verbal test score is 50. (Use a 90% prediction interval.)

    f. Find a 95% confidence interval for the mean final grade for students scoring 35 on the college entrance verbal exam.

9.74 A large supermarket chain has its own store brand for many grocery items. These tend to be priced lower than other brands. For a particular item, the chain wants to study the effect of varying the price of the major competing brand on the sales of the store brand item, while the store and other brand prices are held fixed. The experiment is conducted at one of the chain's stores over a 7-week period, and the results appear in the table.

Week	Major Competitor's Price X, Cents	Store Brand Sales Y
1	37	122
2	32	107
3	29	99
4	35	110
5	33	113
6	31	104
7	35	116

    a. Find the least squares line relating store brand sales, Y, to major competitor's price X.

    b. Plot the data and graph the line as a check on your calculations.

    c. Does X contribute information for the prediction of Y? Use $\alpha = .05$.

    d. Calculate r and r^2 and interpret their values.

    e. Find a 90% confidence interval for mean store brand sales when the competitor's price is 33¢.

    f. Suppose you were to set the competitor's price at 33¢. Find a 90% prediction interval for next week's sales.

9.75 As part of the first-year evaluation of new salespeople, a large food-processing firm projects the second-year sales for each salesperson based on his or her sales for the first year. The data for eight salespeople are given in the table.

First-Year Sales X, Thousands of Dollars	Second-Year Sales Y, Thousands of Dollars
75.2	99.3
91.7	125.7
100.3	136.1
64.2	108.6
81.8	102.0
110.2	153.7
77.3	108.8
80.1	105.4

a. Use the tabulated data to fit a linear prediction model for second-year sales based on the first year's sales. (Assume the data have been adjusted in terms of a base year to discount inflation effects.)
b. Plot the data and graph the line as a check on your calculations.
c. Do the data provide sufficient information to indicate that X contributes information for the prediction of Y? Use $\alpha = .10$.
d. Calculate r^2 and interpret its value.
e. If a salesperson has first-year sales of $90,000, find a 90% prediction interval for the second year's sales.

9.76 In placing a weekly order, a concessionaire that provides services at a baseball stadium must estimate the number of people who will attend games during the coming week. Advanced ticket sales give an indication of expected attendance, so that food requirements may be predicted on the basis of advanced sales. The table gives data collected from 7 weeks of home games.

Hot Dogs Purchased Weekly Y, Thousands	Weekly Advanced Ticket Sales X, Thousands
39.1	54.0
35.9	48.1
20.8	28.8
42.4	62.4
46.0	64.4
40.7	59.5
29.9	42.3

a. Use the tabulated data to find the least squares line.
b. Plot the data and graph the line as a check on your calculations.
c. Do the data provide sufficient information to indicate that advanced ticket sales can help in predicting hot dog demand? Use $\alpha = .01$.
d. Calculate r^2 and interpret its value.
e. Find a 90% confidence interval for the mean number of hot dogs purchased when the advanced ticket sales equal 50,000.
f. If the advanced ticket sales this week equal 55,000, find a 90% prediction interval for the number of hot dogs that will be purchased this week at the game.

9.77 At temperatures approaching absolute zero ($-273°C$), helium exhibits traits that defy many laws of conventional physics. An experiment has been conducted with helium in solid form at various temperatures near absolute zero. The solid helium is placed in a dilution refrigerator along with a solid impure substance, and the fraction (in weight) of the impurities that pass through the solid helium is recorded. (This phenomenon of solids passing directly through solids is known as *quantum tunnelling*.) The data appear in the table at the top of page 560.

Temperature X, °C	Proportion of Impurities Passing through Helium Y	Temperature X, °C	Proportion of Impurities Passing through Helium Y
−262.0	.315	−272.0	.935
−265.0	.202	−272.4	.957
−266.0	.204	−272.7	.906
−267.0	.620	−272.8	.985
−270.0	.715	−272.9	.987

a. Fit a least squares line to the data.
b. Test $H_0: \beta_1 = 0$ against $H_a: \beta_1 < 0$, at the $\alpha = .01$ level of significance.
c. Compute r^2 and interpret your results.
d. Find a 95% prediction interval for the percentage of the solid impurities that pass through solid helium at −273°C. (Note that this value of X is outside the experimental region; therefore, use of the model for prediction may be dangerous!)

9.78   A certain manufacturer evaluates the sales potential of a product in a new marketing area by selecting several stores within the area to sell the product on a trial basis for 1 month. The sales figures for the trial period are then used to project sales for the entire area. [Note: The same number of trial stores are used each time.]

Total Sales during Trial Period X, Hundreds of Dollars	Total Sales for Entire Area during First Month Y, Hundreds of Dollars
16.8	48.2
14.0	46.8
18.3	54.3
22.1	59.7
14.9	48.3
23.2	67.5

a. Use the data in the table to find the least squares line.
b. Plot the data and graph the line as a check on your calculations.
c. Do the data provide sufficient evidence to indicate that total sales during the trial period contribute information for predicting total sales during the first month? Use $\alpha = .10$.
d. Use a 90% prediction interval to predict total sales for the first month for the entire area if the trial sales equal $2,000.

9.79   The management of a manufacturing firm is considering the possibility of setting up its own market research department rather than continuing to use the services of a market research firm. The management wants to know what salary should be paid to a market researcher, based on years of experience. An independent consultant checks with several other firms in the area and obtains the information shown in the table on market researchers.

Annual Salary Y, Thousands of Dollars	Experience X, Years
21.3	2
21.2	1.5
30.0	11
34.1	15
30.4	9
26.9	6

a. Fit a least squares line to the data.
b. Plot the data and graph the line as a check on your calculations.
c. Calculate r and r^2 Explain how these values measure the utility of the model.
d. Estimate the mean annual salary of all market researchers with 8 years of experience. Use a 90% confidence interval.
e. Predict the salary of a market researcher with 7 years of experience using a 90% prediction interval.

9.80 A study was conducted to determine whether there is a linear relationship between the breaking strength, Y, of wooden beams and the specific gravity, X, of the wood. Ten randomly selected beams of the same cross-sectional dimensions were stressed until they broke. The breaking strength and density for each beam is listed in the table.

Beam	Specific Gravity X	Strength Y	Beam	Specific Gravity X	Strength Y
1	.499	11.14	6	.528	12.60
2	.558	12.74	7	.418	11.13
3	.604	13.13	8	.480	11.70
4	.441	11.51	9	.406	11.02
5	.550	12.38	10	.467	11.41

a. Find the least squares prediction equation.
b. Test $H_0: \beta_1 = 0$ against $H_a: \beta_1 \neq 0$. Use $\alpha = .01$.
c. Estimate the mean strength for beams with specific gravity .590 using a 90% confidence interval.

9.81 Although the income tax system is structured so that people with higher incomes should pay a higher percentage of their incomes in taxes, there are many loopholes and tax shelters available for individuals with higher incomes. A sample of seven individual 1985 tax returns gave the data listed in the table.

Individual	Gross Income X, Thousands of Dollars	Taxes Paid Y, Percentage of Total Income
1	35.8	16.7
2	80.2	21.4
3	14.9	15.2
4	7.3	10.1
5	9.1	12.2
6	150.7	19.6
7	25.9	17.3

a. Fit a least squares line to the data.
b. Plot the data and graph the line as a check on your calculations.
c. Calculate r and r^2 and interpret each.
d. Find a 90% confidence interval for the mean percentage of income paid in taxes by individuals with gross incomes of $70,000.

9.82 The data in the table were collected to calibrate a new instrument for measuring intra-ocular pressure. The pressure for each of ten glaucoma patients was measured by the new instrument and by a standard, reliable (but more time-consuming) method.

Patient	Standard Method X	New Instrument Y	Patient	Standard Method X	New Instrument Y
1	20.2	20.0	6	21.8	22.1
2	16.7	17.1	7	19.1	18.9
3	17.1	17.2	8	22.9	22.2
4	26.3	25.1	9	23.5	24.0
5	22.2	22.0	10	17.0	18.1

a. Fit a least squares line to the data.
b. Calculate r and r^2. Interpret each of these quantities.
c. Predict the pressure measured by the new instrument when the standard method gives a reading of 20.0. Use a 90% prediction interval.

9.83 The data in the table give the mileages obtained by a test automobile when using gasolines of varying octane levels.

Mileage Y, Miles per Gallon	Octane X	Mileage Y, Miles per Gallon	Octane X
13.0	89	13.3	89
13.2	93	13.8	95
13.0	87	14.1	100
13.6	90	14.0	98

    a. Calculate r and r^2.

    b. Do the data provide sufficient evidence to indicate a correlation between octane level and miles per gallon for the test automobile? Test using $\alpha = .05$.

9.84 The accompanying figure contains a scatterplot that compares mean beginning salaries of teachers in contiguous school districts with the beginning salaries of individual districts in Florida during the 1979–1980 school year. Also shown in the graph is the least squares line for the data. The correlation coefficient for the data is equal to .71.

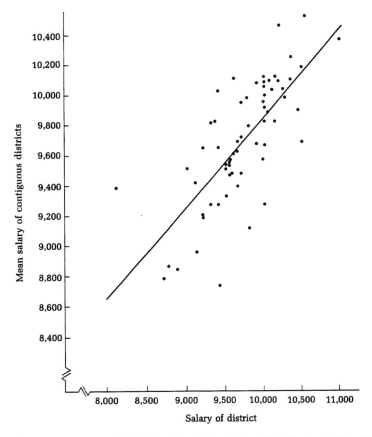

Source: K. M. Matthews and C. T. Holmes, "Implications of Regional Cost Adjustments to School Finance Plans," *Educational Administration Quarterly*, Vol. 20, No. 1, Winter 1984, p. 86. Copyright © 1984. Reprinted by permission of Sage Publications, Inc.

    a. Interpret the scatterplot shown in the figure.

    b. Calculate the coefficient of determination and interpret this value.

*9.85 To compare the literacy rate, Y, to the daily newspaper circulation, X, the sample data shown in the table were obtained.

Nation	Newspaper Circulation per 1,000 Inhabitants X	Literacy Rate Y, Percentage
Albania	52	75
Burma	14	78
Denmark	356	99
Greece	88	95
Italy	132	98
Malawi	2	25
Pakistan	24	23
Saudi Arabia	29	15
Sweden	578	99
Uruguay	337	94
Zambia	20	54

Source: *The World Almanac & Book of Facts 1985* (New York: Newspaper Enterprise Association, Inc., 1984), pp. 515–600. Copyright © Newspaper Enterprise Association, Inc., 1984, New York, New York 10166.

Use a computer to do the following:
a. Draw a scatterplot of the data.
b. Find the least squares line.
c. Determine whether there is sufficient evidence to conclude that literacy rate is linearly related to daily newspaper circulation. Use $\alpha = .10$.
d. Calculate r and r^2. Interpret their values.
e. The Republic of Guatemala has a daily newspaper circulation of 47 per 1,000. Predict Guatemala's literacy rate. Use a 90% prediction interval.

CHAPTER 9 QUIZ

1. A car dealer is interested in modeling the relationship between the number of cars sold by the firm each week and the number of salespeople who work on the showroom floor. The dealer believes the relationship between the two variables can best be described by a straight line. The sample data shown in the table were supplied by the car dealer.
a. Construct a scatterplot for the data.
b. Use the method of least squares to estimate the Y-intercept and the slope of a straight-line model.
c. Plot the least squares line on the scatterplot.
d. Is there sufficient evidence to conclude that X helps to predict Y in a linear manner? Use $\alpha = .05$.

Week of	Number of Cars Sold Y	Number of Salespeople on Duty X
January 30	20	6
June 3	18	6
March 2	10	4
October 26	6	2
February 7	11	3

e. Calculate and interpret r and r^2.

f. Estimate the average number of cars sold in a week for which five salespeople are on duty. Use a 90% confidence interval.

g. Form a 95% prediction interval for the number of cars sold in a week by four salespeople.

CHAPTERS 7–9 CUMULATIVE QUIZ

1. Two companies produce an electronic component that is used in weather satellites. It is desired to compare the mean lifetimes of the two brands of components. A summary of the data collected from two independent samples is given in the table:

Company 1	Company 2
$n_1 = 3$	$n_2 = 4$
$\bar{x}_1 = 1{,}436.7$ hours	$\bar{x}_2 = 1{,}345.5$ hours
$s_1 = 35.9$ hours	$s_2 = 41.6$ hours

Form a 95% confidence interval for the difference between the mean lifetimes of the two brands of components.

2. A company will purchase a mine if it is convinced that the ore contains more than 10% of a certain metal. Fifty samples of ore from the mine are assayed, and the percentage of metal in each sample is recorded. A summary of the data is shown below:

$$\bar{x} = 11.3\% \qquad s = 2.9\%$$

a. Do the data provide sufficient evidence to indicate that the mean percentage of the metal in the ore is larger than 10%? Use $\alpha = .01$.

b. Why would the mining company want a small value of α for this test?

3. It is desired to estimate the mean amount of pollutant emitted per mile during city driving by a certain type of automobile. For how many miles should the pollutant be measured if it is desired to estimate the mean to within .03 with 95% confidence? Assume the standard deviation is approximately .13.

4. The data in the table represent a sample of mathematics achievement test scores and first-semester calculus grades for eight independently selected college students.

Student	Mathematics Achievement Test Scores X	Calculus Grade Y	Student	Mathematics Achievement Test Scores X	Calculus Grade Y
1	68	65	5	83	78
2	70	75	6	60	63
3	54	58	7	97	98
4	95	88	8	64	71

a. Use the method of least squares to estimate the Y-intercept and the slope of a straight-line model relating Y to X.

b. Is there sufficient evidence to conclude that X helps to predict Y in a linear manner? Use $\alpha = .05$.

On Your Own

Many dependent variables in all areas of research are the subject of regression modeling efforts. We list five examples below:

1. Crime rates in various communities
2. Daily maximal temperatures in your town
3. Grade-point averages of students who have completed one academic year at your college
4. Gross National Product of the United States
5. Points scored by your favorite football team in any one game

Choose one of these dependent variables or choose some other dependent variable for which you want to construct a prediction model. There may be a large number of independent variables that should be included in a prediction equation for the dependent variable you choose. List three potentially important independent variables, X_1, X_2, and X_3, that you think might be (individually) strongly related to your dependent variable. Next, obtain ten data values, each of which consists of a measure of your dependent variable, Y, and the corresponding values of X_1, X_2, and X_3.

a. Use the least squares formulas given in this chapter to fit three straight-line models—one for each independent variable—for predicting Y.

b. Interpret the sign of the estimated slope coefficient b_1 in each case, and test the utility of the model by testing $H_0: \beta_1 = 0$ against $H_a: \beta_1 \neq 0$.

c. Calculate the coefficient of determination, r^2, for each model. Which of the independent variables predicts Y best for the ten sampled observations? Is this variable necessarily best in general (that is, for the entire population)? Explain.

Chapter 10

ANALYSIS OF VARIANCE: COMPARING MORE THAN TWO POPULATION MEANS

COMPARING THE MAXIMAL GRIP STRENGTH OF WOMEN ATHLETES

Vivian Heyward and Leslie McCreary (1977) discuss the strength and endurance of women athletes in an article published in *The Research Quarterly*. To determine whether differences exist in the mean grip strengths of women participating in eight different sports, the authors sampled women athletes from each sport and measured their maximal grip strength. The sample sizes, means, and standard deviations for the data are shown in Table 10.1 (page 568).

Using a statistical procedure known as an **analysis of variance**, Heyward and McCreary concluded that the data in Table 10.1 provide sufficient evidence to indicate that there are differences in mean maximal grip strength among all women participating in the eight sports.

In this chapter we will introduce the statistical methods that are used when comparing more than two population means. The sampling procedure we will consider is called a **completely randomized design**, and it is analogous to the independent sampling scheme used when comparing two population means. The method of analysis is called an **analysis of variance.**

Table 10.1

Maximal grip strength data

Sport	Number of Women Sampled	Maximal Grip Strength	
		$\bar{x}$, Kilograms	s, Kilograms
Basketball	7	40.68	5.96
Field hockey	7	39.16	4.84
Golf	6	42.74	4.62
Gymnastics	5	39.14	2.80
Swimming	6	33.38	4.63
Tennis	5	45.44	5.68
Track and field	8	37.26	3.69
Volleyball	6	38.37	5.24

Source: Heyward and McCreary (1977) in *The Research Quarterly for Exercise and Sport,* published by American Alliance for Health, Physical Education, Recreation and Dance.

10.1

THE COMPLETELY RANDOMIZED DESIGN

In Section 8.1 we stated that two samples from two populations are independent if the selection of objects from one population is unrelated to the selection of objects from the other population. This definition may be extended to more than two samples.

Definition 10.1

If three (or more) samples are selected, one from each of three (or more) populations, then the samples are **independent** if the selection of objects from any one population is unrelated to the selection of objects from any of the other populations.

When independent samples are randomly selected, the sampling method is called a **completely randomized design,** or **independent sampling design.**

Definition 10.2

A **completely randomized design,** or **independent sampling design,** is one in which independent random samples are selected from each of the populations of interest.

Consider the maximal grip strength experiment discussed in the introduction. Samples of women athletes were obtained from each of eight sports (basketball, field hockey, golf, etc.). Heyward and McCreary (1977) indicate that a random sample of female basketball players was selected, and then a second random sample of female field hockey players was independently selected, and then a third independent random sample of female golfers was selected, etc. Thus, a completely randomized design was used to collect the data.

Consider another example. Suppose a medical researcher wants to compare the mean lengths of time for patients to respond to treatment by three drugs, A, B, and C. Thirty patients are randomly selected and divided into three equal groups. The ten patients in the first group are treated with drug A, the ten patients in the second group with drug B, and the ten patients in the third group with drug C. The length of time until recovery is measured for each of the sampled patients. Because the patients employed in the experiment are randomly selected and randomly assigned to the three drug categories, this is a completely randomized design.

Note that in this experiment ten sample measurements (recovery times) were randomly and independently chosen from each of three conceptual populations of recovery times for all patients who might possibly use the three drugs.

In this experiment a sample measurement (recovery time) was recorded for each patient, or **experimental unit,** involved in the study. Generally, experimental units are the objects on which the sample measurements are obtained. Many experiments involve the **treatment** of experimental units in different ways and a comparison of the means of the conceptual populations of measurements corresponding to the different treatments. Thus, the three drugs given to the patients in this study are referred to as **treatments.** Since each treatment determines a population (here, the recovery time for patients using that drug), we will commonly call the population a *treatment* and say that we are *comparing treatments*.

10.2

THE ANALYSIS OF VARIANCE FOR A COMPLETELY RANDOMIZED DESIGN

In this section we will discuss the comparison of more than two population (treatment) means using a completely randomized design. The procedure we will use is called an **analysis of variance;** this method is valid when the conditions listed in the box at the top of page 570 are met.

The implications of the second and third conditions are shown in Figure 10.1. Three normally distributed populations with equal variances are pictured. Notice that the populations may have different means (denoted by μ_1, μ_2, and μ_3), and that this is what we want to compare. The comparison will be based upon a completely randomized design.

> **When to Use an Analysis of Variance to
> Compare More Than Two Population Means**
>
> To conduct a valid analysis of variance, as presented in this section, the following conditions must be met:
>
> 1. A completely randomized design must be used to collect the sample data.
> 2. All the populations of interest must have normal distributions.
> 3. All the population variances must be equal.

Figure 10.1
Three normally distributed populations with equal variances

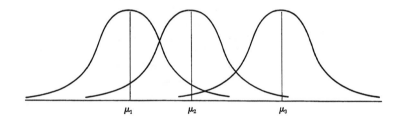

$\mu_1 \qquad \mu_2 \qquad \mu_3$

Notation that is useful when performing an analysis of variance is shown in Table 10.2. This notation may be used when comparing k population means, where k could equal 3, 4, 5, ...* As indicated in Section 10.1, it is often convenient to refer to the populations of interest as *treatments*.

The sample total is the sum of the measurements in that sample. Thus, each sample mean is equal to the sample total divided by the sample size. For example, $\bar{x}_1 = T_1/n_1$.

When performing an analysis of variance, we will test the null hypothesis that the k treatment means are equal,

H_0: $\mu_1 = \mu_2 = \cdots = \mu_k$

against the alternative hypothesis

H_a: At least two of the treatment means differ

To see the principle behind the test we will use, consider the dot diagrams for two samples shown in Figure 10.2. Five observations for each of two samples are shown, and the sample means are indicated by arrows. We think you will agree that these two samples provide a great deal of evidence to indicate a difference between μ_1 and μ_2, the corresponding population means. This is because the *difference between the sam-*

---

* Actually, k could equal 2, but then the independent two-sample t statistic is usually used to analyze the data. The two-sided t test and analysis of variance are equivalent when $k = 2$.

Table 10.2

Notation for a completely randomized design

	Populations (Treatments)				
	1	2	3	$\cdots$	k
Mean	μ_1	μ_2	μ_3	$\cdots$	μ_k
Variance	σ^2	σ^2	σ^2	$\cdots$	σ^2

	Independent Random Samples				
	1	2	3	$\cdots$	k
Sample size	n_1	n_2	n_3	$\cdots$	n_k
Sample total	T_1	T_2	T_3	$\cdots$	T_k
Sample mean	$\bar{x}_1$	$\bar{x}_2$	$\bar{x}_3$	$\cdots$	$\bar{x}_k$
Sample variance	s_1^2	s_2^2	s_3^2	$\cdots$	s_k^2

Total number of measurements $= n = n_1 + n_2 + \cdots + n_k$

Sum of all n measurements $= T_1 + T_2 + \cdots + T_k = \sum X$

Mean of all n measurements $= \bar{X} = \dfrac{\sum X}{n}$

Sum of squares of all n measurements $= \sum X^2$

Figure 10.2
Dot diagrams for two samples (Strong evidence of a difference between population means)

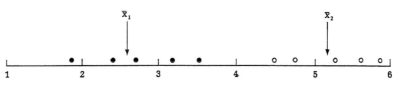

●: Sample 1

○: Sample 2

ple means, $\bar{x}_1$ and $\bar{x}_2$, is large compared to the variability of the measurements within each sample.

Now look at the dot diagrams in Figure 10.3 for two other samples of five measurements each. The two sample means are rather close together, and the variability within each sample is quite large. The data provide little, if any, evidence to indicate a difference between μ_1 and μ_2, because the *difference between the sample means is small compared to the variability of the measurements within each sample.*

Figure 10.3
Dot diagrams for two samples (Little evidence of a difference between population means)

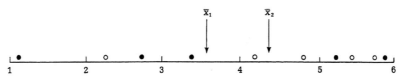

●: Sample 1

○: Sample 2

When deciding whether a difference exists among k population means μ_1, $\mu_2, \ldots, \mu_k$, we will compare the variation among the sample means to the variation within the samples. The variation among the sample means, $\bar{x}_1, \bar{x}_2, \ldots, \bar{x}_k$, is measured by a weighted sum of squares of deviations about the overall mean, $\bar{X}$. It is called the **sum of squares for treatments (SST)** and is given by the expression

$$SST = n_1(\bar{x}_1 - \bar{X})^2 + n_2(\bar{x}_2 - \bar{X})^2 + \cdots + n_k(\bar{x}_k - \bar{X})^2$$

To measure the within-sample variability, we use a pooled measure of the variability within each sample. It is called the **sum of squared errors (SSE)** and is given by

$$SSE = (n_1 - 1)s_1^2 + (n_2 - 1)s_2^2 + \cdots + (n_k - 1)s_k^2$$

Before comparing the variability among treatment means (SST) to the within-sample variability (SSE), each sum of squares must be divided by its degrees of freedom to obtain a **mean square.** The degrees of freedom for treatments equals $(k - 1)$, and the **mean square for treatments (MST)** is

$$MST = \frac{SST}{k - 1}$$

The degrees of freedom for error equals $(n - k)$, and the **mean square for error (MSE)** is

$$MSE = \frac{SSE}{n - k}$$

When the null hypothesis is true, both MST and MSE estimate σ^2, the common population variance. When the alternative hypothesis is true, MSE still estimates σ^2, but MST does not. The larger MST is, compared to MSE, the more evidence to indicate that the alternative hypothesis is true. We will compare MST to MSE by forming their ratio, MST/MSE. This statistic has a sampling distribution that has an **F distribution.**[*] The distribution of $F = $ MST/MSE depends upon the $(k - 1)$ degrees of freedom associated with MST *and* the $(n - k)$ degrees of freedom associated with MSE. We say that there are **$(k - 1)$ numerator degrees of freedom** and **$(n - k)$ denominator degrees of freedom** when describing this F distribution. An F distribution with 7 numerator degrees of freedom and 9 denominator degrees of freedom is shown in Figure 10.4. As you can see, the distribution is skewed to the right, since the value of MST/MSE cannot be less than 0, but has no upper limit.

To conduct an analysis of variance, we need to be able to find F-values corresponding to the upper-tail areas of the F distribution. The upper-tail F-values corresponding to areas of .10, .05, .025, and .01 are given in Appendix Tables IX, X, XI, and XII, respectively. Table X is partially reproduced in Table 10.3. The columns of this table give the numerator degrees of freedom, $(k - 1)$, and the rows give the denominator degrees of freedom, $(n - k)$. Thus, if $k - 1 = 7$ and $n - k = 9$, we look in the

---

[*] The F distribution was first introduced in Section 8.7, an optional section. We will describe the F distribution again here, in case you omitted Section 8.7.

Figure 10.4
An F distribution with
7 numerator degrees of
freedom and 9 denominator
degrees of freedom

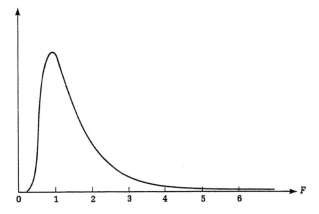

Table 10.3

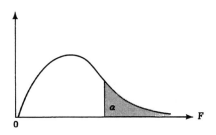

A partial reproduction of
Appendix Table X, $\alpha = .05$

| | | \multicolumn{9}{c|}{Numerator Degrees of Freedom} |
		1	2	3	4	5	6	7	8	9
	1	161.4	199.5	215.7	224.6	230.2	234.0	236.8	238.9	240.5
	2	18.51	19.00	19.16	19.25	19.30	19.33	19.35	19.37	19.38
	3	10.13	9.55	9.28	9.12	9.01	8.94	8.89	8.85	8.81
	4	7.71	6.94	6.59	6.39	6.26	6.16	6.09	6.04	6.00
Denominator Degrees of Freedom	5	6.61	5.79	5.41	5.19	5.05	4.95	4.88	4.82	4.77
	6	5.99	5.14	4.76	4.53	4.39	4.28	4.21	4.15	4.10
	7	5.59	4.74	4.35	4.12	3.97	3.87	3.79	3.73	3.68
	8	5.32	4.46	4.07	3.84	3.69	3.58	3.50	3.44	3.39
	9	5.12	4.26	3.86	3.63	3.48	3.37	3.29	3.23	3.18
	10	4.96	4.10	3.71	3.48	3.33	3.22	3.14	3.07	3.02
	11	4.84	3.98	3.59	3.36	3.20	3.09	3.01	2.95	2.90
	12	4.75	3.89	3.49	3.26	3.11	3.00	2.91	2.85	2.80
	13	4.67	3.81	3.41	3.18	3.03	2.92	2.83	2.77	2.71
	14	4.60	3.74	3.34	3.11	2.96	2.85	2.76	2.70	2.65

Figure 10.5
The *F* distribution with
7 and 9 df

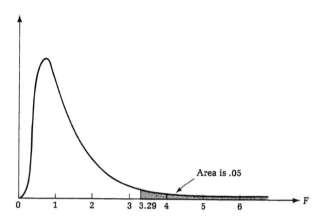

seventh column and ninth row to find the value $F = 3.29$. As shown in Figure 10.5, the tail area to the right of 3.29 is .05 for the *F* distribution with 7 and 9 df.

The complete test of hypothesis for *k* population means using data from a completely randomized design is given in the following box:

**Test to Compare *k* Treatment Means
for a Completely Randomized Design**

H_0: $\mu_1 = \mu_2 = \cdots = \mu_k$

H_a: At least two treatment means differ

Test statistic: $F = MST/MSE$

Rejection region: Reject H_0 if $F >$ Table value.

The table value is based on $(k - 1)$ numerator degrees of freedom and $(n - k)$ denominator degrees of freedom, and it is found in the *F* table with an upper-tail area equal to the value of α.

This testing procedure is called an *analysis of variance* because the *F* statistic compares two sources of variation—the source of variation due to differences among the sample means and the source of variation due to within-sample differences among experimental units. Although the sums of squares, SST and SSE, could be calculated by the formulas given earlier in this section, simpler computing formulas are available. The formulas needed to complete the analysis of variance are given in the next box.

**Calculating Formulas for the Analysis of Variance
in a Completely Randomized Design**

$$\sum X = T_1 + T_2 + \cdots + T_k$$

$$\text{SST} = \frac{T_1^2}{n_1} + \frac{T_2^2}{n_2} + \cdots + \frac{T_k^2}{n_k} - \frac{\left(\sum X\right)^2}{n}$$

$$\text{SSE} = \sum X^2 - \left(\frac{T_1^2}{n_1} + \frac{T_2^2}{n_2} + \cdots + \frac{T_k^2}{n_k}\right)$$

$$\text{MST} = \frac{\text{SST}}{k-1}$$

$$\text{MSE} = \frac{\text{SSE}}{n-k}$$

$$F = \frac{\text{MST}}{\text{MSE}}$$

EXAMPLE 10.1   A sociologist conducted an experiment to compare the mean grade-point averages of first-year college students associated with four socioeconomic groups. Note that in this experiment, treatments (the socioeconomic conditions of the students) were not "applied" to the experimental units but instead identify the four populations of interest to the sociologist. The populations of grade-point averages are normally distributed with equal variances. Independent random samples of grade-point averages at the end of the freshman year were selected from each group. The sample data are shown in Table 10.4. Do the data provide sufficient evidence to indicate a difference in mean grade-point averages for at least two of the four socioeconomic groups? Test at the $\alpha = .05$ level of significance.

Table 10.4

	Group 1	Group 2	Group 3	Group 4
	2.87	3.23	2.61	2.25
	2.16	3.45	3.56	3.13
	3.14	3.67	2.97	2.44
	2.51	2.78	2.33	3.27
	1.80	3.77	3.64	2.81
	3.01		2.67	1.36
	2.16		3.31	2.70
			3.01	2.41
Totals	$T_1 = 17.65$	$T_2 = 16.90$	$T_3 = 24.10$	$T_4 = 20.37$
	$n_1 = 7$	$n_2 = 5$	$n_3 = 8$	$n_4 = 8$

Solution   A completely randomized design has been used to obtain samples from four normally distributed populations with equal variances. We may proceed with an analysis of variance and complete the following test, where μ_1 is the population mean grade-point average for group 1, etc.:

H_0: $\mu_1 = \mu_2 = \mu_3 = \mu_4$

H_a: At least two of the means are not equal

Test statistic: $F = \text{MST/MSE}$

Rejection region: Reject H_0 if $F > 3.01$.

The tabulated F-value that locates the rejection region was found in Table X using $k - 1 = 4 - 1 = 3$ df and $n - k = 28 - 4 = 24$ df.

We will now perform the calculations needed to complete the test:

$$\sum X = T_1 + T_2 + T_3 + T_4 = 17.65 + 16.90 + 24.10 + 20.37 = 79.02$$

$$\sum X^2 = (2.87)^2 + (2.16)^2 + \cdots + (2.41)^2 = 232.2534$$

$$\text{SST} = \frac{T_1^2}{n_1} + \frac{T_2^2}{n_2} + \frac{T_3^2}{n_3} + \frac{T_4^2}{n_4} - \frac{(\sum X)^2}{n}$$

$$= \frac{(17.65)^2}{7} + \frac{(16.90)^2}{5} + \frac{(24.10)^2}{8} + \frac{(20.37)^2}{8} - \frac{(79.02)^2}{28}$$

$$= 226.0936 - 223.0057$$

$$= 3.0879$$

$$\text{SSE} = \sum X^2 - \left(\frac{T_1^2}{n_1} + \frac{T_2^2}{n_2} + \frac{T_3^2}{n_3} + \frac{T_4^2}{n_4} \right)$$

$$= 232.2534 - 226.0936 = 6.1598$$

$$\text{MST} = \frac{\text{SST}}{k - 1} = \frac{3.0879}{3} = 1.0293$$

$$\text{MSE} = \frac{\text{SSE}}{n - k} = \frac{6.1598}{24} = .2567$$

The calculated value of the test statistic is

$$F = \frac{\text{MST}}{\text{MSE}} = \frac{1.0293}{.2567} = 4.01$$

Since $F = 4.01 > 3.01$, we may conclude that there is a difference in the mean grade-point averages for at least two of the socioeconomic groups. We make this conclusion at the .05 level of significance. A natural question that arises is, "Which treatments are different?" An analysis of variance does not answer this question; it can indicate only that differences exist. In the next section we will discuss a method for comparing specific pairs of treatment means.   ∎

The results of an analysis of variance are often summarized in the form of a table called an **ANOVA table.** The general form of an ANOVA table for a completely ran-

Table 10.5

ANOVA table for
a completely
randomized design

Source	df	SS	MS	F
Treatments	$k-1$	SST	MST	MST/MSE
Error	$n-k$	SSE	MSE	

domized design is shown in Table 10.5. The heading **Source** refers to the possible sources of variation; **df** refers to the degrees of freedom associated with each source of variation; **SS** refers to the sums of squares; **MS** refers to the mean squares; and **F** refers to the F statistic.

The analysis of variance for the grade-point average data of Example 10.1 is summarized in the ANOVA table shown in Table 10.6. Notice that the nature of the treatments is indicated.

Table 10.6

ANOVA table for
Example 10.1

Source	df	SS	MS	F
Socioeconomic groups	3	3.0879	1.0293	4.01
Error	24	6.1598	.2567	

In the introductory example we indicated that there was a difference in mean maximal grip strengths among women athletes participating in different sports. An ANOVA table for the data reported in Table 10.1 is given in Table 10.7. (The values in this table were obtained from the raw data, not from the rounded values given in Table 10.1.)

Table 10.7

ANOVA table for maximal
grip strength data

Source	df	SS	MS	F
Sports	7	511.50	73.07	3.15
Error	42	974.88	23.21	

The F statistic is 3.15, and it is significant at the .05 level of significance since the tabulated F-value for $\alpha = .05$ and 7 and 42 df is approximately 2.25. On the basis of this analysis we can conclude that there is a difference in the mean maximal grip strengths for women participating in at least two of the eight sports of interest.

In concluding this discussion of the analysis of variance for a completely randomized design, we should comment on the conditions given earlier for performing a valid analysis of variance. The populations of interest should be normally distributed, but moderate departures from normality have little effect on the significance level of the test. On the other hand, the population variances should be equal, and even moderate departures from equality can affect the significance level of the test. However, the effect is less when the sample sizes are equal. If you are in doubt about whether the conditions of normality or equal variances are met in an experiment, we recommend that you use a nonparametric statistical method to analyze the data. (For example, see the discussion of the Kruskal–Wallis H test in Section 12.3.)

In Example 10.1 we conducted an analysis of variance to compare the mean grade-point averages of four socioeconomic groups. Minitab may be used to produce an ANOVA table for a completely randomized design. With the data for Example 10.1 shown in Table 10.4, the following Minitab printout was obtained.*

```
ANALYSIS OF VARIANCE
SOURCE      DF        SS         MS        F         P
FACTOR       3      3.088      1.029      4.01     0.019
ERROR       24      6.160      0.257
TOTAL       27      9.248
```

The ANOVA table generated by Minitab is essentially the same as the one we gave in Table 10.6. In Table 10.6 we indicated that the treatments were socioeconomic groups. Minitab always labels the treatments source of variability as "FACTOR." Other than this, the only difference is that Minitab gives the total degrees of freedom $(3 + 24 = 27)$ and the total sum of squares $(3.088 + 6.160 = 9.248)$. Although the sum of squares and the mean squares are rounded to three decimal places rather than the four decimal places given in Table 10.6, both F-values are 4.01. Minitab also reports the p-value as 0.019.

Once the ANOVA table is obtained with the aid of Minitab, the interpretation would proceed exactly as in Example 10.1. Recall that we concluded at the .05 level of significance that there is a difference in the mean grade-point averages for at least two of the socioeconomic groups. ∎

Exercises (10.1–10.12)

Learning the Language

10.1 Define the following important terms:
a. Completely randomized design   b. Independent samples

10.2 State what the following symbols represent:
a. SST   c. k e. $n - k$ g. MSE   i. $k - 1$
b. MST   d. F f. SSE   h. n

Using the Tables

10.3 Give the rejection region for testing

$$H_0: \ \mu_1 = \mu_2 = \mu_3$$

against

$$H_a: \ \text{At least two of the means are unequal}$$

for each of the following situations:
a. $n = 20$, $\alpha = .05$ e. $n = 90$, $\alpha = .10$
b. $n = 15$, $\alpha = .05$ f. $n = 75$, $\alpha = .05$
c. $n = 24$, $\alpha = .01$ g. $n = 21$, $\alpha = .01$
d. $n = 24$, $\alpha = .10$ h. $n = 9$, $\alpha = .05$

* Actually, this is only half of the printout. The other half will be discussed in the next section.

Learning the Mechanics

10.4 Independent random samples were selected from three normally distributed populations with common (but unknown) variance, σ^2. The data are shown in the table below:

Sample 1	Sample 2	Sample 3
3.1	5.4	1.1
4.3	3.6	.2
1.2	4.0	3.0
	2.9	

a. Make the appropriate calculations and complete the analysis of variance table shown below:

Source	df	SS	MS	F
Treatments				
Error				

b. Test the hypothesis that the three population means are equal ($\mu_1 = \mu_2 = \mu_3$) against the alternative hypothesis that at least one mean is different from the other two. Test using $\alpha = .05$.

10.5 Independent random samples were selected from four normally distributed populations with common (but unknown) variance, σ^2. The data are shown in the table below:

Sample 1	Sample 2	Sample 3	Sample 4
2	1	2	6
4	1	3	5
	3	4	7
	2	3	6
	3	4	5
			6

a. Make the appropriate calculations and complete the analysis of variance table shown below:

Source	df	SS	MS	F
Treatments				
Error				

b.   Test

$$H_0: \quad \mu_1 = \mu_2 = \mu_3 = \mu_4$$

against

H_a: At least two of the means are different

Use $\alpha = .10$.

Applying the Concepts

In each of the following exercises, specify the conditions required to make the statistical procedure valid.

10.6   Studies conducted at the University at Melbourne (Australia) indicate that there may be a difference in the pain threshold of blonds and brunettes (*Family Weekly*, February 5, 1978). Men and women of various ages were divided into four categories according to hair color: light blond, dark blond, light brunette, and dark brunette. The purpose of the experiment was to determine whether hair color is related to the amount of pain elicited by common types of mishaps and assorted types of traumas. Each person in the experiment was given a pain threshold score based upon his or her performance in a pain sensitivity test (the higher the score, the lower the person's pain tolerance). The results are given in the table.

Light Blond	Dark Blond	Light Brunette	Dark Brunette
62	63	42	32
60	57	50	39
71	52	41	51
55	41	37	30
48	43		35

Source: *Family Weekly*, February 5, 1978. Reprinted by permission of *Family Weekly*, copyright 1978, 641 Lexington Avenue, New York, N.Y. 10022.

a.   Based on the given information, what type of experimental design appears to have been employed?

b.   Is there evidence of a difference among mean pain thresholds for people possessing the four hair colors? Use $\alpha = .01$.

10.7   Some varieties of nematodes (very small roundworms that live in the soil) feed upon the roots of lawn grasses and such crops as strawberries and tomatoes. These farm and garden pests, which are particularly troublesome in warm climates, can be treated by the application of nematicides, but due to the small size of the worms, it is very difficult to measure directly the effectiveness of the nematicides. To compare four nematicides, the yields of equal plots of one variety of tomatoes were collected and weighed. The data (in pounds per plot) are shown in the table. Do the data provide sufficient evidence to indicate a difference in the mean yields of tomatoes per plot for at least two of the four nematicides? Test using $\alpha = .05$.

	Nematicide		
1	2	3	4
18.6	18.7	18.4	19.0
18.2	19.3	19.9	18.5
17.6	18.9	19.7	18.6
		19.1	

10.8 A research psychologist wants to investigate the difference in maze test scores for a strain of laboratory mice trained under different laboratory conditions. The experiment is conducted using eighteen randomly selected mice of this strain, with six receiving no training at all (control group), six trained under condition 1, and six trained under condition 2. Each mouse is then given a test score between 0 and 100, according to its performance in a test maze. The experiment produced the results shown in the table. Is there sufficient evidence to indicate a difference in mean maze test scores for mice trained under the three laboratory conditions? Use $\alpha = .05$.

Control	Condition 1	Condition 2
58	73	53
32	70	74
59	68	72
64	71	62
55	60	58
49	62	61

10.9 An experiment is conducted to determine whether there is a difference in the mean increases in growth produced by five inoculins of growth hormones for plants. The experimental material consists of twenty cuttings of a shrub (all of equal weight), with four cuttings randomly assigned to each of the five inoculins. The results of the experiment are given in the table; all measurements represent increases in weight. Based on this information, can we conclude that the mean increases in weight differ for at least two of the five inoculins of growth hormone? Test at the $\alpha = .05$ level of significance.

		Inoculin		
A	B	C	D	E
15	21	22	10	6
18	13	19	14	11
9	20	24	21	15
16	17	21	13	8

10.10 To compare the attitudes of students preparing for eight occupations, random samples of students independently selected for each occupation were asked questions designed to measure how they envisioned their future relationships with clients. The relative attitudinal autonomy was of particular interest to the researchers, who hypothesized that there would be a difference in the mean attitudes expressed by students in different occupations. A summary of the sample data is given in the table. Higher mean scores indicate a greater feeling of attitudinal autonomy toward potential clients.

Occupation	n	Mean	Standard Deviation
Medicine	80	51.01	7.41
Law	168	54.14	8.96
Education	131	51.14	9.11
Nursing	136	45.08	7.69
Social work	137	43.38	9.31
Librarianship	71	48.31	9.29
Engineering	169	53.62	7.26
Business administration	62	53.19	7.00

Source: P. B. Forsyth and T. J. Danisiewicz, "Toward a Theory of Professionalization," *Work and Occupations*, Vol. 12, No. 1, February 1985, p. 70. Copyright © 1985. Reprinted by permission of Sage Publications, Inc.

The analysis of variance to test for differences among the mean attitude scores for the eight occupations is shown in the following table.

Source	df	SS	MS	F
Occupations	7	15,474.06	2,210.58	31.59
Error	946	66,190.23	69.97	

a. Verify that $F = 31.59$.
b. Is there evidence to support the researchers' hypothesis at the .01 level of significance?

10.11 A study was conducted to investigate human social preening:

> Preening by subjects (defined as hair grooming, clothes straightening, and gazing at self in mirror) was unobtrusively observed and timed in the restrooms of five restaurants and bars located within the university community.*

One hypothesis was that people preen more when developing a relationship than when relations are well established. Subjects involved in the study were later inter-

---

* J. A. Daly, E. Hogg, D. Sacks, M. Smith, and L. Zimring, "Sex and Relationship Affect Social Self-Grooming," *Journal of Nonverbal Behavior*, Vol. 7, No. 3, Spring 1983, p. 185. Copyright 1983 by Human Sciences Press, Inc., 72 Fifth Avenue, New York, New York 10011.

viewed and classified into four types of relationship:

(1) long-term, stable relationships which included married couples and long-term same-sex friends who had no intentions of meeting new people that evening, (2) couples who were dating and who had dated for more than four dates, (3) couples who were in the early stages of the relationship: the date was one of their first four with each other, and (4) individuals and groups who were specifically in the establishment trying to meet new people.*

Using the preening times and categories obtained, the researchers conducted an analysis of variance and concluded that there was a difference in the mean preening times for the four types of relationship. The observed level of significance was stated to be less than .005—that is, $p < .005$.

a. Identify the populations of interest.
b. State the null hypothesis and alternative hypothesis that were tested.
c. Interpret the results of this experiment.

*10.12 An accounting firm that specializes in auditing the financial records of large corporations is interested in evaluating the appropriateness of the fees it charges for its services. As part of its evaluation it wants to compare the costs it incurs in auditing corporations of different sizes. The accounting firm decided to measure the size of its client corporations in terms of their yearly sales. Accordingly, its population of client corporations was divided into three subpopulations:

A: Those with sales over $250 million

B: Those with sales between $100 million and $250 million

C: Those with sales under $100 million

The firm chose random samples of ten corporations from each of the subpopulations and determined from its records the costs (in thousands of dollars) given in the table.

Costs Incurred in Audits

A	B	C
250	100	80
150	150	125
275	75	20
100	200	186
475	55	52
600	80	92
150	110	88
800	160	141
325	132	76
230	233	200

Use a computer package to obtain an analysis of variance table that may be used to test whether the three classes of firms have different mean costs incurred in audits. Use $\alpha = .05$.

* Ibid., p. 186.

10.3

CONFIDENCE INTERVALS FOR ONE OR TWO POPULATION MEANS

Since the completely randomized design involves the selection of independent random samples from normally distributed populations with equal variances, we can find a confidence interval for a single treatment mean using the method of Section 7.5 or for the difference between two treatment means using the method of Section 8.3. We will use the value of MSE that was calculated in the analysis of variance to estimate the common variance, σ^2. The formulas for the confidence intervals of interest are given in the boxes.

Confidence Interval for a Single Treatment Mean

The confidence interval for μ_i, the mean of treatment i, is expressed as

$$\bar{x}_i - t \sqrt{\frac{MSE}{n_i}} \quad \text{to} \quad \bar{x}_i + t \sqrt{\frac{MSE}{n_i}}$$

The value of t is found in Appendix Table VI, using $(n - k)$ degrees of freedom (the degrees of freedom for error in the ANOVA table.)

Confidence Interval for the Difference Between Two Treatment Means

The confidence interval for $\mu_i - \mu_j$, the difference between the means of treatment i and treatment j, is expressed as

$$(\bar{x}_i - \bar{x}_j) - t \sqrt{\frac{MSE}{n_i} + \frac{MSE}{n_j}} \quad \text{to} \quad (\bar{x}_i - \bar{x}_j) + t \sqrt{\frac{MSE}{n_i} + \frac{MSE}{n_j}}$$

The value of t is found in Appendix Table VI, using $(n - k)$ degrees of freedom (the degrees of freedom for error in the ANOVA table).

EXAMPLE 10.2　In Example 10.1 we concluded that there was a difference among the mean grade-point averages of students in four socioeconomic groups. A summary of the data is given in Table 10.8.

Table 10.8

	Group 1	Group 2	Group 3	Group 4
Totals	$T_1 = 17.65$	$T_2 = 16.90$	$T_3 = 24.10$	$T_4 = 20.37$
Sample sizes	$n_1 = 7$	$n_2 = 5$	$n_3 = 8$	$n_4 = 8$

The value of MSE was found to be .2567 with 24 df. Find a 95% confidence interval for μ_1, the mean grade-point average of all possible first-year students who belong to group 1.

Solution   Since the conditions for conducting a valid analysis of variance were met, we may now determine the confidence interval. The form of the interval is

$$\bar{x}_1 - t \sqrt{\frac{MSE}{n_1}} \quad \text{to} \quad \bar{x}_1 + t \sqrt{\frac{MSE}{n_1}}$$

The value of $\bar{x}_1$ is

$$\bar{x}_1 = \frac{T_1}{n_1} = \frac{17.65}{7} = 2.52$$

Using Appendix Table VI, we find the t-value associated with 95% confidence and 24 df to be $t = 2.06$. Substituting the appropriate values into the confidence interval formula, we obtain

$$2.52 - (2.06)\sqrt{\frac{.2567}{7}} \quad \text{to} \quad 2.52 + (2.06)\sqrt{\frac{.2567}{7}}$$

$$2.52 - .39 \quad \text{to} \quad 2.52 + .39$$

$$2.13 \quad \text{to} \quad 2.91$$

Thus, we are 95% confident that the mean grade-point average for all first-year students in group 1 is between 2.13 and 2.91.   ■

EXAMPLE 10.3   Now, suppose that prior to conducting the experiment described in Example 10.1, the sociologist was particularly interested in estimating the difference in mean grade-point averages between all possible first-year students in group 1 and those in group 2. Find a 95% confidence interval for this difference in mean grade-point averages.

Solution   Using the information given in Table 10.8 and some of our calculations from Example 10.2, we have $\bar{x}_1 = 2.52$, $n_1 = 7$, $T_2 = 16.90$, $n_2 = 5$, and MSE = .2567 with 24 df. Calculating $\bar{x}_2$, we get

$$\bar{x}_2 = \frac{T_2}{n_2} = \frac{16.90}{5} = 3.38$$

The general form of the confidence interval for a difference in the means, $\mu_1 - \mu_2$, is

$$(\bar{x}_1 - \bar{x}_2) - t \sqrt{\frac{MSE}{n_1} + \frac{MSE}{n_2}} \quad \text{to} \quad (\bar{x}_1 - \bar{x}_2) + t \sqrt{\frac{MSE}{n_1} + \frac{MSE}{n_2}}$$

From Table VI, the value of t with 24 df for 95% confidence is 2.06. Calculating the interval, we obtain

$$(2.52 - 3.38) - (2.06)\sqrt{\frac{.2567}{7} + \frac{.2567}{5}} \quad \text{to} \quad (2.52 - 3.38) + (2.06)\sqrt{\frac{.2567}{7} + \frac{.2567}{5}}$$

$$-.86 - .61 \quad \text{to} \quad -.86 + .61$$

$$-1.47 \quad \text{to} \quad -.25$$

We can be 95% confident that the difference between μ_1 and μ_2 is between -1.47 and $-.25$. Since the interval contains only negative numbers, we can conclude that the mean grade-point average of all possible first-year students in group 2 exceeds the mean grade-point average of those in group 1. ■

EXAMPLE 10.4   Consider the introductory example to this chapter, which discussed the comparison of the mean maximal grip strengths of women participating in eight sports. Suppose that it is of particular interest to compare the mean grip strengths of women basketball players and women volleyball players (the first and eighth sports). Use the data given in the table to form a 90% confidence interval for the difference in the mean grip strengths of all female basketball players and all female volleyball players. From the analysis of variance table given earlier (Table 10.7), we have MSE = 23.21 with 42 df.

Sport	Sample Size	Sample Mean
Basketball	$n_1 = 7$	$\bar{x}_1 = 40.68$
Volleyball	$n_8 = 6$	$\bar{x}_8 = 38.37$

Solution   From Table VI, we obtain the t-value for 90% confidence and 42 df, which is 1.65. (This is actually a z-value, since the degrees of freedom are so large.) Calculating the interval, we find

$$(\bar{x}_1 - \bar{x}_8) - t\sqrt{\frac{MSE}{n_1} + \frac{MSE}{n_8}} \quad \text{to} \quad (\bar{x}_1 - \bar{x}_8) + t\sqrt{\frac{MSE}{n_1} + \frac{MSE}{n_8}}$$

$$(40.68 - 38.37) - (1.65)\sqrt{\frac{23.21}{7} + \frac{23.21}{6}} \quad \text{to} \quad (40.68) - 38.37) + (1.65)\sqrt{\frac{23.21}{7} + \frac{23.21}{6}}$$

$$2.31 - 4.42 \quad \text{to} \quad 2.31 + 4.42$$

$$-2.11 \quad \text{to} \quad 6.73$$

We are 90% confident that the difference in mean grip strengths for women who play basketball versus women who play volleyball is between -2.11 and 6.73. Since the interval contains 0, we cannot conclude that women in one sport have a greater mean grip strength than women in the other. ■

There are a few important points to be made about forming a confidence interval for the difference in a pair of treatment means. It is advisable to choose the pair (or pairs) of means for which you want to construct confidence intervals before collecting the sample data, so that your choice will not be influenced by the data. It is also a good idea to keep the number of pairs of means as small as possible. The more intervals that are formed, the higher the risk that at least one of the intervals will be incorrect. We will be very confident that any one interval is correct, but we would have less confidence that all of a large number of intervals are correct.

It may be tempting to form a confidence interval for every possible pair of treatment means to determine which treatments are different; however, this is not advisable. Statistical methods (called **multiple comparison procedures**) are available to compare all possible pairs of treatment means. A discussion of these methods can be found in Snedecor and Cochran (1980).

COMPUTER
EXAMPLE 10.2

In Example 10.2 we found a 95% confidence interval for μ_1, the mean grade-point average of all possible first-year students who belong to the first of four socioeconomic groups. The resulting interval was from 2.13 to 2.91.

Minitab may also be used to generate such an interval. In fact, whenever Minitab produces an ANOVA table, it also displays 95% confidence intervals for all the individual treatment means. In Computer Example 10.1 we displayed a Minitab printout of an ANOVA table for the grade-point averages data. After the ANOVA table, the printout showed the following information.

```
                                    INDIVIDUAL 95 PCT CI'S FOR MEAN
                                    BASED ON POOLED STDEV
LEVEL        N      MEAN     STDEV  --------+---------+---------+--------
GROUP 1      7    2.5214    0.5041  (------*-------)
GROUP 2      5    3.3800    0.3948                 (---------*--------)
GROUP 3      8    3.0125    0.4674            (------*-------)
GROUP 4      8    2.5463    0.5955  (.------*------)
                                    --------+---------+---------+--------
POOLED STDEV =     0.5066             2.50      3.00      3.50
```

For each of the four socioeconomic groups, Minitab first gives the sample size ("N"). This is followed by the mean of each sample and the standard deviation of each sample. Below this information, the value of $\sqrt{MSE}$ (which Minitab calls the "POOLED STDEV") is given as 0.5066. Finally, the confidence interval for each mean is displayed at the right. The values of the intervals' endpoints are not noted, and it is difficult to tell at exactly what value an interval begins or ends. Recall that the interval we calculated for group 1 was from 2.13 to 2.91. This agrees with the interval displayed in the printout. ∎

Exercises (10.13–10.17)

Applying the Concepts

In each of the following exercises, specify the conditions required to make the statistical procedure valid.

10.13 Most new products are test marketed in several locations, and different advertising techniques are often used. The accompanying table gives the number of sales for a new product during randomly selected months at each of three locations.

Location		
1	2	3
456	441	501
421	419	467
397	415	520
419	420	493

a. Treat this as a completely randomized design and test to determine whether there is a difference in mean sales for at least two of the three locations. Use $\alpha = .05$.
b. Suppose you want to estimate the difference in the mean sales between locations 1 and 3. Find a 90% confidence interval for $\mu_1 - \mu_3$.
c. Form a 95% confidence interval for μ_1.

10.14 A company that employs a large number of salespeople is interested in learning which of the salespeople sell the most: those who work strictly on commission, those who receive a fixed salary, or those who receive a reduced fixed salary plus a commission. The previous month's records for a sample of salespeople are inspected and the amount of sales (in dollars) is recorded for each, as shown in the table.

Commission	Fixed Salary	Commission Plus Salary
$425	$420	$430
507	448	492
450	437	470
483	432	501
466	444	
492		

a. Do the data provide sufficient evidence to indicate a difference in the mean sales for at least two of the three types of compensation? Use $\alpha = .05$.
b. Use a 90% confidence interval to estimate the mean sales for salespeople who receive a commission plus salary.
c. Use a 90% confidence interval to estimate the difference in mean sales between salespeople on commission plus salary versus those on fixed salary.

10.15 Refer to Exercise 10.10. Random samples of students were independently selected for each of eight occupations. The students expressed their attitudes regarding their feelings of autonomy toward future clients. The sample data are summarized in the next table. Higher mean scores indicate a greater feeling of attitudinal autonomy.

Occupation	n	Mean	Standard Deviation
Medicine	80	51.01	7.41
Law	168	54.14	8.96
Education	131	51.14	9.11
Nursing	136	45.08	7.69
Social work	137	43.38	9.31
Librarianship	71	48.31	9.29
Engineering	169	53.62	7.26
Business administration	62	53.19	7.00

Source: P. B. Forsyth and T. J. Danisiewicz, "Toward a Theory of Professionalization," Work and Occupations, Vol. 12, No. 1, February 1985, p. 70.

The analysis of variance table for the data is given below.

Source	df	SS	MS	F
Occupations	7	15,474.06	2,210.58	31.59
Error	946	66,190.23	69.97	

a. Form a 95% confidence interval for the mean attitudinal score for medical students.
b. Estimate the difference in the mean scores of law students and nursing students using a 99% confidence interval.
c. Estimate the difference in the mean scores of law students and engineering students using a 90% confidence interval.

10.16 Refer to Exercise 10.11. Preening times were compared for people classified into four types of relationship. The mean length of time (in seconds) for each group is given in the table. Assume that each type of relationship contained a sample of eighteen people, and an analysis of variance for a completely randomized design had a mean square error term equal to 1,486.1.

Relationship	Mean Preening Time
Married/close friend	9.8
Established dating	27.9
Early dating	57.8
Seeking new acquaintances	46.8

Source: J. A. Daly et al., "Sex and Relationship Affect Social Self-Grooming," *Journal of Nonverbal Behavior,* Vol. 7, No. 3, Spring 1983, p. 187.

a. Find a 95% confidence interval for the difference in the mean preening times of those classified as Early dating and Established dating.
b. Find a 95% confidence interval for the difference in the mean preening times of those classified as Early dating and Seeking new acquaintances.

*10.17 One of the main selling points for golf balls is their durability. An independent testing laboratory is commissioned to compare the durability of three brands of golf balls. Balls of each type are put into a machine that hits them with the force of a golfer on the course. The number of hits delivered until the outer covering cracks is recorded for each ball, with the results given in the table for ten balls randomly selected from each brand. With the aid of a computer, answer the following.

Brand A		Brand B		Brand C	
310	284	261	197	233	208
235	259	219	207	289	245
279	273	263	221	301	271
306	219	247	244	264	298
237	301	288	228	273	276

a. Is there evidence that the mean durabilities differ for at least two of the three brands? Use $\alpha = .05$.
b. Estimate the difference between the mean durabilities of brands A and C using a 99% confidence interval.
c. Form a 99% confidence interval for the mean durability of brand C.

Chapter Summary

An **analysis of variance** is a hypothesis test for comparing the means of several populations. The test is valid only when all the populations have normal distributions with the same variance and the sampling procedure is a **completely randomized design.** A completely randomized design requires that the samples be random and also that they be **independent,** that is, that the selection of objects from one population be unrelated to the selection from others. The populations are called **treatments** because they often result from varying treatments of one kind of object.

For a completely randomized design with k populations, we use the following notation for the jth population:

μ_j = Population mean $\bar{x}_j$ = Sample mean
σ^2 = Population variance s_j^2 = Sample variance

n_j = Sample size
T_j = Total of all n_j sample measurements
$n = n_1 + n_2 + \cdots + n_k$

The components of an analysis of variance test are summarized in an **ANOVA table:**

Source	df	SS	MS	F
Treatments	$k-1$	SST	MST	MST/MSE
Error	$n-k$	SSE	MSE	

These components are:

Treatments: The k populations, one possible source of variation among the means

Error: Variation among means due to variation *within* each sample

df: Degrees of freedom; $(k-1)$ is the **numerator degrees of freedom** and $(n-k)$ is the **denominator degrees of freedom**

SS: Sum of squares; **SST** is the **sum of squares for treatments,** a measure of variation among means:

$$\text{SST} = \sum[n_j(\bar{x}_j - \bar{x})^2] = \left(\sum \frac{T_j^2}{n_j}\right) - \frac{(\sum x)^2}{n}$$

SSE is the **sum of squared errors,** a measure of variability within each sample:

$$SSE = \sum[(n_j - 1)s_j^2] = \sum X^2 - \left(\sum \frac{T_j^2}{n_j}\right)$$

MS: Mean square; **MST** is the **mean square for treatments:**

$$MST = \frac{SST}{k - 1}$$

MSE is the **mean square for error,** an estimate for σ^2:

$$MSE = \frac{SSE}{n - k}$$

F: $F = MST/MSE$ is the statistic of interest; it has an F distribution for which values are tabulated in Tables IX, X, XI, and XII

The analysis of variance hypothesis test is:

H_0: $\mu_1 = \mu_2 = \cdots = \mu_k$

H_a: At least two population means are different

Test statistic: $F = MST/MSE$

Rejection region: Use Table IX, X, XI, or XII.

For each population we can form a confidence interval for the mean μ_j:

Obtain t from Table VI using $(n - k)$ df for the confidence required.

Calculate $\bar{x}_j$, MSE, and $B = t\sqrt{\dfrac{MSE}{n_j}}$.

Interval: $\bar{x}_j - B$ to $\bar{x}_j + B$

For two populations we can form a confidence interval for the difference between the means, $\mu_i - \mu_j$:

Obtain t from Table VI using $(n - k)$ df for the confidence required.

Calculate $\bar{x}_i$, $\bar{x}_j$, and $B = t\sqrt{\dfrac{MSE}{n_i} + \dfrac{MSE}{n_j}}$.

Interval: $(\bar{x}_i - \bar{x}_j) - B$ to $(\bar{x}_i - \bar{x}_j) + B$

SUPPLEMENTARY EXERCISES (10.18–10.30)

Using the Tables

10.18   A total of twenty-five measurements are independently and randomly sampled from five normally distributed populations with equal variances. For each of the following values of α, give the rejection region for testing

$$H_0: \quad \mu_1 = \mu_2 = \mu_3 = \mu_4 = \mu_5$$

against

H_a: At least two of the means differ

a. $\alpha = .05$ b. $\alpha = .10$ c. $\alpha = .01$

Learning the Mechanics

10.19 Independent random samples were selected from four normally distributed populations with common (but unknown) variance, σ^2. The data are shown in the table.

Sample 1	Sample 2	Sample 3	Sample 4
8	6	9	12
10	9	10	13
9	8	8	10
10	8	11	11
11	7	12	11

a. Do the data provide sufficient evidence to indicate a difference among treatment means? Test using $\alpha = .05$.

b. Form a 90% confidence interval for $\mu_1 - \mu_2$.

c. Form a 95% confidence interval for μ_4.

Applying the Concepts

In each of the following exercises, specify the conditions required to make the statistical procedure valid.

10.20 In recent years, higher wholesale beef prices have resulted in the sale of ground beef with higher fat contents in an attempt to keep retail prices down. Four 1-pound packages of ground beef were randomly selected from the stock of each of four supermarket chains. The fat content (in ounces) was measured for each package, with the results shown in the table.

	Supermarket		
A	B	C	D
4.2	4.5	5.0	3.8
4.0	4.7	4.0	4.0
4.3	4.4	4.3	3.7
4.5	4.4	4.7	3.7

a. What type of experimental design does this represent?

b. Do the data provide evidence at the $\alpha = .05$ level that the mean fat contents differ for at least two of the four supermarket chains?

c. Use a 90% confidence interval to estimate the mean fat content per pound of ground beef at supermarket C.

10.21 The plastic used in optic lenses will wear better if it is treated after casting. Four treatments are to be tested. To look for differences in mean wear among the treatments, twenty-eight castings from a single formulation of the plastic are made and divided randomly among the four treatments (seven castings per treatment). Wear is determined by measuring the increase in "haze" after 200 cycles of abrasion (better wear is indicated by smaller increases). The results are given in the table.

| | Treatment | | |
A	B	C	D
9.16	11.95	11.47	11.35
13.29	15.15	9.54	8.73
12.07	14.75	11.26	10.00
11.97	14.79	13.66	9.75
13.31	15.48	11.18	11.71
12.32	13.47	15.03	12.45
11.78	13.06	14.86	12.38

a. Is there evidence of a difference in mean increases in haze for at least two of the four treatments? Use $\alpha = .05$.

b. Estimate the difference in mean haze increases between treatments B and C using a 99% confidence interval.

c. Find a 90% confidence interval for the mean increase in haze for lenses receiving treatment A.

10.22 A professor wants to find out whether performance in an introductory statistics course depends on the student's year in college. Random samples of the final grade-point averages of students from all four years produced the data summary shown in the table.

Freshman	Sophomore	Junior	Senior
$T_1 = 52.55$	$T_2 = 168.25$	$T_3 = 760.50$	$T_4 = 363.10$
$n_1 = 25$	$n_2 = 75$	$n_3 = 350$	$n_4 = 150$
	$\sum X^2 = 3,144.52$		

a. Do the data provide sufficient evidence to indicate that students tend to perform differently depending on their year in college? Use $\alpha = .05$.

b. Find a 95% confidence interval for the difference in mean grade-point averages between seniors and freshmen.

10.23   A drug company has synthesized three new drugs intended for relief of pain due to ulcers. To determine whether the drugs will be absorbed by the stomach (and hence may be effective), nine pigs are randomly assigned oral doses of the drugs, three pigs per drug. The concentration of the given drug in each pig's stomach lining is later ascertained, and the data are shown in the table.

Drug		
1	2	3
1.70	1.73	1.67
1.72	1.79	1.63
1.81	1.76	1.67

a. Do the data provide sufficient evidence to indicate a difference in the mean concentrations of the three drugs in the stomach lining? Use $\alpha = .01$.

b. Use a 95% confidence interval to estimate the difference in the mean concentrations of drugs 1 and 2. Interpret the interval.

c. Find a 95% confidence interval for the mean concentration of drug 3. Interpret the interval.

10.24   In a nutrition experiment, an investigator studied the effects of different rations on the growth of young rats. Forty rats from the same inbred strain were divided at random into four groups of ten and used for the experiment. Each group was fed a different ration, and after a certain period of time each rat's increase in growth was measured (in grams). The data appear in the table.

Ration A		Ration B		Ration C		Ration D	
100	60	130	90	120	100	150	210
80	60	150	100	160	120	130	180
120	90	140	80	130	100	150	200
110	50	130	100	110	90	100	190
90	60	170	80	150	90	120	220

a. Do the data provide sufficient evidence to indicate a difference in mean increases in growth for rats receiving the different rations? Test using $\alpha = .05$.

b. Estimate the difference in mean increases in growth for rations A and D using a 90% confidence interval.

c. Find a 90% confidence interval for the mean gain for rats on ration D.

10.25   An experimenter believes that weight gains in chickens can be increased by adding small amounts of thyroxine (a hormone) to their diet. To test this theory, fifteen chickens are divided into three groups of five each. Each group is then fed a regular diet to which different amounts of thyroxine are added. The first diet has none, the second diet contains 2 milligrams of thyroxine per kilogram of feed, and the third diet contains 5 milligrams of thyroxine per kilogram of feed. After 8 weeks, each chicken's weight gain is measured (in grams), and the results are given in the table.

	Diet	
1	2	3
500	505	825
620	765	870
685	730	695
440	570	740
645	760	850

a. Do the data provide sufficient evidence to conclude that there is a difference in mean weight gains among the three diets? Use $\alpha = .05$.

b. Estimate the difference in mean weight gains between diets 2 and 3 using a 95% confidence interval.

10.26 Psychologists have studied the effects of the working environment on the quality and quantity of work done. Many businesses have music piped into work areas to improve the environment. An experiment is performed to determine which type of music is best suited for a certain company. Three types of music—country, rock, and classical—are tried, each on four randomly selected days. Each day's productivity is measured by recording the number of items produced. The results appear in the table. Can we conclude from this information that the mean number of items produced differs for at least two of the three types of music? Use $\alpha = .05$.

Country	Rock	Classical
857	791	824
801	753	847
795	781	881
842	776	865

10.27 One indicator of employee morale is how long employees stay with the company. A large corporation has three factories located in similar areas of the country. While the management attempts to maintain uniformity in the supervision, working conditions, employee relations, etc., at the various factories, it realizes that differences may exist. To study employee morale at its factories, employee records from all three factories are randomly selected, and each employee's length of service with the company is recorded. The summarized data are given in the table. Is there evidence of a difference in mean lengths of service at the three factories? Use $\alpha = .05$.

Factory	1	2	3
Number in Sample	15	21	17

$$SST = 421.74$$
$$SSE = 3,574.06$$

10.28 Many companies have experimented with three different forms of the 40-hour work week in order to determine which form will maximize production and minimize expenses. A factory has tested a 5-day week (8 hours per day), a 4-day week (10 hours per day), and a $3\frac{1}{3}$-day week (12 hours per day), with the weekly production results (in thousands of dollars' worth of items produced) shown in the table.

8-Hour Day	10-Hour Day	12-Hour Day
87	75	95
96	82	76
75	90	87
90	80	82
72	73	65
86		

a. Construct an ANOVA summary table for this experiment.
b. Is there evidence of a difference in the mean productivities for the three lengths of work day?
c. Form a 90% confidence interval for the mean weekly productivity with 12-hour days.

10.29 Sixty-two girls participated in an experiment to study isokinetic leg extension strength among track and field athletes. Four groups were compared: throwers (shot put and discus), jumpers (long jump and high jump), middle-distance runners (over 400 meters), and sprinters. The sample means of leg extension strengths are shown in the following table:

Group	n	Mean
Throwers	16	134.53
Jumpers	11	108.89
Middle-distance runners	12	85.85
Sprinters	23	99.18

Source: T. J. Housh, W. G. Thorland, G. D. Tharp, G. O. Johnson, and C. J. Cisar, "Isokinetic Leg Flexion and Extension Strength of Elite Adolescent Female Track and Field Athletes," *Research Quarterly for Exercise and Sport*, Vol. 55, No. 4, December 1984, p. 349.

The leg extension strengths measured for the sixty-two sampled athletes produced the following results:

Source	df	SS
Groups	3	18,927.79
Error	58	20,392.43

a. Is there sufficient evidence to conclude that there is a difference in the mean leg extension strength among the four groups? Use $\alpha = .01$.
b. Find a 95% confidence interval for the mean leg extension strength of throwers.
c. Estimate the difference in the mean leg extension strength of throwers and sprinters using a 99% confidence interval.

*10.30 Three methods have been devised to reduce the time spent in transferring materials from one location to another. With no prior information available on their effectiveness, a study is performed. Each method is tried several times, and the length of time (in hours) to complete the transfer is recorded in the table. Use a computer package to help answer the following.

Method		
A	B	C
8.2	7.9	7.1
7.1	8.1	7.4
7.8	8.3	6.9
8.9	8.5	6.8
8.8	7.6	
	8.5	

a. Is there evidence that the mean times to complete the transfer differ for at least two of the three methods? Use $\alpha = .01$.
b. Form a 95% confidence interval for the mean time to completion for method B.

CHAPTER 10 QUIZ

1. The concentration of a catalyst used in producing grouted sand is thought to affect its strength. An experiment designed to investigate the effects of three concentrations of the catalyst utilized five test specimens of grouted sand per concentration. The strength of the grouted sand was determined by placing the test specimen in a press and applying pressure until the specimen broke. The pressure required to break each specimen (in pounds per square inch) is listed in the table.

Concentration of Catalyst		
35%	40%	45%
5.9	6.8	9.9
8.1	7.9	9.0
5.6	8.4	8.6
6.3	9.3	7.9
7.7	8.2	8.7

a. Construct an ANOVA table for this experiment.
b. Do the data provide sufficient evidence to indicate a difference in the mean strengths of the grouted sand among the three concentrations of catalyst? Test using $\alpha = .05$.
c. Find a 95% confidence interval for the difference in mean strengths of specimens produced with a 35% concentration of the catalyst versus those produced with a 45% concentration.
d. Find a 90% confidence interval for the mean strength of specimens produced with a 40% concentration of the catalyst.

CHAPTERS 8–10 CUMULATIVE QUIZ

1. Three groups of randomly selected students in an introductory sociology class were subjected to different teaching techniques. At the end of the semester, the students were given a comprehensive examination. The scores on this test appear in the table.

Teaching Technique		
1	2	3
67	77	95
85	71	89
71	80	88
79	81	77
79	70	80
71	77	94
68		91

a. Do the data provide sufficient evidence to indicate that at least two of the teaching techniques are associated with different mean test scores? Use $\alpha = .05$.
b. Find a 95% confidence interval for the difference in the mean test scores for teaching techniques 1 and 3.

2. An agronomist wants to determine whether there is a difference in the mean bacteria counts for two plots of land. Independent random samples of soil are taken from each plot, and the results are summarized in the table.

Plot 1	Plot 2
$n_1 = 35$	$n_2 = 35$
$\bar{x}_1 = 13.3$	$\bar{x}_2 = 12.0$
$s_1^2 = 4.3$	$s_2^2 = 4.9$

a. Is there sufficient evidence to conclude that there is a difference in the mean bacteria counts for the two plots of land? Use $\alpha = .10$.

b. How many observations would need to be sampled from each plot of land to estimate the mean difference in bacteria counts to within .5 with 95% confidence?

3. An experiment was conducted to examine the relationship between heart rate and age of exercising women. A random sample of seven women was selected, and each subject jogged on a treadmill for 5 minutes. The increase in heart rate (the difference before and after jogging) was recorded for each. The women's ages and increases in heart rate are shown in the table.

Woman	Age X	Increase in Heart Rate Y
1	18	24
2	25	25
3	40	29
4	63	35
5	55	26
6	29	28
7	37	31

a. Calculate the coefficient of correlation, r, and the coefficient of determination, r^2. Interpret these values.

b. Use the method of least squares to find the Y-intercept and slope for a straight-line relationship between Y and X.

c. Estimate the mean increase in heart rate for women who are 50 years old. Use a 99% confidence interval.

On Your Own

Choose three or more populations that you would like to compare, and select independent random samples of measurements from the populations. For example, you could compare the assessed values of houses in different subdivisions of a city. The sample data could be obtained from records on file at the court house.

Use the data you obtain to test:

H_0: The means of the populations are equal

H_a: At least two of the population means differ

References

Heyward, V. & McCreary, L. "Analysis of the Static Strength and Relative Endurance of Women Athletes," *The Research Quarterly for Exercise and Sport*, 1977, *48*, pp. 703–710.

Snedecor, G. W. & Cochran, W. G. *Statistical Methods*. 7th ed. Ames, Iowa: Iowa State University Press, 1980.

THE CHI-SQUARE TEST AND
THE ANALYSIS OF COUNT DATA

A TELEPHONE SURVEY

When conducting a telephone survey, it is important to obtain a high rate of response. An article in the *Public Opinion Quarterly* (Gunn & Rhodes, 1981) discusses an experiment conducted by the Centers for Disease Control and the Opinion Research Corporation to examine the effects of giving physicians monetary incentives for their participation in a telephone interview. The experiment involved a nationwide sample of physicians: general and family practitioners, internists, pediatricians, and industrial doctors.

The sampled physicians were offered no incentive, $25, or $50 to participate in the telephone survey. The numbers in each group that did and did not complete the telephone interview are shown in Table 11.1. The data in this table are **count data**—that is, the number of physicians in each classification who did or did not complete the interview has been counted and tabulated.

The hypothesis of interest is that the rate of response for the survey depends on the monetary incentive. In this chapter we will present an analysis of count data that will test such a hypothesis.

Table 11.1

Physician participation

	No incentive	$25	$50
Completed interview	79	101	112
Did not complete interview	58	46	33
Totals	137	147	145

Source: Reprinted by permission of Elsevier Science Publishing Co., Inc., from W. J. Gunn and I. N. Rhodes, *Public Opinion Quarterly*, *45*, pp. 109–115. Copyright 1981 by The Trustees of Columbia University.

11.1

ONE-DIMENSIONAL COUNT DATA: THE MULTINOMIAL EXPERIMENT

In Section 5.4 we presented the binomial experiment in which qualitative data are classified into two distinct categories. We will now consider a generalization of this experiment and consider cases in which qualitative data are classified into more than two categories. Such data are usually referred to as **count data,** or **enumerative data.**

For example, we might count (enumerate) the number of consumers who choose each of three brands of coffee, or the number of students who major in each of five different areas. In these examples the data are classified on a single scale—that is, brand of coffee in the first case and college major in the second. We will say that count data classified on a single scale have a **one-dimensional classification,** and in this section we will discuss the analysis of one-dimensional count data.

Many practical experiments result in one-dimensional count data. We will consider a class of experiments with characteristics similar to those of the binomial experiment. This type of experiment, called the **multinomial experiment,** is actually an extension of the binomial experiment, and its characteristics are given in the box.

Characteristics of a Multinomial Experiment

1. The population contains an infinite number of observations, each of which may be classified into exactly one of k categories.
2. The fractions of the population's observations in the k categories are denoted by $\pi_1, \pi_2, \ldots, \pi_k$. Note that

$$\pi_1 + \pi_2 + \cdots + \pi_k = 1$$

3. The experiment consists of sampling n observations from the population and recording the category in which each observation falls. The experiment must be performed in such a way that the following properties hold:
   a. The probability is π_1 that any sampled observation is in the first category, π_2 in the second category, π_3 in the third category, etc.
   b. The probability that any sampled observation is in any category is not affected by what the other sampled observations happen to be.
4. The statistics of interest are the observed numbers of observations $o_1, o_2, \ldots, o_k$ in each of the k categories.

You can see that the properties of the multinomial experiment closely resemble those of the binomial experiment and that, in fact, a binomial experiment is a special case of the multinomial experiment with $k = 2$. As with the binomial experiment, the characteristics of the multinomial experiment are rarely satisfied exactly for practical experiments, because populations of interest usually contain a finite number of observations. Thus, we will consider an experiment to be multinomial if a random sample of n observations is taken from a very large population.

For example, suppose we want to compare the percentage of voters who favor each of three political candidates running for the same elective position. The voting preferences of a random sample of 150 eligible voters are obtained, and the resulting count data are classified according to a single criterion, candidate preference. The data are shown in Table 11.2.

Table 11.2

Voter-preference data

Candidate 1	Candidate 2	Candidate 3
61	53	36

This voter-preference survey satisfies the properties of a multinomial experiment. The experiment consists of randomly sampling $n = 150$ voters from a large population of voters containing unknown proportions π_1 who favor candidate 1, π_2 who favor candidate 2, and π_3 who favor candidate 3. Each voter sampled represents a single trial that can result in one of three outcomes; the voter will favor candidate 1, 2, or 3 with probabilities π_1, π_2, and π_3, respectively (assuming all voters have a preference). The voting preference of any one voter in the sample does not affect the preference of another. And, finally, you can see that the recorded data are the observed numbers of voters in each of the three voter-preference categories.

In the voter-preference survey and in most practical applications of the multinomial experiment, the k outcome probabilities $\pi_1, \pi_2, \ldots, \pi_k$ are unknown and we want to use the survey data to make inferences about their values. The unknown probabilities in the voter-preference survey are

$\pi_1 =$ Proportion of all voters who favor candidate 1

$\pi_2 =$ Proportion of all voters who favor candidate 2

$\pi_3 =$ Proportion of all voters who favor candidate 3

To decide whether the voters have a preference for any of the candidates, we will want to test the null hypothesis that the candidates are equally preferred (that is, $\pi_1 = \pi_2 = \pi_3 = \frac{1}{3}$) against the alternative hypothesis that one candidate is preferred (that is, at least one of the probabilities π_1, π_2, and π_3 differs from $\frac{1}{3}$):

H_0: $\pi_1 = \pi_2 = \pi_3 = \frac{1}{3}$ (no preference)

H_a: At least one of the proportions differs from $\frac{1}{3}$ (a preference exists)

If the null hypothesis is true and $\pi_1 = \pi_2 = \pi_3 = \frac{1}{3}$, then the expected value (mean value) of the number of voters who prefer candidate 1 is given by

$$E_1 = n\pi_1 = (n)\frac{1}{3} = (150)\frac{1}{3} = 50$$

Similarly, $E_2 = E_3 = 50$ if the null hypothesis is true and no preference exists.

The following test statistic measures the degree of disagreement between the observed sample values and the expected values:

$$X^2 = \frac{(o_1 - E_1)^2}{E_1} + \frac{(o_2 - E_2)^2}{E_2} + \frac{(o_3 - E_3)^2}{E_3}$$
$$= \frac{(o_1 - 50)^2}{50} + \frac{(o_2 - 50)^2}{50} + \frac{(o_3 - 50)^2}{50}$$

Note that the farther the observed numbers o_1, o_2, and o_3 are from their expected value (50), the larger X^2 will become. That is, large values of X^2 imply that the null hypothesis is false. We have to know the sampling distribution of X^2 in order to decide whether the data indicate a preference exists.

If the null hypothesis is true, the distribution of X^2 in repeated sampling is approximately a χ^2 **(chi-square) distribution.** [*] This approximation for the sampling distribution of X^2 is adequate as long as the expected number of observations in each of the k categories is at least 5. The χ^2 distribution is characterized by a single parameter, called the **degrees of freedom associated with the distribution.** Several χ^2 distributions with different degrees of freedom are shown in Figure 11.1.

Figure 11.1
Several χ^2 distributions

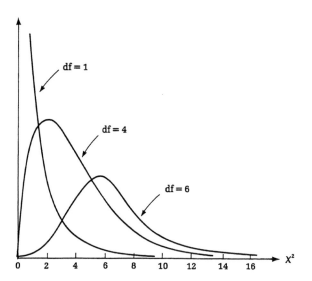

For a multinomial experiment yielding a one-dimensional classification, the degrees of freedom corresponding to the approximate sampling distribution of X^2 will always be $(k - 1)$, that is, 1 less than the number of categories being compared. Because large values of X^2 support the alternative hypothesis, the rejection region for the test will be located in the upper tail of the χ^2 distribution, as shown in Figure 11.2.

[*] The χ^2 distribution was first introduced in Section 7.8, an optional section. We will describe the χ^2 distribution again here, in case you omitted Section 7.8.

Figure 11.2
Location of the rejection region

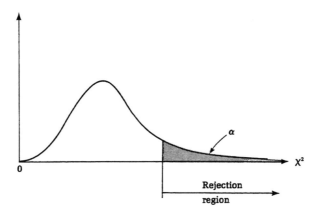

For the voter-preference example, the approximate distribution of the test statistic X^2 is a χ^2 distribution with $k - 1 = 2$ df. To determine how large X^2 must be to reject the null hypothesis, we consult Appendix Table VIIA. A partial reproduction of this table is shown in Table 11.3. To use this table, we must specify the value of α and the appropriate degrees of freedom. We then use the column for a one-tailed test (>) to find the appropriate table value. Thus, for a value of $\alpha = .05$ and 2 df, we can reject the null hypothesis that there is no voter preference among the three candidates if

$$X^2 > 6.00$$

Table 11.3

A partial reproduction of Appendix Table VIIA, χ^2 Values for Rejection Regions for One-Tailed Tests

df	$\alpha = .10$		$\alpha = .05$		$\alpha = .01$	
	One-tailed test, >	One-tailed test, <	One-tailed test, >	One-tailed test, <	One-tailed test, >	One-tailed test, <
1	2.71	0.0158	3.84	0.00393	6.64	0.000157
2	4.61	0.211	6.00	0.103	9.21	0.0201
3	6.25	0.584	7.82	0.352	11.4	0.115
4	7.78	1.0636	9.50	0.711	13.3	0.297
5	9.24	1.61	11.1	1.15	15.1	0.554
6	10.6	2.20	12.6	1.64	16.8	0.872
7	12.0	2.83	14.1	2.17	18.5	1.24
8	13.4	3.49	15.5	2.73	20.1	1.65
9	14.7	4.17	17.0	3.33	21.7	2.09
10	16.0	4.87	18.3	3.94	23.2	2.56

This rejection region is shown in Figure 11.3 (page 606).

To complete the test of hypothesis for the voter-preference survey, we must find the computed value of the test statistic. The calculation is summarized in Table 11.4. The sum of the numbers in the last column gives the value of the test statistic, X^2.

Figure 11.3
Rejection region for
voter-preference survey

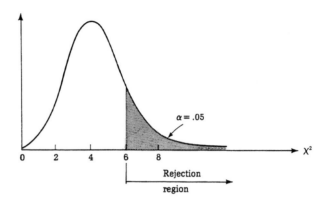

Table 11.4

Calculation of X^2
test statistic

Candidate	Observed o	Expected E	$o - E$	$(o - E)^2$	$\dfrac{(o - E)^2}{E}$
1	61	50	11	121	2.42
2	53	50	3	9	.18
3	36	50	-14	196	3.92
					$X^2 = \overline{6.52}$

Since the computed $X^2 = 6.52$ exceeds the table value of 6.00, we conclude at the $\alpha = .05$ level of significance that a voter preference for one or more of the candidates does exist.

The conditions that must be met to test a hypothesis about $\pi_1, \pi_2, \ldots, \pi_k$ are given in the box. These conditions guarantee that the sampling distribution of X^2 has an approximate χ^2 distribution with $(k - 1)$ df.

When to Test a Hypothesis About $\pi_1, \pi_2, \ldots, \pi_k$

To make a valid test of hypothesis about $\pi_1, \pi_2, \ldots, \pi_k$, the following conditions must be met:

1. A multinomial experiment must be conducted. This is generally satisfied by taking a random sample from the population of interest.
2. For each category, the expected number of observations must be greater than or equal to 5. This means that $E_1 \geq 5, E_2 \geq 5, \ldots, E_k \geq 5$.

The general form of the test of hypothesis about the multinomial probabilities $\pi_1, \pi_2, \ldots, \pi_k$ is given in the next box.

Test of Hypothesis About $\pi_1, \pi_2, \ldots, \pi_k$

H_0: $\pi_1 = \pi_{1,0}, \ \pi_2 = \pi_{2,0}, \ldots, \pi_k = \pi_{k,0}$, where $\pi_{1,0}, \ \pi_{2,0}, \ldots, \pi_{k,0}$ represent the hypothesized values of the multinomial probabilities and satisfy $\pi_{1,0} + \pi_{2,0} + \cdots + \pi_{k,0} = 1$

H_a: At least one of the multinomial probabilities does not equal its hypothesized value

Test statistic:

$$X^2 = \frac{(o_1 - E_1)^2}{E_1} + \frac{(o_2 - E_2)^2}{E_2} + \cdots + \frac{(o_k - E_k)^2}{E_k}$$

where $E_1 = n\pi_{1,0}$, $E_2 = n\pi_{2,0}$, etc. The total sample size is n.

Rejection region: Reject H_0 if $X^2 >$ Table value, where the table value is obtained from Appendix Table VIIA, using $(k-1)$ df.

EXAMPLE 11.1   Suppose an educational television station has broadcast a series of programs that discussed the existing scientific evidence about the physiological and psychological effects of smoking marijuana. Now that the series is finished, the station wants to find out whether the citizens within the viewing area have changed their minds about how possession of marijuana should be considered legally. Before the series ran, it was determined that 7% of the citizens favored legalization, 18% favored decriminalization, 65% favored the existing law (fine or imprisonment), and 10% had no opinion.

A summary of opinions after the series ended of a random sample of 500 people in the viewing area is given in Table 11.5. Test at the $\alpha = .01$ level to determine whether the data indicate a significant difference in the distribution of opinions relative to the proportions prior to the series.

Table 11.5

Distribution of opinions about marijuana possession

Legalization	Decriminalization	Existing Law	No Opinion
39	99	336	26

Solution   Define the proportions after the series was shown to be

π_1 = Proportion of citizens favoring legalization

π_2 = Proportion of citizens favoring decriminalization

π_3 = Proportion of citizens favoring existing laws

π_4 = Proportion of citizens with no opinion

Then the null hypothesis representing no change in the distribution of percentages is

H_0: $\pi_1 = .07, \pi_2 = .18, \pi_3 = .65, \pi_4 = .10$

and the alternative is

H_a: At least one of the proportions differs from its null hypothesized value

Test statistic: $X^2 = \dfrac{(o_1 - E_1)^2}{E_1} + \cdots + \dfrac{(o_4 - E_4)^2}{E_4}$

where

$E_1 = n\pi_{1,0} = 500(.07) = 35$

$E_2 = n\pi_{2,0} = 500(.18) = 90$

$E_3 = n\pi_{3,0} = 500(.65) = 325$

$E_4 = n\pi_{4,0} = 500(.10) = 50$

Since all these values are larger than 5, the χ^2 approximation is appropriate. Also, the citizens in the sample were randomly selected, so the properties of the multinomial experiment are satisfied.

Rejection region: Reject H_0 if $X^2 > 11.4$

The rejection region was found using Appendix Table VIIA, with $\alpha = .01$ and $k - 1 = 3$ df.

The test statistic calculation is shown in Table 11.6.

Table 11.6

Calculation of X^2 for
Example 11.2

Category	o	E	o − E	(o − E)²	$\dfrac{(o-E)^2}{E}$
1	39	35	4	16	.46
2	99	90	9	81	.90
3	336	325	11	121	.37
4	26	50	−24	576	11.52
					13.25

Since the value $X^2 = 13.25$ exceeds the table value of χ^2 (11.4), the data provide strong evidence ($\alpha = .01$) that the opinions on legalization of marijuana have changed since the series was aired. ■

COMPUTER EXAMPLE 11.1   Minitab does not have a single command that will perform the analyses discussed in this section. However, with the aid of Minitab it is possible to perform the calculations that are summarized in Table 11.6. A printout of such a table follows.

ROW	CATEGORY	O	E	(O−E)	(O−E)SQ	VALUE	X−SQRD
1	1	39	35	4	16	0.4571	13.2495
2	2	99	90	9	81	0.9000	
3	3	336	325	11	121	0.3723	
4	4	26	50	−24	576	11.5200	

Other than the number of decimal places displayed and slightly different column headings, the Minitab printout and Table 11.6 are essentially identical. With Minitab the value of the test statistic is labeled "X-SQRD" and it equals 13.2495, which rounds to 13.25, the value given in Table 11.6.

Whether you calculate the value of the test statistic or use Minitab to perform the calculations, the interpretation will be the same. Namely, at the .01 level of significance, we may conclude that the opinions on legalization of marijuana have changed since the series was aired. ∎

Exercises (11.1–11.7)

Using the Tables

11.1 Find the rejection region for a one-dimensional test of hypothesis concerning π_1, $\pi_2, \ldots, \pi_k$ if:

a. $k = 3$ and $\alpha = .10$ c. $k = 4$ and $\alpha = .05$
b. $k = 5$ and $\alpha = .01$ d. $k = 5$ and $\alpha = .05$

Learning the Mechanics

11.2 A multinomial experiment with $k = 3$ cells and $n = 200$ produced the data shown in the table. Do the data provide sufficient evidence to contradict the null hypothesis that $\pi_1 = .25$, $\pi_2 = .25$, and $\pi_3 = .50$? Test using $\alpha = .05$.

Cell		
1	2	3
48	69	83

11.3 A multinomial experiment with four possible outcomes and 100 trials produced the data shown in the table. Do the data provide sufficient evidence to indicate that the four outcomes are not equally likely? Test using $\alpha = .10$.

Outcome			
1	2	3	4
15	30	25	30

Applying the Concepts

11.4 After purchasing a policy from a life insurance company, a person has a certain period of time in which the policy can be cancelled without financial penalty. An insurance company is interested in determining whether those who cancel a policy during this time period are as likely to be in one policy size category as another. Records for 250 people who cancelled policies during this period were selected at random from company files, and the numbers of people cancelling per policy-size category are shown in the table. Is there sufficient evidence to conclude that the cancelled policies are not distributed equally among the five policy-size categories? Use $\alpha = .05$.

Size of Policy (Thousands of Dollars)				
10	15	20	25	30
31	39	67	54	59

11.5 In the game of chess, the first few moves play a crucial role in determining the outcome. Five different opening strategies are highly favored by chess experts. To determine whether one or more of these strategies is preferred by Grand Masters in international competition, a random sample of 100 Grand Masters is taken, and each is asked which of the strategies is preferred. A summary of their responses is shown below. Do the data present sufficient evidence to indicate a preference for one or more of the five strategies? Use $\alpha = .05$.

Strategy:	A	B	C	D	E
Frequency:	17	27	22	15	19

11.6 According to genetic theory, a crossing of red and white snapdragons should produce offspring that are 25% red, 50% pink, and 25% white. An experiment conducted to test the theory produced 30 red, 78 pink, and 36 white offspring in 144 crossings. Do the data provide sufficient evidence to contradict the theory? Use $\alpha = .05$.

*11.7 Supermarket chains often carry products with their own brand labels and usually price them below the nationally known brands. A supermarket conducted a taste test of four brands of ice cream it carries: its own brand (A) and three nationally known brands (B, C, D). A sample of 200 people participated, and they indicated the preferences shown in the table. Is there evidence of a difference in preference for the four brands? Use a computer package to help test the hypotheses of interest, at $\alpha = .05$.

Brand			
A	B	C	D
39	57	55	49

11.2

TWO-DIMENSIONAL COUNT DATA: CONTINGENCY TABLES

In Section 11.1 we introduced the multinomial probability distribution and considered data classified according to a single criterion. We now consider multinomial experiments in which the data are classified according to two factors.

For example, when purchasing a car, many energy-conscious consumers consider the size of the automobile as an important factor in making their choice. Suppose an automobile manufacturer is interested in determining whether there is a relationship between the size and manufacturer of newly purchased cars. One thousand recent buyers of American-made cars are randomly sampled and each purchase is classified with respect to the size and manufacturer of the purchased automobile. The data are summarized in the two-way table shown in Table 11.7. This table is called a **contingency table;** it presents **multinomial count data classified on two scales, or dimensions,** namely automobile size and manufacturer.

Table 11.7

Contingency table for automobile size–manufacturer example

| | | Manufacturer | | | | Totals |
		A	B	C	D	
	Small	157	65	181	10	413
Size	Middle	126	82	142	46	396
	Large	58	45	60	28	191
Totals		341	192	383	84	1,000

Table 11.7 shows that this really is a multinomial experiment with a sample of 1,000 observations and $(3)(4) = 12$ possible categories, or **cells.** The notation we will use to represent the observed cell counts in a contingency table is shown in Table 11.8. Thus, o_{11} represents the number of sampled buyers who purchased manufacturer A's small car, o_{12} represents the number who bought manufacturer B's small car, etc. Note that we have also presented notation for the total of the cell counts for each row and column. The symbol r_1 represents the total number of buyers who purchased a small car, c_1 represents the total number of buyers who purchased a car from manufacturer A, etc. The total number of observations is denoted by n.

Table 11.8

Notation for a contingency table

| | | Manufacturer | | | | Totals |
		A	B	C	D	
	Small	o_{11}	o_{12}	o_{13}	o_{14}	r_1
Size	Middle	o_{21}	o_{22}	o_{23}	o_{24}	r_2
	Large	o_{31}	o_{32}	o_{33}	o_{34}	r_3
Totals		c_1	c_2	c_3	c_4	n

The notation for the actual probabilities that observations will fall in each cell is shown in Table 11.9 (page 612). For instance, π_{23} is the probability that a buyer will buy manufacturer C's mid-sized car. The row and column probabilities are called **marginal probabilities.** Note the symbols for the probabilities. The marginal probability π_1

Table 11.9

Probabilities for a
contingency table

		Manufacturer				Totals
		A	B	C	D	
Size	Small	π_{11}	π_{12}	π_{13}	π_{14}	π_1
	Middle	π_{21}	π_{22}	π_{23}	π_{24}	π_2
	Large	π_{31}	π_{32}	π_{33}	π_{34}	π_3
Totals		π_A	π_B	π_C	π_D	1

represents the probability that a small car is purchased, and the marginal probability π_A is the probability that manufacturer A's car is purchased. Therefore, we have

$$\pi_1 = \pi_{11} + \pi_{12} + \pi_{13} + \pi_{14} \quad \text{and} \quad \pi_A = \pi_{11} + \pi_{21} + \pi_{31}$$

Suppose we want to know whether the two classifications, manufacturer and size, are dependent. That is, if we know which size car a buyer wants, does that information give us a clue about the manufacturer of the car the buyer will choose? In the contingency table analysis, if the two classifications are independent, the probability that an observation is classified in any particular cell of the table is the product of the corresponding marginal probabilities. Thus, under the hypothesis of independence, in Table 11.9, we must have

$$\pi_{11} = \pi_1 \pi_A \qquad \pi_{12} = \pi_1 \pi_B$$

and so forth.

To test the hypothesis of independence, we use the same reasoning employed in the one-dimensional tests of Section 11.1. First, we calculate the expected, or mean, count in each cell assuming the null hypothesis of independence is true. We do this by noting that the expected count in a cell of the table is just the total number of multinomial trials, n, times the cell probability. Recall that o_{11} represents the observed count in the cell located in the first row and first column. Then the expected cell count for the first row and first column is

$$E_{11} = n\pi_{11}$$

or, when the null hypothesis (the classifications are independent) is true,

$$E_{11} = n\pi_1 \pi_A$$

Since these true probabilities are not known, we estimate π_1 and π_A by the sample proportions $p_1 = r_1/n$ and $p_A = c_1/n$. Thus, the estimate of the expected cell count is*

$$E_{11} = n\left(\frac{r_1}{n}\right)\left(\frac{c_1}{n}\right) = \frac{r_1 c_1}{n}$$

---

* Some books denote this estimated expected cell count as $\hat{E}_{11}$. To simplify the notation, we have deleted the "hat," but you should be sure to remember that the calculated quantity is not the actual expected value, but an estimate.

Similarly,

$$E_{12} = \frac{r_1 c_2}{n}$$

$$\vdots$$

$$E_{34} = \frac{r_3 c_4}{n}$$

Using the data in Table 11.7, we find

$$E_{11} = \frac{r_1 c_1}{n} = \frac{(413)(341)}{1,000} = 140.83$$

$$E_{12} = \frac{r_1 c_2}{n} = \frac{(413)(192)}{1,000} = 79.30$$

$$\vdots$$

$$E_{34} = \frac{r_3 c_4}{n} = \frac{(191)(84)}{1,000} = 16.04$$

We can now use the X^2 statistic to compare the observed and estimated expected counts in each cell of the contingency table:

$$X^2 = \frac{(O_{11} - E_{11})^2}{E_{11}} + \frac{(O_{12} - E_{12})^2}{E_{12}} + \cdots + \frac{(O_{34} - E_{34})^2}{E_{34}}$$

The calculation of the test statistic is given in Table 11.10.

Table 11.10

Row, Column	o	E	o − E	(o − E)²	$\frac{(o-E)^2}{E}$
1, 1	157	140.83	16.17	261.47	1.86
1, 2	65	79.30	−14.30	204.49	2.58
1, 3	181	158.18	22.82	520.75	3.29
1, 4	10	34.69	−24.69	609.60	17.57
2, 1	126	135.04	−9.04	81.72	.61
2, 2	82	76.03	5.97	35.64	.47
2, 3	142	151.67	−9.67	93.51	.62
2, 4	46	33.26	12.74	162.31	4.88
3, 1	58	65.13	−7.13	50.84	.78
3, 2	45	36.67	8.33	69.39	1.89
3, 3	60	73.15	−13.15	172.92	2.36
3, 4	28	16.04	11.96	143.04	8.92
				$X^2 =$	45.83

Large values of X^2 imply that the observed and expected counts do not closely agree, and therefore imply that the hypothesis of independence is false. To determine how large X^2 must be before it is too large to be attributed to chance, we make use of the fact that the sampling distribution of X^2 is approximately a χ^2 probability distribution when the classifications are independent.

When testing the null hypothesis of independence in a two-way contingency table, the degrees of freedom will be $(r - 1)(c - 1)$, where r is the number of rows and c is the number of columns in the table.

For the size–manufacturer example, the degrees of freedom for χ^2 are

$$(r - 1)(c - 1) = (3 - 1)(4 - 1) = 6$$

Then, for $\alpha = .05$, we reject the hypothesis of independence when $X^2 > 12.6$ (from Appendix Table VIIA). Since the computed $X^2 = 45.83$ exceeds the value 12.6, we conclude that the size and manufacturer of a car selected by a purchaser are dependent events.

The conditions that must be met to conduct a test of independence for two-dimensional count data are given in the box. These conditions guarantee that the sampling distribution of X^2 has an approximately χ^2 distribution with $(r - 1)(c - 1)$ df.

When to Test for Independence of Two-Dimensional Count Data

To conduct a valid test of hypothesis for independence using data from a contingency table, the following conditions must be met:

1. A multinomial experiment must be conducted. This is generally satisfied by taking a random sample from the population of interest.
2. For each cell (category), the estimated expected number of observations must be greater than or equal to 5. That is, $E_{11} \geq 5$, $E_{12} \geq 5$, etc.

The general form of a contingency table with r rows and c columns is shown in Table 11.11. Such a table is referred to as an **r by c contingency table.** Using this

Table 11.11

General r by c contingency table

		Column 1	Column 2	$\cdots$	Column c	Row Totals
Row	1	O_{11}	O_{12}	$\cdots$	O_{1c}	r_1
	2	O_{21}	O_{22}	$\cdots$	O_{2c}	r_2
	$\vdots$	$\vdots$	$\vdots$		$\vdots$	$\vdots$
	r	O_{r1}	O_{r2}	$\cdots$	O_{rc}	r_r
Column Totals		c_1	c_2	$\cdots$	c_c	n

notation, we give the general form of the contingency table test for independence in the box.

Test for Independence Using a Contingency Table

H_0: The two classifications are independent

H_a: The two classifications are dependent

Test statistic:

$$X^2 = \frac{(o_{11} - E_{11})^2}{E_{11}} + \frac{(o_{12} - E_{12})^2}{E_{12}} + \cdots + \frac{(o_{rc} - E_{rc})^2}{E_{rc}}$$

where $E_{11} = \dfrac{r_1 c_1}{n}$, $E_{12} = \dfrac{r_1 c_2}{n}$, etc.

Rejection region: Reject H_0 if $X^2 >$ Table value, where the rejection region is obtained from Appendix Table VIIA, using $(r - 1)(c - 1)$ df.

EXAMPLE 11.2   A social scientist wants to determine whether the divorce status (divorced or never divorced) among American men is dependent on their religious affiliation (or lack thereof). A random sample of 500 American men who were, or are, married is surveyed, and the results are summarized in Table 11.12. Test to see if there is sufficient evidence to indicate that divorce status is dependent on religious affiliation. Test using $\alpha = .01$.

Table 11.12

Observed cell counts for Example 11.2

		Religious Affiliation					Totals
		A	B	C	D	None	
Status	Divorced	39	19	12	28	18	116
	Never divorced	172	61	44	70	37	384
Totals		211	80	56	98	55	500

Solution   The test of interest is

H_0: The divorce status and religious affiliation of American men are independent

H_a: The divorce status and religious affiliation of American men are dependent

Test statistic: $X^2 = \dfrac{(o_{11} - E_{11})^2}{E_{11}} + \cdots + \dfrac{(o_{25} - E_{25})^2}{E_{25}}$

Rejection region: Reject H_0 if $X^2 > 13.3$.

The rejection region was found using Appendix Table VIIA, with $\alpha = .01$ and $(r - 1)(c - 1) = (1)(4) = 4$ df.

First, we must calculate the estimated expected cell counts under the assumption that the classifications are independent. Thus,

$$E_{11} = \frac{r_1 c_1}{n} = \frac{(116)(211)}{500} = 48.95$$

$$E_{12} = \frac{r_1 c_2}{n} = \frac{(116)(80)}{500} = 18.56$$

$$\vdots$$

$$E_{25} = \frac{r_2 c_5}{n} = \frac{(384)(55)}{500} = 42.24$$

All the observed cell counts and the estimated expected cell counts are shown in Table 11.13. Since all the estimated expected cell counts are greater than 5, and the men were randomly selected from the population of interest, the test for independence is valid.

Table 11.13

Calculation of X^2 for Example 11.2

Row, Column	o	E	$o - E$	$(o - E)^2$	$\dfrac{(o - E)^2}{E}$
1, 1	39	48.95	−9.95	99.00	2.02
1, 2	19	18.56	.44	.19	.01
1, 3	12	12.99	−.99	.98	.08
1, 4	28	22.74	5.26	27.67	1.22
1, 5	18	12.76	5.24	27.46	2.15
2, 1	172	162.05	9.95	99.00	.61
2, 2	61	61.44	−.44	.19	.00
2, 3	44	43.01	.99	.98	.02
2, 4	70	75.26	−5.26	27.67	.37
2, 5	37	42.24	−5.24	27.46	.65
					$X^2 = \overline{7.13}$

As shown in Table 11.13, the calculated value of the test statistic is $X^2 = 7.13$. Since $X^2 = 7.13$ is less than 13.3, we cannot conclude that the divorce status of American men depends on their religious affiliation. Perhaps the two classifications are independent, or perhaps a larger sample would reveal a dependence.   ■

COMPUTER EXAMPLE 11.2   In Example 11.2, we tested to determine whether there is sufficient evidence to conclude that divorce status is dependent on religious affiliation. Using the data in Table 11.12, we calculated the value of the test statistic to be $X^2 = 7.13$. There was insufficient evidence to conclude that divorce status and religious affiliation are dependent.

Minitab may be used to calculate the value of the test statistic when analyzing the data from a contingency table. From the data in Table 11.12, the following printout was obtained:

```
Expected counts are printed below observed counts

              A          B          C          D       NONE      Total
    1        39         19         12         28         18        116
           48.95      18.56      12.99      22.74      12.76

    2       172         61         44         70         37        384
          162.05      61.44      43.01      75.26      42.24

Total      211         80         56         98         55        500

ChiSq =   2.023 +    0.010 +    0.076 +    1.219 +    2.152 +
          0.611 +    0.003 +    0.023 +    0.368 +    0.650 = 7.135
df = 4
```

The observed cell counts that were given in Table 11.12 are on the printout, and under them appear the expected cell counts (rounded to one decimal place). Below the table of observed and expected cell counts, Minitab reports the individual cell contributions to the value of the test statistic, as well as the total value of the statistic. The test statistic is identified as "ChiSq," and its value is given as 7.135. Due to internal rounding by Minitab, this is slightly different from the value of 7.13 that we calculated in Example 11.2. The last information provided on the printout is that there are 4 degrees of freedom associated with this test.

The interpretation of the results is identical to that made in Example 11.2. ■

Exercises (11.8–11.18)

Using the Tables

11.8 Find the rejection region for a test of independence of two directions of classification if the contingency table contains r rows and c columns and:

a. $r = 2$, $c = 2$, $\alpha = .05$
b. $r = 3$, $c = 4$, $\alpha = .01$
c. $r = 3$, $c = 6$, $\alpha = .10$
d. $r = 4$, $c = 4$, $\alpha = .05$

Learning the Mechanics

11.9 Test the null hypothesis of independence of the two directions of classification for the 2 by 2 contingency table shown below. Test using $\alpha = .05$.

		B	
		B_1	B_2
A	A_1	39	75
	A_2	63	51

11.10   Test the null hypothesis that the rows and columns for the 4 by 3 contingency table below are independent. Use $\alpha = .01$.

		Column		
		1	2	3
	1	20	30	50
	2	40	20	40
Row	3	100	50	50
	4	40	0	60

Applying the Concepts

11.11   Many scientists believe that alcoholism is linked to social isolation. One measure of social isolation is marital status—that is, whether a person is married or not. To test the relationship between alcoholism and isolation, 280 adults were selected and each was classified as a diagnosed alcoholic, an undiagnosed alcoholic, or nonalcoholic, and as married or unmarried. A summary of the designations is shown in the table. Can you conclude that there is a relationship between the marital status and alcoholism classifications? Test using $\alpha = .05$.

		Alcoholic Classification		
		Diagnosed	Undiagnosed	Nonalcoholic
Status	Married	21	37	58
	Unmarried	59	63	42

11.12   One criterion used to evaluate employees in the assembly section of a large factory is the number of defective pieces per 1,000 parts produced. The quality control department wants to find out whether there is a relationship between years of experience and defect rate. Since the job is rather repetitious, any improvement due to a learning effect after the training period might be offset by a decrease in the motivation of a worker. A defect rate is calculated for each worker for a yearly evaluation. The results for 100 workers are given in the table. Is there evidence of a relationship between defect rate and years of experience? Use $\alpha = .05$.

		Years of Experience (After Training)		
		1	2–5	6–10
	High	6	9	9
Defect Rate	Average	9	19	23
	Low	7	8	10

11.13 An experimenter wants to determine whether there is a relationship between hair color and eye color. One hundred people are randomly sampled and their eyes and hair are judged to be light or dark. The number of people in each of the four categories is listed in the table. Do the data provide sufficient evidence to indicate a relationship between eye and hair color? Test using $\alpha = .10$.

| | | **Hair** | | **Totals** |
		Light	Dark	
Eyes	Light	31	21	52
	Dark	14	34	48
Totals		45	55	100

11.14 A team of market researchers conducted a study of 200 American citizens to discover what people fear the most. Each person's sex was noted, and then each was asked what was most fearsome: speaking before a group, heights, insects, financial problems, sickness/death, or other. The results of the poll appear in the table. Do the data provide sufficient information to indicate a relationship between sex and greatest fear? Test at the $\alpha = .05$ level.

| | | **Greatest Fear** | | | | | |
		Public speaking	Heights	Insects	Financial problems	Sickness/ death	Other
Sex	Male	21	10	7	23	15	21
	Female	16	22	15	9	18	23

11.15 A study was conducted to help determine who takes advantage of sales and specials at food stores that advertise in newspapers. Randomly selected shoppers were asked whether they usually check the advertisements before shopping and into which of the following brackets their annual income falls: (1) below \$5,000; (2) at least \$5,000 but less than \$15,000; (3) at least \$15,000 but less than \$25,000; (4) at least \$25,000 but less than \$35,000; or (5) at least \$35,000. The data are given in the table. Test to determine whether there is a relationship between income level and checking food store advertisements. Use $\alpha = .10$.

| | **Income Bracket** | | | | |
	(1)	(2)	(3)	(4)	(5)
Read ads	33	62	31	14	6
Do not read ads	3	8	19	15	14

11.16 *Academic inbreeding* refers to the practice of a school's hiring its own graduates as faculty members. Faculty homogeneity, which is considered detrimental to scholarly achievement and innovation, has been a problem in certain fields. One study of nursing schools chose to investigate the relationship between the size of a university's faculty and its level of inbreeding. A sample of nursing faculty was taken, and the results are summarized in the accompanying table. Do the data provide sufficient evidence to indicate a relationship between the size of the faculty and the level of inbreeding? Use $\alpha = .01$.

		Size of Faculty		
		Less than 60	60 to 89	More than 89
Faculty Characteristic	*Inbred*	78	211	610
	Not Inbred	128	263	502

Source: Adapted from Table 4, "Comparison of Inbreeding in Schools of Nursing by Size of Faculty." In M. L. Kornguth and M. H. Miller, "Academic Inbreeding in Nursing: Intentional or Inevitable?" *Journal of Nursing Education*, Vol. 24, No. 1, January 1985, p. 23.

11.17 New admissions to state and county mental hospitals in 1975 were classified according to the primary diagnosis and sex of the patient. Data for four types of mental disorder are summarized in the table. Is there sufficient evidence to indicate a relationship between the sex of the patient and the primary diagnosis for these four diseases? Use $\alpha = .01$.

		Primary Diagnosis			
		Depressive neurosis	Personality disorders	Drug-related disorders	Childhood-related disorders
Sex	*Male*	9,058	16,999	15,373	8,234
	Female	14,036	7,064	5,759	4,723

Source: Adapted from Table 2/7, "New Admissions to State and County Mental Hospitals, by Primary Diagnosis and Sex of Patient: 1975." In Center for Demographic Studies, U.S. Bureau of the Census, *Social Indicators III* (Washington, D.C.: Government Printing Office, 1980), p. 93.

*11.18 An insurance company that sells hospitalization policies wants to know whether there is a relationship between the amount of hospitalization coverage a person has and the length of stay in the hospital. Records are selected at random at a large hospital by hospital personnel, and the information on length of stay and hospitalization coverage is given to the insurance company. The results are summarized in the table. Can you conclude that there is a relationship between length of stay and hospitalization coverage? Use $\alpha = .01$. Use a computer package to help perform the analysis.

		Length of Stay in Hospital (Days)			
		5 or less	6–10	11–15	Over 15
Coverage of Hospitalization Costs	Under 25%	26	30	6	5
	At least 25% but less than 50%	21	30	11	7
	At least 50% but less than 75%	25	25	45	9
	At least 75%	11	32	17	11

11.3

CONTINGENCY TABLES WITH SPECIFIED MARGINAL TOTALS

In many experiments that produce two-dimensional count data, the number of observations for each category of one of the classifications (rows or columns) is set at a specified number before the data are collected. For instance, in the divorce status–religious affiliation example (Example 11.2), the experimenter might decide to sample 100 men from each of the religious affiliation classifications. Thus, the observed data would not represent one random sample from the population of all American men, but would represent individual random samples of men from each of five classifications of religious affiliation.

The analysis of data from this type of experiment is the same as that presented in Section 11.2, but the conditions required for a valid analysis are different, as shown in the box.

When to Test for Independence of Two-Dimensional Count Data with Specified Column Totals*

To conduct a valid test of hypothesis for independence using data from a contingency table with specified column totals, the following conditions must be met:

1. A random sample is selected from each of the populations for which the sample size is determined by a specified column total.
2. The samples are independent.
3. The estimated expected number of observations in each cell must be greater than or equal to 5.

---

* In this text we will always consider the column totals to be specified. If you are presented with a contingency table in which the row totals are specified, simply interchange the rows and columns.

The form of the test for independence is identical whether the column totals are specified or not. For convenience, the information presented in Section 11.2 is repeated in the next box.

Test for Independence Using a Contingency Table with Specified Column Totals

H_0: The two classifications are independent

H_a: The two classifications are dependent

Test statistic:

$$X^2 = \frac{(o_{11} - E_{11})^2}{E_{11}} + \frac{(o_{12} - E_{12})^2}{E_{12}} + \cdots + \frac{(o_{rc} - E_{rc})^2}{E_{rc}}$$

where $E_{11} = \dfrac{r_1 c_1}{n}$, $E_{12} = \dfrac{r_1 c_2}{n}$, etc.

Rejection region: Reject H_0 if $X^2 >$ Table value, where the rejection region is obtained from Appendix Table VIIA, using $(r - 1)(c - 1)$ df.

EXAMPLE 11.3   Suppose three different techniques for teaching elementary calculus are to be investigated. Technique 1 involves using computer-assisted instruction (CAI) in conjunction with lectures, technique 2 involves CAI only, and technique 3 involves lectures only.

Random samples of 100 students are assigned to each of the three teaching techniques, and their final grades are used to compare the methods. Do the data in Table 11.14 provide sufficient evidence to indicate that the distribution of final grades depends on the teaching technique employed? Test at the .10 level of significance.

Table 11.14

Final grades for three teaching techniques

		Technique 1	Technique 2	Technique 3	Totals
	A	15	13	12	40
	B	34	28	35	97
Final Grade	C	40	36	38	114
	D	3	19	6	28
	F	8	4	9	21
Totals		100	100	100	300

Solution We want to test

H_0: The proportions of final grades in the grade categories do not depend on the teaching technique

H_a: The proportions of final grades in the grade categories depend on the teaching technique

Test statistic: $X^2 = \dfrac{(O_{11} - E_{11})^2}{E_{11}} + \cdots + \dfrac{(O_{53} - E_{53})^2}{E_{53}}$

Rejection region: For $\alpha = .10$ and $(r - 1)(c - 1) = (4)(2) = 8$ df, we will reject H_0 if $X^2 > 13.4$.

We calculate the estimated expected counts exactly as in Section 11.2:

$$E_{11} = \frac{r_1 c_1}{n} = \frac{(40)(100)}{300} = 13.33$$

$$E_{12} = \frac{r_1 c_2}{n} = \frac{(40)(100)}{300} = 13.33$$

and so forth. The observed cell counts and estimated expected cell counts are shown in Table 11.15. Since all these values are greater than 5, the χ^2 approximation is appropriate.

Table 11.15

Calculation of X^2 for Example 11.3

Row, Column	o	E	o − E	(o − E)²	$\dfrac{(o - E)^2}{E}$
1, 1	15	13.33	1.67	2.79	.21
1, 2	13	13.33	−.33	.11	.01
1, 3	12	13.33	−1.33	1.77	.13
2, 1	34	32.33	1.67	2.79	.09
2, 2	28	32.33	−4.33	18.75	.58
2, 3	35	32.33	2.67	7.13	.22
3, 1	40	38.00	2.00	4.00	.11
3, 2	36	38.00	−2.00	4.00	.11
3, 3	38	38.00	.00	.00	.00
4, 1	3	9.33	−6.33	40.07	4.29
4, 2	19	9.33	9.67	93.51	10.02
4, 3	6	9.33	−3.33	11.09	1.19
5, 1	8	7.00	1.00	1.00	.14
5, 2	4	7.00	−3.00	9.00	1.29
5, 3	9	7.00	2.00	4.00	.57
				$X^2 =$	$\overline{18.96}$

The calculated value of the test statistic is $X^2 = 18.96$. Since $X^2 = 18.96$ is greater than 13.4, we conclude that the distribution of final grades does depend on the teaching technique used. We could now use the methods of Chapters 7 and 8 to

624 11 THE CHI-SQUARE TEST AND THE ANALYSIS OF COUNT DATA

make inferences about the individual cell probabilities, or to compare cell probabilities between two columns.

Our procedure and conclusion are valid as long as the three samples are random and independent. We emphasize that we can make inferences only about the population from which the students were chosen. For example, if all students were selected from the same university, then we can make our inference only with respect to that university. ■

EXAMPLE 11.4   In the introduction to this chapter, we discussed an experiment that considered the effects of paying physicians for their participation in a telephone interview (Gunn & Rhodes, 1981). The data from that experiment are repeated in Table 11.16.

Table 11.16

Physician participation

	No incentive	$25	$50	Totals
Completed interview	79	101	112	292
Did not complete interview	58	46	33	137
Totals	137	147	145	429

Source: Reprinted by permission of Elsevier North Holland, Inc., from W. J. Gunn and I. N. Rhodes, *Public Opinion Quarterly*, 45, pp. 109–115. Copyright 1981 by The Trustees of Columbia University.

The total number of physicians sampled in each monetary classification was specified at the beginning of the experiment. Do the data provide sufficient evidence to indicate that response rate depends on monetary incentive? Use $\alpha = .01$.

Solution   We want to test

H_0: Response rate does not depend on monetary incentive

H_a: Response rate depends on monetary incentive

Test statistic: $X^2 = \dfrac{(o_{11} - E_{11})^2}{E_{11}} + \cdots + \dfrac{(o_{23} - E_{23})^2}{E_{23}}$

Rejection region: For $\alpha = .01$ and $(r - 1)(c - 1) = (1)(2) = 2$ df, we will reject H_0 if $X^2 > 9.21$.

The estimated expected cell counts are

$$E_{11} = \frac{r_1 c_1}{n} = \frac{(292)(137)}{429} = 93.25$$

$$E_{12} = \frac{r_1 c_2}{n} = \frac{(292)(147)}{429} = 100.06$$

and so forth. Table 11.17 gives all the observed cell counts and estimated expected cell counts. The χ^2 approximation is appropriate since each of the estimated expected cell counts is greater than 5.

Table 11.17

Calculation of X^2 for Example 11.4

Row, Column	o	E	$o - E$	$(o - E)^2$	$\dfrac{(o - E)^2}{E}$
1, 1	79	93.25	−14.25	203.06	2.18
1, 2	101	100.06	.94	.88	.01
1, 3	112	98.69	13.31	177.16	1.79
2, 1	58	43.75	14.25	203.06	4.64
2, 2	46	46.94	−.94	.88	.02
2, 3	33	46.31	−13.31	177.16	3.82
					$X^2 = \overline{12.46}$

The calculated value of the test statistic has been found to be $X^2 = 12.46$. Since $X^2 = 12.46$ is greater than the table value of 9.21, we conclude that response rate does depend on the monetary incentive. We make this conclusion at the .01 level of significance. ∎

COMPUTER EXAMPLE 11.3

Since the analyses are the same for the two types of contingency tables presented in this section and in Section 11.2, the information obtained from Minitab is the same. The following is a Minitab printout for the data considered in Example 11.4:

```
Expected counts are printed below observed counts

      NO INCNT     $25       $50     Total
   1       79      101       112       292
        93.25   100.06     98.69

   2       58       46        33       137
        43.75    46.94     46.31

Total     137      147       145       429

ChiSq =  2.177 +  0.009 +  1.794 +
         4.641 +  0.019 +  3.823 = 12.463
df = 2
```

The calculated value of the test statistic, "ChiSq = 12.463," when rounded, matches the value of 12.46 found in Example 11.4. Recall that, based on this value, we concluded at the .01 level of significance that the response rate to a telephone interview depended on the monetary incentive offered. ∎

Exercises (11.19–11.25)

Learning the Mechanics

11.19 Four independent random samples of 100 observations each were classified into one of three categories. The contingency table, with columns corresponding to the four samples and rows corresponding to the three categories, is shown on page 626. Do the data provide sufficient evidence to indicate that the distributions of observations within the rows are dependent upon the columns? Test using $\alpha = .05$.

		Column (Samples)			
		1	2	3	4
	1	20	30	20	10
Row	2	30	40	40	60
	3	50	30	40	30

Applying the Concepts

11.20 A study was conducted to determine whether a relationship exists between obesity in children and obesity in their parents. Random samples of fifty obese and fifty nonobese children were obtained. For each child, it was determined whether one or both parents were obese. A summary of the data appears in the table. Do the data provide sufficient evidence to indicate that child obesity is dependent upon parental obesity? Use $\alpha = .10$.

		Child		Totals
		Obese	Nonobese	
Parent	Obese	34	29	63
	Nonobese	16	21	37
Total		50	50	100

11.21 An experiment was conducted to compare two methods for operating a group family medical practice. Four hundred patients were randomly assigned to two groups: one group received the conventional direct contact with physicians, while the other group was first screened by a nurse-practitioner and then referred to a doctor only if the doctor's services were deemed necessary. At the conclusion of the experiment, the quality of each person's medical care was rated as satisfactory or unsatisfactory by an impartial medical observer in consultation with the patient. The results of the experiment are shown in the table. Do the data present sufficient evidence to indicate a difference in the proportions of satisfactory ratings for the two methods of patient care? Test using $\alpha = .05$.

		Method		Totals
		Conventional	Nurse-practitioner	
Rating	Satisfactory	148	161	309
	Unsatisfactory	52	39	91
Totals		200	200	400

11.22 A study was conducted to determine whether the treatment a psychiatric patient receives is affected by the patient's social status. Generally, the treatment of a psychiatric patient is classified as follows: psychotherapy, organic treatment (physical–chemical), or no treatment (custodial care in an institution with neither of the two other types of treatment). One hundred psychiatric patients were randomly sampled from each of four social classes, and each was classified according to the type of treatment received. Determine whether the data given in the table support the theory that type of psychiatric treatment and social status are dependent. Test at the $\alpha = .10$ level.

| | | **Social Class** | | | |
		Upper	Upper-middle	Middle-lower	Lower
Treatment	Psychotherapy	78	53	31	17
	Organic	13	27	38	33
	None	9	20	31	50

11.23 Some people would work even if they had no economic need, but does the proportion of Americans committed to working change over time? To help answer this question, 820 people were sampled in 1973–1974 and 842 were sampled in 1976–1977. Each one was asked "If you were to get enough money to live as comfortably as you would like for the rest of your life, would you continue to work or would you stop working?" The results of the survey are shown in the table. Do the data present sufficient evidence to indicate a difference in the proportions of those who would continue to work for the two time periods? Use $\alpha = .01$.

| | | **Time Period** | |
		1973–1974	1976–1977
Commitment to Work	Would continue	549	586
	Would stop	271	256

Source: Adapted from Table 7/1, "Commitment to Work: 1973–74 and 1976–77." In Center for Demographic Studies, U.S. Bureau of the Census, *Social Indicators III* (Washington, D.C.: Government Printing Office, 1980), p. 348.

11.24 A psychological study of the police service was described as follows:

This article describes an exploratory study into occupational stress among three groups of British police officers: (i) probationary constables, (ii) station sergeants, and (iii) relatively senior officers (i.e., inspectors or above). The police officers completed a 45-item Situation Stress Inventory which covered a range of different stressors commonly seen in police work. The inventory gave empirical data on the types of occupational events actually experienced and their subjective impact. Some significant differences emerged between the three groups of police officers. The findings are discussed with reference to practical implications for the British police service.[*]

[*] G. H. Gudjonsson and K. R. C. Adlam, "Occupational Stressors Among British Police Officers," *The Police Journal*, Vol. 58, No. 1, January 1985, p. 73.

Samples of 75 probationers, 33 sergeants, and 79 senior officers participated. The responses of each group of officers to one survey item, "delivering of death messages," are summarized in the accompanying table.* Is there evidence of a difference among the three groups of British policemen with respect to the fractions who find "delivering of death messages" stressful? Use $\alpha = .05$.

		Group		
		Probationers	Sergeants	Senior officers
Response	Stressful	50	14	13
	Unstressful	25	19	66

*11.25 Nausea is a common symptom among postoperative patients. A group of physicians is interested in comparing two new drugs, A and B, for their effectiveness in reducing postoperative nausea. One hundred eighty patients scheduled for surgery at a large hospital are used in the study, with sixty assigned to receive drug A and sixty to receive drug B after their operations. The remaining sixty patients are given a placebo (no drug). Shortly after their operations, the patients are classified according to the degree of nausea felt. The results are given in the table. Is there evidence of a difference among the drugs and the placebo with respect to their effectiveness in reducing postoperative nausea? Test using $\alpha = .05$, with the aid of a computer package.

		Drug A	Drug B	Placebo	**Totals**
Degree of Nausea	None	40	36	30	106
	Slight	10	12	16	38
	Moderate	6	4	8	18
	Severe	4	8	6	18
Totals		60	60	60	180

11.4

A CAUTION

Because the X^2 statistic for testing hypotheses about multinomial probabilities is one of the most widely applied statistical tools, it is also one of the most abused statistical procedures. The user should always be certain that the experiment satisfies the conditions given with each procedure. Furthermore, the user should be certain that the sample is drawn from the correct population—that is, from the population about which the inference is to be made. In Example 11.3, if the experimenter had chosen

* G. H. Gudjonsson and K. R. C. Adlam, "Occupational Stressors Among British Police Officers," *The Police Journal*, Vol. 58, No. 1, January 1985. Adapted from Table 1.

100 students from each of three different colleges, no valid inference could be made about the teaching methods. We would be comparing the three colleges, as well as the teaching methods.

The use of the χ^2 distribution as an approximation to the sampling distribution for X^2 should be avoided when the expected counts are very small. The approximation can become very poor when these expected counts are small, and thus the true α level may be very different from the table value. As a rule of thumb, an expected cell count of at least 5 will mean that the χ^2 probability distribution can be used to determine an approximate critical value.

If the X^2 value does not exceed the established critical value of χ^2, *do not accept the hypothesis of independence*. You would be risking a Type II error (accepting H_0 when it is false), and the probability β of committing such an error is unknown. Therefore, we avoid concluding that two classifications are independent, even when X^2 is small.

Finally, if a contingency table X^2 value *does* exceed the critical value, we must be careful to *avoid* inferring that a causal relationship exists between the classifications. Our alternative hypothesis states that the two classifications are statistically dependent, and *statistical dependence does not imply causality*. Therefore, the existence of a causal relationship cannot be established by a contingency table analysis.

Chapter Summary

The **multinomial experiment** is a generalization of the binomial experiment in which the observations in a population (infinite or very large) are classified into $k \geq 2$ categories. The fraction of the population in category j is π_j, and this is also the probability that any sampled observation is in category j. Sampled observations must be random and independent. We are interested in the number of observations, o_j, in each category; these are referred to as the **count data.** If $\sum o_j = n$ is the total number of observations, the **expected number** in category j is $E_j = n\pi_j$. To test whether there is disagreement between the data and hypothesized values for π_j, we require that each E_j be at least 5.

Test of hypothesis:

H_0: Each π_j is its hypothesized value

H_a: At least one π_j does not equal its hypothesized value

Test statistic: $X^2 = \sum \dfrac{(o_j - E_j)^2}{E_j}$

The distribution of X^2 is approximately a χ^2 distribution.

Rejection region: Use Appendix Table VIIA, with $(k - 1)$ df.

When the categories in a multinomial experiment are determined by two factors, we may be interested in the statistical dependence or independence of the factors. The factors are **dependent** if knowledge of one factor affects the probabilities for the other. Each category determined by the two factors is called a **cell.** Counts and probabilities can be summarized in **contingency tables,** using double subscripts, as shown in the tables.

		c, Column Factors A B ...			Totals
	1	o_{11}	o_{12}	...	r_1
r, Row Factors	2	o_{21}	o_{22}	...	r_2
	⋮	⋮	⋮		
Totals		c_1	c_2		n

		c, Column Factors A B ...			Totals
	1	π_{11}	π_{12}	...	π_1
	2	π_{21}	π_{22}	...	π_2
	⋮	⋮	⋮		
		π_A	π_B		1

The category probabilities can be estimated from the count data by $\pi_1 \approx p_1 = r_1/n$, $\pi_A \approx p_A = c_1/n$, etc. The expected cell counts $E_{ij} = n\pi_{ij}$ are then estimated by $r_i c_j/n$. To test whether the two classification factors are independent, each estimated cell count must be at least 5. If the total number of observations for each category of one of the classifications is specified in advance, we further require that the samples for these categories be independent.

Test of hypothesis:

H_0: The classification factors are statistically independent

H_a: The factors are dependent

Test statistic: $X^2 = \sum \dfrac{(o_{ij} - E_{ij})^2}{E_{ij}}$

The distribution of X^2 is approximately a χ^2 distribution.

Rejection region: Use Appendix Table VIIA, with $(r-1)(c-1)$ df.

Remember: Statistical dependence does not imply a cause–effect relationship.

SUPPLEMENTARY EXERCISES (11.26–11.41)

Learning the Mechanics

11.26 A random sample of 250 observations was classified according to the row and column categories shown in the table.

		Column 1	2	3
	1	20	20	10
Row	2	10	20	70
	3	20	50	30

a. Do the data provide sufficient evidence to conclude that the row and column classifications are dependent? Test using $\alpha = .05$.

b. Would the *analysis* change if the row totals were fixed before the data were collected?

c. Would the *assumptions* required for the analysis to be valid differ if the row totals were fixed? Explain.

11.27 A random sample of 150 observations was classified into five categories, as shown in the table. Do the data provide sufficient evidence to indicate that the categories are not equally likely to occur? Use $\alpha = .10$.

		Category		
1	2	3	4	5
28	35	33	25	29

Applying the Concepts

11.28 A computer used by a 24-hour banking service is supposed to assign each transaction to one of five memory locations at random. At the end of a day's transactions, the count for each of the five memory locations was recorded, as shown in the table. Is there evidence to indicate a difference in the proportions of transactions assigned to the five memory locations? Test using $\alpha = .05$.

		Memory Location		
1	2	3	4	5
90	78	100	72	85

11.29 In a recent poll, 656 people were randomly selected and classified according to whether they were government employees and how they believed the quality of life had changed since 1975. A summary of the responses is shown in the table. Do the data provide sufficient evidence to indicate that government employees perceive the change in quality of life differently from others? Test using $\alpha = .05$.

	Quality of Life		
	Worse	Better	Same
Civil servants	17	31	20
Others	317	191	80

11.30 A restaurateur who owns restaurants in four cities is considering the possibility of building separate dining rooms for nonsmokers. Since this would involve great expense, the restaurateur plans to survey customers at each restaurant, asking: "Would you be more comfortable dining here if there were a separate dining room for nonsmokers only?" Suppose seventy-five people were randomly selected and surveyed at each restaurant, with the results shown in the table. Is there sufficient evidence to indicate that customer preferences are different at the four restaurants? Use $\alpha = .10$.

		Reply		
		Yes	No	Indifferent
Restaurant	1	38	32	5
	2	42	26	7
	3	35	34	6
	4	37	30	8

11.31 Despite a good winning percentage, a certain major league baseball team has not drawn as many fans as expected. In hopes of finding ways to increase attendance, the management plans to interview fans who come to the games, to find out why they come. One question of interest is whether various age groups support the team differently. Suppose the data in the table were collected from randomly selected fans. Can you conclude that there is a relationship between age and number of games attended per year? Use $\alpha = .05$.

		Number of Games Attended per Year		
		1 or 2	3–5	Over 5
Age of Fan	Under 20	78	107	17
	21–30	147	87	13
	31–40	129	86	19
	41–55	55	103	40
	Over 55	23	74	22

11.32 If a company can identify times of day when on-the-job accidents are most likely to occur, extra precautions can be instituted during those times. A random sampling of accident report records over the last year at a particular plant gives the frequency of occurrence of accidents at different times of the day. Can it be concluded from the data given below that the proportions of accidents are different for the four time periods? Use $\alpha = .05$.

Hours:	1–2	3–4	5–6	7–8
Number of accidents:	31	28	45	47

11.33 A sociologist wants to determine whether sons have a tendency to choose the same occupation as their fathers. To find out, 500 males were polled and asked their occupation and their father's. A summary of the number of father–son pairs falling in each occupational category is shown in the table. Do the data provide sufficient evidence to indicate a dependence between a son's choice of occupation and his father's occupation? Test using $\alpha = .05$.

		Son			
		Professional or business	Skilled	Unskilled	Farmer
Father	Professional or business	55	38	7	0
	Skilled	79	71	25	0
	Unskilled	22	75	38	10
	Farmer	15	23	10	32

11.34 Teenage alcoholism is a big problem in the United States. To help determine why teen-agers are turning to alcohol, a survey was conducted to find out whether a teenager's family status has any relationship to the frequency with which he or she consumes alcohol. A random sample of 200 teenagers, from 15 to 19 years old, was asked about their use of alcohol. A summary of the responses is shown in the table. Do the data provide sufficient evidence to indicate a relationship between family status and the use of alcohol? Test using $\alpha = .05$.

		Alcohol Use		
		None	Occasional	Frequent
Family Status	Upper class	4	16	10
	Upper-middle class	11	40	24
	Lower-middle class	9	47	9
	Lower class	6	17	7

11.35 Five candidates have just entered the race for mayor of a large city. To determine whether any of the candidates has an early lead in popularity, 2,000 voters were polled and asked which candidate they preferred. A summary of their responses is shown in the table. Do the data provide sufficient evidence to indicate a difference in preference for the five candidates? Test using $\alpha = .01$.

Candidate				
I	II	III	IV	V
385	493	628	235	259

11.36 A city has three television stations. Each station has its own evening news program from 6:00 to 6:30 P.M. every weekday. An advertising firm wants to know how their audiences compare in size. One hundred people selected at random from the total audience are asked which news program they watch. Do the results given below provide sufficient evidence to indicate that the three stations do not have equal shares of the evening news audience? Use $\alpha = .05$.

Station:	1	2	3
Number of viewers:	35	43	22

11.37 In late 1977, many farmers across the United States went on strike, protesting that the prices of farm products (chiefly grains) were less than the cost of production. Although their main goal was to receive 100% of parity prices for all farm products, a second controversial strike goal was to induce farmers to reduce production, thereby reducing surpluses and boosting prices. A sample survey of 100 farmers was conducted to determine whether a relationship exists between a farmer's decision to participate in the strike and the farmer's opinion of the need for a 50% cutback in production. The results are shown in the table. Is there evidence of a relationship between a farmer's strike position and the stand on reduced production? Use $\alpha = .05$.

		On Strike	
		Yes	No
	Favor	21	7
50% Cutback in Production	Undecided	37	2
	Oppose	22	11

11.38 The Census Bureau has compared how often inpatients and outpatients were served in mental health facilities during 1955 and 1975. Suppose that fifty patients are sampled for each of the two years and classified according to the type of care they received. The data are summarized in the table. is there sufficient evidence to conclude that the proportion of outpatient care depends on the year? Use $\alpha = .05$.

		Year	
		1955	1975
Type of Care	Inpatient	38	14
	Outpatient	12	36

Source: Adapted from Table 2/6, "Inpatient and Outpatient Care Episodes in Mental Health Facilities, by Type of Facility: 1955 and 1975." In Center for Demographic Studies, U.S. Bureau of the Census, *Social Indicators III* (Washington, D.C.: Government Printing Office, 1980), p. 93.

*11.39 A national survey was conducted to determine how the general public views government involvement in domestic projects. Two hundred people from each of three income levels were asked if they thought the government was involved too much, too little, or just enough. Their responses are summarized in the table. Do the data provide sufficient information to indicate a relationship between income and view on government involvement in domestic projects? Test using $\alpha = .05$, with the aid of a computer package.

		Too little	Involvement Just enough	Too much	Totals
	Low	125	48	27	200
Income	Medium	103	58	39	200
	High	72	69	59	200
Totals		300	175	125	600

*11.40 A local bank plans to offer a special service to its young customers. To determine their economic interests, a survey of 100 people under 30 years of age is conducted. Each person is asked to identify his or her top two financial concerns from the six combinations listed in the table. Use the χ^2 test to determine whether the proportions of responses differ for the six pairs of top concerns. Test at $\alpha = .10$, using a computer package to find the value of the test statistic.

First Concern	Second Concern	Number of Responses
Buy a car	Go on a trip	15
Buy a car	Save money	14
Save money	Buy a car	22
Save money	Go on a trip	23
Go on a trip	Buy a car	10
Go on a trip	Save money	16
		100

*11.41 A political scientist wanted to find out whether there is a relationship between income and political affiliation. A random sample of 265 registered voters provided the information on income and political affiliation shown in the table. Do the data provide sufficient evidence to indicate a relationship between political affiliation and annual income? Test using $\alpha = .10$, with the aid of a computer package.

		Annual Income (thousands of dollars)			
		At least 25	At least 16 but less than 25	At least 8 but less than 16	Less than 8
	Republican	50	28	20	12
Affiliation	Democrat	14	35	35	41
	Other	6	7	10	7

CHAPTER 11 QUIZ

1. A random sample of 500 college students from a large university was classified according to athletic involvement and grade-point average. The data are summarized in the table. Do the data provide sufficient evidence to conclude that grade-point averages are dependent on athletic involvement? Test at the .05 level of significance.

| | | Student | |
		Athlete	Nonathlete
GPA	Under 2.5	68	218
	2.5 or above	94	120

2. Independent random samples of 500 Republicans and 500 Democrats were surveyed to ascertain their opinions about the president's economic policy. The data appear in the table. Is there sufficient evidence to indicate that the opinions depend on party affiliation? Test at the .10 level of significance.

| | | | Opinion | |
		Approve	Disapprove	No opinion
Party	Republican	243	217	40
	Democrat	204	268	28

CHAPTERS 9–11 CUMULATIVE QUIZ

1. A tobacco company decides to test market a new brand of cigarettes in three cities, using a different promotional campaign in each city. After the cigarette has been on the market for 2 months, independent random samples of 500 smokers are taken from each city and questioned about their preference in cigarettes. The results of the sample surveys are summarized in the table. Do the data provide sufficient evidence to indicate that the proportion preferring the new brand depends on the city? Test using $\alpha = .05$.

| | | | City | |
		1	2	3
Preference	New brand	23	12	41
	Other brand	477	488	459

2. Five varieties of peas are being tested to compare their mean yields per acre. A field is divided into twenty plots, and each variety of pea is planted in four randomly assigned plots. The yields, in bushels of peas produced per plot, are listed in the table.

Variety of Peas				
1	2	3	4	5
26.2	29.2	29.1	21.3	20.1
24.3	28.1	30.8	22.4	19.3
21.8	27.3	33.9	24.3	19.9
28.1	31.2	32.8	21.8	22.1

a. Is there sufficient evidence to conclude that at least two of the mean yields differ? Use $\alpha = .05$.

b. Form a 95% confidence interval for the difference between the mean yield of variety 1 and the mean yield of variety 3.

3. A veterinarian is studying a daily diet supplement which is believed to increase weight gain in young piglets. To investigate a possible relationship between weight gain and daily dosage of the supplement, the veterinarian randomly selects eight piglets of the same age and weight and feeds them different dosages (pellets) for 1 month. The resulting data appear in the table.

Piglet	Weight Gain Y, pounds	Daily Pellet Dose X
1	15	0
2	19	0
3	20	1
4	22	1
5	25	2
6	22	2
7	26	3
8	28	3

a. Find the least squares estimates of the Y-intercept and slope for a straight-line relationship between Y and X.

b. Is there sufficient evidence to conclude that Y and X are linearly related? Use $\alpha = .05$.

On Your Own

Many researchers rely on surveys to estimate the proportions of experimental units in populations that possess certain specified characteristics. For example, a political scientist may want to estimate the proportion of an electorate in favor of a certain legislative bill. Or, a social scientist may be interested in the proportions of people in a geographical region who fall into certain socioeconomic classifications. Or a psychologist might want to compare the proportions of patients who have different types of psychological disorders.

Choose a specific topic, similar to those described above, that interests you. Clearly define the population of interest, identify data categories of specific interest, and identify the proportions associated with them. Now, *guesstimate* the proportions of the population that you think fall into each category. For instance, you might guess that all the proportions are equal, or that the first proportion is twice as large as the second but equal to the third, etc.

Now, collect actual data by obtaining a random sample from your population of interest. Select a sample size so that all expected cell counts are at least 5 (preferably larger).

Use the count data you have obtained to test the null hypothesis that the true proportions in the population equal your presampling guesstimates of these actual proportions. Would failure to reject this null hypothesis imply that your guesstimates are correct?

Reference

Gunn, W. J., & Rhodes, I. N. "Physician Response Rates to a Telephone Survey: Effects of Monetary Incentive Level," *Public Opinion Quarterly*, 1981, *45*, pp. 109–115.

Chapter 12

NONPARAMETRIC STATISTICS

OCCUPATIONAL TITLES AND PRESTIGE

Wayne J. Villemez and Burton B. Silver (1976) have examined the possibility that the prestige of an occupation may depend on the title used to describe it. This has important implications, considering "the preoccupation of American sociology with status." Table 12.1 (page 640) shows ten occupational categories described in three slightly different ways. Each set of descriptions is accompanied by rankings of the categories by a random sample of 144 college students. In this table, the lower the rank, the higher the rating. Villemez and Silver (1976) concluded that these rankings are not entirely consistent and that title changes may bias the ratings.

In this chapter we will present methods for analyzing data that can be ranked in order of magnitude. These methods will be used when the t and F tests for comparing the means of two or more populations (Chapters 8 and 10) are not appropriate. Recall that these tests require the populations of interest to have normal distributions. If the populations of interest are markedly not normal, the tabulated values of t and F are not meaningful, the correct value of α is not known, and the t and F tests are of little value. The alternatives to the t and F tests presented in this chapter are called **nonparametric tests.**

The nonparametric counterparts of the t and F tests compare the distributions of the sampled populations, rather than specific parameters of these populations (such as the means or variances). For example, nonparametric tests can be used to compare the distribution of the strengths of preferences for a new product to the distributions of the strengths of preferences for the currently popular brands. If it can be inferred that the distribution of the strengths of preferences for the new product lies above (to the right of) that for the other products, as shown in Figure 12.1, the implication is that the new product tends to be more preferred than the currently popular products. Such an inference might lead to a decision to market the product nationally.

Table 12.1

Occupational descriptions and ranks

Occupational Category	Description 1	Rank	Description 2	Rank	Description 3	Rank
1	Building and maintenance	9	Constructing and maintaining	7.5	Construction	6.5
2	Arts and entertainment	3	Culture and information	9	Aesthetics and information	3
3	Transportation	8	Movement of goods and people	7.5	Movement of materials and people	8
4	Extraction	10	Removing materials from the earth	10	Farming and mining	9
5	Health and welfare	6.5	Citizen well-being	5.5	Social, physical, and moral well-being	1
6	Commerce	4	Business	2	Trade	5
7	Legal authority	1	Law and control	5.5	Law enforcement	6.5
8	Finance and records	5	Monetary affairs	3	Records and accounts	4
9	Manufacturing	6.5	Production	4	Production and assembly	10
10	Education and research	2	Knowledge	1	Training and education	2

Figure 12.1
Distributions of strengths of preference measurements (new product is preferred)

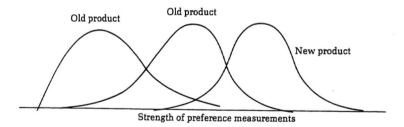

Strength of preference measurements

Many nonparametric methods use **relative ranks** of sample observations, rather than their actual numerical values. These methods are particularly valuable when we are unable to obtain numerical measurements of some phenomena but are able to rank them in comparison to each other. Statistics based on ranks of measurements are called **rank statistics.** In Sections 12.1 and 12.3, we will present rank statistics for comparing two or more distributions using independent samples. In Section 12.2 the **paired difference** design is used to make nonparametric comparisons of two populations. Finally, in Section 12.4, we will present a nonparametric measure of correlation between two variables called **Spearman's rank correlation coefficient.**

12.1

COMPARING TWO POPULATIONS: WILCOXON RANK SUM TEST FOR INDEPENDENT SAMPLES

In Section 8.3 we used the t statistic to compare two population means. One of the conditions that must be met to make valid use of the t statistic is that both populations of interest have normal distributions. If we suspect that this condition is not met, or if we are unable to obtain exact values of the sample measurements, the t statistic is not appropriate. An alternative statistical procedure that can be used to compare two populations is the **Wilcoxon rank sum test** (developed by Frank Wilcoxon).* To use this procedure, we must be able to rank the sample data in order of magnitude.

For example, suppose an experimental psychologist wants to compare reaction times for adult males under the influence of drug A to those under the influence of drug B. Experience has shown that populations of reaction time measurements possess distributions that are skewed to the right, as shown in Figure 12.2. Consequently, a t test should not be used to compare the mean reaction times for the two drugs, because the normality condition is not met.

Suppose the psychologist randomly selects six subjects to receive drug A and seven others to receive drug B. Their reaction times are measured (in seconds) and recorded at the completion of the experiment. The data are shown in Table 12.2 (page 642).

The population of reaction times for drug A is the conceptual set of reaction times that would be obtained by giving drug A to all adult males. Similarly, the population of reaction times for drug B is the set of all reaction times that would be obtained by giving drug B to all adult males. To compare the distributions for populations A and B, we first *rank the sample observations as though they were all drawn from the same population.* That is, we pool the measurements from both samples, and then

Figure 12.2
Typical distribution of reaction times

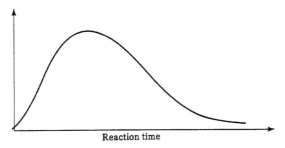

Reaction time

* Another statistic that is used for comparing two populations based on independent random samples is the Mann–Whitney U statistic. The U statistic is a simple function of the rank sums. It can be shown that the Wilcoxon rank sum test and the Mann–Whitney U test are equivalent.

Table 12.2

Drug A	Drug B
1.96	2.11
2.24	2.43
1.71	2.07
2.41	2.71
1.62	2.50
1.93	2.84
	2.88

Reaction times (in seconds) of subjects given drug A and drug B

Table 12.3

Ranks of reaction times

Drug A Reaction time, seconds	Rank	Drug B Reaction time, seconds	Rank
1.96	4	2.11	6
2.24	7	2.43	9
1.71	2	2.07	5
2.41	8	2.71	11
1.62	1	2.50	10
1.93	3	2.84	12
		2.88	13

rank the measurements from the smallest (1.62 receives a rank of 1) to the largest (2.88 is ranked 13). The results of this ranking process are shown in Table 12.3.

If the two populations had identical distributions of reaction times, we would expect the ranks to be randomly mixed between the two samples. If, on the other hand, one population tends to have slower reaction times than the other, we would expect the larger ranks to be mostly in one sample and the smaller ranks mostly in the other. Thus, the test statistic for the Wilcoxon test is based on the totals of the ranks for each of the two samples—that is, on the **rank sums.**

In the reaction times example, we denote the rank sum for drug A by T_A and the rank sum for drug B by T_B. Then

$$T_A = 4 + 7 + 2 + 8 + 1 + 3 = 25$$
$$T_B = 6 + 9 + 5 + 11 + 10 + 12 + 13 = 66$$

The sum of T_A and T_B will always equal $n(n + 1)/2$, where $n = n_1 + n_2$. So, for this example, $n_1 = 6$, $n_2 = 7$, and

$$T_A + T_B = \frac{13(13 + 1)}{2} = 91$$

Since $T_A + T_B$ is a constant, a small value for T_A implies a large value for T_B (and vice versa) and a large difference between T_A and T_B. Therefore, the smaller the value of one of the rank sums, the greater will be the evidence to indicate that the samples were selected from different populations.

When two different sample sizes are obtained, the Wilcoxon test uses the rank sum for the smaller sample. If the sample sizes are equal, either rank sum may be used.

Table 12.4

A partial reproduction of Appendix Table XIII, Critical Values for the Wilcoxon Rank Sum Test

A. $\alpha = .025$ One-Tailed; $\alpha = .05$ Two-Tailed

									n_1									
	3		4		5		6		7		8		9		10	
n_2	T_L	T_U	T_L	T_U	T_L	T_U	T_L	T_U	T_L	T_U	T_L	T_U	T_L	T_U	T_L	T_U
3	5	16	6	18	6	21	7	23	7	26	8	28	8	31	9	33
4	6	18	11	25	12	28	12	32	13	35	14	38	15	41	16	44
5	6	21	12	28	18	37	19	41	20	45	21	49	22	53	24	56
6	7	23	12	32	19	41	26	52	28	56	29	61	31	65	32	70
7	7	26	13	35	20	45	28	56	37	68	39	73	41	78	43	83
8	8	28	14	38	21	49	29	61	39	73	49	87	51	93	54	98
9	8	31	15	41	22	53	31	65	41	78	51	93	63	108	66	114
10	9	33	16	44	24	56	32	70	43	83	54	98	66	114	79	131

Values that locate the rejection region for this test are given in Appendix Table XIII. A partial reproduction of this table is shown in Table 12.4.

To illustrate the use of this table, suppose that $n_1 = 8$, $n_2 = 10$, and a two-tailed test is conducted with $\alpha = .05$. The columns of the table correspond to n_1, the first sample size, and the rows correspond to n_2, the second sample size. The T_L and T_U entries in the table specify the upper and lower boundaries, respectively, of the rejection region for the rank sum being used as the test statistic. Consulting the table for $n_1 = 8$ and $n_2 = 10$, we find that $T_L = 54$ and $T_U = 98$. Since $n_1 < n_2$, the null hypothesis will be rejected if the rank sum of sample 1 (the sample with fewer measurements) is less than or equal to 54 or greater than or equal to 98.

The conditions that must be met for the Wilcoxon rank sum test to be valid are given in the box.

When to Use the Wilcoxon Rank Sum Test

To conduct a valid Wilcoxon rank sum test to compare two populations, the following conditions must be met:

1. Two random samples must be taken—one from each of the two populations of interest.
2. The two samples must be independent.
3. It must be possible to rank the observations in order of magnitude.

[*Note*: The distributions of the populations of interest may have any shape.]

The general form of the Wilcoxon rank sum test is given in the next box.

Wilcoxon Rank Sum Test: Independent Samples

One-Tailed Test

H_0: Two sampled populations have identical distributions

H_a: The distribution for population 1 is shifted to the right of that for population 2

Test statistic: The rank sum T associated with the sample with fewer measurements; if sample sizes are equal, either rank sum can be used

Rejection region: Reject H_0 if $T \geq T_U$, where T_U is the upper value given by Appendix Table XIII for the chosen *one-tailed* α value.

Two-Tailed Test

H_0: Two sampled populations have identical distributions

H_a: The distribution for population 1 is shifted to the left or to the right of that for population 2

Test statistic: The rank sum T associated with the sample with fewer measurements; if sample sizes are equal, either rank sum can be used

Rejection region: Reject H_0 if $T \leq T_L$ or $T \geq T_U$, where T_L is the lower value and T_U is the upper value given by Appendix Table XIII for the chosen *two-tailed* α value.

[*Note:* If the one-tailed alternative is that population 1 is shifted to the *left* of population 2, we reject H_0 if $T \leq T_L$.]

EXAMPLE 12.1   Do the data given in Table 12.3 provide sufficient evidence to indicate that the distribution of reaction times corresponding to drug A lies either to the right or to the left of the distribution corresponding to drug B? Test at the .05 level of significance.

Solution   The test of interest is:

H_0:   The two populations of reaction times corresponding to drug A and drug B have the same distribution

H_a: The distribution of reaction times for drug A is shifted to the right or left of the distribution corresponding to drug B

Test statistic: Since drug A has fewer subjects than drug B, the test statistic is T_A, the rank sum of drug A's reaction times.

Rejection region: Since the test is two-sided, we consult Appendix Table XIIIA for the rejection region corresponding to $\alpha = .05$ and $n_1 = 6$, $n_2 = 7$. We will reject H_0 if $T_A \leq T_L$ or $T_A \geq T_U$. Thus, we will reject H_0 if $T_A \leq 28$ or $T_A \geq 56$.

Since T_A, the rank sum of drug A's reaction times, was calculated earlier to be $T_A = 25$, it is in the rejection region (see Figure 12.3). Therefore, we can conclude that the distributions of reaction times for drugs A and B are not identical. In fact, it appears that drug B tends to be associated with reaction times that are larger (slower) than those associated with drug A, because T_A falls in the lower tail of the rejection region. Figure 12.3 depicts only this side of the two-sided alternative hypothesis. The other would show the distribution for drug A shifted to the right of the distribution for drug B.

Figure 12.3
Alternative hypothesis
and rejection region for
Example 12.1

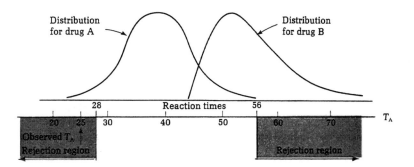

When you apply the Wilcoxon rank sum test in a practical situation, you may encounter one or more ties in the observations. (A tie occurs when two of the sample observations are equal.) The Wilcoxon rank sum test will still be valid if the number of ties is small in comparison with the number of sample measurements and if you *assign to each tied observation the average of the ranks the two observations would have received if the observations had not been tied.* For example, suppose the fourth and fifth smallest observations are tied. Since these observations would have received the ranks 4 and 5, you should assign the average of these ranks, 4.5, to both of them and then proceed with the test in the usual manner.

COMPUTER
EXAMPLE 12.1

Minitab may be used to find the value of the test statistic in conducting the Wilcoxon rank sum test. As reported in the footnote on page 641, the Wilcoxon rank sum test and the Mann–Whitney U test are equivalent methods for comparing two populations based on independent random samples. Minitab uses the Mann–Whitney U test when analyzing such data, but we can still find the value of the Wilcoxon statistic.

Consider the test conducted in Example 12.1. We found that the rank sum of drug A's reaction times was $T_A = 25$. The data previously given in Table 12.2 were analyzed using Minitab, resulting in the printout shown on page 646.

Mann-Whitney Confidence Interval and Test

```
DRUG A     N =   6      MEDIAN =        1.9450
DRUG B     N =   7      MEDIAN =        2.5000
POINT ESTIMATE FOR ETA1-ETA2 IS        -0.4950
96.2  PCT C.I. FOR ETA1-ETA2 IS (   -0.9497,  -0.1099)
W =      25.0
TEST OF ETA1 = ETA2  VS.   ETA1 N.E. ETA2 IS SIGNIFICANT AT  0.0184
```

The "W = 25.0" corresponds to the value $T_A = 25$ that was calculated previously. Notice that the level of significance, the p-value, is also reported for this test. Since the observed significance level of 0.0184 is less than the value of $\alpha = .05$ given in Example 12.1, we conclude that the distributions of reaction times for drug A tend to be smaller than those for drug B.   ∎

Exercises (12.1–12.7)

Using the Tables

12.1   a.   Suppose you want to compare two treatments, A and B, and you want to determine whether the distribution of the population of B measurements is shifted to the right of the distribution of the population of A measurements. If $n_A = 7$, $n_B = 5$, and $\alpha = .05$, give the rejection region for the test.

     b.   Suppose you want to detect a shift in the distributions, either A to the right of B or vice versa. Locate the rejection region for the test, assuming $n_A = 7$, $n_B = 5$, and $\alpha = .05$.

Learning the Mechanics

12.2   Independent random samples were selected from two populations. The data are shown in the table.

Sample from Population 1		Sample from Population 2	
15	14	6	7
16	12	13	5
13	17	8	4
		9	10

     a.   Use the Wilcoxon rank sum test to determine whether the data provide sufficient evidence to indicate a difference (a shift) in the locations of the sampled population distributions. Test using $\alpha = .05$.

     b.   Do the data provide sufficient evidence to indicate that population distribution 1 is shifted to the right of population distribution 2? Use the Wilcoxon rank sum test with $\alpha = .05$.

Applying the Concepts

12.3 A realtor wants to determine whether a difference exists between home prices in two subdivisions. Six homes from subdivision A and eight homes from subdivision B are sampled, and the prices (in thousands of dollars) are recorded in the table.

Subdivision			
A		B	
43	60	57	88
48	39	39	46
42	47	55	41
		52	64

a. Use the two-sample *t* test to compare the population mean home prices for the two subdivisions. What conditions are necessary to validate this procedure? Are they reasonable in this case?

b. Use the Wilcoxon rank sum test to determine whether there is a difference (a shift in location) in the distributions of home prices in the two subdivisions.

12.4 An educational psychologist claims that the order in which test questions are asked affects a student's ability to answer correctly. To investigate this assertion, a professor prepares one set of test questions but arranges the questions in two different orders. On test A they appear in order of increasing difficulty (that is, from easiest to hardest), while on test B the order is reversed. Seven randomly chosen students in the class take test A, and six other students take test B. Each student's score is recorded, as follows:

Test A: 90, 71, 83, 82, 75, 91, 65

Test B: 66, 78, 50, 68, 80, 60

Do the data provide sufficient evidence to indicate a difference in a student's ability to answer the questions between the two tests? Test using $\alpha = .05$.

12.5 A major razor blade manufacturer advertises that its twin-blade disposable razor will "get you a lot more shaves" than any single-blade disposable razor on the market. A rival company, which has been very successful in selling single-blade razors, wants to test this claim. Independent random samples of eight single-blade shavers and eight twin-blade shavers are taken, and the number of shaves that each gives before the razor is disposed of is recorded. The results are shown in the table.

Number of Shaves			
Twin Blades		Single Blades	
8	15	10	13
17	10	6	14
9	6	3	5
11	12	7	7

a. Do the data support the twin-blade manufacturer's claim? Use $\alpha = .05$.

b. Do you think that this experiment was designed in the best possible way? If not, what design might have been better?

12.6 Fourteen rats were used in an experiment aimed at comparing two deprivation schedules, A and B, for their effects on hoarding behavior. An independent sampling design was used, with seven rats randomly assigned to each schedule. At the end of the deprivation period, the rats were permitted free access to food pellets, and the number of pellets hoarded (taken but not eaten) during a given time period was recorded. The data are given in the table. Is there sufficient evidence to indicate that rats on one of the deprivation schedules have a greater tendency to hoard than those on the other schedule? Test using $\alpha = .05$.

Number of Pellets Hoarded

Schedule A		Schedule B	
15	4	5	2
10	9	1	6
5	7	2	3
7		8	

*12.7 In a comparison of visual acuity of deaf and hearing children, eye movement rates were noted for ten deaf and ten hearing children. Test the claim that deaf children have greater visual acuity than hearing children, using the data given in the table. (A larger rate of eye movement indicates greater visual acuity.) Use a computer package to help with this analysis.

Eye Movement Rates

Deaf Children		Hearing Children	
2.75	1.95	1.15	1.23
3.14	2.17	1.65	2.03
3.23	2.45	1.43	1.64
2.30	1.83	1.83	1.96
2.64	2.23	1.75	1.37

12.2

COMPARING TWO POPULATIONS: WILCOXON SIGNED RANK TEST FOR PAIRED SAMPLES

In Section 8.4 we use the t statistic to compare two population means when paired (dependent) samples were obtained. But for that procedure to be valid, the population of differences must be normally distributed. When this condition is not met, we use a nonparametric technique called the **Wilcoxon signed rank test** to analyze paired observations.

For instance, suppose that a company wants to compare the softness of two paper products. One way to compare the products would be to give ten judges samples of the two products and have each judge rate their softness on a scale of 1 to 10 (higher ratings imply greater softness). The results of such an experiment are shown in Table 12.5.

Table 12.5

Softness ratings of paper

Judge	Product A	Product B	Judge	Product A	Product B
1	6	4	6	7	9
2	8	5	7	6	2
3	4	5	8	5	3
4	9	8	9	6	7
5	4	1	10	8	2

Since this is a paired difference experiment, we analyze the differences between the measurements. However, the nonparametric approach requires that we calculate the ranks of the absolute values of the differences between the measurements—that is, the ranks of the differences after removing any minus signs. The differences, absolute values of the differences, and corresponding ranks are shown in Table 12.6. Note that tied absolute differences are assigned the average of the ranks they would receive if they were unequal but successive measurements. After the absolute differences are ranked, the sum of the ranks of the positive differences of the original measurements, T_+, and the sum of the ranks of the negative differences of the original measurements, T_-, are computed (see Table 12.6).

Using the data in Table 12.6, we can test the hypotheses:

H_0: The distributions of the ratings for products A and B are identical

H_a: The distributions of the ratings differ (in location) for the two products

[Note: This is a two-sided alternative and therefore it implies a two-tailed test.]

Test statistic: $T =$ Smaller of the positive and negative rank sums, T_+ and T_-

Table 12.6

Calculations required for Wilcoxon signed rank test

Judge	Product A B	Difference (A − B)	Absolute Value of Difference	Rank of Absolute Value
1	6 4	2	2	5
2	8 5	3	3	7.5
3	4 5	−1	1	2
4	9 8	1	1	2
5	4 1	3	3	7.5
6	7 9	−2	2	5
7	6 2	4	4	9
8	5 3	2	2	5
9	6 7	−1	1	2
10	8 2	6	6	10
			$T_+ =$ Positive rank sum =	46
			$T_- =$ Negative rank sum =	9

The smaller the value of T, the greater will be the evidence to indicate that the two distributions differ in location. The rejection region for T can be determined by consulting Appendix Table XIV. A portion of this table is shown in Table 12.7. This table gives a value for both one-tailed tests and two-tailed tests for each value of n, the number of paired differences. For a two-tailed test with $\alpha = .05$, we will reject H_0 if $T \leq$ Table value. You can see in Table 12.7 that the table value that locates the rejection region for the judges' ratings for $\alpha = .05$ and $n = 10$ pairs of observations is 8. Therefore, the rejection region for the test is

Rejection region: Reject H_0 if $T \leq 8$ for $\alpha = .05$.

Table 12.7

A partial reproduction of Appendix Table XIV, Critical Values for the Wilcoxon Signed Rank Test

One-Tailed	Two-Tailed	$n = 5$	$n = 6$	$n = 7$	$n = 8$	$n = 9$	$n = 10$
$\alpha = .05$	$\alpha = .10$	1	2	4	6	8	11
$\alpha = .025$	$\alpha = .05$		1	2	4	6	8
$\alpha = .01$	$\alpha = .02$			0	2	3	5
$\alpha = .005$	$\alpha = .01$				0	2	3

One-Tailed	Two-Tailed	$n = 11$	$n = 12$	$n = 13$	$n = 14$	$n = 15$	$n = 16$
$\alpha = .05$	$\alpha = .10$	14	17	21	26	30	36
$\alpha = .025$	$\alpha = .05$	11	14	17	21	25	30
$\alpha = .01$	$\alpha = .02$	7	10	13	16	20	24
$\alpha = .005$	$\alpha = .01$	5	7	10	13	16	19

Since the smaller rank sum for the paper softness data, $T_- = 9$, does not fall within the rejection region, the experiment has not provided sufficient evidence to indicate that the two paper products differ with respect to their softness ratings at the $\alpha = .05$ significance level.

Note that if a level of $\alpha = .10$ had been used, the rejection region would have been $T \leq 11$, and we would have rejected H_0. In other words, the samples do provide evidence that the distributions of the softness ratings differ at the $\alpha = .10$ significance level.

When the Wilcoxon signed rank test is applied to a set of data, it is possible that one (or more) of the paired differences may equal 0. The test will continue to be valid if the number of zeros is small in comparison to the number of pairs, but to perform the test you must delete the zeros and reduce the number of differences accordingly. For example, if you have $n = 12$ pairs and two of the differences equal 0, you should delete these pairs, rank the remaining 10 differences, and use $n = 10$ when using Table XIV. Ties in ranks are treated in the same manner as for the Wilcoxon rank sum test for a completely randomized design. Assign to each of the tied ranks the average of the ranks the two observations would have received if the observations had not been tied.

The conditions that must be met for the Wilcoxon signed rank test to be valid are given in the box.

When to Use the Wilcoxon Signed Rank Test

To conduct a valid Wilcoxon signed rank test to compare two populations, the following conditions must be met:

1. A random sample of pairs of observations must be taken.
2. It must be possible to rank the absolute differences of the paired observations.

[*Note:* The population of differences may have any shape.]

The general form of the Wilcoxon signed rank test is given in the next box.

Wilcoxon Signed Rank Test for a Paired Difference Experiment

One-Tailed Test

H_0: Two sampled populations have identical distributions

H_a: The distribution for population A is shifted to the right of that for population B

Test statistic: T_-, the negative rank sum (we assume the differences are computed by subtracting each paired B measurement from the corresponding A measurement)

Rejection region: Reject H_0 if $T_- \leq$ Table value, found in Appendix Table XIV, for the one-tailed significance level α and the number of untied pairs, n

Two-Tailed Test

H_0: Two sampled populations have identical distributions

H_a: The distribution for population A is shifted to the left or to the right of that for population B

Test statistic: T, the smaller of the positive and negative rank sums, T_+ and T_-

Rejection region: Reject H_0 if $T \leq$ Table value, found in Appendix Table XIV, for the two-tailed significance level α and the number of untied pairs, n

[*Note:* If the alternative hypothesis is that the distribution for A is shifted to the left of B, we use T_+ as the test statistic and reject H_0 if $T_+ \leq$ Table value.]

EXAMPLE 12.2   Suppose the police commissioner in a small community must choose between two plans for patrolling the town's streets. Plan A, the less expensive plan, uses voluntary citizen groups to patrol certain high-risk neighborhoods. Plan B would use police patrols instead. As an aid in reaching a decision, both plans are examined by ten crime prevention experts, each of whom is asked to rate the plans on a scale from 1 to 10 (high ratings imply more effective crime prevention). The city will adopt plan B (and hire extra police) only if the data provide evidence that the experts tend to rate plan B as more effective than plan A.

The results of the survey and the preliminary calculations are shown in Table 12.8. Do the data provide evidence at the $\alpha = .05$ level that the distribution of ratings for plan B lies to the right of that for plan A?

Table 12.8

Ratings by ten crime prevention experts

Crime Prevention Expert	Plan A	Plan B	Difference $(A - B)$	Rank of Absolute Difference
1	7	9	−2	4.5
2	4	5	−1	2
3	8	8	0	(Eliminated)
4	9	8	1	2
5	3	6	−3	6
6	6	10	−4	7.5
7	8	9	−1	2
8	10	8	2	4.5
9	9	4	5	9
10	5	9	−4	7.5

Positive rank sum $= T_+ = 15.5$

Solution   The null and alternative hypotheses are

H_0: The two distributions of effectiveness ratings are identical

H_a: The effectiveness ratings of plan B tend to exceed those of plan A

Observe that the alternative hypothesis is one-sided (that is, we want to detect only a shift in the distribution of B ratings to the right of the distribution of A ratings), and therefore it implies a one-tailed test of the null hypothesis (see Figure 12.4). If the alternative hypothesis is true, the B ratings will tend to be larger than their paired A ratings, more negative differences in pairs will occur, T_- will be large, and T_+ will be

Figure 12.4
The alternative hypothesis for Example 12.2: We expect T_+ to be small

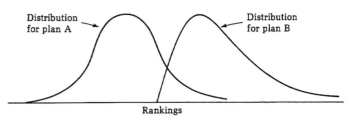

small. Because Table XIV is constructed to give lower-tailed values of T, we will use T_+ as the test statistic and reject H_0 for $T_+ \leq$ Table value.

The differences in ratings for the pairs $(A - B)$ are shown in Table 12.8. Note that one of the differences equals 0. Consequently, we eliminate this pair from the ranking and reduce the number of pairs to $n = 9$. Using this value in Table XIV, we find that for a one-tailed test with $\alpha = .05$ and $n = 9$, the table value is 8. Therefore, the test statistic and rejection region for the test are

Test statistic: T_+, the positive rank sum

Rejection region: Reject H_0 if $T_+ \leq 8$.

Summing the ranks of the positive differences in Table 12.8, we find $T_+ = 15.5$. Since this value exceeds the table value (8), we conclude that this sample provides insufficient evidence at the $\alpha = .05$ level to support the alternative hypothesis. The commissioner *cannot* conclude that the plan utilizing police patrols tends to be rated higher than the plan using citizen volunteers. That is, on the basis of this study, extra police will not be hired. ∎

COMPUTER EXAMPLE 12.2   In Example 12.2 we demonstrated the use of the Wilcoxon signed rank test for paired samples. The value of the test statistic was $T_+ = 15.5$. Based on this result, there was insufficient evidence at the $\alpha = .05$ level of significance to conclude that the plan using police patrols tends to be rated higher than the plan using citizen volunteers.

The following printout shows the results of using Minitab to calculate the value of the test statistic:

```
TEST OF MEDIAN = 0.000000000 VERSUS MEDIAN L.T. 0.000000000

              N FOR    WILCOXON            ESTIMATED
         N    TEST     STATISTIC  P-VALUE   MEDIAN
C3       10    9         15.5      0.221    -1.000
```

The value of 15.5 that Minitab calls the "WILCOXON STATISTIC" corresponds to the $T_+ = 15.5$ calculated in Example 12.2. Again we see that there is insufficient evidence to make a conclusion (reject H_0), because the reported "P-VALUE" of 0.221 is larger than $\alpha = .05$. ∎

Exercises (12.8–12.14)

Using the Tables

12.8 a. Suppose you want to test a hypothesis that two treatments, A and B, are equivalent against the alternative that the responses for A tend to be larger than those for B. If $n = 8$ and $\alpha = .01$, give the rejection region for a Wilcoxon signed rank test.

b. Suppose you want to detect a difference in the locations of the distributions of the responses for A and B. If $n = 7$ and $\alpha = .10$, give the rejection region for the Wilcoxon signed rank test.

Learning the Mechanics

12.9 A random sample of nine pairs of measurements is shown in the table.

Pair	Sample Data from Population 1	Sample Data from Population 2	Pair	Sample Data from Population 1	Sample Data from Population 2
1	8	7	6	8	3
2	10	1	7	4	6
3	6	4	8	9	2
4	10	10	9	8	4
5	7	4			

a. Use the Wilcoxon signed rank test to determine whether the data provide sufficient evidence to indicate that distribution 1 is shifted to the right of distribution 2. Test using $\alpha = .05$.

b. Use the Wilcoxon signed rank test to determine whether the data provide sufficient evidence to indicate that distribution 1 is shifted either to the right or to the left of distribution 2. Test using $\alpha = .05$.

Applying the Concepts

12.10 Hypoglycemia is a condition in which blood sugar is below normal limits. To compare two compounds, X and Y, for treating hypoglycemia, each compound is applied to half the diaphragms of each of seven white mice. Blood glucose uptake (in milligrams per gram of tissue) is measured for each half, producing the results listed in the table.

Mouse	Compound X	Y
1	4.7	5.1
2	3.3	4.6
3	8.5	8.7
4	3.9	3.6
5	7.0	6.1
6	4.7	4.1
7	5.2	5.1

Do the data provide sufficient evidence to indicate that one of the compounds tends to produce higher blood sugar uptake readings than the other? Test using $\alpha = .10$.

12.11 Children completing the sixth grade at a school in a large city have two junior high schools, A and B, to choose from. The school board wants to compare the academic effectiveness of the two schools. Parents of six sets of identical twins have agreed to send one child to school A and the other to school B. Since each set of twins was in the same class at each grade level through the sixth grade, a paired difference design can be employed. Near the end of the ninth grade, the twins are given an achievement test, with the results shown in the table. Test the hypothesis that there is a difference (a shift in location) between the distributions of achievement test scores for the two schools. Use $\alpha = .05$.

Twin Pair	School	
	A	B
1	65	69
2	73	72
3	86	74
4	50	52
5	60	47
6	81	72

12.12 Dental researchers have developed a new material for preventing cavities, a plastic sealant that is applied to the chewing surfaces of the teeth. To determine its effectiveness, the sealant was applied to half of the teeth of each of twelve school-age children. After 5 years, the number of cavities in the coated and the untreated teeth were counted. The results are given in the table. Is there sufficient evidence to indicate that sealant-coated teeth are less prone to cavities than are untreated teeth? Test using $\alpha = .05$.

Child	Coated	Untreated	Child	Coated	Untreated
1	3	3	7	1	5
2	1	3	8	2	0
3	0	2	9	1	6
4	4	5	10	0	0
5	1	0	11	0	3
6	0	1	12	4	3

12.13 A food vending company currently uses vending machines made by two different manufacturers. Before purchasing new machines, the company wants to compare the two types in terms of reliability. The company has the same number of machines of each type. Records for 7 weeks are given in the table, where the data indicate the number of breakdowns per week for each type of machine. Do the data present sufficient evidence to indicate that one of the machine types is less prone to breakdowns than the other? Test using $\alpha = .05$.

Week	Machine Type	
	A	B
1	14	12
2	17	13
3	10	14
4	15	12
5	14	9
6	9	11
7	12	11

*12.14 Twelve sets of identical twins were given psychological tests to determine whether the firstborn tends to be more aggressive than the secondborn. The results are shown in the table, where higher scores indicate greater aggressiveness. Do the data provide sufficient evidence to indicate that the firstborn of twins is more aggressive than the other? Test using $\alpha = .05$. Use a computer package to calculate the value of the test statistic and the associated p-value.

Set	Firstborn	Secondborn	Set	Firstborn	Secondborn
1	86	88	7	77	65
2	71	77	8	91	90
3	77	76	9	70	65
4	68	64	10	71	80
5	91	96	11	88	81
6	72	72	12	87	72

12.3

COMPARING MORE THAN TWO POPULATIONS:
THE KRUSKAL–WALLIS H TEST FOR A
COMPLETELY RANDOMIZED DESIGN

In Chapter 10 we used an analysis of variance and the F test to compare the means of k populations based on random samples from populations that were normally distributed with a common variance, σ^2. We now present a nonparametric technique, called the **Kruskal–Wallis *H* test,** that does not require the populations to be normally distributed.

For example, suppose a health administrator wants to compare the unoccupied bed space for three hospitals located in the same city. Ten different days are randomly selected from the records of each hospital, and the number of empty beds for each day is recorded (see Table 12.9). Because the number of empty beds per day occasionally may be quite large, it is conceivable that the population distributions of data may be skewed to the right and that this type of data may not satisfy the conditions needed for a parametric comparison of the population means. We therefore use a nonparametric analysis and base our comparison on the rank sums for the three sets of sample data. Just as with two independent samples (Section 12.1), the ranks are computed for each observation according to the relative magnitude of the measurements *when the data for all the samples are combined* (see Table 12.9). *Ties are treated as they were for the Wilcoxon rank sum and signed rank tests, by assigning the average value of the ranks to each of the tied observations.* We want to test the hypotheses

H_0: The distributions of the number of empty beds are the same for all three hospitals

H_a: At least two of the distributions differ in location

Table 12.9

Number of available beds

	Hospital 1		Hospital 2		Hospital 3	
	Beds	Rank	Beds	Rank	Beds	Rank
	6	5	34	25	13	9.5
	38	27	28	19	35	26
	3	2	42	30	19	15
	17	13	13	9.5	4	3
	11	8	40	29	29	20
	30	21	31	22	0	1
	15	11	9	7	7	6
	16	12	32	23	33	24
	25	17	39	28	18	14
	5	4	27	18	24	16
		$R_1 = \overline{120}$		$R_2 = \overline{210.5}$		$R_3 = \overline{134.5}$

If we denote the three sample rank sums by R_1, R_2, and R_3, the test statistic is given by

$$H = \frac{12}{n(n+1)} \left[\frac{R_1^2}{n_1} + \frac{R_2^2}{n_2} + \frac{R_3^2}{n_3} \right] - 3(n+1)$$

where n_1 is the number of measurements in the first sample, n_2 is the number in the second, n_3 is the number in the third, and n is the total sample size ($n = n_1 + n_2 + n_3$). For the data in Table 12.9, we have $n_1 = n_2 = n_3 = 10$, and $n = 30$. The rank sums are $R_1 = 120$, $R_2 = 210.5$, and $R_3 = 134.5$. Thus,

$$H = \frac{12}{30(31)} \left[\frac{(120)^2}{10} + \frac{(210.5)^2}{10} + \frac{(134.5)^2}{10} \right] - 3(31)$$

$$= 99.097 - 93 = 6.10$$

If the null hypothesis is true, the distribution of H in repeated sampling is approximately a χ^2 (chi-square) distribution (discussed in Sections 7.8 and 11.1). This approximation for the sampling distribution of H is adequate as long as each of the k sample sizes is at least 5. The χ^2 distribution is characterized by a single parameter, called the **degrees of freedom associated with the distribution.** Several χ^2 distributions with different degrees of freedom are shown in Figure 12.5. The degrees of freedom corresponding to the approximate sampling distribution of H will always be $(k - 1)$, that is, 1 less than the number of populations being compared. Because large values of H support the alternative hypothesis that the populations have different distributions, the rejection region for the test will be located in the upper tail of the χ^2 distribution.

Figure 12.5
Several χ^2 distributions

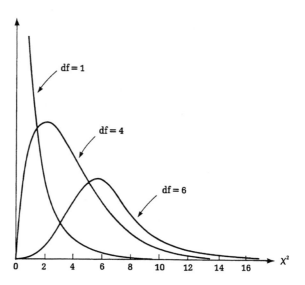

For the data of Table 12.9, the approximate distribution of the test statistic H is a χ^2 distribution with $k - 1 = 2$ df. To determine how large H must be before we will reject the null hypothesis, we consult Appendix Table VIIA, which is partially repro-

Table 12.10

A partial reproduction of Appendix Table VIIA, χ^2 Values for Rejection Regions for One-Tailed Tests

df	$\alpha = .10$		$\alpha = .05$		$\alpha = .01$	
	One-tailed test, >	One-tailed test, <	One-tailed test, >	One-tailed test, <	One-tailed test, >	One-tailed test, <
1	2.71	0.0158	3.84	0.00393	6.64	0.000157
2	4.61	0.211	6.00	0.103	9.21	0.0201
3	6.25	0.584	7.82	0.352	11.4	0.115
4	7.78	1.0636	9.50	0.711	13.3	0.297
5	9.24	1.61	11.1	1.15	15.1	0.554
6	10.6	2.20	12.6	1.64	16.8	0.872
7	12.0	2.83	14.1	2.17	18.5	1.24
8	13.4	3.49	15.5	2.73	20.1	1.65
9	14.7	4.17	17.0	3.33	21.7	2.09
10	16.0	4.87	18.3	3.94	23.2	2.56

duced in Table 12.10. To use this table, find the column corresponding to the desired value of α and a one-tailed test ($>$). The appropriate table value is then found in the row with $(k - 1)$ df. Thus, for $\alpha = .05$ and 2 df, we will reject the null hypothesis that the three distributions are the same if the calculated value of the test statistic is greater than 6.00. Since the calculated value, $H = 6.10$, is greater than the table value of 6.00, we conclude that at least one of the three hospitals tends to have a larger number of empty beds than the others. We have 95% confidence in this decision.

Note that prior to conducting the experiment we might have decided to compare the distributions of the daily number of empty beds for a specific pair of hospitals. The Wilcoxon rank sum test presented in Section 12.1 could be used for this purpose.

The conditions that must be met to conduct a valid Kruskal–Wallis H test are given in the box.

When to Use the Kruskal–Wallis H Test for Comparing k Distributions

To conduct a valid Kruskal–Wallis H test to compare k distributions, the following conditions must be met:

1. The k samples must be random and independent. (A completely randomized design must be used.)

2. There must be 5 or more measurements in each sample.

3. It must be possible to rank the observations.

[*Note:* The distributions of the populations may have any shapes.]

The general form of the Kruskal–Wallis H test is given in the next box.

Kruskal–Wallis H Test for Comparing k Distributions

H_0: The k distributions are identical

H_a: At least two of the k distributions differ in location

Test statistic:

$$H = \frac{12}{n(n+1)} \left[\frac{R_1^2}{n_1} + \cdots + \frac{R_k^2}{n_k} \right] - 3(n+1)$$

where

n_1 = Number of measurements in sample 1, etc.

R_1 = Rank sum for sample 1, etc., where the rank of each measurement is computed according to its relative magnitude in the totality of data for the k samples

n = Total sample size = $n_1 + n_2 + \cdots + n_k$

Rejection region: Reject H_0 if $H >$ Table value, where the table value is obtained from Appendix Table VIIA, using $(k-1)$ df

---

EXAMPLE 12.3   Suppose a dairy farmer wants to compare the amount of milk produced by dairy cattle fed on four different diets. Five cows are randomly assigned and maintained on each of the diets for 3 months. After the 3-month period, each cow's milk production is recorded for 1 week. The numbers given in Table 12.11 are the ranks of the actual productivity data. Do the data provide sufficient evidence to indicate that at least one of the diets tends to achieve greater milk production than the others? Test at the .10 level of significance.

Table 12.11

Ranks for milk production data

Diet 1	Diet 2	Diet 3	Diet 4
1	12	8	14
5	2	9	15
6	17	3	16
7	19	11	4
10	20	13	18
$R_1 = 29$	$R_2 = 70$	$R_3 = 44$	$R_4 = 67$

Solution   The elements of the test are as follows:

H_0: The population distributions of milk production for the four diets are identical

H_a: At least two of the diets have distributions with different locations

Test statistic:

$$H = \frac{12}{n(n+1)}\left[\frac{R_1^2}{n_1} + \cdots + \frac{R_4^2}{n_4}\right] - 3(n+1)$$

$$= \frac{12}{20(21)}\left[\frac{(29)^2}{5} + \frac{(70)^2}{5} + \frac{(44)^2}{5} + \frac{(67)^2}{5}\right] - 3(21)$$

$$= 69.52 - 63 = 6.52$$

Rejection region: Since we are comparing four distributions, there are $4 - 1 = 3$ df associated with the test statistic. Using $\alpha = .10$, we will reject H_0 if $H > 6.25$.

Since 6.52 is greater than 6.25, we have sufficient evidence at the .10 level of significance to conclude that one or more of the diets tends to achieve greater milk productivity than the others. ■

COMPUTER EXAMPLE 12.3   Minitab may be used to calculate the value of the test statistic when analyzing a completely randomized design with the Kruskal–Wallis *H* test. From the data in Table 12.11, which were used to calculate the value of *H* given in Example 12.3, the following Minitab printout was obtained:

```
LEVEL     NOBS    MEDIAN   AVE. RANK    Z VALUE
  1         5      6.000       5.8       -2.05
  2         5     17.000      14.0        1.53
  3         5      9.000       8.8       -0.74
  4         5     15.000      13.4        1.27
OVERALL    20                 10.5
H = 6.520
```

The "H = 6.520" is the value of the test statistic of interest. This is exactly the same value we obtained in Example 12.3. Since Minitab does not report a *p*-value for this test, the value of the test statistic must be compared to the table value given in the rejection region. As before, since 6.520 > 6.250, we conclude that at least one of the diets tends to achieve greater milk production than the rest. This conclusion was reached with $\alpha = .10$. ■

Exercises (12.15–12.21)

Learning the Mechanics

12.15 Independent random samples were selected from four populations, and the results are shown in the table at the top of page 662. Do the data provide sufficient evidence to indicate a difference in location between at least two of the four distributions? Test using the Kruskal–Wallis *H* test with $\alpha = .05$.

Sample 1	Sample 2	Sample 3	Sample 4
8.1	7.8	2.1	4.8
2.3	3.5	3.3	3.9
3.4	5.2	1.0	2.6
5.1	5.0	4.5	6.7
3.6	4.6	0.8	5.2
	7.7		

Applying the Concepts

12.16 Three lists of words, representing three levels of abstractness, are randomly assigned to twenty-one experimental subjects so that seven subjects receive each list. The subjects are asked to respond to each word on their list with as many associated words as possible within a given period. A subject's score is the total number of associates, summing over all words in the list. Scores for each list are given in the table. Do the data provide sufficient evidence to indicate differences among the three lists in the number of word associates that subjects can list? Test using $\alpha = .05$.

List 1	List 2	List 3
48	41	18
43	36	42
39	29	28
57	40	38
21	35	15
47	45	33
58	32	31

12.17 A large charitable fund-raising organization wants to appeal to civic pride to increase contributions during its annual drive. Contribution records for each of four cities are selected randomly. The amount of each contribution selected is given in the table. Is there sufficient evidence to indicate that the contributions in one or more of the cities tend to be larger than the contributions in the others? Use $\alpha = .10$.

Amount Contributed (in dollars)			
City 1	City 2	City 3	City 4
75	65	15	45
20	30	25	30
30	45	10	25
45	50	35	60
25	35	5	55
	70		

12.18 The Environmental Protection Agency wants to determine whether temperature changes in the ocean water near a nuclear power plant will have a significant effect on the animal life in the region. Recently hatched specimens of a certain species of fish are randomly divided into four groups. The groups are placed in separate simulated ocean environments that are identical in every way except for water temperature. Six months later, the specimens are weighed. The results (in ounces) are given in the table. Do the data provide sufficient evidence to indicate that one (or more) of the temperatures tend(s) to produce larger weight increases than the others? Test using $\alpha = .10$.

Weight of Specimen (in ounces)			
38°F	42°F	46°F	50°F
22	15	14	17
24	21	28	18
16	26	21	13
18	16	19	20
19	25	24	21
	17	23	

12.19 An experiment was conducted to compare how long it takes a human to recover from each of three types of influenza—Victoria A, Texas, and Russian. Twenty-one human subjects were selected at random from a group of volunteers and divided into three groups of seven each. Each group was randomly assigned a strain of the virus and the influenza was induced in the subjects. All the subjects were then cared for under identical conditions, and the recovery time (in days) was recorded. The results are given in the table.

Victoria A	Texas	Russian
12	9	7
6	10	3
13	5	7
10	4	5
8	9	6
11	8	4
7	11	8

 a. Do the data provide sufficient evidence to indicate that the recovery times for one (or more) type(s) of influenza tend(s) to be longer than for the other types? Test using $\alpha = .05$.

 b. Do the data provide sufficient evidence to indicate a difference in locations of the distributions of recovery times for the Victoria A and Russian types? Test using $\alpha = .05$.

12.20 Three brands of magnetron tubes (the key components in microwave ovens) were subjected to stressful testing, and the number of hours each operated without repair was recorded. Although these times do not represent typical lifetimes, they do indicate how well the tubes can withstand extreme stress. The data are given in the table.

Brand		
A	B	C
36	49	71
48	33	31
5	60	140
67	2	59
53	55	42

a. Use the F test for a completely randomized design (Chapter 10) to test the hypothesis that the mean lifetime under stress is the same for the three brands. Use $\alpha = .05$. What assumptions are necessary for the validity of this procedure? Is there any reason to doubt these assumptions?

b. Use the Kruskal–Wallis H test to determine whether evidence exists to conclude that the brands of magnetron tubes tend to differ in lifetime under stress. Test using $\alpha = .05$.

*12.21 An experiment was conducted to determine whether a test designed to identify a particular form of mental illness could be easily interpreted with little psychological training. Thirty judges were selected to review the results of 100 tests, half of which were given to disturbed patients and half to normal people. Of the thirty judges chosen, ten were staff members of a mental hospital, ten were trainees at the hospital, and ten were undergraduate psychology majors. The results shown in the table give the percentages of the 100 tests correctly classified by each judge. Do the data provide sufficient evidence to indicate that the three types of judges differ in their ability to identify the mentally disturbed patients? Test using $\alpha = .05$. Use a computer package to find the value of the test statistic.

Staff		Trainees		Undergraduates	
78	76	80	69	65	74
79	86	75	81	70	80
85	88	72	76	74	73
93	84	68	72	78	75
90	81	75	76	68	73

12.4

SPEARMAN'S RANK CORRELATION COEFFICIENT

In Section 9.7 we introduced the notion of measuring the correlation between two variables. We will now present a method of measuring the **correlation between two sets of ranked data.**

Suppose ten new paintings are shown to two art critics and each critic ranks them from 1 (best) to 10 (worst). We want to determine whether the critics' ranks are related. Does a correspondence exist between their ratings? If a painting is rated high by critic 1, is it likely to be rated high by critic 2? Or do high rankings by one critic correspond to low rankings by the other? That is, we want to determine whether the rankings of the critics are **correlated.**

If the rankings are as shown in the "Perfect Agreement" columns of Table 12.12, we immediately notice that the critics agree on the rank of every painting. High ranks correspond to high ranks and low ranks to low ranks. This is an example of **perfect positive correlation** between the ranks.

Table 12.12

Rankings of ten paintings by two critics

Painting	Perfect Agreement		Perfect Disagreement	
	Critic 1	Critic 2	Critic 1	Critic 2
1	4	4	9	2
2	1	1	3	8
3	7	7	5	6
4	5	5	1	10
5	2	2	2	9
6	6	6	10	1
7	8	8	6	5
8	3	3	4	7
9	10	10	8	3
10	9	9	7	4

In contrast, if the rankings appear as shown in the "Perfect Disagreement" columns of Table 12.12, high ranks for one critic correspond to low ranks for the other. This is an example of **perfect negative correlation.**

In practice, you will rarely see perfect positive or negative correlation between the ranks. In fact, it is more likely that the critics' ranks will appear as shown in Table 12.13. Note that these rankings indicate some agreement between the critics but not perfect agreement. Thus, we need a measure of **rank correlation.**

Table 12.13

Rankings of paintings: Less than perfect agreement

Painting	Critic 1	Critic 2	Painting	Critic 1	Critic 2
1	4	5	6	10	9
2	1	2	7	7	7
3	9	10	8	3	3
4	5	6	9	6	4
5	2	1	10	8	8

Spearman's rank correlation, r_s, provides a measure of correlation between ranks. The formula we will use to calculate this measure of correlation is given in the box on page 666.

Spearman's Rank Correlation Coefficient

To calculate r_s, Spearman's rank correlation coefficient, use the following formula:

$$r_s = 1 - \frac{6 \sum d^2}{n(n^2 - 1)}$$

where $\sum d^2$ is the sum of the squared differences between the pairs of ranks and n is the number of pairs of ranks.

Note that if the ranks for the two critics are identical, as in the second and third columns of Table 12.12, the differences between the ranks, d, will all be 0. Thus,

$$r_s = 1 - \frac{6 \sum d^2}{n(n^2 - 1)} = 1 - \frac{6(0)}{10(99)} = 1$$

That is, **perfect positive correlation** between the pairs of ranks is characterized by a Spearman correlation coefficient of $r_s = 1$. When the ranks indicate perfect disagreement, as in the fourth and fifth columns of Table 12.12, we calculate

$$\sum d^2 = 330$$

and

$$r_s = 1 - \frac{6(330)}{10(99)} = 1 - \frac{1,980}{990} = -1$$

Thus, **perfect negative correlation** is indicated by $r_s = -1$.

The data with less than perfect agreement, which were originally given in Table 12.13, are repeated in Table 12.14. We have calculated d, the differences between

Table 12.14

Rankings of paintings:
Less than perfect
agreement

Painting	Critic 1	Critic 2	Difference Between Rank 1 and Rank 2 d	d^2
1	4	5	−1	1
2	1	2	−1	1
3	9	10	−1	1
4	5	6	−1	1
5	2	1	1	1
6	10	9	1	1
7	7	7	0	0
8	3	3	0	0
9	6	4	2	4
10	8	8	0	0
				$\sum d^2 = \overline{10}$

rank 1 and rank 2, and d^2, the squares of these differences. The sum of the squared differences is $\sum d^2 = 10$. Thus, we obtain

$$r_s = 1 - \frac{6 \sum d^2}{n(n^2 - 1)} = 1 - \frac{6(10)}{10(99)} = 1 - \frac{6}{99} = .94$$

The fact that r_s is *close* to 1 indicates that the critics tend to agree, but the agreement is not perfect.

The value of r_s will always fall between -1 and $+1$, with $+1$ indicating perfect positive correlation and -1 indicating perfect negative correlation. The closer r_s falls to $+1$ or -1, the greater the correlation between the ranks. Conversely, the nearer r_s is to 0, the less will be the correlation. We summarize the properties of r_s in the box.

Properties of Spearman's Rank Correlation Coefficient

1. The value of r_s is always between -1 and 1.
2. r_s *positive:* The ranks of the pairs of sample observations tend to increase together.
3. $r_s = 0$: The ranks are not correlated.
4. r_s *negative:* The ranks of one variable tend to decrease as the ranks of the other variable increase.

You will note that the concept of correlation implies that two responses are obtained for each experimental unit. In the art critics example, each painting received two ranks (one for each critic) and the objective of the study was to determine the degree of positive correlation between the two rankings. Rank correlation methods can be used to measure the correlation between any pair of variables. If two variables are measured on each of n experimental units, we rank the measurements associated with each variable separately. *Ties receive the average of the ranks of the tied observations.*[*] Then we calculate the value of r_s for the two rankings. This value will measure the rank correlation between the two variables. We will illustrate the procedure with Example 12.4.

EXAMPLE 12.4   A study is conducted to investigate the relationship between cigarette smoking during pregnancy and the weights of newborn infants. A sample of fifteen women smokers kept accurate records of the number of cigarettes smoked during their pregnancies, and the weights of their children were recorded at birth. The data and preliminary calculations are given in Table 12.15 (page 668). Calculate and interpret Spearman's rank correlation coefficient for the data.

[*] If there are a large number of ties in the ranks for either variable, the formula for r_s that we have been using is *not* appropriate. In this text we will consider only examples with few, or no, ties in the ranks, so that the given formula may be used. If you need to calculate r_s for a problem in which there are many ties in the ranks, refer to Chapter 10 of McClave and Dietrich (1988).

Table 12.15

Data and calculations for
Example 12.4

Woman	Number of Cigarettes per Day	Rank	Baby's Weight Pounds	Rank	d	d^2
1	12	1	7.7	5	-4	16
2	15	2	8.1	9	-7	49
3	35	13	6.9	4	9	81
4	21	7	8.2	10	-3	9
5	20	5.5	8.6	13.5	-8	64
6	17	3	8.3	11.5	-8.5	72.25
7	19	4	9.4	15	-11	121
8	46	15	7.8	6	9	81
9	20	5.5	8.3	11.5	-6	36
10	25	8.5	5.2	1	7.5	56.25
11	39	14	6.4	3	11	121
12	25	8.5	7.9	7	1.5	2.25
13	30	12	8.0	8	4	16
14	27	10	6.1	2	8	64
15	29	11	8.6	13.5	-2.5	6.25
					Total $=$	795

Solution We first ranked the number of cigarettes smoked per day, assigning a 1 to the smallest (12) and a 15 to the largest number (46). Note that the two ties received the averages of their respective ranks. Similarly, we assigned ranks to the fifteen babies' weights. The differences between the ranks of the babies' weights and the ranks of the number of cigarettes smoked per day are shown in the column labeled d in Table 12.15. The squares of the differences are also given. Thus,

$$r_s = 1 - \frac{6 \sum d^2}{n(n^2 - 1)} = 1 - \frac{6(795)}{15(15^2 - 1)} = 1 - 1.42 = -.42$$

This negative correlation coefficient indicates that in this sample, an increase in the number of cigarettes smoked per day is associated with—but is not necessarily the cause of—a decrease in the weight of the newborn infant. ∎

Can the relationship seen in Example 12.4 be generalized from the sample to the population? That is, can we conclude that weights of newborns and the number of cigarettes smoked per day are negatively correlated for the populations of observations for all smoking mothers?

If we define ρ_s as the **Spearman rank correlation coefficient of the population** (that is, the rank correlation coefficient corresponding to the pairs of ranks for number of cigarettes smoked per day by the mother during pregnancy and baby's weight for all births), the question can be answered by conducting the following test:

H_0: $\rho_s = 0$ (no population correlation between ranks)

H_a: $\rho_s < 0$ (negative population correlation between ranks)

Test statistic: r_s, the sample Spearman rank correlation coefficient

To determine a rejection region, we consult Appendix Table XV, which is partially reproduced in Table 12.16. Note that the left-hand column gives values of n, the number of pairs of observations. The entries in the table are values for an upper-tail rejection region, since only positive values are given. Thus, for $n = 15$ and $\alpha = .05$, the value .441 is the boundary of the upper-tail rejection region. Similarly, we expect to see $r_s < -.441$ only 5% of the time if there is really no relationship between the ranks of the variables. The lower-tailed rejection region for $\alpha = .05$ is therefore:

Rejection region: Reject H_0 if $r_s < -.441$.

Table 12.16

A partial reproduction of
Appendix Table XV, Critical
Values of Spearman's Rank
Correlation Coefficient

n	$\alpha = .05$	$\alpha = .025$	$\alpha = .01$	$\alpha = .005$
5	.900	—	—	—
6	.829	.886	.943	—
7	.714	.786	.893	—
8	.643	.738	.833	.881
9	.600	.683	.783	.833
10	.564	.648	.745	.794
11	.523	.623	.736	.818
12	.497	.591	.703	.780
13	.475	.566	.673	.745
14	.457	.545	.646	.716
15	.441	.525	.623	.689

Since the calculated $r_s = -.42$ is not less than $-.441$, we cannot reject H_0 at the $\alpha = .05$ level of significance. That is, this sample of fifteen smoking mothers provides insufficient evidence to conclude that a negative correlation exists between number of cigarettes smoked and the weight of newborns for the populations of measurements corresponding to all smoking mothers. Of course, this does not mean that no relationship exists. A study using a larger sample of smokers and taking other factors into account (father's weight, sex of infant, etc.) would be more likely to indicate whether smoking and the weight of a newborn child are related.

The nonparametric test for rank correlation is valid when the conditions given in the box are met.

When to Use Spearman's Nonparametric Test for Rank Correlation

To perform a valid test of hypothesis about ρ_s, Spearman's rank correlation coefficient for a population, the following conditions must be met:

1. A random sample of pairs of observations must be selected for the two variables of interest.
2. It must be possible to rank the observations for each of the variables.

A summary of Spearman's nonparametric test for correlation is given in the next box.

Spearman's Nonparametric Test for Rank Correlation

<table>
<tr><td>One-Tailed Test</td><td>Two-Tailed Test</td></tr>
<tr><td>

H_0: $\rho_S = 0$

H_a: $\rho_S > 0$
  (or H_a: $\rho_S < 0$)

Test statistic: r_S, the sample rank correlation

Rejection region:
Reject H_0 if $r_S >$ Table value (or $r_S < -$ Table value with H_a: $\rho_S < 0$), where the table value is found in Appendix Table XV, using the column corresponding to the value of α and the row corresponding to n, the number of pairs of observations.

</td><td>

H_0: $\rho_S = 0$

H_a: $\rho_S \neq 0$

Test statistic: r_S, the sample rank correlation

Rejection region:
Reject H_0 if $r_S >$ Table value or $r_S < -$ Table value, where the table value is found in Appendix Table XV, using the column corresponding to the value of $\alpha/2$ and the row corresponding to n, the number of pairs of observations.

</td></tr>
</table>

EXAMPLE 12.5   Manufacturers of perishable foods often use preservatives to retard spoilage, but using too much preservative will change the flavor of the food. Suppose an experiment is conducted using samples of a food product with varying amounts of preservative added. The length of time (in days) until the food shows signs of spoiling and a taste

Table 12.17

Data for Example 12.5

Sample	Time Until Spoilage	Rank	Taste Rating	Rank
1	30	2	4.3	11
2	47	5	3.3	7.5
3	26	1	4.7	12
4	94	11	2.0	3
5	67	7	3.0	6
6	83	10	1.7	2
7	36	3	4.0	10
8	77	9	3.7	9
9	43	4	3.3	7.5
10	109	12	1.3	1
11	56	6	2.7	5
12	70	8	2.3	4

rating are recorded for each sample. The taste rating is the average rating for three tasters, each of whom rates each sample on a scale from 1 (good) to 5 (bad). Twelve sample measurements are shown in Table 12.17. Use a nonparametric test to find out whether the spoilage times and taste ratings are correlated. Use $\alpha = .05$. [*Note:* Tied measurements are assigned the average of the ranks that would be given the measurements if they were different but consecutive.]

Solution The test is two-tailed, with

$$H_0: \quad \rho_s = 0$$
$$H_a: \quad \rho_s \neq 0$$

Test statistic: $\quad r_s = 1 - \dfrac{6 \sum d^2}{n(n^2 - 1)}$

Rejection region: Since the test is two-tailed, we need to halve the α value before consulting Table XV. For $\alpha = .05$, we calculate $\alpha/2 = .025$ and look up the table value corresponding to $n = 12$ pairs of observations. Since this value is .591, we will reject H_0 if $r_s < -.591$ or $r_s > .591$.

The first step in the computation of r_s is to sum the squares of the differences between ranks:

$$\sum d^2 = (2 - 11)^2 + (5 - 7.5)^2 + \cdots + (8 - 4)^2 = 536.5$$

Then

$$r_s = 1 - \frac{6(536.5)}{12(144 - 1)} = -.876$$

Since $-.876 < -.591$, we reject H_0 and conclude that the amount of preservative is associated with the taste ratings of the food. The fact that r_s is negative suggests that the preservative has an adverse effect on the taste. ∎

EXAMPLE 12.6 In the introduction to this chapter, we discussed a study by Villemez and Silver (1976) that investigated whether changes in the descriptions of occupations affect how students view the occupations. Table 12.1 is repeated in Table 12.18 (page 672), which shows ten job categories described in three ways, with a ranking of these categories obtained from a random sample of 144 college students.*

Villemez and Silver (1976) reported Spearman's rank correlation coefficients between the different sets of occupational descriptions. For description 1 versus description 2, $r_s = .59$; for description 1 versus description 3, $r_s = .49$; and for description 2

---

* In this study the *smallest rank* refers to the occupation rated *highest*. Although this method of assigning ranks is opposite to the method we have been using, the numerical value of r_s, the rank correlation, is the same for both methods.

Table 12.18

Occupational descriptions and ranks

Occupational Category	Description 1	Rank	Description 2	Rank	Description 3	Rank
1	Building and maintenance	9	Constructing and maintaining	7.5	Construction	6.5
2	Arts and entertainment	3	Culture and information	9	Aesthetics and information	3
3	Transportation	8	Movement of goods and people	7.5	Movement of materials and people	8
4	Extraction	10	Removing materials from the earth	10	Farming and mining	9
5	Health and welfare	6.5	Citizen well-being	5.5	Social, physical, and moral well-being	1
6	Commerce	4	Business	2	Trade	5
7	Legal authority	1	Law and control	5.5	Law enforcement	6.5
8	Finance and records	5	Monetary affairs	3	Records and accounts	4
9	Manufacturing	6.5	Production	4	Production and assembly	10
10	Education and research	2	Knowledge	1	Training and education	2

versus description 3, $r_s = .33$. Villemez and Silver state:

> These moderate (and non-significant) correlations resulting from using different words suggest at least that situs [category] ranking is not entirely consistent . . . . That is, when rating broad categories of type of work, raters probably do so with sets of specific familiar occupations in mind, which biases their ratings. The changing titles in this study may have brought to mind different occupational sets [Villemez & Silver, 1976, pp. 327–328]. ∎

Exercises (12.22–12.29)

Using the Tables

12.22 *a.* Suppose you want to test H_a: $\rho_s > 0$ and you have $n = 14$ and $\alpha = .01$. Give the rejection region for the test.

b. Suppose you want to test H_a: $\rho_s \neq 0$ and you have $n = 22$ and $\alpha = .05$. Give the rejection region for the test.

Learning the Mechanics

12.23 A random sample of seven pairs of observations are recorded on two variables, X and Y. The data are shown in the table. Use Spearman's nonparametric test for rank correlation to answer the questions.

Pair	Value of X	Value of Y
1	65	58
2	57	61
3	55	58
4	38	23
5	29	34
6	43	38
7	49	37

a. Do the data provide sufficient evidence to conclude that ρ_s, the rank correlation between X and Y, is greater than 0? Test using $\alpha = .05$.

b. Do the data provide sufficient evidence to conclude that $\rho_s \neq 0$? Test using $\alpha = .05$.

Applying the Concepts

12.24 Two expert wine tasters were asked to rank six brands of wine. Their rankings are shown in the table. Do the data present sufficient evidence to indicate a positive correlation in the rankings of the two experts? Use $\alpha = .01$.

Brand	Expert 1	Expert 2
A	6	5
B	5	6
C	1	2
D	3	1
E	2	4
F	4	3

12.25 An experiment was designed to study whether the eye pupil size is related to lying. Eight students were asked to lie in answering some in a series of questions (the number to be answered dishonestly was left to the individual). Pupil size was measured before the questioning, and during questioning the percentage increase in dilation was recorded. Each student was given a deception score (a higher score indicates a larger number of deceptive replies). The results are given in the table at the top of page 674. Can you conclude that the percentage increase in eye pupil size is positively correlated with deception score? Use $\alpha = .05$.

Student	Deception Score	Percentage Increase in Eye Pupil Size	Student	Deception Score	Percentage Increase in Eye Pupil Size
1	87	10	5	43	0
2	63	6	6	89	15
3	95	11	7	33	4
4	50	7	8	55	5

12.26 It is often conjectured that income is one of the primary determinants of social status for an adult male. To investigate this theory, fifteen adult males are chosen at random from a community inhabited primarily by professional people, and their annual gross incomes are noted. Each subject is then asked to complete a questionnaire designed to measure social status within the community. The social status scores (higher scores correspond to higher social status) and gross incomes (in thousands of dollars) for the fifteen adult males are given in the table.

Subject	Social Status	Income	Subject	Social Status	Income
1	92	29.9	9	45	16.0
2	51	18.7	10	72	25.0
3	88	32.0	11	53	17.2
4	65	15.0	12	43	9.7
5	80	26.0	13	87	20.1
6	31	9.0	14	30	15.5
7	38	11.3	15	74	16.5
8	75	22.1			

a. Compute Spearman's rank correlation coefficient for the data.
b. Is there evidence that social status and income are positively correlated? Use $\alpha = .05$.

12.27 Many large businesses send representatives to college campuses to conduct job interviews. To aid the recruiter, one company decides to study the correlation between the strength of an applicant's references (the company requires three references) and the applicant's on-the-job performance. Eight recently hired employees are sampled, and independent evaluations of references and of job performance are made on a scale from 1 to 20. The scores appear in the table.

Employee	References	Job Performance	Employee	References	Job Performance
1	18	20	5	16	14
2	14	13	6	11	18
3	19	16	7	20	15
4	13	9	8	9	12

a. Compute Spearman's rank correlation coefficient for the data.

b. Is there evidence that strength of references and job performance are positively correlated? Use $\alpha = .05$.

12.28 Recreation therapy is a current treatment for the mentally retarded. Psychologists theorize that some types of recreation soothe highly excitable patients. In one experiment, nine mentally retarded patients were monitored for 1 week. The patients were allowed to spend as much time as they wanted in a special recreation room, and the total number of hours was recorded for each. The number of times during the week that highly excitable patients had to be given tranquilizing medication was also recorded. From the data in the table, determine whether there is sufficient evidence to indicate that recreation time is negatively correlated with the number of times a tranquilizer must be given. Test using $\alpha = .05$.

Patient	Recreation Time	Tranquilizers	Patient	Recreation Time	Tranquilizers
1	16	3	6	34	2
2	22	1	7	26	0
3	10	4	8	13	10
4	8	9	9	5	9
5	14	5			

12.29 A large manufacturing firm wants to determine whether a relationship exists between the number of work-hours an employee misses per year and the employee's annual wages (in thousands of dollars). A sample of fifteen employees produced the data in the table. Do the data provide evidence that the work-hours missed are related to annual wages? Use $\alpha = .05$.

Employee	Work-Hours Missed	Annual Wages	Employee	Work-Hours Missed	Annual Wages
1	49	15.8	9	191	10.8
2	36	17.5	10	6	18.8
3	127	11.3	11	63	13.8
4	91	13.2	12	79	12.7
5	72	13.0	13	43	15.1
6	34	14.5	14	57	24.4
7	155	11.8	15	82	13.9
8	11	20.2			

Chapter Summary

Nonparametric tests are tests that compare the distributions of sampled populations rather than specific parameters. They are used when the distributions are not normal so that t and F tests do not apply. Four nonparametric tests were considered in this chapter, all involving rankings of measurements.

I. Wilcoxon Rank Sum Test

Situation: Two independent random samples, one from each of two populations; the observations are ranked in order of magnitude and the average rank is used for any ties that occur

Test of hypothesis:

H_0: Two populations have identical distributions

H_a: (1) Population 1 is shifted to the right of population 2
or   (2) Population 1 is shifted to the left of population 2
or   (3) Population 1 is shifted to the right or left of population 2

Test statistic: T = Sum of ranks associated with the sample with fewer measurements

Rejection region: Use Appendix Table XIII: (1) One-tailed test, $T \geq T_U$; (2) One-tailed test, $T \leq T_L$; (3) Two-tailed test, $T \leq T_L$ or $T \geq T_U$.

II. Wilcoxon Signed Rank Test

Situation: Paired difference experiment; the absolute values of the paired differences are ranked in order of magnitude and 0 differences are deleted; the average rank is used for any ties that occur

Test of hypothesis:

H_0: Two populations have identical distributions

H_a: (1) Population A is shifted to the right of population B
or   (2) Population A is shifted to the left of population B
or   (3) Population A is shifted to the right or left of population B

Test statistic: (1) T_- = Sum of ranks for observations that are negative when paired measurement B is subtracted from corresponding measurement A. (2) T_+ = Sum of ranks for observations that are positive when paired measurement B is subtracted from corresponding measurement A. (3) T = Smaller of T_- and T_+.

Rejection region: Use Appendix Table XIV to obtain table value: (1) One-tailed test, $T_- \leq$ Table value; (2) One-tailed test, $T_+ \leq$ Table value; (3) Two-tailed test, $T \leq$ Table value.

III. Kruskal–Wallis *H* Test

Situation: Completely randomized design on k populations with each sample size at least 5; the observations are ranked in order of magnitude and the average rank is used for any ties that occur

Test of hypothesis:

H_0: The k distributions are identical

H_a: At least two of the k distributions differ in location

Test statistic: $H = \dfrac{12}{n(n+1)} \left[\sum \dfrac{R_j^2}{n_j} \right] - 3(n+1)$

where R_j is the rank sum for sample j, n_j is the number of measurements for sample j, and $n = \sum n_j =$ Total sample size. The distribution of H is approximately the χ^2 distribution.

Rejection region: Use Appendix Table VIIA ($>$), with $(k-1)$ df; reject H_0 if H exceeds the table value.

IV. Spearman's Rank Correlation Coefficient

Situation: n paired rankings; any ties that occur receive the average rank

$$r_s = 1 - \frac{6 \sum d^2}{n(n^2 - 1)}$$

where d is the difference between paired ranks; the value of r_s indicates the strength of the relationship, if any, between the two rankings in the sample of n:

$r_s = 1$:   Perfect agreement

r_s *positive:*   Ranks tend to increase together

$r_s = 0$:   No correlation

r_s *negative:*   One rank tends to increase as the other decreases

$r_s = -1$:   Perfect disagreement

Test of hypothesis:

H_0:   Population correlation coefficient $\rho_s = 0$

H_a: (1) $\rho_s > 0$; (2) $\rho_s < 0$; (3) $\rho_s \neq 0$

Test statistic: r_s

Rejection region: Use Appendix Table XV to obtain a table value: (1) Value corresponds to α, reject H_0 if $r_s >$ Table value; (2) Value corresponds to α, reject H_0 if $r_s < -$Table value; (3) Value corresponds to $\alpha/2$, reject H_0 if $r_s < -$Table value or $r_s >$ Table value.

For all of the tests I–IV, the number of ties in rankings must be small.

SUPPLEMENTARY EXERCISES (12.30–12.60)

Learning the Mechanics

12.30 The data for three independent random samples are shown in the table at the top of page 678. Assume that we know that the sampled populations are *not* normally distributed. Then use an appropriate test to determine whether the data provide sufficient evidence to indicate that at least two of the populations differ in location. Test using $\alpha = .05$.

Sample from Population 1	Sample from Population 2	Sample from Population 3
18	12	87
32	33	53
43	10	65
15	34	50
63	18	64
		77

12.31   A random sample of nine pairs of observations are recorded on two variables, X and Y. The data are shown in the table.

Pair	Value of X	Value of Y	Pair	Value of X	Value of Y
1	19	12	6	29	10
2	27	19	7	16	16
3	15	7	8	22	10
4	35	25	9	16	18
5	13	11			

a. Do the data provide sufficient evidence to indicate that ρ_s, the rank correlation between X and Y, differs from 0? Test using $\alpha = .05$.

b. Do the data provide sufficient evidence to indicate that the distribution for X is shifted to the right of that for Y? Test using $\alpha = .05$.

12.32   Two independent random samples produced the measurements shown in the table. Do the data provide sufficient evidence to conclude that there is a difference in location in the distributions for the sampled populations? Test using $\alpha = .05$.

Sample from Population 1	Sample from Population 2
1.2	1.5
1.9	1.3
0.7	2.9
2.5	1.9
1.0	2.7
1.8	3.5
1.1	

Applying the Concepts

12.33   An experiment was conducted to compare two print types, A and B, to determine whether type A is easier to read than type B. Ten subjects were randomly divided into two groups of five. Each subject received the same material to read, but one group's material was printed in type A and the other's in type B. The times necessary for each subject to read the material (in seconds) are given below:

Type A: 95, 122, 101, 99, 108
Type B: 110, 102, 115, 112, 120

Do the data provide sufficient evidence to indicate that print type A is easier to read than type B? Test using $\alpha = .05$.

12.34 A study was conducted to investigate whether the installation of a traffic light reduced the number of accidents at a busy intersection. The number of accidents per month was sampled for 6 months prior to the installation and 5 months afterward; the data appear in the table. Is there sufficient evidence to conclude that the distribution of the number of monthly accidents prior to installation of the traffic light is shifted to the right of the distribution of the number of monthly accidents after installation of the light? Test using $\alpha = .025$.

Before	After
12	4
5	2
10	7
9	3
14	8
6	

12.35 The time required for humans to respond to a new pain killer was tested in the following manner: Seven randomly chosen subjects received both aspirin and the new drug, with the treatments spaced over time and given in random order. The length of time (in minutes) before the subject felt physical pain relief was recorded for both the aspirin and the new drug. The data appear in the table. Do the data provide sufficient evidence that the distribution of times to relief with aspirin is shifted to the right of the distribution of times to relief with the new drug? Test using $\alpha = .05$.

Subject	Aspirin	New Drug
1	15	7
2	20	14
3	12	13
4	20	11
5	17	10
6	14	16
7	17	11

12.36 A new diet is supposed to be effective in reducing weight. To investigate whether the diet will produce a weight loss within a week, ten people were randomly chosen and put on the diet. Their weights (in pounds) were recorded at the start and end of the week, as listed in the table. Do the data provide sufficient evidence that the distribution of weights before the diet is shifted to the right of the distribution of weights after 1 week on the diet? Test using $\alpha = .05$.

Subject	Before	After	Subject	Before	After
1	115	112	6	166	166
2	123	124	7	185	180
3	155	153	8	172	172
4	220	219	9	245	241
5	215	216	10	184	182

12.37 A manufacturer of household appliances is considering one of two chains of department stores to sell its product in a region of the United States. The manufacturer first wants to compare the product exposure that might be expected from the two chains. Eight locations are selected where both chains have stores and, on a certain day, the shoppers entering each store are counted. The table shows the data. Do the data provide sufficient evidence to indicate that one chain tends to have more customers per day than the other? Test using $\alpha = .05$.

Location	Store A	Store B	Location	Store A	Store B
1	879	1,085	5	2,326	2,778
2	445	325	6	857	992
3	692	848	7	1,250	1,303
4	1,565	1,421	8	773	1,215

12.38 The director of a state highway patrol wondered whether frequent patrolling of highways substantially reduces the number of speeders. Two similar interstate highways were selected for a study, and one was very heavily patrolled while the other was patrolled only occasionally. After 1 month, random samples of 100 cars were chosen on each highway and the number exceeding the speed limit was recorded. This process was repeated on five randomly selected days. The data appear in the table.

Day	Highway 1 Patrolled often	Highway 2 Patrolled seldom
1	35	60
2	40	36
3	25	48
4	38	54
5	47	63

a. Do the data provide evidence to indicate that the heavily patrolled highway tends to have fewer cars (per 100) exceeding the speed limit than the occasionally patrolled highway? Test using $\alpha = .05$.

b. Use the paired t test with $\alpha = .05$ to compare the population mean number of speeders per 100 cars for the two highways. What conditions must be met for this procedure to be valid?

12.39 A drug company has synthesized two new compounds to be used in sleeping pills. The data in the table represent the additional hours of sleep gained by ten patients through the use of the two drugs. Do the data present sufficient evidence to indicate that the distributions of additional hours of sleep differ for the two drugs? Test using $\alpha = .10$.

Patient	Drug A	Drug B	Patient	Drug A	Drug B
1	0.4	0.7	6	2.9	3.4
2	−0.7	−1.6	7	4.0	3.7
3	−0.4	−0.2	8	0.1	0.8
4	−1.4	−1.4	9	3.1	0.0
5	−1.6	−0.2	10	1.9	2.0

12.40 An insurance company wants to determine whether a relationship exists between the number of claims filed by owners of family policies and the annual incomes of the families. A random sample of ten policies is selected, with the results listed in the table. Is there evidence of a relationship between the number of claims filed and the incomes of the family policyholders? Use $\alpha = .10$.

Family	Number of Claims 3-year period	Annual Income Thousands of dollars, averaged over 3 years
1	5	16.5
2	1	12.6
3	9	62.5
4	0	25.6
5	4	15.3
6	7	17.6
7	0	9.3
8	2	21.6
9	6	20.1
10	3	14.5

12.41 The trend among doctors in some parts of the country is to form a group practice. By combining treatment resources, doctors hope to be more efficient. One doctor who recently joined a group family practice wants to compare the distribution of the number of patients treated daily when he was an individual practitioner with the distribution since joining the group practice. Records were checked for 8 days before and 8 days after, with the results shown in the table. Use the Wilcoxon rank sum test to determine whether these samples indicate that the doctor now tends to treat more patients daily than before. Use $\alpha = .05$.

Number of Patients Treated per Day

Before		After	
26	24	28	29
25	20	30	23
27	22	27	28
26	21	31	29

12.42 Each applicant to a certain university is judged by his or her high school grade-point average and a score on a standard aptitude test. The GPAs and aptitude scores for eight randomly selected applicants are shown in the table. Do the data provide sufficient evidence to indicate a positive correlation between high school GPA and the aptitude test score? Use $\alpha = .025$.

Applicant	GPA	Aptitude Test Score	Applicant	GPA	Aptitude Test Score
1	3.25	1,200	5	3.10	1,510
2	2.85	890	6	2.90	950
3	3.01	980	7	2.75	1,010
4	4.00	1,150	8	3.35	1,080

12.43 The coach of a mediocre basketball team has one all-star player who attempts the majority of the team's shots. For each of the last ten games, the star's number of shots and the team's winning margin (negative number implies a loss) are recorded:

Game:	1	2	3	4	5	6	7	8	9	10
Star's shots:	19	15	21	17	25	22	14	11	18	18
Winning margin:	−3	13	−8	−5	−2	16	4	1	−5	−12

Compute Spearman's rank correlation coefficient. Is there evidence of a relationship between the team's performance and the star's number of shots? Use $\alpha = .05$.

12.44 In recent years, many magazines have been forced to raise their prices because of increased postage, printing, and paper costs. These higher prices may mean that some households now subscribe to fewer magazines than they did 3 years ago. Ten households were selected at random, and the number of magazines subscribed to 3 years ago and now was determined. The results are given in the table. Does this sample provide sufficient evidence that households tend to subscribe to fewer magazines now than they did 3 years ago? Use $\alpha = .05$.

Household	Number of Magazine Subscriptions		Household	Number of Magazine Subscriptions	
	3 Years Ago	Now		3 Years Ago	Now
1	8	4	6	6	5
2	3	5	7	4	3
3	6	4	8	2	2
4	3	3	9	9	6
5	10	5	10	8	2

12.45 A hotel had a problem with people reserving rooms for a weekend and then not honoring their reservations (no-shows). As a result, the hotel developed a new plan requiring deposits, in the hope of reducing the number of no-shows. One year after initiating the new policy, the management evaluated its effectiveness. Compare the records given in the table for ten randomly selected nonholiday weekends prior to instituting the new policy and ten randomly selected nonholiday weekends during the year of evaluation. Has the situation improved under the new policy? Test at $\alpha = .05$.

Number of No-Shows			
Before		After	
10	11	4	4
5	8	3	2
3	9	8	5
6	6	5	7
7	5	6	1

12.46 An economist is interested in knowing whether property tax rates differ among three types of school districts—urban, suburban, and rural. A random sample of several districts of each type produced the data in the table (rate is in mills, where 1 mill = $1/1,000). Do the data indicate a difference in the level of property taxes among the three types of school districts? Use $\alpha = .05$.

Urban	Suburban	Rural
4.3	5.9	5.1
5.2	6.7	4.8
6.2	7.6	3.9
5.6	4.9	6.2
5.1	6.3	4.5

12.47 Suppose a company wants to study how personality relates to leadership. Four supervisors with different types of personalities are selected. Several employees are then selected from the group supervised by each, and these employees are asked to rate the leader of their group on a scale from 1 to 20 (20 signifies highly favorable). The resulting data are shown in the table at the top of page 684.

Supervisor			
I	II	III	IV
20	17	16	8
19	11	15	12
20	13	13	10
18	15	18	14
17	14	11	9
	16		10

a. Is there sufficient evidence to indicate that one or more of the supervisors tends to receive higher ratings than the others? Use $\alpha = .05$.

b. Suppose the company is particularly interested in comparing the ratings of the personality types represented by supervisors I and III. Make this comparison using $\alpha = .05$.

12.48 Two fluoride toothpastes and one nonfluoride toothpaste were compared for their effectiveness in preventing cavities. Three randomly selected groups of subjects used the toothpastes for 6 months, and each subject was examined before and after the study to determine how many cavities developed. The data appear in the table. Do the data provide sufficient evidence to indicate that the distributions of the number of new cavities differ in location for at least two of the toothpastes? Use $\alpha = .05$.

Fluoride A		Fluoride B		Nonfluoride	
0	0	2	1	4	3
1	1	0	2	3	4
3	3	3	1	5	5
1	2	3		4	
2	2	0		4	

12.49 Weevils cause millions of dollars worth of damage each year to cotton crops. Three chemicals designed to control weevil populations were applied, one to each of three fields of cotton. After 3 months, ten plots of equal size were randomly selected within each field, and the percentage of cotton plants with weevil damage was recorded for each. Do the data in the table provide sufficient evidence to indicate a difference in location among the distributions of damage rates corresponding to the three treatments? Use $\alpha = .05$.

Chemical A		Chemical B		Chemical C	
10.8	9.8	22.3	20.4	9.8	10.8
15.6	16.7	19.5	23.6	12.3	12.2
19.2	19.0	18.6	21.2	16.2	17.3
17.9	20.3	24.3	19.8	14.1	15.1
18.3	19.4	19.9	22.6	15.3	11.3

12.50 A manufacturer wants to know whether the number of defective items produced by its employees tends to rise as the day progresses. Unknown to the employees, a complete inspection is made of every item produced on a certain day, and the hourly fraction of defectives is recorded. The resulting data are given in the table. Do they provide evidence that the fraction of defectives increases as the day progresses? Test at the $\alpha = .05$ level.

Hour	Fraction Defective	Hour	Fraction Defective
1	.02	5	.06
2	.05	6	.09
3	.03	7	.11
4	.08	8	.10

12.51 For many years, the Girl Scouts of America have sold cookies using various sales techniques. One troop experimented with several techniques and reported the number of boxes of cookies sold per girl, as listed in the table.

Door-to-Door	Telephone	Grocery Store Stand	Department Store Stand
47	63	113	25
93	19	50	36
58	29	68	21
37	24	37	27
62	33	39	18
		77	31

a. Is there sufficient evidence to indicate that one sales method tends to produce more sales per scout than the others? Use $\alpha = .05$.

b. Does the door-to-door approach tend to produce a different number of sales per scout from that at the grocery store stand? Test using $\alpha = .05$.

12.52 A psychologist ranked a random sample of ten children on two subjective scales according to the amounts of paranoid behavior and aggressiveness they exhibit. The rankings according to the two criteria are:

Child:	1	2	3	4	5	6	7	8	9	10
Paranoia:	7	3	6	1	2	4	10	8	5	9
Aggressiveness:	5	1	4	2	8	7	9	3	6	10

Compute Spearman's rank correlation coefficient. Is there evidence of a relationship between aggression and paranoia in children as judged by this psychologist? Use $\alpha = .05$.

12.53 Performance in a personal interview often determines whether a candidate is offered a job. Suppose the personnel director of a company interviewed six job applicants, knowing nothing about their backgrounds, and then rated them on a scale from 1 to 10. Independently, the director's supervisor evaluated their background qualifications on the same scale. The results are shown in the table. Is there evidence that a candidate's qualification score is positively correlated with the interview performance score? Use $\alpha = .05$.

Candidate	Qualifications	Interview Performance
1	10	8
2	8	9
3	9	10
4	4	5
5	5	3
6	6	6

12.54 Two car rental companies have long waged an advertising war. An independent testing agency is hired to compare the number of rentals at a major airport. After 10 days, the agency has gathered the data listed in the table. At this point, can either car rental company claim to be number one at this airport? Use $\alpha = .05$.

Day	Rental Company	
	A	B
1	29	22
2	26	29
3	19	30
4	28	25
5	27	26
6	16	20
7	35	30
8	43	45
9	29	38
10	32	40

12.55 Suppose three health-food diets are compared by placing eight overweight individuals on each diet for 6 weeks. The values in the table represent the weight losses (in pounds) of the twenty-four dieters.

Diet A		Diet B		Diet C	
11	2	0	8	3	9
19	13	4	11	7	10
23	20	19	14	8	16
7	22	15	17	11	5

a. Can we conclude that the distributions of the weight losses differ for the three diets? Use $\alpha = .05$.

b. Compare the weight loss distributions for diets A and B using $\alpha = .10$.

12.56 A researcher asked the staff of a public library to rank fourteen clienteles according to their ideal priority for receiving service. The way in which the professional and the nonprofessional staff members ranked the fourteen groups is shown in the accompanying table. Is there evidence to indicate that the priorities of the professionals and nonprofessionals are positively correlated? Use $\alpha = .05$.

Clientele	Professional Staff's Rank	Nonprofessional Staff's Rank
Children	1	1
Old people	2	2
Teenagers	3	4
Migrants	4	3
The housebound	5	5
The physically handicapped	6	6
The unemployed	7	8
People near or below the poverty line	9.5	7
Local societies	11	9
Local council	8	11
The institutionalized	9.5	12
Teachers	12.5	10
Business and industrial firms	12.5	13
Trade unions	14	14

Source: Ida Vincent, "Staff's Perceptions of Public Library Goals: A Case Study of an Australian Public Library," *The Library Quarterly*, Vol. 54, No. 4, 1984, p. 401. Copyright © 1984 by The University of Chicago Press. Reprinted by permission.

12.57 Refer to Exercise 12.56. The ideal priority rankings of the entire library staff were reported, as well as a second ranking of the fourteen clienteles according to the actual priority that the staff believed the library afforded them. These rankings appear in the table at the top of page 688.

Clientele	Rank of Ideal Priority	Rank of Actual Priority
Children	1	1
Old people	2	2
Teenagers	3	7
Migrants	4	3
The housebound	5	10
The physically handicapped	6	9
The unemployed	7	11.5
People near or below the poverty line	8	13
Local societies	9	5
Local council	10	4
The institutionalized	11	11.5
Teachers	12	8
Business and industrial firms	13	6
Trade unions	14	14

Source: Ida Vincent, "Staff's Perceptions of Public Library Goals: A Case Study of an Australian Public Library," *The Library Quarterly*, Vol. 54, No. 4, 1984, p. 401. Copyright © 1984 by The University of Chicago Press. Reprinted by permission.

a. Calculate Spearman's rank correlation coefficient for the data. Interpret this value.
b. Is there evidence that the two perceptions of priority are positively correlated? Use $\alpha = .05$.

Use a computer package to help solve the following exercises.

*12.58 A savings and loan association is considering three locations in a large city as potential office sites. The company has hired a marketing firm to compare the family incomes in each of the three neighborhoods. The market researchers interview ten households chosen at random in each area to determine the type of job, length of employment, etc., of those in the household who work. This information will enable them to estimate the annual income of each household. The results shown in the table are obtained.

Estimated Annual Income (thousands of dollars)					
Location 1		Location 2		Location 3	
19.3	21.2	24.3	27.2	19.5	23.3
20.5	28.5	30.5	88.5	14.3	28.3
17.1	19.7	35.2	32.9	22.2	21.7
13.3	23.0	57.1	26.2	18.2	25.0
25.5	20.1	33.6	29.0	17.6	20.2

a. Is there evidence that incomes in one (or more) of the locations tend to be higher than in the other locations? Use $\alpha = .05$.

b. Use the Wilcoxon rank sum test to compare the distributions of incomes in locations 1 and 2. Use $\alpha = .05$.

*12.59 Twelve samples of variously priced carpeting were chosen and tested for wear. The cost (in dollars) per square yard and the number of months of wear for each of the twelve samples of carpeting are listed in the table. Do the data provide sufficient evidence to indicate that wearability tends to increase as the price increases? Test using $\alpha = .05$.

Sample	Cost	Amount of Wear	Sample	Cost	Amount of Wear
1	9.95	32.5	7	10.45	25.2
2	8.50	24.8	8	17.95	35.3
3	10.85	25.6	9	12.95	34.6
4	7.99	18.4	10	9.99	29.7
5	15.25	28.3	11	14.85	29.9
6	20.50	20.4	12	9.75	26.3

*12.60 A national clothing store franchise operates two stores in one city—one urban and one suburban. To stock the stores with clothing suited to the clientele, a survey is conducted to determine the customers' incomes. Ten customers in each store are offered significant discounts if they will reveal the annual income of their household. The results (in thousands of dollars) are listed in the table. Is there evidence that customers of one of the stores tend to have higher incomes than customers of the other store? Use $\alpha = .05$.

Store 1		Store 2	
23.8	34.5	16.3	14.3
32.9	21.3	23.2	19.6
17.2	27.1	10.3	13.8
90.3	20.7	28.5	12.6
18.1	29.0	15.0	23.3

CHAPTER 12 QUIZ

1. To compare the popularity of two domestic small cars within a city, a local trade organization obtained the information given in the table (page 690) from two car dealers—one dealer for each of the two makes of car. Is there evidence of a difference in location between the distributions of number of cars sold for the two makes of car? Use the Wilcoxon signed rank test for paired samples at the .05 level of significance.

Month	Car Sales	
	A	B
1	9	17
2	10	20
3	13	15
4	11	12
5	7	18
6	8	14
7	9	8

2. Two anticoagulant drugs were studied to compare their effectiveness in dissolving blood clots. A random sample of five subjects received drug A, and a second random sample received drug B. The length of time required for a small cut to stop bleeding was then recorded, with the results shown in the table.

Clotting Time (seconds)	
Drug A	Drug B
127.5	174.4
130.6	144.3
118.3	111.7
115.5	129.1
180.7	129.0

a. Use the Wilcoxon rank sum test for independent samples to determine whether the data provide sufficient evidence to indicate a difference in mean clotting time between the two drugs. Use $\alpha = .05$.

b. Suggest another method of designing this experiment that could provide more information.

3. The accompanying table lists the number of newspaper copies and radios per 1,000 people in ten countries.

Country	Newspaper Copies	Radios
Czechoslovakia	280	266
Italy	142	230
Kenya	10	114
Norway	391	313
Panama	86	329
Philippines	17	42
Tunisia	21	49
U.S.A.	314	1,695
U.S.S.R.	333	430
Venezuela	91	182

a. Calculate Spearman's rank correlation coefficient and interpret this value.

b. Is there sufficient evidence to conclude that the number of newspaper copies and number of radios are correlated? Use $\alpha = .05$.

CHAPTERS 10–12 CUMULATIVE QUIZ

1. Methods of displaying goods can have an effect on their sales. The manager of a large produce market would like to compare three ways of displaying fruit. Each type of display is used for 5 weeks (total time is 15 weeks), and the sales (in dollars) of the fruit from each display are determined. The results appear in the table.

| Type of Display | | |
A	B	C
$125	$153	$108
137	140	115
110	119	115
119	130	110
141	160	133

a. Conduct an analysis of variance (F test) to determine whether there is a difference in mean sales for at least two of the three types of displays. Use $\alpha = .05$.

b. Use the Kruskal–Wallis H test to determine whether there is sufficient evidence to conclude that the displays tend to differ in the amount of sales. Use $\alpha = .05$.

2. The data in the table present unemployment figures for independent random samples from the work forces in four cities. Is there sufficient evidence to conclude that the distribution of employment status is dependent on the choice of city? Use $\alpha = .01$.

| Employment Status | City | | | |
	1	2	3	4
Full-time	722	643	814	596
Part-time	180	244	100	229
Unemployed	98	113	86	175

On Your Own

In Chapters 10 and 12 we have discussed two methods of analyzing data obtained from a completely randomized design. When the populations of interest are normally distributed, we can employ the analysis of variance described in Chapter 10. If not, we should use the Kruskal–Wallis H test.

Choose three or more populations that you would like to compare, and select independent random samples of measurements from the populations. Analyze the data using both an analysis of variance and the Kruskal–Wallis H test.

How do the results of the two analyses compare? Are the interpretations the same? How can you explain the similarity (or lack of similarity) between the two results?

References

McClave, J. T. & Dietrich, II, F. H. *Statistics*. 4th ed. San Francisco: Dellen, 1988.

Villemez, W. J. & Silver, B. B. "Occupational Situs as Horizontal Social Position: A Reconsideration," *Sociology and Social Research*, 1976, *61*, pp. 320–335.

TABLES

Table I

■■■■■■■■■■

Binomial Sampling Distributions

n	p	.10	.20	.30	.40	π .50	.60	.70	.80	.90	p
2	0	.810	.640	.490	.360	.250	.160	.090	.040	.010	0
	$\frac{1}{2}$	.180	.320	.420	.480	.500	.480	.420	.320	.180	$\frac{1}{2}$
	1	.010	.040	.090	.160	.250	.360	.490	.640	.810	1
3	0	.729	.512	.343	.216	.125	.064	.027	.008	.001	0
	$\frac{1}{3}$	.243	.384	.441	.432	.375	.288	.189	.096	.027	$\frac{1}{3}$
	$\frac{2}{3}$	.027	.096	.189	.288	.375	.432	.441	.384	.243	$\frac{2}{3}$
	1	.001	.008	.027	.064	.125	.216	.343	.512	.729	1
4	0	.656	.410	.240	.130	.062	.026	.008	.002	.00+	0
	$\frac{1}{4}$	.292	.410	.412	.346	.250	.154	.076	.026	.004	$\frac{1}{4}$
	$\frac{2}{4}$	.049	.154	.265	.346	.375	.346	.265	.154	.049	$\frac{2}{4}$
	$\frac{3}{4}$	.004	.026	.076	.154	.250	.346	.412	.410	.292	$\frac{3}{4}$
	1	.00+	.002	.008	.026	.062	.130	.240	.410	.656	1
5	0	.590	.328	.168	.078	.031	.010	.002	.00+	.00+	0
	$\frac{1}{5}$	.328	.410	.360	.259	.156	.077	.028	.006	.00+	$\frac{1}{5}$
	$\frac{2}{5}$	.073	.205	.309	.346	.312	.230	.132	.051	.008	$\frac{2}{5}$
	$\frac{3}{5}$	.008	.051	.132	.230	.312	.346	.309	.205	.073	$\frac{3}{5}$
	$\frac{4}{5}$	.00+	.006	.028	.077	.156	.259	.360	.410	.328	$\frac{4}{5}$
	1	.00+	.00+	.002	.010	.031	.078	.168	.328	.590	1
6	0	.531	.262	.118	.047	.016	.004	.001	.00+	.00+	0
	$\frac{1}{6}$	.354	.393	.303	.187	.094	.037	.010	.002	.00+	$\frac{1}{6}$
	$\frac{2}{6}$	.098	.246	.324	.311	.234	.138	.060	.015	.001	$\frac{2}{6}$
	$\frac{3}{6}$	.015	.082	.185	.276	.312	.276	.185	.082	.015	$\frac{3}{6}$
	$\frac{4}{6}$	.001	.015	.060	.138	.234	.311	.324	.246	.098	$\frac{4}{6}$
	$\frac{5}{6}$	.00+	.002	.010	.037	.094	.187	.303	.393	.354	$\frac{5}{6}$
	1	.00+	.00+	.001	.004	.016	.047	.118	.262	.531	1
7	0	.478	.210	.082	.028	.008	.002	.00+	.00+	.00+	0
	$\frac{1}{7}$	.372	.367	.247	.131	.055	.017	.004	.00+	.00+	$\frac{1}{7}$
	$\frac{2}{7}$	.124	.275	.318	.261	.164	.077	.025	.004	.00+	$\frac{2}{7}$
	$\frac{3}{7}$	.023	.115	.227	.290	.273	.194	.097	.029	.003	$\frac{3}{7}$
	$\frac{4}{7}$	.003	.029	.097	.194	.273	.290	.227	.115	.023	$\frac{4}{7}$
	$\frac{5}{7}$	.00+	.004	.025	.077	.164	.261	.318	.275	.124	$\frac{5}{7}$
	$\frac{6}{7}$	.00+	.00+	.004	.017	.055	.131	.247	.367	.372	$\frac{6}{7}$
	1	.00+	.00+	.00+	.002	.008	.028	.082	.210	.478	1

Table I

Continued

n	p	.10	.20	.30	.40	π .50	.60	.70	.80	.90	p
8	0	.430	.168	.058	.017	.004	.001	.00+	.00+	.00+	0
	$\frac{1}{8}$	.383	.336	.198	.090	.031	.008	.001	.00+	.00+	$\frac{1}{8}$
	$\frac{2}{8}$	.149	.294	.296	.209	.109	.041	.010	.001	.00+	$\frac{2}{8}$
	$\frac{3}{8}$	.033	.147	.254	.279	.219	.124	.047	.009	.00+	$\frac{3}{8}$
	$\frac{4}{8}$	.005	.046	.136	.232	.273	.232	.136	.046	.005	$\frac{4}{8}$
	$\frac{5}{8}$	.00+	.009	.047	.124	.219	.279	.254	.147	.033	$\frac{5}{8}$
	$\frac{6}{8}$	.00+	.001	.010	.041	.109	.209	.296	.294	.149	$\frac{6}{8}$
	$\frac{7}{8}$	.00+	.00+	.001	.008	.031	.090	.198	.336	.383	$\frac{7}{8}$
	1	.00+	.00+	.00+	.001	.004	.017	.058	.168	.430	1
9	0	.387	.134	.040	.010	.002	.00+	.00+	.00+	.00+	0
	$\frac{1}{9}$	.387	.302	.156	.060	.018	.004	.00+	.00+	.00+	$\frac{1}{9}$
	$\frac{2}{9}$	.172	.302	.267	.161	.070	.021	.004	.00+	.00+	$\frac{2}{9}$
	$\frac{3}{9}$	.045	.176	.267	.251	.164	.074	.021	.003	.00+	$\frac{3}{9}$
	$\frac{4}{9}$	.007	.066	.172	.251	.246	.167	.074	.017	.001	$\frac{4}{9}$
	$\frac{5}{9}$	.001	.017	.074	.167	.246	.251	.172	.066	.007	$\frac{5}{9}$
	$\frac{6}{9}$	.00+	.003	.021	.074	.164	.251	.267	.176	.045	$\frac{6}{9}$
	$\frac{7}{9}$	.00+	.00+	.004	.021	.070	.161	.267	.302	.172	$\frac{7}{9}$
	$\frac{8}{9}$	.00+	.00+	.00+	.004	.018	.060	.156	.302	.387	$\frac{8}{9}$
	1	.00+	.00+	.00+	.00+	.002	.010	.040	.134	.387	1
10	0	.349	.107	.028	.006	.001	.00+	.00+	.00+	.00+	0
	$\frac{1}{10}$	.387	.268	.121	.040	.010	.002	.00+	.00+	.00+	$\frac{1}{10}$
	$\frac{2}{10}$	.194	.302	.233	.121	.044	.011	.001	.00+	.00+	$\frac{2}{10}$
	$\frac{3}{10}$	.057	.201	.267	.215	.117	.042	.009	.001	.00+	$\frac{3}{10}$
	$\frac{4}{10}$	.011	.088	.200	.251	.205	.111	.037	.006	.00+	$\frac{4}{10}$
	$\frac{5}{10}$	.001	.026	.103	.201	.246	.201	.103	.026	.001	$\frac{5}{10}$
	$\frac{6}{10}$	.00+	.006	.037	.111	.205	.251	.200	.088	.011	$\frac{6}{10}$
	$\frac{7}{10}$	.00+	.001	.009	.042	.117	.215	.267	.201	.057	$\frac{7}{10}$
	$\frac{8}{10}$	.00+	.00+	.001	.011	.044	.121	.233	.302	.194	$\frac{8}{10}$
	$\frac{9}{10}$	.00+	.00+	.00+	.002	.010	.040	.121	.268	.387	$\frac{9}{10}$
	1	.00+	.00+	.00+	.00+	.001	.006	.028	.107	.349	1

Table I

▂▂▂▂▂▂▂▂▂▂▂▂

Continued

n	p	.10	.20	.30	.40	π .50	.60	.70	.80	.90	p
11	0	.314	.086	.020	.004	.00+	.00+	.00+	.00+	.00+	0
	$1/11$	.384	.236	.093	.027	.005	.001	.00+	.00+	.00+	$1/11$
	$2/11$	.213	.295	.200	.089	.027	.005	.001	.00+	.00+	$2/11$
	$3/11$	.071	.221	.257	.177	.081	.023	.004	.00+	.00+	$3/11$
	$4/11$	.016	.111	.220	.236	.161	.070	.017	.002	.00+	$4/11$
	$5/11$	.002	.039	.132	.221	.226	.147	.057	.010	.00+	$5/11$
	$6/11$	.00+	.010	.057	.147	.226	.221	.132	.039	.002	$6/11$
	$7/11$	.00+	.002	.017	.070	.161	.236	.220	.111	.016	$7/11$
	$8/11$	.00+	.00+	.004	.023	.081	.177	.257	.221	.071	$8/11$
	$9/11$	.00+	.00+	.001	.005	.027	.089	.200	.295	.213	$9/11$
	$10/11$	.00+	.00+	.00+	.001	.005	.027	.093	.236	.384	$10/11$
	1	.00+	.00+	.00+	.00+	.00+	.004	.020	.086	.314	1
12	0	.282	.069	.014	.002	.00+	.00+	.00+	.00+	.00+	0
	$1/12$	.377	.206	.071	.017	.003	.00+	.00+	.00+	.00+	$1/12$
	$2/12$	.230	.283	.168	.064	.016	.002	.00+	.00+	.00+	$2/12$
	$3/12$	.085	.236	.240	.142	.054	.012	.001	.00+	.00+	$3/12$
	$4/12$	.021	.133	.231	.213	.121	.042	.008	.001	.00+	$4/12$
	$5/12$	.004	.053	.158	.227	.193	.101	.029	.003	.00+	$5/12$
	$6/12$	.00+	.016	.079	.177	.226	.177	.079	.016	.00+	$6/12$
	$7/12$	.00+	.003	.029	.101	.193	.227	.158	.053	.004	$7/12$
	$8/12$	.00+	.001	.008	.042	.121	.213	.231	.133	.021	$8/12$
	$9/12$	.00+	.00+	.001	.012	.054	.142	.240	.236	.085	$9/12$
	$10/12$	.00+	.00+	.00+	.002	.016	.064	.168	.283	.230	$10/12$
	$11/12$	.00+	.00+	.00+	.00+	.003	.017	.071	.206	.377	$11/12$
	1	.00+	.00+	.00+	.00+	.00+	.002	.014	.069	.282	1

Table I

Continued

n	p	.10	.20	.30	.40	π .50	.60	.70	.80	.90	p
13	0	.254	.055	.010	.001	.00+	.00+	.00+	.00+	.00+	0
	$1/13$	.367	.179	.054	.011	.002	.00+	.00+	.00+	.00+	$1/13$
	$2/13$	.245	.268	.139	.045	.010	.001	.00+	.00+	.00+	$2/13$
	$3/13$	.100	.246	.218	.111	.035	.006	.001	.00+	.00+	$3/13$
	$4/13$	.028	.154	.234	.184	.087	.024	.003	.00+	.00+	$4/13$
	$5/13$	.006	.069	.180	.221	.157	.066	.014	.001	.00+	$5/13$
	$6/13$	.001	.023	.103	.197	.209	.131	.044	.006	.00+	$6/13$
	$7/13$	.00+	.006	.044	.131	.209	.197	.103	.023	.001	$7/13$
	$8/13$	.00+	.001	.014	.066	.157	.221	.180	.069	.006	$8/13$
	$9/13$	.00+	.00+	.003	.024	.087	.184	.234	.154	.028	$9/13$
	$10/13$	.00+	.00+	.001	.006	.035	.111	.218	.246	.100	$10/13$
	$11/13$	.00+	.00+	.00+	.001	.010	.045	.139	.268	.245	$11/13$
	$12/13$	.00+	.00+	.00+	.00+	.002	.011	.054	.179	.367	$12/13$
	1	.00+	.00+	.00+	.00+	.00+	.001	.010	.055	.254	1
14	0	.229	.044	.007	.001	.00+	.00+	.00+	.00+	.00+	0
	$1/14$	.356	.154	.041	.007	.001	.00+	.00+	.00+	.00+	$1/14$
	$2/14$	.257	.250	.113	.032	.006	.001	.00+	.00+	.00+	$2/14$
	$3/14$	.114	.250	.194	.085	.022	.003	.00+	.00+	.00+	$3/14$
	$4/14$	.035	.172	.229	.155	.061	.014	.001	.00+	.00+	$4/14$
	$5/14$	.008	.086	.196	.207	.122	.041	.007	.00+	.00+	$5/14$
	$6/14$	.001	.032	.126	.207	.183	.092	.023	.002	.00+	$6/14$
	$7/14$	.00+	.009	.062	.157	.209	.157	.062	.009	.00+	$7/14$
	$8/14$	.00+	.002	.023	.092	.183	.207	.126	.032	.001	$8/14$
	$9/14$	.00+	.00+	.007	.041	.122	.207	.196	.086	.008	$9/14$
	$10/14$	.00+	.00+	.001	.014	.061	.155	.229	.172	.035	$10/14$
	$11/14$	.00+	.00+	.00+	.003	.022	.085	.194	.250	.114	$11/14$
	$12/14$	.00+	.00+	.00+	.001	.006	.032	.113	.250	.257	$12/14$
	$13/14$	.00+	.00+	.00+	.00+	.001	.007	.041	.154	.356	$13/14$
	1	.00+	.00+	.00+	.00+	.00+	.001	.007	.044	.229	1

Table I

Continued

n	p	.10	.20	.30	.40	π .50	.60	.70	.80	.90	p
15	0	.206	.035	.005	.00+	.00+	.00+	.00+	.00+	.00+	0
	1/15	.343	.132	.031	.005	.00+	.00+	.00+	.00+	.00+	1/15
	2/15	.267	.231	.092	.022	.003	.00+	.00+	.00+	.00+	2/15
	3/15	.129	.250	.170	.063	.014	.002	.00+	.00+	.00+	3/15
	4/15	.043	.188	.219	.127	.042	.007	.001	.00+	.00+	4/15
	5/15	.010	.103	.206	.186	.092	.024	.003	.00+	.00+	5/15
	6/15	.002	.043	.147	.207	.153	.061	.012	.001	.00+	6/15
	7/15	.00+	.014	.081	.177	.196	.118	.035	.003	.00+	7/15
	8/15	.00+	.003	.035	.118	.196	.177	.081	.014	.00+	8/15
	9/15	.00+	.001	.012	.061	.153	.207	.147	.043	.002	9/15
	10/15	.00+	.00+	.003	.024	.092	.186	.206	.103	.010	10/15
	11/15	.00+	.00+	.001	.007	.042	.127	.219	.188	.043	11/15
	12/15	.00+	.00+	.00+	.002	.014	.063	.170	.250	.129	12/15
	13/15	.00+	.00+	.00+	.00+	.003	.022	.092	.231	.267	13/15
	14/15	.00+	.00+	.00+	.00+	.00+	.005	.031	.132	.343	14/15
	1	.00+	.00+	.00+	.00+	.00+	.00+	.005	.035	.206	1
16	0	.185	.028	.003	.00+	.00+	.00+	.00+	.00+	.00+	0
	1/16	.329	.113	.023	.003	.00+	.00+	.00+	.00+	.00+	1/16
	2/16	.275	.211	.073	.015	.002	.00+	.00+	.00+	.00+	2/16
	3/16	.142	.246	.146	.047	.009	.001	.00+	.00+	.00+	3/16
	4/16	.051	.200	.204	.101	.028	.004	.00+	.00+	.00+	4/16
	5/16	.014	.120	.210	.162	.067	.014	.001	.00+	.00+	5/16
	6/16	.003	.055	.165	.198	.122	.039	.006	.00+	.00+	6/16
	7/16	.00+	.020	.101	.189	.175	.084	.019	.001	.00+	7/16
	8/16	.00+	.006	.049	.142	.196	.142	.049	.006	.00+	8/16
	9/16	.00+	.001	.019	.084	.175	.189	.101	.020	.00+	9/16
	10/16	.00+	.00+	.006	.039	.122	.198	.165	.055	.003	10/16
	11/16	.00+	.00+	.001	.014	.067	.162	.210	.120	.014	11/16
	12/16	.00+	.00+	.00+	.004	.028	.101	.204	.200	.051	12/16
	13/16	.00+	.00+	.00+	.001	.009	.047	.146	.246	.142	13/16
	14/16	.00+	.00+	.00+	.00+	.002	.015	.073	.211	.275	14/16
	15/16	.00+	.00+	.00+	.00+	.00+	.003	.023	.113	.329	15/16
	1	.00+	.00+	.00+	.00+	.00+	.00+	.003	.028	.185	1

Table I

Continued

n	p	.10	.20	.30	.40	π .50	.60	.70	.80	.90	p
17	0	.167	.023	.002	.00+	.00+	.00+	.00+	.00+	.00+	0
	$1/17$	.315	.096	.017	.002	.00+	.00+	.00+	.00+	.00+	$1/17$
	$2/17$	.280	.191	.058	.010	.001	.00+	.00+	.00+	.00+	$2/17$
	$3/17$	.156	.239	.125	.034	.005	.00+	.00+	.00+	.00+	$3/17$
	$4/17$	.060	.209	.187	.080	.018	.002	.00+	.00+	.00+	$4/17$
	$5/17$	.017	.136	.208	.138	.047	.008	.001	.00+	.00+	$5/17$
	$6/17$	.004	.068	.178	.184	.094	.024	.003	.00+	.00+	$6/17$
	$7/17$	.001	.027	.120	.193	.148	.057	.009	.00+	.00+	$7/17$
	$8/17$	.00+	.008	.064	.161	.185	.107	.028	.002	.00+	$8/17$
	$9/17$	.00+	.002	.028	.107	.185	.161	.064	.008	.00+	$9/17$
	$10/17$	.00+	.00+	.009	.057	.148	.193	.120	.027	.001	$10/17$
	$11/17$	.00+	.00+	.003	.024	.094	.184	.178	.068	.004	$11/17$
	$12/17$	.00+	.00+	.001	.008	.047	.138	.208	.136	.017	$12/17$
	$13/17$	.00+	.00+	.00+	.002	.018	.080	.187	.209	.060	$13/17$
	$14/17$	.00+	.00+	.00+	.00+	.005	.034	.125	.239	.156	$14/17$
	$15/17$	.00+	.00+	.00+	.00+	.001	.010	.058	.191	.280	$15/17$
	$16/17$	.00+	.00+	.00+	.00+	.00+	.002	.017	.096	.315	$16/17$
	1	.00+	.00+	.00+	.00+	.00+	.00+	.002	.023	.167	1
18	0	.150	.018	.002	.00+	.00+	.00+	.00+	.00+	.00+	0
	$1/18$	.300	.081	.013	.001	.00+	.00+	.00+	.00+	.00+	$1/18$
	$2/18$	.284	.172	.046	.007	.001	.00+	.00+	.00+	.00+	$2/18$
	$3/18$	.168	.230	.105	.025	.003	.00+	.00+	.00+	.00+	$3/18$
	$4/18$	.070	.215	.168	.061	.012	.001	.00+	.00+	.00+	$4/18$
	$5/18$	.022	.151	.202	.115	.033	.004	.00+	.00+	.00+	$5/18$
	$6/18$	.005	.082	.187	.166	.071	.015	.001	.00+	.00+	$6/18$
	$7/18$	.001	.035	.138	.189	.121	.037	.005	.00+	.00+	$7/18$
	$8/18$	.00+	.012	.081	.173	.167	.077	.015	.001	.00+	$8/18$
	$9/18$	.00+	.003	.039	.128	.185	.128	.039	.003	.00+	$9/18$
	$10/18$	.00+	.001	.015	.077	.167	.173	.081	.012	.00+	$10/18$
	$11/18$	.00+	.00+	.005	.037	.121	.189	.138	.035	.001	$11/18$
	$12/18$	.00+	.00+	.001	.015	.071	.166	.187	.082	.005	$12/18$
	$13/18$	.00+	.00+	.00+	.004	.033	.115	.202	.151	.022	$13/18$
	$14/18$	.00+	.00+	.00+	.001	.012	.061	.168	.215	.070	$14/18$
	$15/18$	.00+	.00+	.00+	.00+	.003	.025	.105	.230	.168	$15/18$
	$16/18$	.00+	.00+	.00+	.00+	.001	.007	.046	.172	.284	$16/18$
	$17/18$	.00+	.00+	.00+	.00+	.00+	.001	.013	.081	.300	$17/18$
	1	.00+	.00+	.00+	.00+	.00+	.00+	.002	.018	.150	1

Table I

■■■■■■■■■■■■■■■■

Continued

n	p	.10	.20	.30	.40	π .50	.60	.70	.80	.90	p
19	0	.135	.014	.001	.00+	.00+	.00+	.00+	.00+	.00+	0
	$1/19$	.285	.068	.009	.001	.00+	.00+	.00+	.00+	.00+	$1/19$
	$2/19$	.285	.154	.036	.005	.00+	.00+	.00+	.00+	.00+	$2/19$
	$3/19$	.180	.218	.087	.017	.002	.00+	.00+	.00+	.00+	$3/19$
	$4/19$	.080	.218	.149	.047	.007	.001	.00+	.00+	.00+	$4/19$
	$5/19$	.027	.164	.192	.093	.022	.002	.00+	.00+	.00+	$5/19$
	$6/19$	.007	.095	.192	.145	.052	.008	.001	.00+	.00+	$6/19$
	$7/19$	.001	,044	.153	.180	.096	.024	.002	.00+	.00+	$7/19$
	$8/19$	.00+	.017	.098	.180	.144	.053	.008	.00+	.00+	$8/19$
	$9/19$	.00+	.005	.051	.146	.176	.098	.022	.001	.00+	$9/19$
	$10/19$	.00+	.001	.022	.098	.176	.146	.051	.005	.00+	$10/19$
	$11/19$	.00+	.00+	.008	.053	.144	.180	.098	.017	.00+	$11/19$
	$12/19$	.00+	.00+	.002	.024	.096	.180	.153	.044	.001	$12/19$
	$13/19$	.00+	.00+	.001	.008	.052	.145	.192	.095	.007	$13/19$
	$14/19$	.00+	.00+	.00+	.002	.022	.093	.192	.164	.027	$14/19$
	$15/19$	.00+	.00+	.00+	.001	.007	.047	.149	.218	.080	$15/19$
	$16/19$	.00+	.00+	.00+	.00+	.002	.017	.087	.218	.180	$16/19$
	$17/19$	.00+	.00+	.00+	.00+	.00+	.005	.036	.154	.285	$17/19$
	$18/19$	.00+	.00+	.00+	.00+	.00+	.001	.009	.068	.285	$18/19$
	1	.00+	.00+	.00+	.00+	.00+	.00+	.001	.014	.135	1
20	0	.122	.012	.001	.00+	.00+	.00+	.00+	.00+	.00+	0
	$1/20$	.270	.058	.007	.00+	.00+	.00+	.00+	.00+	.00+	$1/20$
	$2/20$	.285	.137	.028	.003	.00+	.00+	.00+	.00+	.00+	$2/20$
	$3/20$	.190	.205	.072	.012	.001	.00+	.00+	.00+	.00+	$3/20$
	$4/20$	.090	.218	.130	.035	.005	.00+	.00+	.00+	.00+	$4/20$
	$5/20$	.032	.175	.179	.075	.015	.001	.00+	.00+	.00+	$5/20$
	$6/20$	.009	.109	.192	.124	.037	.005	.00+	.00+	.00+	$6/20$
	$7/20$	.002	.055	.164	.166	.074	.015	.001	.00+	.00+	$7/20$
	$8/20$	.00+	.022	.114	.180	.120	.035	.004	.00+	.00+	$8/20$
	$9/20$	.00+	.007	.065	.160	.160	.071	.012	.00+	.00+	$9/20$
	$10/20$	.00+	.002	.031	.117	.176	.117	.031	.002	.00+	$10/20$
	$11/20$	.00+	.00+	.012	.071	.160	.160	.065	.007	.00+	$11/20$

Table I

Continued

n	p	.10	.20	.30	.40	π .50	.60	.70	.80	.90	p
	$12/20$	.00+	.00+	.004	.035	.120	.180	.114	.022	.00+	$12/20$
	$13/20$	.00+	.00+	.001	.015	.074	.166	.164	.055	.002	$13/20$
	$14/20$	.00+	.00+	.00+	.005	.037	.124	.192	.109	.009	$14/20$
	$15/20$	.00+	.00+	.00+	.001	.015	.075	.179	.175	.032	$15/20$
	$16/20$	.00+	.00+	.00+	.00+	.005	.035	.130	.218	.090	$16/20$
	$17/20$	.00+	.00+	.00+	.00+	.001	.012	.072	.205	.190	$17/20$
	$18/20$	.00+	.00+	.00+	.00+	.00+	.003	.028	.137	.285	$18/20$
	$19/20$	.00+	.00+	.00+	.00+	.00+	.00+	.007	.058	.270	$19/20$
	1	.00+	.00+	.00+	.00+	.00+	.00+	.001	.012	.122	1
21	0	.109	.009	.001	.00+	.00+	.00+	.00+	.00+	.00+	0
	$1/21$	.255	.048	.005	.00+	.00+	.00+	.00+	.00+	.00+	$1/21$
	$2/21$	.284	.121	.022	.002	.00+	.00+	.00+	.00+	.00+	$2/21$
	$3/21$	.200	.192	.058	.009	.001	.00+	.00+	.00+	.00+	$3/21$
	$4/21$	.100	.216	.113	.026	.003	.00+	.00+	.00+	.00+	$4/21$
	$5/21$	.038	.183	.164	.059	.010	.001	.00+	.00+	.00+	$5/21$
	$6/21$	.011	.122	.188	.105	.026	.003	.00+	.00+	.00+	$6/21$
	$7/21$	.003	.065	.172	.149	.055	.009	.00+	.00+	.00+	$7/21$
	$8/21$	.001	.029	.129	.174	.097	.023	.002	.00+	.00+	$8/21$
	$9/21$	.00+	.010	.080	.168	.140	.050	.006	.00+	.00+	$9/21$
	$10/21$	.00+	.003	.041	.134	.168	.089	.018	.001	.00+	$10/21$
	$11/21$	.00+	.001	.018	.089	.168	.134	.041	.003	.00+	$11/21$
	$12/21$	.00+	.00+	.006	.050	.140	.168	.080	.010	.00+	$12/21$
	$13/21$	.00+	.00+	.002	.023	.097	.174	.129	.029	.001	$13/21$
	$14/21$	.00+	.00+	.00+	.009	.055	.149	.172	.065	.003	$14/21$
	$15/21$	.00+	.00+	.00+	.003	.026	.105	.188	.122	.011	$15/21$
	$16/21$	.00+	.00+	.00+	.001	.010	.059	.164	.183	.038	$16/21$
	$17/21$	.00+	.00+	.00+	.00+	.003	.026	.113	.216	.100	$17/21$
	$18/21$	.00+	.00+	.00+	.00+	.001	.009	.058	.192	.200	$18/21$
	$19/21$	.00+	.00+	.00+	.00+	.00+	.002	.022	.121	.284	$19/21$
	$20/21$	.00+	.00+	.00+	.00+	.00+	.00+	.005	.048	.255	$20/21$
	1	.00+	.00+	.00+	.00+	.00+	.00+	.001	.009	.109	1

Table I

███████████████████

Continued

n	p	.10	.20	.30	.40	π .50	.60	.70	.80	.90	p
22	0	.098	.007	.00+	.00+	.00+	.00+	.00+	.00+	.00+	0
	$1/22$	.241	.041	.004	.00+	.00+	.00+	.00+	.00+	.00+	$1/22$
	$2/22$	.281	.107	.017	.001	.00+	.00+	.00+	.00+	.00+	$2/22$
	$3/22$	.208	.178	.047	.006	.00+	.00+	.00+	.00+	.00+	$3/22$
	$4/22$	.110	.211	.096	.019	.002	.00+	.00+	.00+	.00+	$4/22$
	$5/22$	.044	.190	.149	.046	.006	.00+	.00+	.00+	.00+	$5/22$
	$6/22$	.014	134	.181	.086	.018	.001	.00+	.00+	.00+	$6/22$
	$7/22$	.004	.077	.177	.131	.041	.005	.00+	.00+	.00+	$7/22$
	$8/22$	.001	.036	.142	.164	.076	.014	.001	.00+	.00+	$8/22$
	$9/22$	.00+	.014	.095	.170	.119	.034	.003	.00+	.00+	$9/22$
	$10/22$	.00+	.005	.053	.148	.154	.066	.010	.00+	.00+	$10/22$
	$11/22$	.00+	.001	.025	.107	.168	.107	.025	.001	.00+	$11/22$
	$12/22$	.00+	.00+	.010	.066	.154	.148	.053	.005	.00+	$12/22$
	$13/22$	.00+	.00+	.003	.034	.119	.170	.095	.014	.00+	$13/22$
	$14/22$	.00+	.00+	.001	.014	.076	.164	.142	.036	.001	$14/22$
	$15/22$	.00+	.00+	.00+	.005	.041	.131	.177	.077	.004	$15/22$
	$16/22$	.00+	.00+	.00+	.001	.018	.086	.181	.134	.014	$16/22$
	$17/22$	.00+	.00+	.00+	.00+	.006	.046	.149	.190	.044	$17/22$
	$18/22$	.00+	.00+	.00+	.00+	.002	.019	.096	.211	.110	$18/22$
	$19/22$	.00+	.00+	.00+	.00+	.00+	.006	.047	.178	.208	$19/22$
	$20/22$	.00+	.00+	.00+	.00+	.00+	.001	.017	.107	.281	$20/22$
	$21/22$	.00+	.00+	.00+	.00+	.00+	.00+	.004	.041	.241	$21/22$
	1	.00+	.00+	.00+	.00+	.00+	.00+	.00+	.007	.098	1
23	0	.089	.006	.00+	.00+	.00+	.00+	.00+	.00+	.00+	0
	$1/23$	.226	.034	.003	.00+	.00+	.00+	.00+	.00+	.00+	$1/23$
	$2/23$	.277	.093	.013	.001	.00+	.00+	.00+	.00+	.00+	$2/23$
	$3/23$	.215	.163	.038	.004	.00+	.00+	.00+	.00+	.00+	$3/23$
	$4/23$	.120	.204	.082	.014	.001	.00+	.00+	.00+	.00+	$4/23$
	$5/23$	.051	.194	.133	.035	.004	.00+	.00+	.00+	.00+	$5/23$
	$6/23$	.017	.145	.171	.070	.012	.001	.00+	.00+	.00+	$6/23$
	$7/23$	.005	.088	.178	.113	.029	.003	.00+	.00+	.00+	$7/23$
	$8/23$	.001	.044	.153	.151	.058	.009	.00+	.00+	.00+	$8/23$
	$9/23$	.00+	.018	.109	.168	.097	.022	.002	.00+	.00+	$9/23$
	$10/23$	.00+	.006	.065	.157	.136	.046	.005	.00+	.00+	$10/23$
	$11/23$	.00+	.002	.033	.123	.161	.082	.014	.00+	.00+	$11/23$
	$12/23$	.00+	.00+	.014	.082	.161	.123	.033	.002	.00+	$12/23$

Table I

████████████████████

Continued

n	p	.10	.20	.30	.40	π .50	.60	.70	.80	.90	p
	$^{13}/_{23}$	.00+	.00+	.005	.046	.136	.157	.065	.006	.00+	$^{13}/_{23}$
	$^{14}/_{23}$	.00+	.00+	.002	.022	.097	.168	.109	.018	.00+	$^{14}/_{23}$
	$^{15}/_{23}$	.00+	.00+	.00+	.009	.058	.151	.153	.044	.001	$^{15}/_{23}$
	$^{16}/_{23}$	.00+	.00+	.00+	.003	.029	.113	.178	.088	.005	$^{16}/_{23}$
	$^{17}/_{23}$	.00+	.00+	.00+	.001	.012	.070	.171	.145	.017	$^{17}/_{23}$
	$^{18}/_{23}$	.00+	.00+	.00+	.00+	.004	.035	.133	.194	.051	$^{18}/_{23}$
	$^{19}/_{23}$	.00+	.00+	.00+	.00+	.001	.014	.082	.204	.120	$^{19}/_{23}$
	$^{20}/_{23}$	.00+	.00+	.00+	.00+	.00+	.004	.038	.163	.215	$^{20}/_{23}$
	$^{21}/_{23}$	.00+	.00+	.00+	.00+	.00+	.001	.013	.093	.277	$^{21}/_{23}$
	$^{22}/_{23}$	.00+	.00+	.00+	.00+	.00+	.00+	.003	.034	.226	$^{22}/_{23}$
	1	.00+	.00+	.00+	.00+	.00+	.00+	.00+	.006	.089	1
24	0	.080	.005	.00+	.00+	.00+	.00+	.00+	.00+	.00+	0
	$^{1}/_{24}$	.213	.028	.002	.00+	.00+	.00+	.00+	.00+	.00+	$^{1}/_{24}$
	$^{2}/_{24}$	.272	.081	.010	.001	.00+	.00+	.00+	.00+	.00+	$^{2}/_{24}$
	$^{3}/_{24}$	.221	.149	.031	.003	.00+	.00+	.00+	.00+	.00+	$^{3}/_{24}$
	$^{4}/_{24}$	.129	.196	.069	.010	.001	.00+	.00+	.00+	.00+	$^{4}/_{24}$
	$^{5}/_{24}$	.057	.196	.118	.027	.003	.00+	.00+	.00+	.00+	$^{5}/_{24}$
	$^{6}/_{24}$	.020	.155	.160	.056	.008	.00+	.00+	.00+	.00+	$^{6}/_{24}$
	$^{7}/_{24}$	.006	.100	.176	.096	.021	.002	.00+	.00+	.00+	$^{7}/_{24}$
	$^{8}/_{24}$	.001	.053	.160	.136	.044	.005	.00+	.00+	.00+	$^{8}/_{24}$
	$^{9}/_{24}$	.00+	.024	.122	.161	.078	.014	.001	.00+	.00+	$^{9}/_{24}$
	$^{10}/_{24}$	.00+	.009	.079	.161	.117	.032	.003	.00+	.00+	$^{10}/_{24}$
	$^{11}/_{24}$	.00+	.003	.043	.137	.149	.061	.008	.00+	.00+	$^{11}/_{24}$
	$^{12}/_{24}$	.00+	.001	.020	.099	.161	.099	.020	.001	.00+	$^{12}/_{24}$
	$^{13}/_{24}$	.00+	.00+	.008	.061	.149	.137	.043	.003	.00+	$^{13}/_{24}$
	$^{14}/_{24}$	.00+	.00+	.003	.032	.117	.161	.079	.009	.00+	$^{14}/_{24}$
	$^{15}/_{24}$	.00+	.00+	.001	.014	.078	.161	.122	.024	.00+	$^{15}/_{24}$
	$^{16}/_{24}$	.00+	.00+	.00+	.005	.044	.136	.160	.053	.001	$^{16}/_{24}$
	$^{17}/_{24}$	.00+	.00+	.00+	.002	.021	.096	.176	.100	.006	$^{17}/_{24}$
	$^{18}/_{24}$	.00+	.00+	.00+	.00+	.008	.056	.160	.155	.020	$^{18}/_{24}$
	$^{19}/_{24}$	.00+	.00+	.00+	.00+	.003	.027	.118	.196	.057	$^{19}/_{24}$
	$^{20}/_{24}$	.00+	.00+	.00+	.00+	.001	.010	.069	.196	.129	$^{20}/_{24}$
	$^{21}/_{24}$	.00+	.00+	.00+	.00+	.00+	.003	.031	.149	.221	$^{21}/_{24}$
	$^{22}/_{24}$	.00+	.00+	.00+	.00+	.00+	.001	.010	.081	.272	$^{22}/_{24}$
	$^{23}/_{24}$	.00+	.00+	.00+	.00+	.00+	.00+	.002	.028	.213	$^{23}/_{24}$
	1	.00+	.00+	.00+	.00+	.00+	.00+	.00+	.005	.080	1

Table I

━━━━━━━━━━━━━

Continued

n	p	.10	.20	.30	.40	π .50	.60	.70	.80	.90	p
25	0	.072	.004	.00+	.00+	.00+	.00+	.00+	.00+	.00+	0
	1/25	.199	.024	.001	.00+	.00+	.00+	.00+	.00+	.00+	1/25
	2/25	.266	.071	.007	.00+	.00+	.00+	.00+	.00+	.00+	2/25
	3/25	.226	.136	.024	.002	.00+	.00+	.00+	.00+	.00+	3/25
	4/25	.138	.187	.057	.007	.00+	.00+	.00+	.00+	.00+	4/25
	5/25	.065	.196	.103	.020	.002	.00+	.00+	.00+	.00+	5/25
	6/25	.024	.163	.147	.044	.005	.00+	.00+	.00+	.00+	6/25
	7/25	.007	.111	.171	.080	.014	.001	.00+	.00+	.00+	7/25
	8/25	.002	.062	.165	.120	.032	.003	.00+	.00+	.00+	8/25
	9/25	.00+	.029	.134	.151	.061	.009	.00+	.00+	.00+	9/25
	10/25	.00+	.012	.092	.161	.097	.021	.001	.00+	.00+	10/25
	11/25	.00+	.004	.054	.147	.133	.043	.004	.00+	.00+	11/25
	12/25	.00+	.001	.027	.114	.155	.076	.011	.00+	.00+	12/25
	13/25	.00+	.00+	.011	.076	.155	.114	.027	.001	.00+	13/25
	14/25	.00+	.00+	.004	.043	.133	.147	.054	.004	.00+	14/25
	15/25	.00+	.00+	.001	.021	.097	.161	.092	.012	.00+	15/25
	16/25	.00+	.00+	.00+	.009	.061	.151	.134	.029	.00+	16/25
	17/25	.00+	.00+	.00+	.003	.032	.120	.165	.062	.002	17/25
	18/25	.00+	.00+	.00+	.001	.014	.080	.171	.111	.007	18/25
	19/25	.00+	.00+	.00+	.00+	.005	.044	.147	.163	.024	19/25
	20/25	.00+	.00+	.00+	.00+	.002	.020	.103	.196	.065	20/25
	21/25	.00+	.00+	.00+	.00+	.00+	.007	.057	.187	.138	21/25
	22/25	.00+	.00+	.00+	.00+	.00+	.002	.024	.136	.226	22/25
	23/25	.00+	.00+	.00+	.00+	.00+	.00+	.007	.071	.266	23/25
	24/25	.00+	.00+	.00+	.00+	.00+	.00+	.001	.024	.199	24/25
	1	.00+	.00+	.00+	.00+	.00+	.00+	.00+	.004	.072	1

Table II

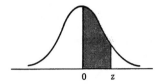

0   z

Normal Curve Areas

z	.00	.01	.02	.03	.04	.05	.06	.07	.08	.09
.0	.000	.004	.008	.012	.016	.020	.024	.028	.032	.036
0.1	.040	.044	.048	.052	.056	.060	.064	.067	.071	.075
0.2	.079	.083	.087	.091	.095	.099	.103	.106	.110	.114
0.3	.118	.122	.126	.129	.133	.137	.141	.144	.148	.152
0.4	.155	.159	.163	.166	.170	.174	.177	.181	.184	.188
0.5	.191	.195	.198	.202	.205	.209	.212	.216	.219	.222
0.6	.226	.229	.232	.236	.239	.242	.245	.249	.252	.255
0.7	.258	.261	.264	.267	.270	.273	.276	.279	.282	.285
0.8	.288	.291	.294	.297	.300	.302	.305	.308	.311	.313
0.9	.316	.319	.321	.324	.326	.329	.331	.334	.336	.339
1.0	.341	.344	.346	.348	.351	.353	.355	.358	.360	.362
1.1	.364	.367	.369	.371	.373	.375	.377	.379	.381	.383
1.2	.385	.387	.389	.391	.393	.394	.396	.398	.400	.401
1.3	.403	.405	.407	.408	.410	.411	.413	.415	.416	.418
1.4	.419	.421	.422	.424	.425	.426	.428	.429	.431	.432
1.5	.433	.434	.436	.437	.438	.439	.441	.442	.443	.444
1.6	.445	.446	.447	.448	.449	.451	.452	.453	.454	.454
1.7	.455	.456	.457	.458	.459	.460	.461	.462	.462	.463
1.8	.464	.465	.466	.466	.467	.468	.469	.469	.470	.471
1.9	.471	.472	.473	.473	.474	.474	.475	.476	.476	.477
2.0	.477	.478	.478	.479	.479	.480	.480	.481	.481	.482
2.1	.482	.483	.483	.483	.484	.484	.485	.485	.485	.486
2.2	.486	.486	.487	.487	.487	.488	.488	.488	.489	.489
2.3	.489	.490	.490	.490	.490	.491	.491	.491	.491	.492
2.4	.492	.492	.492	.492	.493	.493	.493	.493	.493	.494
2.5	.494	.494	.494	.494	.494	.495	.495	.495	.495	.495
2.6	.495	.495	.496	.496	.496	.496	.496	.496	.496	.496
2.7	.497	.497	.497	.497	.497	.497	.497	.497	.497	.497
2.8	.497	.498	.498	.498	.498	.498	.498	.498	.498	.498
2.9	.498	.498	.498	.498	.498	.498	.498	.499	.499	.499
3.0	.499	.499	.499	.499	.499	.499	.499	.499	.499	.499
3.1	.499	.499	.499	.499	.499	.499	.499	.499	.499	.499
3.2	.499	.499	.499	.499	.499	.499	.499	.499	.499	.499
3.3	.500	.500	.500	.500	.500	.500	.500	.500	.500	.500

Table III

z-Values for Rejection
Regions for Large-Sample
Tests of Hypotheses

Value of α	Type of Alternative Hypothesis, H_a		
	$>$	$<$	$\neq$
$\alpha = .10$	Reject H_0 if $z > 1.28$	Reject H_0 if $z < -1.28$	Reject H_0 if $z > 1.65$ or $z < -1.65$
$\alpha = .05$	Reject H_0 if $z > 1.65$	Reject H_0 if $z < -1.65$	Reject H_0 if $z > 1.96$ or $z < -1.96$
$\alpha = .01$	Reject H_0 if $z > 2.33$	Reject H_0 if $z < -2.33$	Reject H_0 if $z > 2.58$ or $z < -2.58$

Table IV

z-Scores for Large-Sample
Confidence Intervals

	Confidence Desired		
	90%	95%	99%
Value of z	1.65	1.96	2.58

Table V

t-Values for Rejection
Regions for Small-Sample
Tests of Hypotheses

df	$\alpha = .10$ One-tailed test	$\alpha = .10$ Two-tailed test	$\alpha = .05$ One-tailed test	$\alpha = .05$ Two-tailed test	$\alpha = .01$ One-tailed test	$\alpha = .01$ Two-tailed test
1	3.08	6.31	6.31	12.71	31.82	63.66
2	1.89	2.92	2.92	4.30	6.97	9.92
3	1.64	2.35	2.35	3.18	4.54	5.84
4	1.53	2.13	2.13	2.78	3.75	4.60
5	1.48	2.02	2.02	2.57	3.37	4.03
6	1.44	1.94	1.94	2.45	3.14	3.71
7	1.42	1.89	1.89	2.36	3.00	3.50
8	1.40	1.86	1.86	2.31	2.90	3.36
9	1.38	1.83	1.83	2.26	2.82	3.25
10	1.37	1.81	1.81	2.23	2.76	3.17
11	1.36	1.80	1.80	2.20	2.72	3.11
12	1.36	1.78	1.78	2.18	2.68	3.05
13	1.35	1.77	1.77	2.16	2.65	3.01
14	1.35	1.76	1.76	2.14	2.62	2.98
15	1.34	1.75	1.75	2.13	2.60	2.95
16	1.34	1.75	1.75	2.12	2.58	2.92
17	1.33	1.74	1.74	2.11	2.57	2.90
18	1.33	1.73	1.73	2.10	2.55	2.88
19	1.33	1.73	1.73	2.09	2.54	2.86
20	1.33	1.72	1.72	2.09	2.53	2.85
21	1.32	1.72	1.72	2.08	2.52	2.83
22	1.32	1.72	1.72	2.07	2.51	2.82
23	1.32	1.71	1.71	2.07	2.50	2.81
24	1.32	1.71	1.71	2.06	2.49	2.80
25	1.32	1.71	1.71	2.06	2.49	2.79
26	1.32	1.71	1.71	2.06	2.48	2.78
27	1.31	1.70	1.70	2.05	2.47	2.77
28	1.31	1.70	1.70	2.05	2.47	2.76
z	1.28	1.65	1.65	1.96	2.33	2.58

Table VI

t-Values for
Confidence Intervals

df	Confidence Level		
	90%	95%	99%
1	6.31	12.71	63.66
2	2.92	4.30	9.92
3	2.35	3.18	5.84
4	2.13	2.78	4.60
5	2.02	2.57	4.03
6	1.94	2.45	3.71
7	1.89	2.36	3.50
8	1.86	2.31	3.36
9	1.83	2.26	3.25
10	1.81	2.23	3.17
11	1.80	2.20	3.11
12	1.78	2.18	3.05
13	1.77	2.16	3.01
14	1.76	2.14	2.98
15	1.75	2.13	2.95
16	1.75	2.12	2.92
17	1.74	2.11	2.90
18	1.73	2.10	2.88
19	1.73	2.09	2.86
20	1.72	2.09	2.85
21	1.72	2.08	2.83
22	1.72	2.07	2.82
23	1.71	2.07	2.81
24	1.71	2.06	2.80
25	1.71	2.06	2.79
26	1.71	2.06	2.78
27	1.70	2.05	2.77
28	1.70	2.05	2.76
z	1.65	1.96	2.58

Table VII

χ^2-Values for
Rejection Regions
A. One-Tailed Tests

df	$\alpha = .10$ One-tailed test, >	$\alpha = .10$ One-tailed test, <	$\alpha = .05$ One-tailed test, >	$\alpha = .05$ One-tailed test, <	$\alpha = .01$ One-tailed test, >	$\alpha = .01$ One-tailed test, <
1	2.71	0.0158	3.84	0.00393	6.64	0.000157
2	4.61	0.211	6.00	0.103	9.21	0.0201
3	6.25	0.584	7.82	0.352	11.4	0.115
4	7.78	1.0636	9.50	0.711	13.3	0.297
5	9.24	1.61	11.1	1.15	15.1	0.554
6	10.6	2.20	12.6	1.64	16.8	0.872
7	12.0	2.83	14.1	2.17	18.5	1.24
8	13.4	3.49	15.5	2.73	20.1	1.65
9	14.7	4.17	17.0	3.33	21.7	2.09
10	16.0	4.87	18.3	3.94	23.2	2.56
11	17.2	5.58	19.7	4.58	24.7	3.05
12	18.6	6.30	21.0	5.23	26.2	3.57
13	19.8	7.04	22.4	5.90	27.7	4.11
14	21.1	7.79	23.7	6.57	29.1	4.66
15	22.3	8.55	25.0	7.26	30.6	5.23
16	23.5	9.31	26.3	7.96	32.0	5.81
17	24.8	10.1	27.6	8.67	33.4	6.41
18	26.0	10.9	28.9	9.39	34.8	7.01
19	27.2	11.7	30.1	10.1	36.2	7.63
20	28.4	12.4	31.4	10.9	37.6	8.26
21	29.6	13.2	32.7	11.6	39.0	8.90
22	30.8	14.0	33.9	12.3	40.3	9.54
23	32.0	14.9	35.2	13.1	41.6	10.2
24	33.2	15.7	36.4	13.9	43.0	10.9
25	34.4	16.5	37.7	14.6	44.3	11.5
26	35.6	17.3	38.9	15.4	45.6	12.2
27	36.7	18.1	40.1	16.2	47.0	12.9
28	37.9	18.9	41.3	16.9	48.3	13.6
29	39.1	19.8	42.6	17.7	49.6	14.3
30	40.3	20.6	43.8	18.5	50.9	15.0
40	51.8	29.1	55.8	26.5	63.7	22.2
50	63.2	37.7	67.5	34.8	76.2	29.7
60	74.4	46.5	79.1	43.2	88.4	37.5
70	85.5	55.3	90.5	51.8	100.0	45.4
80	96.6	64.3	102.0	60.4	112.0	53.5
90	108.0	73.3	113.0	69.1	124.0	61.8
100	114.0	82.4	124.0	77.9	136.0	70.1

APPENDIX TABLES

Table VII

df	$\alpha = .10$ Smaller value	$\alpha = .10$ Larger value	$\alpha = .05$ Smaller value	$\alpha = .05$ Larger value	$\alpha = .01$ Smaller value	$\alpha = .01$ Larger value
1	0.00393	3.84	0.000982	5.02	0.0000393	7.88
2	0.103	6.00	0.0506	7.38	0.0100	10.6
3	0.352	7.82	0.216	9.35	0.0717	12.9
4	0.711	9.50	0.484	11.1	0.207	14.9
5	1.15	11.1	0.831	12.8	0.412	16.8
6	1.64	12.6	1.24	14.5	0.676	18.6
7	2.17	14.1	1.69	16.0	0.990	20.3
8	2.73	15.5	2.18	17.5	1.34	22.0
9	3.33	17.0	2.70	19.0	1.73	23.6
10	3.94	18.3	3.25	20.5	2.16	25.2
11	4.58	19.7	3.82	21.9	2.60	26.8
12	5.23	21.0	4.40	23.3	3.07	28.3
13	5.90	22.4	5.01	24.7	3.57	29.8
14	6.57	23.7	5.63	26.1	4.07	31.3
15	7.26	25.0	6.26	27.5	4.60	32.8
16	7.96	26.3	6.91	28.9	5.14	34.3
17	8.67	27.6	7.56	30.2	5.70	35.7
18	9.39	28.9	8.23	31.5	6.26	37.2
19	10.1	30.1	8.91	32.9	6.84	38.6
20	10.9	31.4	9.59	34.2	7.43	40.0
21	11.6	32.7	10.3	35.5	8.03	41.4
22	12.3	33.9	11.0	36.8	8.64	42.8
23	13.1	35.2	11.7	38.1	9.26	44.2
24	13.9	36.4	12.4	39.4	9.89	45.6
25	14.6	37.7	13.1	40.7	10.5	46.9
26	15.4	38.9	13.8	41.9	11.2	48.3
27	16.2	40.1	14.6	43.2	11.8	49.7
28	16.9	41.3	15.3	44.5	12.5	51.0
29	17.7	42.6	16.1	45.7	13.1	52.3
30	18.5	43.8	16.8	47.0	13.8	53.7
40	26.5	55.8	24.4	59.3	20.7	66.8
50	34.8	67.5	32.4	71.4	28.0	79.5
60	43.2	79.1	40.5	83.3	35.5	92.0
70	51.8	90.5	48.8	95.0	43.3	104.0
80	60.4	102.0	57.2	107.0	51.2	116.0
90	69.1	113.0	65.7	118.0	59.2	128.0
100	77.9	124.0	74.2	130.0	67.3	140.0

χ^2-Values for Rejection Regions
B. Two-Tailed Tests

Table VIII

df	90% confidence Smaller value	90% confidence Larger value	95% confidence Smaller value	95% confidence Larger value	99% confidence Smaller value	99% confidence Larger value
1	0.00393	3.84	0.000982	5.02	0.0000393	7.88
2	0.103	6.00	0.0506	7.38	0.0100	10.6
3	0.352	7.82	0.216	9.35	0.0717	12.9
4	0.711	9.50	0.484	11.1	0.207	14.9
5	1.15	11.1	0.831	12.8	0.412	16.8
6	1.64	12.6	1.24	14.5	0.676	18.6
7	2.17	14.1	1.69	16.0	0.990	20.3
8	2.73	15.5	2.18	17.5	1.34	22.0
9	3.33	17.0	2.70	19.0	1.73	23.6
10	3.94	18.3	3.25	20.5	2.16	25.2
11	4.58	19.7	3.82	21.9	2.60	26.8
12	5.23	21.0	4.40	23.3	3.07	28.3
13	5.90	22.4	5.01	24.7	3.57	29.8
14	6.57	23.7	5.63	26.1	4.07	31.3
15	7.26	25.0	6.26	27.5	4.60	32.8
16	7.96	26.3	6.91	28.9	5.14	34.3
17	8.67	27.6	7.56	30.2	5.70	35.7
18	9.39	28.9	8.23	31.5	6.26	37.2
19	10.1	30.1	8.91	32.9	6.84	38.6
20	10.9	31.4	9.59	34.2	7.43	40.0
21	11.6	32.7	10.3	35.5	8.03	41.4
22	12.3	33.9	11.0	36.8	8.64	42.8
23	13.1	35.2	11.7	38.1	9.26	44.2
24	13.9	36.4	12.4	39.4	9.89	45.6
25	14.6	37.7	13.1	40.7	10.5	46.9
26	15.4	38.9	13.8	41.9	11.2	48.3
27	16.2	40.1	14.6	43.2	11.8	49.7
28	16.9	41.3	15.3	44.5	12.5	51.0
29	17.7	42.6	16.1	45.7	13.1	52.3
30	18.5	43.8	16.8	47.0	13.8	53.7
40	26.5	55.8	24.4	59.3	20.7	66.8
50	34.8	67.5	32.4	71.4	28.0	79.5
60	43.2	79.1	40.5	83.3	35.5	92.0
70	51.8	90.5	48.8	95.0	43.3	104.0
80	60.4	102.0	57.2	107.0	51.2	116.0
90	69.1	113.0	65.7	118.0	59.2	128.0
100	77.9	124.0	74.2	130.0	67.3	140.0

χ^2-Values for
Confidence Intervals

Table IX

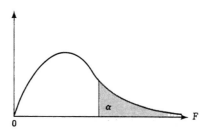

Percentage Points of the
F Distribution, $\alpha = .10$

		Numerator Degrees of Freedom							
		1	2	3	4	5	6	7	8	9
Denominator Degrees of Freedom	1	39.86	49.50	53.59	55.83	57.24	58.20	58.91	59.44	59.86
	2	8.53	9.00	9.16	9.24	9.29	9.33	9.35	9.37	9.38
	3	5.54	5.46	5.39	5.34	5.31	5.28	5.27	5.25	5.24
	4	4.54	4.32	4.19	4.11	4.05	4.01	3.98	3.95	3.94
	5	4.06	3.78	3.62	3.52	3.45	3.40	3.37	3.34	3.32
	6	3.78	3.46	3.29	3.18	3.11	3.05	3.01	2.98	2.96
	7	3.59	3.26	3.07	2.96	2.88	2.83	2.78	2.75	2.72
	8	3.46	3.11	2.92	2.81	2.73	2.67	2.62	2.59	2.56
	9	3.36	3.01	2.81	2.69	2.61	2.55	2.51	2.47	2.44
	10	3.29	2.92	2.73	2.61	2.52	2.46	2.41	2.38	2.35
	11	3.23	2.86	2.66	2.54	2.45	2.39	2.34	2.30	2.27
	12	3.18	2.81	2.61	2.48	2.39	2.33	2.28	2.24	2.21
	13	3.14	2.76	2.56	2.43	2.35	2.28	2.23	2.20	2.16
	14	3.10	2.73	2.52	2.39	2.31	2.24	2.19	2.15	2.12
	15	3.07	2.70	2.49	2.36	2.27	2.21	2.16	2.12	2.09
	16	3.05	2.67	2.46	2.33	2.24	2.18	2.13	2.09	2.06
	17	3.03	2.64	2.44	2.31	2.22	2.15	2.10	2.06	2.03
	18	3.01	2.62	2.42	2.29	2.20	2.13	2.08	2.04	2.00
	19	2.99	2.61	2.40	2.27	2.18	2.11	2.06	2.02	1.98
	20	2.97	2.59	2.38	2.25	2.16	2.09	2.04	2.00	1.96
	21	2.96	2.57	2.36	2.23	2.14	2.08	2.02	1.98	1.95
	22	2.95	2.56	2.35	2.22	2.13	2.06	2.01	1.97	1.93
	23	2.94	2.55	2.34	2.21	2.11	2.05	1.99	1.95	1.92
	24	2.93	2.54	2.33	2.19	2.10	2.04	1.98	1.94	1.91
	25	2.92	2.53	2.32	2.18	2.09	2.02	1.97	1.93	1.89
	26	2.91	2.52	2.31	2.17	2.08	2.01	1.96	1.92	1.88
	27	2.90	2.51	2.30	2.17	2.07	2.00	1.95	1.91	1.87
	28	2.89	2.50	2.29	2.16	2.06	2.00	1.94	1.90	1.87
	29	2.89	2.50	2.28	2.15	2.06	1.99	1.93	1.89	1.86
	30	2.88	2.49	2.28	2.14	2.05	1.98	1.93	1.88	1.85
	40	2.84	2.44	2.23	2.09	2.00	1.93	1.87	1.83	1.79
	60	2.79	2.39	2.18	2.04	1.95	1.87	1.82	1.77	1.74
	120	2.75	2.35	2.13	1.99	1.90	1.82	1.77	1.72	1.68
	∞	2.71	2.30	2.08	1.94	1.85	1.77	1.72	1.67	1.63

		\multicolumn{10}{c}{**Numerator Degrees of Freedom**}
		10	12	15	20	24	30	40	60	120	∞
	1	60.19	60.71	61.22	61.74	62.00	62.26	62.53	62.79	63.06	63.33
	2	9.39	9.41	9.42	9.44	9.45	9.46	9.47	9.47	9.48	9.49
	3	5.23	5.22	5.20	5.18	5.18	5.17	5.16	5.15	5.14	5.13
	4	3.92	3.90	3.87	3.84	3.83	3.82	3.80	3.79	3.78	3.76
	5	3.30	3.27	3.24	3.21	3.19	3.17	3.16	3.14	3.12	3.10
	6	2.94	2.90	2.87	2.84	2.82	2.80	2.78	2.76	2.74	2.72
	7	2.70	2.67	2.63	2.59	2.58	2.56	2.54	2.51	2.49	2.47
	8	2.54	2.50	2.46	2.42	2.40	2.38	2.36	2.34	2.32	2.29
	9	2.42	2.38	2.34	2.30	2.28	2.25	2.23	2.21	2.18	2.16
	10	2.32	2.28	2.24	2.20	2.18	2.16	2.13	2.11	2.08	2.06
	11	2.25	2.21	2.17	2.12	2.10	2.08	2.05	2.03	2.00	1.97
	12	2.19	2.15	2.10	2.06	2.04	2.01	1.99	1.96	1.93	1.90
	13	2.14	2.10	2.05	2.01	1.98	1.96	1.93	1.90	1.88	1.85
	14	2.10	2.05	2.01	1.96	1.94	1.91	1.89	1.86	1.83	1.80
	15	2.06	2.02	1.97	1.92	1.90	1.87	1.85	1.82	1.79	1.76
	16	2.03	1.99	1.94	1.89	1.87	1.84	1.81	1.78	1.75	1.72
	17	2.00	1.96	1.91	1.86	1.84	1.81	1.78	1.75	1.72	1.69
	18	1.98	1.93	1.89	1.84	1.81	1.78	1.75	1.72	1.69	1.66
	19	1.96	1.91	1.86	1.81	1.79	1.76	1.73	1.70	1.67	1.63
Denominator Degrees of Freedom	20	1.94	1.89	1.84	1.79	1.77	1.74	1.71	1.68	1.64	1.61
	21	1.92	1.87	1.83	1.78	1.75	1.72	1.69	1.66	1.62	1.59
	22	1.90	1.86	1.81	1.76	1.73	1.70	1.67	1.64	1.60	1.57
	23	1.89	1.84	1.80	1.74	1.72	1.69	1.66	1.62	1.59	1.55
	24	1.88	1.83	1.78	1.73	1.70	1.67	1.64	1.61	1.57	1.53
	25	1.87	1.82	1.77	1.72	1.69	1.66	1.63	1.59	1.56	1.52
	26	1.86	1.81	1.76	1.71	1.68	1.65	1.61	1.58	1.54	1.50
	27	1.85	1.80	1.75	1.70	1.67	1.64	1.60	1.57	1.53	1.49
	28	1.84	1.79	1.74	1.69	1.66	1.63	1.59	1.56	1.52	1.48
	29	1.83	1.78	1.73	1.68	1.65	1.62	1.58	1.55	1.51	1.47
	30	1.82	1.77	1.72	1.67	1.64	1.61	1.57	1.54	1.50	1.46
	40	1.76	1.71	1.66	1.61	1.57	1.54	1.51	1.47	1.42	1.38
	60	1.71	1.66	1.60	1.54	1.51	1.48	1.44	1.40	1.35	1.29
	120	1.65	1.60	1.55	1.48	1.45	1.41	1.37	1.32	1.26	1.19
	∞	1.60	1.55	1.49	1.42	1.38	1.34	1.30	1.24	1.17	1.00

Source: From M. Merrington and C. M. Thompson, "Tables of Percentage Points of the Inverted Beta (*F*)-Distribution," *Biometrika*, 1943, *33*, 73–88. Reproduced by permission of the *Biometrika* Trustees.

Table X

Percentage Points of the F Distribution, α = .05

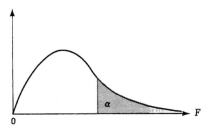

		Numerator Degrees of Freedom							
	1	2	3	4	5	6	7	8	9
1	161.4	199.5	215.7	224.6	230.2	234.0	236.8	238.9	240.5
2	18.51	19.00	19.16	19.25	19.30	19.33	19.35	19.37	19.38
3	10.13	9.55	9.28	9.12	9.01	8.94	8.89	8.85	8.81
4	7.71	6.94	6.59	6.39	6.26	6.16	6.09	6.04	6.00
5	6.61	5.79	5.41	5.19	5.05	4.95	4.88	4.82	4.77
6	5.99	5.14	4.76	4.53	4.39	4.28	4.21	4.15	4.10
7	5.59	4.74	4.35	4.12	3.97	3.87	3.79	3.73	3.68
8	5.32	4.46	4.07	3.84	3.69	3.58	3.50	3.44	3.39
9	5.12	4.26	3.86	3.63	3.48	3.37	3.29	3.23	3.18
10	4.96	4.10	3.71	3.48	3.33	3.22	3.14	3.07	3.02
11	4.84	3.98	3.59	3.36	3.20	3.09	3.01	2.95	2.90
12	4.75	3.89	3.49	3.26	3.11	3.00	2.91	2.85	2.80
13	4.67	3.81	3.41	3.18	3.03	2.92	2.83	2.77	2.71
14	4.60	3.74	3.34	3.11	2.96	2.85	2.76	2.70	2.65
15	4.54	3.68	3.29	3.06	2.90	2.79	2.71	2.64	2.59
16	4.49	3.63	3.24	3.01	2.85	2.74	2.66	2.59	2.54
17	4.45	3.59	3.20	2.96	2.81	2.70	2.61	2.55	2.49
18	4.41	3.55	3.16	2.93	2.77	2.66	2.58	2.51	2.46
19	4.38	3.52	3.13	2.90	2.74	2.63	2.54	2.48	2.42
20	4.35	3.49	3.10	2.87	2.71	2.60	2.51	2.45	2.39
21	4.32	3.47	3.07	2.84	2.68	2.57	2.49	2.42	2.37
22	4.30	3.44	3.05	2.82	2.66	2.55	2.46	2.40	2.34
23	4.28	3.42	3.03	2.80	2.64	2.53	2.44	2.37	2.32
24	4.26	3.40	3.01	2.78	2.62	2.51	2.42	2.36	2.30
25	4.24	3.39	2.99	2.76	2.60	2.49	2.40	2.34	2.28
26	4.23	3.37	2.98	2.74	2.59	2.47	2.39	2.32	2.27
27	4.21	3.35	2.96	2.73	2.57	2.46	2.37	2.31	2.25
28	4.20	3.34	2.95	2.71	2.56	2.45	2.36	2.29	2.24
29	4.18	3.33	2.93	2.70	2.55	2.43	2.35	2.28	2.22
30	4.17	3.32	2.92	2.69	2.53	2.42	2.33	2.27	2.21
40	4.08	3.23	2.84	2.61	2.45	2.34	2.25	2.18	2.12
60	4.00	3.15	2.76	2.53	2.37	2.25	2.17	2.10	2.04
120	3.92	3.07	2.68	2.45	2.29	2.17	2.09	2.02	1.96
∞	3.84	3.00	2.60	2.37	2.21	2.10	2.01	1.94	1.88

Denominator Degrees of Freedom

		Numerator Degrees of Freedom								
	10	12	15	20	24	30	40	60	120	∞
1	241.9	243.9	245.9	248.0	249.1	250.1	251.1	252.2	253.3	254.3
2	19.40	19.41	19.43	19.45	19.45	19.46	19.47	19.48	19.49	19.50
3	8.79	8.74	8.70	8.66	8.64	8.62	8.59	8.57	8.55	8.53
4	5.96	5.91	5.86	5.80	5.77	5.75	5.72	5.69	5.66	5.63
5	4.74	4.68	4.62	4.56	4.53	4.50	4.46	4.43	4.40	4.36
6	4.06	4.00	3.94	3.87	3.84	3.81	3.77	3.74	3.70	3.67
7	3.64	3.57	3.51	3.44	3.41	3.38	3.34	3.30	3.27	3.23
8	3.35	3.28	3.22	3.15	3.12	3.08	3.04	3.01	2.97	2.93
9	3.14	3.07	3.01	2.94	2.90	2.86	2.83	2.79	2.75	2.71
10	2.98	2.91	2.85	2.77	2.74	2.70	2.66	2.62	2.58	2.54
11	2.85	2.79	2.72	2.65	2.61	2.57	2.53	2.49	2.45	2.40
12	2.75	2.69	2.62	2.54	2.51	2.47	2.43	2.38	2.34	2.30
13	2.67	2.60	2.53	2.46	2.42	2.38	2.34	2.30	2.25	2.21
14	2.60	2.53	2.46	2.39	2.35	2.31	2.27	2.22	2.18	2.13
15	2.54	2.48	2.40	2.33	2.29	2.25	2.20	2.16	2.11	2.07
16	2.49	2.42	2.35	2.28	2.24	2.19	2.15	2.11	2.06	2.01
17	2.45	2.38	2.31	2.23	2.19	2.15	2.10	2.06	2.01	1.96
18	2.41	2.34	2.27	2.19	2.15	2.11	2.06	2.02	1.97	1.92
19	2.38	2.31	2.23	2.16	2.11	2.07	2.03	1.98	1.93	1.88
20	2.35	2.28	2.20	2.12	2.08	2.04	1.99	1.95	1.90	1.84
21	2.32	2.25	2.18	2.10	2.05	2.01	1.96	1.92	1.87	1.81
22	2.30	2.23	2.15	2.07	2.03	1.98	1.94	1.89	1.84	1.78
23	2.27	2.20	2.13	2.05	2.01	1.96	1.91	1.86	1.81	1.76
24	2.25	2.18	2.11	2.03	1.98	1.94	1.89	1.84	1.79	1.73
25	2.24	2.16	2.09	2.01	1.96	1.92	1.87	1.82	1.77	1.71
26	2.22	2.15	2.07	1.99	1.95	1.90	1.85	1.80	1.75	1.69
27	2.20	2.13	2.06	1.97	1.93	1.88	1.84	1.79	1.73	1.67
28	2.19	2.12	2.04	1.96	1.91	1.87	1.82	1.77	1.71	1.65
29	2.18	2.10	2.03	1.94	1.90	1.85	1.81	1.75	1.70	1.64
30	2.16	2.09	2.01	1.93	1.89	1.84	1.79	1.74	1.68	1.62
40	2.08	2.00	1.92	1.84	1.79	1.74	1.69	1.64	1.58	1.51
60	1.99	1.92	1.84	1.75	1.70	1.65	1.59	1.53	1.47	1.39
120	1.91	1.83	1.75	1.66	1.61	1.55	1.50	1.43	1.35	1.25
∞	1.83	1.75	1.67	1.57	1.52	1.46	1.39	1.32	1.22	1.00

Denominator Degrees of Freedom

Source: From M. Merrington and C. M. Thompson, "Tables of Percentage Points of the Inverted Beta (F)-Distribution," Biometrika, 1943, 33, 73–88. Reproduced by permission of the Biometrika Trustees.

Table XI

Percentage Points of the F Distribution, $\alpha = .025$

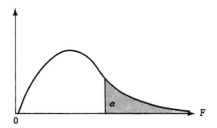

					Numerator Degrees of Freedom				
	1	**2**	**3**	**4**	**5**	**6**	**7**	**8**	**9**
1	647.8	799.5	864.2	899.6	921.8	937.1	948.2	956.7	963.3
2	38.51	39.00	39.17	39.25	39.30	39.33	39.36	39.37	39.39
3	17.44	16.04	15.44	15.10	14.88	14.73	14.62	14.54	14.47
4	12.22	10.65	9.98	9.60	9.36	9.20	9.07	8.98	8.90
5	10.01	8.43	7.76	7.39	7.15	6.98	6.85	6.76	6.68
6	8.81	7.26	6.60	6.23	5.99	5.82	5.70	5.60	5.52
7	8.07	6.54	5.89	5.52	5.29	5.12	4.99	4.90	4.82
8	7.57	6.06	5.42	5.05	4.82	4.65	4.53	4.43	4.36
9	7.21	5.71	5.08	4.72	4.48	4.32	4.20	4.10	4.03
10	6.94	5.46	4.83	4.47	4.24	4.07	3.95	3.85	3.78
11	6.72	5.26	4.63	4.28	4.04	3.88	3.76	3.66	3.59
12	6.55	5.10	4.47	4.12	3.89	3.73	3.61	3.51	3.44
13	6.41	4.97	4.35	4.00	3.77	3.60	3.48	3.39	3.31
14	6.30	4.86	4.24	3.89	3.66	3.50	3.38	3.29	3.21
15	6.20	4.77	4.15	3.80	3.58	3.41	3.29	3.20	3.12
16	6.12	4.69	4.08	3.73	3.50	3.34	3.22	3.12	3.05
17	6.04	4.62	4.01	3.66	3.44	3.28	3.16	3.06	2.98
18	5.98	4.56	3.95	3.61	3.38	3.22	3.10	3.01	2.93
19	5.92	4.51	3.90	3.56	3.33	3.17	3.05	2.96	2.88
20	5.87	4.46	3.86	3.51	3.29	3.13	3.01	2.91	2.84
21	5.83	4.42	3.82	3.48	3.25	3.09	2.97	2.87	2.80
22	5.79	4.38	3.78	3.44	3.22	3.05	2.93	2.84	2.76
23	5.75	4.35	3.75	3.41	3.18	3.02	2.90	2.81	2.73
24	5.72	4.32	3.72	3.38	3.15	2.99	2.87	2.78	2.70
25	5.69	4.29	3.69	3.35	3.13	2.97	2.85	2.75	2.68
26	5.66	4.27	3.67	3.33	3.10	2.94	2.82	2.73	2.65
27	5.63	4.24	3.65	3.31	3.08	2.92	2.80	2.71	2.63
28	5.61	4.22	3.63	3.29	3.06	2.90	2.78	2.69	2.61
29	5.59	4.20	3.61	3.27	3.04	2.88	2.76	2.67	2.59
30	5.57	4.18	3.59	3.25	3.03	2.87	2.75	2.65	2.57
40	5.42	4.05	3.46	3.13	2.90	2.74	2.62	2.53	2.45
60	5.29	3.93	3.34	3.01	2.79	2.63	2.51	2.41	2.33
120	5.15	3.80	3.23	2.89	2.67	2.52	2.39	2.30	2.22
∞	5.02	3.69	3.12	2.79	2.57	2.41	2.29	2.19	2.11

Denominator Degrees of Freedom

		Numerator Degrees of Freedom								
	10	12	15	20	24	30	40	60	120	∞
1	968.6	976.7	984.9	993.1	997.2	1001	1006	1010	1014	1018
2	39.40	39.41	39.43	39.45	39.46	39.46	39.47	39.48	39.49	39.50
3	14.42	14.34	14.25	14.17	14.12	14.08	14.04	13.99	13.95	13.90
4	8.84	8.75	8.66	8.56	8.51	8.46	8.41	8.36	8.31	8.26
5	6.62	6.52	6.43	6.33	6.28	6.23	6.18	6.12	6.07	6.02
6	5.46	5.37	5.27	5.17	5.12	5.07	5.01	4.96	4.90	4.85
7	4.76	4.67	4.57	4.47	4.42	4.36	4.31	4.25	4.20	4.14
8	4.30	4.20	4.10	4.00	3.95	3.89	3.84	3.78	3.73	3.67
9	3.96	3.87	3.77	3.67	3.61	3.56	3.51	3.45	3.39	3.33
10	3.72	3.62	3.52	3.42	3.37	3.31	3.26	3.20	3.14	3.08
11	3.53	3.43	3.33	3.23	3.17	3.12	3.06	3.00	2.94	2.88
12	3.37	3.28	3.18	3.07	3.02	2.96	2.91	2.85	2.79	2.72
13	3.25	3.15	3.05	2.95	2.89	2.84	2.78	2.72	2.66	2.60
14	3.15	3.05	2.95	2.84	2.79	2.73	2.67	2.61	2.55	2.49
15	3.06	2.96	2.86	2.76	2.70	2.64	2.59	2.52	2.46	2.40
16	2.99	2.89	2.79	2.68	2.63	2.57	2.51	2.45	2.38	2.32
17	2.92	2.82	2.72	2.62	2.56	2.50	2.44	2.38	2.32	2.25
18	2.87	2.77	2.67	2.56	2.50	2.44	2.38	2.32	2.26	2.19
19	2.82	2.72	2.62	2.51	2.45	2.39	2.33	2.27	2.20	2.13
20	2.77	2.68	2.57	2.46	2.41	2.35	2.29	2.22	2.16	2.09
21	2.73	2.64	2.53	2.42	2.37	2.31	2.25	2.18	2.11	2.04
22	2.70	2.60	2.50	2.39	2.33	2.27	2.21	2.14	2.08	2.00
23	2.67	2.57	2.47	2.36	2.30	2.24	2.18	2.11	2.04	1.97
24	2.64	2.54	2.44	2.33	2.27	2.21	2.15	2.08	2.01	1.94
25	2.61	2.51	2.41	2.30	2.24	2.18	2.12	2.05	1.98	1.91
26	2.59	2.49	2.39	2.28	2.22	2.16	2.09	2.03	1.95	1.88
27	2.57	2.47	2.36	2.25	2.19	2.13	2.07	2.00	1.93	1.85
28	2.55	2.45	2.34	2.23	2.17	2.11	2.05	1.98	1.91	1.83
29	2.53	2.43	2.32	2.21	2.15	2.09	2.03	1.96	1.89	1.81
30	2.51	2.41	2.31	2.20	2.14	2.07	2.01	1.94	1.87	1.79
40	2.39	2.29	2.18	2.07	2.01	1.94	1.88	1.80	1.72	1.64
60	2.27	2.17	2.06	1.94	1.88	1.82	1.74	1.67	1.58	1.48
120	2.16	2.05	1.94	1.82	1.76	1.69	1.61	1.53	1.43	1.31
∞	2.05	1.94	1.83	1.71	1.64	1.57	1.48	1.39	1.27	1.00

Denominator Degrees of Freedom

Source: From M. Merrington and C. M. Thompson, "Tables of Percentage Points of the Inverted Beta (*F*)-Distribution," *Biometrika,* 1943, *33,* 73–88. Reproduced by permission of the *Biometrika* Trustees.

Table XII

Percentage Points of the F Distribution, $\alpha = .01$

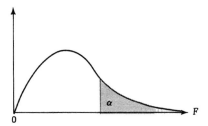

		Numerator Degrees of Freedom							
	1	**2**	**3**	**4**	**5**	**6**	**7**	**8**	**9**
1	4,052	4,999.5	5,403	5,625	5,764	5,859	5,928	5,982	6,022
2	98.50	99.00	99.17	99.25	99.30	99.33	99.36	99.37	99.39
3	34.12	30.82	29.46	28.71	28.24	27.91	27.67	27.49	27.35
4	21.20	18.00	16.69	15.98	15.52	15.21	14.98	14.80	14.66
5	16.26	13.27	12.06	11.39	10.97	10.67	10.46	10.29	10.16
6	13.75	10.92	9.78	9.15	8.75	8.47	8.26	8.10	7.98
7	12.25	9.55	8.45	7.85	7.46	7.19	6.99	6.84	6.72
8	11.26	8.65	7.59	7.01	6.63	6.37	6.18	6.03	5.91
9	10.56	8.02	6.99	6.42	6.06	5.80	5.61	5.47	5.35
10	10.04	7.56	6.55	5.99	5.64	5.39	5.20	5.06	4.94
11	9.65	7.21	6.22	5.67	5.32	5.07	4.89	4.74	4.63
12	9.33	6.93	5.95	5.41	5.06	4.82	4.64	4.50	4.39
13	9.07	6.70	5.74	5.21	4.86	4.62	4.44	4.30	4.19
14	8.86	6.51	5.56	5.04	4.69	4.46	4.28	4.14	4.03
15	8.68	6.36	5.42	4.89	4.56	4.32	4.14	4.00	3.89
16	8.53	6.23	5.29	4.77	4.44	4.20	4.03	3.89	3.78
17	8.40	6.11	5.18	4.67	4.34	4.10	3.93	3.79	3.68
18	8.29	6.01	5.09	4.58	4.25	4.01	3.84	3.71	3.60
19	8.18	5.93	5.01	4.50	4.17	3.94	3.77	3.63	3.52
20	8.10	5.85	4.94	4.43	4.10	3.87	3.70	3.56	3.46
21	8.02	5.78	4.87	4.37	4.04	3.81	3.64	3.51	3.40
22	7.95	5.72	4.82	4.31	3.99	3.76	3.59	3.45	3.35
23	7.88	5.66	4.76	4.26	3.94	3.71	3.54	3.41	3.30
24	7.82	5.61	4.72	4.22	3.90	3.67	3.50	3.36	3.26
25	7.77	5.57	4.68	4.18	3.85	3.63	3.46	3.32	3.22
26	7.72	5.53	4.64	4.14	3.82	3.59	3.42	3.29	3.18
27	7.68	5.49	4.60	4.11	3.78	3.56	3.39	3.26	3.15
28	7.64	5.45	4.57	4.07	3.75	3.53	3.36	3.23	3.12
29	7.60	5.42	4.54	4.04	3.73	3.50	3.33	3.20	3.09
30	7.56	5.39	4.51	4.02	3.70	3.47	3.30	3.17	3.07
40	7.31	5.18	4.31	3.83	3.51	3.29	3.12	2.99	2.89
60	7.08	4.98	4.13	3.65	3.34	3.12	2.95	2.82	2.72
120	6.85	4.79	3.95	3.48	3.17	2.96	2.79	2.66	2.56
∞	6.63	4.61	3.78	3.32	3.02	2.80	2.64	2.51	2.41

					Numerator Degrees of Freedom					
		10	12	15	20	24	30	40	60	120	∞
	1	6,056	6,106	6,157	6,209	6,235	6,261	6,287	6,313	6,339	6,366
	2	99.40	99.42	99.43	99.45	99.46	99.47	99.47	99.48	99.49	99.50
	3	27.23	27.05	26.87	26.69	26.60	26.50	26.41	26.32	26.22	26.13
	4	14.55	14.37	14.20	14.02	13.93	13.84	13.75	13.65	13.56	13.46
	5	10.05	9.89	9.72	9.55	9.47	9.38	9.29	9.20	9.11	9.02
	6	7.87	7.72	7.56	7.40	7.31	7.23	7.14	7.06	6.97	6.88
	7	6.62	6.47	6.31	6.16	6.07	5.99	5.91	5.82	5.74	5.65
	8	5.81	5.67	5.52	5.36	5.28	5.20	5.12	5.03	4.95	4.86
	9	5.26	5.11	4.96	4.81	4.73	4.65	4.57	4.48	4.40	4.31
	10	4.85	4.71	4.56	4.41	4.33	4.25	4.17	4.08	4.00	3.91
	11	4.54	4.40	4.25	4.10	4.02	3.94	3.86	3.78	3.69	3.60
	12	4.30	4.16	4.01	3.86	3.78	3.70	3.62	3.54	3.45	3.36
	13	4.10	3.96	3.82	3.66	3.59	3.51	3.43	3.34	3.25	3.17
Denominator Degrees of Freedom	14	3.94	3.80	3.66	3.51	3.43	3.35	3.27	3.18	3.09	3.00
	15	3.80	3.67	3.52	3.37	3.29	3.21	3.13	3.05	2.96	2.87
	16	3.69	3.55	3.41	3.26	3.18	3.10	3.02	2.93	2.84	2.75
	17	3.59	3.46	3.31	3.16	3.08	3.00	2.92	2.83	2.75	2.65
	18	3.51	3.37	3.23	3.08	3.00	2.92	2.84	2.75	2.66	2.57
	19	3.43	3.30	3.15	3.00	2.92	2.84	2.76	2.67	2.58	2.49
	20	3.37	3.23	3.09	2.94	2.86	2.78	2.69	2.61	2.52	2.42
	21	3.31	3.17	3.03	2.88	2.80	2.72	2.64	2.55	2.46	2.36
	22	3.26	3.12	2.98	2.83	2.75	2.67	2.58	2.50	2.40	2.31
	23	3.21	3.07	2.93	2.78	2.70	2.62	2.54	2.45	2.35	2.26
	24	3.17	3.03	2.89	2.74	2.66	2.58	2.49	2.40	2.31	2.21
	25	3.13	2.99	2.85	2.70	2.62	2.54	2.45	2.36	2.27	2.17
	26	3.09	2.96	2.81	2.66	2.58	2.50	2.42	2.33	2.23	2.13
	27	3.06	2.93	2.78	2.63	2.55	2.47	2.38	2.29	2.20	2.10
	28	3.03	2.90	2.75	2.60	2.52	2.44	2.35	2.26	2.17	2.06
	29	3.00	2.87	2.73	2.57	2.49	2.41	2.33	2.23	2.14	2.03
	30	2.98	2.84	2.70	2.55	2.47	2.39	2.30	2.21	2.11	2.01
	40	2.80	2.66	2.52	2.37	2.29	2.20	2.11	2.02	1.92	1.80
	60	2.63	2.50	2.35	2.20	2.12	2.03	1.94	1.84	1.73	1.60
	120	2.47	2.34	2.19	2.03	1.95	1.86	1.76	1.66	1.53	1.38
	∞	2.32	2.18	2.04	1.88	1.79	1.70	1.59	1.47	1.32	1.00

Source: From M. Merrington and C. M. Thompson, "Tables of Percentage Points of the Inverted Beta (*F*)-Distribution," *Biometrika*, 1943, *33*, 73–88. Reproduced by permission of the *Biometrika* Trustees.

Table XIII

Critical Values for the Wilcoxon Rank Sum Test

Test statistic is rank sum associated with smaller sample (if equal sample sizes, either rank sum can be used).

A. $\alpha = .025$ One-Tailed; $\alpha = .05$ Two-Tailed

	n_1 3		4		5		6		7		8		9		10	
n_2	T_L	T_U	T_L	T_U	T_L	T_U	T_L	T_U	T_L	T_U	T_L	T_U	T_L	T_U	T_L	T_U
3	5	16	6	18	6	21	7	23	7	26	8	28	8	31	9	33
4	6	18	11	25	12	28	12	32	13	35	14	38	15	41	16	44
5	6	21	12	28	18	37	19	41	20	45	21	49	22	53	24	56
6	7	23	12	32	19	41	26	52	28	56	29	61	31	65	32	70
7	7	26	13	35	20	45	28	56	37	68	39	73	41	78	43	83
8	8	28	14	38	21	49	29	61	39	73	49	87	51	93	54	98
9	8	31	15	41	22	53	31	65	41	78	51	93	63	108	66	114
10	9	33	16	44	24	56	32	70	43	83	54	98	66	114	79	131

B. $\alpha = .05$ One-Tailed; $\alpha = .10$ Two-Tailed

	n_1 3		4		5		6		7		8		9		10	
n_2	T_L	T_U	T_L	T_U	T_L	T_U	T_L	T_U	T_L	T_U	T_L	T_U	T_L	T_U	T_L	T_U
3	6	15	7	17	7	20	8	22	9	24	9	27	10	29	11	31
4	7	17	12	24	13	27	14	30	15	33	16	36	17	39	18	42
5	7	20	13	27	19	36	20	40	22	43	24	46	25	50	26	54
6	8	22	14	30	20	40	28	50	30	54	32	58	33	63	35	67
7	9	24	15	33	22	43	30	54	39	66	41	71	43	76	46	80
8	9	27	16	36	24	46	32	58	41	71	52	84	54	90	57	95
9	10	29	17	39	25	50	33	63	43	76	54	90	66	105	69	111
10	11	31	18	42	26	54	35	67	46	80	57	95	69	111	83	127

Source: From F. Wilcoxon and R. A. Wilcox, "Some Rapid Approximate Statistical Procedures," 1964, 20–23. Reproduced with the permission of American Cyanamid Company.

Table XIV

Critical Values for the
Wilcoxon Signed Rank Test

One-Tailed	Two-Tailed	$n = 5$	$n = 6$	$n = 7$	$n = 8$	$n = 9$	$n = 10$
$\alpha = .05$	$\alpha = .10$	1	2	4	6	8	11
$\alpha = .025$	$\alpha = .05$		1	2	4	6	8
$\alpha = .01$	$\alpha = .02$			0	2	3	5
$\alpha = .005$	$\alpha = .01$				0	2	3
		$n = 11$	$n = 12$	$n = 13$	$n = 14$	$n = 15$	$n = 16$
$\alpha = .05$	$\alpha = .10$	14	17	21	26	30	36
$\alpha = .025$	$\alpha = .05$	11	14	17	21	25	30
$\alpha = .01$	$\alpha = .02$	7	10	13	16	20	24
$\alpha = .005$	$\alpha = .01$	5	7	10	13	16	19
		$n = 17$	$n = 18$	$n = 19$	$n = 20$	$n = 21$	$n = 22$
$\alpha = .05$	$\alpha = .10$	41	47	54	60	68	75
$\alpha = .025$	$\alpha = .05$	35	40	46	52	59	66
$\alpha = .01$	$\alpha = .02$	28	33	38	43	49	56
$\alpha = .005$	$\alpha = .01$	23	28	32	37	43	49
		$n = 23$	$n = 24$	$n = 25$	$n = 26$	$n = 27$	$n = 28$
$\alpha = .05$	$\alpha = .10$	83	92	101	110	120	130
$\alpha = .025$	$\alpha = .05$	73	81	90	98	107	117
$\alpha = .01$	$\alpha = .02$	62	69	77	85	93	102
$\alpha = .005$	$\alpha = .01$	55	61	68	76	84	92
		$n = 29$	$n = 30$	$n = 31$	$n = 32$	$n = 33$	$n = 34$
$\alpha = .05$	$\alpha = .10$	141	152	163	175	188	201
$\alpha = .025$	$\alpha = .05$	127	137	148	159	171	183
$\alpha = .01$	$\alpha = .02$	111	120	130	141	151	162
$\alpha = .005$	$\alpha = .01$	100	109	118	128	138	149
		$n = 35$	$n = 36$	$n = 37$	$n = 38$	$n = 39$	
$\alpha = .05$	$\alpha = .10$	214	228	242	256	271	
$\alpha = .025$	$\alpha = .05$	195	208	222	235	250	
$\alpha = .01$	$\alpha = .02$	174	186	198	211	224	
$\alpha = .005$	$\alpha = .01$	160	171	183	195	208	
		$n = 40$	$n = 41$	$n = 42$	$n = 43$	$n = 44$	$n = 45$
$\alpha = .05$	$\alpha = .10$	287	303	319	336	353	371
$\alpha = .025$	$\alpha = .05$	264	279	295	311	327	344
$\alpha = .01$	$\alpha = .02$	238	252	267	281	297	313
$\alpha = .005$	$\alpha = .01$	221	234	248	262	277	292
		$n = 46$	$n = 47$	$n = 48$	$n = 49$	$n = 50$	
$\alpha = .05$	$\alpha = .10$	389	408	427	446	466	
$\alpha = .025$	$\alpha = .05$	361	379	397	415	434	
$\alpha = .01$	$\alpha = .02$	329	345	362	380	398	
$\alpha = .005$	$\alpha = .01$	307	323	339	356	373	

Source: From F. Wilcoxon and R. A. Wilcox, "Some Approximate Statistical Procedures," 1964, 28. Reproduced with the permission of American Cyanamid Company.

Table XV

The α values correspond to a one-tailed test of H_0: $\rho_s = 0$. The value should be doubled for two-tailed tests.

Critical Values of
Spearman's Rank
Correlation Coefficient

n	α = .05	α = .025	α = .01	α = .005
5	.900	—	—	—
6	.829	.886	.943	—
7	.714	.786	.893	—
8	.643	.738	.833	.881
9	.600	.683	.783	.833
10	.564	.648	.745	.794
11	.523	.623	.736	.818
12	.497	.591	.703	.780
13	.475	.566	.673	.745
14	.457	.545	.646	.716
15	.441	.525	.623	.689
16	.425	.507	.601	.666
17	.412	.490	.582	.645
18	.399	.476	.564	.625
19	.388	.462	.549	.608
20	.377	.450	.534	.591
21	.368	.438	.521	.576
22	.359	.428	.508	.562
23	.351	.418	.496	.549
24	.343	.409	.485	.537
25	.336	.400	.475	.526
26	.329	.392	.465	.515
27	.323	.385	.456	.505
28	.317	.377	.448	.496
29	.311	.370	.440	.487
30	.305	.364	.432	.478

Source: From E. G. Olds, "Distribution of Sums of Squares of Rank Differences for Small Samples," *Annals of Mathematical Statistics*, 1938, 9. Reproduced with the kind permission of the Institute of Mathematical Statistics.

ANSWERS TO SELECTED EXERCISES

CHAPTER 2

2.2 **a.** Quantitative **b.** Qualitative **c.** Qualitative **d.** Quantitative **e.** Quantitative
    f. Quantitative **g.** Qualitative **h.** Qualitative **i.** Quantitative **j.** Qualitative

2.3 **a.** Qualitative **b.** Quantitative **c.** Qualitative **d.** Qualitative **e.** Qualitative
    f. Quantitative **g.** Qualitative **h.** Quantitative **i.** Quantitative **j.** Quantitative

2.4 **a.** Qualitative **b.** Qualitative **c.** Qualitative **d.** Quantitative
    e. Quantitative **f.** Quantitative **g.** Qualitative **h.** Quantitative

2.5 **a.** Quantitative **b.** Qualitative **c.** Quantitative **d.** Qualitative
    e. Qualitative **f.** Quantitative **g.** Quantitative

2.8 **a.**

Category	Frequency	Relative Frequency	Percentage
Red	5	.05	5%
Green	20	.20	20%
Yellow	50	.50	50%
Blue	25	.25	25%
Totals	100	1.00	100%

2.9 **a.**

Category	Frequency	Relative Frequency	Percentage
A	20	.1	10%
B	40	.2	20%
C	20	.1	10%
D	40	.2	20%
E	80	.4	40%
Totals	200	1.0	100%

2.10 a.

Number	Frequency
1	6
2	6
3	2
4	5
5	2
6	9
Total	30

2.11 a.–d.

Letter	Frequency	Relative Frequency	Percentage
D	17	$^{17}/_{40} = .425$	42.5%
R	3	$^{3}/_{40} = .075$	7.5%
S	13	$^{13}/_{40} = .325$	32.5%
T	7	$^{7}/_{40} = .175$	17.5%
Totals	40	$^{40}/_{40} = 1.000$	100.0%

2.12 a.

Number	Frequency
0	5
1	15
2	10
3	6
4	12
5	2
Total	50

2.14 b. 21 **2.17 b.** 10.7% **2.18 b.** 32.7% **2.19 a.** White and others **b.** 100% − 21.8% = 78.2%
2.21 b. 5.0, 12.7 **c.** 6.4, 10.9 **2.22 b.** 10 **2.25 b.** 1.8 million

2.29 a.–b.

Category	Frequency
1.15 to 2.15	1
2.15 to 3.15	3
3.15 to 4.15	1
4.15 to 5.15	1
5.15 to 6.15	6
6.15 to 7.15	3
7.15 to 8.15	1
8.15 to 9.15	7
9.15 to 10.15	1
Total	24

2.31 a.

Category	Frequency	Relative Frequency
10.5 to 15.5	1	.05
15.5 to 20.5	2	.10
20.5 to 25.5	4	.20
25.5 to 30.5	6	.30
30.5 to 35.5	4	.20
35.5 to 40.5	3	.15
Totals	20	1.00

2.34 a. 7.5 to 9.5 **b.** .15 **c.** .20 **d.** 20 students

2.35 a.

Category	Frequency	Relative Frequency
0.45 to 3.95	7	.35
3.95 to 7.45	8	.40
7.45 to 10.95	2	.10
10.95 to 14.45	1	.05
14.45 to 17.95	0	.00
17.95 to 21.45	2	.10
Totals	20	1.00

c. .15 **2.38 b.** $^{17}/_{50} = .34$

2.41 Graph *b* has a stretched vertical axis. The vertical axis in graph *c* does not start at 0.

2.45 a.

Category	Frequency	Relative Frequency	Percentage
X	90	.45	45%
Y	50	.25	25%
Z	60	.30	30%
Totals	200	1.00	100%

2.47 a.

Type of Crime	Frequency	Relative Frequency	Percentage	**c.** Vandalism
Burglary	35	.070	7%	
Larceny	155	.310	31%	
Vandalism	310	.620	62%	
Totals	500	1.000	100%	

2.49 b. $^{19}/_{30}$ **c.** No; a histogram should be used.   **2.50 d.** 7   **2.51 b.** 48   **c.** 5
2.52 b. Yes   **c.** The $15,000 policy   **2.54** Red, .25; pink, .50; white, .25   **2.55 b.** 59   **c.** 24
2.57 It is impossible to make a statement of reliability.
2.58 a. Personal income tax, sales tax   **b.** Education K–12, health and welfare

CHAPTER 3

3.3 a. 5   **b.** 14   **c.** 2.8   **3.4 a.** 4.8, 5.5, 6   **b.** 1.2, 1, 0   **c.** 0.8, 1, 1   **3.5 a.** 5   **b.** 7   **c.** 5   **d.** 0.2
3.6 a. −0.4, −2, −2   **b.** 6, 7, 7   **c.** 20.9, 9, 9   **3.8 c.** 81.2, 83, 83
3.9 a. Skewed to the right   **b.** Skewed to the left   **c.** Skewed to the right   **d.** The shape would depend on the week
e. Skewed to the left   **f.** Skewed to the left
3.10 a. 19,325, 16,000, 16,000   **b.** 15,563, 16,000, 16,000; mean   **3.11 a.** 78.2, 84, 88   **b.** Median   **c.** 83.1, 86, 88; mean
3.12 a. No   **b.** Yes; 110   **3.13 a.** Mean   **b.** Median   **3.14** 69 inches   **3.15 a.** 3 hours   **b.** 2 hours   **d.** No
3.20 a. 14   **b.** 54   **c.** 5   **d.** 3.7   **e.** 1.9   **3.21 a.** 4, 2.6, 1.6   **b.** 4, 3.1, 1.8   **c.** 10, 9.2, 3.0
3.22 a. 10.16   **b.** 0.01   **c.** 26.05   **d.** 51.11   **3.23 a.** 10, 11.6, 3.4   **b.** 7, 5.2, 2.3   **c.** 99, 1,226.8, 35.0
3.24 Data set b is most variable; a is least variable.   **3.25 a.** 23.6, 9.6   **b.** 23.65, 9.58   **c.** 4.86, 3.09
3.26 0.110, 0.00028, .017
3.27 a. 54.9, 95.0, 9.7   **b.** $\bar{x} - 2s = 35.5$, $\bar{x} + 2s = 74.3$; 29 measurements
c. $\bar{x} - 3s = 25.8$, $\bar{x} + 3s = 84.0$; 30 measurements
3.28 First professor
3.29 a. Sample 1   **b.** Range is 10 for both samples.   **c.** Sample 1 variance = 26.4, sample 2 variance = 10
d. Variance
3.30 a. 341.4, 798.9   **b.** $^{25}/_{26}$ **c.** $^{25}/_{26}$
3.31 a. 2,097.4, 3,506.9   **b.** $\bar{x} - 2s = -4,916.4$, $\bar{x} + 2s = 9,111.2$ **c.** 17
3.33 a. 1   **b.** 2   **c.** −0.5   **d.** 1.33   **e.** −2   **3.34 a.** 10%, 50%   **b.** 90%, 1%   **c.** 25%, 89%
3.35 a. $z = 1.7$ **b.** $z = 1.2$ **c.** $z = 2.86$; this value is relatively largest
3.36 a. Approx. 17.0, approx. 28.0, approx. 38.5   **b.** Approx. 14.0, approx. 17.0, approx. 30.5, approx. 38.5
3.37 a. $z = -1.56$ **b.** $z = -2.5$; this value is relatively smallest   **c.** $z = -1.33$
3.38 a. $z = -2.7$; this value is relatively farthest from the mean   **b.** $z = 2.6$ **c.** $z = -1.6$
3.39 a. (i) 0.75, −1.75   (ii) 1.11, −1.11   (iii) 1.9, −1.43   (iv) 2.18, −0.76   (v) 1.30, −1.74   (vi) 2.26, −0.57   **b.** Yes
3.41 b. Student with a 680 on the SAT. The z-score is higher.   **3.43 a.** −3   **b.** 90 or above   **3.44 b.** 94   **c.** 0

3.45 b. 130   **c.** 1,965   **3.46 b.** 6.7, 2.4   **c.** 69   **d.** 70   **3.47 a.** 6   **b.** $s \approx 1.5$ **c.** 3.1, 1.8
3.48 a. 6   **b.** $s \approx 1.5$ **c.** 3.2, 1.8   **3.50** He will not buy the land.   **3.51** The number is near 50.
3.52 a. No   **b.** 70 to 130   **3.53 a.** 26.9   **b.** 10.8   **c.** 0, $\bar{x} - 3s$ is less than 0   **d.** 59, $\bar{x} + 3s = 59.3$
3.54 a. No; $z = 3.11$ (42 was the largest number.)   **b.** Skewed
3.55 a. 34.35, 17.60   **b.** 29 of 31   **3.56 a.** 180, 69.6   **b.** 18 of 19
3.58 a. -3.5, rare   **b.** 3.5, rare   **c.** -1.7, not rare   **d.** 0.5, not rare   **e.** -5, rare
3.59 $z = -2.5$; conclude that the production line is not operating correctly
3.60 a. Almost all are between $90 and $240   **b.** Yes; $z = -3$ **c.** $z = -1.4$; not rare
3.61 a. Almost none   **b.** Claim is probably wrong   **c.** $z = -1.2$; not rare   **3.62 a.** 7.6 to 27.6   **b.** Yes; $z = 1.48$
3.63 a. Yes; $z = -.88$ **b.** Yes; $z = 2.86$ **3.64 a.** 34, 100   **b.** 47, 52.5, 78
3.72 a. 3, 2, 7, 7, 2.6   **b.** 8, 9, 10, 18.7, 4.3   **c.** 2, 1.5, 5, 4.4, 2.1   **d.** 4.8, 3.5, 8, 12.9, 3.6
3.73 a. 3, 7.93, 2.82   **b.** 4.24, 3.90, 1.97   **c.** 1.29, 8.90, 2.98   **3.74 a.** -2 **b.** 2   **c.** -2.5 **d.** 1.67
3.75 a. Most measurements are between 65 and 85.   **b.** Almost all measurements are between 60 and 90.
    c. Yes; $z = 3.4$ **3.76** 10, 32, 5.7   **3.77** Variance would decrease in both cases.
3.78 a. 30.0, 4.8, 2.2   **3.79 a.** 13   **b.** 4.5, 2.1   **3.80 a.** 11, $s \approx 2.75$ **b.** 14.7, 11.2, 3.3   **c.** All the data
3.81 a. 58.2, 65.4, 8.1   **b.** Most in the interval $\bar{x} - 2s$ to $\bar{x} + 2s$; almost all in the interval $\bar{x} - 3s$ to $\bar{x} + 3s$ **c.** 24, 25
3.82 a. 22, $s \approx 5.5$ **b.** 33.9, 5.8   **c.** All the measurements
3.83 Almost all days would have between 53 and 113 jobs submitted.   **3.84** 58 minutes
3.85 a. 18.2, 13.6, 3.7   **b.** 20.2, 13.6, 3.7   **3.86 a.** 98.7, 330.2, 18.2   **b.** 49.3, 82.6, 9.1
3.87 Few boxes would contain more than 16.25 ounces.   **3.88** 12.6, 24, 52.7, 7.3   **3.89** No
3.90 3,075 to 3,175 would contain most audience sizes; 3,050 to 3,200 would contain almost all the audience sizes
3.91 a. 23.8, 6.6, 2.6   **b.** Most in the interval $\bar{x} - 2s$ to $\bar{x} + 2s$; almost all in the interval $\bar{x} - 3s$ to $\bar{x} + 3s$ **c.** 24, 25
3.92 a. -0.4, -3, -1.8, 0.8, 2.4, 0   **c.** 70 could be the lowest, and 97 is one of the highest
3.93 a. -3 **b.** Yes   **c.** No longer much evidence; $z = -1.5$ is not rare   **3.94** No; $z = 1.90$
3.95 a. Mathematics   **b.** English   **c.** Mathematics; compare z-scores   **3.96 a.** 60.47, 195.70, 13.99   **b.** 29
3.97 a. 58.2, 64.3, 69.3   **3.98 a.** $\bar{x} = 58,229.6$, $s = 123,557.4$ **b.** 3.76, -0.46 **c.** 6,800, $z = -0.42$
3.99 a. $z = 0.66$ **b.** $\mu + 3\sigma = 64.9\%$ (Actually, the maximum was 53%, in Upper Volta)
3.100 a. 0% to 56%   **b.** No; $z = 3.5$ **3.101 a.** $1,268.6, $942.5   **b.** $2,292.8, $430.6

CHAPTER 4

4.3   a. 840   **b.** 117,600   **c.** 90   **d.** 120   **e.** 30,240   **f.** 25

4.4   c. 30   **d.**

	First Draw	Second Draw		First Draw	Second Draw		First Draw	Second Draw
1.	1	2	11.	3	1	21.	5	1
2.	1	3	12.	3	2	22.	5	2
3.	1	4	13.	3	4	23.	5	3
4.	1	5	14.	3	5	24.	5	4
5.	1	6	15.	3	6	25.	5	6
6.	2	1	16.	4	1	26.	6	1
7.	2	3	17.	4	2	27.	6	2
8.	2	4	18.	4	3	28.	6	3
9.	2	5	19.	4	5	29.	6	4
10.	2	6	20.	4	6	30.	6	5

4.5 c. 24 **d.**

	First Letter	Second Letter	Third Letter		First Letter	Second Letter	Third Letter
1.	A	B	C	13.	C	A	B
2.	A	B	D	14.	C	A	D
3.	A	C	B	15.	C	B	A
4.	A	C	D	16.	C	B	D
5.	A	D	B	17.	C	D	A
6.	A	D	C	18.	C	D	B
7.	B	A	C	19.	D	A	B
8.	B	A	D	20.	D	A	C
9.	B	C	A	21.	D	B	A
10.	B	C	D	22.	D	B	C
11.	B	D	A	23.	D	C	A
12.	B	D	C	24.	D	C	B

4.6 c. 20 **d.**

	First Marble	Second Marble		First Marble	Second Marble
1.	R1	R2	11.	G1	G2
2.	R1	G1	12.	G1	O
3.	R1	G2	13.	G2	R1
4.	R1	O	14.	G2	R2
5.	R2	R1	15.	G2	G1
6.	R2	G1	16.	G2	O
7.	R2	G2	17.	O	R1
8.	R2	O	18.	O	R2
9.	G1	R1	19.	O	G1
10.	G1	R2	20.	O	G2

4.7

	Liked Best	Second Best		Liked Best	Second Best
1.	G	S	4.	S	W
2.	G	W	5.	W	S
3.	S	G	6.	W	G

4.8 We use H = High jump, T = Triple jump, L = Long jump, and D = 100-yard dash to denote the events.

	First Event	Second Event	Third Event		First Event	Second Event	Third Event		First Event	Second Event	Third Event
1.	H	T	L	9.	T	L	H	17.	L	D	H
2.	H	T	D	10.	T	L	D	18.	L	D	T
3.	H	L	T	11.	T	D	H	19.	D	H	T
4.	H	L	D	12.	T	D	L	20.	D	H	L
5.	H	D	T	13.	L	H	T	21.	D	T	H
6.	H	D	L	14.	L	H	D	22.	D	T	L
7.	T	H	L	15.	L	T	H	23.	D	L	H
8.	T	H	D	16.	L	T	D	24.	D	L	T

4.9 Let 1, 2, 3, 4, 5 represent the top five best sellers.

	First Book	Second Book		First Book	Second Book		First Book	Second Book		First Book	Second Book
1.	1	2	6.	2	3	11.	3	4	16.	4	5
2.	1	3	7.	2	4	12.	3	5	17.	5	1
3.	1	4	8.	2	5	13.	4	1	18.	5	2
4.	1	5	9.	3	1	14.	4	2	19.	5	3
5.	2	1	10.	3	2	15.	4	3	20.	5	4

4.10 We use S1 = Spinach and S2 = String beans to differentiate between these.

	First Vegetable	Second Vegetable		First Vegetable	Second Vegetable		First Vegetable	Second Vegetable
1.	A	B	11.	C	A	21.	S1	A
2.	A	C	12.	C	B	22.	S1	B
3.	A	P	13.	C	P	23.	S1	C
4.	A	S1	14.	C	S1	24.	S1	P
5.	A	S2	15.	C	S2	25.	S1	S2
6.	B	A	16.	P	A	26.	S2	A
7.	B	C	17.	P	B	27.	S2	B
8.	B	P	18.	P	C	28.	S2	C
9.	B	S1	19.	P	S1	29.	S2	P
10.	B	S2	20.	P	S2	30.	S2	S1

4.11 a. 742,560   **4.12 a.** 24,024   **4.13 a.** 1,004,000   **b.** (1,004,000) × (1,003,999)

4.16 a. $1/30$

b. A: 1, 2, 3, 4, 5, 6, 7, 8, 9, 11, 12, 13, 16, 17, 18, 21, 22, 26
B: 3, 4, 5, 8, 9, 10, 13, 14, 15, 16, 17, 18, 19, 20, 21, 22, 23, 24, 25, 26, 27, 28, 29, 30
C: 1, 3, 5, 12, 13, 15, 22, 24, 25
D: 1, 2, 3, 4, 5, 8, 10, 11, 12, 13, 14, 15, 17, 20, 21, 22, 23, 24, 25, 27, 29

c. $P(A) = {}^{18}/_{30} = {}^3/_5$, $P(B) = {}^{24}/_{30} = {}^4/_5$, $P(C) = {}^9/_{30} = {}^3/_{10}$, $P(D) = {}^{21}/_{30} = {}^7/_{10}$

4.17 a. A: 1, 2, 3, 4, 5, 6, 7, 8, 9, 11, 13, 14, 15, 17, 19, 20, 21, 23
B: 2, 5, 8, 11, 19, 21
C: 3, 4, 9, 10, 13, 14, 15, 16, 17, 18, 23, 24
D: Contains no samples.

b. $P(A) = {}^{18}/_{24} = {}^3/_4$, $P(B) = {}^6/_{24} = {}^1/_4$, $P(C) = {}^{12}/_{24} = {}^1/_2$, $P(D) = 0$

4.18 a. A: 1, 5, 11, 15
B: 1, 2, 3, 5, 6, 7, 9, 10, 11, 13, 14, 15
C: 17, 18, 19, 20
D: 4, 8, 12, 16, 17, 18, 19, 20

b. $P(A) = {}^4/_{20} = {}^1/_5$, $P(B) = {}^{12}/_{20} = {}^3/_5$, $P(C) = {}^4/_{20} = {}^1/_5$, $P(D) = {}^8/_{20} = {}^2/_5$

4.19 $P(A) = {}^{20}/_{72} = {}^5/_{18}$, $P(B) = {}^{12}/_{72} = {}^1/_6$, $P(C) = {}^{56}/_{72} = {}^7/_9$

4.20 $P(A) = {}^8/_{20} = {}^2/_5$, $P(B) = {}^{10}/_{20} = {}^1/_2$, $P(C) = {}^{14}/_{20} = {}^7/_{10}$, $P(D) = {}^4/_{20} = {}^1/_5$, $P(E) = {}^4/_{20} = {}^1/_5$

4.21 a. $1/6$ **b.** $2/6$

4.22 a. A: 1, 5

B: 1, 2, 3, 4, 5, 6, 7, 8, 9, 10, 13, 14, 17, 18

C: 2, 3, 4, 6, 7, 8, 9, 10, 13, 14, 17, 18

b. $P(A) = \frac{2}{20} = \frac{1}{10}$, $P(B) = \frac{14}{20} = \frac{7}{10}$, $P(C) = \frac{12}{20} = \frac{3}{5}$

4.23 $P(A) = \frac{2}{12} = \frac{1}{6}$, $P(B) = \frac{8}{12} = \frac{2}{3}$, $P(C) = \frac{10}{12} = \frac{5}{6}$

4.24 $P(A) = \frac{18}{38} = \frac{9}{19}$, $P(B) = \frac{18}{38} = \frac{9}{19}$, $P(C) = \frac{18}{38} = \frac{9}{19}$, $P(D) = \frac{2}{38} = \frac{1}{19}$

4.25 $P(A) = \frac{60}{504} = \frac{5}{42}$, $P(B) = \frac{24}{504} = \frac{1}{21}$ **4.26** $P(A) = \frac{6}{120} = \frac{1}{20}$, $P(B) = \frac{6}{120} = \frac{1}{20}$

4.27 $\frac{2,520}{1,860,480} \approx .001$ **4.28 a.** $\frac{2}{30} \approx .067$ **b.** $\frac{12}{30} = .40$ **4.29** $\frac{60}{504} \approx .119$

4.30 a. .0540   **b.** .9460   **c.** .0029   **d.** .8949   **4.32 b.** $\frac{6}{720} \approx .008$

4.33 $P(P) = \frac{30}{56} \approx .54$, $P(D) = \frac{2}{56} \approx .04$ **c.** No   **d.** Yes   **4.34** $P(4 \text{ aces}) = \frac{24}{6,497,400} \approx .000004$

4.35 b. $\frac{3,321}{112,892} \approx .03$ **c.** Yes   **4.36 b.** $\frac{89,700}{99,990,000} \approx .0009$ **c.** Yes   **d.** $\frac{94,080,300}{99,990,000} \approx .94$ **e.** No

4.37 a. $\frac{40,320}{1,814,400} \approx .02$ **4.38 a.** $\frac{117,600}{970,200} \approx .12$ **4.39 a.** $\frac{5,814}{970,200} \approx .006$

4.43 b. $P(E) = \frac{10}{12} = \frac{5}{6}$, $P(F) = \frac{8}{12} = \frac{2}{3}$, $P(E \cup F) = \frac{10}{12} = \frac{5}{6}$, $P(E \cap F) = \frac{8}{12} = \frac{2}{3}$, $P(F^c) = \frac{4}{12} = \frac{1}{3}$

c. $P(E \cup F) = \frac{10}{12} = \frac{5}{6}$, $P(F^c) = \frac{4}{12} = \frac{1}{3}$

4.44 b. $P(E) = \frac{8}{20} = \frac{2}{5}$, $P(F) = \frac{2}{20} = \frac{1}{10}$, $P(E \cup F) = \frac{10}{20} = \frac{1}{2}$, $P(E \cap F) = 0$, $P(E^c) = \frac{12}{20} = \frac{3}{5}$

c. $P(E \cup F) = \frac{10}{20} = \frac{1}{2}$, $P(E^c) = \frac{12}{20} = \frac{3}{5}$

4.45 a. $E \cap F$: 2, 5, 7, 8   **b.** $E \cup F$: Contains all twelve samples   **c.** E^c: 9, 12

d. $P(E) = \frac{10}{12} = \frac{5}{6}$, $P(F) = \frac{6}{12} = \frac{1}{2}$, $P(E \cap F) = \frac{4}{12} = \frac{1}{3}$, $P(E \cup F) = 1$, $P(E^c) = \frac{2}{12} = \frac{1}{6}$

e. $P(E \cup F) = 1$, $P(E^c) = \frac{2}{12} = \frac{1}{6}$

4.46 $P(A \cap B) = 0$, $P(B \cap C) = 0$, $P(C^c) = .40$, $P(A \cup C) = .85$, $P(B \cup C) = .75$

4.47 $P(A) = \frac{18}{38} = \frac{9}{19}$, $P(B) = \frac{18}{38} = \frac{9}{19}$, $P(A \cap B) = \frac{8}{38} = \frac{4}{19}$, $P(A \cup B) = \frac{28}{38} = \frac{14}{19}$, $P(C) = \frac{18}{38} = \frac{9}{19}$

4.51 $P(B|A) = \frac{1}{3}$, $P(B|A^c) = 0$, $P(B|C) = \frac{1}{9}$, $P(A|C) = \frac{1}{9}$, $P(C|A^c) = \frac{2}{3}$

4.52 a. 0   **b.** 0   **c.** No; $P(A|B) \neq P(A)$

4.53 a. $P(A|B) = \frac{1}{2}$, $P(B|A) = \frac{1}{3}$, $P(A \cap B) = \frac{1}{6}$ **b.** $P(A \cap B) = \frac{1}{5}$, $P(B|A) = \frac{2}{5}$

4.54 a. $P(E) = .16$, $P(F) = .64$, $P(E \cap F) = .16$ **b.** $P(E|F) = \frac{.16}{.64} = \frac{1}{4}$, $P(F|E) = \frac{.16}{.16} = 1$

4.55 a. $\frac{24}{50} = \frac{12}{25}$ **b.** $\frac{16}{50} = \frac{8}{25}$ **c.** $\frac{32}{50} = \frac{16}{25}$ **d.** $\frac{8}{50} = \frac{4}{25}$ **e.** $\frac{16}{50} = \frac{8}{25}$ **f.** $\frac{1}{2}$

g. No; $P(A) \neq P(A|B)$

4.56 a. .10   **b.** .67   **c.** .56   **4.57 a.** .30   **b.** $\frac{.10}{.52} \approx .19$ **c.** $\frac{.22}{.30} \approx .73$ **d.** $\frac{.40}{.48} \approx .83$

4.58 a. .15   **b.** .50   **4.59** .891

4.60 a. Yes; $P(A) = \frac{26}{52} = \frac{1}{2}$ and $P(A|B) = \frac{6}{12} = \frac{1}{2}$

b. Yes; $P(C) = \frac{26}{52} = \frac{1}{2}$ and $P(C|B) = \frac{6}{12} = \frac{1}{2}$

4.61 a. .028   **4.62 a.** .40   **b.** .30   **c.** $\frac{.15}{.40} = .375$ **d.** No; $P(A) = .40$ and $P(A|B) = .375$

4.63 a. 30   **b.** 3,024   **c.** 2,730   **d.** 1,860,480

4.64 c. 30

d.

	First Jelly Bean	Second Jelly Bean		First Jelly Bean	Second Jelly Bean		First Jelly Bean	Second Jelly Bean
1.	R1	R2	11.	R3	R1	21.	O2	R1
2.	R1	R3	12.	R3	R2	22.	O2	R2
3.	R1	O1	13.	R3	O1	23.	O2	R3
4.	R1	O2	14.	R3	O2	24.	O2	O1
5.	R1	B	15.	R3	B	25.	O2	B
6.	R2	R1	16.	O1	R1	26.	B	R1
7.	R2	R3	17.	O1	R2	27.	B	R2
8.	R2	O1	18.	O1	R3	28.	B	R3
9.	R2	O2	19.	O1	O2	29.	B	O1
10.	R2	B	20.	O1	B	30.	B	O2

e. A: 1, 2, 6, 7, 11, 12

B: 5, 10, 15, 20, 25, 26, 27, 28, 29, 30

C: 3, 4, 8, 9, 13, 14, 16, 17, 18, 19, 20, 21, 22, 23, 24, 25, 29, 30

D: Contains no samples

f. $P(A) = \frac{6}{30} = \frac{1}{5}$, $P(B) = \frac{10}{30} = \frac{1}{3}$, $P(C) = \frac{18}{30} = \frac{3}{5}$, $P(D) = 0$

4.65 $P(A) = \dfrac{6}{15,600} \approx .0002$, $P(B) = \dfrac{1,716}{15,600} = .11$, $P(C) = \dfrac{13,800}{15,600} \approx .88$

4.66 a. $P(G) = \dfrac{2,730}{6,840} \approx .40$, $P(W) = \dfrac{60}{6,840} \approx .009$ **b.** No; not a rare event **c.** Yes; rare event

4.67 a. $\frac{12}{30} = \frac{2}{5}$ **b.** $\frac{10}{30} = \frac{1}{3}$ **c.** $\frac{18}{30} = \frac{3}{5}$ **4.68** $P(A) = \frac{6}{20} = \frac{3}{10}$, $P(B) = \frac{18}{20} = \frac{9}{10}$, $P(C) = \frac{2}{20} = \frac{1}{10}$

4.69 a. $\dfrac{5,146}{41,641,670} \approx .0001$ **b.** Yes; rare event **c.** $\dfrac{71,703,853}{83,283,340} \approx .86$ **d.** No; not a rare event

4.70 a. 10,000 **b.** Sample space **c.** $P(A) = \dfrac{3,000}{10,000} = .30$, $P(B) = \dfrac{3,500}{10,000} = .35$, $P(C) = \dfrac{5,500}{10,000} = .55$, $P(D) = \dfrac{1,200}{10,000} = .12$

4.71 $P(A) = \frac{2}{30} = \frac{1}{15}$, $P(B) = \frac{12}{30} = \frac{2}{5}$, $P(C) = \frac{28}{30} = \frac{14}{15}$

4.72 $P(A) = \frac{175}{700} = .25$, $P(B) = \frac{665}{700} = .95$, $P(C) = 0$, $P(D) = \frac{455}{700} = .65$

4.73 a. $\dfrac{1,190}{489,300} \approx .002$ **b.** Yes; rare event **c.** Yes **4.74 a.** $\frac{174}{870} = .20$ **b.** $\frac{552}{870} \approx .63$ **c.** $\frac{30}{870} \approx .03$

4.75 $\dfrac{2,520}{6,720} = .375$ **4.76 a.** $\dfrac{1}{(26)(25)\cdots(17)}$ **b.** Yes; extremely rare event **4.77 a.** $\frac{174}{870} = .20$ **b.** $\frac{552}{870} \approx .63$

4.78 $\dfrac{360}{1,680} \approx .214$ **4.79 a.** $\dfrac{2,520}{30,240} \approx .083$ **b.** $\dfrac{5,040}{151,200} \approx .033$ **4.80** $\dfrac{990}{1,320} = .75$ **4.81** $\frac{42}{90} \approx .47$ **4.82** $\frac{2}{20} = .10$

4.83 a. $\dfrac{3,024}{5,040} = .60$ **b.** $\dfrac{360}{5,040} \approx .07$ **4.84 a.** 30,240 **b.** $\dfrac{2,520}{30,240} \approx .083$ **c.** $\dfrac{15,120}{30,240} = .50$

4.85 a. $\dfrac{(150)(149)(148)(147)(146)}{(1,000)(999)(998)(997)(996)} \approx .00007$

4.86 a. Both, .64; one, .32; neither, .04 **b.** Both, .72; one, .22; neither, .06 **c.** Model in part b is more realistic.

4.88 a. $P(A) = .50$, $P(B) = .50$, $P(C) = .30$, $P(D) = .15$, $P(E) = .20$, $P(F) = .35$ **b.** $P(A \cup B) = 1$ **c.** $P(B \cap C) = .05$

d. $P(A \cap F) = .05$ **e.** $P(A \cap B) = 0$ **f.** $P(C \cup D) = .45$

4.89 a. $P(A) = .15$, $P(B) = .85$, $P(C) = .70$, $P(D) = .15$, $P(E) = .05$, $P(F) = .10$ **b.** $P(A \cup B) = 1$ **c.** $P(B \cap F) = .08$
   d. $P(C \cup D) = .85$ **e.** $P(E \cap F) = 0$ **f.** $P(F \cup B) = .87$
4.90 .79

CHAPTER 5

5.3 a. Sum of probabilities cannot equal 1.1.   **b.** Negative probability is not possible.
5.4 a. $\frac{1}{6}$ **b.** $\frac{4}{6} = \frac{2}{3}$ **c.** $\frac{5}{6}$ **5.5 a.** $-4, 0, 1, 3$ **b.** 1   **c.** .9   **d.** 0   **e.** .8

5.6 a.

M	1	2
P(M)	$^{12}/_{24} = \frac{1}{2}$	$^{12}/_{24} = \frac{1}{2}$

b.

L	2	10
P(L)	$^{6}/_{24} = \frac{1}{4}$	$^{18}/_{24} = \frac{3}{4}$

c.

s^2	1.0	24.3	28.0	30.3
P(s^2)	$^{6}/_{24} = \frac{1}{4}$	$^{6}/_{24} = \frac{1}{4}$	$^{6}/_{24} = \frac{1}{4}$	$^{6}/_{24} = \frac{1}{4}$

5.7 a.

$\bar{x}$	1.5	2.0	2.5	3.5	4.0	4.5	5.0	6.5
P($\bar{x}$)	$\frac{1}{10}$	$\frac{1}{10}$	$\frac{1}{10}$	$\frac{1}{10}$	$\frac{2}{10} = \frac{1}{5}$	$\frac{2}{10} = \frac{1}{5}$	$\frac{1}{10}$	$\frac{1}{10}$

b.

R	1	2	3	4	5	6
P(R)	$\frac{3}{10}$	$\frac{1}{10}$	$\frac{1}{10}$	$\frac{2}{10} = \frac{1}{5}$	$\frac{2}{10} = \frac{1}{5}$	$\frac{1}{10}$

c.

X	1	2	3	6	7
P(X)	$\frac{1}{5}$	$\frac{1}{5}$	$\frac{1}{5}$	$\frac{1}{5}$	$\frac{1}{5}$

5.8 a. .048   **b.** .524
5.10 a. Distribution I   **b.** For distribution I, $P(X > 5) = .08$; for distribution II, $P(X > 5) = .73$.
   c. For distribution I, $P(X \leq 3) = .61$; for distribution II, $P(X \leq 3) = .08$.

5.11 a.

X	0	1	2
P(X)	$\frac{1}{6}$	$\frac{4}{6} = \frac{2}{3}$	$\frac{1}{6}$

b. $\frac{1}{6}$ **c.** $\frac{1}{6}$

5.12 a.

X	0	1	2	3
P(X)	$\frac{1}{20}$	$\frac{9}{20}$	$\frac{9}{20}$	$\frac{1}{20}$

b. $\frac{1}{20}$ **c.** $^{19}/_{20}$

5.13 a. Yes; $\frac{21}{15{,}504} \approx .001$ **b.** 3   **c.** $\frac{12{,}298}{15{,}504}$ **5.14 a.** $\frac{588{,}120}{742{,}560} \approx .79$ **b.** No; $\frac{1{,}560}{742{,}560} \approx .002$

5.15 a. $\frac{18{,}000}{24{,}024} \approx .75$ **b.** $\frac{24}{24{,}024} \approx .001$ **5.20 a.** 10   **5.21 a.** 1.5

5.22 b. (1) 5.5, (2) 2.0, (3) 7.2   **5.23** Yes, $\mu_X = \$11{,}500$
5.24 a. \$79   **b.** No, the average cost is over \$70 (the normal fare).

5.25 a.

X	−1	5
P(X)	$7/8$	$1/8$

b. −.25   **c.** No, the mean gain is negative—a loss.   **5.26** 1.29   **5.27** 2.72

5.28 a. 114.4   **5.29 a.** $\dfrac{3{,}071}{39} \approx 78.7$ **5.32 a.** 140   **b.** 11.8   **5.33 a.** 4.8   **b.** 12.56   **c.** 3.5

5.34 a. 0, 3.96, 1.99   **c.** $\mu_X - 2\sigma_X = -3.98$; $\mu_X + 2\sigma_X = 3.98$; probability that X is in this interval is .90

5.35 c. (1) $\mu_X = 1.0$, $\sigma_X^2 = .6$; (2) $\mu_X = 1.0$, $\sigma_X^2 = .2$ **5.36 a.** $81.25   **b.** 195.3125   **c.** .90

5.37 a. 1.52   **b.** 1.4896   **c.** 1.22   **5.38** Yes; 60 is 2.5 standard deviations above the mean.

5.39 $\mu_X = 1$, $\sigma_X = \sqrt{.96} \approx .98$, $\mu_X - 2\sigma_X = -.96$, $\mu_X + 2\sigma_X = 2.96$; probability that X is in this interval is .92

5.40 a. $\sqrt{3.1616} \approx 1.78$ **b.** 1

5.41 a. 1,000.64   **b.** 31.6

   d. Depending on the economy, etc., order either $\mu_X - 2\sigma_X \approx 51$ or $\mu_X + 2\sigma_X \approx 178$ bicycles

5.42 a. 9.27   **b.** 4.29   **c.** $2/39 \approx .05$

5.45 a. Yes   **b.** No   **c.** No   **d.** No   **e.** Yes

5.46 a. .2, .008, .089   **b.** .8, .008, .089   **c.** .5, .0025, .05   **d.** .2, .0016, .04   **e.** .2, .0032, .057   **f.** .01, .0000198, .0044

5.47 a. .194   **b.** .403   **c.** .813   **d.** .00+   **e.** .998   **f.** .00+

5.48 a.

p	0	$1/5$	$2/5$	$3/5$	$4/5$	1
P(p)	.590	.328	.073	.008	.00+	.00+

b.

p	0	$1/5$	$2/5$	$3/5$	$4/5$	1
P(p)	.031	.156	.312	.312	.156	.031

c.

p	0	$1/5$	$2/5$	$3/5$	$4/5$	1
P(p)	.00+	.00+	.008	.073	.328	.590

5.49 a. .5, .025, .16   **c.** .978   **5.50 a.** .9, .0036, .06   **c.** .966

5.51 a. Yes   **b.** No   **c.** For part a: $\mu_p = .2$, $\sigma_p^2 = .008$, $\sigma_p = .089$ **5.52 a.** .6   **b.** .00+   **c.** Yes

5.53 a. .590   **b.** .410   **5.54** .098   **5.55** .021   **5.56 a.** .001   **b.** .595

5.57 a. .76   **b.** Between $\mu_p - 2\sigma_p = .70$ and $\mu_p + 2\sigma_p = .82$

5.58 a. .15, .000255, .016   **b.** $\mu_p - 3\sigma_p = .102$, $\mu_p + 3\sigma_p = .198$; probability is approximately 1

5.59 a. .009   **b.** Yes   **5.60 a.** .009   **b.** Yes   **c.** .153

5.61 a. .35, .0011, .034   **b.** 2.35; yes   **c.** −1.47; no   **d.** −2.94; yes   **5.62 a.** .009

5.63 $\mu_p = .001$, $\sigma_p = .0014$, $z = 5$ for $p = 4/500$; very improbable   **5.64 a.** .2   **b.** .2   **c.** .195   **d.** No

5.65 a. .00+   **b.** Conclude the drug was effective.

5.66 a. .12, .000264, .016   **b.** No; $z = -2.66$ for $p = 31/400$, which is improbable.   **c.** Yes   **5.67 a.** .273   **b.** No

5.68 a. It is not likely; $z = -2.58$ for $p = 80/250$ **b.** Yes

5.69 a. 16.2, 18.76, 4.3   **b.** .4   **c.** $\mu_X - 2\sigma_X = 7.6$, $\mu_X + 2\sigma_X = 24.8$ **d.** 1

5.70 a.

	First Number	Second Number		First Number	Second Number		First Number	Second Number
1.	1	5	5.	5	9	9.	9	11
2.	1	9	6.	5	11	10.	11	1
3.	1	11	7.	9	1	11.	11	5
4.	5	1	8.	9	5	12.	11	9

b. $\bar{x}$	3	5	6	7	8	10
$P(\bar{x})$	$^2/_{12} = ^1/_6$	$^2/_{12} = ^1/_6$	$^2/_{12} = ^1/_6$	$^2/_{12} = ^1/_6$	$^2/_{12} = ^1/_6$	$^2/_{12} = ^1/_6$

c. R	2	4	6	8	10
$P(R)$	$^2/_{12} = ^1/_6$	$^4/_{12} = ^1/_3$	$^2/_{12} = ^1/_6$	$^2/_{12} = ^1/_6$	$^2/_{12} = ^1/_6$

d. 6.5, 4.9, 2.2   **e.** 5.7, 7.2, 2.7

5.71 b. .3, .0028, .053   **c.** $z = 2.45$; rare event
5.72 a. .124   **b.** .244   **c.** .756   **d.** .975   **e.** .985   **f.** .6   **g.** .012   **h.** .110   **i.** .036
5.73 22,250 square feet   **5.74 a.** .53   **b.** \$1,240   **c.** \$240   **d.** 312,400; \$558.9   **5.75 a.** .349   **b.** .930
5.76 a. .026   **b.** .375   **5.77 a.** .125   **b.** .056   **5.78** .651   **5.79 a.** .328   **b.** .205
5.80 a. Company A, 4.6; company B, 3.7   **b.** Company A, \$46,000; company B, \$55,500
　c. Variances: company A, 1.34; company B, 1.21; standard deviations: company A, 1.2; company B, 1.1
　d. Company A, .95; company B, .95
5.81 a. .657   **b.** .027   **5.82** .292   **5.83 a.** .2, .0064   **b.** .618   **c.** .005   **5.84 a.** .005   **5.85** .059
5.86 a. .021   **b.** .412   **5.87 a.** .50, .05   **b.** $\mu_p - 3\sigma_p = .35$, $\mu_p + 3\sigma_p = .65$

CHAPTER 6

6.5 For each part, $\mu_{\bar{x}} = 50$; the values of $\sigma_{\bar{x}}$ are: **a.** 2.65   **b.** 1.67   **c.** .84   **d.** 1   **e.** .26   **f.** .42
6.6 a. 10, 2   **b.** 20, 1   **c.** 50, 30   **d.** 30, 20
6.9 a. Smallest value = 98.94, largest value = 101.06   **b.** $3\sigma_{\bar{x}} = 1.06$ **c.** No
6.10 a. Number of breakdowns per 8-hour shift; $\mu = 1.5$, $\sigma = 1.1$ **b.** 1.5, .28
　c. Not necessarily; $n = 15$ is not a large sample size.
6.11 a. Approximately normal   **b.** 6, .35   **c.** No; 10 parts per billion is 11.43 standard errors above $\mu_{\bar{x}}$.
6.12 a. Normal
　b. The percentage is large. Most people would have an energy requirement within 2 standard deviations of the mean.
6.13 a. $\mu = 3.3$, $\sigma = 2.6$; describe the lengths of time a customer must wait   **b.** Approximately normal   **c.** 3.3, .26
6.14 a. Approximately normal   **b.** 2,460, 54.8   **6.17 a.** 0   **b.** 3.64   **c.** 1.74   **d.** −.79   **e.** .79   **f.** −3.16
6.18 a. −2.03   **b.** 7.60   **c.** −4.05   **d.** −1.01   **e.** −2.53   **f.** 2.87
6.19 a. .341   **b.** .045   **c.** .341   **d.** .982   **e.** .888   **f.** .308
6.20 a. .500   **b.** .878   **c.** .014   **d.** .363   **e.** .926   **f.** .307
6.21 a. .500   **b.** .627   **c.** .002   **d.** .020   **6.22 a.** .997   **b.** .082   **c.** .512   **d.** .987
6.23 a. 1.3, .24   **b.** Yes; $n = 50$ **c.** .106   **d.** .006   **6.24 a.** Approximately 0   **b.** .982   **c.** $\sigma_{\bar{x}}$ would decrease
6.25 a. .117   **b.** Decrease   **c.** No   **6.26 a.** .011   **b.** .063   **c.** .998   **6.27** Area is .023 or 2.3%
6.28 a. .071   **b.** .019   **6.29 a.** .005   **b.** .022   **6.32 a.** −1.95   **b.** 0   **c.** 1.95   **d.** 2.54   **e.** .39   **f.** −2.93
6.33 a. 9.06   **b.** −3.62   **c.** 1.58   **d.** 2.49   **e.** −1.58   **f.** −.45
6.34 a. No   **b.** Yes   **c.** Yes   **d.** No   **e.** Yes   **f.** Yes
6.35 a. Yes   **b.** Yes   **c.** No   **d.** No   **e.** Yes   **f.** No
6.36 a. .477   **b.** .477   **c.** .008   **d.** .096   **e.** .989   **f.** .785
6.37 a. .750   **b.** .006   **c.** .994   **d.** .797   **e.** .228   **f.** .015
6.38 a. .905   **b.** .052   **c.** .740   **d.** .989   **6.39 a.** .500   **b.** .940   **c.** .012   **d.** .001   **6.40 a.** .984   **b.** .111
6.41 .075   **6.42 a.** .977   **b.** .782   **6.43 a.** .040   **b.** .788   **6.44** .003   **6.45** .026

6.46 a. Yes; $\mu_p - 3\sigma_p = .10 > 0$ and $\mu_p + 3\sigma_p = .40 < 1$ **b.** .001 **6.47** $\bar{x} = 54.8$ and $\bar{x} = 53.6$
6.48 $\bar{x} = 6.65$ and $\bar{x} = 6.50$ **6.49** $p = .22$ **6.50** $p = .54$ and $p = .51$ **6.51 a.** .012
6.52 a. Distribution is approximately normal, with $\mu_{\bar{x}} = 100$ and $\sigma_{\bar{x}} = 1.06$ **b.** .002 **6.53 a.** .001
6.54 a. Approximately 0 **b.** No **6.55 a.** .104 **b.** No **6.56 a.** .062 **b.** No **6.57** Yes; $P(p \geq {}^{233}\!/_{300}) = .014$
6.58 a. 1.89 **b.** −3.14 **c.** 0 **d.** −1.89 **e.** 2.52 **f.** .63
6.59 a. −3.50 **b.** .25 **c.** −8.87 **d.** 2.62 **e.** 1.37 **f.** −1.87
6.60 a. 120, 2.34 **b.** Approximately normal **c.** .195 **d.** .884 **e.** .976 **f.** .125
6.61 a. .987 **b.** .709 **c.** .046 **d.** .369 **6.62 a.** .966 **b.** .907 **6.63 a.** .099 **b.** .042 **c.** .953
6.64 .989 **6.65 a.** .500 **b.** .002 **6.66 a.** .988 **b.** .576 **6.67 a.** .008 **b.** Yes
6.68 a. .2, .01 **b.** $P(p \geq .25) \approx 0$ **6.69** .026
6.70 a. Approximately normal with $\mu_{\bar{x}} = 45$, $\sigma_{\bar{x}} = .258$ **b.** .851 **c.** Approximately 0
6.71 a. Approximately normal with $\mu_{\bar{x}} = 250$, $\sigma_{\bar{x}} = 6.364$ **b.** .442 **c.** .009
6.72 a. Approximately normal with $\mu_{\bar{x}} = 400$, $\sigma_{\bar{x}} = 11.859$ **b.** .017, .500 **6.73 a.** .104 **b.** No **c.** .009; yes
6.74 .010 **6.75 a.** .017 **b.** The average distance jogged is more than 2.5 miles.

CHAPTER 7

7.6 **a.** Reject H_0 if $z > 2.58$ or $z < -2.58$. **b.** Reject H_0 if $z > 1.65$ or $z < -1.65$.
c. Reject H_0 if $z > 1.96$ or $z < -1.96$.
7.7 **a.** Reject H_0 if $z > 1.65$. **b.** Reject H_0 if $z > 1.96$ or $z < -1.96$. **c.** Reject H_0 if $z < -1.65$.
7.8 **a.** Reject H_0 if $z > 1.28$. **b.** Reject H_0 if $z > 1.65$. **7.9 a.** $z = 1.38$ **b.** $z = 1.39$ **c.** $z = -3.16$
7.10 **a.** $z = 1.57$ **b.** $z = 2.83$ **c.** $z = -2.03$ **7.11 a.** $z = 3.86$ **b.** $z = -1.08$ **c.** $z = -1.59$
7.12 **a. and b.** $\bar{x} = 1.03$, $s = .58$, $z = -1.81$ **7.15 a.** $H_0: \mu = 43$, $H_a: \mu > 43$ **b.** $z = 2.92$; yes
7.16 $z = -4.34$; yes **7.17** $z = 3.54$; yes **7.18** $z = -0.18$; no **7.19** $z = -3.20$; yes
7.20 **a.** $H_0: \mu = 122{,}000$, $H_a: \mu < 122{,}000$ **b.** $z = -2.07$; yes **7.21 a.** .055 **b.** .022 **c.** .005
7.22 **a.** c **b.** b, c **c.** a, b, c **7.23 a.** p-value = .117 **b.** p-value = .022; this provides more support
7.24 .006 **7.25 a.** .002 **b.** Yes; .002 < .01 **7.26 a.** .034 **7.28 a.** .27 **b.** No **7.29 a.** .019
7.30 **a.** 1.96 **b.** 1.65 **c.** 2.58 **7.32 a.** 72.4 to 77.6 **b.** 72.0 to 78.0 **c.** 71.0 to 79.0 **d.** Width increases
7.33 **a.** 38.5 to 41.5 **b.** 39.2 to 40.8 **c.** 49.1 to 50.9 **d.** 48.8 to 51.2
7.34 $\bar{x} = 7.8$, $s = 9.225$ **a.** 4.4 to 11.2 **b.** 5.6 to 10.0
7.37 **a.** 12.8 to 14.4 **b.** 13.2 to 14.0 **c.** Width is decreased by half. **7.38** 18.1 to 25.9 hours
7.39 8.94 to 10.26 hours **7.40** .6659 to .7741 second **7.41** 31.78 to 33.42 hours
7.42 **a.** 24.0 to 28.0 years **b.** 753.4 to 1,030.6 acres **7.43 a.** 2.10 to 2.70 **b.** 3.77 to 4.23
7.44 18.115 to 21.085 **7.45 a.** 4.617 to 5.623
7.48 **a.** Reject H_0 if $t > 2.09$ to $t < -2.09$. **b.** Reject H_0 if $t > 2.49$. **c.** Reject H_0 if $t > 1.38$.
d. Reject H_0 if $t < -2.68$. **e.** Reject H_0 if $t > 1.75$ or $t < -1.75$. **f.** Reject H_0 if $t < -1.89$.
7.49 **a.** 1.76 **b.** 4.03 **c.** 2.95 **d.** 2.05 **e.** 1.70
7.50 **a.** $t = -1.79$ **b.** $t = -1.79$ **c.** 2.94 to 6.66 **d.** 3.37 to 6.23
7.51 **a.** $\bar{x} = 5.0$, $s = 1.8$ **b.** $t = 3.54$ **c.** $t = 3.54$ **d.** $t = 3.54$ **e.** 3.11 to 6.89 **f.** 2.04 to 7.96
7.52 51.13 to 54.07 beats per minute **7.53 a.** $t = -2.01$; yes **b.** Type II error **c.** Type I error
7.54 19.733 to 19.987 ounces **7.55 b.** $\bar{x} = 4.68$, $s^2 = .667$; $t = -2.25$; no **7.56** 9.10 to 13.50 inches
7.57 $t = 3.57$; yes **7.58** $t = -4.19$; yes **7.59** $t = -3.63$; yes **7.60 a.** $t = -2.20$; no **7.61** 10.562% to 15.618%
7.62 **a.** 5.55 to 7.45 **7.63 a.** $t = .26$; no **7.64 a.** 2.64 to 10.12 **b.** −.26 to 13.02
7.65 **a.** .33 to .57 **b.** .35 to .55 **c.** $z = -.8$ **d.** $z = -.8$
7.66 **a.** $z = 3.33$ **b.** $z = 3.33$ **c.** $z = 1.67$ **d.** .84 to .96 **e.** .82 to .98 **7.67 b.** $z = -.99$ **7.68** .04 to .10
7.69 .58 to .68 **7.70** $z = -1.51$; no **7.71** $z = 2.5$; yes **7.72** .035 to .075 **7.73** .04 to .08 **7.74** $z = -1.24$; no

7.75 $z = 98$; no   **7.76** $z = -2.97$; yes   **7.77** .154 to .218   **7.78** $z = 1.48$; no   **7.79** .760 to .800

7.80 a. 683   **b.** 6,147   **c.** 4,734   **d.** 484   **7.81 a.** 5,991   **b.** 240   **c.** 2,305   **d.** 726   **7.82** 577

7.83 a. 1,430   **b.** 1,702   **7.84** 39   **7.85 a.** 69   **b.** Decrease   **7.86** 666   **7.87** 62   **7.88** 135   **7.89** 1,702

7.90 55   **7.91** 69   **7.92** 457   **7.93 a.** 2,401   **b.** 385 (rounding up from 384.16)   **7.94** 322

7.97 a. Reject H_0 if $\chi^2 < 8.91$ or $\chi^2 > 32.9$.   **b.** Reject H_0 if $\chi^2 > 43.0$.   **c.** Reject H_0 if $\chi^2 > 14.7$.

    d. Reject H_0 if $\chi^2 < 3.57$.   **e.** Reject H_0 if $\chi^2 < 1.64$ or $\chi^2 > 12.6$.   **f.** Reject H_0 if $\chi^2 < 3.33$.

7.98 a. 5.63, 26.1   **b.** 6.57, 23.7   **c.** 4.07, 31.3   **7.99 b.** $\chi^2 = 17.64$ **c.** $\chi^2 = 17.64$ **d.** 1.857 to 24.810

7.100 a. $\chi^2 = 436.59$ **c.** 3.521 to 5.604   **7.101** 1.64 to 7.19   **7.102** $\chi^2 = 4.5$; no   **7.103** $\chi^2 = 54$; yes

7.104 $\chi^2 = 133.90$; no   **7.105** .0014 to .0053

7.106 a. Reject H_0 if $z < -1.65$.   **b.** 2.58   **c.** 1.65   **d.** Reject H_0 if $z > 2.58$ or $z < -2.58$.

7.107 a. Reject H_0 if $\chi^2 < 13.9$.   **b.** 1.71   **c.** Reject H_0 if $t < -2.49$.   **d.** Reject H_0 if $\chi^2 < 13.9$ or $\chi^2 > 36.4$.

    e. 12.4, 39.4   **f.** Reject H_0 if $t > 1.71$.   **g.** 2.80

7.108 a. 65.10 to 80.10   **b.** $t = -1.71$ **c.** $t = -1.71$ **d.** 60.19 to 85.01   **e.** 161

7.109 a. $z = -1.78$ **b.** $z = -1.78$ **c.** .23 to .35   **d.** .21 to .37   **e.** 549

7.110 a. 8.08 to 8.32   **b.** $z = -1.67$ **c.** $z = -3.35$

7.111 a. 34.13 to 71.86   **b.** 28.51 to 92.00   **c.** $\chi^2 = 63.48$ **d.** $\chi^2 = 63.48$ **7.112** $z = -2.99$; yes

7.113 $t = 3.51$; no   **7.114 a.** $t = 1.56$; no   **7.115** $z = 6.13$; yes   **7.116** $z = -3.79$; yes

7.117 a. $t = 3.82$; yes   **b.** Not a random sample from all years   **7.118 a.** $z = 1.41$; no   **b.** Small

7.119 Small   **7.120 a.** 10.83 to 13.57 days   **b.** 459   **7.121 a.** $z = 1.07$; no   **b.** 65

7.122 a. 13.72 to 14.48 credit hours   **b.** $219.52 to $231.68   **7.123 a.** .038 to .112   **b.** $z = 1.62$; yes

7.124 3.27 to 3.93 hours   **7.125** .82 to .90   **7.126 a.** 47.9 to 60.1 inches   **b.** Measure of maximum snowfall

7.127 119   **7.128** .52 to .66   **7.129** .030 to .220   **7.130** .80 to .90   **7.131 a.** $z = -1.45$; yes   **7.132** .54 to .58

7.133 a. $z = -3.03$; yes   **b.** Not random   **7.134** .2215 to .3465   **7.135 a.** 2.75 to 2.85   **b.** 8,950

7.136 a. .623 to .677   **b.** 1,549   **7.137** .174 to .226   **7.138** .0047 to .0173   **7.139** $\chi^2 = 3.81$; no

7.140 $\chi^2 = 22.47$; no   **7.141 a.** 25.41 to 26.63   **b.** $t = -2.77$ **7.142 a.** $z = -1.88$ **b.** .03   **c.** 2.57 to 4.00

CHAPTER 8

8.2 Independent   **8.3** Independent   **8.4** Dependent   **8.6** Dependent   **8.7** Dependent   **8.8** Independent

8.10 a. 1.96   **b.** Reject H_0 if $z > 2.58$ or $z < -2.58$.   **c.** Reject H_0 if $z > 1.65$.   **d.** 1.65   **e.** Reject H_0 if $z < -2.33$.

8.11 a. $-.73$ to $-.27$ **b.** $z = -4.33$ **8.12 a.** -6.16 to -2.04 **b.** -7.31 to $-.89$ **c.** $z = -3.29$

8.13 a. $z = 1.67$ **b.** $z = 1.67$ **c.** $-.36$ to 4.56   **8.15 b.** $z = -3.76$; yes   **c.** -4.89 to $-.91$ **d.** Narrower

8.16 a. $z = 2.83$; yes   **b.** 9.2 to 50.8   **8.17** -5.36 to 12.16   **8.18** $z = 2.69$; yes   **8.19** $z = 3.43$; yes

8.20 $z = -2.09$; yes   **8.21 a.** $z = -4.22$; yes   **b.** -1.53 to $-.67$ **8.22** $z = 2.76$; yes

8.24 a. $z = 4.52$; yes   **b.** Yes   **8.25 a.** $z = -15.67$; yes   **c.** -4.30 to -2.34

8.27 a. Reject H_0 if $t > 2.49$.   **b.** Reject H_0 if $t > 1.71$ or $t < -1.71$.   **c.** 2.79   **d.** Reject H_0 if $t < -1.71$.   **e.** 2.06

8.28 a. 2.47, 3.55, .69, .34   **b.** .56   **c.** -1.98 to $-.18$ **d.** $t = -2.24$ **e.** -2.70 to .54   **f.** $t = -2.24$

8.29 a. $t = 1.58$ **b.** $-.87$ to 6.47   **c.** $-.23$ to 5.83   **d.** $t = 1.58$ **8.30** $t = .82$; no   **8.31** $t = -2.91$

8.32 $t = -1.56$; no   **8.33 a.** $t = 2.10$; yes   **b.** .7 to 7.3   **8.34** 4.372 to 8.808   **8.35** $t = 1.17$; no

8.36 a. $t = .61$ **b.** -4.46 to 9.26   **8.37 b.** $t = 1.53$

8.39 a. 2.20   **b.** 1.80   **c.** Reject H_0 if $t < -2.72$.   **d.** Reject H_0 if $t > 2.20$ or $t < -2.20$.

8.40 a. $\bar{x}_D = 2$, $s_D = 2.1$ **b.** $\mu_D = \mu_1 - \mu_2$ **c.** $-.2$ to 4.2   **d.** $t = 2.33$ **e.** .27 to 3.73   **f.** $t = 2.33$

8.41 a. $t = -6.87$; yes   **b.** -5.32 to -3.08 **8.42** 3.03 to 6.97   **8.43 a.** $t = 3.73$; yes   **b.** 1.23 to 4.11

8.44 $t = 3.66$; no   **8.45** $t = 2.98$; yes   **8.46 a.** $t = 7.68$, yes   **b.** .28 to .56

8.47 a. $t = .40$; no   **b.** -1.88 to 2.72   **d.** No!   **8.48 a.** No   **b.** $t = 5.83$; yes   **c.** 1.40 to 3.00

8.49 a. $t = -4.96$; yes   **b.** -11.20 to -3.80 **8.51 a.** $t = -2.58$ **8.52** $-$1,413.45 to $4,445.11

8.54 **a.** 1.65 **b.** Reject H_0 if $z > 2.33$. **c.** 2.58 **d.** Reject H_0 if $z > 1.96$ or $z < -1.96$. **e.** Reject H_0 if $z < -1.28$.
8.55 **a.** $-.02$ to $.24$ **b.** $-.176$ to $-.004$ **c.** $-.12$ to $.22$
8.56 **a.** $z = -3.00$ **b.** $z = -9.02$ **c.** $-.15$ to $-.05$ **d.** $-.19$ to $-.01$ **e.** $z = -3.00$
8.57 $z = -1.67$; no **8.58 a.** $z = 4$; yes **b.** $.12$ to $.28$ **8.59** $-.59$ to $-.13$ **8.60** $z = -.86$; no
8.61 **a.** $.088$ to $.108$ **b.** $z = 17.98$ **8.62 a.** $.135$ to $.205$ **b.** $z = 9.18$; yes **8.63** $-.202$ to $-.058$
8.64 $z = -2.14$; yes **8.66 a.** $n_1 = n_2 = 158$ **b.** $n_1 = n_2 = 54$ **c.** $n_1 = n_2 = 149$
8.67 **a.** $n_1 = n_2 = 29,954$ **b.** $n_1 = n_2 = 545$ **c.** $n_1 = n_2 = 273$ **8.68** $n_1 = n_2 = 769$ **8.69** $n_1 = n_2 = 137$
8.70 $n_1 = n_2 = 4,802$ **8.71** $n_1 = n_2 = 58$ **8.72** $n_1 = n_2 = 349$ **8.73** $n_1 = n_2 = 38$ **8.74** $n_1 = n_2 = 1,182$
8.76 **a.** Reject H_0 if $F > 3.02$. **b.** Reject H_0 if $F < .32$ or $F > 3.02$. **c.** Reject H_0 if $F > .19$. **d.** $F_1 = 3.02$, $F_2 = 3.14$
e. Reject H_0 if $F > 4.62$. Reject H_0 if $F < .34$ or $F > 4.62$. Reject H_0 if $F < .22$. $F_1 = 4.62$, $F_2 = 2.90$
8.77 **a.** $F = .40$ **b.** $.17$ to 1.13 **c.** $F = .40$
8.78 **a.** 2.05, $.41$ **b.** $.55$ to 27.05 **c.** $.34$ to 38.80 **d.** $F = 5.00$ **e.** $F = 5.00$ **8.79** 1.02 to 4.37
8.80 **a.** $F = .90$ **b.** $.28$ to 2.86 **8.81** $F = .25$ **8.82** $F = .38$; yes **8.83** $.01$ to $.96$
8.84 **a.** 2.58 **b.** Reject H_0 if $z > 1.28$.
8.85 **a.** Reject H_0 if $t > 2.13$ or $t < -2.13$. **b.** 1.75 **c.** $F_1 = 4.53$, $F_2 = 4.90$ **d.** Reject H_0 if $F < .16$.
e. Reject H_0 if $F < .27$ or $F > 3.50$.
8.86 **a.** 2.12 **b.** Reject H_0 if $t < -2.58$. **8.87 a.** Reject H_0 if $z > 1.65$. **b.** 1.96
8.88 **a.** $t = .78$ **b.** -6.5 to 11.5 **c.** $n_1 = n_2 = 225$ **8.89 a.** $.092$ to $.68$ **b.** $F = .26$ **c.** No
8.90 **a.** 3.49 to 4.11 **b.** $z = 20.08$ **c.** $n_1 = n_2 = 348$ **8.91 a.** $z = 2.56$ **b.** $.03$ to $.21$ **c.** $n_1 = n_2 = 18,942$
8.92 **a.** $t = 5.73$ **b.** 1.96 to 5.64 **8.93** **a.** $t = 1.06$ **b.** $F = 1.21$ **c.** $.38$ to 3.85 **8.94** $t = -4.02$
8.95 $z = .90$; no **8.96** **a.** -3.45 to -1.95 **b.** $n_1 = n_2 = 113$ **8.97** $n_1 = n_2 = 545$ **8.98** $z = 2.57$; yes
8.99 $n_1 = n_2 = 197$ **8.100** $t = -.70$; no **8.101** $z = -4.03$; yes **8.102** $t = -1.61$; no **8.103** $n_1 = n_2 = 193$
8.104 **a.** $t = 2.27$; yes **c.** $.20$ to 1.74 **8.105** $t = 6.14$; yes **8.106** $-.003$ to $.065$ **8.107** $z = -1.30$; no
8.108 $-\$256.26$ to $-\$97.74$ **8.109** $t = -2.12$; no **8.110** $z = -2.61$; yes **8.111** $F = 2.79$; no
8.112 **a.** $-\$3,159.51$ to $\$876.18$ **8.113 a.** $-.13$ to $.05$ **b.** $n_1 = n_2 = 1,184$
8.114 **a.** $t = -2.84$; yes **b.** -76.9 to -8.7 **8.115** $z = -2.93$; yes **8.116** $-.13$ to $.03$
8.117 **a.** $t = -1.38$; yes **b.** $F = .55$ **8.118** $n_1 = n_2 = 70$ **8.119** $t = -2.00$; yes **8.120** $F = 1.74$; no
8.121 $z = -54$; no **8.122 a.** $-.333$ to $-.267$ **b.** $-.332$ to $-.268$ **c.** The two intervals are almost identical.
8.123 **a.** 0.51 to $.109$ **b.** $n_1 = n_2 = 3,156$ **8.124 a.** Dependent **b.** 2.65 to 3.51 **8.125 a.** -7.06 to 2.06

CHAPTER 9

9.10 **a.** $2, 3$ **b.** $-1, 1$ **c.** $3, -2$ **d.** $5, 0$ **e.** $-3, 4$ **f.** $-2, -1$ **9.11 b.** $10, -20,000$ **9.12 b.** $-2, 250$
9.21 **a.** $100, -1,250$ **b.** -12.5 **c.** 35 **d.** $\hat{Y} = 35 - 12.5X$ **9.22 a.** $21, 91, 18, 78, 6$ **b.** $105, 90$ **c.** $0, .857$
9.23 **a.** $2, -1.2$ **c.** $.8$ **d.** 3.8 **9.24** **a.** $\hat{Y} = 57.912 - .811X$ **c.** $9.252 \approx 9$ **9.25 a.** $\hat{Y} = 133.088 - .265X$
9.26 **a.** $\hat{Y} = 288 - 9.28X$ **b.** $\$148.80$ **9.27 a.** $b_0 = 23.452$, $b_1 = 2.811$ **b.** 33.993 **9.30** 6.08
9.31 **a.** $68, 84$ **b.** $.286$ **9.32** $.533$ **9.33 a.** $3, 2, 200$ **b.** 0 **9.34 a.** $\hat{Y} = -45.78 + 1.562X$ **c.** 4.901
9.35 **a.** $-.246, 1.315$ **b.** 2.487 **9.36** 392.03 **9.37** 11.08 **9.38 a.** $t = .84$ **b.** $-.96$ to 2.24 **c.** $t = .84$
9.39 **a.** $t = -2.97$ **b.** -10.57 to -2.23
9.40 **a.** $6.404, -.617$ **b.** $.489$ **c.** -1.052 to $-.182$ **d.** $t = -6.53$ **e.** $t = -6.53$
9.41 **a.** $t = 6.35$; yes **b.** $.036$ to $.184$ **9.42 a.** $\hat{Y} = 18.890 + .011X$ **b.** $t = .81$; no **c.** $-.016$ to $.038$
9.43 **a.** $Y = \beta_0 + \beta_1 X + \varepsilon$, $\hat{Y} = 70.837 - .282X$ **b.** $t = .49$; no **9.44 b.** $29.423, -1.196$ **d.** $t = -4.30$
9.45 $t = 2.56$; no **9.46** $\$4.19$ to $\$7.09$ **9.47 a.** $t = -7.23$; yes **b.** -11.79 to -6.76
9.48 **a.** $t = 6.12$; yes **b.** 1.626 to 3.996 **9.51** $.94, .89$ **9.52** $.96, .92$ **9.53** $-.95, .90$
9.54 **a.** $-.69, .47$ **b.** $t = -.58$; no **9.55** $.917, .842$ **9.55 a.** $t = -.79$; no **b.** $-.37, .13$
9.57 **a.** $.995$ **b.** $.990$ **9.58** $.994, .989$ **9.59** $-.778, .621$ **9.60** $.704, .496$

9.63 a. 2.26 to 4.14   **b.** 1.24 to 5.16   **c.** −.54 to 4.54   **d.** −2.20 to 1.40
9.64 a. $\hat{Y} = -1.4 + .8X$ **b.** .05   **c.** 2.06 to 2.34   **d.** 2.12 to 3.08   **e.** .98 to 1.82   **f.** 2.92 to 3.08
9.65 a. $t = 6.97$; yes   **b.** 17.20 to 25.22   **c.** 13.32 to 29.10
9.66 a. $\hat{Y} = 4.554 + 2.671X$ **c.** .97, .93   **d.** 11.91 to 14.29   **e.** 12.14 to 13.00
9.67 a. $\hat{Y} = 1,082.402 - 31.809X$ **b.** $t = -17.08$; yes
    d. Prediction interval: 580.03 to 630.51; confidence interval: 595.74 to 614.80
9.68 a. $\hat{Y} = 40.926 - .134X$ **b.** $t = -4.19$; yes   **c.** 30.35 to 32.17   **d.** 30.93 to 35.35
9.71 a. $\hat{Y} = 27.78 - .6X$ **b.** .389   **c.** $t = -6.80$; yes   **d.** −.85, .72   **e.** 25.35 to 28.05   **f.** 26.39 to 27.01
9.72 a. −.12, .02   **c.** $t = -.35$; no
9.73 a. $\hat{Y} = 40.784 + .766X$ **c.** $t = 4.38$; yes   **d.** .362 to 1.170   **e.** 62.05 to 96.11   **f.** 59.83 to 75.35
9.74 a. $\hat{Y} = 22.325 + 2.650X$ **c.** $t = 6.28$; yes   **d.** .94, .89   **e.** 107.62 to 111.94   **f.** 103.68 to 115.88
9.75 a. $\hat{Y} = 18.224 + 1.166X$ **c.** $t = 4.97$; yes   **d.** .80   **e.** 104.11 to 142.29
9.76 a. $\hat{Y} = 2.084 + .668X$ **c.** $t = 16.28$; yes   **d.** .98   **e.** 34.51 to 36.47   **f.** 36.06 to 41.60
9.77 a. $\hat{Y} = -20.044 - .077X$ **b.** $t = -8.76$ **c.** .87   **d.** .712 to 1.242
9.78 a. $\hat{Y} = 16.283 + 2.078X$ **c.** $t = 6.67$; yes   **d.** 51.69 to 63.99
9.79 a. $\hat{Y} = 20.108 + .972X$ **c.** .98, .96   **d.** 26.90 to 28.86   **e.** 24.34 to 29.48
9.80 a. $\hat{Y} = 6.514 + 10.829X$ **b.** $t = 6.34$ **c.** 12.54 to 13.26
9.81 a. $\hat{Y} = 13.482 + .056X$ **c.** .74, .55   **d.** 14.93 to 19.87
9.82 a. $\hat{Y} = 2.758 + .866X$ **b.** .99, .97   **c.** 19.08 to 20.08   **9.83 a.** .89, .79   **b.** $t = 4.82$; yes
9.85 b. $\hat{Y} = 52.030 + .112X$ **c.** $t = 2.47$; yes   **d.** .64, .40   **e.** 4.4 to 100.0

CHAPTER 10

10.3 a. Reject H_0 if $F > 3.59$.   **b.** Reject H_0 if $F > 3.89$.   **c.** Reject H_0 if $F > 5.78$.
    d. Reject H_0 if $F > 2.57$.   **e.** Reject H_0 if $F > 2.39$.   **f.** Reject H_0 if $F > 3.15$.
    g. Reject H_0 if $F > 6.01$.   **h.** Reject H_0 if $F > 5.14$.

10.4 a.

Source	df	SS	MS	F
Treatments	2	11.075	5.538	3.15
Error	7	12.301	1.757	

10.5 a.

Source	df	SS	MS	F
Treatments	3	43.978	14.659	17.64
Error	14	11.633	.831	

10.6 a. Completely randomized

b. Yes;

Source	df	SS	MS	F
Hair colors	3	1,360.726	453.575	6.79
Error	15	1,001.800	66.787	

10.7 No;

Source	df	SS	MS	F
Nematicides	3	2.348	.783	3.20
Error	9	2.201	.245	

10.8 Yes;

Source	df	SS	MS	F
Conditions	2	673	336.5	4.55
Error	15	1,109.5	73.967	

10.9 Yes;

Source	df	SS	MS	F
Inoculins	4	292.8	73.2	5.29
Error	15	207.75	13.85	

10.10 b. Yes

10.13 a.

Source	df	SS	MS	F
Locations	2	13,728.667	6,864.333	16.95
Error	9	3,644.25	404.917	

b. −98.04 to −45.96   **c.** 400.51 to 445.99

10.14 a. No;

Source	df	SS	MS	F
Forms of compensation	2	4,195.35	2,097.675	3.17
Error	12	7,945.05	662.088	

b. 450.35 to 496.15   **c.** 6.33 to 67.77

10.15 a. 49.177 to 52.843   **b.** 6.571 to 11.549   **c.** −.984 to 2.024   **10.16 a.** 4.71 to 55.09   **b.** −14.19 to 36.19

10.17 a. Yes;

Source	df	SS	MS	F
Brands	2	6,323.267	3,161.633	3.53
Error	27	24,188.2	895.859	

b. −32.58 to 41.58   **c.** 239.6 to 292.0

10.18 a. Reject H_0 if $F > 2.87$.   **b.** Reject H_0 if $F > 2.25$.   **c.** Reject H_0 if $F > 4.43$.

10.19 a. Yes;

Source	df	SS	MS	F
Treatments	3	36.95	12.317	7.70
Error	16	25.6	1.6	

b. .6 to 3.4   **c.** 10.20 to 12.60

10.20 a. Completely randomized

b. Yes;

Source	df	SS	MS	F
Supermarkets	3	1.308	.436	6.32
Error	12	.83	.069	

c. 4.27 to 4.73

10.21 a. Yes;

Source	df	SS	MS	F
Treatments	3	36.750	12.250	4.88
Error	24	60.282	2.512	

b. −.71 to 4.03   **c.** 10.96 to 13.01

10.22 a. Yes;

Source	df	SS	MS	F
Years	3	6.951	2.317	11.03
Error	596	125.217	.210	

b. .13 to .51

10.23 a. No;

Source	df	SS	MS	F
Drugs	2	.0185	.0092	5.75
Error	6	.0097	.0016	

b. −.097 to .063   **c.** 1.60 to 1.72

10.24 a. Yes;

Source	df	SS	MS	F
Rations	3	34,867.5	11,622.5	12.22
Error	36	34,230.0	950.8	

b. −105.8 to −60.2   **c.** 148.9 to 181.1

10.25 a. Yes;

Source	df	SS	MS	F
Diets	2	120,280	60,140	5.85
Error	12	123,270	10,272.5	

b. −269.74 to 9.74

10.26 Yes;

Source	df	SS	MS	F
Music types	2	12,698	6,349	10.66
Error	9	5,358.25	595.361	

10.27 No;

Source	df	SS	MS	F
Factories	2	421.74	210.87	2.95
Error	50	3,574.06	71.481	

10.28 a.

Source	df	SS	MS	F
Hours	2	57.604	28.802	.34
Error	13	1,109.333	85.333	

b. No   **c.** 73.69 to 88.31

10.29 a. $F = 17.94$; yes   **b.** 125.342 to 143.718   **c.** 19.601 to 51.099

10.30 a. Yes;

Source	df	SS	MS	F
Methods	2	3.579	1.790	7.02
Error	13	3.057	.255	

b. 7.70 to 8.60

CHAPTER 11

11.1 a. Reject H_0 if $X^2 > 4.61$.   **b.** Reject H_0 if $X^2 > 13.3$.   **c.** Reject H_0 if $X^2 > 7.82$.   **d.** Reject H_0 if $X^2 > 9.50$.
11.2 $X^2 = 10.19$; yes   **11.3** $X^2 = 6$; no   **11.4** $X^2 = 17.36$; yes   **11.5** $X^2 = 4.4$; no   **11.6** $X^2 = 1.5$; no
11.7 $X^2 = 3.92$; no

11.8 **a.** Reject H_0 if $X^2 > 3.84$. **b.** Reject H_0 if $X^2 > 16.8$. **c.** Reject H_0 if $X^2 > 16.0$. **d.** Reject H_0 if $X^2 > 17.0$.
11.9 $X^2 = 10.22$ **11.10** $X^2 = 66.25$ **11.11** $X^2 = 19.72$; yes **11.12** $X^2 = 1.28$; no **11.13** $X^2 = 9.35$; yes
11.14 $X^2 = 14.41$; yes **11.15** $X^2 = 43.72$; yes **11.16** $X^2 = 28.30$; yes **11.17** $X^2 = 6,813.39$; yes
11.18 $X^2 = 40.70$; yes **11.19** $X^2 = 28.51$; yes **11.20** $X^2 = 1.07$; no **11.21** $X^2 = 2.40$; no **11.22** $X^2 = 93.64$; yes
11.23 $X^2 = 1.34$; no **11.24** $X^2 = 40.08$; yes **11.25** $X^2 = 5.57$; no **11.26** **a.** $X^2 = 54.14$; yes **b.** No **c.** Yes
11.27 $X^2 = 2.13$; no **11.28** $X^2 = 5.51$; no **11.29** $X^2 = 23.089$; yes **11.30** $X^2 = 2.60$; no **11.31** $X^2 = 103.08$; yes
11.32 $X^2 = 7.38$; no **11.33** $X^2 = 180.87$; yes **11.34** $X^2 = 8.66$; no **11.35** $X^2 = 269.91$; yes **11.36** $X^2 = 6.74$; yes
11.37 $X^2 = 9.50$; yes **11.38** $X^2 = 23.08$; yes **11.39** $X^2 = 30.51$; yes **11.40** $X^2 = 7.40$; no **11.41** $X^2 = 42.54$; yes

---

CHAPTER 12

12.1 **a.** Reject H_0 if $T \geq 43$. **b.** Reject H_0 if $T \leq 20$ or $T \geq 45$. **12.2** **a.** $T = 67.5$ **b.** $T = 67.5$; yes
12.3 **a.** $t = -1.26$ **b.** $T = 37.5$ **12.4** $T = 29$; no **12.5** **a.** $T = 82$; no **12.6** $T = 69.5$; yes **12.7** $T = 150.5$
12.8 **a.** Reject H_0 if $T_- \leq 2$. **b.** Reject H_0 if $T \leq 4$. **12.9** **a.** $T_- = 2.5$; yes **b.** $T_- = 2.5$; yes **12.10** $T_- = 13$; no
12.11 $T_- = 5$ **12.12** $T_+ = 11$; yes **12.13** $T_- = 8$; no **12.14** $T_- = 24.5$ **12.15** $H = 8.01$; yes
12.16 $H = 7.15$; no **12.17** $H = 7.97$; yes **12.18** $H = 2.03$; no **12.19** **a.** $H = 6.66$; yes **b.** $T = 71$; yes
12.20 **a.** $F = 1.33$ **b.** $H = 1.22$ **12.21** $H = 15.33$; yes
12.22 **a.** Reject H_0 if $r_s > .646$. **b.** Reject H_0 if $r_s > .428$ or $r_s < -.428$. **12.23** **a.** $r_s = .866$; yes **b.** $r_s = .866$; yes
12.24 $r_s = .657$; no **12.25** $r_s = .881$; yes **12.26** **a.** $r_s = .861$ **b.** Yes **12.27** **a.** $r_s = .452$ **b.** No
12.28 $r_s = -.804$; yes **12.29** $r_s = -.854$; yes **12.30** $H = 9.86$ **12.31** **a.** $r_s = .400$; no **b.** $T_- = 1.5$; yes
12.32 $T = 55.5$; no **12.33** $T = 21$; no **12.34** $T = 19$; yes **12.35** $T_- = 3$; yes **12.36** $T_- = 4$; yes
12.37 $T_+ = 6$; no **12.38** **a.** $T_+ = 1$; yes **b.** $t = -2.96$ **12.39** $T_+ = 19.5$; no **12.40** $r_s = .421$; no
12.41 $T = 41.5$ **12.42** $r_s = .643$; no **12.43** $r_s = -.148$; no **12.44** $T_- = 3.5$; yes **12.45** $T = 132.5$; yes
12.46 $H = 4.94$; no **12.47** **a.** $H = 14.61$; yes **b.** $T = 38.5$ **12.48** $H = 14.27$; yes **12.49** $H = 19.47$; yes
12.50 $r_s = .929$; yes **12.51** **a.** $H = 12.79$; yes **b.** $T = 28.5$; no **12.52** $r_s = .479$; no **12.53** $r_s = .771$; no
12.54 $T_+ = 17.5$; no **12.55** **a.** $H = 3.26$; no **b.** $T = 78$ **12.56** $r_s = .92$; yes **12.57** **a.** $r_s = .47$ **b.** Yes
12.58 **a.** $H = 16.39$; yes **b.** $T = 59$ **12.59** $r_s = .364$; no **12.60** $T = 139$; yes

INDEX

	α = .10		α = .05		α = .01	
df	One-tailed test	Two-tailed test	One-tailed test	Two-tailed test	One-tailed test	Two-tailed test
1	3.08	6.31	6.31	12.71	31.82	63.66
2	1.89	2.92	2.92	4.30	6.97	9.92
3	1.64	2.35	2.35	3.18	4.54	5.84
4	1.53	2.13	2.13	2.78	3.75	4.60
5	1.48	2.02	2.02	2.57	3.37	4.03
6	1.44	1.94	1.94	2.45	3.14	3.71
7	1.42	1.89	1.89	2.36	3.00	3.50
8	1.40	1.86	1.86	2.31	2.90	3.36
9	1.38	1.83	1.83	2.26	2.82	3.25
10	1.37	1.81	1.81	2.23	2.76	3.17
11	1.36	1.80	1.80	2.20	2.72	3.11
12	1.36	1.78	1.78	2.18	2.68	3.05
13	1.35	1.77	1.77	2.16	2.65	3.01
14	1.35	1.76	1.76	2.14	2.62	2.98
15	1.34	1.75	1.75	2.13	2.60	2.95
16	1.34	1.75	1.75	2.12	2.58	2.92
17	1.33	1.74	1.74	2.11	2.57	2.90
18	1.33	1.73	1.73	2.10	2.55	2.88
19	1.33	1.73	1.73	2.09	2.54	2.86
20	1.33	1.72	1.72	2.09	2.53	2.85
21	1.32	1.72	1.72	2.08	2.52	2.83
22	1.32	1.72	1.72	2.07	2.51	2.82
23	1.32	1.71	1.71	2.07	2.50	2.81
24	1.32	1.71	1.71	2.06	2.49	2.80
25	1.32	1.71	1.71	2.06	2.49	2.79
26	1.32	1.71	1.71	2.06	2.48	2.78
27	1.31	1.70	1.70	2.05	2.47	2.77
28	1.31	1.70	1.70	2.05	2.47	2.76
z	1.28	1.65	1.65	1.96	2.33	2.58

t Values for Rejection Regions for Small-Sample Tests of Hypotheses